Tutto s'accoglie in lei, e fuor di quella
É defettivo ciò ch'é lì perfetto.

DANTE, *Paradiso* XXXIII, vv. 104–105

For Leopoldine

Contents

Preface

Even a cursory look at the table of contents of this book will reveal that the material is arranged in an unusual way, starting immediately with the application of techniques (Part I) which are justified and explained in greater detail in later chapters (Parts II and III). Indeed, the book is not meant to be read sequentially and I have not attempted to force the material into a sequential exposition. Rather, I have organized it as a series of modules through which each reader can fashion an individual pathway according to background, needs and interests.

The widespread availability of computers and symbolic manipulation packages has made mathematics relevant to many more people than in the past. A serious problem, however, remains the high "cost of entry" of mathematics. While some might argue that in a perfect world people would spend the time required to learn all the mathematics they might need before applying it, this is a utopia certainly, and probably also a misguided ideal as any user of mathematics – having gone through many cycles of learning and applying – can testify.

Hopefully, the modular structure of the book will serve a variety of users addressing different needs ranging from a quick impression of a mathematical tool to a fairly deep understanding of its basis and "inner workings." The many cross-references, detailed index and table of contents will render possible a non-systematic "navigation" through the material.

Knowledge is a multi-dimensional network. Most ideas, facts and techniques can be approached from different directions, and are connected to each other by a web of relations. While each reader at first may resonate more readily with one or the other approach, an appreciation of the various interconnections will ultimately provide a deeper understanding of each topic and facilitate its mastery.

A network permits a multitude of one-dimensional paths, which are what most of the existing books and courses of instruction provide. While of course legitimate, each one of them is to some extent the fruit of an arbitrary selection. Here the choice is the opposite: by starting, as it were, from the end point of each path and following it "upstream," a reader may find a reasonably efficient trajectory suitable for his/her knowledge, need and experience.

By the same token, an instructor using this book would not be forced into a fixed trajectory through the material, but can structure the course in many different ways.

The mathematician's tools are theorems and proofs, but these seldom are the aspects "users" of mathematics – the intended audience of this book – care about. On the other hand, the correct use of mathematical methods requires an appreciation of the underlying theory, both to enhance their usefulness and to avoid pitfalls. Here I have chosen to state theorems with a reasonable degree of precision, but to outline few proofs with a level of rigor that would satisfy the mathematician. Rather, where possible within the bounds imposed by concision of exposition, I have tried to make the results plausible with heuristic arguments which both justify the hypotheses necessary to derive the results and illustrate why they are true. Some simple proofs, or proofs which are desirable to illuminate a particular result, have however been included.

One significant benefit of the avoidance of hard proofs is that theorems can be stated in greater generality than would otherwise be possible. This feature opens up "vistas" that greatly enhance the import of many results and help illuminate their mutual relationship.

Given the intent of this book, it seemed out of place to include end-of-chapter problems. Problem sets are however available on the book's web page at www.cambridge.org/prosperetti.

To the reader

This book is not meant to be read sequentially. The material is organized according to a modular structure with abundant cross-referencing and indexing to permit a variety of pathways through it.

Each chapter in Part I, Applications, is devoted to a particular technique: Fourier series, Fourier transform, etc. The chapters open with a section summarizing very briefly the basic relations and proceeds directly to show on a variety of examples how they are applied and how they "work."

A fairly detailed exposition of the essential background of the various techniques is given in the chapters of Part II, Essential Tools. Other chapters here describe general concepts (e.g., Green's functions and analytic functions) that occur repeatedly elsewhere. The last chapter on matrices and finite-dimensional linear spaces is included mostly to introduce Part III, Some Advanced Tools. Here the general theory of linear spaces, generalized functions and linear operators provides a unified foundation to the various techniques of Parts I and II.

The book starts with some general remarks and introductory material in Part 0. Here the first chapter summarizes the basic equations of classical field theory to establish a connection between specific physical problems and the many examples of Part I in which, by and large, no explicit reference to physics is made. The last section of this chapter provides a very elementary introduction to the basic idea of eigenfunction expansion. Readers not familiar with this powerful point of view are strongly encouraged to read it. Since the book is meant to be useful to readers with a variety of backgrounds and mathematical experience, the second chapter of Part 0 treats in a very elementary way some basic notions and techniques which are repeatedly encountered in the applications that follow and which are treated in much greater depth in later chapters.

In many places it proves necessary to invoke general mathematical notions on differentiation, integration etc. which, for convenience and ease of reference, are collected in the Appendix.

Tables

PART 0

General Remarks and Basic Concepts

1 The Classical Field Equations

It is a remarkable aspect of the "unreasonable effectiveness of mathematics in the natural sciences" (Wigner 1960) that a handful of equations are sufficient to describe mathematically a vast number of physically disparate phenomena, at least at some level of approximation. Key reasons are the isotropy and uniformity of space-time (at least locally), the attendant conservation laws,[1] and the useful range of applicability of linear approximations to constitutive relations.

After a very much abbreviated survey of the principal properties of vector fields, we present a summary of these fundamental equations and associated boundary conditions, and then describe several physical contexts in which they arise. The initial chapters of a book on any specific discipline give a far better derivation of the governing equations for that discipline than space constraints permit here. Our purpose is, firstly, to remind the reader of the meaning accorded to the various symbols in any specific application and of the physics that they describe and, secondly, to show the similarity among different phenomena.

The final section of this chapter is a very simple-minded description of the method of eigenfunction expansion systematically used in many of the applications treated in this book. The starting point is an analogy with vectors and matrices in finite-dimensional spaces and the approach is purposely very elementary; a "real" theory is to be found in Part III of the book.

1.1 Vector fields

Throughout most of the book the symbol u is used to denote the unknown function. Although mathematically unnecessary, a distinction is made between the time variable t and the space variables denoted, in Cartesian coordinates, by $\mathbf{x} = (x, y, z)$; vectors are shown in bold face.

The gradient ∇u of a scalar function u is a vector with Cartesian components

$$\nabla u = \left(\frac{\partial u}{\partial x}, \frac{\partial u}{\partial y}, \frac{\partial u}{\partial z}\right) = \frac{\partial u}{\partial x}\mathbf{i} + \frac{\partial u}{\partial y}\mathbf{j} + \frac{\partial u}{\partial z}\mathbf{k}, \qquad (1.1.1)$$

---

[1] Noether's theorem asserts that any differentiable symmetry of the action (i.e., the integral of the Lagrangian) of a physical system has a corresponding conservation law (see e.g. Lanczos 1970, p. 401; Goldstein 1980, p. 588; Weinberg 1995, vol. 1, p. 307; Srednicki 2007, p. 133).

in which $\mathbf{i}$, $\mathbf{j}$ and $\mathbf{k}$ are the unit vectors in the x-, y- and z-directions respectively. Physically, $(\nabla u) \cdot d\mathbf{x}$ represents the increment of u when its spatial argument is changed from $\mathbf{x} = (x, y, z)$ to $\mathbf{x} + d\mathbf{x} = (x + dx, y + dy, z + dz)$. The same meaning is attached to $(\nabla u) \cdot d\mathbf{x}$ in any other coordinate system (ξ, η, ζ) and, for this reason, the component of the gradient in the ξ-direction, for example, is not necessarily given by $\partial u / \partial \xi$;[2] the form of the gradient in cylindrical and spherical coordinate systems is given in Tables 6.4 p. 148 and 7.3 p. 173, respectively.

The divergence of a vector $\mathbf{A}$ with Cartesian components (A_x, A_y, A_z) is

$$\nabla \cdot \mathbf{A} = \frac{\partial A_x}{\partial x} + \frac{\partial A_y}{\partial y} + \frac{\partial A_z}{\partial z}. \tag{1.1.2}$$

The physical meaning of $\nabla \cdot \mathbf{A}$ is made evident by the divergence theorem (pp. 400 and 590) which implies that

$$(\nabla \cdot \mathbf{A})(\mathbf{x}) = \lim_{\Delta V \to 0} \frac{1}{\Delta V} \oint_S \mathbf{A} \cdot \mathbf{n} \, dS, \tag{1.1.3}$$

in which S is the boundary of a small volume ΔV centered at $\mathbf{x}$, $\mathbf{n}$ is the outwardly directed unit normal and the limit is understood in the sense that all dimensions of ΔV tend to zero approximately at the same rate. Thus $(\Delta V) \nabla \cdot \mathbf{A}$ represents the net flux of $\mathbf{A}$ out of the elementary volume ΔV.

A third important differential vector operator is the curl:

$$\nabla \times \mathbf{A} = \begin{vmatrix} \mathbf{i} & \mathbf{j} & \mathbf{k} \\ \frac{\partial}{\partial x} & \frac{\partial}{\partial y} & \frac{\partial}{\partial z} \\ A_x & A_y & A_z \end{vmatrix} = \left(\frac{\partial A_z}{\partial y} - \frac{\partial A_y}{\partial z} \right) \mathbf{i} + \left(\frac{\partial A_x}{\partial z} - \frac{\partial A_z}{\partial x} \right) \mathbf{j} + \left(\frac{\partial A_y}{\partial x} - \frac{\partial A_x}{\partial y} \right) \mathbf{k}.$$
$$\tag{1.1.4}$$

Similarly to (1.1.3), we can get some insight into the physical meaning of this quantity from Stokes's theorem (p. 401) which permits us to write

$$\mathbf{n} \cdot (\nabla \times \mathbf{A}) = \lim_{\Delta S \to 0} \frac{1}{\Delta S} \oint_L \mathbf{A} \cdot \mathbf{t} \, d\ell. \tag{1.1.5}$$

Here L is a small planar loop enclosing an area ΔS which has a unit normal $\mathbf{n}$; $\mathbf{t}$ is the unit vector tangent to L oriented in the direction of the fingers of the right hand when the thumb points along $\mathbf{n}$; $(\Delta S)\mathbf{n} \cdot (\nabla \times \mathbf{A})$ represents therefore the circulation of $\mathbf{A}$ around the small loop encircling the area ΔS.

Just as in the case of the gradient, although the physical meaning associated with divergence and curl is the same in all coordinate systems, the forms shown in (1.1.2) and (1.1.4) are only valid in Cartesian coordinates because this is the only system in which the unit vectors along the coordinate lines are constant. In a general orthogonal system $\{\mathbf{e}_1, \mathbf{e}_2, \mathbf{e}_3\}$ a vector field is represented as

$$\mathbf{A} = A_1(\mathbf{x})\mathbf{e}_1 + A_2(\mathbf{x})\mathbf{e}_2 + A_3(\mathbf{x})\mathbf{e}_3 = \sum_{k=1}^{3} A_k \mathbf{e}_k. \tag{1.1.6}$$

---

[2] $\partial u / \partial \xi$ is the component of the gradient in the ξ-direction only when $d\xi$ represents the actual distance between the points having coordinates (ξ, η, ζ) and $(\xi + d\xi, \eta, \zeta)$. Evidently, this cannot be the case, for example, if ξ is an angular coordinate.

Then, using the standard rules of vector calculus to express $\nabla \cdot [f \, \mathbf{V}]$ and $\nabla \times [f \, \mathbf{V}]$, we would have

$$\nabla \cdot \mathbf{A} = \sum_{k=1}^{3} [(\nabla A_k) \cdot \mathbf{e}_k + A_k \nabla \cdot \mathbf{e}_k], \qquad \nabla \times \mathbf{A} = \sum_{k=1}^{3} [(\nabla A_k) \times \mathbf{e}_k + A_k \nabla \times \mathbf{e}_k].$$

(1.1.7)

Expressions for the divergence and the curl of a vector field in cylindrical and spherical coordinate systems are shown in Tables 6.4 p. 148 and 7.3 p. 173, respectively.

In the following a central role is played by the Laplace operator, or *Laplacian*,

$$\nabla^2 u = \nabla \cdot (\nabla u) = \frac{\partial^2 u}{\partial x^2} + \frac{\partial^2 u}{\partial y^2} + \frac{\partial^2 u}{\partial z^2}.$$

(1.1.8)

While the form $\nabla \cdot (\nabla u)$ is valid in all coordinate systems, it follows from the previous considerations that the last form is only valid in Cartesian coordinates; the form of this operator in cylindrical and spherical coordinate systems is shown in Tables 6.4 p. 148 and 7.3 p. 173, respectively. For a vector field $\mathbf{A}$, in Cartesian coordinates we have

$$\nabla^2 \mathbf{A} = (\nabla^2 A_x)\mathbf{i} + (\nabla^2 A_y)\mathbf{j} + (\nabla^2 A_z)\mathbf{k},$$

(1.1.9)

but in a general coordinate system the Laplacian of a vector is defined using the relation

$$\nabla^2 \mathbf{A} = \nabla(\nabla \cdot \mathbf{A}) - \nabla \times \nabla \times \mathbf{A},$$

(1.1.10)

which is an identity in Cartesian coordinates and provides a consistent definition of $\nabla^2 \mathbf{A}$ in non-Cartesian coordinate systems.

Two standard vector identities are

$$\nabla \times (\nabla u) = 0, \qquad \nabla \cdot (\nabla \times \mathbf{A}) = 0.$$

(1.1.11)

Furthermore, the following two implications are well known:

$$\nabla \times \mathbf{C} = 0 \quad \Longleftrightarrow \quad \mathbf{C} = \nabla u,$$

(1.1.12)

$$\nabla \cdot \mathbf{B} = 0 \quad \Longleftrightarrow \quad \mathbf{B} = \nabla \times \mathbf{A}.$$

(1.1.13)

In (1.1.12) u is the *scalar potential*[3] of the field $\mathbf{C}$ and, in (1.1.13), $\mathbf{A}$ is the *vector potential* of the field $\mathbf{B}$. Since the equation $\nabla \cdot \mathbf{B} = 0$ implies that only two of the three scalars that constitute the field $\mathbf{B}$ are independent, the same must be true of $\mathbf{A}$. Thus, (1.1.13) must be insufficient to uniquely specify the vector potential and, indeed, there is the freedom to add various subsidiary conditions to make $\mathbf{A}$ unique without affecting the physical field $\mathbf{B} = \nabla \times \mathbf{A}$. These conditions, examples of which will be found in §1.5, are known as *gauge conditions*.

For a general vector field $\mathbf{V}$ we have the *Helmholtz* or *Hodge decomposition* (see e.g. Morse & Feshbach 1953, p. 53)

$$\mathbf{V} = \nabla u + \nabla \times \mathbf{A},$$

(1.1.14)

---

[3] The relation (1.1.12) holds in a simply connected domain. In a multiply connected domain the local condition $\nabla \times \mathbf{C} = 0$ must be supplemented by the global requirement that the circulation of $\mathbf{C}$ around any closed path vanish.

with $\nabla^2 u = \nabla \cdot \mathbf{V}$. The two terms, ∇u and $\nabla \times \mathbf{A}$, are the *longitudinal* and *transverse* parts of $\mathbf{V}$.

The addition of the gradient of a scalar function to the vector potential $\mathbf{A}$ does not affect the fields $\mathbf{B}$ or $\mathbf{V}$ and, because of this freedom, in some situations it is possible to assume without loss of generality particularly convenient forms for the vector potential. For example, if the problem is planar, one may take $\mathbf{A} = \nabla \times (\psi \mathbf{k})$ in which $\mathbf{k}$ is a constant vector orthogonal to the plane and ψ a scalar function. With axial symmetry, one may take $\mathbf{A} = \nabla \times (\psi \mathbf{e}_\phi)$ in which $\mathbf{e}_\phi$ is a unit vector in the direction of the angle of rotation around the symmetry axis. In other cases a useful representation can be given in the form (see e.g Chandrasekhar 1961, p. 622)

$$\mathbf{A} = \nabla \times (\psi \mathbf{x}) + \nabla \times \nabla \times (\chi \mathbf{x}), \qquad (1.1.15)$$

in which ψ and χ are the two *defining scalar* functions. The two terms in this expression are the *toroidal* and *poloidal* components of $\mathbf{A}$, respectively; it is evident that only the second one possesses a radial component. We return to this decomposition in much greater detail in §14.4 and show its usefulness in an example in §7.18.

1.2 The fundamental equations

This book deals with the practice (Part I) and theory (Parts II and III) of solution methods of some fundamental equations which are the most important examples of the three fundamental groups of elliptic, hyperbolic and parabolic equations (see e.g. Courant & Hilbert 1953; Garabedian 1964; Renardy & Rogers 1993 and many others). Here we summarize these equations but, first, we establish some terminology related to boundary and initial conditions.

1.2.1 Boundary conditions

There is a standard terminology to refer to the types of space-time boundary conditions (which would be initial or "final" conditions for the time variable) normally associated with second-order equations:

- *Dirichlet data*: The function is prescribed (e.g. equal to 0) on the boundary;
- *Neumann data*: The normal gradient of the function is prescribed (e.g. equal to 0) on the boundary. With this type of boundary condition, if only derivatives of the unknown function appear in the governing equation, the solution cannot be unique as any function differing by a constant from a solution of the problem is also a solution;
- *Cauchy data*: Both the function and its normal gradient are prescribed on the boundary;
- *Mixed data*: A linear combination of the unknown function and its normal gradient is prescribed on the boundary;

- Conditions may also be of one type (e.g., Dirichlet) on part of the boundary and of another type (e.g. Neumann) on other parts. For example, Neumann conditions might be specified on the finite part of the boundaries, and the Dirichlet-type condition $u \to 0$ at infinity may be added to make the solution unique.

The nature of the boundary conditions to be associated with each specific equation type is quite important and is one aspect of the distinction between properly and improperly posed problems. Although we do not deal with the general theory, we illustrate this point with an example in §4.7.

1.2.2 Elliptic equations

Generally speaking, elliptic equations describe a situation of equilibrium established a long time after the beginning of a process: all forces have equilibrated, all disturbances have either damped out or propagated to infinity and so forth. The most fundamental equation of this type is the *Laplace equation*

$$\nabla^2 u = 0. \tag{1.2.1}$$

Functions satisfying this equation in some spatial region Ω are termed *harmonic* in Ω. The non-homogeneous form of this equation is known as the *Poisson equation*:[4]

$$-\nabla^2 u = f, \tag{1.2.2}$$

in which the function $f(\mathbf{x})$, representing distributed sources of the field u, is given.

Another equation of the same mathematical type which is frequently encountered is the *Helmholtz equation* or *reduced wave equation*

$$\nabla^2 u + k^2 u = -f(\mathbf{x}), \tag{1.2.3}$$

in which both f and the (real or complex, positive or negative) parameter k^2 are given. As will be seen in several examples, when the parameter k^2 is real and positive, there are some very significant qualitative differences between the solutions of (1.2.1) or (1.2.2) and those of (1.2.3).

It will be clear from the physical considerations described in the following sections that the boundary conditions to be associated with these equations cannot be of the Cauchy type. An attempt to prescribe such conditions would result in an ill-behaved and physically unacceptable solution (see example in §4.7). Appropriate boundary conditions are of the Dirichlet, Neumann, or mixed type.

There are important types of higher-order equations of the same mathematical type, chief among them the *biharmonic equation*:

$$\nabla^4 u = \nabla^2 \left(\nabla^2 u \right) = 0, \tag{1.2.4}$$

---

[4] A physical justification for the minus sign is that an integration by parts of $-u\nabla^2 u$ causes the appearance of the non-negative-definite quantity $|\nabla u|^2$ which often has the physical meaning of an energy density.

which is efficiently split into the Laplace–Poisson system (§7.7)

$$\nabla^2 v = 0, \qquad \nabla^2 u = v. \tag{1.2.5}$$

1.2.3 Hyperbolic equations

Hyperbolic equations generally describe propagation with a finite speed. The prototypical equation of this type is the scalar *wave equation*

$$\frac{1}{c^2}\frac{\partial^2 u}{\partial t^2} - \nabla^2 u = F(\mathbf{x}, t), \tag{1.2.6}$$

in which the constant c is the speed of propagation of the wave and F, representing distributed sources of the field associated with the wave, is given. By referring, for instance, to a mechanical problem in which u might represent a displacement caused by the wave, it is intuitively clear that, on the "time boundary" $t = 0$, conditions of the Cauchy type, namely

$$u(\mathbf{x}, t = 0) = u_0(\mathbf{x}), \qquad \left.\frac{\partial u}{\partial t}\right|_{t=0} = v(\mathbf{x}), \tag{1.2.7}$$

would be appropriate. On the spatial boundaries through which the wave enters the domain of interest either Dirichlet, Neumann or mixed conditions can be applied.

Particularly simple solutions of the homogeneous wave equation have the form

$$u(\mathbf{x}, t) = \Phi(\mathbf{e} \cdot \mathbf{x} - ct), \tag{1.2.8}$$

where $\mathbf{e}$ is a unit vector and Φ an arbitrary function admitting a (conventional or distributional) double derivative. It is evident that u as given by this formula has a constant value on the planes $\mathbf{e} \cdot \mathbf{x} - ct = $ const., which justifies the denomination *plane wave* given to solutions of this type. The normal to this family of planes is parallel to $\mathbf{e}$ and identifies the direction of propagation. In one space dimension only two such waves are possible, one propagating to the left and one to the right so that the most general solution of (1.2.7) has the well-known d'Alembert form

$$u(x, t) = \Phi_+(x - ct) + \Phi_-(x + ct). \tag{1.2.9}$$

In spaces of higher dimensionality a much greater variety of plane waves is possible and any arbitrary superposition of them is also a solution of the equation (see John 1955).

When there is a steady boundary or volume source maintaining monochromatic waves, i.e. waves of a single frequency $\omega/2\pi$ (ω being the angular frequency), and all transients have died out, one expects the solution to have the form

$$u(\mathbf{x}, t) = v(\mathbf{x}) \cos[\omega t + \psi(\mathbf{x})], \tag{1.2.10}$$

i.e., to consist of monochromatic waves of the same frequency with a spatially modulated amplitude v and phase ψ. It is convenient to express this functional form as $u(\mathbf{x}, t) = w(\mathbf{x})e^{-i\omega t}$ allowing w to be complex and tacitly agreeing to take the real part of the complex expression. When this functional form is substituted into (1.2.6) with F assumed to have the form $F(\mathbf{x}, t) = f(\mathbf{x})e^{-i\omega t}$, w is found to satisfy

$$\nabla^2 w + \frac{\omega^2}{c^2} w = -f, \tag{1.2.11}$$

i.e., the Helmholtz equation (1.2.3) with $k^2 = \omega^2/c^2$; in this context, k is the *wave number*, equal to 2π divided by the wavelength of the monochromatic wave. The same association between the wave and Helmholtz equations is encountered if (1.2.6) is solved by the Fourier transform in the time variable.

If there is no agent to sustain waves and all those initially present have propagated away from the region of interest, the solution of (1.2.6) becomes independent of time and the equation reduces then to the Laplace (1.2.1) or Poisson (1.2.2) forms.

In some cases the effect of damping modifies (1.2.6) to the form

$$\frac{1}{c^2}\frac{\partial^2 u}{\partial t^2} + \frac{2a}{c^2}\frac{\partial u}{\partial t} - \nabla^2 u = F(\mathbf{x}, t), \tag{1.2.12}$$

with $a \geq 0$, which is known as the *telegrapher equation*. An example is given in §1.5.

In electromagnetism and elasticity the propagating quantities are vectors rather than scalars. In these cases it is preferable to write the wave equation using the expression (1.1.10) for the Laplacian to put the equation in the coordinate-invariant form

$$\frac{1}{c^2}\frac{\partial^2 \mathbf{A}}{\partial t^2} + \nabla \times \nabla \times \mathbf{A} = \mathbf{F}, \tag{1.2.13}$$

where it has been assumed that $\nabla \cdot \mathbf{A} = 0$ as often happens, or can be caused to happen by means of a suitable gauge transformation. In this case, plane wave solutions of the homogeneous equation have the form

$$\mathbf{A} = (\mathbf{b} \times \mathbf{e})\, \Psi\, (\mathbf{e} \cdot \mathbf{x} - ct)\,, \tag{1.2.14}$$

in which $\mathbf{b}$ is an arbitrary constant vector and Ψ a function. When $\mathbf{A}$ is a vector potential, the quantity with a physical meaning is $\nabla \times \mathbf{A} = [\mathbf{b} - (\mathbf{b} \cdot \mathbf{e})\mathbf{e}]\Psi'$, which is perpendicular to $\mathbf{e}$. These waves are therefore polarized in the direction perpendicular to the direction of propagation and, for this reason, are termed *transverse*.

1.2.4 Parabolic equations

The prototypical parabolic equation is the *diffusion equation*

$$\frac{\partial u}{\partial t} = D\nabla^2 u + f(\mathbf{x}, t), \tag{1.2.15}$$

in which the constant D, with dimensions of (length)^2/time, is the *diffusivity* or diffusion coefficient of the quantity u and $f(\mathbf{x}, t)$ is given. The introduction of a scaled time $t_* = Dt$ with $f_* = f/D$ permits us to consider the equation with $D = 1$, which we will often do without explicitly mentioning that a scaled time variable is being used. For these equations initial conditions

$$u(\mathbf{x}, t = 0) = u_0(\mathbf{x}) \tag{1.2.16}$$

are appropriate, together with Dirichlet, Neumann or mixed conditions on the spatial boundaries.

For steady conditions the diffusion equation degenerates to the Laplace (1.2.1) or Poisson (1.2.2) forms. If (1.2.15) is solved by a Fourier or Laplace transform in time, the transformed function satisfies the Helmholtz equation (1.2.3) with k^2 negative or complex (see §5.5).

In diffusion problems it is possible that the source f, rather than being prescribed a priori, is a function of u describing, e.g., the disappearance of u due to a chemical reaction. When this dependence is linear we have

$$\frac{\partial u}{\partial t} = D\nabla^2 u - \alpha u. \tag{1.2.17}$$

If $\alpha = $ const. the simple substitution $u = e^{-\alpha t}v$ brings the equation into the standard form (1.2.15), written for v, with $f = 0$.

1.3 Diffusion

The most elementary balance relation states that the rate of change of the total amount of some quantity distributed with a density U within a fixed volume V equals the net transport through the boundary S of V plus the contribution of the sources q inside the volume:

$$\frac{\mathrm{d}}{\mathrm{d}t}\int_V U\,\mathrm{d}V = -\oint_S \mathbf{Q}\cdot\mathbf{n}\,\mathrm{d}S + \int_V q\,\mathrm{d}V, \tag{1.3.1}$$

in which $\mathbf{Q}$ is the flux of U and $\mathbf{n}$ the outward unit normal. By applying the divergence theorem (p. 400), in view of the arbitrariness of the control volume V, the previous equation implies that, almost everywhere in space,[5]

$$\frac{\partial U}{\partial t} = -\nabla\cdot\mathbf{Q} + q. \tag{1.3.2}$$

In heat transfer (1.3.1) derives from the first principle of thermodynamics for an incompressible medium: U is the enthalpy per unit volume, $\mathbf{Q}$ the heat flux and q the internal heat generation rate per unit volume. If the medium has constant properties, $U = \rho c_p T$, with ρ the density, c_p the specific heat and T the temperature. The heat flux $\mathbf{Q}$ must depend on T in such a way that it vanishes when T is spatially uniform. For an isotropic medium the simplest functional form satisfying this condition is embodied in *Fourier's law of conduction*:

$$\mathbf{Q} = -k\nabla T, \tag{1.3.3}$$

in which the thermal conductivity k must be positive for heat to flow away from the hotter regions. A relation of this type may be seen as a truncated Taylor series expansion of a more

---

[5] The qualification "almost everywhere" has the technical meaning "aside from a set of zero measure"; see p. 682.

general constitutive relation. This is a typical mechanism through which the divergence of a flux gives rise to the Laplacian operator. Upon substitution into (1.3.2) we have

$$\frac{\partial T}{\partial t} = \frac{k}{\rho c_p} \nabla^2 T + \frac{q}{\rho c_p}, \tag{1.3.4}$$

which has the form (1.2.15) with $D = k/\rho c_p$. Because of this derivation one often refers to this latter equation as to the heat, or conduction, equation.

In steady conditions and in the absence of sources, (1.3.4) reduces to $\nabla^2 T = 0$ so that the temperature field is harmonic and $\oint \mathbf{n} \cdot \nabla T \, dS = 0$ from (1.1.3). The underlying physical picture permits us to give a physically compelling justification of the *maximum principle* for harmonic functions, according to which a harmonic function defined in a region Ω cannot attain a local maximum or minimum inside Ω but only on the boundary (see p. 76). Indeed, if T had e.g. a maximum at a certain point interior to Ω, on a sufficiently small surface surrounding this point heat would flow away so that $\mathbf{n} \cdot \nabla T$ would be negative everywhere and the integral could not vanish.

Solution of the diffusion equation clearly requires that an initial temperature distribution $T(\mathbf{x}, t = 0)$ be known. In physical terms, it is evident that the solution will be affected by conditions at the spatial boundaries: the imposition of a prescribed temperature (Dirichlet condition) or a prescribed heat flux (Neumann condition) will certainly affect the spatial and temporal evolution of the temperature field. If the surface of the medium is in contact with a moving fluid capable of removing heat at a rate $h(T - T_\infty)$, where h is a heat transfer coefficient and T_∞ a constant ambient temperature, continuity of heat flux at the surface results in

$$-k\mathbf{n} \cdot \nabla T = h(T - T_\infty), \tag{1.3.5}$$

which is a condition of the mixed type for the unknown $u = T - T_\infty$. A similar condition is approximately valid if the surface of the medium exchanges radiant energy with its surroundings. According to the Stefan–Boltzmann law, a surface emits radiant energy at a rate $\varepsilon \sigma T^4$ per unit area, where ε is the surface emissivity and σ is the Stefan–Boltzmann constant. The incident energy absorbed from the surroundings at temperature T_∞ is $\alpha \sigma T_\infty^4$, with α the surface absorptivity. From Kirchhoff's law, $\alpha = \varepsilon$ and, therefore, continuity of heat flux across the surface requires that

$$-k\mathbf{n} \cdot \nabla T = \varepsilon \sigma (T^4 - T_\infty^4) = \varepsilon \sigma \frac{T^4 - T_\infty^4}{T - T_\infty} (T - T_\infty) \simeq 4\varepsilon \sigma T_\infty^3 (T - T_\infty), \tag{1.3.6}$$

provided the temperature difference is not too large. For this reason, the denomination *radiation condition* is encountered in the heat transfer literature to denote conditions of the mixed type.

Another physical process resulting in an equation of the type (1.3.1) is the diffusion of a solute in a solvent. In this case we may take the mass density ρ_d of the diffusing substance as the conserved quantity U in (1.3.1). The flux $\mathbf{Q}$ is given by Fick's law of diffusion as $\mathbf{Q} = -\rho \mathscr{D} \nabla(\rho_d/\rho)$, in which ρ is the total mass density of the solute–solvent mixture and $\mathscr{D}$ the mass diffusivity. When both these quantities are approximately constant $\nabla \cdot \mathbf{Q} \simeq -\rho \mathscr{D} \nabla^2 \rho_d$ and the standard form (1.2.15) is recovered. In this case the source term f may represent the generation or disappearance of solute due e.g. to chemical reaction. If this mechanism is proportional to the amount of solute already present, we have an instance of the occurrence of an equation of the form (1.2.17).

In the evaporation of a liquid into a gas, the vapor density at the liquid surface is set by the equality of the chemical potentials in the two phases, which results in a well-defined vapor density, i.e., a Dirichlet condition. The same condition is found in the diffusion of gas in a liquid in contact with the free gas, where the dissolved gas concentration at the free surface is set by Henry's law, another manifestation of the equality of the chemical potentials. In other cases a gas may have very little solubility in a liquid or a solid, in which case $-\mathscr{D}\,\mathbf{n}\cdot\nabla\rho_d \simeq 0$, i.e., a Neumann condition.

1.4 Fluid mechanics

With $q = 0$, the balance relation (1.3.2) reflects mass conservation for a fluid if U is identified with ρ, the mass density, and $\mathbf{Q}$ with the mass flux $\rho\mathbf{u}$, in which $\mathbf{u}$ is the velocity field:

$$\frac{\partial\rho}{\partial t} + \nabla\cdot(\rho\mathbf{u}) = 0. \tag{1.4.1}$$

When compressibility effects are small, $\rho \simeq$ const. and (1.4.1) reduces to

$$\nabla\cdot\mathbf{u} = 0, \tag{1.4.2}$$

which expresses the mathematical model of an incompressible fluid.

If the fluid is in *irrotational motion*, i.e. $\nabla\times\mathbf{u} = 0$, then by (1.1.13) the velocity field admits a scalar potential, usually denoted by φ, so that $\mathbf{u} = \nabla\varphi$. Substitution into (1.4.2) leads then to the Laplace equation (1.2.1). One method of solving this equation is to imagine that the flow is driven by fluid sources or sinks located outside the physical domain of interest (e.g., inside a solid body). In this case the equation for φ takes the Poisson form $\nabla^2\varphi = q$ with q the strength of these sources.

If the fluid is in contact with a surface in motion with a velocity $\mathbf{v}$ into which it cannot penetrate, it is evidently necessary that $\mathbf{n}\cdot\mathbf{u} = \mathbf{n}\cdot\nabla\varphi = \mathbf{n}\cdot\mathbf{v}$, which is a Neumann boundary condition.

For small-amplitude motion the acceleration of a fluid element is approximately given by $\partial\mathbf{u}/\partial t$ and the forces per unit volume are the divergence of the stress tensor, $\nabla\cdot\boldsymbol{\sigma}$, and a body force per unit mass $\mathbf{f}$. Newton's equation then becomes, approximately,[6]

$$\rho\frac{\partial\mathbf{u}}{\partial t} = \nabla\cdot\boldsymbol{\sigma} + \rho\mathbf{f}. \tag{1.4.3}$$

---

[6] The approximation consists in taking $\partial\mathbf{u}/\partial t$ as the acceleration of a fluid particle in place of the true expression $\partial\mathbf{u}/\partial t + (\mathbf{u}\cdot\nabla)\mathbf{u}$. The term quadratic in $\mathbf{u}$ is negligible if the velocity is small. For an incompressible fluid this equation can be obtained from the general balance relations (1.3.1) or (1.3.2) by setting $U = \rho\mathbf{u}$, $\mathbf{Q} = \boldsymbol{\sigma}$ and $q = \rho\mathbf{f}$. The equation that one finds in so doing is the same as (1.4.3) by virtue of the incompressibility condition (1.4.2).

When ρ is constant and the stress in the fluid is only due to pressure forces so that $\nabla \cdot \boldsymbol{\sigma} = -\nabla p$, an initially irrotational state of motion remains irrotational for all time (see e.g. Landau & Lifshitz 1987a, chapter 1; Batchelor 1999, chapter 6). With ρ constant, if $\mathbf{f}$ is also constant, the equation can then be integrated to give

$$\frac{\partial \varphi}{\partial t} + \frac{p}{\rho} - \mathbf{x} \cdot \mathbf{f} = F(t), \tag{1.4.4}$$

with $F(t)$ arbitrary. This is the linearized form of the *Bernoulli integral* and is applicable to the small-amplitude motion of incompressible fluids when viscous effects are insignificant (see e.g. Landau & Lifshitz 1987a, chapter 1; Batchelor 1999, chapter 6). Due to the assumption of incompressibility, here p is not the absolute thermodynamic pressure, but the pressure in excess of an arbitrary reference level. For example, when the fluid is separated by a free surface from a medium with negligible dynamical effects, the pressure in this medium can be taken as the reference value and a balance of normal stresses across the free surface leads then to the boundary condition

$$\frac{\partial^2 \varphi}{\partial t^2} + K\mathbf{n} \cdot \nabla \varphi = 0, \tag{1.4.5}$$

with K a suitable constant related, e.g., to gravity or surface tension; an example of the derivation of this equation in the case of surface waves will be found in §6.8. This relation becomes a condition of the mixed type for monochromatic waves or when a Fourier or Laplace time transform are used to solve the problem in the case of a more general time dependence.

Small density perturbations in a fluid are the realm of acoustics. By setting $\mathrm{d}\rho = \mathrm{d}p/c^2$, where $c^2 = \partial p/\partial \rho$ (with the derivative taken at constant entropy) and neglecting the term $\mathbf{u}\mathrm{d}\rho$ which is quadratic in the perturbation, the equation of continuity (1.4.1) becomes

$$\frac{1}{c^2}\frac{\partial p}{\partial t} + \rho \nabla \cdot \mathbf{u} = 0, \tag{1.4.6}$$

where now ρ is to be interpreted as the undisturbed density. When, as before, the stress reduces to pressure, the momentum equation (1.4.3) is

$$\rho \frac{\partial \mathbf{u}}{\partial t} = -\nabla p, \tag{1.4.7}$$

in which we have assumed $\mathbf{f}$ to be a constant and we write ∇p in place of $\nabla(p - \rho \mathbf{x} \cdot \mathbf{f})$ so that p, strictly speaking, is the pressure in excess of the static pressure $-\rho \mathbf{x} \cdot \mathbf{f}$. The velocity $\mathbf{u}$ can be eliminated between (1.4.6) and (1.4.7) by differentiating the first equation with respect to time and taking the divergence of the second one. The result is the wave equation (1.2.6):

$$\frac{\partial^2 p}{\partial t^2} - c^2 \nabla^2 p = 0 \tag{1.4.8}$$

where we see that the parameter c has the meaning of speed of sound in the medium.

Taking the divergence of (1.4.6) and the time derivative of (1.4.7) leads to the same equation for $\mathbf{u}$.

For a broad class of fluids the stress tensor $\boldsymbol{\sigma} = \{\sigma_{ij}\}$ has the Newtonian form

$$\sigma_{ij} = -p\delta_{ij} + \mu\left(\frac{\partial u_i}{\partial x_j} + \frac{\partial u_j}{\partial x_i}\right), \tag{1.4.9}$$

in which the parameter μ is the viscosity coefficient. With this constitutive relation for the stress, the momentum equation (1.4.3) for an incompressible fluid in slow, viscosity-dominated motion takes the form of the time-dependent *Stokes equation*:

$$\rho\frac{\partial \mathbf{u}}{\partial t} = -\boldsymbol{\nabla}p + \mu\nabla^2\mathbf{u} + \rho\mathbf{f}, \tag{1.4.10}$$

in which we have assumed that μ is a constant. Since the fluid is assumed incompressible, (1.4.2) applies and the divergence of this equation gives

$$\nabla^2 p = \rho\boldsymbol{\nabla}\cdot\mathbf{f}, \tag{1.4.11}$$

i.e., the Poisson equation. The curl of (1.4.10), on the other hand, gives

$$\frac{\partial}{\partial t}(\boldsymbol{\nabla}\times\mathbf{u}) = \frac{\mu}{\rho}\nabla^2(\boldsymbol{\nabla}\times\mathbf{u}) + \boldsymbol{\nabla}\times\mathbf{f}, \tag{1.4.12}$$

which is a non-homogeneous vector diffusion equation for $\boldsymbol{\nabla}\times\mathbf{u}$, the *vorticity* of the fluid, usually denoted by $\boldsymbol{\omega}$. By (1.1.13), the incompressibility constraint (1.4.2) implies the existence of a vector potential for the velocity field, $\mathbf{u} = \boldsymbol{\nabla}\times\mathbf{A}$, on which one often imposes the subsidiary condition $\boldsymbol{\nabla}\cdot\mathbf{A} = 0$ (cf. end of §1.1). With this,

$$\boldsymbol{\omega} = \boldsymbol{\nabla}\times\mathbf{u} = \boldsymbol{\nabla}\times(\boldsymbol{\nabla}\times\mathbf{A}) = \boldsymbol{\nabla}(\boldsymbol{\nabla}\cdot\mathbf{A}) - \nabla^2\mathbf{A} = -\nabla^2\mathbf{A}. \tag{1.4.13}$$

For steady conditions and an irrotational force field, substitution into (1.4.12) gives the biharmonic equation (1.2.4) for $\mathbf{A}$.

A rather extreme case of viscosity-dominated flow takes place in a porous medium where the porosity (i.e., the fraction of the volume available to the fluid flow) is small, wall effects very strong and velocities correspondingly low. An empirical relation governing the flow in these conditions is *Darcy's law*, according to which

$$\frac{\mu}{\kappa}\mathbf{u} = -\boldsymbol{\nabla}p + \rho\mathbf{f}, \tag{1.4.14}$$

in which κ is the permeability of the medium and $\mathbf{u}$, p and ρ are to be interpreted as volume- or ensemble-averaged fields. If μ/κ is a constant, and the average flow incompressible, the divergence of this relation gives yet another instance of the Poisson equation.

1.5 Electromagnetism

In the Gauss system of units Maxwell's equations are (see e.g. Jackson 1998, chapter 6)

$$\boldsymbol{\nabla}\cdot\mathbf{E} = \frac{4\pi}{\varepsilon}\rho, \qquad \boldsymbol{\nabla}\times\mathbf{H} \quad \frac{\varepsilon}{c}\frac{\partial\mathbf{E}}{\partial t} = \frac{4\pi}{c}\mathbf{J}, \tag{1.5.1}$$

$$\mathbf{\nabla}\cdot\mathbf{H} = 0, \qquad \mathbf{\nabla}\times\mathbf{E} + \frac{\mu}{c}\frac{\partial\mathbf{H}}{\partial t} = 0. \tag{1.5.2}$$

Here $\mathbf{E}$ and $\mathbf{H}$ are the electric and magnetic fields, ρ and $\mathbf{J}$ are the charge and current densities and c is the speed of light in vacuum; ε and μ are the electric permittivity and magnetic permeability of the medium, which we have assumed to be constant. Combining the time derivative of the first of (1.5.1) with the divergence of the second one, we find the continuity equation for charge which is identical with (1.4.1) with $\rho\mathbf{u}$ replaced by $\mathbf{J}$.

On the basis of the first of (1.5.2), by (1.1.13), we can introduce a vector potential $\mathbf{A}$:

$$\mathbf{H} = \frac{1}{\mu}\mathbf{\nabla}\times\mathbf{A}, \tag{1.5.3}$$

and, after substitution into the second one of (1.5.2), a scalar potential V in view of (1.1.12):

$$\mathbf{E} + \frac{1}{c}\frac{\partial\mathbf{A}}{\partial t} = -\mathbf{\nabla}V. \tag{1.5.4}$$

In terms of these potentials, equations (1.5.1) become

$$\nabla^2 V + \frac{1}{c}\frac{\partial}{\partial t}\mathbf{\nabla}\cdot\mathbf{A} = -\frac{4\pi}{\varepsilon}\rho, \quad \mathbf{\nabla}(\mathbf{\nabla}\cdot\mathbf{A}) - \nabla^2\mathbf{A} + \frac{\varepsilon\mu}{c}\frac{\partial}{\partial t}\left(\mathbf{\nabla}V + \frac{1}{c}\frac{\partial\mathbf{A}}{\partial t}\right) = \frac{4\pi\mu}{c}\mathbf{J}. \tag{1.5.5}$$

As noted in connection with (1.1.13), we are at liberty to impose one gauge constraint on $\mathbf{A}$. The *Lorentz gauge condition* prescribes

$$\frac{\varepsilon\mu}{c}\frac{\partial V}{\partial t} + \mathbf{\nabla}\cdot\mathbf{A} = 0, \tag{1.5.6}$$

with which the two equations become

$$\frac{\partial^2 V}{\partial t^2} - \frac{c^2}{\varepsilon\mu}\nabla^2 V = \frac{4\pi c^2}{\varepsilon^2\mu}\rho, \quad \frac{\partial^2\mathbf{A}}{\partial t^2} - \frac{c^2}{\varepsilon\mu}\nabla^2\mathbf{A} = \frac{4\pi c}{\varepsilon}\mathbf{J}, \tag{1.5.7}$$

i.e., uncoupled non-homogeneous wave equations with speed of propagation $c/\sqrt{\varepsilon\mu}$. The *Coulomb gauge* consists in prescribing $\mathbf{\nabla}\cdot\mathbf{A} = 0$, with which (1.5.5) become

$$\nabla^2 V = -\frac{4\pi}{\varepsilon}\rho, \quad \frac{\partial^2\mathbf{A}}{\partial t^2} - \frac{c^2}{\varepsilon\mu}\nabla^2\mathbf{A} = \frac{4\pi c}{\varepsilon}\mathbf{J} - c\frac{\partial}{\partial t}\mathbf{\nabla}V. \tag{1.5.8}$$

For *electrostatic problems* there is no magnetic field and time derivatives vanish so that, with both gauges, or directly from the second one of (1.5.2) and the first one of (1.5.1), we have the Poisson equation $\nabla^2 V = -4\pi\rho/\varepsilon$. In electrostatics conductors have a uniform potential and the problem then is to determine the spatial distribution of the potential given the potential of each conductor, i.e., Dirichlet-type boundary conditions. In this case, the normal component of the field, $\mathbf{n}\cdot\mathbf{\nabla}V$, is proportional to the local charge density at the surface of the conductor. At the interface between two dielectrics the normal electric displacement field $\mathbf{D} = \varepsilon\mathbf{E}$ is continuous, so that $\varepsilon_1\mathbf{n}\cdot\mathbf{\nabla}V_1 = \varepsilon_1\mathbf{n}\cdot\mathbf{\nabla}V_2$. If $\varepsilon_1 \gg \varepsilon_2$, we have approximately $\mathbf{n}\cdot\mathbf{\nabla}V_1 = 0$, which is a Neumann condition.

For *magnetostatic problems* $\mathbf{E} = 0$ and the vector potential satisfies the vector Poisson equation $\nabla^2\mathbf{A} = -(4\pi\mu/c)\mathbf{J}$. At the interface between two magnetic materials, in the absence of surface currents, the tangential component of $\mathbf{H}$ and the normal component of the magnetic induction $\mathbf{B} = \mathbf{H}/\mu$ are continuous; the latter condition is $\mathbf{n} \cdot \mathbf{H}_1/\mu_1 = \mathbf{n} \cdot \mathbf{H}_2/\mu_2$. In this case the nature of the relation between the potential and the physical field leads to less straightforward boundary conditions in general. However, in a region of space where $\mathbf{J} = 0$, the second one of (1.5.1) permits the introduction of a magnetic scalar potential Φ in terms of which $\mathbf{H} = -\nabla\Phi$. In this case, if $\mu_1 \ll \mu_2$ we have, approximately, the Neumann condition $\mathbf{n} \cdot \nabla\Phi_1 = 0$. For other aspects of this situation see e.g. Jackson 1998, chapter 5.

The Lorentz gauge condition (1.5.6) does not entirely remove the arbitrariness inherent in the potentials as the transformation

$$\mathbf{A} = \mathbf{A}' + \nabla\Lambda, \qquad V = V' - \frac{1}{c}\frac{\partial\Lambda}{\partial t}, \tag{1.5.9}$$

with Λ an arbitrary solution of the homogeneous wave equation, preserves both the gauge condition and the form of the physical fields as given by (1.5.7) as is readily checked. Using this freedom, we can show that the Lorentz gauge can be reduced to the Coulomb gauge when there are no free charges, $\rho = 0$. If we set

$$\mathbf{A} = \mathbf{A}' - c\nabla \int^t V(\mathbf{x}, \tau)\,d\tau, \tag{1.5.10}$$

substitution into (1.5.6) and use of the wave equation satisfied by V shows that $\partial(\nabla \cdot \mathbf{A}')/\partial t = 0$ so that $\nabla \cdot \mathbf{A}'$ is independent of time and can be taken to vanish. Furthermore (1.5.4) shows that $\mathbf{E} = -(1/c)\partial\mathbf{A}'/\partial t$. In other words, both fields can be described solely in terms of $\mathbf{A}'$ and we can take $V = 0$ identically. The remaining three components of the vector potential $\mathbf{A}'$ can be taken to satisfy the Coulomb gauge $\nabla \cdot \mathbf{A}' = 0$. This choice is particularly advantageous in the case of electromagnetic waves in free space and of magnetic fields inside conductors, where there can be no free charges, although there is a current given by $\mathbf{J} = \sigma\mathbf{E} = -(\sigma/c)\partial\mathbf{A}'/\partial t$, with σ the electrical conductivity of the material. With this relation the second one of (1.5.8) gives the *telegrapher equation*

$$\frac{\partial^2\mathbf{A}}{\partial t^2} + \frac{4\pi\sigma}{\varepsilon}\frac{\partial\mathbf{A}}{\partial t} - \frac{c^2}{\varepsilon\mu}\nabla^2\mathbf{A} = 0. \tag{1.5.11}$$

1.6 Linear elasticity

Similarly to (1.4.3), the momentum equation of linear elasticity is

$$\rho\frac{\partial^2\mathbf{u}}{\partial t^2} = \nabla \cdot \boldsymbol{\sigma} + \mathbf{f}, \tag{1.6.1}$$

in which $\boldsymbol{\sigma}$ is the stress tensor, assumed to vanish when all the particles of the medium are in their equilibrium position, and $\mathbf{u}$ is the displacement of a particle from this equilibrium position (rather than the particle velocity as in §1.4). In a homogeneous isotropic linear elastic medium the constitutive relations connecting stress and deformation are

$$\sigma_{ij} = 2\mu e_{ij} + \lambda(\nabla \cdot \mathbf{u})\delta_{ij}, \qquad e_{ij} = \frac{1}{2\mu}\left[\sigma_{ij} - \frac{\lambda}{2\mu + 3\lambda}(\mathrm{Tr}\,\boldsymbol{\sigma})\delta_{ij}\right], \tag{1.6.2}$$

in which $\text{Tr}\,\boldsymbol{\sigma} = \sigma_{11} + \sigma_{22} + \sigma_{33}$ is the trace of $\boldsymbol{\sigma}$, $e_{ij} = \frac{1}{2}\left(\partial u_i/\partial x_j + \partial u_j/\partial x_i\right)$ is the deformation tensor, and λ and μ are the *Lamé coefficients*, which we take to be constants.

Substitution of (1.6.2) into (1.6.1) gives

$$\rho\frac{\partial^2 \mathbf{u}}{\partial t^2} = -\mu\boldsymbol{\nabla}\times\boldsymbol{\nabla}\times\mathbf{u} + (\lambda + 2\mu)\boldsymbol{\nabla}(\boldsymbol{\nabla}\cdot\mathbf{u}) + \mathbf{f}. \qquad (1.6.3)$$

By taking the divergence and the curl of this expression we immediately see the possibility of two types of waves. In the first case we have the scalar wave equation

$$\frac{\partial^2}{\partial t^2}(\boldsymbol{\nabla}\cdot\mathbf{u}) = \frac{\lambda + 2\mu}{\rho}\nabla^2(\boldsymbol{\nabla}\cdot\mathbf{u}) + \boldsymbol{\nabla}\cdot\mathbf{f}. \qquad (1.6.4)$$

These *longitudinal waves* propagate with the speed $c_L^2 = (\lambda + 2\mu)/\rho$. The curl of (1.6.3) gives the vector wave equation

$$\frac{\partial^2}{\partial t^2}(\boldsymbol{\nabla}\times\mathbf{u}) = -\frac{\mu}{\rho}\boldsymbol{\nabla}\times\boldsymbol{\nabla}\times(\boldsymbol{\nabla}\times\mathbf{u}) + \boldsymbol{\nabla}\times\mathbf{f}, \qquad (1.6.5)$$

so that these *transverse waves* have the slower speed $c_T^2 = \mu/\rho < c_L^2$.

For static conditions in the absence of forces the governing equation is simply $\boldsymbol{\nabla}\cdot\boldsymbol{\sigma} = 0$. There are several representations of the stress, or of the strain, which simplify the solution of the equation in particular conditions (see e.g. Gurtin 1984, p. 54; Barber 1992; Hetnarski & Ignaczak 2004, p. 107). For example, for a body in a state of plane stress in the (x, y) plane, $\boldsymbol{\sigma}\cdot\mathbf{n} = 0$ if $\mathbf{n}$ is normal to the plane and the remaining stress components can be expressed in terms of the *Airy stress function* φ:

$$\sigma_{xx} = \frac{\partial^2 \varphi}{\partial y^2}, \qquad \sigma_{yy} = \frac{\partial^2 \varphi}{\partial x^2}, \qquad \sigma_{xy} = -\frac{\partial^2 \varphi}{\partial x\partial y}. \qquad (1.6.6)$$

In the presence of conservative forces derived from a potential V, the stress function satisfies the equation

$$\nabla^4\varphi = -(1 - \nu)\,\nabla^2 V, \qquad (1.6.7)$$

with $\nu = \lambda/2(\lambda + \mu)$ the Poisson ratio; this becomes the biharmonic equation (1.2.4) in the absence of body forces.

Other static solutions can be generated by means of the *strain potential*

$$2\mu\mathbf{u} = \boldsymbol{\nabla}\phi, \qquad (1.6.8)$$

which satisfies a Poisson equation $\nabla^2\phi = C$, with C an arbitrary constant, of the *Galerkin vector* $\mathbf{F}$

$$2\mu\mathbf{u} = 2(1 - \nu)\nabla^2\mathbf{F} - \boldsymbol{\nabla}(\boldsymbol{\nabla}\cdot\mathbf{F}), \qquad (1.6.9)$$

which satisfies the vector biharmonic equation $\nabla^4\mathbf{F} = 0$, and others (see e.g. Barber 1992).

The momentum equation (1.6.3) takes special forms in the case of plates or beams, i.e., when one or two dimensions of the elastic body are much smaller than the remaining ones and much smaller than other relevant length scales, such as the wavelength of elastic waves. In this case one can average the momentum equation (1.6.1) over the cross section of the body and, for transverse motion, the variable $\mathbf{u}$ becomes the displacement of a reference surface, or line, in the direction normal to its equilibrium position. The cross-sectional averages of the stresses can be expressed in terms of this variable and one finds, for the transverse vibrations of a beam (see e.g. Landau & Lifshitz 1987b; Gurtin 1984; Hetnarski & Ignaczak 2004),

$$\frac{\partial^2 u}{\partial t^2} + \frac{EI}{\rho S}\frac{\partial^4 u}{\partial x^4} = 0, \tag{1.6.10}$$

in which $E = (3\lambda + 2\mu)\mu/(\lambda + \mu)$ is Young's modulus, I the moment of inertia of the cross section and S its area. The waves are now dispersive as is immediately apparent by assuming a solution proportional to $e^{(ikx-\omega t)}$, which can only exist if $c(\omega) = \omega/k = (EI/\rho S)^{1/4}\omega$. Similarly, for a plate,

$$\frac{\partial^2 u}{\partial t^2} + \frac{h^2}{\rho}\frac{(\lambda+\mu)\mu}{3(3\lambda+2\mu)}\nabla_2^2 u = 0, \tag{1.6.11}$$

in which h is the thickness of the plate and ∇_2^2 the two-dimensional Laplacian in the plane the plate occupies when in equilibrium. Typical boundary conditions for problems of this type are specified in §3.9 p. 80.

The limit of an extremely thin rod or plate is a string, or membrane, respectively. In this case we must relinquish the assumption of zero stress in the equilibrium state as, otherwise, there would be no resistance to small deformation. We thus assume that, at equilibrium, the two sides of an ideal cut across a membrane would pull on each other with a tension T per unit length and find that the transverse vibrations of the membrane are governed by the scalar two-dimensional wave equation with wave speed $c^2 = T/m$, in which m is the mass per unit area of the membrane. Similarly, for a string, one finds the one-dimensional wave equation with the wave speed $c^2 = T/m$, except that now m is the mass per unit length and T has the dimensions of a force, rather than force per unit length.

1.7 Quantum mechanics

The fundamental equation of non-relativistic quantum mechanics for a single particle of mass m is the Schrödinger equation (see e.g. Landau & Lifshitz 1981; Sakurai 1994):

$$i\hbar\frac{\partial\psi}{\partial t} = -\frac{\hbar^2}{2m}\nabla^2\psi + V(\mathbf{x})\psi, \tag{1.7.1}$$

in which $\hbar$ is Planck's constant divided by 2π, V is the potential of the external forces and ψ is the wave function. For stationary states (i.e., steady conditions), ψ has the exponential time dependence $\psi(\mathbf{x}, t) = u(\mathbf{x})e^{-iEt/\hbar}$ and the equation becomes

$$-\frac{\hbar^2}{2m}\nabla^2 u + V(\mathbf{x})u = Eu, \tag{1.7.2}$$

in which the particle energy E can be given (e.g., in a scattering problem), or be an eigenvalue of the operator in the left-hand side. For a free particle $V = $ const. and this reduces to the Helmholtz equation (1.2.3).

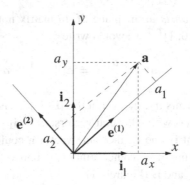

Fig. 1.1   Unit vectors in the plane.

Generally speaking, quantum mechanical problems can be attacked by the methods described in this book which, actually, have received a great impulse from these applications. However, in many cases, the detailed solution of quantum mechanical problems requires specialized techniques which are more efficiently left to the specific literature (see e.g. Landau & Lifshitz 1981; Sakurai 1994). Accordingly, we shall only consider a few examples in §3.8 and §4.3.

1.8 Eigenfunction expansions: a simple-minded introduction

The point of view that informs this book is that *functions are vectors in a suitable space*, and operations carried out on them, such as differentiation and integration, are analogous to the stretching and rotation that a matrix multiplying an ordinary vector imparts to it. These words are not to be understood metaphorically, but in a very literal sense. The practical usefulness of the idea lies in the fact that, once it is fully appreciated, the geometric intuition that is available in ordinary two- or three-dimensional spaces can be carried over to confer a great deal of transparency to the mathematical manipulations used to solve the problems of Part I of this book. Here we explain the idea by analogy with finite-dimensional spaces: we start with two dimensions, the plane, go on to N dimensions and then, heuristically, to infinite dimensions. These considerations are amplified in a file that can be downloaded from the book's web site cambridge.org/prosperetti. A proper, if synthetic, mathematical treatment is offered in Chapters 19 and 21.

1.8.1 Two dimensions

In an ordinary two-dimensional space we deal with vectors $\mathbf{a} = a_x \mathbf{i}_1 + a_y \mathbf{i}_2$ where $\mathbf{i}_1$ and $\mathbf{i}_2$ are unit orthogonal vectors in the plane directed along the coordinate axes (Figure 1.1). Since linear combinations of these vectors enable us to express any vector, we say that they constitute a *basis* in the plane. If, in matrix notation, we represent these unit vectors as $[1, 0]^T$ and $[0, 1]^T$,[7] we would write

$$\mathbf{a} = \begin{vmatrix} a_x \\ a_y \end{vmatrix} = a_x \begin{vmatrix} 1 \\ 0 \end{vmatrix} + a_y \begin{vmatrix} 0 \\ 1 \end{vmatrix} = a_x \mathbf{i}_1 + a_y \mathbf{i}_2. \tag{1.8.1}$$

The components a_x, a_y of $\mathbf{a}$ in the $\{\mathbf{i}_1, \mathbf{i}_2\}$ basis are found by projecting on – i.e., taking scalar products with – the unit vectors: $a_x = \mathbf{i}_1 \cdot \mathbf{a}$, $a_y = \mathbf{i}_2 \cdot \mathbf{a}$.

In place of $\mathbf{i}_1$ and $\mathbf{i}_2$, the same vector $\mathbf{a}$ could equally well be represented by using a basis consisting of two other orthogonal unit vectors $\mathbf{e}^{(1)}$ and $\mathbf{e}^{(2)}$ obtained by an arbitrary rotation of $\mathbf{i}_1$ and $\mathbf{i}_2$ (Figure 1.1):

$$\mathbf{a} = a_1 \mathbf{e}^{(1)} + a_2 \mathbf{e}^{(2)}. \tag{1.8.2}$$

If such a representation is known, a_x, a_y can be found very readily by another projection, or scalar product, operation:

$$a_x = \mathbf{i}_1 \cdot \mathbf{a} = a_1 (\mathbf{i}_1 \cdot \mathbf{e}^{(1)}) + a_2 (\mathbf{i}_1 \cdot \mathbf{e}^{(2)}), \tag{1.8.3}$$

and similarly for a_y. A first point to note is the simplicity that is afforded by the use of pairs of orthogonal vectors of length 1 as bases in the plane. Clearly there is an infinity of such bases: a particular one may be preferable to others in a specific situation, but simple scalar products, or *projections*, is all that is needed to convert from one representation to another one.

Suppose that the vector $\mathbf{a}$ undergoes a linear transformation to $\mathbf{b} = \mathsf{M}\mathbf{a}$, where M is a 2×2 matrix, and that we are interested in finding $\mathbf{a}$ given $\mathbf{b}$, i.e., in solving the equation $\mathsf{M}\mathbf{a} = \mathbf{b}$. We might form the scalar products $\mathbf{i}_1 \cdot (\mathsf{M}\mathbf{a}) = \mathbf{i}_1 \cdot \mathbf{b} = b_x$, i.e.

$$\mathbf{i}_1 \cdot \mathsf{M} \left(a_x \mathbf{i}_1 + a_y \mathbf{i}_2 \right) = (\mathbf{i}_1 \cdot \mathsf{M}\mathbf{i}_1) \, a_x + (\mathbf{i}_1 \cdot \mathsf{M}\mathbf{i}_2) \, a_y = \mathbf{i}_1 \cdot \mathbf{b}, \tag{1.8.4}$$

and similarly with $\mathbf{i}_2$, and end up with a linear algebraic system:

$$m_{11} a_x + m_{12} a_y = b_x, \qquad m_{21} a_x + m_{22} a_y = b_y, \tag{1.8.5}$$

with $m_{ij} = \mathbf{i}_i \cdot \mathsf{M}\mathbf{i}_j$. The "difficulty" in solving these equations is that the two unknowns are coupled because M applied to $\mathbf{i}_1$, for example, produces a vector which is a "mixture" of $\mathbf{i}_1$ and $\mathbf{i}_2$, i.e., $m_{ij} \neq 0$ for $i \neq j$.

This remark suggests that it might be profitable to use another orthogonal basis of unit vectors $\{\mathbf{e}^{(1)}, \mathbf{e}^{(2)}\}$ to represent the vector $\mathbf{a}$ as in (1.8.2) in the hope that, with the new basis, the coupling would disappear. Once the components a_1 and a_2 in the new basis have been determined, the components a_x, a_y in the original $\{\mathbf{i}_1, \mathbf{i}_2\}$ basis can be recovered by taking projections as in (1.8.3).

From the previous derivation of (1.8.5) we see that the coupling between the equations would be avoided if (1) $\mathsf{M}\mathbf{e}^{(j)}$ were parallel to $\mathbf{e}^{(j)}$, without any component in the other direction, and, (2) if $\mathbf{e}^{(i)} \cdot \mathbf{e}^{(j)} = 0$ for $i \neq j$. The first condition is met by choosing the new

---

[7] The superscript T denotes the transpose, which turns the row vectors $[1, 0]$ and $[0, 1]$ into the corresponding column vectors.

vectors to satisfy the relations

$$\mathsf{M}\mathbf{e}^{(1)} = \lambda_1 \mathbf{e}^{(1)}, \qquad \mathsf{M}\mathbf{e}^{(2)} = \lambda_2 \mathbf{e}^{(2)}, \tag{1.8.6}$$

or, in other words, to be *eigenvectors* of M. After this step, whether the other condition holds depends on the nature of the matrix M. For sufficiently nice matrices (e.g., symmetric) it does hold (§18.7), and we assume here and in the rest of this section that this is the case. Thus we represent $\mathbf{a}$ as in (1.8.2) and substitute into $\mathsf{M}\mathbf{a} = \mathbf{b}$ to find

$$\mathsf{M}\mathbf{a} = \lambda_1 a_1 \mathbf{e}^{(1)} + \lambda_2 a_2 \mathbf{e}^{(2)} = \mathbf{b}. \tag{1.8.7}$$

Upon taking scalar products and using the fact that $\mathbf{e}^{(1)} \cdot \mathbf{e}^{(2)} = 0$, $\mathbf{e}^{(1)} \cdot \mathbf{e}^{(1)} = \mathbf{e}^{(2)} \cdot \mathbf{e}^{(2)} = 1$, we find $\lambda_1 a_1 = b_1, \lambda_2 a_2 = b_2$. The coupling has disappeared from the resulting equations as expected. Of course this method is not efficient for as simple a problem as this one, but the power of the approach that we advocate lies in the fact that this example – simple as it is – already contains in essence the same machinery that will enable us to solve a large number of boundary value problems.

1.8.2 N dimensions

The same general procedure goes through with no modifications in N dimensions. Suppose that we wanted to solve numerically in the interval $0 < x < L$ the differential equation

$$-\frac{d^2 u}{dx^2} = g(x), \qquad u(0) = u(L) = 0. \tag{1.8.8}$$

We would divide the interval $[0, L]$ into $N + 1$ equal parts of length h by means of points $x_0 = 0$, $x_1 = h$, $x_2 = 2h$, $\ldots, x_N = Nh$, $x_{N+1} = L$ and store the corresponding values of $u(x)$ in a vector; in effect, we would adopt a "coordinate representation" by writing

$$\mathbf{u} \equiv \begin{vmatrix} u(x_1) \\ u(x_2) \\ \vdots \\ u(x_N) \end{vmatrix} = u(x_1)\mathbf{i}_1 + u(x_2)\mathbf{i}_2 + \cdots + u(x_N)\mathbf{i}_N, \tag{1.8.9}$$

with $\mathbf{i}_1 = [1, 0, \ldots, 0]^T$, $\mathbf{i}_2 = [0, 1, \ldots, 0]^T$, $\ldots$, $\mathbf{i}_N = [0, 0, \ldots, 1]^T$. Upon using the standard finite-difference approximation for the second derivative (see e.g. Abramowitz & Stegun 1964, p. 877; Isaacson & Keller 1966, p. 288)

$$u''(x_j) \simeq \frac{u_{j+1} - 2u_j + u_{j-1}}{h^2}, \tag{1.8.10}$$

in which $u_j = u(x_j)$, the differential equation approximately becomes the linear algebraic system

$$
\mathbf{M} \begin{vmatrix} u_1 \\ u_2 \\ \vdots \\ u_{N-1} \\ u_N \end{vmatrix} \simeq \begin{vmatrix} g_1 \\ g_2 \\ \vdots \\ g_{N-1} \\ g_N \end{vmatrix} \equiv \mathbf{g}, \tag{1.8.11}
$$

where the operator (namely, the matrix) $\mathbf{M}$ is given by

$$
\mathbf{M} = \frac{1}{h^2} \begin{vmatrix} 2 & -1 & 0 & 0 & 0 & \cdots & 0 \\ -1 & 2 & -1 & 0 & 0 & \cdots & 0 \\ 0 & -1 & 2 & -1 & 0 & \cdots & 0 \\ & & & \cdots & & & \\ 0 & & \cdots & -1 & 2 & -1 & 0 \\ 0 & & \cdots & 0 & -1 & 2 & -1 \\ 0 & & \cdots & 0 & 0 & -1 & 2 \end{vmatrix}. \tag{1.8.12}
$$

Just as in the previous case, it is desirable to eliminate the coupling among the different u_j's by replacing the basis vectors $\{\mathbf{i}_1, \ldots, \mathbf{i}_N\}$ by a more convenient set consisting of the unit eigenvectors $\mathbf{e}^{(k)}$ of $\mathbf{M}$ which satisfy

$$
\mathbf{M}\mathbf{e}^{(k)} = \lambda_k \mathbf{e}^{(k)}, \qquad k = 1, 2, \ldots, N. \tag{1.8.13}
$$

Without worrying about how this problem is actually solved, let us simply state the result, namely[8]

$$
\mathbf{e}^{(k)} = \sqrt{\frac{2}{N+1}} \begin{vmatrix} \sin k\frac{\pi}{N+1} \\ \sin k\frac{2\pi}{N+1} \\ \vdots \\ \sin\left(k\frac{(N-1)\pi}{N+1}\right) \\ \sin\left(k\frac{N\pi}{N+1}\right) \end{vmatrix}. \tag{1.8.14}
$$

For example, if $N = 3$, we have the 3 eigenvectors

$$
\mathbf{e}^{(1)} = \frac{1}{\sqrt{2}} \begin{vmatrix} \sin\frac{\pi}{4} \\ \sin\frac{2\pi}{4} \\ \sin\frac{3\pi}{4} \end{vmatrix}, \quad \mathbf{e}^{(2)} = \frac{1}{\sqrt{2}} \begin{vmatrix} \sin\left(2\frac{\pi}{4}\right) \\ \sin\left(2\frac{2\pi}{4}\right) \\ \sin\left(2\frac{3\pi}{4}\right), \end{vmatrix} \quad \mathbf{e}^{(3)} = \frac{1}{\sqrt{2}} \begin{vmatrix} \sin\left(3\frac{\pi}{4}\right) \\ \sin\left(3\frac{2\pi}{4}\right) \\ \sin\left(3\frac{3\pi}{4}\right) \end{vmatrix}.
$$
$$\tag{1.8.15}$$

---

[8] Here is a quick derivation. Let us drop the superscript k for convenience and denote the components of $\mathbf{e}^{(k)}$ by $e_1, e_2, \ldots, e_N$. The generic equation of the system (1.8.13) is $-e_{j-1} + (2 - h^2\lambda)e_j - e_{j+1} = 0$. This is a finite difference equation with constant coefficients the solution of which can be found by substituting $\exp(ij\nu)$ (i is the imaginary unit) just as in the case of ordinary differential equations with constant coefficients (§5.3). The result is $\lambda = 2(1 - \cos \nu)$ which, for given λ, gives two roots $\pm\nu$. Again just as in the case of ordinary differential equations we set $e_j = \alpha \cos(j\nu) + \beta \sin(j\nu)$. A quick way to take the next step is to think of the end nodes $j = 0$ and $j = N + 1$ where the components of the eigenvectors should vanish. For $j = 0$ we find $e_0 = \alpha = 0$, so that $\alpha = 0$ and $e_j = \beta \sin(j\nu)$. For e_{N+1} to vanish without having the trivial solution it is necessary that $(N + 1)\nu = k\pi$, from which we have $\lambda_k = (2/h^2)[1 - \cos k\pi/(N + 1)]$, which is the same as (1.8.16).

If the eigenvalue equation (1.8.13) is satisfied by $\mathbf{e}^{(k)}$ it is also satisfied by any multiple of it; we have chosen the multiplicative constant $\sqrt{2/(N+1)}$ in such a way that each vector has length 1, namely $\mathbf{e}^{(k)} \cdot \mathbf{e}^{(k)} = 1$. It can be verified that different eigenvectors are orthogonal, so that $\mathbf{e}^{(j)} \cdot \mathbf{e}^{(k)} = 0$ if $j \neq k$, and it may be checked that allowing the index k to take values greater than N would not give different vectors. For example, $\mathbf{e}^{(N+1)} = 0$, $\mathbf{e}^{(N+2)} = -\mathbf{e}^N$, etc. so that, indeed, there are only N independent eigenvectors.

The corresponding eigenvalues are

$$\lambda_k = \frac{4}{h^2} \sin^2 \frac{k\pi}{2(N+1)}. \tag{1.8.16}$$

For $N = 3$, for example, $h^2\lambda_1 = 4 \sin^2 (\pi/8) = 2 - \sqrt{2}$, $h^2\lambda_2 = 4 \sin^2 (2\pi/8) = 2$, $h^2\lambda_3 = 4 \sin^2 (3\pi/8) = 2 + \sqrt{2}$. We note for future reference that, since $\sin \alpha \simeq \alpha$ when α is small, for $k \ll N$ we find

$$\lambda_k \simeq \frac{k^2\pi^2}{h^2(N+1)^2} = \frac{k^2\pi^2}{L^2}. \tag{1.8.17}$$

The next step is to express the vector of unknowns $\mathbf{u}$ as in (1.8.2) rather than as in (1.8.9), namely as

$$\mathbf{u} = \sum_{k=1}^{N} U_k \mathbf{e}^{(k)}. \tag{1.8.18}$$

Substitute this expansion into (1.8.11) and use the eigenvalue property (1.8.13) of the $\mathbf{e}^{(k)}$'s to find

$$\sum_{k=1}^{N} \lambda_k U_k \mathbf{e}^{(k)} = \mathbf{g}. \tag{1.8.19}$$

In the present N-dimensional case the scalar product $\mathbf{a} \cdot \mathbf{b}$ between two vectors $\mathbf{a}$ and $\mathbf{b}$ is defined by

$$\mathbf{a} \cdot \mathbf{b} = \sum_{j=1}^{N} a_j b_j. \tag{1.8.20}$$

Upon taking the scalar products of (1.8.19) in succession with $\mathbf{e}^{(1)}$, $\mathbf{e}^{(2)}$, ..., $\mathbf{e}^{(N)}$ we find

$$\lambda_1 U_1 = \mathbf{e}^{(1)} \cdot \mathbf{g}, \qquad \ldots, \qquad \lambda_N U_N = \mathbf{e}^{(N)} \cdot \mathbf{g}. \tag{1.8.21}$$

With the U_k's determined in this way, (1.8.18) already permits us to reconstruct $\mathbf{u}$ but, if we want a representation in the form (1.8.9), we can easily find its components:

$$u_j = \mathbf{i}_j \cdot \mathbf{u} = \sum_{k=1}^{N} U_k \left(\mathbf{i}_j \cdot \mathbf{e}^{(k)} \right) = \sqrt{\frac{2}{N+1}} \sum_{k=1}^{N} U_k \sin \frac{\pi jk}{N+1} \tag{1.8.22}$$

with, from (1.8.21), $U_k = \mathbf{e}^{(k)} \cdot \mathbf{g}/\lambda_k$.

The reader should observe how, in this derivation, we have followed precisely one by one the steps used for the simple two-dimensional case which, as advertised before, is seen to contain the essence of the method, at least in the finite-dimensional case.

1.8.3 Functions and derivatives

Let us now see how the procedure can be extended when we wish to solve the differential equation (1.8.8) exactly, rather than with a discretized approximation as before. The fact that it is possible to generate numerically an accurate approximation to an analytic solution suggests that it should also be possible to establish a parallel between the techniques used to generate the approximate solution of the previous section and the methods that prove useful to find the exact analytic solution.

Writing $u(x)$ is similar to writing (1.8.9), namely expressing u through its values at points x. The lesson of the previous example is that this may not be the most efficient way to proceed. Rather, we will express u as the superposition

$$u(x) = \sum_k U_k\, e_k(x), \tag{1.8.23}$$

where now the e_k's are the eigenvectors (or eigenfunctions) of the operator that appears in our problem, namely $\mathsf{M} = -\mathrm{d}^2/\mathrm{d}x^2$, rather than of the matrix (1.8.12). Now the eigenvalue equation (1.8.13) is a differential equation:

$$-\frac{\mathrm{d}^2 e_k}{\mathrm{d}x^2} = \lambda_k e_k, \tag{1.8.24}$$

the solution of which is

$$e_k(x) = A_k \cos \sqrt{\lambda_k}\,x + B_k \sin \sqrt{\lambda_k}\,x. \tag{1.8.25}$$

We can easily determine that $A_k = 0$ by noting that, for u as given by (1.8.23) to satisfy $u(0) = 0$, it is necessary that $e_k(0) = 0$ for all k's. The other condition is that $u(L) = 0$, which similarly requires that $e_k(L) = 0$, a requirement which can be met non-trivially only if $\sqrt{\lambda_k}\,L = k\pi$, or

$$\lambda_k = \frac{k^2\pi^2}{L^2}, \tag{1.8.26}$$

just as in (1.8.17). The explanation of the correspondence is that a discretized solution can approximate the exact solution only when there is a sufficiently fine discretization which, in the case of (1.8.16), means $k \ll N$.

In the previous N-dimensional case we had found that the index k could not take values greater than N, but there is no such limitation here: the summation (1.8.23) has an infinite number of terms and, therefore, we should worry at least about convergence issues; we do this in detail in chapters 8 and 9.

In the finite-dimensional case we have seen the major role played by the scalar product. We need to understand what this notion means when, as now, we deal with functions rather than finite-dimensional vectors. In keeping with the heuristic point of view of this section we may proceed as follows. A graphical interpretation of the scalar product (1.8.20) is the area of the rectangles shown in the left diagram of Figure 1.2. When we are interested in a problem defined for $0 < x < L$, this representation is not the best as, if we wish to refine the discretization by increasing the number of points, the interval on the horizontal axis

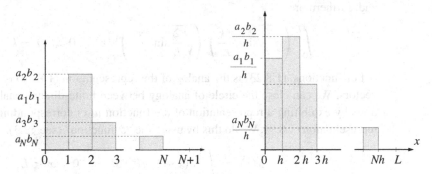

Fig. 1.2 Geometrically, the scalar product $\mathbf{a} \cdot \mathbf{b} = \sum_{j=1}^{N} a_j b_j$ equals the shaded area in the left figure; by rewriting it as $\mathbf{a} \cdot \mathbf{b} = \sum_{j=1}^{N} (a_j / \sqrt{h})(b_j \sqrt{h}) h$ it can equally be represented by the shaded area in the figure on the right.

necessary to draw the rectangles increases also. We can avoid this if we rewrite (1.8.20) as (Figure 1.2, right)

$$\mathbf{a} \cdot \mathbf{b} = \sum_{j=1}^{N} a_j b_j = \sum_{j-1}^{N} \frac{a_j}{\sqrt{h}} \frac{b_j}{\sqrt{h}} h. \tag{1.8.27}$$

Now the total length on the x-axis is constant and equal to L: as we increase the number of points the rectangles become skinnier and higher. Suppose now that

$$\lim_{N \to \infty} \frac{a_j}{\sqrt{h}} = a(x_j), \qquad \lim_{N \to \infty} \frac{b_j}{\sqrt{h}} = b(x_j), \tag{1.8.28}$$

where $a(x)$ and $b(x)$ are ordinary functions. The graphical representation of figure 1.2 suggests that, as $N \to \infty$, the sum of the areas of the rectangles will become an integral (cf. the definition of the Riemann integral on p. 687):

$$\mathbf{a} \cdot \mathbf{b} = \sum_{j=1}^{N} \frac{a_j}{\sqrt{h}} \frac{b_j}{\sqrt{h}} h \to \int_0^L a(x) b(x) \, dx. \tag{1.8.29}$$

For all this to make sense, we must verify that (1.8.28) is true. It is easiest to consider the eigenvectors (1.8.14); for the j-th element of $\mathbf{e}^{(k)}$ we have

$$\frac{e_j^{(k)}}{\sqrt{h}} = \sqrt{\frac{2}{h(N+1)}} \sin \frac{\pi k(jh)}{(N+1)h} = \sqrt{\frac{2}{L}} \sin \frac{\pi k x_j}{L}, \tag{1.8.30}$$

which agrees with our result (1.8.25) if we take $B_k = \sqrt{2/L}$ (recall that $A_k = 0$). As a further check, since the discrete $\mathbf{e}^{(k)}$ has length 1, with this determination of B_k we should find that the eigenfunctions $e_k(x)$ have "length" 1 if the scalar product is defined as in (1.8.29); sure enough

$$\int_0^L \left(\sqrt{\frac{2}{L}} \sin \frac{\pi k}{L} \right)^2 dx = 1 \qquad k = 1, 2, \dots, K, \dots, \tag{1.8.31}$$

and, furthermore,

$$\int_0^L \left(\sqrt{\frac{2}{L}} \sin \frac{\pi k}{L} \right) \left(\sqrt{\frac{2}{L}} \sin \frac{\pi j}{L} \right) dx = 0 \qquad j \neq k. \tag{1.8.32}$$

For functions, (1.8.23) is the analog of the representation (1.8.18) of finite-dimensional vectors. We can close the circle of analogy between finite-dimensional vectors and functions by exhibiting a representation of the function $u(x)$ corresponding to the coordinate representation (1.8.9). We do this by using the "δ-function" (see §2.3):

$$u(x) = \int_0^L \delta(x - \xi) u(\xi) \, d\xi \qquad 0 < x < L. \tag{1.8.33}$$

The "δ-spike" at ξ is in a sense similar to the unit vector $\mathbf{i}_k = [0, \ldots, 1, 0, \ldots, 0]^T$ which has a "spike" in the interval $x_{k-1} < x < x_k$. Of course, this is only a superficial analogy, but it is suggestive nevertheless.

We have set up a close formal correspondence between the procedure for finite-dimensional vectors and matrices, and for functions and operators of a more general nature. It is this framework that will be systematically exploited in many of the applications of Part I of this book. The heuristic arguments given here are further expanded in §2.1 and put on a more solid mathematical basis in Chapters 19 and 21.

Some Simple Preliminaries

This chapter collects in a simplified form some ideas and techniques extensively used in Part I of the book. This material will be revisited in greater detail in later chapters, but the brief summary given here may be helpful to readers who do not have the time or the inclination to tackle the more extensive treatments.

§2.1 continues the considerations of the final section of the previous chapter and further explains the fundamental idea underlying the method of eigenfunction expansions. While this method may be seen as an extension of the elementary "separation of variables" procedure (cf. §3.10), the geometric view advocated here provides a powerful aid to intuition and should greatly facilitate an understanding of "what is really going on" in most of the applications of Part I; the basis of the method is given in some detail in Chapters 19 and 21 of Part III.

§2.2 is a reminder of a useful method to solve linear non-homogeneous ordinary differential equations. Here the solution to some equations that frequently arise in the applications of Part I is derived. A more general way in which this technique may be understood is through its connection with Green's functions. This powerful idea is explained in very simple terms in §2.4 and, in greater detail, in Chapters 15 and 16.

Green's functions make use of the notion of the so-called "δ-function," the principal properties of which are summarized in §2.3. A proper theory for this and other generalized functions is presented in Chapter 20. Finally in §2.5 we briefly explain the power series method for the solution of ordinary differential equations.

2.1 The method of eigenfunction expansions

In (1.8.29) p. 25 of the final section of the previous chapter we introduced the scalar product between two functions. From now on we denote the scalar product of two functions f and g by (f, g) and, slightly generalizing the earlier definition, we define it by

$$(f, g) = \int_a^b \overline{f}(x)g(x)s(x) \, \mathrm{d}x, \tag{2.1.1}$$

where $s(x) > 0$ in $a \leq x \leq b$ is given. Here and throughout this book the overline denotes the complex conjugate and its use follows from the arguments of the previous section applied to the ordinary scalar product between two finite-dimensional complex, rather than real, vectors. The basic motivation is that we want the "length" $(f, f)^{1/2}$ of the "vector" corresponding to f to be a real number. We refer to this length as the *norm* of f and we denote it by $\|f\|$:[1]

$$\|f\| = \sqrt{(f, f)}. \tag{2.1.2}$$

Let us now summarize the framework introduced and motivated in §1.8. We are given a set of eigenfunctions, or *basis vectors*, $e_k(x)$ of a linear differential operator acting over an interval $a < x < b$, e.g. as in (1.8.24).[2] As vectors, these functions have norm ("length") 1 in the sense that

$$(e_k, e_k) = \int_a^b |e_k(x)|^2 s(x)\, dx \equiv \|e_k\|^2 = 1. \tag{2.1.3}$$

We assume also that the e_k are orthogonal to each other according to the scalar product (2.1.1) so that, concisely,

$$(e_j, e_k) = \int_a^b \bar{e}_j(x) e_k(x) s(x)\, dx = \delta_{jk}, \tag{2.1.4}$$

where the Kronecker symbol δ_{jk} equals 1 if $j = k$ and 0 otherwise. As before, this orthogonality property cannot be enforced, but will depend on the operator of which the e_k are eigenfunctions; the eigenfunctions of a very large number of practically important linear operators enjoy this property (see e.g. §15.2 and §21.2).

In the cases that we will consider, a function $f(x)$ can be expressed as an infinite superposition of these eigenfunctions

$$f(x) = \sum_{k=0}^{\infty} f_k e_k(x), \tag{2.1.5}$$

in which the constants f_k, which can be interpreted as "components" of the "vector" f along the "direction" of e_k, are given by[3]

$$f_k = (e_k, f) = \int_a^b \bar{e}_k(x) f(x) s(x)\, dx. \tag{2.1.6}$$

It is evident that two different functions $f_1(x)$ and $f_2(x)$ will have different coefficients.[4] The norm of f can be readily expressed in terms of these coefficients by taking the scalar product and exchanging summation and integration (where legitimate):

$$
\begin{aligned}
(f, f) &= \int_a^b |f|^2 s(x)\, dx = \int_a^b \overline{\left(\sum_{j=0}^{\infty} f_j e_j(x) \right)} \left(\sum_{k=0}^{\infty} f_k e_k(x) \right) s(x)\, dx \\
&= \sum_{j=0}^{\infty} \sum_{k=0}^{\infty} \bar{f}_j f_k \int_a^b \bar{e}_j(x) e_k(x) s(x)\, dx = \sum_{k=0}^{\infty} |f_k|^2,
\end{aligned}
\tag{2.1.7}
$$

---

[1] This notion is generalized and treated in detail in §19.2.

[2] Linearity of an operator L consists in the property that, if $u_{1,2}(x)$ are two functions and $a_{1,2}$ are two constants, then $\mathsf{L}(a_1 u_1 + a_2 u_2) = a_1 \mathsf{L} u_1 + a_2 \mathsf{L} u_2$.

[3] The actual meaning of the equality (2.1.5) should be made more precise e.g. when f is discontinuous, nor is it true that the integral (2.1.6) exists for any f; these issues are treated in detail for the case of the Fourier series in Chapter 9 and, in some greater generality, in Chapter 15.

[4] If, that is, the two functions differ on a set of positive measure; see p. 546 and §A.2.

in which the last step follows from (2.1.4). This is known as *Parseval's equality* and bears a striking resemblance to the way in which the length of an ordinary vector is calculated from its components along mutually orthogonal axes. Here and in much of what follows we have proceeded formally without justifying the term-by-term multiplication of the two series or the exchange of the order of integration and summation; these aspects are addressed later.

Suppose now that we deal with a function of two variables, e.g. $u = u(x, t)$. We can argue as follows: If t_1 and t_2 are two different values of t, regarded as function of x, $u(x, t_1)$ is a different function from $u(x, t_2)$. Therefore, the two functions $u(x, t_1)$ and $u(x, t_2)$ will have different coefficients u_k or, equivalently, the coefficients u_k will depend on t:

$$u_k(t) = \int_a^b \bar{e}_k(x) u(x, t) s(x)\, dx \qquad u(x, t) = \sum_{k=0}^{\infty} u_k(t) e_k(x). \qquad (2.1.8)$$

This remark is used in the following way. Suppose that we wish to solve

$$\frac{\partial u}{\partial t} - \frac{\partial^2 u}{\partial x^2} = g(x, t) \qquad (2.1.9)$$

in which $g(x, t)$ is given and, at $t = 0$, u satisfies the initial condition $u(x, 0) = U_0(x)$; there will be boundary conditions that we do not need to specify for the present purposes. Suppose also that

$$-\frac{d^2 e_k}{dx^2} = \lambda_k e_k. \qquad (2.1.10)$$

We take the expansion (2.1.8) for u and substitute it into (2.1.9), interchange the orders of summation and differentiation[5] and use (2.1.10) to find

$$\sum_{k=0}^{\infty} \left(\frac{du_k}{dt} + \lambda_k u_k \right) e_k(x) = g(x, t). \qquad (2.1.11)$$

Now we take the scalar product of this equation with the e_j taking in succession $j = 1, 2, \ldots$ to find, after interchanging summation and integration,

$$\sum_{k=0}^{\infty} \left(\frac{du_k}{dt} + \lambda_k u_k \right) \int_a^b \bar{e}_j(x)\, e_k(x) s(x)\, dx = \int_a^b \bar{e}_j(x)\, g(x, t) s(x)\, dx.$$

All the integrals in the left-hand side vanish except the one with $j = k$ which equals 1. The result of the x-integration in the right-hand side is a function of t only, which we may denote by $g_j(t)$. Thus, we have reduced the partial differential equation (2.1.9) to the set of uncoupled ordinary differential equations

$$\frac{du_j}{dt} + \lambda_j u_j = \int_a^b \bar{e}_j(x)\, g(x, t) s(x)\, dx \equiv g_j(t), \qquad (2.1.12)$$

[5] Again, our procedure is purely formal. It is by no means obvious that this interchange or the subsequent manipulations are legitimate (see e.g. pp. 227 and 227). As shown in §3.1, in some cases it may be preferable to take the scalar product of (2.1.9) by e_k and integrate by parts to cause the coefficients u_k to appear without having to differentiate term by term.

which are readily solved (see (2.2.31) p. 34):

$$u_j(t) = u_j(0)e^{-\lambda_j t} + \int_0^t e^{-\lambda_j(t-\tau)} g_j(\tau) \, d\tau. \tag{2.1.13}$$

The initial condition $u_j(0)$ is calculated from the integral in (2.1.8) by replacing $u(x, t)$ by the initial data $U_0(x)$ so that $u_j(0) = (e_j, U_0)$.

The reader familiar with the method of separation of variables will realize that, while the two approaches would be equivalent if $g(x, t) = 0$, strict separation of variables would be inapplicable if $g \neq 0$ although, of course, procedures to deal with this case can be developed. Furthermore, the point worth stressing again is that the geometric view afforded by the present approach confers a much greater transparency to the various manipulations.

In place of (2.1.8), in some cases it is more convenient to use an expansion of u of the form

$$u(x, t) = \sum_{k=1}^{\infty} c_k(t) v_k(x), \tag{2.1.14}$$

in which the v_k are eigenfunctions of the operator, but do not have "length" 1,[6]

$$\|v_k\|^2 = \int_a^b |v_k(x)|^2 s(x) \, dx \neq 1. \tag{2.1.15}$$

In this case the expression (2.1.8) for the coefficients of the expansion is slightly modified to

$$c_k(t) = \frac{1}{\|v_k\|^2} \int_a^b \overline{v}_k(x) u(x, t) s(x) \, dx. \tag{2.1.16}$$

By noting that $e_k(x) = v_k(x)/\|v_k(x)\|$ has norm 1, the equivalence between the two representations is immediate.

The solution procedure of all the problems of Chapter 3 and of most of the problems of Chapters 6 and 7 follows this pattern. The underlying theory for eigenfunctions of ordinary differential operators is covered in Chapter 15 and the general theory in Chapters 19 and 21.

A comment on the relation of this approach to the method of separation of variables will be found in §3.10.

2.2 Variation of parameters

In many of the problems that we deal with we are faced with linear non-homogeneous second-order ordinary differential equations of the form

$$u'' + a(x)u'(x) + b(x)u = f(x), \tag{2.2.1}$$

in which $u' = du/dx$, $u'' = d^2u/dx^2$ and the functions $a(x)$, $b(x)$ and $f(x)$ are assumed given and such that the problem is solvable. Let $v_1(x)$ and $v_2(x)$ be any two linearly independent solutions of the homogeneous equation, so that

$$v_j'' + a(x)v_j'(x) + b(x)v_j = 0, \qquad j = 1, 2. \tag{2.2.2}$$

---

[6] Eigenvectors and eigenfunctions are defined only up to an arbitrary multiplicative constant; see e.g. p. 23 and §15.2 and §18.6.

The solution of (2.2.1) will be required to satisfy suitable boundary conditions, but no specific boundary conditions need to be imposed on v_1 and v_2.

We seek the solution of (2.2.1) in the form

$$u(x) = A_1(x)v_1(x) + A_2(x)v_2(x) \qquad (2.2.3)$$

where the "parameters" (in reality, functions) A_1 and A_2 are to be suitably determined. We note at the outset that, since we have replaced a single unknown function, $u(x)$, by two unknown functions, A_1 and A_2, the problem has become under-determined and we will be at liberty to impose additional constraints that might simplify the solution procedure. Upon differentiating (2.2.3) once we have

$$u'(x) = A_1(x)v_1'(x) + A_2(x)v_2'(x) + \left[A_1'(x)v_1(x) + A_2'(x)v_2(x) \right] \qquad (2.2.4)$$

and, differentiating once more,

$$\begin{aligned} u''(x) &= A_1(x)v_1''(x) + A_2(x)v_2''(x) + A_1'(x)v_1'(x) + A_2'(x)v_2'(x) \\ &\quad + \frac{d}{dx} \left[A_1'(x)v_1(x) + A_2'(x)v_2(x) \right] . \end{aligned} \qquad (2.2.5)$$

Upon substitution into the original equation (2.2.1), some terms reproduce the homogeneous equation (2.2.2) multiplied by A_1 and A_2 and therefore vanish. What remains is

$$\begin{aligned} A_1'(x)v_1'(x) + A_2'(x)v_2'(x) &+ \frac{d}{dx} \left[A_1'(x)v_1(x) + A_2'(x)v_2(x) \right] \\ &+ a(x) \left[A_1'(x)v_1(x) + A_2'(x)v_2(x) \right] = f(x). \end{aligned} \qquad (2.2.6)$$

The indicated differentiation of A_1' and A_2' would introduce derivatives of second order and therefore would not result in any progress with respect to the original problem (2.2.1). However, since we can impose additional constraints on A_1 and A_2, we can avoid this difficulty by requiring satisfaction of the subsidiary condition

$$A_1'(x)v_1(x) + A_2'(x)v_2(x) = 0, \qquad (2.2.7)$$

with which (2.2.6) reduces to

$$A_1'(x)v_1'(x) + A_2'(x)v_2'(x) = f(x). \qquad (2.2.8)$$

The two relations (2.2.7) and (2.2.8) form an algebraic system for A_1' and A_2' which is readily solved to find

$$A_1'(x) = -\frac{v_2(x)}{W(v_1, v_2; x)} f(x), \qquad A_2'(x) = \frac{v_1(x)}{W(v_1, v_2; x)} f(x), \qquad (2.2.9)$$

where

$$W(v_1, v_2; x) = \begin{vmatrix} v_1(x) & v_2(x) \\ v_1'(x) & v_2'(x) \end{vmatrix} = v_1(x)v_2'(x) - v_1'(x)v_2(x) \qquad (2.2.10)$$

is the *Wronskian* of the two solutions of the homogeneous equation v_1 and v_2. These two equations are readily integrated with a result that we write in the form

$$A_1 = \alpha_1 - \int^x \frac{v_2(\xi)}{W(v_1, v_2; \xi)} f(\xi) \, d\xi, \qquad (2.2.11)$$

$$A_2 = \alpha_2 + \int^x \frac{v_1(\xi)}{W(v_1, v_2; \xi)} f(\xi) \, d\xi, \tag{2.2.12}$$

in which α_1 and α_2 are constants. In principle this form is redundant as these constants can be adjusted by adjusting the lower limits of integration, which we have left unspecified. Nevertheless, when it comes to imposing boundary conditions on u, it is sometimes convenient to write the solution in this form.

Upon substitution of (2.2.11) and (2.2.12) into (2.2.3) we may write the solution of (2.2.1) as

$$u(x) = \left[\alpha_1 - \int^x \frac{v_2(\xi)}{W(v_1, v_2; \xi)} f(\xi) \, d\xi \right] v_1(x)$$

$$+ \left[\alpha_2 + \int^x \frac{v_1(\xi)}{W(v_1, v_2; \xi)} f(\xi) \, d\xi \right] v_2(x). \tag{2.2.13}$$

It is evident that (2.2.13) has the familiar form of the sum of the *the general solution* u_h of the homogeneous equation,

$$u_h = \alpha_1 v_1(x) + \alpha_2 v_2(x), \tag{2.2.14}$$

and of *one particular solution* u_p of the non-homogeneous equation:

$$u_p(x) = \int^x \frac{v_1(\xi)v_2(x) - v_1(x)v_2(\xi)}{W(v_1, v_2; \xi)} f(\xi) \, d\xi. \tag{2.2.15}$$

In §2.4 and, more fully, in Chapter 15, we will see the relation between this form of the solution and the Green's function for the problem (2.2.1).

Among the many properties of linear equations that make their solution particularly simple is the fact that, as Abel showed, the Wronskian can be conveniently expressed in terms of the coefficient $a(x)$ of the differential equation written in the form (2.2.1). Indeed, a simple direct calculation shows that

$$\frac{dW}{dx} = -a(x) \, W, \tag{2.2.16}$$

so that

$$W(v_1, v_2; x) = W_0 \exp\left(- \int^x a(\xi) \, d\xi \right), \tag{2.2.17}$$

in which the dependence on the two particular solutions v_1 and v_2 is only through the constant W_0. As a simple example, for the equation

$$v'' + k^2 v = 0, \tag{2.2.18}$$

we expect the Wronskian of any two solutions to be a constant, as $a = 0$. If we were to take $v_1 = \cos kx$, $v_2 = \sin kx$, we would find $W_0 = k$; if we were to take $v_1 = \cos kx$, $v_2 = \exp ikx$, $W_0 = ik$, etc.

Example 2.2.1 As an example of the application of (2.2.13), let us consider an equation that arises in Chapter 7 in the solution of problems in spherical coordinates. Since here the independent variable is the radial distance from the origin, we write the equation in terms of a variable r rather than x:

$$u''(r) + \frac{2}{r}u'(r) - \frac{\ell(\ell+1)}{r^2}u(r) = f(r) \tag{2.2.19}$$

with $\ell \geq 0$. It is obvious by inspection[7] that the homogeneous solutions are powers of r; upon setting $v \propto r^\lambda$ one readily finds the two solutions $v_1 = r^\ell$ and $v_2 = r^{-\ell-1}$, from which $W = -(2\ell+1)/r^2$. The formula (2.2.13) gives then

$$u(r) = \left[\alpha_1 + \frac{1}{2\ell+1}\int^r s^{-\ell+1}f(s)\,ds\right]r^\ell$$
$$+ \left[\alpha_2 - \frac{1}{2\ell+1}\int^r s^{\ell+2}f(s)\,ds\right]r^{-\ell-1}. \tag{2.2.20}$$

If, for example, the solution is required to be regular at $r = 0$, we must ensure that the quantity multiplying $r^{-\ell-1}$ vanish for $r = 0$. This can be accomplished by taking $\alpha_2 = 0$ and extending the second integral from 0 to r so that[8]

$$u(r) = \left[\alpha_1 + \frac{1}{2\ell+1}\int^r s^{-\ell+1}f(s)\,ds\right]r^\ell - \frac{r^{-\ell-1}}{2\ell+1}\int_0^r s^{\ell+2}f(s)\,ds. \tag{2.2.21}$$

If the interval of interest is $0 < r < R$ and the boundary condition at $r = R$ specifies e.g. that $u(r = R) = u_0$, it may prove convenient to take R as the integration limit in the remaining integral finding

$$u_0 = \alpha_1 R^\ell - \frac{R^{-\ell-1}}{2\ell+1}\int_0^R s^{\ell+2}f(s)\,ds, \tag{2.2.22}$$

which determines α_1. In another problem the solution may be required to vanish at infinity. We would then take $\alpha_1 = 0$ and extend the first integral in (2.2.20) from r to ∞:

$$u(r) = -\frac{r^\ell}{2\ell+1}\int_r^\infty s^{-\ell+1}f(s)\,ds + \left[\alpha_2 - \frac{1}{2\ell+1}\int^r s^{\ell+2}f(s)\,ds\right]r^{-\ell-1}. \tag{2.2.23}$$

The other constant, and the remaining integration limit, are determined as before from a boundary condition e.g. at $r = R$. If the domain of interest is a finite interval $0 < R < r < S$ bounded away from 0, in principle neither term exhibits singularities; the constants and the integration limits are determined by imposing the given boundary conditions at $r = R$ and $r = S$.   □

Example 2.2.2 In several examples of the following chapters we encounter the equation

$$u'' - h^2 u = -f(x). \tag{2.2.24}$$

---

[7] This is an equation of the Euler type and the solutions shown follow directly from the standard method by which this class of equations is solved; see e.g. Ince (1926), p. 141; Bender & Orszag (2005), p. 12 and many others.
[8] Provided, of course, that f is such that the integrand vanishes sufficiently quickly as $r \to 0$.

In an unbounded region it is often convenient to take $v_1 = e^{-hx}$ and $v_2 = e^{hx}$, with which $W = 2h$ and

$$u(x) = \left(\alpha_1 + \frac{1}{2h} \int^x e^{h\xi} f(\xi) \, d\xi \right) e^{-hx} + \left(\alpha_2 - \frac{1}{2h} \int^x e^{-h\xi} f(\xi) \, d\xi \right) e^{hx}.$$
(2.2.25)

In a bounded region the two solutions $v_1 = \cosh hx$ and $v_2 = \sinh hx$ may prove more convenient. In this case $W = h$ and

$$u(x) = \left(\alpha_1 + \frac{1}{h} \int^x \sinh h\xi \ f(\xi) \, d\xi \right) \cosh hx$$

$$+ \left(\alpha_2 - \frac{1}{h} \int^x \cosh h\xi \ f(\xi) \, d\xi \right) \sinh hx. \qquad (2.2.26)$$

For the related equation

$$u'' + b^2 u = -f(x), \qquad (2.2.27)$$

we can take $v_1 = \cos bx$, $v_2 = \sin bx$ and proceed as before or, more simply, set $b = ih$ in the previous formulae remembering that

$$\cosh i\alpha = \cos \alpha, \qquad \sinh i\alpha = i \sin \alpha, \qquad (2.2.28)$$

and rename the constants to find

$$u(x) = \left(\alpha_1 + \frac{1}{b} \int^x \sin b\xi \ f(\xi) \, d\xi \right) \cos bx$$

$$+ \left(\alpha_2 - \frac{1}{b} \int^x \cos b\xi \ f(\xi) \, d\xi \right) \sin bx. \qquad (2.2.29)$$

$\square$

The general approach of variation of parameters can be applied to equations of any order; in particular, for the first-order equation

$$u' + A(x)u = f(x), \qquad u(a) = u_a, \qquad (2.2.30)$$

it leads to

$$u(x) = u_a \exp \left(- \int_a^x A(\xi) \, d\xi \right) + \int_a^x f(\xi) \exp \left(- \int_\xi^x A(\eta) \, d\eta \right) d\xi, \qquad (2.2.31)$$

in which the first term is the general solution of the homogeneous equation and the second term a particular solution of the non-homogeneous one.

An adaptation of the method of variation of parameters to linear non-homogeneous finite-difference equations is shown in §5.3 p. 133.

The existence of the simple expression (2.2.17) for the Wronskian permits us to generate a second independent solution of the homogeneous differential equation (2.2.2) from a given solution. Suppose $v_1(x)$ is known. Then the expression (2.2.10) for W can be considered as a first-order differential equation for v_2:

$$v_2' - \frac{v_1'}{v_1} v_2 = \frac{W}{v_1} \qquad (2.2.32)$$

a particular solution of which is, according to the second term of (2.2.31),

$$v_2(x) = v_1(x) \int \frac{W(\xi)}{v_1^2(\xi)} \, d\xi. \tag{2.2.33}$$

2.3 The "δ-function"

In elementary treatments, the "δ-function" $\delta(x)$ is introduced as the mathematical entity which, for any function $f(x)$ continuous at 0, has the property that

$$\int_{-\infty}^{\infty} \delta(x) f(x) \, dx = f(0), \tag{2.3.1}$$

or, more generally, that

$$\int_a^b \delta(x - x_0) f(x) \, dx = \begin{cases} f(x_0) & \text{if } a < x_0 < b \\ 0 & \text{otherwise} \end{cases}. \tag{2.3.2}$$

These are really symbolic expressions as there is no ordinary function for which they hold. The original heuristic introduction of the δ has led to the development of the notion of generalized functions, or *distributions*, which are dealt with in Chapter 20. At an intuitive level, the δ can be seen as the limit of a family of functions more and more peaked at the origin (or at x_0) whence the common statement that "$\delta(x)$ vanishes everywhere except at 0 where it is infinite" in such a way that

$$\int_{-\infty}^{\infty} \delta(x) = 1. \tag{2.3.3}$$

Here the integration limits $-\infty$ and ∞ can be replaced by any finite negative and positive number, respectively.

A beautiful feature of the theory of distributions is that treating the δ as an ordinary function seldom leads to error – if coupled with a modicum of common sense. For example, if δ is multiplied by a function $\alpha(x)$

$$\int_{-\infty}^{\infty} [\alpha(x)\delta(x)] \, f(x) \, dx = \int_{-\infty}^{\infty} \delta(x) \alpha(x) f(x) \, dx = \alpha(0) f(0), \tag{2.3.4}$$

which shows that

$$\alpha(x) \delta(x) = \alpha(0) \delta(x), \tag{2.3.5}$$

in the sense that both expressions give the same result when multiplied by $f(x)$ and integrated; thus, in particular, $x \, \delta(x) = 0$. However, it does not make sense to multiply δ by itself as no proper meaning can be attributed to $\delta^2(x)$.

Many manipulations involving the δ can be carried out applying the rules of ordinary calculus in a formal way. For example, if $\lambda > 0$,

$$\int_{-\infty}^{\infty} \delta(\lambda x) \, f(x) \, dx = \frac{1}{\lambda} \int_{-\infty}^{\infty} \delta(y) \, f\left(\frac{y}{\lambda}\right) \, dy = \frac{1}{\lambda} f(0). \tag{2.3.6}$$

If $\lambda < 0$, the integration limits are interchanged, which introduces a minus sign, and we conclude that

$$\delta(\lambda x) = \frac{\delta(x)}{|\lambda|} \tag{2.3.7}$$

from which, in particular, $\delta(-x) = \delta(x)$ so that $\delta(x)$ is "even."

A crucial property of generalized functions is that they can be differentiated even when they would not possess a derivative according to the rules of ordinary calculus. For example, the derivative of the the Heaviside or unit step function $H(x)$, defined by

$$H(x) = \begin{cases} 0 & x < 0 \\ 1 & x > 0 \end{cases}, \tag{2.3.8}$$

is calculated by formally applying the rule of integration by parts according to the chain of relations[9]

$$\int_{-\infty}^{\infty} H'(x) f(x) \, dx = -\int_{-\infty}^{\infty} H(x) f'(x) \, dx = -\int_{0}^{\infty} f'(x) \, dx = f(0), \tag{2.3.9}$$

from which, comparing with (2.3.1), we deduce that

$$H'(x) = \delta(x). \tag{2.3.10}$$

Similarly

$$\int_{-\infty}^{\infty} \delta'(x) f(x) \, dx = -\int_{-\infty}^{\infty} \delta(x) f'(x) \, dx = -f'(0) \tag{2.3.11}$$

and so on.

Use of the step function and of the differentiation formula (2.3.10) enables us to express the derivative of functions which would not possess an ordinary one; for example

$$\frac{d|x|}{dx} = \frac{d}{dx} [H(x)x - H(-x)x] = H(x) - H(-x) + 2x\delta(x) = \operatorname{sgn} x, \tag{2.3.12}$$

where we have used the fact pointed out earlier that $x\delta(x) = 0$ and $\operatorname{sgn} x$ is the *sign* or *signum* function defined by

$$\operatorname{sgn} x = \begin{cases} -1 & x < 0 \\ 1 & x > 0 \end{cases} = H(x) - H(-x). \tag{2.3.13}$$

With this result we have e.g. $de^{|x|}/dx = e^{|x|} d|x|/dx = (\operatorname{sgn} x) e^{|x|}$; furthermore $d^2 |x|/dx^2 = 2\delta(x)$.

The ordinary rule for the derivative of a product applies so that

$$\frac{d}{dx} [\alpha(x)\delta(x)] = \alpha'(x)\delta(x) + \alpha(x)\delta'(x). \tag{2.3.14}$$

Thus, differentiating both sides of (2.3.5), we have

$$\alpha'(x)\delta(x) + \alpha(x)\delta'(x) = \alpha(0)\delta'(x) \tag{2.3.15}$$

---

[9] We assume that f and f' vanish at infinity.

Table 2.1

Some properties of the "δ-function". In the next to the last line, the sum is extended to all the simple roots of $g(x) = 0$; $\delta(g(x))$ is undefined if g has multiple roots.

$$\int_{-\infty}^{\infty} \delta(x - x_0) f(x)\,dx = f(x_0)$$

$$\int_a^b \delta(x - x_0) f(x)\,dx = [H(b - x_0) - H(a - x_0)] f(x_0)$$

$$\int_a^b \delta(x - a) f(x)\,dx = f(a) \qquad \int_a^b \delta(x - b) f(x)\,dx = f(b)$$

$$\int_{-\infty}^{\infty} \delta'(x - x_0) f(x)\,dx = -f'(x_0) \qquad \int_{-\infty}^{\infty} \delta''(x - x_0) f(x)\,dx = f''(x_0)$$

$$\delta(\lambda x - \beta) = \frac{\delta(x - \beta/\lambda)}{|\lambda|} \qquad \delta(\lambda x) = \frac{\delta(x)}{|\lambda|}, \quad \delta(-x) = \delta(x)$$

$$\alpha(x)\,\delta(x - x_0) = \alpha(x_0)\,\delta(x - x_0) \qquad x\,\delta(x) = 0$$

$$\alpha(x)\delta'(x - x_0) = \alpha(x_0)\delta'(x - x_0) \\ -\alpha'(x_0)\delta(x - x_0) \qquad\qquad x\,\delta'(x) = -\delta(x)$$

$$H'(x - x_0) = \delta(x - x_0) \qquad d\,|x|/dx = \operatorname{sgn} x$$

$$\delta(g(x)) = \sum_i \frac{\delta(x - x_i)}{|g'(x_i)|} \qquad \delta(x^2 - a^2) = \frac{\delta(x - a) + \delta(x + a)}{2a}$$

$$\int_{-\infty}^{\infty} e^{\pm ikx}\,dk = \int_{-\infty}^{\infty} \cos kx\,dk = 2\int_0^{\infty} \cos kx\,dk = 2\pi\,\delta(x)$$

which, with (2.3.5), shows that

$$\alpha(x)\delta'(x) = \alpha(0)\delta'(x) - \alpha'(0)\delta(x). \tag{2.3.16}$$

Similar relations for higher-order derivatives are obtained in the same way.

We note for future reference that the action of the δ can be compactly expressed in terms of the Heaviside function generalizing (2.3.9) as

$$\int_{-\infty}^b \delta(x - x_0) f(x)\,dx = H(b - x_0) f(x_0) \tag{2.3.17}$$

or

$$\int_a^b \delta(x - x_0) f(x)\,dx = [H(b - x_0) - H(a - x_0)] f(x_0). \tag{2.3.18}$$

In the use of these formulae the question arises of how to interpret them when $x_0 = a$ or $x_0 = b$. The answer is that one is at liberty to define the value of this integral in any "reasonable" way, e.g. as $\frac{1}{2} f(b)$, or $f(b)$, provided this is done consistently throughout a calculation (see e.g. Friedman 1956, p. 154; Saichev & Woyczyński 1997, p. 58; Hoskins

1999, pp. 53, 107). The most appropriate choice in any particular instance may rest on the physics of the problem at hand. For example, if the δ represents a point force localized at one of the end points of the domain of interest, it may make physical sense to include its full effect by stipulating that

$$\int_a^b \delta(x - b) f(x) \, dx = f(b), \qquad \int_a^b \delta(x - a) f(x) \, dx = f(a). \qquad (2.3.19)$$

This convention proves convenient in our applications (see also p. 578). These and other properties are summarized in Table 2.1. The heuristic arguments given here are put on a more solid footing in Chapter 20.

While it is not legitimate to multiply $\delta(x)$ by itself, an unambiguous meaning can be attributed to $\delta(x) \, \delta(y)$:

$$\int_\Omega \delta(x) \, \delta(y) \, f(x, y) \, d\Omega = f(0, 0), \qquad (2.3.20)$$

provided the origin $(0,0)$ is interior to the integration domain Ω and $f(0, 0)$ is well defined; similarly

$$\int_\Omega \delta(x - x_0) \, \delta(y - y_0) \, f(x, y) \, d\Omega = f(x_0, y_0), \qquad (2.3.21)$$

provided the point $\mathbf{x}_0 = (x_0, y_0)$ is interior to the integration domain. We write

$$\delta(\mathbf{x} - \mathbf{x}_0) = \delta(x - x_0) \, \delta(y - y_0), \qquad (2.3.22)$$

and similar in higher-dimensional spaces.

In extending the δ to more than one space dimension, we encounter the question of how to deal with a change of variables. For example, upon passing from Cartesian to plane polar coordinates, the point $\mathbf{x}_0$ becomes $\mathbf{x}_0 = (r_0, \theta_0)$ while $dx \, dy$ becomes $r \, dr \, d\theta$; the result of the integration, however, must still be $f(\mathbf{x}_0) = f(r_0, \theta_0)$. A moment's thought shows that, to achieve this result, it is necessary to set

$$\delta(x - x_0) \, \delta(y - y_0) = \frac{1}{r}\delta(r - r_0) \, \delta(\theta - \theta_0) = \frac{1}{r_0}\delta(r - r_0) \, \delta(\theta - \theta_0), \qquad (2.3.23)$$

to cancel the factor r_0 that would otherwise result from the integration. This intuitive argument is given a more solid basis in §20.7; Tables 6.3 p. 148 and 7.4 p. 173 show the proper form of $\delta(\mathbf{x} - \mathbf{x}_0)$ in cylindrical and polar coordinates. Special situations arise at the singular points of the mapping which defines the change of variables, for example $r = 0$ in the previous example. Here we recognize that the variable θ is undefined at the origin and we therefore set

$$\delta(x) \, \delta(y) = \frac{1}{2\pi r}\delta(r), \qquad (2.3.24)$$

so that, upon integrating e.g. over a disk of radius R,

$$\int_{|\mathbf{x}| \leq R} \delta(\mathbf{x}) \, f(\mathbf{x}) \, dS = \int_0^R r \, dr \int_0^{2\pi} d\theta \, \frac{\delta(r)}{2\pi r} \, f = f(\mathbf{0}). \qquad (2.3.25)$$

2.4 The idea of Green's functions

We devote two chapters, 15 and 16, to theory and applications of Green's functions. However, since the solution of the problems of the chapters that follow eventually is found in the form of a Green's function combined with the given data, it may be worth while to explain the basic concept so that at least this structure can be recognized for what it is.

The essential idea is to note that the response of a *linear* system to a distributed input can be represented as the superposition of the effect of elementary inputs at each point of the system. As an example, suppose that a unit weight is attached to the point ξ_1 of a string taut between $x = 0$ and $x = L$. Under the action of this weight, the string will deform, taking a shape that we denote by $u(x) = G(x, \xi_1)$. If the magnitude of the weight is not unity but $f(\xi_1)\,d\xi$, and the deformation is small enough for the system to behave linearly, the shape of the string will be $u(x) = G(x, \xi_1)\,[f(\xi_1)\,d\xi]$. If now a weight $f(\xi_2)\,d\xi$ is attached at the point ξ_2, and the hypothesis of linearity still holds, the shape of the string will be $u(x) = [G(x, \xi_1)\,f(\xi_1) + f(\xi_2)\,G(x, \xi_2)\,f(\xi_1)]\,d\xi$, and so on. In the limit of an infinity of infinitesimal weights, the shape of the string will be

$$u(x) = \int_0^L G(x, \xi)\,f(\xi)\,d\xi. \tag{2.4.1}$$

In order for the string to be pinned at the end points, evidently,

$$G(0, \xi) = G(L, \xi) = 0 \qquad \text{for all } \xi. \tag{2.4.2}$$

The function $G(x, \xi)$ appearing here is the Green's function for this problem.[10]

The mathematical problem determining the shape of the string has a structure of the type

$$Lu(x) = f(x), \tag{2.4.3}$$

where the action of the linear operator L on the local string shape $u(x)$ expresses the balance of internal forces which counteract the external load $f(x)$; $u(x)$ given by (2.4.1) must be the solution of this equation, which is only possible if

$$L_x \int_0^L G(x, \xi)\,f(\xi)\,d\xi = \int_0^L L_x\,G(x, \xi)\,f(\xi)\,d\xi = f(x), \tag{2.4.4}$$

where we write L_x to stress the fact that the operator, being applied to $u(x)$, acts on the first one of the two variables on which G depends. It is therefore evident that the Green's function must satisfy

$$L_x G(x, \xi) = \delta(x - \xi), \tag{2.4.5}$$

together with the boundary conditions (2.4.2). For the sake of concreteness, let us look at a couple of specific examples (even though neither one refers to strings).

---

[10] As this example suggests, every linear problem will have its own Green's function. For this reason the common usage is to violate the grammatical rule that forbids the simultaneous use of the genitive and the determinate article. Similarly to "Bessel function", the denomination "Green function" might suggest a specific type of function, which Green's functions definitely are not. This consideration justifies the commonly used reference to "the Green's function" for each specific problem.

Example 2.4.1 We return to the problem of Example 2.2.2

$$-u'' + k^2 u = f(x), \qquad u \to 0 \text{ for } |x| \to \infty \tag{2.4.6}$$

over the entire line $-\infty < x < \infty$; we assume that the given function f is such that a solution with the required properties exists. In order to make the solution (2.2.25) satisfy the boundary conditions at infinity we take $\alpha_1 = \alpha_2 = 0$ and we extend the first integral from $-\infty$ to x and the second one from x to ∞; the result is

$$u(x) = \frac{1}{2k} e^{-kx} \int_{-\infty}^{x} e^{k\xi} f(\xi) \, d\xi + \frac{1}{2k} e^{kx} \int_{x}^{\infty} e^{-k\xi} f(\xi) \, d\xi \tag{2.4.7}$$

which can be written equivalently and more compactly as

$$u(x) = \frac{1}{2k} \int_{-\infty}^{\infty} e^{-k|x-\xi|} f(\xi) \, d\xi. \tag{2.4.8}$$

This formula suggests that

$$G(x, \xi) = \frac{1}{2k} e^{-k|x-\xi|} \tag{2.4.9}$$

is the Green's function for this problem. Using the earlier relation $d^2 |x|/dx^2 = 2\delta(x)$ we readily find that

$$-\frac{\partial^2 G}{\partial x^2} + k^2 G = \delta(x - \xi), \tag{2.4.10}$$

which is what we expect from (2.4.5). We have already encountered another instance of a Green's function in (2.2.15).   □

The same line of reasoning can be extended to more than one space dimension, to space/time-dependent problems and to boundary-value problems. For example, one may think of a membrane oscillating due to the time dependence of the shape of its frame, to the temperature of a solid responding to a time-dependent heat input on the boundary, and so on. The examples of the following chapters will provide abundant illustrations.

The concept of the Green's function can be approached also in a different way, which is the line of attack that we follow in Chapter 15 and which is best explained with another example.

Example 2.4.2 Consider the problem

$$-i\frac{du}{dx} = \lambda s(x)u + f(x), \qquad u(0) = u(b), \tag{2.4.11}$$

where λ is a real or complex number and it is assumed that $s(x) > 0$. The boundary condition may be interpreted as the requirement that u be periodic with period b. We multiply the equation by a function $G(x, y)$ and integrate over $[0, b]$, using integration by parts in the left-hand side and the periodicity condition $u(0) = u(b)$; the result is

$$-i[G(b, y) - G(0, y)]u(0) - \int_{0}^{b} \left(-i\frac{\partial G}{\partial x} + \lambda s G\right) u(x) \, dx = \int_{0}^{b} G(x, y) f(x) \, dx. \tag{2.4.12}$$

If G satisfies the equation

$$-i\frac{\partial G}{\partial x} + \lambda s(x)G(x, y) = \delta(y - x), \qquad (2.4.13)$$

the first integration can be carried out and we find

$$u(y) = -i[G(b, y) - G(0, y)]u(0) - \int_0^b G(x, y)f(x)\,dx. \qquad (2.4.14)$$

This is not yet an explicit expression for u as it contains the unknown constant $u(0)$ but, since the boundary condition on G is so far unspecified, we choose it so that this unknown term disappears:

$$G(0, y) = G(b, y). \qquad (2.4.15)$$

With (2.4.13), this relation completes the specification of the problem for G. Upon rewriting (2.4.13) in the form

$$\frac{\partial}{\partial x}\left\{e^{i\lambda S(x)}G(x, y)\right\} = e^{i\lambda S(y)}\delta(x - y), \qquad S(x) = \int_0^x s(\xi)\,d\xi, \qquad (2.4.16)$$

where use has been made of (2.3.5) in the right-hand side, and integrating between 0 and x, we find, using (2.3.17),

$$G(x, y) = G(0, y)e^{-i\lambda S(x)} + ie^{i\lambda[S(y)-S(x)]}H(y - x). \qquad (2.4.17)$$

We now impose (2.4.15) noting that since, in the present problem, the variable is restricted to lie between 0 and b, we may take $H(y) = 1$ and $H(y - b) = 0$. In this way we determine $G(0, y)$:

$$G(0, y) = -i\frac{e^{-i\lambda S(y)}}{1 - e^{-i\lambda S(b)}}, \qquad (2.4.18)$$

so that, in conclusion

$$G(x, y) = ie^{i\lambda[S(y)-S(x)]}\left[H(y - x) - \frac{1}{1 - e^{-i\lambda S(b)}}\right]. \qquad (2.4.19)$$

Upon substitution into (2.4.14) the result is

$$u(y) = ie^{i\lambda S(y)}\int_0^y e^{-i\lambda S(x)}f(x)\,dx - i\frac{e^{i\lambda S(y)}}{1 - e^{-i\lambda S(b)}}\int_0^b e^{-i\lambda S(x)}f(x)\,dx, \qquad (2.4.20)$$

and coincides with that obtained from (2.2.31) by imposing the periodicity condition directly. Again, for this simple problem, use of the Green's function is not particularly efficient. The point here is only to demonstrate how boundary conditions on G can be determined so as to eliminate unknown features of the solution. We return to this example on p. 375 when we introduce singular Sturm–Liouville problems.   □

For practical purposes the value of the Green's function is to permit to write down the general solution of a linear problem irrespective of the specific form of the data. Thus, the task of solving (2.4.6) or (2.4.11) does not need to be undertaken anew for every different $f(x)$: whatever f, the required solution is given by (2.4.8) or (2.4.20). On a more theoretical level, the availability of a specific form for the solution of a problem permits us to draw several important conclusions on the nature and properties of the solutions of the problem. We will see several examples in Chapters 15 and 16.

2.5 Power series solution of ordinary differential equations

A standard method for generating solutions of ordinary differential equations, due to Frobenius, relies on the use of suitable power series. More often than not this approach is more useful for the investigation of the nature and properties of the solutions than it is for actually finding them, whether by analytical evaluation or numerical computation. Nevertheless, the method is valuable as shown later in sections 12.2 and 13.2 for the Bessel and Legendre functions, respectively. We begin with two examples, and give a concise summary of the theory at the end.

The main properties of power series are covered in §8.4 and §17.4. Further considerations of the method can be found in many books, e.g., Morse & Feshbach 1953, chapter 5, Coddington & Levinson 1955, chapter 4 and Bender & Orszag 2005, chapter 3.

Example 2.5.1 [Ordinary point] Let us consider the equation

$$\frac{d^2u}{dx^2} + \omega^2 u = 0. \tag{2.5.1}$$

We assume that $x = 0$ is an *ordinary point*, namely a point in the neighborhood of which the solution can be represented by a power series of the form

$$u(x) = \sum_{n=0}^{\infty} a_n x^n, \tag{2.5.2}$$

and show how Frobenius's method reproduces the general solution $u = a_0 \cos \omega x + a_1 \sin \omega x$. We substitute the power series into the equation and differentiate term by term, which is legitimate for power series (§8.4); the result is

$$\sum_{n=2}^{\infty} n(n-1)a_n x^{n-2} + \omega^2 \sum_{n=0}^{\infty} a_n x^n = 0. \tag{2.5.3}$$

On the left-hand side we have allowed the summation index to start from $n = 2$ as the product $n(n-1)$ causes the $n = 0$ and $n = 1$ terms to vanish. Let us now set $n' = n - 2$ in the first series to find

$$\sum_{n'=0}^{\infty} (n'+2)(n'+1)a_{n'+2} x^{n'} + \omega^2 \sum_{n=0}^{\infty} a_n x^n = 0, \tag{2.5.4}$$

or, upon renaming the summation index,

$$\sum_{n=0}^{\infty} \left[(n+2)(n+1)a_{n+2} + \omega^2 a_n \right] x^n = 0. \tag{2.5.5}$$

For a power series to vanish identically in a finite region it is necessary that all its coefficients be equal to 0 (Theorem 17.4.2 p. 433), and we therefore have the recurrence relation

$$a_{n+2} = -\frac{\omega^2}{(n+2)(n+1)} a_n. \tag{2.5.6}$$

We may choose a_0 and a_1 in an arbitrary manner, which corresponds to the two arbitrary integration constants for an equation of the second order; after this step, all other coefficients are determined by the recurrence relation. For example, for the even coefficients we find

$$a_{2k+2} = -\frac{\omega^2}{(2k+2)(2k+1)} a_{2k}, \tag{2.5.7}$$

so that

$$a_2 = -\frac{\omega^2}{2 \cdot 1} a_0, \qquad a_4 = -\frac{\omega^2}{4 \cdot 3} a_2 = \frac{\omega^4}{4 \cdot 3 \cdot 2 \cdot 1} a_0, \tag{2.5.8}$$

and so on; thus

$$a_{2k} = \frac{(-1)^k \omega^{2k}}{(2k)!} a_0. \tag{2.5.9}$$

In this way we have generated the series

$$a_0 \sum_{k=0}^{\infty} \frac{(-1)^k (\omega x)^{2k}}{(2k)!} = a_0 \cos \omega x. \tag{2.5.10}$$

Reasoning similarly on the odd coefficients gives rise to the other solution $a_1 \sin \omega x$. $\quad\square$

Before proceeding, let us remark on how to deal with the recurrence relations that are typically generated by this method. Suppose that we have a recurrence of the form

$$\alpha_{n+1} = \lambda(n+a)\alpha_n, \tag{2.5.11}$$

i.e.

$$\alpha_1 = \lambda a \alpha_0, \qquad \alpha_2 = \lambda(1+a)\alpha_1 = \lambda^2 a(1+a)\alpha_0, \tag{2.5.12}$$

and so on. It is evident (and it can readily be proven by induction) that α_n will contain the factor $a(a+1)(a+2)\cdots(a+n-1)$ which can be represented compactly as[11]

$$a(a+1)(a+2)\cdots(a+n-1) = \frac{\Gamma(n+a)}{\Gamma(a)}, \tag{2.5.13}$$

by invoking the fundamental property $\Gamma(a+1) = a\Gamma(a)$ of the Gamma function (p. 442). Thus, the solution of the recurrence relation (2.5.11) is

$$\alpha_n = \lambda^n \frac{\Gamma(n+a)}{\Gamma(a)} \alpha_0. \tag{2.5.14}$$

---

[11] The notation $(a)_n = a(a+1)(a+2)\cdots(a+n-1)$ is also in use for this factor; $(a)_n$ is referred to as the Pochhammer symbol:

The argument is readily extended to show that the solution of the more general recurrence relation

$$\alpha_{n+1} = \lambda \frac{(n+a)\cdots(n+b)}{(n+c)\cdots(n+d)}\alpha_n, \tag{2.5.15}$$

is given by

$$\alpha_n = \lambda^n \frac{\Gamma(n+a)}{\Gamma(a)}\cdots\frac{\Gamma(n+b)}{\Gamma(b)}\frac{\Gamma(c)}{\Gamma(n+c)}\cdots\frac{\Gamma(d)}{\Gamma(n+d)}\alpha_0. \tag{2.5.16}$$

It is evident from (2.5.13) that, when one of the constants $a, \ldots, b$ in the numerator equals a negative integer, the recurrence terminates and the series reduces to a polynomial. When the same happens for one of the constants $c, \ldots, d$ in the denominator, on the other hand, the formula becomes inapplicable; we address this special case later.

In the previous example the method worked because the solution possessed an ordinary power series expansion centered at $x = 0$, but it would have failed had this not been the case. One of the properties of linear equations that makes them simpler than non-linear ones is that, if the equation is written so that the highest-order derivative has coefficient 1, the only *possible* singularities of the solution are those of the coefficients.[12] In the most common cases the singularity is isolated and therefore it can be a pole, a branch point or an essential singularity (p. 435). In the former two cases the singular point is termed *regular* or *Fuchsian*; criteria for identifying points of this type will be given later. At a regular singular point the solution may be regular, have a pole, or a branch point. We illustrate this case with an important example.

Since, for the power-series method to work, the solution must be analytic in a neighborhood of the center of the power series, from here on we allow the independent variable to be complex and denote it by z.

Example 2.5.2 [Regular singular point] Consider the Bessel equation (Chapter 12)

$$\frac{d^2u}{dz^2} + \frac{1}{z}\frac{du}{dz} + \left(1 - \frac{\nu^2}{z^2}\right)u = 0, \tag{2.5.17}$$

with ν^2 a real or complex constant. The only possible singularity of the solution at a finite distance from the origin is a pole or a branch point at the origin itself, $z = 0$, which is the only finite singularity of the coefficients. Since, in general, we cannot expect a power series of the form (2.5.2) to be able to represent the solution, we modify it to the Frobenius form

$$u = z^\beta \sum_{n=0}^{\infty} a_n z^n, \tag{2.5.18}$$

where the *indicial exponent* β needs to be determined as part of the procedure and we can, without loss of generality, assume that $a_0 \neq 0$ as, if not, it can be made so by a redefinition

---

[12] That this is not true for non-linear equations is demonstrated by the elementary example $du/dx = -u^2$. The solution of the equation is $u = (x - c)^{-1}$ in which the constant c depends on the particular solution being sought through its value at the starting point of the integration. In this case, therefore, the location c of the singularity depends on the particular solution chosen.

of β, as shown below. Upon substituting into the differential equation we find

$$\sum_{n=0}^{\infty} \left[(n+\beta)^2 - v^2\right] a_n z^{n+\beta-2} + \sum_{n=0}^{\infty} a_n z^{n+\beta} = 0, \qquad (2.5.19)$$

or, with $n' = n - 2$,

$$\sum_{n'=-2}^{\infty} \left[(n'+2+\beta)^2 - v^2\right] a_{n'+2} z^{n'+\beta} + \sum_{n=0}^{\infty} a_n z^{n+\beta} = 0, \qquad (2.5.20)$$

which may be rewritten as

$$(\beta^2 - v^2) a_0 z^\beta + \left[(1+\beta)^2 - v^2\right] a_1 z^{\beta+1}$$

$$+ \sum_{n=0}^{\infty} \left\{ \left[(n+2+\beta)^2 - v^2\right] a_{n+2} + a_n \right\} z^{n+\beta} = 0. \qquad (2.5.21)$$

Since, as mentioned before, we are at liberty to require that $a_0 \neq 0$, in order to set to 0 the coefficient of z^β we must choose β such that

$$\beta^2 - v^2 = 0, \qquad (2.5.22)$$

which is the indicial equation for this case. Thus, we have the two possibilities $\beta = \pm v$. With this choice for β, in order to cause the second term to vanish we must take $a_1 = 0$,[13] and, furthermore,

$$a_{n+2} = -\frac{a_n}{(n+2+\beta)^2 - v^2} = -\frac{a_n}{(n+2)(n+2\pm 2v)}. \qquad (2.5.23)$$

With $a_1 = 0$, this relation implies that all the odd coefficients vanish. As for the even ones, we can determine them from (2.5.16) by interpreting a_{2k} as α_k and writing

$$a_{2(k+1)} = -\frac{a_{2k}}{2^2(k+1)(k+1\pm v)}. \qquad (2.5.24)$$

The result is

$$a_{2k} = \frac{(-1)^k}{2^{2k}} \frac{\Gamma(1)}{\Gamma(k+1)} \frac{\Gamma(1\pm v)}{\Gamma(k+1\pm v)} a_0 = \frac{(-1)^k}{2^{2k}k!} \frac{\Gamma(1\pm v)}{\Gamma(k+1\pm v)} a_0, \qquad (2.5.25)$$

where we have used $\Gamma(k+1) = k!$, $\Gamma(1) = 1$. The usual standardization of the Bessel functions of the first kind consists in taking $a_0 = 2^{\mp v}/\Gamma(1\pm v)$ so that

$$a_{2k} = \frac{(-1)^k}{2^{2k\pm v}k!} \frac{1}{\Gamma(k+1\pm v)}, \qquad (2.5.26)$$

and we have the two power series

$$u = \left(\frac{z}{2}\right)^{\pm v} \sum_{k=0}^{\infty} \frac{(-1)^k}{k!\Gamma(k+1\pm v)} \left(\frac{z}{2}\right)^{2k}, \qquad (2.5.27)$$

---

[13] If we take $a_0 = 0$, then we must take $\beta = -1 \pm v$ to kill the second term and the power series (2.5.18) begins as $z^{-1\pm v}(a_1 z + \cdots) = z^{\pm v}(a_1 + \cdots)$ which has the same form as the series found by taking $a_0 \neq 0$.

which define the Bessel functions of the first kind $J_{\pm\nu}(z)$. The ratio test (p. 218) shows that the series converge for any finite z.

When $\nu \neq 0$, the two series start with different powers and define therefore two linearly independent functions, but this is not so if $\nu = 0$. Another special case, as noted earlier after (2.5.16), arises when $\nu = N$, a non-negative integer, as (2.5.27) gives then a solution by taking the upper sign, namely $+N$, but becomes invalid if we take $-N$. In either case, however, (2.5.27) does furnish one solution of the equation. We can generate a second solution by using the relation (2.2.33) p. 35. From (2.2.17) p. 32 the Wronskian in this case is $W \propto 1/z$ and therefore, up to an arbitrary multiplicative constant, (2.2.33) becomes

$$u_2(z) = J_N(z) \int \frac{d\zeta}{\zeta J_N^2(\zeta)}. \tag{2.5.28}$$

The denominator of the integrand has a pole of order $2N + 1$ at $\zeta = 0$ and therefore (p. 435) it can be expanded in an ascending power series starting with ζ^{-2N-1}. Upon term-by-term integration of this series we then find a result of the form

$$u_2(z) = J_N(z) \left[A \log z + z^{-2N} \sum_{k=0}^{\infty} b_k z^k \right] \tag{2.5.29}$$

where A and the b_k are constants determined by the expansion.

It is instructive to investigate more closely the case $\nu = N$. If we write the recurrence relation (2.5.24) for $\beta = -N$ in the form

$$2^2(k+1)(k+1-N)a_{2(k+1)} = -a_{2k}, \tag{2.5.30}$$

we see that, when $k + 1 = N$, a_{2N} remains undetermined, while this relation requires that $a_{2(N-1)} = 0$ and as a consequence, going back in k, all a_{2k} with $k \leq N - 1$ must vanish, including a_0. Thus, in this case, we find a solution of the form $z^{-N}(a_{2N}z^{2N} + \cdots) = z^N(a_{2N} + \cdots)$, i.e., in essence, the same solution found by taking $\beta = +N$.   □

We can now generalize the lesson learned in this example by stating a theorem in two parts, preceded by a definition:

Definition 2.5.1 [Regular singular point] The point z_0 is a *regular*, or *Fuchsian*, *singular point* for the equation

$$\frac{d^2u}{dz^2} + \frac{P(z)}{z - z_0}\frac{du}{dz} + \frac{Q(z)}{(z - z_0)^2}u = 0, \tag{2.5.31}$$

if $P(z_0)$ and $Q(z_0)$ are finite (possibly 0); otherwise the singular point is termed *irregular*.

Example 2.5.3 [Point at infinity] In order to study the nature of the possible singularity at the point at infinity we make the transformation $\zeta = 1/z$ and study the behavior of the coefficients of the equation in the neighborhood of $\zeta = 0$. Carrying out this change of variable on the equation

$$\frac{d^2u}{dz^2} + p(z)\frac{du}{dz} + q(z)u = 0, \tag{2.5.32}$$

we find

$$\frac{d^2 u}{d\zeta^2} + \left[\frac{2}{\zeta} - \frac{p(1/\zeta)}{\zeta^2}\right]\frac{du}{dz} + \frac{q(1/\zeta)}{\zeta^4} u = 0. \tag{2.5.33}$$

The condition for $\zeta = 0$ to be a regular singular point is therefore that the limits

$$\lim_{\zeta \to 0}\left(2 - \frac{p(1/\zeta)}{\zeta}\right) = p_0, \qquad \lim_{\zeta \to 0}\frac{q(1/\zeta)}{\zeta^2} = q_0, \tag{2.5.34}$$

exist and be finite (possibly 0). ☐

Theorem 2.5.1 [Fuchs – Part I] *A necessary condition for the second-order equation (2.5.31) to have, in the neighborhood of a point z_0, two linearly independent solutions in the form of Frobenius series, namely*

$$u_{1,2} = (z - z_0)^{\beta_{1,2}} \sum_{n=0}^{\infty} a_n^{(1,2)}(z - z_0)^n, \tag{2.5.35}$$

is that z_0 be a regular singular point and that the difference between the two roots $\beta_{1,2}$ of the indicial equation

$$\beta(\beta - 1) + P(z_0)\beta + Q(z_0) = 0 \quad \text{with} \quad \text{Re}\,\beta_1 \geq \text{Re}\,\beta_2, \tag{2.5.36}$$

be not an integer.

The condition $\beta_1 - \beta_2 \neq$ integer barely misses being also sufficient as we see shortly.

If z_0 is an irregular singular point, the solution may have an essential singularity there. An example is the point at infinity for the Bessel equation for which

$$2 - \frac{p(1/\zeta)}{\zeta} = 1, \qquad \frac{q(1/\zeta)}{\zeta^2} = \frac{1}{\zeta^2} - \nu^2, \tag{2.5.37}$$

and therefore $q_1(1/\zeta)/\zeta^2$ has no limit at $\zeta = 0$. Indeed, the Bessel functions have an essential singularity at infinity and an attempt to generate a solution of the Frobenius form would lead to $a_0 = a_1 = 0$, $a_{n+2} \propto a_n$, which is just the trivial solution. An even simpler example are equations with constant coefficients, for which the limits (2.5.34) evidently do not exist. Indeed, the solution of such equations in general is the sum of exponentials and, since $\zeta = 0$ is an essential singularity for functions of the type $\exp(c/\zeta)$ for any $c \neq 0$, these solutions have an essential singularity at infinity.

Theorem 2.5.2 [Fuchs – Part II] *When z_0 is a regular singular point for the equation (2.5.31) and the solutions $\beta_{1,2}$ (with $\text{Re}\,\beta_1 \geq \text{Re}\,\beta_2$) of the indicial equation (2.5.36) are such that $\beta_1 - \beta_2 = N$, a non-negative integer, one solution has the Fuchs form*

$$u_1 = (z - z_0)^{\beta_1} \sum_{n=0}^{\infty} a_n^{(1)}(z - z_0)^n, \tag{2.5.38}$$

while the other one has the form

$$u_2 = u_1 \left[A \log(z - z_0) + (z - z_0)^{\beta_2 - \beta_1} \sum_{n=0}^{\infty} a_n^{(2)} (z - z_0)^n \right], \qquad (2.5.39)$$

unless the constant A happens to vanish. In this latter case, both solutions have the Fuchs form. The constant A vanishes when

$$\sum_{k=0}^{N-1} [(\beta_2 + k) P_{N-k} + Q_{N-k}] a_k^{(2)} = 0, \qquad (2.5.40)$$

in which the constants are defined by

$$P(z) = \sum_{n=0}^{\infty} P_n (z - z_0)^n, \qquad Q(z) = \sum_{n=0}^{\infty} Q_n (z - z_0)^n. \qquad (2.5.41)$$

An equation of the form (2.5.31) with $P(z) = P_0$ and $Q(z) = Q_0$, both constants, has a regular singular point at z_0 and at infinity as is readily checked. The general solution of this equation is elementary and has the form

$$u = a_0 (z - z_0)^{\beta_1} + a_1 (z - z_0)^{\beta_2}, \qquad (2.5.42)$$

where $\beta_{1,2}$ are the solutions of the indicial equation (2.5.36).

The most general equation with three regular singular points at z_0, z_1 and infinity has the form

$$\frac{d^2 u}{dz^2} + \frac{P_0 + P_1 z}{(z - z_0)(z - z_1)} \frac{du}{dz} + \frac{Q_0 + Q_1 z + Q_2 z^2}{(z - z_0)^2 (z - z_1)^2} u = 0. \qquad (2.5.43)$$

The substitution $Z = (z - z_0)/(z_1 - z_0)$ maps the two finite singular points to $Z = 0$ and $Z = 1$, and a suitable renaming of the constants brings this equation to the standard form

$$Z(1 - Z)\frac{d^2 u}{dZ^2} + [c - (a + b_1)Z]\frac{du}{dZ} - abu = 0, \qquad (2.5.44)$$

with the constants a, b, c dependent on the positions of the singularities and on the coefficients P_j, Q_k of the original equation. This is the *hypergeometric equation* which has been studied in great detail (see e.g. Erdélyi *et al.* 1954, vol. 1, chapter 2; Whittaker & Watson 1927, chapter 14; Morse & Feshbach 1953, chapter 5; Lebedev 1965, chapter 9). Since any three singular points z_0, z_1, z_2 can be mapped to 0, 1 and infinity by the transformation $Z = (z - z_0)(z_2 - z_1)/[(z_1 - z_0)(z_2 - z)]$, the most general second-order equation with three distinct regular singular points can be reduced to the form (2.5.44), which explains the importance of this equation. Many of the special functions of applied mathematics can be expressed in terms of solutions of the hypergeometric equation and its numerous generalizations.

When c is not a negative integer, one of the solutions found by applying the power series method to (2.5.44) is the *hypergeometric function* defined by

$$F(a, b; c; Z) = \frac{\Gamma(c)}{\Gamma(a)\Gamma(b)} \sum_{n=0}^{\infty} \frac{\Gamma(a + n)\Gamma(b + n)}{n!\,\Gamma(c + n)} Z^n. \qquad (2.5.45)$$

The series is regular at $Z = 0$, but the ratio test (p. 218) shows that its radius of convergence extends only to $|Z| = 1$ (see also p. 331). For particular values of the parameters a, b, c the hypergeometric series reduces to other functions; examples are

$$F(a, b; b; Z) = (1 - Z)^{-a}, \qquad F(1, 1; 2; Z) = -\frac{\log(1 - Z)}{Z}, \tag{2.5.46}$$

$$F\left(\frac{1}{2}, \frac{1}{2}; \frac{3}{2}; \sin^2 Z\right) = \frac{Z}{\sin Z}, \qquad F\left(\frac{1}{2}, \frac{1}{2}; \frac{3}{2}; Z^2\right) = \frac{\sin^{-1} Z}{Z}, \tag{2.5.47}$$

and many others. In particular, when $a = 1$, $F(1, b; b; Z) = (1 - Z)^{-1} = \sum_{n=0}^{\infty} Z^n$ is just the geometric series whatever b, which justifies the name of this function.

PART I

Applications

3

Fourier Series: Applications

The Fourier series is arguably the most basic tool to deal with partial differential equations in many commonly encountered bounded domains. This series is the simplest embodiment of the general framework of eigenfunction expansions outlined in §2.1, to which the reader may wish to refer to gain a deeper intuitive understanding of the approach and related manipulations. Furthermore, for the reason explained on p. 364, questions of convergence of other series expansions are reducible ultimately to the convergence of the Fourier series.

We start by presenting a summary of the basic relations and then proceed directly to show several applications. The underlying theory is covered in Chapter 9. In the examples of §3.2.2, 3.2.5 and 3.9.2 the series is not used in its standard form but is adapted to deal with different boundary conditions. In spite of the superficial similarity with the Fourier series, these cases are really instances of the general Sturm–Liouville theory of Chapter 15 rather than applications of the Fourier series proper.

3.1 Summary of useful relations

The Fourier series can represent functions defined on a finite interval. We start by considering an interval of length 2π, which in applications is often $0 \leq x < 2\pi$ or $-\pi < x \leq \pi$. For such an interval, the Fourier series in *exponential form* is

$$f(x) = \sum_{n=-\infty}^{\infty} f_n \, e_n(x), \qquad e_n(x) = \frac{\exp(inx)}{\sqrt{2\pi}}. \tag{3.1.1}$$

The coefficients f_n are given by the *scalar products* (e_n, f) of e_n and f defined as

$$f_n = (e_n, f) \equiv \int_0^{2\pi} \frac{\exp(-in\xi)}{\sqrt{2\pi}} f(\xi) \, d\xi \quad \text{or} \quad \int_{-\pi}^{\pi} \frac{\exp(-in\xi)}{\sqrt{2\pi}} f(\xi) \, d\xi, \tag{3.1.2}$$

depending on whether the interval of interest is $0 \leq x < 2\pi$ or $-\pi < x \leq \pi$. Since a Fourier series, even if divergent, can be integrated term by term resulting in a uniformly convergent series (Theorem 9.8.1 p. 256), these expressions for f_n are readily proved when f is integrable. It is sufficient to multiply (3.1.1) by $\bar{e}_m(x)$ and integrate term by term, using the orthonormality relation $(e_m, e_n) = \delta_{mn}$, with δ_{mn} the Kronecker symbol (p. 28).

Table 3.1	Behavior of the Fourier coefficients under some operations; $X = 0$ if the interval is $0 \leq x < 2\pi$, while $X = -\pi$ if the interval is $-\pi < x \leq \pi$.

$f(x) = \sum\limits_{n=-\infty}^{\infty} f_n \dfrac{\exp(inx)}{\sqrt{2\pi}}$	$f_n = \int_X^{X+2\pi} \dfrac{\exp(-in\xi)}{\sqrt{2\pi}} f(\xi)\,d\xi$
$\dfrac{d^p f}{dx^p}$	$(in)^p f_n$
$f^{(-1)}(x) \equiv \int^x f(y)\,dy$	$(in)^{-1} f_n$ for $n \neq 0$ $f_0^{(-1)}$ arbitrary
$\int_X^{X+2\pi} f(y)\,g(x-y)\,dy$	$(2\pi)^{1/2} f_n g_n$
$f(x)\,g(x)$	$(2\pi)^{-1/2} \sum_{m=-\infty}^{\infty} f_m g_{n-m}$
$\delta(x - x_0)$	$\delta_n = \exp(-inx_0)/\sqrt{2\pi}$

A crucial property that the Fourier series shares with many other series expansions is uniqueness: if two functions have the same Fourier coefficients, then they are point-wise equal, and vice versa.[1]

If f is discontinuous at x, and $f(x \pm 0)$ denote the limits of f as x is approached from the right or from the left, (3.1.1) is modified to

$$\frac{1}{2}[f(x+0) + f(x-0)] = \sum_{n=-\infty}^{\infty} f_n\, e_n(x). \tag{3.1.3}$$

Since, for brevity, we will not write separate relations for discontinuous functions in the following, the reader is invited to keep this relation in mind.

The series in the right-hand sides of (3.1.1) or (3.1.3) are periodic functions of x and it is best to think of the 2π-interval as being "wrapped around" a disk of radius 1 so as not to have a beginning or an end (see Figure 9.2 p. 245). This remark explains the equivalence of the two x-intervals that we have mentioned. In this picture, a function f such that $f(0) \neq f(2\pi)$, or $f(-\pi) \neq f(\pi)$, must be considered discontinuous even if it is continuous at every other point. In such cases, the series (3.1.1) will have convergence difficulties because of the behavior at the end points of the interval (see §9.10 p. 261). We will see later (p. 57) how to deal with this problem in some cases.

Series and integral expansions prove useful when the action of specific operators on the transformed function is simpler than that on the original function. The rules that govern this action are called *operational rules*, and some such rules for the Fourier series are summarized in Table 3.1 for functions having the degree of continuity required in each case. The expressions in the right column of the table provide the Fourier coefficients of

---

[1] Except possibly for a set of measure zero (see p. 546); this concern is immaterial for continuous functions.

the functions in the left column. For example, using the first entry, we can write down directly the Fourier expansion of the (continuous) second derivative of a function:

$$\frac{d^2 f}{dx^2} = \sum_{-\infty}^{\infty} (in)^2 f_n e_n(x). \tag{3.1.4}$$

Use of this relation presupposes that it is legitimate to differentiate the series term by term. In this connection it is useful to remember that a uniformly convergent series can be differentiated term by term (p. 227), and that the Fourier series converges uniformly in any interval *interior* to an interval in which f is continuous (p. 247; see also §9.4 p. 248).

If the intervals of interest are $0 \leq x < 2L$, or $-L < x \leq L$, the proper set of basis functions is (§9.9 p. 258)

$$e_n(x) = \frac{\exp(in\pi x/L)}{\sqrt{2L}}, \tag{3.1.5}$$

and the Fourier coefficients are given by

$$f_n = \int_0^{2L} \frac{e^{-in\pi \xi/L}}{\sqrt{2L}} f(\xi) \, d\xi \quad \text{or} \quad \int_{-L}^{L} \frac{e^{-in\pi \xi/L}}{\sqrt{2L}} f(\xi) \, d\xi. \tag{3.1.6}$$

In some cases use of the series in real or *trigonometric form* is more convenient. For an interval $-L < x < L$ it takes the form (§9.6 p. 252)

$$f(x) = \frac{1}{2} A_0 + \sum_{n=1}^{\infty} \left[A_n \cos \frac{n\pi x}{L} + B_n \sin \frac{n\pi x}{L} \right], \tag{3.1.7}$$

$$A_n = \frac{1}{L} \int_{-L}^{L} \cos \frac{n\pi x}{L} f(x) \, dx, \qquad B_n = \frac{1}{L} \int_{-L}^{L} \sin \frac{n\pi x}{L} f(x) \, dx. \tag{3.1.8}$$

The same formulae hold for the interval $0 < x < 2L$ except that the integrals now extend from 0 to $2L$. The expressions for the coefficients are derived by multiplying both sides of (3.1.7) by $\cos n\pi x/L$ or $\sin n\pi x/L$, integrating, and using the relations

$$\int_{-L}^{L} \left(\cos \frac{k\pi x}{L} \right) \left(\cos \frac{n\pi x}{L} \right) dx = \int_{-L}^{L} \left(\sin \frac{k\pi x}{L} \right) \left(\sin \frac{n\pi x}{L} \right) dx = L \delta_{kn}, \tag{3.1.9}$$

$$\int_{-L}^{L} \left(\sin \frac{k\pi x}{L} \right) \left(\cos \frac{n\pi x}{L} \right) dx = 0, \tag{3.1.10}$$

which also hold when the integrations extend between 0 and $2L$.

These expressions simplify if the function is even or odd. In the former case $f(-x) = f(x)$, all the sine integrals vanish and (3.1.7) becomes the *Fourier cosine series*

$$f(x) = \frac{1}{2} A_0 + \sum_{n=1}^{\infty} A_n \cos \frac{n\pi x}{L}, \qquad A_n = \frac{2}{L} \int_0^{L} \cos \frac{n\pi x}{L} f(x) \, dx. \tag{3.1.11}$$

If the function is odd, $f(-x) = -f(x)$, all the cosine integrals vanish and (3.1.7) becomes the *Fourier sine series*

$$f(x) = \sum_{n=1}^{\infty} B_n \sin \frac{n\pi x}{L}, \qquad B_n = \frac{2}{L} \int_0^{L} \sin \frac{n\pi x}{L} f(x) \, dx. \tag{3.1.12}$$

The corresponding expressions for a general interval $a < x \leq b$ are given in §9.9 p. 258.

Table 3.2	Some useful Fourier series; see p. 236 for derivations and additional examples

$$\sum_{n=1}^{\infty} \frac{\sin nx}{n} = \frac{\pi - x}{2} \qquad\qquad 0 < x < 2\pi$$

$$\sum_{n=1}^{\infty} \frac{\cos nx}{n} = \frac{1}{2} \log \frac{1}{2(1 - \cos x)} \qquad\qquad 0 < x < 2\pi$$

$$\sum_{n=1}^{\infty} (-1)^{n-1} \frac{\sin nx}{n} = \frac{x}{2} \qquad\qquad -\pi < x < \pi$$

$$\sum_{n=1}^{\infty} (-1)^{n-1} \frac{\cos nx}{n} = \frac{1}{2} \log \left(2 \cos \frac{x}{2} \right) \qquad\qquad -\pi < x < \pi$$

$$\sum_{n=0}^{\infty} \frac{\sin(2n + 1)x}{2n + 1} = \frac{\pi}{4} \operatorname{sgn} x \qquad\qquad -\pi < x < \pi$$

$$\sum_{n=0}^{\infty} \frac{\cos(2n + 1)x}{2n + 1} = \frac{1}{2} \log \left| \cot \frac{x}{2} \right| \qquad\qquad -\pi < x < \pi$$

$$\sum_{n=0}^{\infty} (-1)^n \frac{\sin(2n + 1)x}{2n + 1} = \frac{1}{2} \log \tan \frac{2x + \pi}{4} \qquad\qquad -\pi/2 < x < \pi/2$$

$$1 + 2 \sum_{k=1}^{\infty} \cos k(x - \xi) = 2\pi \delta(x - \xi) \qquad\qquad -\pi \le x, \xi \le \pi$$

$$\sum_{k=1}^{\infty} k \sin kx = -\pi \delta'(x) \qquad\qquad -\pi \le x \le \pi$$

$$\sum_{n=1}^{\infty} \sin nx = \frac{1}{2} \cot \frac{x}{2} \qquad\qquad 0 < x < 2\pi$$

To deal with a problem posed over the interval $0 < x < L$, one can effect a change of variable mapping the original interval onto $(-L', L')$, with $L' = L/2$, and use the full series (3.1.7), but it is often much more convenient to extend the function in an even or odd way over the range $(-L, 0)$ and use either the cosine or the sine series. Whether one or the other extension is more suitable depends on the conditions at the end points. If $f(0) = 0$, (3.1.12) shows that an odd extension is appropriate while, for $f(0) \ne 0$, an even extension will preserve continuity of the function. In either case, however, these extensions will in general introduce a discontinuity of the first derivative and, therefore, prevent a second term-by-term differentiation (at least, in the ordinary sense). There are several ways to deal with this problem. The one closest in spirit to our general approach is to expand the derivatives of f rather than f itself. For example,

$$\frac{d^2 f}{dx^2} = \frac{1}{2}a_0 + \sum_{n=1}^{\infty} a_n \cos \frac{n\pi x}{L}, \qquad a_n = \frac{2}{L} \int_0^L \cos \frac{n\pi x}{L} \frac{d^2 f}{dx^2} \, dx. \qquad (3.1.13)$$

Integrating by parts the coefficient twice (legitimate if df/dx and f are absolutely continuous, p. 685), we find

$$a_n = \frac{2}{L} \left[(-1)^n f'(L) - f'(0) \right] - \left(\frac{n\pi}{L}\right)^2 A_n, \qquad (3.1.14)$$

with A_n as in (3.1.11);[2] here we have used the fact that $\cos n\pi = (-1)^n$. Proceeding similarly for the sine series we would set

$$\frac{d^2 f}{dx^2} = \sum_{n=1}^{\infty} b_n \sin \frac{n\pi x}{L}, \qquad b_n = \frac{2}{L} \int_0^L \sin \frac{n\pi x}{L} \frac{d^2 f}{dx^2} \, dx, \qquad (3.1.15)$$

to find

$$b_n = \frac{2n\pi}{L^2} \left[f(0) - (-1)^n f(L) \right] - \left(\frac{n\pi}{L}\right)^2 B_n. \qquad (3.1.16)$$

The appearance of f and f' at the end points will enable us to deal with problems with general boundary conditions (see e.g. §3.2.3).

When the boundary conditions are such that no boundary terms arise in (3.1.14) or (3.1.16), these formulae coincide with the result of term-by-term differentiation of the expansion for f. In these cases, it is more efficient to justify the use of formulae such as (3.1.4) on the basis of (3.1.14) or (3.1.16) rather than by proving the legitimacy of term-by-term differentition.

When, as is often the case, f depends on another variable and differentiation with respect to this variable is required this approach needs to be modified. For example, if $f = f(x,t)$ and we use the Fourier sine series expansion (3.1.12), provided $\partial f/\partial t$ is integrable, we can proceed by setting

$$\frac{\partial f}{\partial t} = \sum_{n=1}^{\infty} C_n(t) \sin \frac{n\pi x}{L} \qquad (3.1.17)$$

with

$$C_n(t) = \frac{2}{L} \int_0^L \sin \frac{n\pi x}{L} \frac{\partial f}{\partial t} \, dx = \frac{d}{dt} \left(\frac{2}{L} \int_0^L \sin \frac{n\pi x}{L} f(x,t) \, dx \right) = \frac{dB_n}{dt}. \qquad (3.1.18)$$

The interchange of t-differentiation and integration is legitimate under relatively mild conditions on f, e.g. continuity of $\partial f/\partial t$ (§A.4.5 p. 691). Another tool that can be used in this case are the considerations following Theorem 8.4.1 on p. 229.

---

[2] Upon substituting (3.1.14) back into (3.1.13), the terms proportional to $f'(0)$ and $f'(L)$ give rise to series which do not converge in the normal sense but converge in the sense of distributions to $\delta(x)$ and $\delta(L - x)$, respectively (§20.11 and in particular (20.11.7) p. 601). These terms arise because of the discontinuity in the derivative mentioned before.

We will set out in some detail the logic of the procedure and the arguments used in the application of the Fourier series in the examples of the next section; in the sections that follow we proceed more rapidly.

3.2 The diffusion equation

We start with examples involving the one-dimensional diffusion equation

$$\frac{\partial u}{\partial t} - D\frac{\partial^2 u}{\partial x^2} = q(x, t), \tag{3.2.1}$$

in which the constant D has the physical meaning of diffusivity (see §1.3) and $q(x, t)$ is a prescribed function. In heat conduction, u would be proportional to the temperature, $-\partial u/\partial x$ to the heat flux, and q would be a distributed heat source. The Fourier series is appropriate when the interval of interest is finite, and we take here $0 < x < L$; the infinite or semi-infinite cases are treated by transform methods in Chapters 4 and 5. To complete the specification of the problem we need to impose initial and boundary conditions; we consider several cases.

3.2.1 Homogeneous boundary conditions

The simplest case is that of homogeneous boundary conditions. Let us suppose for example that the end points $x = 0$ and $x = L$ are insulated so that we have the Neumann conditions

$$\frac{\partial u}{\partial x} = 0 \qquad \text{for} \qquad x = 0, L. \tag{3.2.2}$$

We assume that an initial temperature distribution is prescribed:

$$u(x, 0) = f(x) \tag{3.2.3}$$

with f integrable; consistency with (3.2.2) requires that $f'(0) = f'(L) = 0$.

Since $\partial u/\partial x = 0$ at $x = 0$, the even extension $u(-x, t) = u(x, t)$ gives rise to a continuous function defined for $-L < x < L$, and this consideration suggests the use of the Fourier cosine series expansion (3.1.11):

$$u(x, t) = \frac{A_0(t)}{2} + \sum_{n=1}^{\infty} A_n(t) \cos \frac{n\pi x}{L}. \tag{3.2.4}$$

As explained on p. 29, now $A_n = A_n(t)$ as, insofar as the x-dependence is concerned, $u(x, t)$ really denotes a collection of functions, a different one for each value of t.

From (3.1.14), the Fourier coefficients of $\partial^2 u/\partial x^2$ are simply $-(n\pi/L)^2 A_n$ since the boundary term vanishes in view of the boundary conditions (3.2.2). For the t derivative, we tentatively use (3.1.18), reserving until the end of the calculation a check of the legitimacy of the step leading to this relation. The reliability of this procedure ultimately rests on a uniqueness theorem: if we find – by whatever means – a solution of the problem, and the solution is unique, this must be *the* correct solution. Upon substituting the expansions of $\partial^2 u/\partial x^2$ and $\partial u/\partial t$ into (3.2.1) we have

$$\frac{1}{2}\frac{dA_0}{dt} + \sum_{n=1}^{\infty}\left[\frac{dA_n}{dt} + \left(\frac{\pi n}{L}\right)^2 DA_n\right]\cos\frac{\pi nx}{L} = q(x,t). \qquad (3.2.5)$$

Because of the uniqueness of the Fourier expansion, if this equality is to hold, the coefficients of the Fourier series in the left-hand side must equal those of the Fourier expansion of $q(x,t)$. We can expand q and then equate coefficients, but we steer closer to the geometric idea of projection advocated in §1.8 and §2.1 if we get to the same result by taking the scalar product of this equation with the generic cosine function using

$$\left(\cos\frac{\pi kx}{L},\cos\frac{\pi nx}{L}\right) = \int_0^L \cos\frac{\pi kx}{L}\cos\frac{\pi nx}{L}\,dx = \frac{L}{2}\delta_{kn}, \qquad (3.2.6)$$

also valid for $k = 0$ or $n = 0$; the result is

$$\frac{dA_0}{dt} = \frac{2}{L}\int_0^L q(x,t)\,dx \equiv q_0(t), \qquad (3.2.7)$$

$$\frac{dA_k}{dt} + \left(\frac{\pi k}{L}\right)^2 DA_k = \frac{2}{L}\int_0^L \cos\frac{\pi kx}{L}q(x,t)\,dx \equiv q_k(t). \qquad (3.2.8)$$

The solution of these equations for general q is given in (2.1.13) p. 30. For simplicity, let us take here $q = 0$ henceforth, with which

$$A_0(t) = A_0(0), \qquad A_k(t) = A_k(0)\exp\left[-\left(\frac{\pi k}{L}\right)^2 Dt\right], \qquad (3.2.9)$$

so that

$$u(x,t) = \frac{A_0(0)}{2} + \frac{2}{L}\sum_{n=1}^{\infty}A_n(0)\cos\frac{\pi nx}{L}\exp\left[-\left(\frac{\pi n}{L}\right)^2 Dt\right]. \qquad (3.2.10)$$

To determine the integration constants $A_k(0)$ we substitute this result into the initial condition (3.2.3) to find

$$\frac{A_0(0)}{2} + \frac{2}{L}\sum_{n=1}^{\infty}A_n(0)\cos\frac{\pi nx}{L} = f(x). \qquad (3.2.11)$$

We again take scalar products with the result

$$A_0(0) = \frac{2}{L}\int_0^L f(\xi)\,d\xi, \qquad A_k(0) = \frac{2}{L}\int_0^L \cos\frac{\pi k\xi}{L}f(\xi)\,d\xi. \qquad (3.2.12)$$

Upon substitution into (3.2.10), if f is such that the order of summation and integration can be interchanged[3], the result may be written as

$$u(x, t) = \frac{1}{L} \int_0^L d\xi \, f(\xi) \left\{ 1 + \sum_{n=1}^{\infty} \left(\cos \frac{\pi n(x - \xi)}{L} + \cos \frac{\pi n(x + \xi)}{L} \right) \right.$$

$$\left. \times \exp\left[-\left(\frac{\pi n}{L}\right)^2 Dt \right] \right\}. \qquad (3.2.13)$$

In this form, the solution may be interpreted in terms of an "image" temperature distribution over $-L < x < 0$ which evolves in the same way as the real distribution so as to maintain the condition $\partial u / \partial x = 0$ at $x = 0$. An infinity of images of this pair of temperature distributions also evolves over each $2L$ interval in such a way that $\partial u / \partial x = 0$ at all points $x = nL$ with $-\infty < n < \infty$. With

$$G(x, t; \xi) = \frac{1}{L} \left\{ 1 + \sum_{n=1}^{\infty} \left(\cos \frac{\pi n(x - \xi)}{L} + \cos \frac{\pi n(x + \xi)}{L} \right) \exp\left[-\left(\frac{\pi n}{L}\right)^2 Dt \right] \right\}$$

$$= \frac{1}{L} \left\{ 1 + 2 \sum_{n=1}^{\infty} \cos \frac{\pi n x}{L} \cos \frac{\pi n \xi}{L} \exp\left[-\left(\frac{\pi n}{L}\right)^2 Dt \right] \right\} \qquad (3.2.14)$$

(3.2.13) becomes

$$u(x, t) = \int_0^L f(\xi) \, G(x, t; \xi) \, d\xi . \qquad (3.2.15)$$

This expression has the structure of (2.4.1) p. 39 and shows that G is the Green's function for this problem.

The result (3.2.13) contains some interesting physics. In the first place, as $t \to \infty$, $u(x, t) \to A_0/2 = (1/L) \int_0^L f(x) \, dx$, so that the temperature becomes uniform and equal to the mean value of the initial temperature, as would be expected from the principle of conservation of energy. Before this asymptotic stage is reached, the dominant term of the series corresponds to $n = 1$, which decays proportionally to $\exp - (4\pi^2 Dt/L^2)$. Thus, the asymptotic uniform temperature state $u = A_0/2$ is reached after a time of the order of

$$t_{diff} \sim L^2/D \qquad (3.2.16)$$

which is thus revealed as the characteristic time for diffusion. If the initial temperature distribution contains features with a length scale $\ell \simeq L/\bar{n}$, these features will change little until $L/\bar{n} \sim \sqrt{Dt}$, which shows that the characteristic length for diffusion over a time t is of the order

$$\ell_{diff} \sim \sqrt{Dt}. \qquad (3.2.17)$$

While a shortcut to the justification of the various steps taken to obtain our solution is provided by the considerations following (3.1.13), it may be of some interest to examine them directly and see how the results of Chapter 8 might be applied. We will not work out

---

[3] For the legitimacy of this operation see pp. 226 and 694; see also §9.8 p. 256.

the details, but give a sketch of the procedure. If $f(x)$ is absolutely integrable, $|A_n(0)| \leq (2/L) \int_0^L |f(x)| \, dx \equiv A$ and, therefore,

$$\left| A_n(0) \cos \frac{\pi n x}{L} \exp\left[-\left(\frac{\pi n}{L} \right)^2 Dt \right] \right| \leq A \exp\left(-n^2 bt \right) \qquad (3.2.18)$$

with $b = \pi^2 D / L^2$ for brevity. Thus, for $t \geq \varepsilon > 0$, each term of the series (3.2.10) is smaller than $A \exp\left(-bn^2 \varepsilon \right)$ and, since the numerical series $\sum_n \exp\left(-bn^2 \varepsilon \right)$ is convergent for any $\varepsilon > 0$ (see (8.7.38) p. 239), the series (3.2.13) is uniformly convergent in x and t to a continuous function (Theorem 8.3.1 p. 226). Uniform convergence in x also follows from the bound on $A_n(0)$ and the exponential t-decay of the coefficients (§9.3). Thanks to the exponential decay in time, the same conclusion holds for all the derivatives, and we therefore conclude that use of (3.1.18) was legitimate. From uniform convergence we also deduce that the interchange of summation and integration leading to (3.2.13) was legitimate (§8.3 and §9.8). What remains is to prove that the initial condition is satisfied. Let us suppose for simplicity that f is absolutely integrable and continuous with $f(0) = f(L)$ although, with a little more work, this hypothesis can be relaxed. Then its Fourier series converges absolutely and uniformly (Theorem 9.4.1 p. 248) and, therefore,

$$|f(x) - u(x,t)| \leq \frac{2}{L} \sum_{n=1}^{\infty} |A_n(0)| \left(1 - \exp -\frac{\pi^2 n^2 Dt}{L^2} \right)$$

$$\leq \frac{2}{L} \left(1 - \exp -\frac{\pi^2 Dt}{L^2} \right) \sum_{n=1}^{\infty} |A_n(0)|. \qquad (3.2.19)$$

If f' has bounded variation, the coefficients $A_n(0)$ are of order n^{-2} or smaller for large n (Theorem 9.5.1 p. 251), the series converges and this difference tends to 0 for $t \to 0$. More directly, one can appeal to the relation (20.11.7) p. 601, valid for $-L \leq x, \xi \leq L$,

$$\frac{1}{L} \left[1 + \sum_{n=1}^{\infty} \left(\cos \frac{\pi n(x - \xi)}{L} + \cos \frac{\pi n(x + \xi)}{L} \right) \right] = \delta(x - \xi) + \delta(x + \xi), \quad (3.2.20)$$

to prove that the initial condition is satisfied at points of continuity of f.

The same problem with Dirichlet boundary conditions is solved by means of the Laplace transform in §5.5.2 p. 138.

3.2.2 Other homogeneous boundary conditions

If, in place of (3.2.2), the boundary conditions had been of the Dirichlet type, namely $u(0, t) = u(L, t) = 0$, we would proceed similarly using an odd extension and a Fourier sine series. This problem is solved by means of the Laplace transform in §5.5.2 p. 138.

For the case in which there is a Dirichlet condition at one end and a Neumann one at the other, e.g.

$$\left. \frac{\partial u}{\partial x} \right|_{x=0} = 0, \qquad u(L, t) = 0, \qquad (3.2.21)$$

we can proceed by "manufacturing" a suitable set of eigenfunctions of the operator d^2/dx^2 by solving

$$\frac{d^2 e_k}{dx^2} = -\lambda_k^2 e_k, \qquad \frac{de_k}{dx}\bigg|_{x=0} = 0, \qquad e_k(L) = 0. \tag{3.2.22}$$

Up to an arbitrary multiplicative constant[4] the solution is

$$e_k = \cos\left(k + \frac{1}{2}\right)\frac{\pi x}{L}, \qquad \lambda_k = \left(k + \frac{1}{2}\right)\frac{\pi}{L}, \tag{3.2.23}$$

so that the first step toward the solution of the problem is to set, in place of (3.2.4),

$$u(x,t) = \sum_{k=0}^{\infty} D_k(t) \cos\left(k + \frac{1}{2}\right)\frac{\pi x}{L}. \tag{3.2.24}$$

We would then proceed as before expanding derivatives, taking scalar products and so on. The legitimacy of the procedure is guaranteed by the Sturm–Liouville theory of Chapter 15.

An interesting way to look at (3.2.24) is the following. Suppose that the problem were of the same type as the first one solved in this section, but posed on an interval of length $2L$. We would start by setting

$$u(x,t) = \frac{A_0(t)}{2} + \sum_{n=1}^{\infty} A_n(t) \cos\frac{n\pi x}{2L}. \tag{3.2.25}$$

Now we observe that, at $x = L$, the argument of the cosines is $n\pi/2$ so that they vanish for n odd. Thus, we can satisfy $u(L, t) = 0$ if in (3.2.25) we set all the even coefficients to zero, $A_{2m} = 0$. In this way we recover (3.2.24).

3.2.3 Non-homogeneous boundary conditions

In the previous case the condition $\partial u/\partial x = 0$ at $x = 0$ enabled us to smoothly continue in an even way the function u over the interval $-L < x < 0$ and the solution procedure was rather straightforward. If the boundary conditions are not homogeneous (i.e., non-zero), the situation is slightly more complicated as we show in this example. Consider[5]

$$\frac{\partial u}{\partial t} - \frac{\partial^2 u}{\partial x^2} = 0 \tag{3.2.26}$$

in the interval $0 < x < \pi$ with the boundary and initial conditions

$$u(0, t) = f(t), \qquad u(\pi, t) = 0, \qquad u(x, 0) = 0. \tag{3.2.27}$$

For consistency of the initial data we assume that $f(0) = 0$. This problem may model a bar or slab with a prescribed time-dependent temperature at the left end while the right end is

---

[4] Eigenfunctions are defined only up to an arbitrary multiplicative constant; see e.g. p. 23, §15.2 and §18.6.

[5] Setting $t' = \pi^2 Dt/L^2$ and $x' = \pi x/L$ the diffusivity can be eliminated from the equation and the range $0 < x < L$ can be transformed to $0 < x' < \pi$. We assume that this has been done, but continue to use the symbols x and t for the dimensionless position and time.

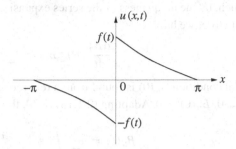

Fig. 3.1 The function $u(x, t)$ defined for $0 < x < \pi$ extended in an odd way to $-\pi < x < 0$.

kept at the reference temperature. The situation is different from the previous one in that the boundary condition at $x = 0$ is not of a type that permits a straightforward even or odd extension of the problem and use of a standard Fourier series expansion. Indeed, if we were to extend the function in an odd way (Figure 3.1) and use a sine series directly, as remarked after (3.1.12), we would find $u(0, t) = 0$ and it would be impossible to fulfill the boundary condition at the left end (Figure 3.1). If we were to use a cosine series, we would find $u(0, t) \neq 0$, but we would not know how to "apportion" $f(t)$ among the different Fourier coefficients. Furthermore, although an extended even u would be continuous at $x = 0$, its derivative would not be and this would cause convergence problems.

As demonstrated at the end of §3.1 p. 57, in this situation it is best to expand $\partial^2 u / \partial x^2$ in place of u itself. In choosing whether to use a sine or a cosine series for this purpose we should consider the known boundary conditions. If we were to use a cosine series, (3.1.14) shows that, in the process of integration by parts, the values of $\partial u / \partial x$ at the end points would arise, which are unknown. On the other hand, use of the sine series expansion (3.1.15) and integration by parts would, according to (3.1.16) p. 57, involve the values of u at the ends which are known. Thus we set

$$\frac{\partial^2 u}{\partial x^2} = \sum_{n=1}^{\infty} b_n(t) \sin nx, \tag{3.2.28}$$

and use (3.1.16) to find, in view of (3.2.27),

$$b_n(t) = \frac{2n}{\pi} f(t) - n^2 B_n(t), \tag{3.2.29}$$

where the $B_n(t)$ are the coefficients of the sine expansion of u as in (3.1.12). Upon substituting (3.2.28) with (3.2.29) into the diffusion equation (3.2.26) and using (3.1.18) for the time derivative we find

$$\sum_{n=1}^{\infty} \frac{d B_n}{dt} \sin nx = \sum_{n=1}^{\infty} \left[\frac{2n}{\pi} f(t) - n^2 B_n(t) \right] \sin nx, \tag{3.2.30}$$

from which, by the uniqueness of the series expansion, or by taking scalar products of both sides as before, we have

$$\frac{dB_n}{dt} + n^2 B_n = \frac{2n}{\pi} f(t). \tag{3.2.31}$$

The initial condition $B_n(0)$ is found from a relation similar to (3.2.12) and therefore, since $u(x, 0) = 0$, $B_n(0) = 0$. Adapting (2.1.13) p. 30, the solution of this equation is therefore

$$B_n(t) = \frac{2n}{\pi} \int_0^t e^{-n^2(t-\tau)} f(\tau) \, d\tau. \tag{3.2.32}$$

Substituting into a sine expansion for u of the form (3.1.12) and interchanging summation and integration we find

$$u(x, t) = \int_0^t f(\tau) \left[\frac{2}{\pi} \sum_{n=1}^{\infty} n \exp[-n^2(t - \tau)] \sin nx \right] d\tau, \tag{3.2.33}$$

in which the quantity in brackets is recognized as the Green's function for this problem. If the boundary condition at $x = 0$ were of the form $\partial u / \partial x = g(t)$, the proper procedure would be to expand $\partial^2 u / \partial x^2$ in a cosine series so as to to pick up the boundary data from the integration by parts according to (3.1.14). Another example of this type can be found in the second part of §4.6 starting on p. 107.

The series in (3.2.33) will converge very slowly for $t - \tau$ small. Furthermore, at least superficially, (3.2.33) looks like a sine series and one may wonder whether the condition $u(0, t) = f(t)$ is truly satisfied. This is a typical case in which use of the Poisson summation formula (8.7.33) proves highly beneficial. For this purpose, let us consider the function

$$F(s; \theta, x) = \frac{2}{\pi} \exp(-\theta s^2) \cos sx, \tag{3.2.34}$$

with $\theta = t - \tau$, in terms of which (3.2.33) may be written as

$$u(x, t) = -\frac{\partial}{\partial x} \int_0^t f(t - \theta) \left[\sum_{n=1}^{\infty} F(n; \theta, x) \right] d\theta. \tag{3.2.35}$$

According to the formula (8.7.33), we have

$$\sum_{n=1}^{\infty} F(n; \theta, x) = -\frac{1}{2} F(0; \theta, x) + \int_0^{\infty} F(\xi; \theta, x) \, d\xi$$

$$+ 2 \sum_{k=1}^{\infty} \int_0^{\infty} F(\xi; \theta, x) \cos 2k\pi \xi \, d\xi. \tag{3.2.36}$$

Since $F(0; \theta, x)$ is actually independent of x, this term will not contribute upon taking the x-derivative indicated in (3.2.35) and can be disregarded. The remaining integrals are

$$I_k = \frac{1}{\pi} \int_0^{\infty} \exp(-\theta s^2) \cos(2\pi k \pm x) s \, ds = \frac{1}{\sqrt{4\pi t}} \exp -\frac{(2\pi k \pm x)^2}{4\theta}. \tag{3.2.37}$$

For θ or x small, the dominant terms are evidently I_0 and I_1 so that

$$\sum_{n=1}^{\infty} F(n; \theta, x) \simeq \frac{1}{\sqrt{\pi \theta}} \exp\left(-\frac{x^2}{4\theta}\right) + \frac{1}{\sqrt{\pi \theta}} \exp\left(-\frac{(2\pi - x)^2}{4\theta}\right). \tag{3.2.38}$$

Keeping the first one only, for small θ we have

$$u(x, t) \simeq \frac{x}{2\sqrt{\pi}} \int_0^t f(t - \theta) \exp\left(-\frac{x^2}{4\theta}\right) \frac{d\tau}{\theta^{3/2}}, \tag{3.2.39}$$

which, upon setting $\xi = x/[2\sqrt{t - \theta}]$, becomes

$$u(x, t) = \frac{2}{\sqrt{\pi}} \int_{x/2\sqrt{t}}^{\infty} f\left(t - \frac{x^2}{4D\xi^2}\right) e^{-\xi^2} d\xi, \tag{3.2.40}$$

from which it is evident that $u(0, t) = f(t)$ given that

$$\frac{2}{\sqrt{\pi}} \int_0^{\infty} e^{-\xi^2} d\xi = 1. \tag{3.2.41}$$

This is precisely the same expression found in §4.11 for the problem of a semi-infinite bar with a prescribed temperature $f(t)$ at $x = 0$. This result shows that, as expected, whether the bar is finite or infinite is irrelevant for very small times. The next term I_1 starts to become important only at times $\theta \sim \pi^2$ or, in dimensional form, $\theta \sim L^2/D$. Thus, for the effect of this and higher-order terms to be felt, one has to wait for a time of the order of the diffusion time (3.2.16) over the length of the bar, as could be expected.

3.2.4 A general case

If the problem were

$$\frac{\partial u}{\partial t} - D \frac{\partial^2 u}{\partial x^2} = q(x, t), \tag{3.2.42}$$

with, for example,

$$u(x, 0) = f(x), \qquad u(0, t) = g_0(t), \qquad u(L, t) = g_1(t), \tag{3.2.43}$$

we can use linearity to decompose it into four pieces. We would write $u = w + u_1 + u_2 + u_3$ with

$$\frac{\partial w}{\partial t} - D \frac{\partial^2 w}{\partial x^2} = q(x, t), \tag{3.2.44}$$

$$w(x, 0) = 0, \qquad w(0, t) = w(L, t) = 0, \tag{3.2.45}$$

and

$$\frac{\partial u_j}{\partial t} - D \frac{\partial^2 u_j}{\partial x^2} = 0, \qquad j = 1, 2, 3, \tag{3.2.46}$$

with

$$u_1(x, 0) = f(x), \qquad u_1(0, t) = u_2(L, t) = 0, \tag{3.2.47}$$

while u_2 and u_3 would have zero initial conditions, with u_2 carrying the boundary condition at 0 and u_3 that at L. Each one of these problems can be solved by the methods illustrated before. Linearity can often be used to advantage in this way to decompose a non-homogeneous problem into the sum of simpler sub-problems.

3.2.5 Mixed boundary conditions

Another typical situation is a mixed boundary condition of the form e.g.

$$u(L, t) + R \left. \frac{\partial u}{\partial x} \right|_{x=L} = 0, \tag{3.2.48}$$

with R a constant. Let us consider the initial-value problem (3.2.3) for this case for the homogeneous diffusion equation (3.2.26); we take the boundary condition $u(0, t) = 0$ at $x = 0$.

If, on the basis of the left boundary condition, we were to try a Fourier sine expansion as in (3.1.15), we would have from (3.1.16)

$$b_n(t) = - \left(\frac{n\pi}{L} \right)^2 B_n(t) - (-1)^n \frac{2n\pi}{L^2} u(L, t) \tag{3.2.49}$$

and we would be faced with the non-trivial task of determining $u(L, t)$. The way to proceed is to use a set of eigenfunctions of the operator $\partial^2 / \partial x^2$ subject to boundary conditions more appropriate for this case. Since we plan to expand u on this set of eigenfunctions, we solve the eigenvalue problem

$$\frac{d^2 e_n}{dx^2} = -\lambda_n e_n, \tag{3.2.50}$$

imposing the boundary conditions $e_n(0) = 0$ and (3.2.48). This problem is solved in Example 15.2.1 p. 363, continued in Example 15.3.1 p. 370. It is shown there that there is an infinity of eigenvalues and orthogonal eigenfunctions

$$e_n(x) = a_n \sin \sqrt{\lambda_n} x, \qquad \frac{\tan \sqrt{\lambda_n} L}{\sqrt{\lambda_n}} = -R, \tag{3.2.51}$$

with a_n a normalization constant given in (15.3.28). We then set

$$u(x, t) = \sum_{n=1}^{\infty} B_n(t) e_n(x), \tag{3.2.52}$$

and proceed as in §3.2.1 using the orthogonality and normalization of the e_n to impose the initial condition.[6]

---

[6] Unlike (3.2.23), here we fix the arbitrary multiplicative constant of the eigenfunctions so that $(e_n, e_n) = 1$. As remarked more than once, there is nothing deep here – it is purely a matter of convenience.

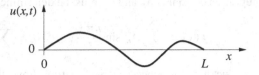

$u(x,t)$

Fig. 3.2   A vibrating string pinned at the ends, §3.3.

3.3 Waves on a string

The displacement $u(x, t)$ of a taut string pinned at $x = 0$ and $x = L$ (Figure 3.2) is governed by the wave equation (p. 18)

$$\frac{\partial^2 u}{\partial t^2} - c^2 \frac{\partial^2 u}{\partial x^2} = 0, \tag{3.3.1}$$

in which c is the wave speed, subject to the boundary conditions

$$u(0, t) = u(L, t) = 0. \tag{3.3.2}$$

We will study this problem for the initial conditions

$$u(x, 0) = f(x), \qquad \left.\frac{\partial u}{\partial t}\right|_{t=0} = 0, \tag{3.3.3}$$

with $f(0) = f(L) = 0$. The conditions (3.3.2) suggest that the function u can be extended smoothly in an odd way to the range $-L < x < 0$. This suggestion can be strengthened if we look at the relations (3.1.13) and (3.1.15) for the expansion of $\partial^2 u/\partial x^2$. The cosine series expansion would bring in the unknown derivatives at the end points, while the sine series expansion only involves $u(0, t)$ and $u(L, t)$ which are known to vanish. Hence we set

$$u(x, t) = \sum_{n=1}^{\infty} B_n(t) \sin \frac{n\pi x}{L}. \tag{3.3.4}$$

By (3.1.16), in view of (3.3.2), the coefficients of $\partial^2 u/\partial x^2$ are simply $-(n\pi c/L)^2 B_n$. For the time derivative we use the analog of (3.1.18). Substitution into (3.3.1) gives then

$$\sum_{n=1}^{\infty} \left[\frac{d^2 B_n}{dt^2} + \left(\frac{n\pi c}{L}\right)^2 B_n \right] \sin \frac{n\pi x}{L} = 0. \tag{3.3.5}$$

Upon taking scalar products (or invoking the linear independence of the sines), we have

$$\frac{d^2 B_n}{dt^2} + \left(\frac{n\pi c}{L}\right)^2 B_n = 0, \tag{3.3.6}$$

with the solution

$$B_n(t) = A_n \cos \frac{n\pi ct}{L} + C_n \sin \frac{n\pi ct}{L}. \tag{3.3.7}$$

The integration constants A_n and C_n must be determined from the initial conditions (3.3.3):

$$f(x) = \sum_{n=1}^{\infty} A_n \sin \frac{n\pi x}{L}, \qquad 0 = \sum_{n=1}^{\infty} C_n \frac{n\pi c}{L} \sin \frac{n\pi x}{L}. \qquad (3.3.8)$$

Scalar product with the generic cosine function shows then that $C_n = 0$ for all n while, upon using the relation (3.2.6), scalar product with the sines gives

$$A_k = \frac{2}{L} \int_0^L \sin \frac{k\pi x}{L} f(x) \, dx . \qquad (3.3.9)$$

Substitution into (3.3.4) and rearrangement give

$$
\begin{aligned}
u(x, t) &= \frac{1}{2} \sum_{n=1}^{\infty} A_n \left[\sin \frac{n\pi (ct + x)}{L} - \sin \frac{n\pi (ct - x)}{L} \right] \\
&= \sum_{n=1}^{\infty} A_n \sin \frac{n\pi x}{L} \cos \frac{n\pi ct}{L} . \qquad (3.3.10)
\end{aligned}
$$

By writing A_n explicitly according to its definition (3.3.9), the solution of the problem can be written similarly to (3.2.15), which permits us to readily identify the Green's function for this case.

Comparison of the first form of (3.3.10) with the first relation of (3.3.8) shows that

$$u(x, t) = \frac{1}{2} [f(ct + x) - f(ct - x)], \qquad (3.3.11)$$

which is a special case of the d'Alembert's general solution of the wave equation (p. 8). Use of the solution in the form (3.3.11) requires extending the function f periodically over the entire range $-\infty < x < \infty$ so that $f(x + NL) = f(x)$ for any positive or negative integer N. In the form (3.3.10) the solution shows that each monochromatic wave contained in the initial condition splits into two waves travelling in opposite directions which interfere destructively at the end points 0 and L. The second form (3.3.10) shows that the solution of the problem may also be thought of as consisting of the superposition of an infinity of *standing waves*, namely states of motion in which every degree of freedom of the system (in this example, each point x) evolves with time in the same way (i.e. proportionally to $\cos(n\pi ct/L)$), but with different amplitudes (i.e. $A_n \sin(n\pi x/L)$).

It is shown in Chapter 20 that, unlike the Laplace, Poisson or diffusion equations, the wave equation admits discontinuous solutions. For this reason, the form (3.3.10) of the solution is not exactly equivalent to (3.3.11) as, at points of discontinuity, the Fourier series will converge to the average value according to (3.1.3), while (3.3.11) will respect the proper discontinuity. More on these matters will be found in Chapter 20.

3.4 Poisson equation in a square

Consider the Poisson equation

$$\frac{\partial^2 u}{\partial x^2} + \frac{\partial^2 u}{\partial y^2} = -f(x, y), \qquad (3.4.1)$$

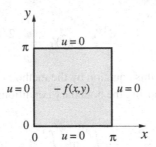

Fig. 3.3   A square membrane subject to a distributed force, §3.4

in the square $0 < x < \pi$, $0 < y < \pi$ (Figure 3.3) subject to homogeneous Dirichlet conditions:

$$u(0, y) = u(\pi, y) = u(x, 0) = u(x, \pi) = 0. \tag{3.4.2}$$

This may be interpreted as the problem of finding the deformation of a membrane fixed to a planar rigid square frame and subject to a force $f(x, y)$ at each point. We will find the solution in three different ways.

By an argument which parallels that made after Eq. (3.2.3), the boundary conditions now suggest an expansion of u in a Fourier sine series in x:

$$u(x, y) = \sum_{n=1}^{\infty} Y_n(y) \sin nx. \tag{3.4.3}$$

The coefficients Y_n depend on y by the same argument used in the previous examples to justify the time dependence of the coefficients. A point to note is that, if the boundary conditions at $y = 0$, π are to be satisfied *for every* x, it is evidently necessary that[7]

$$Y_n(0) = Y_n(\pi) = 0. \tag{3.4.4}$$

After using (3.1.16) to express the Fourier coefficients of $\partial^2 u/\partial y^2$ in terms of those of u and substituting into the field equation (3.4.1) we find

$$\sum_{n=1}^{\infty} \left[\frac{d^2 Y_n}{dy^2} - n^2 Y_n \right] \sin nx = -f(x, y). \tag{3.4.5}$$

We take the scalar product of both sides with the generic sine function $\sin kx$. Upon using the scalar product relation

$$(\sin kx, \sin nx) \equiv \int_0^\pi \sin kx \sin nx \, dx = \frac{\pi}{2} \delta_{kn}, \tag{3.4.6}$$

---

[7] More formally, these relations follow by writing (3.4.3) at $x = 0$ and π and taking scalar products with $\sin nx$.

we find

$$\frac{\mathrm{d}^2 Y_k}{\mathrm{d}y^2} - k^2 Y_k = -\frac{2}{\pi} \int_0^\pi \sin kx \, f(x, y) \, \mathrm{d}x. \tag{3.4.7}$$

The solution of this equation by the method of variation of parameters is given by (2.2.26) p. 34. After imposing the boundary conditions (3.4.4), the result may be compactly written as

$$Y_k(y) = \frac{2}{\pi k \sinh \pi k} \int_0^\pi \mathrm{d}\xi \int_0^\pi \mathrm{d}\eta \, \sinh(ky_<) \, \sinh[k(\pi - y_>)] \, \sin(k\xi) f(\xi, \eta), \tag{3.4.8}$$

where $y_< = \min(y, \eta)$, $y_> = \max(y, \eta)$. The integration over ξ arises from the scalar product while that over η comes from the application of the method of variation of parameters. In conclusion

$$u = \frac{2}{\pi} \sum_{n=1}^\infty \frac{\sin nx}{n \sinh n\pi} \int_0^\pi \mathrm{d}\xi \int_0^\pi \mathrm{d}\eta \, \sinh(ny_<) \, \sinh[n(\pi - y_>)] \, \sin(n\xi) f(\xi, \eta). \tag{3.4.9}$$

Once again we recognize a typical Green's function structure (§2.4) with the solution given by

$$u = \int_0^\pi \mathrm{d}\xi \int_0^\pi \mathrm{d}\eta \, G(x, y; \xi, \eta) \, f(\xi, \eta), \tag{3.4.10}$$

with

$$G(x, y; \xi, \eta) = \frac{2}{\pi} \sum_{n=1}^\infty \frac{\sin nx \, \sinh n(\pi - y)}{n \sinh n\pi} \sinh n\eta \, \sin n\xi. \tag{3.4.11}$$

The second method of solution proceeds in exactly the same way, except that we now expand u in a Fourier sine series in y:

$$u(x, y) = \sum_{n=1}^\infty X_n(x) \sin ny \tag{3.4.12}$$

with $X_n(0) = X_n(\pi) = 0$ to satisfy the boundary conditions in x. The calculations are the same as for the previous case and we find

$$X_k(y) = \frac{2}{\pi k \sinh \pi k} \int_0^\pi \mathrm{d}\xi \int_0^\pi \mathrm{d}\eta \, \sinh(kx_<) \sinh[k(\pi - x_>)] \, \sin(k\eta) \, f(\xi, \eta), \tag{3.4.13}$$

in which $x_< = \min(x, \xi)$, $x_> = \max(x, \xi)$ and now the integration over ξ arises from the use of variation of parameters and that over η from the scalar product. The final result for u takes on the same form as (3.4.9) with the roles of x and y interchanged.

As will be shown shortly, the series (3.4.3) and (3.4.12), with the coefficients given in (3.4.8) and (3.4.13), respectively, define the same function $u(x, y)$. The choice between the two expansions is a practical matter depending, for example, on the speed of convergence of one or the other series at the specific point(s) (x, y) of interest.

The third method of solution starts from the observation that the boundary conditions (3.4.4) on the coefficients $Y_n(y)$ of the expansion (3.4.3) suggest that they themselves be expanded in a Fourier series in y:

$$Y_n(y) = \sum_{k=1}^{\infty} A_{nk} \sin ky. \tag{3.4.14}$$

The coefficients A_{nk} are now constants but carry a double index because one index tells us which function $\sin ky$ the coefficient multiplies, while the other one tells us which of the Y_n is being expanded. Upon substituting into Eq. (3.4.1) and taking a double scalar product in both x and y, we find

$$A_{nk} = \frac{4}{\pi^2} \frac{1}{n^2 + k^2} \int_0^{\pi} d\xi \int_0^{\pi} d\eta \, f(\xi, \eta) \sin n\xi \, \sin k\eta. \tag{3.4.15}$$

The use of a double series in x and y reduces the original differential equation to a simple algebraic equation for the A_{nk}'s, but the price to pay is that the answer is in the form of a *double* series, rather than a single one. In practice, this would be a serious drawback if actual numerical values of $u(x, y)$ were to be found directly from this expression.

In order to reconcile this third solution with the second one, we write

$$
\begin{aligned}
u(x, y) &= \frac{4}{\pi^2} \sum_{k=1}^{\infty} \sum_{n=1}^{\infty} \frac{\sin nx \, \sin ky}{n^2 + k^2} \int_0^{\pi} d\xi \int_0^{\pi} d\eta \, f(\xi, \eta) \sin n\xi \, \sin k\eta \\
&= \frac{2}{\pi^2} \sum_{k=1}^{\infty} \sin ky \int_0^{\pi} d\xi \int_0^{\pi} d\eta \, f(\xi, \eta) \sin k\eta \\
&\quad \times \sum_{n=1}^{\infty} \frac{\cos n(x - \xi) - \cos n(x + \xi)}{n^2 + k^2}.
\end{aligned} \tag{3.4.16}
$$

The last series can be evaluated in closed form using the result (8.7.37) p. 239 obtained in §8.7 as an example of the application of the Poisson summation formula:

$$\sum_{n=1}^{\infty} \frac{\cos n(x \pm \xi)}{n^2 + k^2} = \frac{\pi}{2k} \frac{\cosh k[\pi - |x \pm \xi|]}{\sinh \pi k} - \frac{1}{2k^2}, \tag{3.4.17}$$

to find, if $x > \xi$,

$$\sum_{n=1}^{\infty} \frac{\cos n(x - \xi) - \cos n(x + \xi)}{n^2 + k^2} = \frac{\pi}{k} \frac{\sinh k(\pi - x) \, \sinh k\xi}{\sinh k\pi}, \tag{3.4.18}$$

and the same result with x and ξ interchanged if $x < \xi$. Here use has been made of the relation

$$\cosh a - \cosh b = 2 \sinh \frac{a + b}{2} \sinh \frac{a - b}{2}. \tag{3.4.19}$$

With the result (3.4.18), the double series (3.4.16) is reduced to a single series with the coefficients X_k given in (3.4.13); a parallel development leads to (3.4.8). These arguments prove the identity of the two expansions (3.4.3) and (3.4.12).

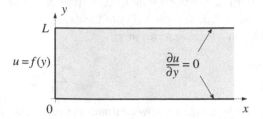

A semi-infinite waveguide for the Helmholtz equation, §3.5.

We can now take a brief look at the steps that we have taken and consider their legitimacy directly rather than via the arguments given in connection with (3.1.16). For large k the hyperbolic functions in (3.4.8) behave as $e^{-n(y_> - y_<)}/n$ with a similar behaviour for (3.4.13). This fact suggests (which can be proven rigorously) that both the series expansions (3.4.3) and (3.4.12) converge uniformly for $y > 0$ and $x > 0$, so that it was indeed legitimate to differentiate term by term and to interchange summation and integration (pp. 226 and 227).

3.5 Helmholtz equation in a semi-infinite strip

Consider now the two-dimensional Helmholtz equation

$$\frac{\partial^2 u}{\partial x^2} + \frac{\partial^2 u}{\partial y^2} + h^2 u = 0, \tag{3.5.1}$$

in the semi-infinite strip $0 < x < \infty$, $0 < y < L$ (Figure 3.4). We take homogeneous Neumann conditions along the strip edges, a Dirichlet condition at $x = 0$

$$\left. \frac{\partial u}{\partial y} \right|_{y=0,L} = 0 \qquad u(0, y) = f(y), \tag{3.5.2}$$

and require the solution to be bounded at infinity. A typical physical counterpart would be an acoustical waveguide with rigid walls and a source of monochromatic waves $f(y)e^{-i\omega t}$ at the left end $x = 0$; in this picture the constant $h = \omega/c$, with c the speed of sound, is the wavenumber, and the space- and time-dependent wave would be given by $u(x, y)e^{-i\omega t}$. An interesting feature of this problem concerns the behavior of the solution at infinity, as we will see.

The boundary conditions suggest the use of a Fourier cosine series in y:

$$u(x, y) = \frac{1}{2}A_0(x) + \sum_{n=1}^{\infty} A_n(x) \cos \frac{n\pi y}{L}, \tag{3.5.3}$$

and we know that the values of $A_n(0)$ will be determined from the condition at $x = 0$:

$$f(y) = \frac{1}{2}A_0(0) + \sum_{n=1}^{\infty} A_n(0) \cos \frac{n\pi y}{L}. \tag{3.5.4}$$

The calculation is as in the derivation of (3.2.12) which provides an expression for the $A_n(0)$ in this case as well. Substitution into the equation leads to

$$\frac{d^2 A_n}{dx^2} + \left[h^2 - \left(\frac{n\pi}{L}\right)^2\right] A_n = 0. \tag{3.5.5}$$

If $h < n\pi/L$, the unique solution bounded at infinity is

$$A_n(x) = A_n(0) e^{-\sqrt{(n\pi/L)^2 - h^2}\, x}. \tag{3.5.6}$$

If, however, $n\pi/L < h$, all we can say is that

$$A_n(x) = C_n e^{i\sqrt{h^2 - (n\pi/L)^2}\, x} + B_n e^{-i\sqrt{h^2 - (n\pi/L)^2}\, x} \tag{3.5.7}$$

and, since both terms are bounded at infinity, we are left with two unknowns and a single boundary condition (3.5.4). This ambiguity is not one of mathematics, but of physics and, to resolve it, we need to return to the complete physical picture.

If we reintroduce the time dependence $e^{-i\omega t}$, we see that the first term of (3.5.7) corresponds to a wave component propagating to the right toward infinity, while the second one describes waves propagating to the left from infinity, which would evidently be a perfectly possible physical situation. Since the present problem specification however states that the source of waves is at $x = 0$, left-propagating waves cannot be present and we must take $B_n = 0$. This type of argument and the conclusion to which it leads is known as *radiation condition* and is characteristic of wave problems in infinite domains. We will encounter other examples in §4.2 p. 90 and in §7.13 p. 196.

In conclusion, for large x, in addition to the longitudinal mode $\frac{1}{2}A_0(0)e^{ihx}$, the solution only contains transverse modes such that $n\pi/L < h$. In particular, if $f(y)$ is such that such terms do not appear (or are very small), essentially nothing other than the $n = 0$ mode will propagate far from the source. In the theory of waveguides, π/L is known as the cut-off wave number and the corresponding frequency $c/(2L)$ as the *cut-off frequency* of the waveguide. The non-propagating modes are known as *evanescent waves*.

3.6 Laplace equation in a disk

Let us consider now the Laplace equation in a disk of radius R (Figure 3.5). Upon adopting polar coordinates in the plane centered at the center of the disk, the equation to be solved is (Table 6.4 p. 148)

$$\frac{1}{r}\frac{\partial}{\partial r}\left(r\frac{\partial u}{\partial r}\right) + \frac{1}{r^2}\frac{\partial^2 u}{\partial \theta^2} = 0, \tag{3.6.1}$$

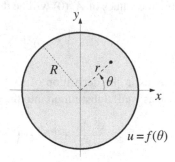

Fig. 3.5 Laplace equation in a disk of radius R, §3.6.

in the circle $0 < r < R$. The range of the angular variable θ is $0 \le \theta < 2\pi$. We will study the Dirichlet problem for the equation (3.6.1), namely we associate to the field equation the boundary condition

$$u(R, \theta) = f(\theta), \qquad (3.6.2)$$

where the function f is given. In addition, we require that the solution be regular (i.e., non-singular) inside the disk. This problem might correspond, for instance, to a force-free circular membrane attached to a frame distorted according to the function (3.6.2). Alternatively, we may think of a cylinder infinite in the direction normal to the page with a prescribed temperature on the boundary independent of the axial position.

The first step is to expand u in a Fourier series in the angular variable. The formulae are more compact if we use the exponential form, as then we do not need to carry separate summations for the sine and cosine terms:

$$u = \sum_{-\infty}^{\infty} S_n(r)\, e_n(\theta) \qquad \text{where} \qquad e_n(\theta) = \frac{1}{\sqrt{2\pi}} \exp in\theta. \qquad (3.6.3)$$

We substitute into (3.6.1) and take the scalar product with the generic e_k to find

$$\frac{1}{r} \frac{d}{dr}\left(r \frac{dS_k}{dr}\right) - \frac{k^2}{r^2} S_k = 0. \qquad (3.6.4)$$

The solution for $k = 0$ is

$$S_0 = A_0 + B_0 \log r, \qquad (3.6.5)$$

while, for $k \ne 0$,

$$S_k = A_k\, r^{|k|} + B_k\, r^{-|k|}. \qquad (3.6.6)$$

It is convenient to express this solution in terms of $r^{\pm|k|}$ rather than $r^{\pm k}$ as, in this way, the coefficients A_k are associated with functions which are regular at $r = 0$ and the B_k's to functions that are singular at the origin irrespective of the sign of k.

The condition of regularity at the origin requires that all the B_k's vanish so that we are left with

$$u = \sum_{-\infty}^{\infty} A_n \, r^{|n|} \, e_n(\theta). \tag{3.6.7}$$

The constants A_n are now determined from the boundary condition (3.6.2) as

$$f(\theta) = \sum_{-\infty}^{\infty} A_n \, R^{|n|} \, e_n(\theta). \tag{3.6.8}$$

By taking scalar products we find

$$A_k = R^{-|k|} \, (e_k, f) = R^{-|k|} \int_0^{2\pi} \bar{e}_k(\eta) \, f(\eta) \, d\eta. \tag{3.6.9}$$

Upon substituting into (3.6.7) and interchanging integration and summation we find[8]

$$u = \frac{1}{2\pi} \int_0^{2\pi} \left[\sum_{-\infty}^{\infty} \left(\frac{r}{R} \right)^{|n|} \exp in(\theta - \eta) \right] f(\eta) \, d\eta. \tag{3.6.10}$$

If the term $n = 0$ is separated and the remaining series between $-\infty$ and -1, and between 1 and ∞ are combined, the series reduces to the series (8.7.19) evaluated in closed form on p. 236; adapting that result we find

$$u(r, \theta) = \frac{1}{2\pi} \int_0^{2\pi} \frac{R^2 - r^2}{R^2 - 2rR \cos(\theta - \eta) + r^2} \, f(\eta) \, d\eta \qquad 0 \le r < R. \tag{3.6.11}$$

This is *Poisson's integral formula for the circle*; in Example 17.10.7 p. 484 we derive this result in a different way by using conformal mapping. The fraction inside the integral, divided by 2π, is the Green's function for this problem.

If we had been interested in the exterior problem, i.e. for $R < r < \infty$, we would have discarded the coefficients A_n in place of the B_n but would have otherwise followed the same procedure. Analogous formulae can be derived for the Neumann problem, in which $\partial u / \partial r$ is assigned on the boundary rather than u itself.

We can use the result (3.6.11) to prove, in this special case, some interesting properties of harmonic functions, i.e., functions which satisfy the Laplace equation. These properties are actually valid in general in any number of dimensions.

In the first place, if we set $r = 0$, we find

$$u(0, \theta) = \frac{1}{2\pi} \int_0^{2\pi} f(\eta) \, d\eta = \frac{1}{L} \oint u(\mathbf{x}_s) \, ds, \tag{3.6.12}$$

where $L = 2\pi R$, s is the arc length, $\mathbf{x}_s$ is the generic point on the circle, and the integration is around the circle. This is obviously true whatever the radius R.[9] Hence we deduce the

---

[8] (3.6.7) is essentially a power series for which this interchange is legitimate inside the radius of convergence $r = R$ (p. 229). It remains to check that (3.6.10) or (3.6.11) satisfy the boundary condition; we omit this step.

[9] Provided, of course, that the function satisfies Laplace's equation throughout the disk, otherwise Poisson's formula does not apply.

mean value theorem (see also p. 431): The value of a harmonic function at a point is equal to its average value calculated over any circle centered at that point. In a higher-dimensional space, the same statement holds with the word "circle" replaced by "(hyper-)sphere". This is a rather remarkable property which shows how "special" harmonic functions are.

Secondly, the solution of (3.6.1) satisfying $f = 1$ is evidently $u = 1$, so that

$$1 = \frac{1}{2\pi} \int_0^{2\pi} \frac{R^2 - r^2}{R^2 - 2rR\cos(\theta - \eta) + r^2} \, d\eta, \tag{3.6.13}$$

which can also be proven directly. Therefore, since the integrand in (3.6.13) is positive (or zero for $r = R$), we find from (3.6.11):

$$u(r, \theta) \le (\max\ f)\ \frac{1}{2\pi} \int_0^{2\pi} \frac{R^2 - r^2}{R^2 - 2rR\cos(\theta - \eta) + r^2} \, d\eta = \max\ f \tag{3.6.14}$$

and, in the same way,

$$u(r, \theta) \ge \min\ f. \tag{3.6.15}$$

It is obvious from the derivation that the equal signs only hold if $f = $ constant but, in this case, u would also equal the same constant. We have thus proven the *maximum principle* for harmonic functions in this special case: unless u is a constant, it cannot take on a maximum or minimum value inside the domain, but only on the boundary. The simple physical interpretation of this principle given on p. 11 indicates that the result is true not only for a circle, but in arbitrarily shaped domains in an arbitrary number of space dimensions. A less obvious consequence is that there can be no equilibrium position for a point charge in the interior of an electrostatic or gravitational field as, at such a position, the potential would be a minimum.

A uniqueness theorem for the solution of the Laplace equation with Dirichlet boundary conditions is also an immediate consequence of the maximum/minimum principle: given two solutions u_1 and u_2, their difference satisfies the Laplace equation and therefore its maximum and minimum must be on the boundary where, if the boundary conditions are identical, $u_1 - u_2 = 0$.

Since the function multiplying $f(\eta)$ in (3.6.11) is infinitely differentiable in the interior of the domain, we also see that u is infinitely differentiable *inside* the domain independently of the nature of the boundary data f.[10]

As a final point, let us set $R = 1$ in (3.6.8) and substitute into (3.6.7) to find

$$u(r, \theta) = \sum_{-\infty}^{\infty} r^{|n|}(e_n, f)\, e_n(\theta), \tag{3.6.16}$$

so that, as $r \to 1$

$$u(1, \theta) = f(\theta) = \frac{1}{2\pi} \lim_{r \to 1-} \int_0^{2\pi} f(\eta)\, \frac{1 - r^2}{1 - 2r\cos(\theta - \eta) + r^2} \, d\eta, \tag{3.6.17}$$

---

[10] This result is similar to that discussed in §20.8 in connection with the relation between distributions and analytic functions.

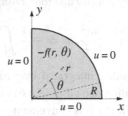

Poisson equation in a circular sector of radius R, §3.7.

whenever the limit exists. By the definition of the Abel (or Abel–Poisson) sum (see §8.5), it is clear that the integral represents the Abel–Poisson sum of the Fourier series for f. It can be shown that, for any integrable function f, the Abel-Poisson sum of the Fourier series converges almost everywhere, to $\frac{1}{2}[f(x+0) + f(x-0)]$ at points of simple discontinuity and to $f(x)$ at points of continuity. We thus deduce that, when applied to continuous functions,

$$\lim_{r \to 1-} \frac{1-r^2}{1 - 2r\cos\theta + r^2} = 2\pi\,\delta(\theta). \tag{3.6.18}$$

3.7 The Poisson equation in a sector

Let us consider the two-dimensional Poisson equation

$$\frac{1}{r}\frac{\partial}{\partial r}\left(r\frac{\partial u}{\partial r}\right) + \frac{1}{r^2}\frac{\partial^2 u}{\partial \theta^2} = -f(r,\theta) \tag{3.7.1}$$

in the sector $0 < r < R$, $0 < \theta < \frac{1}{2}\pi$ (Figure 3.6), subject to homogeneous boundary conditions

$$u(r,0) = u\left(r, \frac{1}{2}\pi\right) = u(R,\theta) = 0. \tag{3.7.2}$$

In considering a Fourier expansion in θ, the condition for $\theta = 0$ suggests to use a sine series $\sin n\theta$. In order to satisfy the condition at $\theta = \frac{1}{2}\pi$ we need $\sin(n/2)\pi = 0$, which can only be satisfied if $n = 2k$ is even. Thus we let

$$u(r,\theta) = \sum_{k=1}^{\infty} R_k(r) \sin 2k\theta. \tag{3.7.3}$$

Exactly the same set of basis functions would be found by taking $L = \frac{1}{2}\pi$ in the relation (3.1.12). With this step we are led to

$$\sum_{k=1}^{\infty} \left[\frac{1}{r}\frac{d}{dr}\left(r\frac{dR_k}{dr}\right) - \frac{4^2}{r^2}R_k\right] \sin 2k\theta = -f(r,\theta). \tag{3.7.4}$$

Upon multiplying by $\sin 2n\theta$ and integrating between 0 and $\frac{1}{2}\pi$ again using (3.4.6), we find

$$\frac{1}{r}\frac{d}{dr}\left(r\frac{dR_n}{dr}\right) - \frac{4n^2}{r^2}R_n = -\frac{4}{\pi}\int_0^{\pi/2} f(r,\theta)\sin 2n\theta \; d\theta \equiv -f_n(r). \qquad (3.7.5)$$

This equation is readily integrated by variation of parameters using the solutions of the homogeneous equation shown in (3.6.6) of an earlier example. In order to write the result in a more compact form, it is convenient to set

$$r_> = \max(r,s), \qquad r_< = \min(r,s), \qquad\qquad (3.7.6)$$

in terms of which we find

$$R_n(r) = \frac{1}{4|n|}\int_0^R \left[\left(\frac{r_<}{r_>}\right)^{2|n|} - \left(\frac{sr}{R^2}\right)^{2|n|}\right] f_n(s)\; ds. \qquad (3.7.7)$$

The series that we thus obtain can be summed in closed form by the methods of §8.7.

In this case the choice of the family of functions suitable for the expansions was evident by inspection. In a sector with an aperture different from $\frac{1}{2}\pi$ we can use the general formulae of §3.1 to find them.

3.8 The quantum particle in a box: eigenvalues of the Laplacian

Let us now consider the stationary-state Schrödinger equation for a particle of mass m confined in a cubic box of side L by infinitely high potential barriers (Figure 3.7):

$$-\frac{\hbar^2}{2m}\nabla^2 u = Eu, \qquad\qquad (3.8.1)$$

or, with $k^2 = 2mE/\hbar^2$,

$$\nabla^2 u + k^2 u = 0. \qquad\qquad (3.8.2)$$

The infinitely high potential on the boundary of the box requires that

$$u(0,y,z) = u(x,0,z) = u(x,y,0) = 0,$$
$$u(L,y,z) = u(x,L,z) = u(x,y,L) = 0. \qquad (3.8.3)$$

This mathematical formulation also describes the vibrations of an elastic solid enclosed by rigid walls, or the standing sound waves in a cubic room with rigid walls. In these cases the constant k would be interpreted as the wave number, equal to ω/c, where ω is the angular frequency and c the speed of the elastic or acoustic waves.

Mathematically, it is evident that the problem embodied in (3.8.2) and (3.8.3) is an eigenvalue problem for the Laplacian. As with all linear eigenvalue problems, and unlike any of the examples considered so far, the trivial solution $u = 0$ satisfies all the conditions.

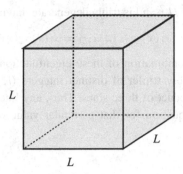

Fig. 3.7 Quantum particle, or the Helmholtz equation, in a box, §3.8.

Just as in the case of the square membrane of §3.4, the boundary conditions suggest a multiple Fourier sine series expansion:

$$u(x, y, z) = \sum_{j=1}^{\infty} \sum_{m=1}^{\infty} \sum_{n=1}^{\infty} a_{jmn} \sin \frac{j\pi x}{L} \sin \frac{m\pi y}{L} \sin \frac{n\pi z}{L}. \tag{3.8.4}$$

Upon substitution into the equation we find

$$\sum_{j=1}^{\infty} \sum_{m=1}^{\infty} \sum_{n=1}^{\infty} a_{kmn} \left[k^2 - (j^2 + m^2 + n^2) \frac{\pi^2}{L^2} \right] \sin \frac{j\pi x}{L} \sin \frac{m\pi y}{L} \sin \frac{n\pi z}{L} = 0. \tag{3.8.5}$$

By virtue of the usual scalar product argument, or simply by the linear independence of the products of sines corresponding to different triplets of integers, we conclude that either all the a_{jmn} vanish, which brings us back to the trivial solution, or they all vanish except those corresponding to a small set of integers j, m, n such that

$$k^2 = (j^2 + m^2 + n^2) \frac{\pi^2}{L^2}. \tag{3.8.6}$$

In the quantum mechanical context, this relation tells us that the energy levels of the system are

$$E_{jmn} = \frac{\pi^2 \hbar^2}{2mL^2} (j^2 + m^2 + n^2). \tag{3.8.7}$$

In the acoustic problem, (3.8.6) identifies the wave numbers of the possible standing waves supported by the system.

The smallest eigenvalue, corresponding to the fundamental quantum state (or lowest frequency), is found for $j = m = n = 1$ and is

$$u_{111} = a_{111} s_1(x) s_1(y) s_1(z), \qquad k_{111}^2 = \frac{3\pi^2}{L^2}, \tag{3.8.8}$$

where we write $s_1(x) = \sin(\pi x / L)$ etc. for brevity. The next higher state is obtained when one of the three integers equals 2 while the other ones equal 1; the corresponding eigenvalue is $k_{211}^2 = 6\pi^2/L^2$ and is triply degenerate, having eigenfunctions

$$u_{211} = s_2(x)\,s_1(y)\,s_1(z), \quad u_{121} = s_1(x)\,s_2(y)\,s_1(z), \quad u_{112} = s_1(x)\,s_1(y)\,s_2(z). \quad (3.8.9)$$

Any linear combination of these eigenfunctions is a possible state of the system. More generally, every triplet of distinct integers (i, j, k) can be assigned in $3! = 6$ different ways to a product of three sines. Thus, any linear combination of these six functions will satisfy the equation for that particular value of k_{ijk}^2. All these eigenvalues are six-fold degenerate.

3.9 Elastic vibrations

The string problem of §3.3 is one example of elastic vibrations. Here we consider the vibrations of a beam, which are governed by (1.6.10) p. 18:

$$\frac{\partial^4 u}{\partial x^4} + \frac{1}{a^2}\frac{\partial^2 u}{\partial t^2} = p(x, t), \quad (3.9.1)$$

where $a^2 = EI/(\rho S)$, $p = P/(EI)$ with E Young's modulus, I the moment of inertia of the cross section of area S and ρ the mass density.

For boundary conditions at the end points of the beam $x = 0$ and $x = L$ involving u, the position, and $\partial^2 u/\partial x^2$, proportional to the bending moment acting on the end point, a consideration of (3.1.16) shows that the appropriate expansion is the sine expansion (3.1.12); indeed, repeated application of (3.1.16) gives

$$\int_0^L \frac{\partial^4 u}{\partial x^4}\sin n'x\,dx = n'\left[\frac{\partial^2 u}{\partial x^2}\bigg|_{x=0} - (-1)^n\frac{\partial^2 u}{\partial x^2}\bigg|_{x=\pi}\right] - n'^2\int_0^L\frac{\partial^2 u}{\partial x^2}\sin n'x\,dx$$

$$= n'\left\{\left[n'^2 u - \frac{\partial^2 u}{\partial x^2}\right]_{x=0} - (-1)^n\left[n'^2 u - \frac{\partial^2 u}{\partial x^2}\right]_{x=\pi}\right\} + n'^4\frac{L}{2}B_n, \quad (3.9.2)$$

where we have set $n' = n\pi/L$ for brevity. In other cases the boundary conditions might specify $\partial u/\partial x$, i.e., the slope of the beam, and $\partial^3 u/\partial x^3$, proportional to the force. In such cases a cosine expansion would be useful. However, it is clear that neither the sine nor the cosine expansion can deal with a situation in which both even- and odd-order derivatives are specified; we consider an example of this type in §3.9.2.

3.9.1 Freely supported beam

For a freely supported beam we prescribe the position of the end points, $u(0, t) = u(L, t) = 0$, and require that the bending moments acting on the end points vanish so that $\partial^2 u/\partial x^2 = 0$ at $x = 0$ and $x = L$.

We begin with an initial-value problem:

$$u(x, 0) = f(x), \quad \frac{\partial u}{\partial t}\bigg|_{t=0} = g(x). \quad (3.9.3)$$

If the beam is not subjected to distributed forces, we see from (3.9.2) that the equation for the coefficients $B_n(t)$ is

$$\frac{d^2 B_n}{dt^2} + \omega_n^2 B_n = 0, \qquad \omega_n^2 = \frac{a^2 n^4 \pi^4}{L^4}. \tag{3.9.4}$$

If f_n and g_n denote the Fourier sine coefficients of f and g, we have then

$$B_n = f_n \cos \omega_n t + \frac{g_n}{\omega_n} \sin \omega_n t, \tag{3.9.5}$$

and

$$u(x, t) = \frac{2}{L} \sum_{n=1}^{\infty} \sin \frac{n\pi x}{L} \cos \frac{n^2 \pi^2 at}{L^2} \int_0^L f(\xi) \sin \frac{n\pi \xi}{L} d\xi$$

$$+ \frac{2L}{\pi^2 a} \sum_{n=1}^{\infty} \frac{1}{n^2} \sin \frac{n\pi x}{L} \sin \frac{n^2 \pi^2 at}{L^2} \int_0^L g(\xi) \sin \frac{n\pi \xi}{L} d\xi. \tag{3.9.6}$$

If, for example, the beam is initially at rest and is imparted an initial velocity by an impulse applied at x_0, we have $f = 0$, $g(x) = g_0 \delta(x - x_0)$ and

$$u(x, t) = \frac{2L}{\pi^2 a} g_0 \sum_{n=1}^{\infty} \frac{1}{n^2} \sin \frac{n\pi x}{L} \sin \frac{n\pi x_0}{L} \sin \frac{n^2 \pi^2 at}{L^2}. \tag{3.9.7}$$

For the case of a distributed forcing we consider the same boundary conditions as before, zero initial velocity and displacement, but non-zero forcing $p(x, t)$ in (3.9.1). We are led to

$$\frac{d^2 B_n}{dt^2} + \omega_n^2 B_n = p_n(t), \qquad p_n(t) = \frac{2}{L} \int_0^L p(x, t) \sin \frac{n\pi x}{L} dx. \tag{3.9.8}$$

The solution is readily written down from (2.2.29) p. 34 and is

$$B_n = \frac{1}{\omega_n} \int_0^t p_n(\tau) \sin \omega_n(t - \tau) d\tau. \tag{3.9.9}$$

For the case of excitation at an end point, e.g. $x = 0$, we take $p(x, t) = 0$, $u(0, t) = h(t)$ and all the other initial and boundary conditions zero. In this case use of (3.9.2) leads to

$$\frac{d^2 B_n}{dt^2} + \omega_n^2 B_n = -\left(\frac{n\pi}{L}\right)^3 h(t), \tag{3.9.10}$$

the solution of which is similar to (3.9.9).

3.9.2 Cantilevered beam

Let us consider now the free vibrations of a cantilevered beam, namely a beam fixed normally to a wall at $x = 0$ but otherwise free to execute transverse vibrations. The boundary conditions appropriate for this case are

$$u(0, t) = 0, \qquad \frac{\partial u}{\partial x}(0, t) = 0, \qquad \frac{\partial^2 u}{\partial x^2}(L, t) = \frac{\partial^3 u}{\partial x^3}(L, t) = 0. \tag{3.9.11}$$

We study the initial-value problem $u(x, 0) = f(x)$, $(\partial u/\partial t)(x, 0) = 0$.

Similarly to the example of §3.2.5, neither the sine nor the cosine series can deal with these boundary conditions without introducing serious difficulties. Let us therefore find a more suitable set of eigenfunctions by solving the eigenvalue problem

$$\frac{d^4 e_n}{dx^4} = \lambda_n^4 e_n \tag{3.9.12}$$

where, without loss of generality, we write the eigenvalue as λ_n^4 to simplify the subsequent formulae. Since we will ultimately expand u on this set of eigenfunctions, we will impose on them the boundary conditions (3.9.11). The solutions of (3.9.12) have the form

$$e_n = A_n \cos \lambda_n x + B_n \cosh \lambda_n x + C_n \sin \lambda_n x + D_n \sinh \lambda_n x. \tag{3.9.13}$$

The first two boundary conditions require $B_n = -A_n$ and $D_n = -C_n$. Upon imposing the remaining conditions we find

$$e_n = B_n \left[\cosh \lambda_n x - \cos \lambda_n x - \frac{\cosh \lambda_n L + \cos \lambda_n L}{\sinh \lambda_n L + \sin \lambda_n L} (\sinh \lambda_n x - \sin \lambda_n x) \right] \tag{3.9.14}$$

with B_n a normalization constant and the eigenvalues determined from the characteristic equation

$$\cosh \lambda_n L \cos \lambda_n L = -1. \tag{3.9.15}$$

It is not difficult to show by means of a graphical argument that this equation has an infinity of roots which, for large order, are approximately given by $\lambda_n L \simeq (n + 1/2)\pi$. We can then write $u(x, t) = \sum_{n=1}^{\infty} U_n(t) e_n(x)$ and proceed as usual exploiting the orthogonality of the eigenfunctions.

3.10 A comment on the method of separation of variables

The reader will have been exposed to the so-called method of separation of variables which is, in essence, a simplified and less powerful form of the eigenfunction expansion approach systematically used in our examples. To illustrate this point, let us consider again the first example of §3.2.

According to the method of separation of variables, to solve the diffusion equation (3.2.26), one would first look for elementary solutions of the form $X_n(x) T_n(t)$. Substitution into the equation and rearrangement gives

$$\frac{1}{DT_n} \frac{dT_n}{dt} = \frac{1}{X_n} \frac{d^2 X_n}{dx^2}. \tag{3.10.1}$$

At this point it is argued that, since the two variables on either side of the equation are independent from each other, the only way in which this equation can be satisfied for all x

and all t is if each side equals one and the same constant so that[11]

$$\frac{1}{X_n}\frac{d^2 X_n}{dx^2} = -\mu_n^2, \qquad \frac{1}{DT_n}\frac{dT_n}{dt} = -\mu_n^2. \tag{3.10.2}$$

The solution of the first equation satisfying the boundary condition $dX_n/dx = 0$ at $x = 0$ is (omitting an irrelevant multiplicative integration constant) $X_n = \cos \mu_n x$ which, upon imposing $dX_n/dx = 0$ at $x = L$, gives $\mu_n = 2\pi n/L$, with which the solution for T_n becomes the same as that for A_n given in (3.2.9).

At this point, it is postulated that the solution of the actual problem of interest can be expressed as a superposition of the elementary solutions just found, which leads essentially to (3.2.10). The constants $A_n(0)$ are determined by integration – i.e., by taking a scalar product – just as before.

The first point to observe is that, in this method, the X_n's are determined by solving (3.10.2), which is just an eigenvalue equation for the operator d^2/dx^2, rather than d/dx. In this sense, separation of variables is also based on an eigenfunction expansion. Secondly, and more importantly, although separation of variables can be applied to non-homogeneous equations, such as those of §3.2.1, §3.4 and §3.7, it obviously fails in the case of non-homogeneous boundary conditions, such as those of §3.2.3, while the eigenfunction expansion can be applied in a straightforward way as we have seen.

3.11 Other applications

Additional applications of the Fourier series will be found elsewhere in connection with other topics:

1. Applications to infinite or semi-infinite strips are shown in conjunction with the Fourier transform in §4.6 and §4.13;
2. Applications to problems in cylindrical geometry are given in §6.2, 6.3, 6.4 and 6.5;
3. Applications to problems in spherical geometry are given in §7.2.

---

[11] Writing the constant as $-\mu_n^2$ simplifies the form of the solution, but exactly the same result would be found if one were to write it as ν_n since, at this stage, we do not know whether it will turn out to be a real or complex number; cf. the footnote on p. 242.

4

Fourier Transform: Applications

The theory of the Fourier transform is outlined in Chapter 10. Over the interval $-\infty < x < \infty$ the exponential version of the transform is appropriate, while over the range $0 < x < \infty$ the sine or cosine transforms may be useful. We start with several applications of the exponential transform; §4.10 to §4.13 show examples of the use of the sine and cosine transforms.

4.1 Useful formulae for the exponential transform

The exponential Fourier transform is useful to solve linear problems when one or more coordinates range over the entire real line $(-\infty, \infty)$. The transform of a function $f(x)$ is defined by*

$$\tilde{f}(k) \equiv \mathscr{F}\{f\} = \frac{1}{\sqrt{2\pi}} \int_{-\infty}^{\infty} e^{ikx} f(x) \, dx, \qquad (4.1.1)$$

with inverse

$$f(x) \equiv \mathscr{F}^{-1}\{\tilde{f}\} = \frac{1}{\sqrt{2\pi}} \int_{-\infty}^{\infty} e^{-ikx} \tilde{f}(k) \, dk. \qquad (4.1.2)$$

For other definitions in common use see p. 267.

At a point of discontinuity this relation is modified to

$$\frac{1}{\sqrt{2\pi}} \int_{-\infty}^{\infty} e^{-ikx} \tilde{f}(k) \, dk = \frac{1}{2} [f(x+0) + f(x-0)]. \qquad (4.1.3)$$

Separating real and imaginary parts in the right-hand side of (4.1.2) we see that, when f is real,

$$\overline{\tilde{f}}(k) = \tilde{f}(-k) \qquad (4.1.4)$$

where the overline denotes the complex conjugate.

The generalized Parseval relation is

$$(f, g) \equiv \int_{-\infty}^{\infty} \overline{f}(x) \, g(x) \, dx = \int_{-\infty}^{\infty} \overline{\tilde{f}}(k) \, \tilde{g}(k) \, dk \equiv (\tilde{f}, \tilde{g}), \qquad (4.1.5)$$

from which, in particular,

$$\| f \|^2 \equiv \int_{-\infty}^{\infty} |f(x)|^2 \, dx = \int_{-\infty}^{\infty} |\tilde{f}(k)|^2 \, dk \equiv \| \tilde{f} \|^2. \qquad (4.1.6)$$

*For other definitions in common use see p. 267.

Table 4.1	Some operational rules for the exponential Fourier transform.

$$f(x) = (2\pi)^{-1/2} \int_{-\infty}^{\infty} e^{-ikx} \tilde{f}(k)\,dk \qquad \tilde{f}(k) = (2\pi)^{-1/2} \int_{-\infty}^{\infty} e^{ikx} f(x)\,dx$$

1	$d^n f/dx^n$	$(-ik)^n \tilde{f}(k)$
2	$\int_{-\infty}^{x} f(\xi)\,d\xi$	$(-ik)^{-1} \tilde{f}(k)$
3	$(2\pi)^{-1/2} \int_{-\infty}^{\infty} f(x-\xi) g(\xi)\,d\xi$	$\tilde{f}(k)\tilde{g}(k)$
4	$x^n f(x)$	$(-id/dk)^n \tilde{f}(k)$
5	$e^{iax} f(x)$	$\tilde{f}(k+a)$
6	$f(ax-b)$	$\exp(ibk/a)\tilde{f}(k/a)/|a|$
7	$f(x) g(x)$	$(2\pi)^{-1/2} \int_{-\infty}^{\infty} \tilde{f}(h)\tilde{g}(k-h)\,dh$

$$8 \qquad \int_{-\infty}^{\infty} |f(x)|^2\,dx = \int_{-\infty}^{\infty} |\tilde{f}(k)|^2\,dk$$

$$9 \qquad \int_{-\infty}^{\infty} \overline{f}(x)\, g(x)\,dx = \int_{-\infty}^{\infty} \overline{\tilde{f}}(k)\,\tilde{g}(k)\,dk$$

The Fourier transform is advantageous when operations on $f(x)$ are replaced by simpler ones on $\tilde{f}(k)$. The correspondence between these operations is established by *operational rules* which are derived in §10.3 and collected in Table 4.1. We see from rule 1 of this table, for example, that

$$\mathscr{F}\left\{\frac{df}{dx}\right\} = -ik\tilde{f}(k), \qquad \mathscr{F}\left\{\frac{d^2 f}{dx^2}\right\} = (-ik)^2 \tilde{f}(k), \qquad (4.1.7)$$

and so forth. Particularly useful is the convolution rule 3

$$\frac{1}{\sqrt{2\pi}} \int_{-\infty}^{\infty} f(x-\xi) g(\xi)\,d\xi = \frac{1}{\sqrt{2\pi}} \int_{-\infty}^{\infty} f(\xi) g(x-\xi)\,d\xi = \tilde{f}(k)\tilde{g}(k), \quad (4.1.8)$$

which, as will be seen in several examples, enables us to solve many problems in general terms without committing ourselves to specific data.

A short list of exponential Fourier transforms is given in Table 4.2. More extended tables giving the exponential transform are available (e.g. Erdélyi *et al.* 1954, vol. 1). Integral tables (e.g. Gradshteyn *et al.* 2007) show integrals of functions $f(x)$ multiplied by $\sin x$ or $\cos x$ over the half-interval $(0, \infty)$. In many cases these tables can be used to calculate the

| Table 4.2 | Some exponential Fourier transform pairs; $H(x)$ is the unit step function, $H_n(x)$ is a Hermite polynomial. |

$f(x) = (2\pi)^{-1/2} \int_{-\infty}^{\infty} e^{-ikx} \tilde{f}(k) \, dk$	$\tilde{f}(k) = (2\pi)^{-1/2} \int_{-\infty}^{\infty} e^{ikx} f(x) \, dx$
$(x^2 + a^2)^{-1}$	$\sqrt{\pi/2} \, e^{-a|k|}/a \quad (a > 0)$
$x^{-1} \sin(ax) \quad (a > 0)$	$\sqrt{\pi/2} \, H(a - |k|) = \begin{cases} \sqrt{\pi/2} & \text{if } |k| < a \\ 0 & \text{if } a < |k| \end{cases}$
$\exp(-a|x|) \quad (\operatorname{Re} a > 0)$	$\sqrt{2/\pi} \, a/(k^2 + a^2)$
$\exp(-ax^2) \quad (\operatorname{Re} a > 0)$	$\exp(-k^2/4a)/\sqrt{2a}$
$\cos(ax^2)$	$\cos[(k^2/4a) - \pi/4]/\sqrt{2a}$
$\sin(ax^2)$	$\cos[(k^2/4a) + \pi/4]/\sqrt{2a}$
$|x|^{-s}, 0 < \operatorname{Re} s < 1$	$\sqrt{2/\pi} \, \Gamma(1 - s) \, |k|^{s-1} \sin(\pi s/2)$
$\exp(-a|x|)/\sqrt{|x|} \quad (a > 0)$	$\sqrt{1 + a/(a^2 + k^2)^{1/2}}$
$(a^2 - x^2)^{-1/2} H(a - |x|)$	$\sqrt{\pi/2} \, J_0(ak)$
$\sin[b\sqrt{a^2 + x^2}]/\sqrt{a^2 + x^2} \quad (b > 0)$	$\sqrt{\pi/2} \, J_0(a\sqrt{b^2 - k^2}) H(b - |k|)$
$H(a - |x|) \cos[b\sqrt{a^2 - x^2}]/\sqrt{a^2 - x^2}$	$\sqrt{\pi/2} \, J_0(a\sqrt{b^2 + k^2})$
$H(a - |x|) \cosh[b\sqrt{a^2 - x^2}]/\sqrt{a^2 - x^2}$	$\sqrt{\pi/2} \, J_0(a\sqrt{k^2 - b^2})$
$e^{-x^2/2} H_n(x)$	$i^n e^{-k^2/2} H_n(k)$

exponential transform observing that (4.1.1) can be rewritten as[1]

$$\tilde{f}(k) = \frac{1}{\sqrt{2\pi}} \int_0^{\infty} [f(x) + f(-x)] \cos kx \, dx + \frac{i}{\sqrt{2\pi}} \int_0^{\infty} [f(x) - f(-x)] \sin kx \, dx.$$

$$(4.1.9)$$

The tables can also be used for the inversion since the iterated transform satisfies the relation

$$\mathscr{F}^2\{f\} \equiv \mathscr{F}\{\mathscr{F}\{f\}\} = \mathscr{F}\left\{\tilde{f}(k)\right\} = f(-x) \qquad (4.1.10)$$

---

[1] To obtain this expression rewrite the original integral in (4.1.1) as the sum of integrals over $(-\infty, 0)$ and $(0, \infty)$, change x to $-x$ in the first one, and use the parity of the sine and cosine.

| Table 4.3 | Some exponential Fourier transforms of tempered distributions; $\operatorname{sgn} x = 2H(x) - 1$ is the sign function defined in (2.3.13) p. 36, C is an arbitrary constant. |

$d(x)$	$\tilde{d}(k)$
$\delta(x - a)$	$(2\pi)^{-1/2} \exp(ika)$
$\delta^{(n)}(x)$	$(2\pi)^{-1/2}(-ik)^n$
$\exp(-iax)$	$(2\pi)^{1/2}\delta(k - a)$
1	$(2\pi)^{1/2}\delta(k)$
$H(x)$	$(2\pi)^{-1/2}[i/k + \pi\delta(k)]$
$\operatorname{sgn} x$	$(2/\pi)^{1/2}i/k$
$x^n \ (n \geq 0)$	$(2\pi)^{1/2}(-i)^n\delta^{(n)}(k)$
$x^n \operatorname{sgn} x \ (n \geq 0)$	$(2/\pi)^{1/2} n! \, (-ik)^{-n-1}$
$x^{-n} \ (n > 0)$	$(\pi/2)^{1/2}i^n k^{n-1} (\operatorname{sgn} k)/(n - 1)!$
$x^{-n} \operatorname{sgn} x$	$(2/\pi)^{1/2}(ik)^{n-1}(C - \log|k|)/(n - 1)!$
$|x|^\lambda \ (\lambda \neq -1, -3, \ldots)$	$-(2/\pi)^{1/2}[\sin(\tfrac{1}{2}\pi\lambda)]\Gamma(\lambda + 1)|k|^{-\lambda-1}$
$|x|^\lambda \operatorname{sgn} x \ (\lambda \neq -2, -4, \ldots)$	$(2/\pi)^{1/2}i[\cos(\tfrac{1}{2}\pi\lambda)]\,\Gamma(\lambda + 1)\,|k|^{-\lambda-1}(\operatorname{sgn} k)$
$|x|^\lambda \log|x|$	$(2/\pi)^{1/2}\cos(\pi\lambda/2)\Gamma(\lambda + 1)|k|^{-\lambda-1}$ $[(\log|k| - \psi(\lambda))\tan(\pi\lambda/2) - \pi/2]$
$|x|^\lambda \log|x|(\operatorname{sgn} x)$	$(2/\pi)^{1/2}i\sin(\pi\lambda/2)\Gamma(\lambda + 1)|k|^{-\lambda-1}$ $[(\psi(\lambda) - \log|k|)\cot(\pi\lambda/2) + \pi/2](\operatorname{sgn} k)$
$x^n \log|x|$	$-(\pi/2)^{1/2}i^n n! k^{-n-1}(\operatorname{sgn} k)$
$x^n \log|x|(\operatorname{sgn} x)$	$(2/\pi)^{1/2}n! i^{n-1} k^{-n-1}[\log|k| - \psi(n)]$
$x^{-n} \log|x|$	$(\pi/2)^{1/2}i^n k^{n-1}[\psi(n - 1) - \log|k|](\operatorname{sgn} k)/(n - 1)!$
$x^{-n} \log|x|(\operatorname{sgn} x)$	$(2\pi)^{-1/2}(ik)^{n-1}[(\psi(n - 1) - \log|k|)^2 + C]/(n - 1)!$

so that

$$f(x) = \mathscr{F}\left\{\tilde{f}\right\}(-x). \qquad (4.1.11)$$

In words: to invert the transform read x for k in the function to be inverted in the left column, and read $-x$ for k in the function in the right column. In some cases a deeper understanding of the inversion is required to choose the desired solution among several possible ones (see e.g. §4.2). Generally speaking, the standard method to invert the transform is contour integration (see §17.9.1 p. 461), in particular with the aid of Jordan's lemma.

4.1.1 Fourier transform of distributions

The power of the Fourier transform is considerably enhanced by its applicability to (tempered) distributions (§20.10 p. 595).

The Fourier transform of distributions satisfies most of the same properties as that of functions listed in Table 4.1 (Table 20.3 p. 598). One important exception concerns relations involving products, both ordinary and of the convolution type. The reason is that in general the Fourier transform of a distribution is another distribution, and neither the ordinary nor the convolution product of two distributions is necessarily defined (p. 576 and §20.9). On the other hand, the product of a distribution by a sufficiently smooth ordinary function is well defined and, in this case, the rules of Table 4.1 which concern multiplication apply. Table 4.3 shows some useful Fourier transforms of distributions; the scope of the table can be enlarged by using the rules in Table 4.1 and the relation

$$H(x) = \frac{1}{2}(1 + \operatorname{sgn} x). \tag{4.1.12}$$

4.1.2 Higher-dimensional spaces

The Fourier transform can also be defined for a function of two variables x and y:

$$\tilde{f}(h, k) = \frac{1}{2\pi} \int_{-\infty}^{\infty} dx \int_{-\infty}^{\infty} dy \, \exp\left[i(hx + ky)\right] f(x, y), \tag{4.1.13}$$

and, more generally, for d variables;

$$\tilde{f}(\mathbf{k}) = \frac{1}{(2\pi)^{d/2}} \int d^d x \, \exp\left(i\mathbf{k} \cdot \mathbf{x}\right) f(\mathbf{x}), \tag{4.1.14}$$

in which $\mathbf{x}$ and $\mathbf{k}$ are vectors in d-dimensional spaces and the integral is extended to the entire space. The inversion formula is

$$f(\mathbf{x}) = \frac{1}{(2\pi)^{d/2}} \int d^d k \, \exp\left(-i\mathbf{k} \cdot \mathbf{x}\right) \tilde{f}(\mathbf{k}). \tag{4.1.15}$$

Parseval's relations (4.1.5) and (4.1.6) hold for this case as well with

$$\|f\|^2 \equiv \int d^d x \, |f(\mathbf{x})|^2 = \int d^d k \, |\tilde{f}(\mathbf{k})|^2 \equiv \|\tilde{f}\|^2. \tag{4.1.16}$$

Some multi-dimensional Fourier transforms of regular functions and distributions are shown in Table 4.4; additional results of this type can be found in Lighthill 1958 p. 43, Jones 1966 p. 469 and Gelfand & Shilov 1964, vol. 1.

4.1.3 Asymptotic behavior

When the transformation or the inversion cannot be carried out analytically, asymptotic formulae for large x or large k may be useful. For ordinary functions, some considerations on this point are presented in §10.8 p. 281.

Table 4.4

Some exponential Fourier transforms of tempered distributions in d-dimensional space. Notation: $r = \sqrt{\mathbf{x} \cdot \mathbf{x}}$, $k = \sqrt{\mathbf{k} \cdot \mathbf{k}}$, $C_\lambda = 2^{\lambda + d/2} |\Gamma(\tfrac{1}{2}(\lambda + d)) / \Gamma(-\tfrac{1}{2}\lambda)|$, $\Omega_d = 2\pi^{d/2} / \Gamma(d/2)$, $m = 0, 1, \ldots$

$d(\mathbf{x}) = d(x_1, \ldots, x_d)$	$\tilde{d}(\mathbf{k}) = \tilde{d}(k_1, \ldots, k_d)$
$\delta(\mathbf{x} - \mathbf{a})$	$(2\pi)^{-d/2} \exp(i\mathbf{k} \cdot \mathbf{a})$
$\exp(-i\mathbf{a} \cdot \mathbf{x})$	$(2\pi)^{d/2} \delta(\mathbf{k} - \mathbf{a})$
1	$(2\pi)^{d/2} \delta(\mathbf{k})$
r^λ $\lambda \neq 2m, -d - 2m$	$C_\lambda k^{-\lambda - d}$
$r^\lambda \log r$ $\lambda \neq 2m, -d - 2m$	$(dC_\lambda / d\lambda) k^{-\lambda - d} + C_\lambda k^{-\lambda - d} \log k$
$\delta(r - a)$, for $d \geq 1$	$(a/2\pi)^{d/2} \Omega_{d-1} k^{1-d/2} J_{d/2-1}(ak)$
$\delta(r - a)$, for $d = 3$	$(2/\pi)^{1/2} a (\sin ak)/k$
$\left(\dfrac{1}{a} \dfrac{d}{da}\right) \dfrac{\delta(r - a)}{a}$	$(2/\pi)^{1/2} (2\pi)^{-d/2} \Omega_d (\sin ak)/k$
$\exp(-ar^2)$	$(2a)^{-d/2} \exp(-k^2/4a)$

The asymptotic behavior of the Fourier transform of distributions is dominated by points in the neighborhood of which the behavior is similar to that of any of the following distributions

$$|x - X|^\lambda, \qquad |x - X|^\lambda \, \text{sgn}\,(x - X), \qquad |x - X|^\lambda \log |x - X|,$$
$$|x - X|^\lambda \, \text{sgn}\,(x - X) \log |x - X|, \qquad \delta^{(p)}(x - X), \qquad (4.1.17)$$

the Fourier transforms of which are given in Table 4.3; see p. 608 for details.

We now consider several examples. Our procedure will be mostly formal: we will not worry about convergence of the integrals or interchange of the order of integration.[2] A more rigorous way to tackle these problems is described, e.g., in Titchmarsh (1948), chapter 10. In principle, it may be advisable to check that a solution obtained by formal methods is the desired one. In practice, even though at times some intermediate steps may not be easily justifiable in terms of classical analysis, they often are from the point of view of distributions and the final result is correct.

---

[2] For classical conditions on the latter operation see p. 690.

4.2 One-dimensional Helmholtz equation

Ordinary differential equations over the line $-\infty < x < \infty$ can often be solved by methods other than the Fourier transform, but the consideration of a few examples permits us to demonstrate the method in a relatively simple setting and to introduce some subtle aspects of the inversion which are encountered in problems of propagation.

Before turning to the Helmholtz equation, let us begin with the simpler problem already solved by variation of parameters in Example 2.2.2 p. 33, namely

$$-\frac{d^2u}{dx^2} + h^2u = f(x), \qquad u \to 0 \qquad \text{for} \qquad |x| \to \infty, \tag{4.2.1}$$

with $h > 0$. We take the Fourier transform of both sides:

$$\mathscr{F}\left\{-\frac{d^2u}{dx^2} + h^2u\right\} = \mathscr{F}\{f\} \tag{4.2.2}$$

and use the operational rule for differentiation (rule 1 in Table 4.1) to find

$$(k^2 + h^2)\tilde{u} = \tilde{f}. \tag{4.2.3}$$

The formal solution of the problem in the original x-domain is found by applying the convolution theorem (rule 3 in Table 4.1) to the product of $1/(k^2 + h^2)$ and $\tilde{f}$, and is therefore

$$u(x) = \frac{1}{\sqrt{2\pi}} \int_{-\infty}^{\infty} f(x - \xi)\, \mathscr{F}^{-1}\left\{\frac{1}{h^2 + k^2}\right\}(\xi)\, d\xi. \tag{4.2.4}$$

The inverse of $(h^2 + k^2)^{-1}$ can be readily found by contour integration as shown in Example 17.9.2 p. 463; using the result given in Table 4.2, we have

$$u(x) = \frac{1}{2h} \int_{-\infty}^{\infty} f(x - \xi)\, \exp\left(-h|\xi|\right)\, d\xi. \tag{4.2.5}$$

This coincides with the solution given earlier provided the integration constants are adjusted as required by the boundary conditions at infinity.

An unexpected difficulty is encountered when we try to solve by the same method a superficially very similar problem, the one-dimensional Helmholtz equation:

$$\frac{d^2u}{dx^2} + h^2u = f(x), \qquad u \to 0 \qquad \text{for} \qquad |x| \to \infty, \tag{4.2.6}$$

again with $h > 0$. Proceeding as before we are led to

$$u(x) = \frac{1}{\sqrt{2\pi}} \int_{-\infty}^{\infty} f(x - \xi)\, \mathscr{F}^{-1}\left\{\frac{1}{h^2 - k^2}\right\}(\xi)\, d\xi, \tag{4.2.7}$$

where

$$\mathscr{F}^{-1}\left\{\frac{1}{h^2 - k^2}\right\}(x) = \frac{1}{\sqrt{2\pi}} \int_{-\infty}^{\infty} \frac{1}{2h}\left(\frac{1}{k + h} - \frac{1}{k - h}\right) e^{-ikx}\, dk. \tag{4.2.8}$$

The difficulty arises because the integrand has non-integrable singularities at $k = \pm h$, which can be regularized in different ways, each one of which produces a different result. The same indeterminacy, which would also be found by variation of parameters, was encountered in §3.5 and is not rooted in mathematics but in physics: each different solution is the correct answer to a different physical problem, and we must therefore turn to physics to solve our dilemma.

A typical context in which (4.2.6) might arise is the propagation of monochromatic waves of angular frequency ω in an infinite non-dispersive medium with wave speed $c = \omega/h$. Including the time dependence $e^{-i\omega t}$, the complete solution to such a problem would then be $u(x)\,e^{-i\omega t}$. Waves can be incident from infinity, as in a scattering problem, or propagate toward infinity, as in a radiation problem. If the problem we are trying to solve is the propagation of waves generated by a distributed spatial source having a local instantaneous amplitude $f(x)\,e^{-i\omega t}$, we are dealing with a radiation problem. Physics then requires that $u(x)\,e^{-i\omega t}$ represent only outgoing waves, propagating toward $x \to \infty$ for $x > 0$ and toward $x \to -\infty$ for $x < 0$.

This requirement, called a *radiation condition*, can be enforced in several equivalent ways (see e.g. Whitham 1974, p. 447; Lighthill 1978, p. 267). One possible approach is to introduce a small damping in the wave equation which is ultimately made to go to zero. Thus, if (4.2.6) arises by considering the monochromatic solutions of the equation

$$\frac{\partial^2 v}{\partial x^2} - \frac{1}{c^2}\frac{\partial^2 v}{\partial t^2} = f(x)e^{-i\omega t}, \tag{4.2.9}$$

as in (1.2.11) p. 9, we consider instead the telegrapher equation (1.2.12) p. 9

$$\frac{\partial^2 v}{\partial x^2} - \frac{1}{c^2}\frac{\partial^2 v}{\partial t^2} - 2\frac{\varepsilon}{c}\frac{\partial v}{\partial t} = f(x)e^{-i\omega t}, \tag{4.2.10}$$

which, upon setting $v = ue^{-i\omega t}$ with $\omega = ch$, becomes

$$\frac{d^2 u}{dx^2} + \left(h^2 + 2i\varepsilon h\right)u = f(x). \tag{4.2.11}$$

Thus, the singularity originally at $k = h$ has moved to $k \simeq h + i\varepsilon$ while the one originally at $k = -h$ has moved to $k \simeq -h - i\varepsilon$ (Figure 4.1, left). After this argument, the integral can be evaluated in several ways which, although superficially different, are all closely related.

A first method is to use contour integration with the aid of Jordan's lemma (p. 461). In the exponential e^{-ikx} we let $k = Re^{i\theta}$ and consider the behavior of $\exp[-iRe^{i\theta}x]$ for $R \to \infty$. If $x > 0$, for this exponential to tend to 0 in the limit it is necessary that $\operatorname{Im} e^{i\theta} < 0$, so that we need to close the integration path by a large semicircle in the lower half-plane, and conversely for $x < 0$. Thus, for $x > 0$, only the pole at $k \simeq -h - i\varepsilon$ contributes and we find

$$\int_{-\infty}^{\infty} \frac{e^{-ikx}}{k^2 - (h + i\varepsilon)^2}\,dk = -2\pi i e^{-i(-h-i\varepsilon)x} \to -2\pi i e^{ihx} \tag{4.2.12}$$

where the minus sign is due to the fact that the path is traversed in the negative direction; similarly, for $x < 0$,

$$\int_{-\infty}^{\infty} \frac{e^{-ikx}}{k^2 - (h + i\varepsilon)^2}\,dk = -2\pi i e^{-i(h+i\varepsilon)x} \to -2\pi i e^{-ihx} \tag{4.2.13}$$

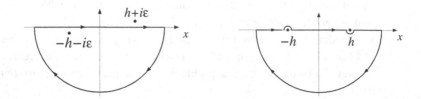

Two equivalent ways of handling the poles in the integral (4.2.8) so that the radiation condition is satisfied; for $x > 0$ the integration path is closed in the lower half plane to ensure the applicability of Jordan's lemma.

with the minus sign because the fraction with the relevant pole is preceded by a minus sign. Thus, finally,

$$\mathscr{F}^{-1}\left\{\frac{1}{h^2 - k^2}\right\}(x) = -\sqrt{2\pi}\, i\, e^{ih|x|} \tag{4.2.14}$$

and (4.2.7) becomes

$$u(x) = \frac{1}{2ih}\int_{-\infty}^{\infty} f(x - \xi)\, e^{ih|\xi|}\, d\xi. \tag{4.2.15}$$

In particular, if $f(x) = \delta(x - x')$, we find the expression of the Green's function appropriate for this problem, namely

$$G(x; x') = \frac{e^{ih|x-x'|}}{2ih}. \tag{4.2.16}$$

An equivalent way is to calculate the inversion integral (4.2.8) as a Cauchy principal value:

$$\mathrm{PV}\int_{-\infty}^{\infty} \frac{e^{-ikx}}{k \pm h}\, dk = \lim_{\varepsilon \to 0}\left(\int_{-\infty}^{\pm h - \varepsilon} + \int_{\pm h + \varepsilon}^{\infty}\right)\frac{e^{-ikx}}{k \pm h}\, dk, \tag{4.2.17}$$

by adding small semi-circles going around the singular points. Whether these semi-circles should be above or below the poles can be decided by considering the radiation condition as before (Figure 4.1, right).

Once we know how the poles are displaced by the radiation condition, a quicker argument consists in using the convolution theorem with the entries for $(x \pm i0)^{-1}$ in Table 4.3 together with the translation rule 6 of Table 4.1:

$$\mathscr{F}\left\{\frac{1}{k \pm h \pm i0}\right\}(x) = e^{\mp ihx}\,\mathscr{F}\left\{\frac{1}{k \pm i0}\right\}(x) = \sqrt{2\pi}\,(\mp i)\, H(\mp x)\, e^{\mp ihx}. \tag{4.2.18}$$

Therefore, from rule (4.1.11),

$$\mathscr{F}^{-1}\left\{\frac{1}{k \pm h \pm i0}\right\}(x) = \mathscr{F}\left\{\frac{1}{k \pm h \pm i0}\right\}(-x) = \sqrt{2\pi}\,(\mp i)\, H(\pm x)\, e^{\pm ihx}. \tag{4.2.19}$$

With these expressions we have, for example for $x > 0$,

$$\mathscr{F}^{-1}\left\{\frac{1}{h^2 - k^2}\right\} = \frac{1}{2h}\mathscr{F}^{-1}\left\{\frac{1}{k + h + i0} - \frac{1}{k - h - i0}\right\} = \frac{\sqrt{2\pi}}{2hi}H(x)\,e^{ihx}, \quad (4.2.20)$$

and similarly for $x < 0$; combining the two results we recover the previous expression.

Yet another method is to use Plemelj's formulae (p. 574), according to which

$$\frac{1}{X \pm i0} = \mathrm{PV}\frac{1}{X} \mp i\pi\delta(X). \quad (4.2.21)$$

Following this route we have

$$\int_{-\infty}^{\infty}\left(\frac{1}{k + h + i0} - \frac{1}{k - h - i0}\right)e^{-ikx}\,dk = -i\pi\left(e^{ihx} + e^{-ihx}\right)$$

$$+\mathrm{PV}\int_{-\infty}^{\infty}\left(\frac{1}{k + h} - \frac{1}{k - h}\right)e^{-ikx}\,dk. \quad (4.2.22)$$

To evaluate the last integral we let $\eta = k \pm h$ to find

$$\mathrm{PV}\int_{-\infty}^{\infty}\frac{e^{-ikx}}{k \pm h}\,dk = 2ie^{\pm ihx}\int_{0}^{\infty}\frac{\sin \eta x}{\eta}\,d\eta = -i\pi e^{\pm ihx}\mathrm{sgn}\,x \quad (4.2.23)$$

where we have used the fact that the principal value of the integral of $(\cos \eta x)/x$ vanishes because of the odd parity of the integrand.[3] Upon combining this result with (4.2.22) we recover (4.2.14).

4.3 Schrödinger equation with a constant force

The steady-state Schrödinger equation for a particle of energy E in a potential $-FX$, corresponding to a constant force field F (i.e., a particle in "free fall") is

$$\frac{d^2U}{dX^2} + \frac{2m}{\hbar^2}(E + FX)U = 0, \quad (4.3.1)$$

which, upon setting

$$x = -\left(\frac{2mF}{\hbar^2}\right)^{1/3}\left(X + \frac{E}{F}\right), \qquad U(X; E) = u(x), \quad (4.3.2)$$

becomes *Airy's equation*

$$-\frac{d^2u}{dx^2} + xu = 0. \quad (4.3.3)$$

With a different normalization, the same equation also describes the diffraction of light near a caustic surface.

---

[3] The remaining integral is calculated in (17.9.22) p. 464.

Airy's equation is the simplest second-order linear differential equation with a turning point, i.e., a point where the character of the solution changes from oscillatory, for $x < 0$, to damped exponential, for $x > 0$. In the quantum mechanical setting, this change reflects the behavior of the wave function in the regions of space where the energy of the system is higher or lower than the potential energy, respectively.

Application of the Fourier transform gives

$$\left(k^2 - i\frac{d}{dk}\right)\tilde{u} = 0 \quad \text{from which} \quad \tilde{u} = C\exp\left(\frac{i}{3}k^3\right), \qquad (4.3.4)$$

with C an integration constant; the inverse is

$$u = \frac{C}{\sqrt{2\pi}}\int_{-\infty}^{\infty}\exp\left(-\frac{i}{3}k^3 - ikx\right)dk = \sqrt{2\pi}\,C\,\mathrm{Ai}(x) \qquad (4.3.5)$$

where Ai is the Airy function of the first kind

$$\mathrm{Ai}(x) = \frac{1}{\pi}\int_0^{\infty}\cos\left(\frac{1}{3}k^3 + kx\right)dk, \qquad (4.3.6)$$

related to Bessel functions of order 1/3 and argument $\xi = \frac{2}{3}x^{3/2}$:

$$\mathrm{Ai}(x) = \frac{1}{3}\sqrt{x}\left[I_{-1/3}(\xi) - I_{1/3}(\xi)\right] = \pi^{-1}\sqrt{\frac{x}{3}}K_{1/3}(\xi), \qquad (4.3.7)$$

$$\mathrm{Ai}(-x) = \frac{1}{3}\sqrt{x}\left[J_{-1/3}(\xi) + J_{1/3}(\xi)\right]. \qquad (4.3.8)$$

At first sight it may appear perplexing that we have found only one solution starting from the second-order equation (4.3.3). The explanation is that the other solution of this equation is proportional to the Airy function of the second kind, $\mathrm{Bi}(x)$, which increases faster than e^x for $x > 0$ and, therefore, does not admit a Fourier transform. This feature – namely the inability to reproduce solutions which do not admit a transform – is common to all transform techniques.

The integration constant C is determined from problem-specific considerations. For example, in the quantum mechanical application, the energy spectrum is continuous and one would impose the normalization

$$\int_{-\infty}^{\infty}\overline{U}(X; E)\,U(X; E')\,dX = \delta(E - E'). \qquad (4.3.9)$$

Given the change of variables (4.3.2), $U(X; E') = u\left(x + \mu(E - E')/F\right)$ with $\mu = (2mF/\hbar^2)^{1/3}$. Thus, the normalization integral (4.3.9) is

$$\int_{-\infty}^{\infty}\overline{u}(x)\,u\left(x + \mu(E - E')/F\right)dx = \mu\delta(E - E'). \qquad (4.3.10)$$

To calculate this integral we can use the generalized Parseval identity (rule 9 of Table 4.1) coupled with rule 3:

$$\mathscr{F}\{u\left(x + \mu(E - E')/F\right)\} = e^{i\mu(E-E')k/F}\,\tilde{u} = e^{i\mu(E-E')k/F}\,C\,e^{(i/3)k^3}, \qquad (4.3.11)$$

and, therefore,

$$\int_{-\infty}^{\infty} \overline{\mathscr{F}\{u(x)\}}\, \mathscr{F}\{u\left(x + \mu(E - E')/F\right)\}\, dk = \int_{-\infty}^{\infty} |C|^2 e^{i\mu(E-E')k/F}\, dk$$

$$= 2\pi |C|^2 \frac{F}{\mu} \delta(E - E') = \mu \delta(E - E'), \tag{4.3.12}$$

from which $2\pi |C|^2 (F/\mu) = \mu$ or $|C| = \mu/\sqrt{2\pi F}$.

4.4 Diffusion in an infinite medium

We now consider several variants of the problem of diffusion in an infinite medium in one, two and three spatial dimensions. The most general version of the problem is one in which the equation contains a forcing term q, corresponding to a distributed source, and there is an initial distribution of the diffusing quantity (e.g. temperature):

$$\frac{\partial u}{\partial t} = D\nabla^2 u + q(\mathbf{x}, t), \tag{4.4.1}$$

$$u(\mathbf{x}, 0) = f(\mathbf{x}), \qquad u \to 0 \quad \text{for} \quad |\mathbf{x}| \to \infty. \tag{4.4.2}$$

As noted before in §3.2.4 p. 65, due to the linearity of the problem, we can split the solution into two parts, $u = u_h + u_f$, in which u_h satisfies the homogeneous equation and the initial condition:

$$\frac{\partial u_h}{\partial t} = D\nabla^2 u_h \tag{4.4.3}$$

$$u_h(\mathbf{x}, 0) = f(\mathbf{x}), \qquad u_h \to 0 \quad \text{for} \quad |\mathbf{x}| \to \infty, \tag{4.4.4}$$

while u_f satisfies the forced, non-homogeneous equation with a homogeneous initial condition:

$$\frac{\partial u_f}{\partial t} = D\nabla^2 u_f + q(\mathbf{x}, t) \tag{4.4.5}$$

subject to

$$u_f(\mathbf{x}, 0) = 0, \qquad u_f \to 0 \quad \text{for} \quad |\mathbf{x}| \to \infty. \tag{4.4.6}$$

Upon integrating (4.4.1) over a volume V bounded by a closed surface S and using the divergence theorem we recover the integral balance relation (1.3.1) p. 10:

$$\frac{d}{dt} \int_V u(\mathbf{x}, t)\, d^3 x = \oint_S D(\nabla u) \cdot \mathbf{n}\, dS + \int_V q(\mathbf{x}, t)\, d^3 x. \tag{4.4.7}$$

In particular, if $q = 0$, V is the entire space and u decays sufficiently rapidly at infinity, the right-hand side vanishes and we have the conservation relation

$$\int_V u_h(\mathbf{x}, t)\, d^3 x = \int_V f(\mathbf{x})\, d^3 x. \tag{4.4.8}$$

We begin with the one-dimensional problem; we will continue with the three-dimensional one and we will recover the solution to the two-dimensional problem by applying the so-called method of descent; we will also demonstrate the use of the similarity method. But we first note an interesting fact which can be proved easily by direct substitution: if $v(x, t)$ is a solution of the one-dimensional version of (4.4.3), so is

$$\frac{e^{-x^2/4Dt}}{\sqrt{Dt}} v\left(\frac{x}{Dt}, -\frac{1}{Dt}\right). \tag{4.4.9}$$

4.4.1 One space dimension

To solve the initial-value homogeneous problem for u_h, we apply the $\mathscr{F}$ operator to (4.4.3) and (4.4.4) to find, using the differentiation rule 1 of Table 4.1,

$$\frac{\partial \tilde{u}_h}{\partial t} = -Dk^2 \tilde{u}_h, \qquad \tilde{u}_h(k, 0) = \tilde{f}(k). \tag{4.4.10}$$

The solution is

$$\tilde{u}_h = \tilde{f} \exp(-Dk^2 t), \tag{4.4.11}$$

which can be inverted with the aid of the convolution theorem (rule 3)

$$u_h(x, t) = \frac{1}{\sqrt{2\pi}} \int_{-\infty}^{\infty} f(x - \xi) \mathscr{F}^{-1}\left\{\exp(-Dk^2 t)\right\}(\xi) \, d\xi. \tag{4.4.12}$$

Upon using an entry of Table 4.2 to invert $\exp(-Dk^2 t)$, we find

$$u_h(x, t) = \frac{1}{2\sqrt{\pi Dt}} \int_{-\infty}^{\infty} f(\xi) \exp\left[-\frac{(x - \xi)^2}{4Dt}\right] d\xi, \tag{4.4.13}$$

which can also be written switching the arguments ξ and $(x - \xi)$ as in (4.1.8). Recall that D has dimensions of $(\text{length})^2/\text{time}$. We thus see that $\sqrt{Dt}$ has dimension length and once again we recognize the role of the diffusion length $\sqrt{2Dt}$ already discussed on p. 60. Upon integrating (4.4.13) over the real line and interchanging the order of integration in the right-hand side (p. 690) we have

$$\int_{-\infty}^{\infty} u_h \, dx = \int_{-\infty}^{\infty} f(x) \, dx, \tag{4.4.14}$$

which is the one-dimensional version of (4.4.8). By considering an initial condition localized at x_0, i.e., by setting $f(x) = \delta(x - x_0)$, we find the Green's function $G_h(x, t; x_0)$ of the problem (§16.2 p. 410)

$$G_h(x, t; x_0) = \frac{H(t)}{2\sqrt{\pi Dt}} \exp\left[-\frac{(x - x_0)^2}{4Dt}\right], \tag{4.4.15}$$

where we have introduced the unit step function $H(t)$ for later reference without affecting G_h which is only defined for $t > 0$.

Let us now consider the problem for u_f. Upon taking the Fourier transform with respect to x we have

$$\frac{\partial \tilde{u}_f}{\partial t} = -Dk^2 \tilde{u}_f + \tilde{q}(k, t), \qquad \tilde{u}_f(k, 0) = 0, \qquad (4.4.16)$$

with solution

$$\tilde{u}_f = \int_0^t \tilde{q}(k, \tau) \exp\left[-k^2 D(t - \tau)\right] d\tau, \qquad (4.4.17)$$

which can be inverted with the help of the convolution theorem and an entry of Table 4.2:

$$
\begin{aligned}
u_f &= \int_0^t d\tau \int_{-\infty}^{\infty} q(x - \xi, \tau) \frac{1}{2\sqrt{\pi D(t - \tau)}} \exp\left[-\frac{\xi^2}{4D(t - \tau)}\right] d\xi \\
&= \int_0^t d\tau \int_{-\infty}^{\infty} q(x - \xi, t - \tau) \frac{1}{2\sqrt{\pi D\tau}} \exp\left[-\frac{\xi^2}{4D\tau}\right] d\xi \\
&= \int_0^t d\tau \int_{-\infty}^{\infty} \frac{q(\xi, t - \tau)}{2\sqrt{\pi D\tau}} \exp\left[-\frac{(x - \xi)^2}{4D\tau}\right] d\xi. \qquad (4.4.18)
\end{aligned}
$$

If we set $q(x, t) = \delta(x - x_0)\,\delta(t - t_0)$ we recover the Green's function for the non-homogeneous equation (see Example 16.2.2 p. 413)

$$G_f(x, t; x_0, t_0) = \frac{H(t - t_0)}{2\sqrt{\pi D(t - t_0)}} \exp\left[-\frac{(x - x_0)^2}{4D(t - t_0)}\right]. \qquad (4.4.19)$$

We see here that G_f is just G_h translated in time, in agreement with the considerations of §16.2.

Similarity solution

The problem for G_h (and G_f) can also be solved by using similarity considerations. For convenience, with a simple translation, we can take $x_0 = 0$ and $t_0 = 0$. A priori we can say that G_h can depend only on x, t and D since there are no intrinsic length or time scales in the statement of the problem. Furthermore, from the initial condition $G_h(x, 0; 0) = \delta(x)$, we see that G_h must have the dimensions of $\delta(x)$, i.e., an inverse length.[4] We thus conclude that G_h must have the form $G_h = F(x, t, D)/x$ or $G_h = F_1(x, t, D)/\sqrt{Dt}$ with F or F_1 dimensionless. Furthermore, the argument of F or F_1 must be dimensionless, and therefore it can only be a function of $x/\sqrt{Dt}$. Let us therefore set[5]

$$G_h(x, t; 0) = \frac{1}{\sqrt{Dt}} F(\eta) \qquad \text{with} \qquad \eta = \frac{x}{2\sqrt{Dt}}. \qquad (4.4.20)$$

---

[4] Since the integral of δ over x gives the dimensionless number 1, δ itself must have dimensions of $(\text{length})^{-1}$.

[5] There is no loss of generality with this choice as, for example, $F(\eta)/\sqrt{Dt} = 2\eta F(\eta)/x = 2F_1(\eta)/x$ with $F_1 = \eta F(\eta)$. Similarly, we could take $\eta_1 = \eta^2$ or $\eta_2 = 1/\eta$ and so forth; the only effect would be to change the appearance of the ordinary differential equation (4.4.22), but there would be no change in the final solution for G_h.

Upon substitution into the one-dimensional diffusion equation we find

$$\frac{d^2 F}{d\eta^2} + 2\eta \frac{dF}{d\eta} + 2F = 0, \tag{4.4.21}$$

or

$$\frac{d}{d\eta}\left(\frac{dF}{d\eta} + 2\eta F\right) = 0. \tag{4.4.22}$$

One integration with the condition $F \to 0$ for $\eta \to \infty$ gives $dF/d\eta + 2F = 0$, from which $F = A e^{-\eta^2}$. The integration constant can be determined by using the conservation relation (4.4.14) to find $A = (2\sqrt{\pi})^{-1}$, thus recovering (4.4.15). The fact that the form of $\sqrt{Dt}\, G_h$ remains unaltered – i.e., self-similar – upon replacing x by λx and t by $\lambda^2 t$ (with λ an arbitrary positive constant) justifies the name *similarity solution* given to this procedure and *similarity variable* given to η.

4.4.2 Three space dimensions

In three space dimensions the problem can be solved by use of the three-dimensional Fourier transform. Proceeding as before we find

$$\frac{\partial \tilde{u}_h}{\partial t} = -Dk^2 \tilde{u}_h, \qquad k^2 = k_x^2 + k_y^2 + k_z^2, \tag{4.4.23}$$

with initial condition $\tilde{u}_h(\mathbf{k}, 0) = \tilde{f}(\mathbf{k})$. With the new meaning of the symbol k^2, the solution is again given by (4.4.11):

$$\tilde{u}_h = \tilde{f} \exp(-Dk^2 t) = \tilde{f} \exp(-Dk_x^2 t) \exp(-Dk_y^2 t) \exp(-Dk_z^2 t). \tag{4.4.24}$$

Use of the convolution theorem now gives

$$u_h(x, t) = \frac{1}{(2\pi)^{3/2}} \int_{-\infty}^{\infty} d\xi \int_{-\infty}^{\infty} d\eta \int_{-\infty}^{\infty} d\zeta \, f(x - \xi, y - \eta, z - \zeta)$$
$$\times \mathscr{F}^{-1}\left\{\exp(-Dk_x^2 t)\right\}(\xi)\, \mathscr{F}^{-1}\left\{\exp(-Dk_y^2 t)\right\}(\eta)\, \mathscr{F}^{-1}\left\{\exp(-Dk_z^2 t)\right\}(\zeta) \tag{4.4.25}$$

i.e.

$$u_h = \frac{1}{(4\pi Dt)^{3/2}} \int_{-\infty}^{\infty} d\xi \int_{-\infty}^{\infty} d\eta \int_{-\infty}^{\infty} d\zeta \, f(x - \xi, y - \eta, z - \zeta)$$
$$\times \exp\left[-\frac{\xi^2 + \eta^2 + \zeta^2}{4Dt}\right]. \tag{4.4.26}$$

In this case, the starting point for the similarity solution for the Green's function is

$$G(x, y, z, t; 0, 0, 0) = (Dt)^{-3/2} F(\eta), \qquad \eta = \frac{1}{2}\left(\frac{x^2 + y^2 + z^2}{Dt}\right)^{1/2} = \frac{r}{2\sqrt{Dt}}$$

and we find

$$\eta \frac{d^2 F}{d\eta^2} + 2\left(1 + \eta^2\right)\frac{dF}{d\eta} + 6\eta F = 0, \tag{4.4.27}$$

or

$$\eta \frac{d}{d\eta}\left(\frac{dF}{d\eta} + 2\eta F\right) + 2\left(\frac{dF}{d\eta} + 2\eta F\right) = 0. \tag{4.4.28}$$

One integration gives $dF/d\eta + 2\eta F = B/\eta^2$ and we must take $B = 0$ for regularity at the origin. A second integration produces $F = A\exp(-\eta^2)$, with the constant A determined as before from the conservation relation (4.4.8).

The solution for u_f proceeds along similar steps and is

$$u_f = \int_0^t \frac{d\tau}{[4\pi D(t-\tau)]^{3/2}} \int_{-\infty}^{\infty} d\xi \int_{-\infty}^{\infty} d\eta \int_{-\infty}^{\infty} d\zeta \, q(x-\xi, y-\eta, z-\zeta, \tau)$$
$$\exp\left[-\frac{\xi^2 + \eta^2 + \zeta^2}{4D(t-\tau)}\right]. \tag{4.4.29}$$

As an application of this result, let us consider the effect of a point source moving along the line $x = a$, $z = 0$ with the constant velocity V. In this case we have $q = q_0\delta(x - a)\delta(z)\delta(y - Vt)$. The spatial integrations in (4.4.29) can be carried out with the result

$$u_f = q_0 \int_0^t \exp\left[-\frac{(x-a)^2 + (y-Vt)^2 + z^2}{4D(t-\tau)}\right] \frac{d\tau}{[4\pi D(t-\tau)]^{3/2}}. \tag{4.4.30}$$

4.4.3 Two space dimensions: Method of descent

In two space dimensions the problem can be solved just as before with a double Fourier transform or, more simply, it can be "guessed" by noting from (4.4.13) and (4.4.26) that each space dimension "carries" a factor of the form $\exp\left(-\xi^2\right)/(4\pi Dt)^{1/2}$ which is convolved with the initial data. But we can use this problem to demonstrate the so-called *method of descent* which, essentially, amounts to starting from the three-dimensional solution and carrying out an integration over the unnecessary dimension.

The procedure is based on the simple observation that a two-dimensional problem can be considered as a three-dimensional one in which the data or the solution do not depend on the third coordinate, say z. In particular, with regard to our three-dimensional solution u_h given by (4.4.26), the initial condition f would not depend on the third variable z. In other words, for each value of z, the value of f is the same or, alternatively, f is constant on each line parallel to the z-axis. With this observation, it is obvious that (4.4.26) also gives a solution of the two-dimensional problem, with a simplification possible in that the integration over ζ can be carried out explicitly. Since

$$\int_{-\infty}^{\infty} \exp\left(-\frac{\zeta^2}{4Dt}\right) d\zeta = 2\sqrt{\pi Dt}, \tag{4.4.31}$$

we have

$$u_h = \frac{1}{4\pi Dt} \int_{-\infty}^{\infty} d\xi \int_{-\infty}^{\infty} d\eta \, f(x - \xi, y - \eta) \exp\left[-\frac{\xi^2 + \eta^2}{4Dt}\right]. \tag{4.4.32}$$

The same approach can be followed to calculate the Green's function and u_f with the results

$$G(x, y, t; 0, 0) = \frac{1}{4\pi Dt} \exp\left[-\frac{x^2 + y^2}{4Dt}\right], \tag{4.4.33}$$

and

$$u_f = \int_0^t \frac{d\tau}{4\pi D\tau} \int_{-\infty}^{\infty} d\xi \int_{-\infty}^{\infty} d\eta \, q(x - \xi, y - \eta, t - \tau) \exp\left[-\frac{\xi^2 + \eta^2}{4D\tau}\right]. \tag{4.4.34}$$

4.5 Wave equation

As another application of the exponential Fourier transform we use it to solve the wave equation

$$\frac{1}{c^2} \frac{\partial^2 u}{\partial t^2} - \nabla^2 u = 0, \tag{4.5.1}$$

in several situations.

4.5.1 One space dimension

In one space dimension the wave equation is

$$\frac{1}{c^2} \frac{\partial^2 u}{\partial t^2} - \frac{\partial^2 u}{\partial x^2} = 0, \qquad -\infty < x < \infty. \tag{4.5.2}$$

We consider the general initial conditions

$$u(x, 0) = f(x), \qquad \left.\frac{\partial u}{\partial t}\right|_{t=0} = g(x), \tag{4.5.3}$$

and look for solutions bounded at infinity. Upon taking the Fourier transform we have

$$\frac{\partial^2 \tilde{u}}{\partial t^2} + k^2 c^2 \tilde{u} = 0, \qquad \tilde{u}(k, 0) = \tilde{f}(k), \qquad \left.\frac{\partial \tilde{u}}{\partial t}\right|_{t=0} = \tilde{g}(k). \tag{4.5.4}$$

The solution is readily found in the form

$$\tilde{u} = \frac{1}{2}\left(\tilde{f} + \frac{\tilde{g}}{ick}\right) e^{ikct} + \frac{1}{2}\left(\tilde{f} - \frac{\tilde{g}}{ick}\right) e^{-ikct}. \tag{4.5.5}$$

With the aid of rule 6 in Table 4.1 we have

$$\mathscr{F}^{-1}\left\{\tilde{f}e^{\pm ikct}\right\}(x) = \mathscr{F}^{-1}\left\{\tilde{f}\right\}(x \mp ct) = f(x \mp ct), \tag{4.5.6}$$

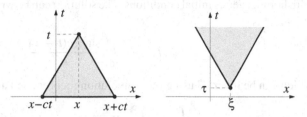

Fig. 4.2 Domain of dependence of the solution of the wave equation at position x and time t (left); domain of influence of the value of the forcing at (ξ, τ).

while, with rules 6 and 2,

$$\mathcal{F}^{-1}\left\{\frac{\tilde{g}}{ik}e^{\pm ikct}\right\}(x) = \mathcal{F}^{-1}\left\{\frac{\tilde{g}}{ik}\right\}(x \mp ct) = -\int_{-\infty}^{x \mp ct} g(\xi)\,\mathrm{d}\xi. \tag{4.5.7}$$

Upon combining the results, we recover the d'Alembert solution (1.2.9) p. 8 consisting of two counter-propagating waves:

$$u(x,t) = \Phi_-(x-ct) + \Phi_+(x+ct), \quad \Phi_\pm = \frac{1}{2}\left[f(x \pm ct) \pm \int_{-\infty}^{x \pm ct} g(\xi)\,\mathrm{d}\xi\right]. \tag{4.5.8}$$

For the first term to remain constant with increasing t, x has to increase, which shows that Φ_- describes a perturbation propagating in the positive x-direction, and conversely for Φ_+. A physically more meaningful alternative form can be found by using the relation $-\int_{-\infty}^{x-ct} + \int_{-\infty}^{x+ct} = \int_{x-ct}^{x+ct}$, namely

$$u(x,t) = \frac{1}{2}[f(x-ct) + f(x+ct)] + \frac{1}{2c}\int_{x-ct}^{x+ct} g(\xi)\,\mathrm{d}\xi. \tag{4.5.9}$$

It is clear here that, at position x and time t, the solution only depends on the initial data in the finite *domain of dependence* $[x - ct, x + ct]$ (Figure 4.2).

It is worth pointing out that, once the solution is put in the form (4.5.9), we can allow f to be discontinuous and therefore, in particular, not differentiable in the usual sense. Nevertheless f and g may be differentiable as distributions and, in this case, (4.5.9) represents the distributional solution of the wave equation.

Following the same logic that led to the decomposition of the non-homogeneous diffusion equation in §4.4 into two problems, the homogeneous equation with non-homogeneous initial data and the non-homogeneous equation with homogeneous initial data, we consider now a problem of the latter type for the wave equation:

$$\frac{1}{c^2}\frac{\partial^2 u}{\partial t^2} - \frac{\partial^2 u}{\partial x^2} = F(x,t), \tag{4.5.10}$$

subject to $u(x,0) = 0$, $(\partial u/\partial t)(x,0) = 0$. The transformed equation is

$$\frac{1}{c^2}\frac{\partial^2 \tilde{u}}{\partial t^2} + k^2\tilde{u} = \tilde{F}(k,t) \tag{4.5.11}$$

with homogeneous initial conditions. The solution can be written down by adapting (2.2.29) p. 34 and is

$$\tilde{u}(k, t) = c \int_0^t \frac{\sin kc(t - \tau)}{k} \tilde{F}(k, \tau) \, d\tau, \tag{4.5.12}$$

which can be inverted using the convolution theorem and an entry of Table 4.2:

$$u(x, t) = \frac{c}{2} \int_0^t d\tau \int_{-\infty}^{\infty} H\left(c(t - \tau) - |x - \xi|\right) F(\xi, \tau) \, d\xi, \tag{4.5.13}$$

or, since the step function will only be non-zero for $x - c(t - \tau) < \xi < x + c(t - \tau)$,

$$u(x, t) = \frac{c}{2} \int_0^t d\tau \int_{x - c(t - \tau)}^{x + c(t - \tau)} F(\xi, \tau) \, d\xi. \tag{4.5.14}$$

Thus, the value of the forcing at (ξ, τ) can only influence the solution for $t > \tau$ in the interval $x - c(t - \tau) < \xi < x + c(t - \tau)$; these conditions define the *domain of influence* of the data (Figure 4.2). Reciprocally, the solution of (x, t) only depends on the values of F in $0 < \tau < t$, $x - c(t - \tau) < \xi < x + c(t - \tau)$, which defines the domain of dependence of the solution as found before.

4.5.2 Three space dimensions

Before considering the general three-dimensional case, let us focus on the special case of a spherically symmetric wave. By using the simple identity

$$\frac{1}{r^2} \frac{\partial}{\partial r} \left(r^2 \frac{\partial u}{\partial r} \right) = \frac{1}{r} \frac{\partial^2}{\partial r^2}(ru), \qquad r = \sqrt{x^2 + y^2 + z^2}, \tag{4.5.15}$$

the spherically symmetric wave equation may be written as

$$\frac{1}{c^2} \frac{\partial^2}{\partial t^2}(ru) - \frac{\partial^2}{\partial r^2}(ru) = 0, \tag{4.5.16}$$

which, formally, is a one-dimensional wave equation for the variable ru. Thus, from (4.5.8), we have the solution

$$u(r, t) = \frac{1}{r} [\Phi_+(r + ct) + \Phi_-(r - ct)]. \tag{4.5.17}$$

As in the one-dimensional case (4.5.8), the first term describes waves propagating inward toward the origin, while the second term describes waves propagating away from the origin. If the problem concerns waves emitted by a localized source, there will be no waves propagating inward from infinity and we must therefore take $\Phi_+ = 0$. In the case of standing waves regular at the origin, it is necessary that $\Phi_+(ct) = -\Phi_-(-ct)$.

Let us now consider the general case:

$$\frac{1}{c^2} \frac{\partial^2 u}{\partial t^2} - \left(\frac{\partial^2 u}{\partial x^2} + \frac{\partial^2 u}{\partial y^2} + \frac{\partial^2 u}{\partial z^2} \right) = 0, \qquad -\infty < x, y, z < \infty. \tag{4.5.18}$$

It is sufficient to solve the initial-value problem

$$u(x, y, z, 0) = 0, \qquad \frac{\partial u}{\partial t}(x, y, z, 0) = g(x, y, z), \qquad (4.5.19)$$

because $v(x, y, z, t) = \partial u/\partial t$ is also a solution of the equation satisfying initial data of the form

$$v(x, y, z, 0) = g(x, y, z), \qquad \frac{\partial v}{\partial t}(x, y, z, 0) = \frac{\partial^2 u}{\partial t^2} = c^2 \nabla^2 u(x, y, z, 0) = 0.$$
$$(4.5.20)$$

Thus, with the solution of (4.5.19) in hand, we can immediately write down the solution of (4.5.20) and, by combining the two, we can find the solution of the general problem with non-zero initial values for both u and $\partial u/\partial t$.

Upon taking the three-dimensional Fourier transform according to (4.1.13) we have

$$\frac{\partial^2 \tilde{u}}{\partial t^2} + c^2 k^2 \tilde{u} = 0 \qquad \text{where} \qquad k^2 = \mathbf{k} \cdot \mathbf{k} = k_x^2 + k_y^2 + k_z^2. \qquad (4.5.21)$$

The solution satisfying $(\partial \tilde{u}/\partial t)(\mathbf{k}, 0) = \tilde{g}(\mathbf{k})$ is

$$\tilde{u} = \frac{\sin kct}{ck} \tilde{g}(\mathbf{k}) \qquad (4.5.22)$$

and therefore, from the convolution theorem,

$$u = \frac{1}{(2\pi)^{3/2}} \int_{-\infty}^{\infty} d\xi \int_{-\infty}^{\infty} d\eta \int_{-\infty}^{\infty} d\zeta \, g(\mathbf{x} - \boldsymbol{\xi}) \mathscr{F}^{-1} \left\{ \frac{\sin kct}{ck} \right\}. \qquad (4.5.23)$$

The inverse transform appearing here is

$$\mathscr{F}^{-1} \left\{ \frac{\sin kct}{ck} \right\} = \frac{1}{(2\pi)^{3/2}} \int \frac{\sin kct}{ck} e^{-i\mathbf{k}\cdot\boldsymbol{\xi}} \, d^3 k. \qquad (4.5.24)$$

We introduce polar coordinates in k-space taking the k_z-axis along $\boldsymbol{\xi}$ so that $\mathbf{k} \cdot \boldsymbol{\xi} = k\rho \cos\theta$, where $\rho = |\boldsymbol{\xi}|$. Then, after a trivial integration over the second angular variable,

$$\mathscr{F}^{-1} \left\{ \frac{\sin kct}{ck} \right\} = \frac{1}{\sqrt{2\pi}} \int_0^{\infty} k^2 \, dk \frac{\sin kct}{ck} \int_{-1}^{1} e^{-ik\rho \cos\theta} d(\cos\theta)$$

$$= \frac{1}{\sqrt{2\pi}} \int_0^{\infty} \frac{\sin kct}{kc} \left(2\frac{\sin k\rho}{k\rho} \right) k^2 \, dk = \sqrt{\frac{2}{\pi}} \int_0^{\infty} \frac{\sin k\rho \sin kct}{c\rho} \, dk$$

$$= \frac{1}{\sqrt{2\pi} c\rho} \int_0^{\infty} [\cos k(r-ct) - \cos k(r+ct)] \, dk = \sqrt{\frac{\pi}{2}} \frac{\delta(\rho - ct)}{c\rho} \qquad (4.5.25)$$

where in the last step we have used the Fourier integral representation of the delta function (Table 4.3 p. 87 or Table 2.1 p. 37) and we have omitted a term $\delta(\rho + ct)$ which cannot give a non-zero contribution since both ρ and t are inherently positive. Upon substitution into (4.5.23) we have

$$u = \frac{1}{4\pi c} \int g(\mathbf{x} - \boldsymbol{\xi}) \frac{\delta(|\boldsymbol{\xi}| - ct)}{|\boldsymbol{\xi}|} d^3 \xi, \qquad (4.5.26)$$

or, because of the symmetry of the convolution product (cf. (4.1.8)),

$$u = \frac{1}{4\pi c} \int g(\boldsymbol{\xi}) \frac{\delta(|\mathbf{x} - \boldsymbol{\xi}| - ct)}{|\mathbf{x} - \boldsymbol{\xi}|} d^3\xi. \tag{4.5.27}$$

From (4.5.17), $\delta(|\mathbf{x} - \boldsymbol{\xi}| - ct)/|\mathbf{x} - \boldsymbol{\xi}|$ may be interpreted as an elementary outward-propagating spherical wave originating at $\boldsymbol{\xi}$. Thus we see that the disturbance u at $(\mathbf{x}, t)$ is the superposition of elementary waves of strength $g(\boldsymbol{\xi})$ emitted at position $\boldsymbol{\xi}$ at the initial instant, all reaching $\mathbf{x}$ at the same time $t = |\mathbf{x} - \boldsymbol{\xi}|/c$. Upon passing to spherical coordinates in (4.5.26) writing $d^3\xi = \rho^2 \, d\rho \, d\Omega$, with $d\Omega$ the element of solid angle, we have

$$u = \frac{t}{4\pi} \int g(\mathbf{x} - \boldsymbol{\xi}) \delta(|\boldsymbol{\xi}| - ct) \, d\rho \, d\Omega = \frac{t}{4\pi} \int_{|\boldsymbol{\xi}|=ct} g(\mathbf{x} - \boldsymbol{\xi}) \, d\Omega. \tag{4.5.28}$$

The last integral is just the average of $g(\mathbf{x} - \boldsymbol{\xi})$ over the sphere of radius ct surrounding the point $\mathbf{x}$. If the initial data g are confined to a finite region $\mathscr{R}$ in space, the solution at $\mathbf{x}$ will be zero until the first instant at which this sphere intersects $\mathscr{R}$, and will go back to zero when ct is so large as to entirely contain $\mathscr{R}$. Thus, three-dimensional waves have no "tail."

In order to write the solution of the wave equation (4.5.18) corresponding to the more general initial conditions

$$u(\mathbf{x}, 0) = f(\mathbf{x}), \qquad \frac{\partial u}{\partial t}(\mathbf{x}, 0) = g(\mathbf{x}), \tag{4.5.29}$$

we use the remark made above in connection with (4.5.20) and find

$$u = \frac{\partial}{\partial t}\left[\frac{t}{4\pi} \int_{|\boldsymbol{\xi}|=ct} f(\mathbf{x} - \boldsymbol{\xi}) \, d\Omega \right] + \frac{t}{4\pi} \int_{|\boldsymbol{\xi}|=ct} g(\mathbf{x} - \boldsymbol{\xi}) \, d\Omega. \tag{4.5.30}$$

Once again we note that this solution makes sense provided only that the indicated integrals exist, even if the result of the integration does not possess the second partial derivatives that appear in the wave equation (4.5.18).

4.5.3 Two space dimensions

The solution of the wave equation in two space dimensions

$$\frac{1}{c^2} \frac{\partial^2 u}{\partial t^2} - \left(\frac{\partial^2 u}{\partial x^2} + \frac{\partial^2 u}{\partial y^2} \right) = 0, \qquad -\infty < x < \infty, \tag{4.5.31}$$

can be found by the method of descent as shown e.g. in Whitham (1974) p. 234 or in Courant & Hilbert (1953), vol. 2 p. 686. Here however we proceed directly to demonstrate a way to deal with non-convergent integrals with the help of the δ-function and its derivatives.

Upon taking the two-dimensional Fourier transform according to (4.1.13) we have

$$\frac{\partial^2 \tilde{u}}{\partial t^2} + c^2(h^2 + k^2)\tilde{u} = 0. \tag{4.5.32}$$

Let us suppose that $u(x, y, 0) = f(x, y)$, while $(\partial u/\partial t)(x, y, 0) = 0$. Then we have

$$\tilde{u} = \tilde{f}(h, k) \cos ctk, \qquad k = \sqrt{h^2 + k^2}, \tag{4.5.33}$$

and therefore

$$u(x, y, t) = \frac{1}{2\pi} \int_{-\infty}^{\infty} d\xi \int_{-\infty}^{\infty} d\eta \, f(x - \xi, y - \eta) \mathscr{F}^{-1}\{\cos ctk\}(\xi, \eta). \qquad (4.5.34)$$

With $\lambda = ct$, the inverse Fourier transform of the last term is

$$I \equiv \mathscr{F}^{-1}\{\cos \lambda \sqrt{h^2 + k^2}\} = \frac{1}{2\pi} \int_{-\infty}^{\infty} dh \int_{-\infty}^{\infty} dk \, \cos \lambda k \, e^{-i\mathbf{k}\cdot\boldsymbol{\xi}}. \qquad (4.5.35)$$

As before we pass to polar coordinates in k-space letting $\mathbf{k} \cdot \boldsymbol{\xi} = k\rho \cos \theta$, with $\rho = |\boldsymbol{\xi}|$ and find

$$I = \frac{1}{2\pi} \int_0^{\infty} \rho \, d\rho \int_0^{2\pi} d\theta \, \cos \lambda k \, e^{-ik\rho \cos \theta} = \int_0^{\infty} \cos \lambda k \, J_0(\rho r) \, k \, dk, \qquad (4.5.36)$$

where we have used the standard integral representation of the Bessel function J_0 (p. 308). The last integral is not convergent, but we may write

$$I = \frac{\partial}{\partial \lambda} \int_0^{\infty} \sin \lambda \rho \, J_0(\rho r) \, d\rho = \frac{\partial}{\partial \lambda} \frac{H(\lambda - r)}{\sqrt{\lambda^2 - r^2}} \qquad (4.5.37)$$

and, after substitution into (4.5.34),

$$u(x, y, t) = \frac{1}{2\pi c} \frac{\partial}{\partial t} \int \int_{\xi^2 + \eta^2 \le c^2 t^2} \frac{f(x - \xi, y - \eta)}{\sqrt{c^2 t^2 - \xi^2 - \eta^2}} \, d\xi \, d\eta. \qquad (4.5.38)$$

With the more general initial condition (4.5.29), by the same method we find

$$u(x, y, t) = \frac{1}{2\pi c} \frac{\partial}{\partial t} \int \int_{\xi^2 + \eta^2 \le c^2 t^2} \frac{f(x - \xi, y - \eta)}{\sqrt{c^2 t^2 - \xi^2 - \eta^2}} \, d\xi \, d\eta$$
$$+ \frac{1}{2\pi c} \int \int_{\xi^2 + \eta^2 \le c^2 t^2} \frac{g(x - \xi, y - \eta)}{\sqrt{c^2 t^2 - \xi^2 - \eta^2}} \, d\xi \, d\eta. \qquad (4.5.39)$$

Contrary to the three-dimensional case, we see that a two-dimensional wave has a "tail," in the sense that the perturbation never vanishes no matter how large t is. The reason is clear once one envisages, as it is proper, this two-dimensional problem embedded in three-dimensional space: an initial two-dimensional disturbance effectively extends to infinity in the third dimension. Therefore, for all times, the sphere of radius ct centered at $\mathbf{x}$ intersects the region where the initial disturbance is non-zero.

4.5.4 The non-homogeneous problem

Let us return to the three-dimensional case in the presence of a source term:

$$\frac{1}{c^2} \frac{\partial^2 u}{\partial t^2} - \nabla^2 u = F(\mathbf{x}, t). \qquad (4.5.40)$$

As before in §4.5.1 it will be sufficient to solve this problem subject to $u(\mathbf{x}, 0) = 0$, $(\partial u / \partial t)(\mathbf{x}, 0) = 0$ since we already have the solution for general initial conditions. After

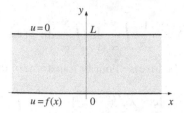

Fig. 4.3   Laplace equation in an infinite strip, §4.6

taking the Fourier transform, we can use (2.2.29) p. 34 to solve the resulting equation in t by variation of parameters finding

$$\tilde{u}(\mathbf{k}, t) = c \int_0^t \tilde{F}(\mathbf{k}, \tau) \frac{\sin kc(t - \tau)}{k} \, d\tau \,. \tag{4.5.41}$$

We now use the convolution theorem and the earlier result (4.5.25) to express u:

$$u = \frac{c}{4\pi} \int d^3\xi \int_0^t d\tau \, F(\mathbf{x} - \boldsymbol{\xi}, \tau) \frac{\delta(|\boldsymbol{\xi}| - c(t - \tau))}{|\boldsymbol{\xi}|}. \tag{4.5.42}$$

By noting that $\delta(|\boldsymbol{\xi}| - c(t - \tau)) = \delta(c(\tau - (t - |\boldsymbol{\xi}|/c))) = c^{-1}\delta(\tau - (t - |\boldsymbol{\xi}|/c))$ we can execute the τ-integration to find

$$u = \frac{1}{4\pi} \int \frac{F(\mathbf{x} - \boldsymbol{\xi}, t - |\boldsymbol{\xi}|/c)}{|\boldsymbol{\xi}|} d^3\xi = \frac{1}{4\pi} \int \frac{F(\boldsymbol{\xi}, t - |\mathbf{x} - \boldsymbol{\xi}|/c)}{|\mathbf{x} - \boldsymbol{\xi}|} d^3\xi. \tag{4.5.43}$$

An interesting special case is that of a point source $4\pi q$ moving with a velocity $\mathbf{v}(t)$. In this case $F = q\delta(\mathbf{x} - \mathbf{X}(t))$ with $\dot{\mathbf{X}} = \mathbf{v}$ so that

$$u = q \int \frac{\delta(\mathbf{x} - \boldsymbol{\xi} - \mathbf{X}(t - |\boldsymbol{\xi}|/c))}{|\boldsymbol{\xi}|} d^3\xi. \tag{4.5.44}$$

In order to evaluate the integral it is convenient to set $\mathbf{y} = \boldsymbol{\xi} + \mathbf{X}(t - |\boldsymbol{\xi}|/c)$, with which we find

$$\frac{\partial y_i}{\partial \xi_j} = \delta_{ij} - \frac{\xi_j}{c|\boldsymbol{\xi}|} v_i = \delta_{ij} + \frac{X_j - y_j}{c|\boldsymbol{\xi}|} v_i. \tag{4.5.45}$$

With this expression, the Jacobian of the transformation is readily evaluated and is found to be

$$\frac{\partial(y_1, y_2, y_3)}{\partial(\xi_1, \xi_2, \xi_3)} = 1 + \frac{(\mathbf{X} - \mathbf{y}) \cdot \mathbf{v}}{c|\mathbf{X} - \mathbf{y}|}. \tag{4.5.46}$$

The solution is then

$$u = \frac{qc}{c|\mathbf{x} - \mathbf{X}(t')| + [\mathbf{X}(t') - \mathbf{x}] \cdot \mathbf{v}(t')}, \qquad t' = t - \frac{|\mathbf{x} - \mathbf{X}(t')|}{c} \tag{4.5.47}$$

with t' the *retarded time*. In electromagnetic theory, in which this problem arises with u having the physical meaning of a potential, this expression is known as the *Liénard–Wiechert potential* (see e.g. Jackson 1998, chapter 14).

4.6 Laplace equation in a strip and in a half-plane

Let us consider the two-dimensional Laplace equation

$$\frac{\partial^2 u}{\partial x^2} + \frac{\partial^2 u}{\partial y^2} = 0, \qquad (4.6.1)$$

in the infinite strip

$$-\infty < x < \infty, \qquad 0 < y < L, \qquad (4.6.2)$$

subject to the Dirichlet conditions

$$u(x, 0) = f(x), \qquad u(x, L) = 0, \qquad u \to 0 \quad |x| \to \infty, \qquad (4.6.3)$$

in which $f(x)$ is given. We recover the solution in the half-plane $y > 0$ by letting $L \to \infty$ in the final result.

Upon taking the Fourier transform with respect to the variable x and using the operational formula for the derivative (rule 1 of table 4.1), the problem is reduced to

$$\frac{\partial^2 \tilde{u}}{\partial y^2} - k^2 \tilde{u} = 0, \quad \tilde{u}(k, 0) = \tilde{f}(k), \qquad \tilde{u}(k, L) = 0. \qquad (4.6.4)$$

The solution is

$$\tilde{u}(k, y) = \frac{\sinh k(L - y)}{\sinh kL} \tilde{f}(k). \qquad (4.6.5)$$

The inverse Fourier transform of the fraction can be readily calculated by the method of residues (§17.9.1 p. 461) to find

$$\mathscr{F}^{-1}\left\{ \frac{\sinh k(L - y)}{\sinh kL} \right\} = \frac{\sqrt{2\pi}}{2L} \frac{\sin(\pi y/L)}{\cosh(\pi x/L) - \cos(\pi y/L)}, \qquad (4.6.6)$$

so that, upon application of the convolution theorem (rule 3 of Table 4.1), we have

$$u(x, y) = \frac{1}{2L} \int_{-\infty}^{\infty} \frac{\sin(\pi y/L)}{\cosh[\pi(x - \xi)/L] - \cos(\pi y/L)} f(\xi) \, d\xi. \qquad (4.6.7)$$

Here we recognize the fraction divided by $2L$ as the Green's function of the problem. In Example 17.10.8 p. 484 this result is derived by conformal mapping methods.

Upon taking the limit of L large, with x and y fixed, the arguments of the hyperbolic and trigonometric functions become small and can be approximated by the first terms of their Taylor series. In this way it is easy to see that[6]

$$\lim_{L \to \infty} \frac{1}{2L} \frac{\sin(\pi y/L)}{\cosh(\pi x/L) - \cos(\pi y/L)} = \frac{1}{\pi} \frac{y}{x^2 + y^2} \qquad (4.6.8)$$

---

[6] By solving the problem in the half-plane from the very beginning, in place of (4.6.5) one finds $\tilde{u}(k, y) = e^{-ky} \tilde{f}(k)$ and the result follows by the same procedure.

so that the solution for the Laplace equation in the half-plane, with Dirichlet conditions, is

$$u(x, y) = \frac{1}{\pi} \int_{-\infty}^{\infty} \frac{y}{(x - \xi)^2 + y^2} f(\xi) \, d\xi. \tag{4.6.9}$$

Direct differentiation under the integral sign readily shows that this expression satisfies the Laplace equation. It is less trivial to show that the boundary condition (4.6.3) is satisfied. However we show in Example 20.6.1 p. 584 that (see also p. 437)

$$\lim_{y \to 0} \frac{1}{\pi} \frac{y}{(x - \xi)^2 + y^2} = \delta(x - \xi), \tag{4.6.10}$$

so that (4.6.3) is indeed satisfied. Similar arguments can be applied to (4.6.7).

The same problem could be solved by using a Fourier series in y. The procedure is similar to that used in §3.2.3 p. 62: we expand $\partial^2 u / \partial y^2$ in a sine series in y and integrate by parts twice to find, according to (3.1.16) p. 57,

$$\frac{\partial^2 u}{\partial y^2} = \sum_{n=1}^{\infty} \left[\frac{2n\pi}{L^2} f(x) - \left(\frac{n\pi}{L} \right)^2 B_n(x) \right] \sin \frac{n\pi y}{L}, \tag{4.6.11}$$

where the B_n's are the coefficients of the Fourier sine expansion of u. As before, substitution into the Laplace equation then gives the following equation for these coefficients

$$\frac{d^2 B_n}{dx^2} - \left(\frac{n\pi}{L} \right)^2 B_n = -\frac{2n\pi}{L^2} f(x), \tag{4.6.12}$$

which can be solved by variation of parameters to find

$$B_n = \frac{1}{L} \int_{-\infty}^{\infty} \exp \left[-\frac{n\pi |x - \xi|}{L} \right] f(\xi) \, d\xi, \tag{4.6.13}$$

with which

$$u(x, y) = \int_{-\infty}^{\infty} d\xi \, f(\xi) \frac{1}{L} \sum_{n=1}^{\infty} \exp \left[-\frac{n\pi |x - \xi|}{L} \right] \sin \frac{n\pi y}{L}. \tag{4.6.14}$$

Upon setting $r = e^{-\pi |x-\xi|/L}$, the series is seen to be the same as the one summed in (8.7.19) p. 236; in this way, the previous result (4.6.7) is recovered.

4.7 Example of an ill-posed problem

The reader will have noticed that in all the examples involving the Laplace or Poisson equations we have prescribed either the function (Dirichlet data) or its normal derivative (Neumann data) on the entire boundary of the domain, including infinity. For problems involving the wave equation, on the other hand, we have prescribed initial data on the function and its time derivative (Cauchy data) on the line $t = 0$.[7] Although this procedure

---

[7] Reference e.g. to Figure 4.2 graphically demonstrates that an initial condition on $\partial u / \partial t$ is a condition on the normal derivative on the boundary $t = 0$ and, therefore, it amounts to a Neumann condition.

is intuitively correct, from the mathematical point of view a variable is a variable, whether time or space, and one may wonder why we never attempted to solve the wave equation prescribing Dirichlet data at some "final time" T, similar e.g. to the condition $u(x, L) = 0$ used in the previous section or, conversely, the Laplace equation prescribing both the value of the function and of its normal derivative on one part of the boundary only, e.g., in the problem of the previous section, on the line $y = 0$.

For the mathematical reasons of our procedure and the difference between hyperbolic (wave equation) and elliptic (Laplace and Poisson) equations the reader is referred to standard texts on partial differential equations (see e.g. Garabedian 1964, chapter 3; Zauderer 1989, chapter 3; Stakgold 1997, p. 467; and many others). Here we want to show what happens if one tries to solve an elliptic or a hyperbolic equation with the "wrong" type of data. To present a unified analysis, we write y in place of ct for the wave equation and consider the two equations

$$\frac{\partial^2 u_\pm}{\partial x^2} \pm \frac{\partial^2 u_\pm}{\partial y^2} = 0, \qquad -\infty < x < \infty, \ 0 < y < \infty \qquad (4.7.1)$$

in the half-plane $y > 0$. Cauchy data assign values to both the function and its normal derivative on part of the boundary and therefore, in this case, they would have the form e.g.

$$u_\pm(x, 0) = f(x), \qquad \frac{\partial u_\pm}{\partial y}\bigg|_{y=0} = g(x), \qquad (4.7.2)$$

with no condition imposed on the rest of the boundary, i.e. $|x| \to \infty$ and $y \to \infty$. Typical Dirichlet data would be instead

$$u_\pm(x, 0) = f(x), \qquad u_\pm(x, y) \to 0 \quad \text{for} \quad |x|, \ y \to \infty, \qquad (4.7.3)$$

with no specification of the value of the normal derivative.

After taking the Fourier transform of (4.7.1) with respect to x we find

$$\frac{\partial^2 \tilde{u}_\pm}{\partial y^2} \mp k^2 \tilde{u}_\pm = 0, \qquad (4.7.4)$$

the general solutions of which may be written

$$\tilde{u}_+ = \tilde{A} \cosh(ky) + \tilde{B} \sinh(ky), \qquad \tilde{u}_- = \tilde{A} \cos(ky) + \tilde{B} \sin(ky). \qquad (4.7.5)$$

If we were to impose the Cauchy data (4.7.2) on $\tilde{u}_+$, we would have

$$\tilde{u}_+ = \tilde{f} \cosh(ky) + \frac{\tilde{g}}{k} \sinh(ky) = \frac{1}{2}\left(\tilde{f} + \frac{\tilde{g}}{k}\right)e^{ky} + \frac{1}{2}\left(\tilde{f} - \frac{\tilde{g}}{k}\right)e^{-ky}. \qquad (4.7.6)$$

It is evident that, in general, the solution blows up for large y unless g happened to be such that $\tilde{g} = \mp k\tilde{f}$ for $k > 0$, or $k < 0$, respectively. More troubling is the fact that the divergence *increases* with increasing $|k|$: however small the term $\tilde{f} + \tilde{g}/k$, for example, the magnitude of $\tilde{u}_+$ can be made arbitrarily large by considering sufficiently large k, which is clearly a concern as inverting the Fourier transform involves the entire range $-\infty < k < \infty$.

This divergence of the solution is a manifestation of the so-called *non-continuous dependence on the data* which is one aspect of problems designated as *ill-posed* in the sense of Hadamard. The idea is that, if one of piece of data of the problem, e.g. $f(x)$, is modified by an infinitesimal amount $\varepsilon(x)$, provided $\tilde{\varepsilon}$ contains sufficiently large values of $|k|$, the difference between the original and modified solution can become arbitrarily large no matter how small $|\varepsilon(x)|$ is. Note that this concept is quite different from the notion of *sensitive dependence on the data* which arises in the theory of dynamical systems: sensitive dependence means that the difference between two solutions quickly (exponentially) increases as a variable (e.g., time) is changed. Non-continuous dependence, on the other hand, means that for a *fixed value* of the variable the difference can be made arbitrarily large by changing the data by a small amount. If, in order to fix the problem, $\tilde{g}$ is chosen equal to $\tilde{f}/k$ so that u_+ is well behaved as $y \to \infty$, (4.7.6) becomes $\tilde{u}_+ = \tilde{f}(k)\, e^{-|k|x}$ which is equivalent to solving the equation subject to the Dirichlet data (4.7.3).

Conversely, if one tried to impose the Dirichlet data (4.7.3) on the solution of the wave equation $\tilde{u}_-$, one would have

$$\tilde{u}_- = \tilde{f}(k)\cos(ky) + \tilde{B}(k)\sin(ky). \qquad (4.7.7)$$

By using the first entry in Table 4.3 p. 87 we see that

$$\mathscr{F}^{-1}\{\cos ky\} = \sqrt{\frac{\pi}{2}}\,[\delta(x+y) + \delta(x-y)]\,, \quad \mathscr{F}^{-1}\{\sin ky\} = i\sqrt{\frac{\pi}{2}}\,[\delta(x+y) - \delta(x-y)]$$

$$(4.7.8)$$

so that, by appealing to the convolution theorem, we invert the transform with the result

$$u_-(x,y) = \frac{1}{2}\,[f(x+y) + f(x-y)] + \frac{i}{2}\,[B(x+y) - B(x-y)]\,. \qquad (4.7.9)$$

Even if f where to tend to 0 for large values of its argument, $f(x \pm y)$ will not go to zero along the lines $x \pm y = $ const. thus violating the boundedness condition (4.7.3) no matter how B is chosen. Thus, as posed, the problem admits no solution, which is another feature of ill-posed problems.

4.8 The Hilbert transform and dispersion relations

For a physical system which is invariant under time translation, a general linear relationship between an input $X(t)$ and an output $Y(t)$ may be written as

$$Y(t) = \int_{-\infty}^{\infty} G(t-\tau)\, X(\tau)\, d\tau\,. \qquad (4.8.1)$$

This is the form of the solution, for example, if the relation between Y and X is given by

$$\mathsf{L}Y = X \qquad (4.8.2)$$

with L a linear operator not explicitly dependent on time; in this case, G would be the Green's function for the operator L (§15.5 p. 388).

Some reasonable physical assumptions put strong constraints on the function G. In the first place, to satisfy causality, the behavior of X at τ cannot influence that of Y for $t < \tau$ and, therefore, $G(t)$ must vanish for negative values of its variable. If an impulsive input

$X \propto \delta(t - t_0)$ results in a finite output, G cannot have singularities.[8] Furthermore, one expects the behavior of X in the remote past to have little effect on Y, which requires that $G(t) \to 0$ for $t \to \infty$. The power of the output is typically proportional to $|Y|^2$ and, if the total energy is to be finite when the total input energy is finite, then G admits a convolution with itself which, in particular, makes it square integrable.[9] Thus we expect that G should admit a Fourier transform:

$$\tilde{G}(\omega) = \frac{1}{\sqrt{2\pi}} \int_0^\infty G(t) e^{i\omega t} \, d\omega \qquad (4.8.3)$$

which is also square integrable by the Parseval relation (4.1.6). If we allow ω to be complex, $\omega = \omega' + i\omega''$, it is obvious that the integral (4.8.3) exists not only for real ω, but also for $\omega'' > 0$ and, therefore, $\tilde{G}(\omega)$ is an analytic function of ω in the upper half-plane of this variable (see Theorem 10.6.1 p. 276), although it may not be analytic on the real line where it might, for example, have branch points (of such a nature, however, as to preserve square integrability).

Consider the Hilbert transform of $\tilde{G}$ (§20.14 p. 616)

$$\check{G}(\Omega) = \frac{1}{\pi} \mathrm{PV} \int_{-\infty}^\infty \frac{\tilde{G}(\omega)}{\omega - \Omega} \, d\omega = \lim_{R \to \infty} \lim_{\varepsilon \to 0} \frac{1}{\pi} \left(\int_{-R}^{-\varepsilon} + \int_\varepsilon^R \right) \frac{\tilde{G}(\omega)}{\omega - \Omega} \, d\omega. \qquad (4.8.4)$$

Since $\tilde{G}$ is analytic in the upper half-plane, we have from Plemelj's formulae (4.2.21)[10]

$$\mathrm{PV} \frac{1}{\omega - \Omega} - i\pi\delta(\omega - \Omega) = \frac{1}{\omega - \Omega + i0} = 0 \qquad (4.8.5)$$

with the last equality a consequence of the fact that the pole $\omega = \Omega - i0$ is below the real axis so that the integrand is analytic inside the integration domain. Thus we can replace the principal value of the fraction by $i\pi\delta(\omega - \Omega)$ to find

$$\check{G}(\Omega) = \frac{1}{\pi} \mathrm{PV} \int_{-\infty}^\infty \frac{\tilde{G}(\omega)}{\omega - \Omega} \, d\omega = i\tilde{G}(\Omega) \qquad (4.8.6)$$

or, upon writing $\tilde{G} = \tilde{G}_R + i\tilde{G}_I$ and separating real and imaginary parts,

$$\tilde{G}_R(\Omega) = \frac{1}{\pi} \mathrm{PV} \int_{-\infty}^\infty \frac{\tilde{G}_I(\omega)}{\omega - \Omega} \, d\omega \,, \qquad \tilde{G}_I(\Omega) = -\frac{1}{\pi} \mathrm{PV} \int_{-\infty}^\infty \frac{\tilde{G}_R(\omega)}{\omega - \Omega} \, d\omega. \qquad (4.8.7)$$

Thus, the (strong) constraint of causality and the (weaker) ones of regularity have the consequence that the real and imaginary parts of $\tilde{G}$ are not independent and the full function can be reconstructed from either one of them.

---

[8] There are some exceptions. For example, the electric permittivity of metals has a simple pole at zero frequency (see e.g. Jackson 1998, p. 309). This pole can be treated with an additional indentation of the integration contour thus arriving at a modified form of (4.8.7) below.

[9] By interchanging the orders of integration (p. 690) we may write $\int_{-\infty}^\infty |Y(t)|^2 \, dt = \int_{-\infty}^\infty \overline{X}(\tau) \, d\tau \int_{-\infty}^\infty X(\theta) \, d\theta \int_{-\infty}^\infty \overline{G}(\eta) G(\eta - (\theta - \tau)) \, d\eta$.

[10] For a proof without recourse to the Plemelj formulae one may note that the integral over a closed contour similar to that shown in Figure 17.18 p. 464 (but indented at $\omega = \Omega$ rather than 0) vanishes by the analyticity of the integrand inside the contour. From this observation, since the large semi-circle does not contribute, the integral along the real line equals minus the integral around the small semi-circle. To evaluate the latter integral one replaces $\tilde{G}(\omega)$ in the integrand by $\tilde{G}(\Omega) + [\tilde{G}(\omega) - \tilde{G}(\Omega)]$. The integral of the first term can be calculated directly, and that of the second one vanishes as the radius of the semi-circle vanishes.

These results can be elaborated further if $G(t)$ is real because then, from (4.1.4), $\tilde{G}(\omega) = \tilde{G}(-\omega)$ so that $\tilde{\overline{G}}_R(\omega) = \tilde{G}_R(-\omega)$ and $\tilde{\overline{G}}_I(\omega) = -\tilde{G}_I(-\omega)$. With the aid of these relations the integrals from $-\infty$ to ∞ can be rewritten as integrals from 0 to ∞ with the result

$$\tilde{G}_R(\Omega) = \frac{2}{\pi} \mathrm{PV} \int_0^\infty \frac{\omega \tilde{G}_I(\omega)}{\omega^2 - \Omega^2}\, d\omega\,, \quad \tilde{G}_I(\Omega) = -\frac{2}{\pi}\Omega \mathrm{PV} \int_0^\infty \frac{\tilde{G}_R(\omega)}{\omega^2 - \Omega^2}\, d\omega. \quad (4.8.8)$$

These are the *Kramers–Kronig dispersion relations*.

The imaginary part of $\tilde{G}$ is often related to the dissipative part of the relation between the input X and the response Y. For example, if the operator L in (4.8.2) is a linear combination of time derivatives, $\tilde{G}_I$ arises from the derivatives of odd order which are not invariant under time reversal and, therefore, describe irreversible processes. Thus, the Kramers–Kronig relations establish a connection between the dissipative response of the system and its in-phase response, and vice versa, which form the basis of the *fluctuation-dissipation theorem* (see e.g. Chaikin & Lubensky 1995, p. 397; Pécseli 2000, p. 41).

4.9 Fredholm integral equation of the convolution type

An interesting application of the convolution rule is to the solution of Fredholm integral equations of the form

$$u(x) = f(x) + \frac{\lambda}{\sqrt{2\pi}} \int_{-\infty}^{\infty} K(x - \xi)u(\xi)\, d\xi \qquad (4.9.1)$$

where the functions f and K and the parameter λ are given. Upon taking the Fourier transform this equation becomes $\tilde{u}(k) = \tilde{f}(k) + \lambda\, \tilde{K}(k)\, \tilde{u}(k)$ so that

$$u(x) = \mathscr{F}^{-1}\left\{ \frac{\tilde{f}}{1 - \lambda\tilde{K}} \right\} = f + \lambda\mathscr{F}^{-1}\left\{ \frac{\tilde{K}}{1 - \lambda\tilde{K}}\, \tilde{f} \right\}. \qquad (4.9.2)$$

If K admits an ordinary Fourier transform, $|\tilde{K}| \to 0$ at infinity and the inverse transform of $1/(1 - \lambda\tilde{K})$ does not exist in the ordinary sense; the last form shown avoids this problem (cf. §5.6 p. 144 for a similar situation with the Laplace transform). Attention must be paid to the possibility that $1 - \lambda\tilde{K}$ might have real zeros, which would introduce δ-functions, and possibly their derivatives, in the division leading to (4.9.2).[11]

Example 4.9.1 An example is the equation

$$u(x) = f(x) + \lambda \int_{-\infty}^{\infty} e^{-|x-\xi|}u(\xi)\, d\xi \qquad (4.9.3)$$

for which we find

$$\left(k^2 + 1 - 2\lambda\right)\tilde{u} = (k^2 + 1)\tilde{f} = \left(k^2 + 1 - 2\lambda\right)\tilde{f} + 2\lambda\tilde{f}. \qquad (4.9.4)$$

---

[11] As shown on p. 610, the solution e.g. of an equation such as $(k - k_0)\tilde{u} = \tilde{f}$ is $\tilde{u} = \tilde{f}/(k - k_0) + \delta(k - k_0)$.

If $1 - 2\lambda > 0$ the solution is

$$
\begin{aligned}
u(x) &= f(x) + \sqrt{\frac{2}{\pi}}\lambda \int_{-\infty}^{\infty} f(x-\xi)\mathscr{F}^{-1}\left\{(k^2+1-2\lambda)^{-1}\right\}\,d\xi \\
&= f(x) + \frac{\lambda}{\sqrt{1-2\lambda}} \int_{-\infty}^{\infty} f(x-\xi)\exp\left(-\sqrt{1-2\lambda}\,|\xi|\right)\,d\xi.
\end{aligned}
\tag{4.9.5}
$$

If, however, $1 - 2\lambda = -B^2 < 0$, writing $\tilde{u} = \tilde{f} + 2\lambda\tilde{f}/(k^2 - B^2)$ is meaningless in the ordinary sense unless $\tilde{f}$ happens to vanish at least like $k \pm B$ at $k = \mp B$. Other than in this case, the integral must be regularized and a term $C_1\delta(k - B) + C_2\delta(k + B)$, with C_1 and C_2 arbitrary constants, must be added to $\tilde{u}$ (see p. 610). By using (4.1.11) together with the translation rule (item 6 in Table 4.1 p. 85), and the transform of x^{-1} (Table 4.3 p. 87) we find*

$$
\begin{aligned}
\mathscr{F}^{-1}\left\{(k^2-B^2)^{-1}\right\} &= \frac{1}{2B}\mathscr{F}^{-1}\left\{(k-B)^{-1}-(k+B)^{-1}\right\} \\
&= -\sqrt{\frac{\pi}{2}}(\operatorname{sgn}x)\frac{\sin Bx}{B} = -\sqrt{\frac{\pi}{2}}\frac{\sin B|x|}{B}
\end{aligned}
\tag{4.9.6}
$$

so that

$$
\begin{aligned}
u &= f - \frac{1}{2B}\int_{-\infty}^{\infty} f(x-\xi)\sin B|\xi|\,d\xi + C_1 e^{-iBx} + C_2 e^{iBx} \tag{4.9.7} \\
&= f - \frac{1}{2B}\int_{0}^{\infty}[f(x+\xi)+f(x-\xi)]\sin B\xi\,d\xi + C_1 e^{-iBx} + C_2 e^{iBx}.
\end{aligned}
$$

Since here we are concerned with the general solution, unlike the case of §4.2 here we must retain the contribution of both poles.

The case $\lambda = 1/2$ is special since the general solution of the homogeneous equation $k^2\tilde{v}(k) = 0$ is $C_1\delta(k) + C_2\delta'(k)$. Furthermore, $\mathscr{F}^{-1}\{k^{-2}\} = -\sqrt{\pi/2}\,|x|$ so that, upon renaming the arbitrary constants,

$$
u = f - \frac{1}{2}\int_{0}^{\infty}[f(x+\xi)+f(x-\xi)]\xi\,d\xi + C_1 + C_2 x.
\tag{4.9.8}
$$

The integral can be found by taking the formal limit $B \to 0$ in the previous expression, but not so the last two terms. □

4.10 Useful formulae for the sine and cosine transforms

The Fourier cosine and sine transforms, the theory of which is covered in §10.5, are useful for semi-infinite domains. The definitions of these transforms are

$$
\tilde{f}_c(k) \equiv \mathscr{F}_c\{f\} = \sqrt{\frac{2}{\pi}}\int_{0}^{\infty}\cos kx\, f(x)\,dx,
\tag{4.10.1}
$$

---

* Since here we are concerned with the general solution, unlike the case of §4.2, we must retain the contribution of both poles.

| Table 4.5 | Operational rules for the Fourier cosine transform. |

$f(x) = (2/\pi)^{1/2} \int_0^\infty \cos(kx)\,\tilde{f}_c(k)\,dk$	$\tilde{f}_c(k) = (2/\pi)^{1/2} \int_0^\infty \cos(kx)\,f(x)\,dx$
1 $f(ax)$ $(a > 0)$	$\tilde{f}_c(k/a)/a$
2 f'	$k\,\tilde{f}_s - (2/\pi)^{1/2} f(0)$
3 f''	$-k^2 \tilde{f}_c - (2/\pi)^{1/2} f'(0)$
4 $x^{2n} f(x)$	$(-1)^n (d^{2n}/dk^{2n})\tilde{f}_c$
5 $x^{2n+1} f(x)$	$(-1)^n (d^{2n+1}/dk^{2n+1})\tilde{f}_s$
6 $2a\cos(bx)\,f(ax)\,(a,\,b > 0)$	$\tilde{f}_c[(k+b)/a] + \tilde{f}_c[(k-b)/a]$
7 $2a\sin(bx)\,f(ax)\,(a,\,b > 0)$	$\tilde{f}_s[(k+b)/a] - \tilde{f}_s[(k-b)/a]$

$$8 \qquad \int_0^\infty f(x)g(x)\,dx = \int_0^\infty \tilde{f}_c(k)\tilde{g}_c(k)\,dk$$

$$9 \qquad \int_0^\infty [f(x)]^2\,dx = \int_0^\infty [\tilde{f}_c(k)]^2\,dk$$

$$\tilde{f}_s(k) \equiv \mathscr{F}_s\{f\} = \sqrt{\frac{2}{\pi}} \int_0^\infty \sin kx\, f(x)\, dx , \qquad (4.10.2)$$

and the inverses, at points of continuity, are

$$f(x) = \mathscr{F}_c^{-1}\{\tilde{f}_c\} = \sqrt{\frac{2}{\pi}} \int_0^\infty \cos kx\, \tilde{f}_c(k)\, dk, \qquad (4.10.3)$$

$$f(x) = \mathscr{F}_s^{-1}\{\tilde{f}_s\} = \sqrt{\frac{2}{\pi}} \int_0^\infty \sin kx\, \tilde{f}_s(k)\, dk. \qquad (4.10.4)$$

The corresponding relations at a point of discontinuity are given in (10.5.3) and (10.5.6) p. 273. It is evident from the definitions that

$$\tilde{f}_c(-k) = \tilde{f}_c(k), \qquad \tilde{f}_s(-k) = -\tilde{f}_s(k). \qquad (4.10.5)$$

While the sine and cosine transforms share with the Laplace transform applicability to the half-infinite range, in those cases in which either transform can be used, the Fourier ones are preferable because inversion is easier.

Some operational rules are summarized in Tables 4.5 and 4.6 respectively. For the present purposes, the most important ones are those governing the second derivatives

$$\mathscr{F}_c\{f''\} = -k^2 \tilde{f}_c - (2/\pi)^{1/2} f'(0), \qquad \mathscr{F}_s\{f''\} = -k^2 \tilde{f}_s + (2/\pi)^{1/2} kf(0), \qquad (4.10.6)$$

Table 4.6	Operational rules for the Fourier sine transform.	

$$f(x) = (2/\pi)^{1/2} \int_0^\infty \sin(kx)\, \tilde{f}_s(k)\, dk \qquad\qquad \tilde{f}_s(k) = (2/\pi)^{1/2} \int_0^\infty \sin(kx)\, f(x)\, dx$$

1	$f(ax)\ \ (a > 0)$	$\tilde{f}_s(k/a)/a$
2	f'	$-k\,\tilde{f}_c$
3	f''	$-k^2 \tilde{f}_s + (2/\pi)^{1/2}\, k f(0)$
4	$x^{2n} f(x)$	$(-1)^n (d^{2n}/dk^{2n}) \tilde{f}_s$
5	$x^{2n+1} f(x)$	$(-1)^{n+1} (d^{2n+1}/dk^{2n+1}) \tilde{f}_c$
6	$2a \cos(bx) f(ax)\ (a, b > 0)$	$\tilde{f}_s[(k+b)/a] + \tilde{f}_s[(k-b)/a]$
7	$2a \sin(bx) f(ax)\ (a, b > 0)$	$\tilde{f}_c[(k-b)/a] - \tilde{f}_c[(k+b)/a]$
8	$\displaystyle\int_0^\infty f(x)g(x)\, dx = \int_0^\infty \tilde{f}_s(k)\tilde{g}_s(k)\, dk$	
9	$\displaystyle\int_0^\infty [f(x)]^2\, dx = \int_0^\infty [\tilde{f}_s(k)]^2\, dk$	

which show that the cosine and sine transforms are suitable, respectively, for Neumann data at $x = 0$ (prescribed derivative) and Dirichlet data (prescribed function). Relations (4.10.6) are similar to (3.1.14) and (3.1.16) p. 57 valid for the Fourier series expansion of second derivatives.

The operational rules concerning derivatives of odd order (and, in particular, of the first) show that a cosine transform is turned into a sine one and vice versa. Thus, these transforms are not normally useful for equations in which both even and odd derivatives appear. In such cases, the Laplace transform may be more suitable. Other useful relations are

$$\mathscr{F}_s^{-1}\left\{\frac{d\tilde{f}}{dk}\right\} = -x\mathscr{F}_c^{-1}\{\tilde{f}(k)\}, \qquad \mathscr{F}_c^{-1}\left\{\frac{d\tilde{f}}{dk}\right\} = x\mathscr{F}_s^{-1}\{\tilde{f}(k)\} - \sqrt{\frac{2}{\pi}}\tilde{f}(0)\,. \quad (4.10.7)$$

There is no convolution theorem for the sine and cosine transforms. However, as will be seen in some examples, the use of suitable parity considerations sometimes permits to turn the inversion of a sine or cosine transform into that of an exponential transform to which the convolution theorem (4.1.8) can be applied.

Some results on the asymptotic behavior of the sine and cosine transforms are given at the end of §10.8 on p. 281.

4.11 Diffusion in a semi-infinite medium

Let us consider the one-dimensional diffusion equation in a semi-infinite medium

$$\frac{\partial u}{\partial t} = D\frac{\partial^2 u}{\partial x^2}, \qquad 0 < x < \infty, \tag{4.11.1}$$

subject to Neumann and initial conditions

$$-\frac{\partial u}{\partial x}\bigg|_{x=0} = q(t), \qquad u(x,0) = 0, \tag{4.11.2}$$

with $u \to 0$ at infinity. Upon taking the Fourier cosine transform we find

$$\frac{\partial \tilde{u}_c}{\partial t} = D\left[-k^2\tilde{u}_c + \sqrt{\frac{2}{\pi}}\,q(t)\right] \tag{4.11.3}$$

which is readily solved with the result

$$\tilde{u}_c = \sqrt{\frac{2}{\pi}}D \int_0^t q(\tau)\,\exp\left[-k^2 D(t-\tau)\right]\,\mathrm{d}\tau. \tag{4.11.4}$$

The inverse is

$$u(x,t) = \sqrt{\frac{D}{\pi}} \int_0^t q(\tau)\,\exp\left[-\frac{x^2}{4D(t-\tau)}\right]\frac{\mathrm{d}\tau}{\sqrt{t-\tau}}, \tag{4.11.5}$$

and in particular, at $x = 0$,

$$u(0,t) = \sqrt{\frac{D}{\pi}} \int_0^t \frac{q(\tau)}{\sqrt{t-\tau}}\,\mathrm{d}\tau. \tag{4.11.6}$$

If we were interested in finding the $q(t)$ which would result in a prescribed $u(0,t) = f(t)$, this would be an integral equation for q which happens to be of the particularly simple Volterra convolution type; this equation is solved in §5.6 p. 144.

The question can also be answered by solving the diffusion equation (4.11.1) with Dirichlet conditions, which we do next. Let us now denote the unknown function by v, and require then that it satisfy the Dirichlet boundary condition

$$v(0,t) = f(t), \tag{4.11.7}$$

again with homogeneous conditions at $t = 0$ and at infinity. As shown below, we can determine the solution by using the Fourier sine transform, but it is easier to note that v is readily obtainable from the previous expression for u by observing that, if u satisfies (4.11.1), so does $v = \partial u/\partial x$. But then a Neumann condition for u is a Dirichlet condition for v, so that the required solution can be found by simply taking the derivative of (4.11.5) and replacing $-q$ by f; the result is

$$v(x,t) = \frac{x}{2\sqrt{\pi D}} \int_0^t f(\tau)\,\exp\left[-\frac{x^2}{4D(t-\tau)}\right]\frac{\mathrm{d}\tau}{(t-\tau)^{3/2}}. \tag{4.11.8}$$

That this expression satisfies the boundary condition can be established with the change of variable $\xi = x/[2\sqrt{D(t-\tau)}]$, with which we find

$$v(x,t) = \frac{2}{\sqrt{\pi}} \int_{x/2\sqrt{Dt}}^{\infty} f\left(t - \frac{x^2}{4D\xi^2}\right) e^{-\xi^2} \, d\xi \tag{4.11.9}$$

from which we directly find $v(0,t) = f(t)$ since $(2/\sqrt{\pi}) \int_0^{\infty} e^{-\xi^2} \, d\xi = 1.$[12] This conclusion can also be expressed by writing

$$\lim_{x \to 0} \frac{x}{2\sqrt{\pi D(t-\tau)^3}} \exp\left[-\frac{x^2}{4D(t-\tau)}\right] = \delta(t-\tau). \tag{4.11.10}$$

The same argument shows that (4.11.5) is indeed the solution of the diffusion equation satisfying the Neumann condition (4.11.2).

The problem posed before in connection with (4.11.6), namely the determination of the flux $q(t)$ to be imposed at $x = 0$ so that the function would have a certain behavior $f(t)$ at $x = 0$, can be solved by taking the derivative of v and evaluating it at $x = 0$:

$$q(t) = -\left.\frac{\partial v}{\partial x}\right|_{x=0} = -\frac{1}{2\sqrt{\pi D}} \int_0^t \frac{f(\tau)}{(t-\tau)^{3/2}} \, d\tau, \tag{4.11.11}$$

provided $f(t)$ is such that the integral exists. Substitution into (4.11.6), interchange of the order of integration (p. 690), and use of the beta function relation (17.5.38) p. 445 proves that this is indeed the correct solution.

The same result (4.11.11) can be found directly by using the sine transform. Upon applying it, we find

$$\frac{\partial \tilde{u}_s}{\partial t} = D\left[-k^2 \tilde{u}_s + \sqrt{\frac{2}{\pi}} k f(t)\right] \tag{4.11.12}$$

which is solved by

$$\tilde{u}_s = \sqrt{\frac{2}{\pi}} D \int_0^t f(\tau) k \, \exp\left[-k^2 D(t-\tau)\right] d\tau$$

$$= -\frac{1}{\sqrt{2\pi}} \int_0^t f(\tau) \frac{\partial}{\partial k} \exp\left[-k^2 D(t-\tau)\right] \frac{d\tau}{t-\tau}. \tag{4.11.13}$$

Upon using the first of (4.10.7) we have

$$\mathscr{F}_s^{-1}\left\{\frac{\partial}{\partial k} \exp\left[-k^2 D(t-\tau)\right]\right\} = -x\mathscr{F}_c^{-1}\left\{\exp\left[-k^2 D(t-\tau)\right]\right\}$$

$$= \frac{x}{\sqrt{2D(t-\tau)}} \exp\left[-\frac{x^2}{4D(t-\tau)}\right] \tag{4.11.14}$$

and (4.11.11) follows.

---

[12] This result can be derived by making the change of variable $t = \xi^2$ which renders the integral proportional to that defining $\Gamma(1/2) = \sqrt{\pi}$ (see (17.5.19) p. 442). An ingenious and more direct way is to consider the squared integral $\left(\int_{-\infty}^{\infty} e^{-\xi^2} \, d\xi\right) \left(\int_{-\infty}^{\infty} e^{-\eta^2} \, d\eta\right)$ and regard it as an area integral. Upon adopting polar coordinates (ρ, θ) in the (ξ, η) plane $d\xi \, d\eta = \rho \, d\rho \, d\theta$ and the necessary integrations are elementary.

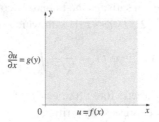

Fig. 4.4 Laplace equation in a quadrant, §4.12.

The case of a mixed boundary condition is treated by means of the Laplace transform in §5.5.3 p. 140.

4.12 Laplace equation in a quadrant

Consider the Laplace equation

$$\frac{\partial^2 u}{\partial x^2} + \frac{\partial^2 u}{\partial y^2} = 0 \tag{4.12.1}$$

in the quadrant $0 < x,\ y < \infty$ (Figure 4.4) subject to boundedness at infinity and

$$u(x,0) = f(x), \qquad \left.\frac{\partial u}{\partial x}\right|_{x=0,y} = g(y). \tag{4.12.2}$$

As noted in §3.2.4 p. 65, by linearity, one could write $u = v + w$, with v and w both harmonic, both vanishing at infinity, and such that

$$v(x,0) = f(x), \qquad \left.\frac{\partial v}{\partial x}\right|_{x=0,y} = 0, \tag{4.12.3}$$

$$w(x,0) = 0, \qquad \left.\frac{\partial w}{\partial x}\right|_{x=0,y} = g(y). \tag{4.12.4}$$

Rather than relying on this observation, let us use a double transform; we will find that the same decomposition arises naturally.

We start with a Fourier cosine transform in x:

$$\frac{\partial^2 \tilde{u}_c}{\partial y^2} - k^2 \tilde{u}_c = \sqrt{\frac{2}{\pi}}\, g(y). \tag{4.12.5}$$

We could solve this equation e.g. by variation of parameters, but we can also take a Fourier sine transform:

$$\tilde{u}_{sc} = \sqrt{\frac{2}{\pi}} \int_0^\infty \tilde{u}_c(k,y) \sin hy \, dy \tag{4.12.6}$$

with the result

$$\tilde{u}_{sc} = \sqrt{\frac{2}{\pi}} \left[\frac{h\,\tilde{f}_c(k)}{h^2 + k^2} - \frac{\tilde{g}_s(h)}{h^2 + k^2} \right] \equiv \tilde{v}_{sc} + \tilde{w}_{sc}. \tag{4.12.7}$$

To invert the first term we need to calculate

$$v(x, y) = \left(\frac{2}{\pi}\right)^{3/2} \int_0^\infty dk\, \tilde{f}_c(k) \cos kx \int_0^\infty \frac{h}{h^2 + k^2} \sin hy\, dh. \tag{4.12.8}$$

We may manipulate the inner integral so as to be able to use one of the entries of Table 4.2:

$$\sqrt{\frac{2}{\pi}} \int_0^\infty \frac{h \sin hy}{h^2 + k^2}\, dh = -\sqrt{\frac{2}{\pi}} \frac{\partial}{\partial y} \int_0^\infty \frac{\cos hy}{h^2 + k^2}\, dh$$

$$= -\frac{\partial}{\partial y} \operatorname{Re} \frac{1}{\sqrt{2\pi}} \int_{-\infty}^\infty \frac{e^{-ihy}}{h^2 + k^2}\, dh = \sqrt{\frac{\pi}{2}} e^{-ky}. \tag{4.12.9}$$

Then

$$v(x, y) = \sqrt{\frac{2}{\pi}} \int_0^\infty \tilde{f}_c(k) e^{-ky} \cos kx\, dk. \tag{4.12.10}$$

Since, from (4.10.5), $\tilde{f}_c(-k) = \tilde{f}_c(k)$, if we define $f(-x) = f(x)$, $\tilde{f}_c$ equals the exponential transform $\tilde{f}$ (cf. (4.1.9 p. 86)) and we may write

$$v(x, y) = \frac{1}{\sqrt{2\pi}} \int_{-\infty}^\infty \tilde{f}(k) e^{-|k|y}\, e^{-ikx}\, dk. \tag{4.12.11}$$

The advantage is that, at this point, we can use the convolution theorem for the exponential transform, provided of course that we remember the even extension of f. Thus

$$v(x, y) = \frac{1}{\sqrt{2\pi}} \int_{-\infty}^\infty f(\xi) \mathscr{F}^{-1}\{e^{-|k|y}\}(x - \xi)\, d\xi. \tag{4.12.12}$$

In the last term we can use the same result (4.12.9) used before so that

$$v(x, y) = \frac{1}{\pi} \int_{-\infty}^\infty \frac{y}{(x - \xi)^2 + y^2} f(\xi)\, d\xi. \tag{4.12.13}$$

At this point we break up the range of integration into a positive and a negative part, and remember the stipulation we made on the extension of $f(x)$ to negative values of x; the final result is then

$$v(x, y) = \frac{1}{\pi} \int_0^\infty \left[\frac{y}{(x - \xi)^2 + y^2} + \frac{y}{(x + \xi)^2 + y^2} \right] f(\xi)\, d\xi \tag{4.12.14}$$

a result which could actually have been written down directly from (4.6.9) by using the even extension of f.

For the other term w, the intermediate result corresponding to (4.12.10) is

$$w(x, y) = -\sqrt{\frac{2}{\pi}} \int_0^\infty \tilde{g}_s(h) e^{-hx} \frac{\sin hy}{h}\, dh \tag{4.12.15}$$

from which, since $\tilde{g}_s(-h) = -\tilde{g}_s(h)$,

$$\frac{\partial w}{\partial x} = \sqrt{\frac{2}{\pi}} \int_0^\infty \tilde{g}_s(h) e^{-hx} \sin hy \, dh = \frac{1}{\sqrt{2\pi}} \int_{-\infty}^\infty \tilde{g}_s(h) e^{-|h|x} \sin hy \, dh. \quad (4.12.16)$$

If we extend $g(y)$ to make an odd fucntion, i.e. $g(-y) = -g(y)$, and use the fact that, for such a function, $\tilde{g}_s = \text{Im} \, \tilde{g}$ while $\text{Re} \, \tilde{g} = 0$ (cf. (4.1.9) p. 86), we can then write

$$\frac{\partial w}{\partial x} = \text{Re} \frac{1}{\sqrt{2\pi}} \int_{-\infty}^\infty \tilde{g}(h) e^{-|h|x} e^{-ihy} \, dh \quad (4.12.17)$$

which, aside from the interchange of x and y, is the same as (4.12.11); thus we have, noting the parity of $g(y)$,

$$\begin{aligned}
\frac{\partial w}{\partial x} &= \frac{1}{\pi} \int_{-\infty}^\infty \frac{x}{x^2 + (y - \eta)^2} g(\eta) \, d\eta \\
&= \frac{1}{\pi} \int_0^\infty \left[\frac{x}{x^2 + (y - \eta)^2} - \frac{x}{x^2 + (y + \eta)^2} \right] g(\eta) \, d\eta. \quad (4.12.18)
\end{aligned}$$

We now integrate over x:

$$w(x, y) - w(0, y) = \frac{1}{2\pi} \int_0^\infty \log \left[\frac{x^2 + (\eta - y)^2}{(\eta - y)^2} \frac{(\eta + y)^2}{x^2 + (\eta + y)^2} \right] g(\eta) \, d\eta. \quad (4.12.19)$$

As $x \to \infty$ $w(x, y) \to 0$ and therefore

$$-w(0, y) = \frac{1}{2\pi} \int_0^\infty g(\eta) \log \frac{(\eta + y)^2}{(\eta - y)^2} \, d\eta \quad (4.12.20)$$

so that the final result is

$$w(x, y) = \frac{1}{2\pi} \int_0^\infty g(\eta) \log \frac{x^2 + (\eta - y)^2}{x^2 + (\eta + y)^2} \, d\eta. \quad (4.12.21)$$

4.13 Laplace equation in a semi-infinite strip

Let us consider the semi-infinite strip $0 < x < \infty$, $0 < y < L$ (Figure 4.5) and solve the Laplace equation imposing homogeneous Dirichlet conditions on the short vertical side, $u(0, y) = 0$, and on the upper horizontal side, $u(x, L) = 0$; on the lower horizontal boundary we set $u(x, 0) = f(x)$. This problem is solvable by means of a Fourier series in y, but here we use the Fourier sine transform as suggested by the second of (4.10.6). By applying the transform we find

$$\frac{\partial^2 \tilde{u}_s}{\partial y^2} - k^2 \tilde{u}_s(k, y) = 0. \quad (4.13.1)$$

The solution of this equation satisfying the transformed boundary conditions at $y = 0$ and $y = L$ is

$$\tilde{u}_s(k, y) = \frac{\sinh k(L - y)}{\sinh kL} \tilde{f}_s(k). \quad (4.13.2)$$

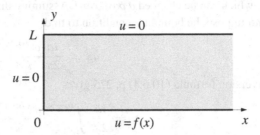

Laplace equation in a semi-infinite strip, §4.13.

One way to deal with the inversion of this result has been shown in connection with (4.12.16); here we use an alternative and somewhat more formal procedure. Let us write the inversion integral of (4.13.2) substituting for $\tilde{f}_s(k)$ its definition; for those f's for which an interchange of the order of the integrations is legitimate (p. 690), we have

$$u(x, y) = \frac{2}{\pi} \int_0^\infty f(\xi)\, d\xi \int_0^\infty \frac{\sinh k(L-y)}{\sinh kL} \sin kx \sin k\xi\, dk$$

$$= \frac{\sin(\pi y/L)}{2L} \int_0^\infty \left[\frac{1}{\cosh X_- + \cos Y} - \frac{1}{\cosh X_+ + \cos Y} \right] f(\xi)\, d\xi \qquad (4.13.3)$$

where we have put $X_\pm = \pi(x \pm \xi)/L$ and $Y = \pi(L-y)/L$ for brevity. This result can also be found from the infinite-strip expression (4.6.7) by using an odd extension of the function $f(x)$, in keeping with the boundary condition $u(0, y) = 0$.

4.14 One-sided transform

In §10.6 we have introduced the one-sided Fourier transform

$$\tilde{f}_+(\kappa) = \frac{1}{\sqrt{2\pi}} \int_0^\infty f(x)\, e^{i\kappa x}\, dx \qquad (4.14.1)$$

with $\kappa = k + ih$, defined for h such that $e^{-hx} f(x)$ is integrable.

As an application of this transform let us consider the problem

$$\frac{\partial^2 u}{\partial x \partial y} = u \qquad (4.14.2)$$

in the quadrant $0 < x,\ y < \infty$, with $u(x, 0) = u(0, y) = a$, a constant. Use of the one-sided transform e.g. in x and integration by parts gives

$$\frac{1}{\sqrt{2\pi}} \int_0^\infty \frac{\partial^2 u}{\partial x \partial y}\, e^{i\kappa x}\, dx = \frac{1}{\sqrt{2\pi}} \left[e^{i\kappa x} \frac{\partial u}{\partial y} \right]_0^\infty - i\kappa \frac{\partial \tilde{u}_+}{\partial y}. \qquad (4.14.3)$$

For the integrated term the contribution of the upper limit vanishes; as for the lower limit, the substitution of a double Taylor series expansion of u near $x = 0$ suggests that $\partial u/\partial y = 0$ at

$x = 0$, which can be checked *a posteriori*. Assuming this for the moment, we can integrate in y and impose the boundary condition to find

$$\tilde{u}_+ = \frac{ia}{2\pi} \frac{e^{iy/\kappa}}{\kappa}. \tag{4.14.4}$$

The inversion formula (10.6.4) p. 275 gives

$$u = \frac{ia}{2\pi} \int_{ia-\infty}^{ia+\infty} \frac{e^{iy/\kappa - i\kappa x}}{\kappa} \, d\kappa. \tag{4.14.5}$$

If we put $\kappa = i\sigma$, the integral becomes

$$u = \frac{a}{2\pi i} \int_{a-i\infty}^{a+i\infty} \frac{e^{y/\sigma}}{\sigma} e^{\sigma x} \, d\sigma \tag{4.14.6}$$

which is the inverse Laplace transform of $ae^{y/\sigma}/\sigma$ and equals $aI_0(2\sqrt{xy})$ (see Table 5.1 p. 124).

4.15 Other applications

Two other applications of the exponential Fourier transform can be found in §6.4 and §7.11 to the determination of the fundamental solution of the Poisson equation in cylindrical coordinates and to the solution of the Laplace equation in the presence of a conical boundary.

An application of the Fourier sine transform to the solution of the Laplace equation in a half-space is described in §6.7 p. 163.

5

Laplace Transform: Applications

The Laplace transform, the essential theory of which is covered in Chapter 11, is a useful tool for the solution of the initial-value problem for ordinary and partial differential equations and for the solution of certain types of integral equations. Since very often the transform is applied with respect to a time-like variable, we will use the letter t to indicate this variable.

The Laplace transform is applicable to functions that vanish for negative values of t (i.e., the "past") and, as $t \to \infty$, do not grow faster than e^{ct} for some real constant c (see §11.1). Like the Fourier sine and cosine transforms, it is a one-sided transform in the sense that it is defined over the half line $0 < t < \infty$. Its inversion, however, is usually more difficult than that of the Fourier transforms which, therefore, are preferable when they are applicable.

As pointed out in the examples that follow, in some cases solution techniques other than the Laplace transform are available, and may be easier to use. In other cases, such as initial value problems for partial differential equations, some differential equations with delay and integral equations of the convolution type, the Laplace transform is the most suitable tool available.

5.1 Summary of useful relations

The Laplace transform of a function $u(t)$ is defined by

$$\mathcal{L}\{u\}(s) \equiv \hat{u}(s) = \int_0^\infty e^{-st} u(t) \, dt. \tag{5.1.1}$$

In general, the integral converges only to the right of some smallest value of $\mathrm{Re}\, s$ which is called the *abscissa of convergence* σ_c. Applying the Laplace transform to a problem consists in multiplying the relations specifying the problem by e^{-st} and integrating over t from 0 to ∞. A short list of Laplace transform pairs is given in Table 5.1.

The inversion operation, denoted by $\mathcal{L}^{-1}\{\hat{u}\}(t)$, is given by

$$\mathcal{L}^{-1}\{\hat{u}\}(t) = \frac{1}{2\pi i} \int_{a-i\infty}^{a+i\infty} \hat{u}(s) \, e^{st} \, ds = u(t), \tag{5.1.2}$$

where a is a real constant to the right of all singularities of $\hat{u}(s)$. As discussed in §11.3, a convenient procedure to determine the inverse is the use of contour integration. It is prudent to remember that $\mathcal{L}^{-1}\{\hat{u}\}(t)$ is usually discontinuous at $t = 0$ so that $\lim_{t \to 0+} \mathcal{L}^{-1}\{\hat{u}\}(t) \neq (2\pi i)^{-1} \int_{a-i\infty}^{a+i\infty} \hat{u}(s) \, ds$.

Table 5.1 — Some Laplace transform pairs.

$\hat{f}(s)$	$f(t)$
s^{-1}	$H(t)$
s^{-n}	$t^{n-1}/(n-1)!,\, n=1,2,\dots$
$s^{-1/2}$	$(\pi t)^{-1/2}$
$s^{-\beta}$	$t^{\beta-1}/\Gamma(\beta),\, \beta>0$
$1/(s^2+a^2)$	$\sin(at)/a$
$s/(s^2+a^2)$	$\cos(at)$
$[\sqrt{s}\,(s-a^2)]^{-1}$	$\exp(a^2t)\,\mathrm{erf}(a\sqrt{t})/a$
$[\sqrt{s}\,(\sqrt{s}+a)]^{-1}$	$\exp(a^2t)\,\mathrm{erfc}(a\sqrt{t})$
$s^{-1}\exp(-as)$	$H(t-a)$
$\sqrt{\pi/s}\,\exp(-a^2/s)$	$\cos(2a\sqrt{t})/\sqrt{t}$
$a\sqrt{\pi/s^3}\,\exp(-a^2/s)$	$\sin(2a\sqrt{t})$
$s^{-\nu-1}\exp(-a/s)\quad(\mathrm{Re}\,\nu>-1)$	$(t/a)^{\nu/2}J_\nu(2\sqrt{at})$
$\sqrt{\pi/s}\,\exp(-2\sqrt{as})$	$\exp(a/t)/\sqrt{t}$
$(s^2+a^2)^{-1/2}$	$J_0(at)$
$[\sqrt{s^2+a^2}-s]^\nu/\sqrt{s^2+a^2}$	$a^\nu J_\nu(at),\ \mathrm{Re}\,\nu>-1$
$(s^2+a^2)^{-\nu-1/2}$	$\sqrt{\pi}\,t^\nu J_\nu(at)/[(2a)^\nu\Gamma(\nu+1/2)]$
$e^{-\beta\sqrt{s}}/\sqrt{s}$	$[\exp(-\beta^2/4t)]/\sqrt{\pi t}$
$e^{-as},1$	$\delta(t-a),\delta(t)$
se^{-as},s	$\delta'(t-a),\delta'(t)$

When $\hat{f}=p(s)/q(s)$, where $p(s)$ is an entire function (i.e., with no finite singularities) and $q(s)$ has only a finite number of simple zeros at $s_1,s_2,\dots$, we have Heaviside's expansion theorem (p. 292):

$$f(t)=\mathscr{L}^{-1}\left\{\frac{p(s)}{q(s)}\right\}=\sum_k\frac{p(s_k)}{q'(s_k)}\exp(s_kt) \tag{5.1.3}$$

in which $q'(s_k)$ is the derivative of $q(s)$ evaluated at $s=s_k$. This relation is also applicable when the number of simple zeros is infinite provided the series converges. A zero of $q(s)$

| Table 5.2 | Some Laplace transforms of distributions; C denotes an arbitrary constant and $\psi(\alpha)$ is the ψ function defined on p. 445. |

$d(t)$	$\hat{d}(s) = \mathscr{L}\{d\}$
$\delta(t)$	1
$\delta^{(m)}(t)$	s^m
$\delta(t-a)$	e^{-as}
$\delta^{(m)}(t-a)$	$s^m e^{-as}$
$t^\alpha H(t)\ (\alpha \neq -1, -2, \ldots)$	$\Gamma(\alpha+1)/s^{\alpha+1}$
$t^{-m-1} H(t)$	$(-s)^m (C - \gamma - \log s)/m!$
$t^\alpha H(t) \log t\ (\alpha \neq -1, -2 \ldots)$	$\Gamma(\alpha+1)[\psi(\alpha) - \log s]/s^{\alpha+1}$

order 2 at s_k gives to the sum (5.1.3) the contribution

$$2\frac{p'(s_k) + tp(s_k)}{q''(s_k)} e^{s_k t}. \tag{5.1.4}$$

For zeros of higher order see p. 292.

These results are valid in the sense of ordinary functions provided that, in the semi-plane to the left of the abscissa of convergence, p/q tends to zero sufficiently fast as $|s| \to \infty$ that Jordan's lemma (p. 291) is applicable. If this condition is not satisfied for p/q, but it is for $s^{-N} p/q$ for some integer N, then the inverse of p/q is a distribution defined by

$$\mathscr{L}^{-1}\left\{\frac{p}{q}\right\} = = \frac{d^N}{dt^N}\left[\mathscr{L}^{-1}\left\{s^{-N}\frac{p}{q}\right\}\right] \tag{5.1.5}$$

in which the derivative is understood in the distributional sense (see p. 607). A short list of the Laplace transform of distributions is given in Table 5.2.

To invert the transforms, in the examples that follow we mostly use the entries in Table 5.1 and the operational rules in Table 5.3 rather than explicit calculations. The need for the actual calculation of a Laplace transform or its inverse is obviated in many cases by the abundant resources available in book form (see e.g. Erdélyi *et al.* 1954, Doetsch 1974, Debnath 1995, Zwillinger 2002) or on the web. Standard integral tables (e.g. Gradshteyn *et al.* 2007) can be used for the direct transform, while the inversion usually requires specific tables.

5.1.1 Operational rules

The scope of the tables can be considerably expanded and the manipulation of both the direct and inverse Laplace transforms considerably simplified by the operational rules of Table 5.3 (see Table 11.1 p. 291 for additional ones).

| Table 5.3 | Some operational rules for the Laplace transform; more specialized rules are given in Table 11.1 p. 291. | |

	$f(t)$	$\hat{f}(s) = \int_0^\infty e^{-st} f(t)\, dt$
1	df/dt	$s\hat{f} - f(0+)$
2	$d^n f/dt^n$	$s^n \hat{f} - s^{n-1} f(0+) - s^{n-2} f'(0+)$ $- \cdots - f^{(n-1)}(0+)$
3	$\int_0^t f(\tau)\, d\tau$	$\hat{f}/s$
4	$\int_0^t f(\tau)\, g(t-\tau)\, dz$	$\hat{f}(s)\, \hat{g}(s)$
5	$t^n f(t)$	$(-1)^n (d^n \hat{f}/ds^n)$
6	$f(t)/t$	$\int_s^\infty \hat{f}(\sigma)\, d\sigma$
7	$e^{at} f(t)$	$\hat{f}(s-a)$
8	$[f(t/c)]/c$	$\hat{f}(cs)$
9	$f(t-a),\ f(t) = 0$ for $t < 0$	$e^{-as} \hat{f}(s)$
10	$\sum_{j=1}^n [q(\lambda_j)/p'(\lambda_j)] \exp(\lambda_j t)$	$q(s)/p(s)$, see (5.1.3)
11	$\int_0^t [f(t)/t]\, dt$	$s^{-1} \int_s^\infty \hat{f}(\sigma)\, d\sigma$
12	$f(t^2)$	$\pi^{-1/2} \int_0^\infty \exp\left(-s^2/4u^2\right) \hat{f}(u^2)\, du$
13	$t^{-3/2} \int_0^\infty \tau \exp[-(\tau^2/4t)]\, f(\tau)\, d\tau$	$2\sqrt{\pi}\, \hat{f}(\sqrt{s})$

As a first example, from rule 7 of Table 5.3 we have

$$\mathscr{L}^{-1}\{(s-a)^{-\beta}\}(t) = e^{at}\, \mathscr{L}^{-1}\{s^{-\beta}\}(t) = e^{at}\, t^{\beta-1}/\Gamma(\beta). \qquad (5.1.6)$$

From rule 1 we find

$$
\begin{aligned}
\mathscr{L}^{-1}\left\{\frac{s}{(s^2+a^2)^{3/2}}\right\}(t) &= \frac{d}{dt}\mathscr{L}^{-1}\left\{\frac{1}{(s^2+a^2)^{3/2}}\right\}(t) + \mathscr{L}^{-1}\left\{\frac{1}{(s^2+a^2)^{3/2}}\right\}(0+) \\
&= \frac{1}{a}\frac{d}{dt}[t J_1(at)] + 0 \qquad (5.1.7)
\end{aligned}
$$

where we have used $J_1(0) = 0$.

As an example of the interesting rule 13, consider, with $\sigma = \sqrt{s}$

$$\mathcal{L}^{-1}\left\{\frac{\sqrt{s}}{s-a^2}\right\}(t) = \frac{1}{2\sqrt{\pi}\,t^{3/2}}\int_0^\infty \tau\, e^{-(\tau^2/4t)}\frac{1}{2}\mathcal{L}^{-1}\left\{\frac{1}{\sigma+a}+\frac{1}{\sigma-a}\right\}d\tau$$

$$= \frac{1}{4\sqrt{\pi}\,t^{3/2}}\int_0^\infty \tau\,\exp\left(-\frac{\tau^2}{4t}\right)\left(e^{a\tau}+e^{-a\tau}\right)d\tau$$

$$= \frac{1}{\sqrt{\pi t}} + a e^{a^2 t}\,\mathrm{erf}(a\sqrt{t}) \tag{5.1.8}$$

where in the first step we have decomposed in partial fractions $\sigma/(\sigma^2 - a^2)$. By combining rules 2 and 5 of Table 5.3 we find

$$\mathcal{L}\left\{t\frac{d^2u}{dt^2}\right\} = -\frac{d}{ds}\mathcal{L}\left\{\frac{d^2u}{dt^2}\right\} = -\frac{d}{ds}\left[s^2\hat{u} - su(0+) - u'(0+)\right]$$

$$= -s^2\frac{d\hat{u}}{ds} - 2s\hat{u} + u(0+). \tag{5.1.9}$$

Just as in the case of the Fourier transform, the convolution theorem (rule 4 in Table 5.3)

$$\mathcal{L}^{-1}\{\tilde{f}(s)\,\tilde{g}(s)\} = \int_0^t f(t-\tau)\,g(\tau)\,d\tau = \int_0^t f(\tau)\,g(t-\tau)\,d\tau \tag{5.1.10}$$

is very useful to find the general solution of a problem leaving the precise functional form of some data unspecified. In other words, use of the convolution theorem often enables us to determine the Green's function for the problem.

The operational rules are particularly convenient when used in conjunction with generalized functions. For example

$$\mathcal{L}^{-1}\left\{\frac{s}{s+a}\hat{f}(s)\right\} = \int_0^t \mathcal{L}^{-1}\{s\}(t-\tau)\,\mathcal{L}^{-1}\left\{\frac{\hat{f}}{s+a}\right\}(\tau)\,d\tau$$

$$= \int_0^t \delta'(t-\tau)\int_0^\tau e^{-a(\tau-\theta)}f(\theta)\,d\theta$$

$$= \int_0^t \delta(t-\tau)\frac{\partial}{\partial\tau}\int_0^\tau e^{-a(\tau-\theta)}f(\theta)\,d\theta = \frac{d}{dt}\int_0^t e^{-a(t-\theta)}f(\theta)\,d\theta. \tag{5.1.11}$$

5.1.2 Small and large t

As described in §11.4, some results are available to determine the behavior of the original function for small and large values of t. These results are particularly useful as it is frequently the case that a complete analytic inversion of the transform is not possible.

Broadly speaking, the behavior of $\hat{f}(s)$ for large s is connected to that of $f(t)$ for small t and vice versa. More precisely, for Re $\lambda > 0$, (Theorems 11.4.1 p. 296 and 11.4.3 p. 298):

$$\lim_{t\to 0+} t^{1-\lambda}f(t) = A \implies \lim_{s\to\infty} s^\lambda \hat{f}(s) = A\Gamma(\lambda), \tag{5.1.12}$$

$$\lim_{t\to\infty} t^{1-\lambda}f(t) = A \implies \lim_{s\to 0+} s^\lambda \hat{f}(s) = A\,\Gamma(\lambda). \tag{5.1.13}$$

Watson's lemma (p. 296) is an extension of (5.1.12) and states that, if the behavior of $f(t)$ for $t \to 0$ is given by the asymptotic series

$$f(t) \sim \sum_{n=1}^{N} a_n t^{\lambda_n - 1} \qquad \text{with} \qquad 0 < \operatorname{Re} \lambda_1 < \operatorname{Re} \lambda_2 < \cdots \qquad (5.1.14)$$

then, as $s \to \infty$, uniformly in $0 \le |\arg s| < \frac{1}{2}\pi$,

$$\hat{f}(s) \sim \sum_{n=1}^{\infty} \Gamma(\lambda_n) a_n s^{-\lambda_n}. \qquad (5.1.15)$$

One must be careful in applying the converse of this result as (5.1.14) does not necessarily follow from (5.1.15): it must be known in advance that $f(t)$ has an asymptotic expansion of the form (5.1.14) before the constants a_n can be deduced from (5.1.15). A similar caution is needed with the converses of (5.1.12) and (5.1.13) (see §11.4).

If $\hat{f} = p(s)/q(s)$, where $p(s)$ is an entire function and $q(s)$ has only a finite or infinite number of simple poles, from Heaviside's expansion theorem (5.1.3), the leading-order behavior of $f(t)$ for $t \to \infty$ is given by

$$f(t) \simeq \frac{p(s_1)}{q'(s_1)} \exp(s_1 t) \qquad (5.1.16)$$

where s_1 is the pole with largest real part. If $\operatorname{Re} s_1 > 0$, $f(t)$ grows indefinitely, while it decays to zero if $\operatorname{Re} s_1 < 0$. If $\operatorname{Re} s_1 = 0$, so that the pole is on the imaginary axis, $f(t)$ remains bounded. If the dominant pole has order 2, the corresponding contribution is given by (5.1.4). In this case a pole on the imaginary axis will give a growing oscillating contribution proportional to $t\, e^{s_1 t}$. For the general case of a pole of order m see (11.3.3) p. 292.

If $\hat{f}$ has a branch point at s_b, in the neighborhood of which it is represented asymptotically by

$$\hat{f}(s) \sim \sum_{n=0}^{\infty} a_n (s - s_b)^{\lambda_n}, \qquad s \to s_b, \qquad \operatorname{Re} \lambda_0 < \operatorname{Re} \lambda_1 < \cdots \to \infty \qquad (5.1.17)$$

then (§11.4.2 p. 297)

$$f(t) \sim e^{s_b t} \sum_{n=0}^{\infty} \frac{a_n}{\Gamma(-\lambda_n)} t^{-\lambda_n - 1}, \qquad t \to \infty. \qquad (5.1.18)$$

Several other results of this type and further specifications on the applicability of those summarized here will be found in §11.4 p. 295.

5.2 Ordinary differential equations

Since the Laplace transform is applicable to functions defined in the range $0 < t < \infty$, it is a natural tool for the solution of time-dependent problems with conditions prescribed at $t = 0$. It is clear from the setting of the problem that such initial conditions are to be understood as limits for $t \to 0+$, rather than values *at* $t = 0$ (see comments in connection with Equation (11.1.8) p. 287).

5.2.1 Homogeneous equation with constant coefficients

Several approaches are available to solve the initial-value problem for ordinary differential equations with constant coefficients such as

$$u'' + c_1 u' + c_0 u = 0 \qquad u(0) = u_0, \qquad u'(0) = v_0. \tag{5.2.1}$$

As an application of the Laplace transform technique, we use rules 1 and 2 of Table 5.3 for the transformation of derivatives to find

$$s^2 \hat{u} - su_0 - v_0 + c_1 \left(s\hat{u} - u_0\right) + c_2 \hat{u} = 0 \tag{5.2.2}$$

from which

$$\hat{u} = \frac{s + c_1}{q(s)} u_0 + \frac{v_0}{q(s)} \tag{5.2.3}$$

where

$$q(s) = s^2 + c_1 s + c_2 = (s - \lambda_1)(s - \lambda_2), \qquad \lambda_{1,2} = \frac{1}{2}\left(-c_1 \pm \sqrt{c_1^2 - 4c_2}\right). \tag{5.2.4}$$

If we let $G(t) = \mathcal{L}^{-1}\{1/q(s)\}$ and apply rule 1 of Table 5.3 we have

$$u(t) = u_0 \left[\frac{dG}{dt} + G(0)\right] + (c_1 u_0 + v_0)G(t). \tag{5.2.5}$$

If $\lambda_1 \neq \lambda_2$, we apply (5.1.3) with the result

$$G(t) = \left(e^{\lambda_1 t} - e^{\lambda_2 t}\right)/(\lambda_1 - \lambda_2) \tag{5.2.6}$$

where we have used $c_1 = -(\lambda_1 + \lambda_2)$ to express $q'(\lambda_1)$ and $q'(\lambda_2)$. Use of the same relation puts (5.2.5) into the explicit form

$$u(t) = \frac{\lambda_2 e^{\lambda_1 t} - \lambda_1 e^{\lambda_2 t}}{\lambda_2 - \lambda_1} u_0 + \frac{e^{\lambda_1 t} - e^{\lambda_2 t}}{\lambda_1 - \lambda_2} v_0 \tag{5.2.7}$$

from which

$$u'(t) = \frac{\lambda_1 \lambda_2}{\lambda_2 - \lambda_1} \left(e^{\lambda_1 t} - e^{\lambda_2 t}\right) u_0 + \frac{\lambda_2 e^{\lambda_2 t} - \lambda_1 e^{\lambda_1 t}}{\lambda_2 - \lambda_1} v_0. \tag{5.2.8}$$

It is readily seen that these expressions satisfy the initial conditions and that, when the roots are complex, $\lambda_{1,2} = -b \pm i\omega$, the expected solution in terms of damped sinusoidal functions is recovered.

The result (5.2.7) embodies the well-known rule that the solution of a homogeneous differential equation with constant coefficients is to be found by assuming solutions of the form $\exp(\lambda t)$ and combining them linearly so as to satisfy the prescribed conditions.

The parameters λ are determined by requiring that $\exp(\lambda t)$ satisfy the equation, which is evidently the same as finding the zeros of $q(s)$ defined in (5.2.4).

The case of coincident roots can be handled either by noting that $(s - \lambda)^{-2} = (\partial/\partial\lambda)$ $(s - \lambda)^{-1}$, and that $\mathscr{L}^{-1}\{(s - \lambda)^{-1}\} = \exp(\lambda t)$, or by taking the limit $\lambda_2 \to \lambda_1$ in (5.2.7). With this latter approach, upon letting $\lambda_2 = \lambda_1 + \varepsilon$, we find

$$u(t) = \left(1 - \lambda_1 \frac{e^{\varepsilon t} - 1}{\varepsilon}\right) e^{\lambda_1 t} u_0 + \frac{e^{\varepsilon t} - 1}{\varepsilon} e^{\lambda_1 t} v_0 \rightarrow (1 - \lambda_1 t) e^{\lambda_1 t} u_0 + t e^{\lambda_1 t} v_0.$$

(5.2.9)

To establish the connection with the matrix-based solution procedure of §18.12 p. 530, let us set $v = u'$ and rewrite the original equation as the first-order system

$$\frac{d}{dt} \begin{vmatrix} u \\ v \end{vmatrix} = \begin{vmatrix} 0 & 1 \\ -c_2 & -c_1 \end{vmatrix} \begin{vmatrix} u \\ v \end{vmatrix}, \qquad \begin{vmatrix} u(0) \\ v(0) \end{vmatrix} = \begin{vmatrix} u_0 \\ v_0 \end{vmatrix}.$$

(5.2.10)

Our solution (5.2.5), (5.2.8) can then be expressed in the form

$$\begin{vmatrix} u(t) \\ v(t) \end{vmatrix} = \begin{vmatrix} A_{11}(t) & A_{12}(t) \\ A_{21}(t) & A_{22}(t) \end{vmatrix} \begin{vmatrix} u_0 \\ v_0 \end{vmatrix}$$

(5.2.11)

where the functions A_{ij} can be read off (5.2.7) and (5.2.8). The matrix in the right-hand side of (5.2.11) is the *fundamental matrix* of the system, whose elements all satisfy the original differential equation and the initial conditions $A_{ij}(0) = \delta_{ij}$.

The Laplace transform can be applied to systems of ordinary differential equations with constant coefficients such as the present problem in the form (5.2.10). This step transforms the differential system into a more easily solvable algebraic one. The procedure may be slightly more convenient than elimination of all but one independent variable by successive differentiation, but is essentially equivalent to it. Since the transformed solution will be the ratio of two polynomials, we immediately conclude from (5.1.16) that it will be asymptotically stable (i.e., $u(t) \to 0$ for $t \to \infty$) if and only if the real part of all the zeros of the denominator is negative.

5.2.2 Non-homogeneous equations with constant coefficients

Let us now use the Laplace transform for the non-homogeneous equation

$$u^{(n)} + c_1 u^{(n-1)} + c_2 u^{(n-2)} + \cdots + c_{n-1} u' + c_n u = f(t)$$

(5.2.12)

in which $u^{(k)}$ denotes the k-th order derivative of u. Since we have just shown in the previous example how to find the general solution of the homogeneous equation, let us focus here on the particular solution satisfying homogeneous initial conditions:

$$u(0) = u'(0) = u''(0) = \cdots = u^{(n-1)}(0).$$

(5.2.13)

Upon taking the Laplace transform we find $\hat{u} = \hat{f}(s)/q(s)$, with $q(s) = s^n + c_1 s^{n-1} + \cdots + c_n$. The convolution theorem leads us to

$$u(t) = \int_0^t f(\tau) G(t - \tau) \, d\tau$$

(5.2.14)

where $G = \mathcal{L}^{-1}\{1/q(s)\}$ is the Green's function (§15.5 p. 388) given, according to
(5.1.3), by

$$G(t) = \sum_{j=1}^{n} \frac{e^{\lambda_j t}}{q'(\lambda_j)} \tag{5.2.15}$$

assuming that $q = 0$ does not have repeated roots. It is evident from (5.2.14) that $u(0) = 0$.
By direct differentiation of (5.2.14) it is found that, for all the derivatives of $u(t)$ to vanish
for $t = 0$ up to order $n - 1$ so that the initial conditions (5.2.13) are satisfied, it is necessary
that all the derivatives of G up to order $n - 2$ vanish for $t = 0$, i.e.

$$\sum_{j=1}^{n} \frac{\lambda_j^k}{q'(\lambda_j)} = 0 \qquad k = 0, 1, \ldots, n - 2. \tag{5.2.16}$$

Since, for $s \to \infty$, $1/q(s) = s^{-n}$, we conclude from (5.1.14) that $\mathcal{L}^{-1}\{1/q\} = t^{n-1}/$
$(n - 1)!$ for $t \to 0$,[1] which proves that all derivatives of the sum of exponentials up to order
$n - 2$ do indeed vanish.

The form (5.2.14) of the solution shows that $u(t)$ is continuous even if the forcing $f(t)$
is not. In the particular case $f(t) = \delta(t - t_0)$, the solution is

$$u(t) = \begin{cases} 0 & t < t_0 \\ G(t - t_0) & t > t_0 \end{cases} \tag{5.2.17}$$

which justifies the alternative denomination *impulse response* given to the Green's
function G.

The transform can be applied to a forced linear constant-coefficient system

$$B\frac{d\mathbf{u}}{dt} + A\mathbf{u} = \mathbf{f}(t) \tag{5.2.18}$$

with A and B constant $N \times N$ matrices, $\mathbf{u}^T = |u_1(t)\, u_2(t) \ldots u_N(t)|$ the vector of
unknowns and $\mathbf{f}^T = |f_1(t)\, f_2(t) \ldots f_N(t)|$ a given time-dependent vector; the standard
row-by-column product of matrix theory is implied in (5.2.18) and initial conditions of the
form $\mathbf{u}(0) = \mathbf{v}$ are prescribed. Any higher-order equation or system with constant coefficients can be written in this way by introducing auxiliary variables as done before with
(5.2.10). Upon taking the Laplace transform we find

$$(Bs + A)\,\hat{\mathbf{u}} = \hat{\mathbf{f}} + B\mathbf{v}. \tag{5.2.19}$$

This is an algebraic system for $\hat{u}_j$, the components of $\hat{\mathbf{u}}$, which can be solved to find

$$\hat{\mathbf{u}}(s) = \hat{G}(s)\left[\hat{\mathbf{f}}(s) + B\mathbf{v}\right]. \tag{5.2.20}$$

Cramer's rule (p. 499) shows that the elements $\hat{G}_{ij}$ of the Green's matrix G are given by
$\hat{G}_{ij} = (-1)^{i+j} \det(C_{ji})/\det(C)$ where $C = Bs + A$ and the matrices C_{ji} are obtained
from C by deleting the jth row and the ith column. Evidently $\det(C) = s^N \det(B) +$

---

[1] The application of the converse of (5.1.12) is legitimate as it is evident from (5.2.15) that G possesses an
expansion of the form (5.1.14).

$O(s^{N-1})$, while $\det(C_{ji})$ is a polynomial of degree at most $N - 1$. Thus, if $\det(B) \neq 0$, the inversion can be effected by using (5.1.3) and the result will have the form

$$\mathbf{u}(t) = \int_0^t G(t - \tau)\,\mathbf{f}(\tau)\,d\tau + G(t)B\mathbf{v}. \tag{5.2.21}$$

If $\det B = 0$, however, the degree of the polynomial $\det C_{ij}$ may be equal to or higher than that of $\det C$ and the Laplace transform of G is not invertible in terms of ordinary functions; we deal then with an *anomalous system*. The use of the generalized functions δ, δ' etc. as demonstrated in (5.1.11) p. 127 is particularly convenient in such cases. Alternative procedures are available, however, such as the reduction of the system to an ordinary non-singular form by elimination of some variables (see p. 532) or others (see e.g. Doetsch 1974, p. 115; Davies 2002, p. 71).

5.2.3 Non-constant coefficients

By (5.1.10), the Laplace transform of the ordinary product of two functions is a convolution of their transforms and, therefore, differential equations with non-constant coefficients lend themselves to solution by Laplace transform methods only in special cases, e.g. when the coefficients are polynomials in t. As an example, let us consider the Bessel equation of real order v:

$$t^2\frac{d^2u}{dt^2} + t\frac{du}{dt} + \left(t^2 - v^2\right)u = 0, \qquad 0 < t < \infty. \tag{5.2.22}$$

If the Laplace transform is applied directly, by operational rule 5 of Table 5.3, the multiplication by t^2 gives rise to a second-order derivative with respect to the conjugate variable s and not much is gained. However, the transformation $u(t) = t^{-v}v(t)$ (in which, without loss of generality, we take $v \geq 0$) puts the equation in the form

$$t\frac{d^2v}{dt^2} - (2v - 1)\frac{dv}{dt} + tv = 0. \tag{5.2.23}$$

Upon using (5.1.9) and rules 1 and 5 the transformed equation is

$$(1 + s^2)\frac{d\hat{v}}{ds} + (2v + 1)s\hat{v} = 2vv(0) \tag{5.2.24}$$

which can be rewritten as

$$\frac{d}{ds}\left[(s^2 + 1)^{v+1/2}\hat{v}\right] = 2vv(0)(s^2 + 1)^{v-1/2}. \tag{5.2.25}$$

Note that, although this is a first-order equation, the general solution still depends on two arbitrary constants, $v(0)$ plus an integration constant for $\hat{v}$. On the basis of (5.1.12), if we are interested in a solution u regular at $t = 0$ we cannot allow $\hat{u}$ to diverge as $s \to \infty$ and we must therefore take $v(0) = 0$; with this assumption, the solution of (5.2.25) is

$$\hat{v}(s) = C\,(s^2 + 1)^{-(v+1/2)} \tag{5.2.26}$$

which, for large s, behaves like $Cs^{-(2v+1)}$. From (5.1.14) we therefore expect $v(t) \simeq Ct^{2v}/\Gamma(2v + 1)$ for $t \to 0$. The choice $C = 2^v\Gamma(v + 1/2)/\sqrt{\pi}$ recovers the known behavior of $J_\nu \simeq (t/2)^\nu / \Gamma(\nu + 1)$ (Table 6.5 p. 151) in agreement with one of the entries of Table 5.1.[2]

In general, an equation of the form

$$\sum_{k=0}^{N} \sum_{j=0}^{M} a_{jk} t^j \frac{d^k u}{dt^k} = f(t) \tag{5.2.27}$$

with a_{jk} constants, is transformed to

$$\sum_{k=0}^{N} \sum_{j=0}^{M} b_{jk} s^k \frac{d^j \hat{u}}{ds^j} = \hat{f}_1(s) \tag{5.2.28}$$

with the b_{jk} other constants given by a linear combinations of the a_{jk}; the right-hand side $\hat{f}_1(s)$ will contain, in addition to the transform of f, terms arising from the initial conditions imposed on u as in (5.2.24). It will be noted that the order N of the highest t-derivative in (5.2.27) has become the order of the highest power of s in (5.2.28), while the highest power t^M has become the highest s-derivative. If $M < N$, the transformed equation is of order smaller than the original one, but not so if $M > N$. In this latter case the application of the Laplace transform may not be useful unless the transformed equation has some special form which facilitates its solution. Since an equation of order M has M linearly independent solutions, when $M < N$ some of the solutions of original equation (5.2.27) may be "lost." In particular, it may not be possible to impose all the given initial conditions on the Laplace transformed solution. This phenomenon signals the fact that not all solutions of the original equation (5.2.27) admit a Laplace transform.

5.3 Difference equations

The Laplace transform can be used to solve difference equations with constant coefficients. We illustrate the procedure with an example. Additional aspects are covered e.g. in Smith 1966, chapter 9,

We start with the homogeneous problem for the second-order difference equation

$$u_{n+2} - 2bu_{n+1} + cu_n = 0, \tag{5.3.1}$$

with $u_0 = 1$ and u_1 given. Define a function $u(t)$ equal to u_n in each interval $n < t < n + 1$. In terms of this function the difference equation becomes

$$u(t + 2) - 2bu(t + 1) + cu(t) = 0. \tag{5.3.2}$$

---

[2] Use of Watson's lemma in reverse is justified from the fact that the existence of an asymptotic expansion for small t can be established directly from the differential equation for v. In the evaluation of C the duplication formula (17.5.28) p. 443 for the Gamma function proves useful.

The function $u(t)$ may be written as

$$u(t) = \sum_{n=0}^{\infty} [H(t-n) - H(t-(n+1))] u_n, \tag{5.3.3}$$

with a Laplace transform given by

$$\hat{u}(s) = \sum_{n=0}^{\infty} u_n \int_n^{n+1} e^{-st} \, dt = \sum_{n=0}^{\infty} u_n \frac{e^{-ns} - e^{-(n+1)s}}{s} = \frac{1-e^{-s}}{s} \sum_{n=0}^{\infty} u_n e^{-ns}. \tag{5.3.4}$$

We note that

$$\int_0^{\infty} u(t+k) e^{-st} \, ds = e^{ks} \int_k^{\infty} u(t') e^{-st'} \, dt' = e^{ks} \left[\hat{u}(s) - \int_0^k u(t') e^{-st'} \, dt' \right], \tag{5.3.5}$$

so that, for $k=1$, we have from (5.3.3)

$$\int_0^1 u(t') e^{-st'} \, dt' = u_0 \int_0^1 e^{-st'} \, dt' = u_0 \frac{1-e^{-s}}{s} \tag{5.3.6}$$

and, for $k=2$,

$$\int_0^2 u(t') e^{-st'} \, dt' = (u_0 + u_1 e^{-s}) \frac{1-e^{-s}}{s}. \tag{5.3.7}$$

Thus, upon taking the Laplace transform of (5.3.2) and recalling that we have assumed $u_0 = 1$, we find

$$e^{2s} \left[\hat{u} - \frac{1-e^{-s}}{s} (1 + u_1 e^{-s}) \right] - 2be^s \left[\hat{u} - \frac{1-e^{-s}}{s} \right] + c\hat{u} = 0, \tag{5.3.8}$$

from which

$$\hat{u} = \frac{e^s - 1}{s} \frac{e^s - 2b + u_1}{e^{2s} - 2be^s + c} = \frac{e^s - 1}{s} \frac{1}{\lambda_1 - \lambda_2} \left[\frac{\lambda_1 - 2b + u_1}{e^s - \lambda_1} - \frac{\lambda_2 - 2b + u_1}{e^s - \lambda_2} \right], \tag{5.3.9}$$

in which $\lambda_{1,2} = b \pm \sqrt{b^2 - c}$ are the roots of the denominator. This step leads us to consider the inversion of fractions of the form $(e^s - 1)/[s(e^s - \lambda)]$. This objective is readily attained by inspection of (5.3.4), where we see that, if we replace u_n by λ^n, we find

$$\mathscr{L} \left\{ \sum_{n=0}^{\infty} [H(t-n) - H(t-(n+1))] \lambda^n \right\} = \frac{1-e^{-s}}{s} \sum_{k=0}^{\infty} (\lambda e^{-s})^k.$$

Let us suppose that $\operatorname{Re} s$ is so large that $|\lambda e^{-s}| < 1$. Then the series can be summed finding

$$\mathscr{L} \left\{ \sum_{n=0}^{\infty} [H(t-n) - H(t-(n+1))] \lambda^n \right\} = \frac{1-e^{-s}}{s} \frac{1}{1 - \lambda e^{-s}} = \frac{1}{s} \frac{e^s - 1}{e^s - \lambda}. \tag{5.3.10}$$

With this result we can now invert (5.3.9) to find

$$u^h(t) = \sum_{n=0}^{\infty} [H(t-n) - H(t-(n+1))] \frac{(\lambda_1 - 2b + u_1)\lambda_1^n - (\lambda_2 - 2b + u_1)\lambda_2^n}{\lambda_1 - \lambda_2}$$

(5.3.11)

or

$$u_n^h = \frac{\lambda_1 - 2b + u_1}{\lambda_1 - \lambda_2} \lambda_1^n - \frac{\lambda_2 - 2b + u_1}{\lambda_1 - \lambda_2} \lambda_2^n.$$

(5.3.12)

This result demonstrates the general rule that the solutions of a homogeneous finite difference equation with constant coefficients are powers. Thus, the situation is similar to that encountered with homogeneous differential equations with constant coefficients the solutions of which are linear combinations of terms of the form $e^{\mu x}$ comparable to $\lambda^n = \exp(n \log \lambda)$. To treat the case of a double root of the denominator we would use the fact that

$$\frac{1}{s} \frac{e^s - 1}{(e^s - \lambda)^2} = \frac{\partial}{\partial \lambda} \cdot \frac{1}{s} \frac{e^s - 1}{e^s - \lambda},$$

(5.3.13)

so that

$$\mathcal{L}^{-1} \left\{ \frac{1}{s} \frac{e^s - 1}{(e^s - \lambda)^2} \right\} = \sum_{n=1}^{\infty} [H(t-n) - H(t-(n+1))] n\lambda^{n-1}.$$

(5.3.14)

We can solve the non-homogeneous problem

$$u_{n+2} - 2bu_{n+1} + cu_n = f_n,$$

(5.3.15)

with f_n given, by adapting the method of variation of parameters of §2.2 p. 30. Let

$$u_n = A_n \lambda_1^n + B_n \lambda_2^n$$

(5.3.16)

where $\lambda_{1,2}$ satisfy $\lambda_{1,2}^2 - 2b\lambda_{1,2} + c = 0$ as before, and impose the subsidiary condition

$$(A_{n+1} - A_n) \lambda_1^{n+1} + (B_{n+1} - B_n) \lambda_2^{n+1} = 0.$$

(5.3.17)

Upon substituting (5.3.16) into (5.3.15) and repeatedly using (5.3.17), in view of the fact that $\lambda_{1,2}^n$ satisfy the homogeneous equation, we are led to

$$(A_{n+1} - A_n) \lambda_1^{n+2} + (B_{n+1} - B_n) \lambda_2^{n+2} = f_n.$$

(5.3.18)

The system (5.3.17), (5.3.18) is readily solved:

$$A_{n+1} - A_n = \frac{f_n}{(\lambda_1 - \lambda_2)\lambda_1^{n+1}}, \qquad B_{n+1} - B_n = -\frac{f_n}{(\lambda_1 - \lambda_2)\lambda_2^{n+1}},$$

(5.3.19)

from which

$$A_n = A_0 + \frac{1}{\lambda_1 - \lambda_2} \sum_0^{n-1} \frac{f_n}{\lambda_1^{n+1}}, \qquad B_n = B_0 - \frac{1}{\lambda_1 - \lambda_2} \sum_0^{n-1} \frac{f_n}{\lambda_2^{n+1}},$$

(5.3.20)

with A_0, B_0 arbitrary. If the solution of the homogeneous is chosen as in (5.3.12), it already satisfies the initial conditions and it is therefore appropriate to take $A_0 = B_0 = 0$. With this choice A_1 and B_1 are non-zero, but their contribution to the solution, $A_1\lambda_1 + B_1\lambda_2$, vanishes due to the subsidiary condition (5.3.17).

5.4 Differential-difference equations

Differential-difference equations involve a function and its derivatives evaluated at different instants of time. A fairly general linear example of such an equation having differential order N and difference order M may be written as

$$\sum_{i=0}^{M}\sum_{j=0}^{N} a_{ij} u^{(j)}(t - c_i) = f(t) \tag{5.4.1}$$

in which $u^{(j)}$ denotes the j-th order derivative of u, f is given, and the c_i are constants with $c_0 = 0$. If $M = 0$ (5.4.1) reduces to an ordinary linear differential equation, while if $N = 0$ it is a difference equation. To introduce some further terminology let us consider the special case

$$a_0 \frac{d}{dt} u(t) + a_1 \frac{d}{dt} u(t - c) + b_0 u(t) + b_1 u(t - c) = f(t). \tag{5.4.2}$$

If $a_0 \neq 0$ while $a_1 = 0$, the equation is of the *retarded type*; if both a_0 and a_1 are non-zero, it is of the *neutral type*, while if $a_0 = 0$ and $a_1 \neq 0$, it is of the *advanced type*.

Let us illustrate the application of the Laplace transform to a simple equation of the retarded type

$$\frac{d}{dt} u(t) + b_0 u(t) + b_1 u(t - c) = f(t) \qquad c < t \tag{5.4.3}$$

with $c > 0$ and constant coefficients b_0 and b_1. In order to have a completely specified solution, it is necessary to prescribe

$$u(t) = g(t), \qquad 0 \le t \le c. \tag{5.4.4}$$

For purposes of orientation, let us consider the homogeneous case and let us assume that c is small. Then, upon a Taylor series expansion, we find

$$\frac{1}{2} b_1 c^2 \frac{d^2 u}{dt^2} + (1 - b_1 c)\frac{du}{dt} + (b_0 + b_1)u = 0 \tag{5.4.5}$$

where all the arguments are t. For $c = 0$ $u \propto e^{-\sigma t}$ with $\sigma = b_0 + b_1$. At the next order in c, the character of the solution does not change but $\sigma \simeq (b_0 + b_1)(1 + b_1 c)$. At order c^2 this solution persists but another exponential solution with $\sigma \simeq 2(1 - b_1 c)/(b_1 c^2)$ appears. If b_1 is positive, this solution is heavily damped, while it rapidly grows if $b_1 < 0$. As more and more terms in the expansion are retained, more and more exponential modes arise.

To apply the Laplace transform, we assume for simplicity that $g(t)$ is continuous with its first derivative so that, from (5.4.3), du/dt is continuous. Upon multiplying (5.4.3) by

e^{-st}, integrating between c and ∞ and rearranging we find

$$\left(s + b_0 + b_1 e^{-cs}\right) \hat{u}_1(s) = \hat{f}_1(s) + e^{-cs} \left[g(c) - b_1 \int_0^c e^{-st} g(t)\, dt \right] \equiv \hat{F}(s) \quad (5.4.6)$$

where

$$\hat{u}_1(s) = \int_c^\infty e^{-st} u(t)\, dt, \qquad \hat{f}_1(s) = \int_c^\infty e^{-st} f(t)\, dt. \qquad (5.4.7)$$

It is evident that $\hat{u}_1(s)$ can be interpreted as the Laplace transform of a function $u_1(t)$ which equals 0 for $0 \leq t < c$ and $u(t)$ for $c \leq t$. Applying the inversion theorem (5.1.2) we then deduce that

$$u_1(t) = \frac{1}{2\pi i} \int_{a-i\infty}^{a+i\infty} \frac{\hat{F}(s)}{s + b_0 + b_1 e^{-cs}} e^{st}\, ds \qquad c < t \qquad (5.4.8)$$

provided all the singularities of the integrand have a finite real part. In view of the hypotheses on $g(t)$, if $f(t)$ admits a Laplace transform, $\hat{F}$ will satisfy this condition. Therefore we need to investigate the roots of the denominator. For this purpose let us set $s = \sigma + i\tau$. Equating to zero the real and imaginary parts of the denominator we find $\sigma + b_0 = -b_1 e^{-c\sigma} \cos \tau$, $\tau = b_1 e^{-c\sigma} \sin \tau$. Squaring and adding gives

$$\tau^2 = b_1^2 e^{-2c\sigma} - (\sigma + b_0)^2 \qquad (5.4.9)$$

which represents a curve in the s plane on which all the roots of the denominator are located. Since the right-hand side becomes negative for σ positive and sufficiently large, the extent of the curve in the positive half-plane $\sigma = \mathrm{Re}\, s > 0$ is limited and, therefore, no singularities of the integrand in (5.4.8) can exist beyond a finite maximum positive value of σ dependent on c, b_0 and b_1. Thus a finite a for use in (5.4.8) can be found and the inversion formula can be applied. Upon taking the ratio of $\sigma + b_0 = -b_1 e^{-c\sigma} \cos \tau$ and $\tau = b_1 e^{-c\sigma} \sin \tau$ we find $\tau \cot \tau = -(\sigma + b_0)$, which has a countably infinite number of solutions. Thus, the denominator of the integrand in (5.4.8) has an infinite number of simple poles occurring in pairs $s_{\pm n} = \sigma_n \pm i\tau_n$ with $\sigma_n \to -\infty$. The inversion can be carried out by closing the contour in the left half-plane (see §11.3) and the solution will be a sum of exponentials most of which decay with increasing t. It is easy to see that, for small c, the leading terms are those found before with the Taylor series argument, while all the other modes are strongly damped.

If the previous heuristic approach is followed to $O(c)$ for an equation of the advanced type such as

$$\frac{d}{dt} u(t - c) + b_0 u(t) + b_1 u(t - c) = 0 \qquad c < t \qquad (5.4.10)$$

one finds two solutions proportional to $e^{-\sigma t}$, one with $\sigma \simeq b_0 + b_1$ as before, while for the other one $\dot{\sigma} \simeq -1/c$. With c positive, this solution will be very rapidly growing. It is found that the denominator of the integrand corresponding to (5.4.8) has an infinite number of roots with arbitrarily large positive real part so that the Laplace transform cannot be inverted. The solution grows faster than exponentially and does not admit a Laplace transform. Finally, for an equation of the neutral type,

$$\frac{\mathrm{d}}{\mathrm{d}t}u(t) + a_1\frac{\mathrm{d}}{\mathrm{d}t}u(t-c) + b_0 u(t) + b_1 u(t-c) = f(t) \qquad c < t \qquad (5.4.11)$$

the zeros of the denominator are asymptotically located on a vertical line. The inversion therefore cannot be carried out by contour integration and the nature of the solution is significantly altered.

5.5 Diffusion equation

Let us consider now several initial-value problems for the one-dimensional diffusion equation

$$\frac{\partial^2 u}{\partial x^2} = \frac{\partial u}{\partial t}, \qquad u(x,0) = f(x), \qquad (5.5.1)$$

with suitable boundary conditions. Upon taking the Laplace transform we have

$$\frac{\partial^2 \hat{u}}{\partial x^2} - s\hat{u} = -f(x). \qquad (5.5.2)$$

5.5.1 Infinite medium

We consider the domain $-\infty < x < \infty$ and require that $u \to 0$ for $|x| \to 0$. This is the same problem solved in §4.4.1 p. 96 by means of the Fourier transform. The pertinent solution of (5.5.2) is readily obtained by specializing (2.2.25) p. 34 and can be written in compact form as

$$\hat{u} = \frac{1}{2}\int_{-\infty}^{\infty} \frac{\exp[-\sqrt{s}\,(x_> - x_<)]}{\sqrt{s}} f(\xi)\,\mathrm{d}\xi \qquad (5.5.3)$$

where $x_> = \max(x,\xi)$, $x_< = \min(x,\xi)$. The function of s can be inverted by using one of the entries in Table 5.1 with the result

$$u(x,t) = \frac{1}{2\sqrt{\pi t}}\int_{-\infty}^{\infty} \exp\left[-\frac{(x-\xi)^2}{4t}\right] f(\xi)\,\mathrm{d}\xi, \qquad (5.5.4)$$

which is the same result (4.4.13) p. 96 obtained in the previous chapter. The use of transform tables renders the two solution procedures quite comparable. However, if the inverse transforms had to be actually calculated, inversion of the Fourier transform would have been easier than that of the Laplace transform. The method works similarly for the other diffusion problems solved earlier by application of the Fourier transforms.

5.5.2 Finite interval

The diffusion problems solved by means of a Fourier series in a spatial variable in Chapter 3 can also be solved by means of the Laplace transform and it is interesting to reveal the relation between the solutions obtained in the two ways.

With this aim in mind, we apply the Laplace transform to a variant of the problem of §3.2 p. 58, in which the diffusion equation is considered in the finite domain $0 < x < L$ subject to homogeneous Dirichlet (rather than Neumann) boundary conditions, $u(0, t) = u(L, t) = 0$. The relevant solution of (5.5.1) in this case can be found by adapting (2.2.26) p. 34 and is

$$\hat{u}(s, t) = \int_0^L \frac{\sinh \sqrt{s}(L - x_>) \; \sinh \sqrt{s} x_<}{\sqrt{s} \; \sinh \sqrt{s} L} f(\xi) \, d\xi, \tag{5.5.5}$$

where $x_> = \max(x, \xi)$, $x_< = \min(x, \xi)$ as before. This form looks significantly simpler than the Fourier series result (3.2.13), but we still need to determine the Green's function

$$G(x_>, x_<, t) = \mathscr{L}^{-1} \left\{ \frac{\sinh \sqrt{s}(L - x_>) \; \sinh \sqrt{s} x_<}{\sqrt{s} \; \sinh \sqrt{s} L} \right\} \tag{5.5.6}$$

after which the solution will be given by

$$u(x, t) = \int_0^L G(x_>, x_<, t) f(\xi) \, d\xi. \tag{5.5.7}$$

In order to bring out more clearly the relation with the Fourier series solution, it is useful to consider the inversion of the Laplace transform. We refer to §11.3 p. 291 for the procedure and to §17.9.1 p. 461 for the evaluation of integrals by the method of residues.

In spite of the presence of square roots, $s = 0$ is not a branch point for (5.5.6) as the function is even in its argument. Furthermore, for large s,

$$\frac{\sinh \sqrt{s}(L - x_>) \; \sinh \sqrt{s} x_<}{\sqrt{s} \; \sinh \sqrt{s} L} \simeq \frac{\exp -\sqrt{s} \, (x_> - x_<)}{\sqrt{s}} \tag{5.5.8}$$

which tends to 0 for $|s| \to \infty$ by the definition of $x_>$ and $x_<$. Hence Jordan's lemma applies and we can close the inversion contour by a semi-circle in the left half-plane. Since $\sinh i\alpha = i \sin \alpha$, the integrand has poles at

$$\sqrt{s_k} L = ik\pi \qquad \text{or} \qquad s_k = -\left(\frac{k\pi}{L}\right)^2, \qquad k = 1, 2, \dots \tag{5.5.9}$$

while, at $s = 0$, the function is regular. With the help of the residue theorem, if we consider, for example, $x_> = x$, $x_< = \xi$, we thus have

$$
\begin{aligned}
G(x, \xi, t) &= \sum_{k=1}^{\infty} \text{Res} \left[\frac{\sinh \sqrt{s}(L - x) \ \sinh \sqrt{s}\xi}{\sqrt{s} \ \sinh \sqrt{s}L} e^{st}, s = s_k \right] \\
&= \sum_{k=1}^{\infty} \frac{\sinh \sqrt{s_k}(L - x) \ \sinh \sqrt{s_k}\xi}{\sqrt{s_k} \ (L/2\sqrt{s_k}) \cosh \sqrt{s_k}L} e^{s_k t} \\
&= \sum_{k=1}^{\infty} (-1)^{k+1} \frac{2}{L} \sin\left(k\pi \frac{L-x}{L}\right) \sin\left(k\pi \frac{\xi}{L}\right) \exp\left[-\left(\frac{k\pi}{L}\right)^2 t\right] \\
&= \frac{2}{L} \sum_{k=1}^{\infty} \sin\left(k\pi \frac{x}{L}\right) \sin\left(k\pi \frac{\xi}{L}\right) \exp\left[-\left(\frac{k\pi}{L}\right)^2 t\right], \quad (5.5.10)
\end{aligned}
$$

which is actually symmetric in the two spatial variables and is therefore valid also for $x_> = \xi$, $x_< = x$. With this result, (5.5.7) has the form that would be found by attacking the problem by means of a Fourier sine series.

5.5.3 Semi-infinite medium, mixed boundary condition

We have seen in Chapter 4 that the Fourier sine and cosine transforms offer a convenient technique for the solution of the diffusion equation in a semi-infinite medium with Dirichlet or Neumann boundary conditions. When the boundary condition is mixed, i.e.

$$
\left[u + h\frac{\partial u}{\partial x} \right]_{x=0} = 0, \quad (5.5.11)
$$

with h a constant, one can either generalize the Fourier transforms (see e.g. Sneddon 1951, p. 173) or, more conveniently, use the Laplace transform.

The solution of the transformed diffusion equation (5.5.2) subject to $\hat{u} \to 0$ at infinity and the transformed version of (5.5.11) is readily found by adapting (2.2.25) p. 34 and is

$$
\begin{aligned}
\hat{u}(x, s) &= \frac{1}{2\sqrt{s}} \int_0^{\infty} f(\xi) \left[e^{-\sqrt{s}\,(x+\xi)} + e^{-\sqrt{s}\,(x_> - x_<)} \right] d\xi \\
&+ \frac{1}{h - \sqrt{s}} \int_0^{\infty} f(\xi) e^{-\sqrt{s}\,(x+\xi)} d\xi \quad (5.5.12)
\end{aligned}
$$

in which $x_> = \max(x, \xi)$, $x_< = \min(x, \xi)$. The inverse of the terms in the first integral is given in Table 5.1. In the second integral let $\sqrt{s} = \sigma$ and $X = x + \xi$. Applying operational rules 8 and 10 (Table 5.3) we have

$$
\begin{aligned}
\mathcal{L}^{-1}\left\{ (h - \sigma)^{-1} e^{-X\sigma} \right\} &= -e^{-hX} \mathcal{L}^{-1}\left\{ (\sigma - h)^{-1} e^{-X(\sigma - h)} \right\}(\tau) \\
&= -e^{h(\tau - X)} \mathcal{L}^{-1}\left\{ e^{-X\sigma}/\sigma \right\}(\tau) = -e^{h(\tau - X)} H(\tau - X) \quad (5.5.13)
\end{aligned}
$$

where H is the step function. Upon substituting into rule 12 we have

$$\mathcal{L}^{-1}\left\{\frac{e^{-\sqrt{s}\,X}}{h-\sqrt{s}}\right\} = \frac{e^{-hX}}{2\sqrt{\pi}\,t^{3/2}}\int_X^\infty d\tau\,\tau\,\exp\left(-\frac{\tau^2}{4t}+h\tau\right)$$

$$= \frac{e^{h^2 t - hX}}{2t}\,\mathrm{erfc}\left(\frac{X}{2\sqrt{t}}-h\sqrt{t}\right) \qquad (5.5.14)$$

in which erfc is the complementary error function defined by

$$\mathrm{erfc}\,x = \frac{2}{\sqrt{\pi}}\int_x^\infty e^{-\xi^2}\,d\xi. \qquad (5.5.15)$$

The final result is therefore

$$u(x,t) = \int_0^\infty f(\xi)\left[\frac{e^{-(x+\xi)^2/4t}+e^{-(x-\xi)^2/4t}}{2\sqrt{\pi t}}\right.$$

$$\left. -\frac{e^{h^2 t - h(x+\xi)}}{2t}\,\mathrm{erfc}\left(\frac{x+\xi}{2\sqrt{t}}-h\sqrt{t}\right)\right]d\xi. \qquad (5.5.16)$$

5.5.4 Duhamel's theorem

Duhamel's theorem addresses the solution of diffusion equation problems of the form

$$\frac{\partial u}{\partial t} - \nabla^2 u - c(\mathbf{x})u = S(\mathbf{x},t), \qquad u(\mathbf{x},0) = f(\mathbf{x}), \qquad (5.5.17)$$

subject to time-dependent mixed boundary conditions:

$$a(\mathbf{x})u + b(\mathbf{x})\mathbf{n}\cdot\nabla u = B(\mathbf{x},t). \qquad (5.5.18)$$

Here $\mathbf{n}$ is the unit normal directed out of the spatial domain Ω of interest, $a(\mathbf{x})$, $b(\mathbf{x})$ and $c(\mathbf{x})$ are given functions of position but not of time, while $S(\mathbf{x},t)$ and $B(\mathbf{x},t)$ can depend on both position and time.

The Laplace transform is particularly effective for problems of this type since, operating on time, it leads to results valid irrespective of the specific geometry or dimensionality of the problem.

The first step in the derivation of the result is to use the linearity of the problem to decompose it into the sum of two sub-problems, one of which carries the inhomogeneity of the boundary conditions and of the initial data and the other one the inhomogeneity of the differential equation. The first sub-problem is

$$\frac{\partial u_1}{\partial t} - \nabla^2 u_1 - c(\mathbf{x})u_1 = 0, \qquad (5.5.19)$$

subject to the boundary and initial conditions

$$a(\mathbf{x})u_1 + b(\mathbf{x})\mathbf{n}\cdot\nabla u_1 = B(\mathbf{x},t), \qquad u_1(\mathbf{x},0) = f(\mathbf{x}). \qquad (5.5.20)$$

The second sub-problem has homogeneous boundary and initial conditions, but a non-homogeneous forcing term in the equation:

$$\frac{\partial u_2}{\partial t} - \nabla^2 u_2 - c(\mathbf{x}) u_2 = S(\mathbf{x}, t) \tag{5.5.21}$$

$$a(\mathbf{x}) u_2 + b(\mathbf{x}) \mathbf{n} \cdot \nabla u_2 = 0, \qquad u_2(\mathbf{x}, 0) = 0. \tag{5.5.22}$$

It is evident that the solution of the original problem can be reconstructed from that of the two sub-problems simply by superposition, $u = u_1 + u_2$ (see e.g. Sneddon 1951, p. 164).

To solve the first sub-problem we take the Laplace transform of (5.5.19) and (5.5.20) to find

$$s\hat{u}_1 - f - [\nabla^2 + c(\mathbf{x})]\hat{u}_1 = 0, \qquad a\hat{u}_1 + b\mathbf{n} \cdot \nabla\hat{u}_1 = \hat{B}. \tag{5.5.23}$$

Solving for $\hat{u}_1$ is facilitated by introducing an auxiliary problem for a function v_1:

$$\frac{\partial v_1}{\partial t} - [\nabla^2 + c(\mathbf{x})]v_1 = 0, \qquad a(\mathbf{x}) v_1 + b(\mathbf{x})\mathbf{n} \cdot \nabla v_1 = B(\mathbf{x}, \tau) \tag{5.5.24}$$

with $v_1 = v_1(\mathbf{x}, t, \tau)$ and τ regarded as a parameter; the initial condition is the same as the original one, namely $v_1(\mathbf{x}, 0, \tau) = f(\mathbf{x})$. The solution of this problem is simpler than that of the original one as the boundary data B do not depend on the time variable t, but only on the parameter τ.

The transform $\hat{v}_1(\mathbf{x}, s, \tau)$ of v_1 with respect to t satisfies

$$s\hat{v}_1 - f - [\nabla^2 + c(\mathbf{x})]\hat{v}_1 = 0, \qquad a\hat{v}_1 + b\mathbf{n} \cdot \nabla\hat{v}_1 = \frac{1}{s}B(\mathbf{x}, \tau). \tag{5.5.25}$$

Since $\hat{v}_1$ also depends on τ, we can define an iterated Laplace transform by multiplying by $e^{-s\tau}$ and integrating with respect to this variable as well. We denote this iterated transform of v_1 by

$$\hat{V}_1 = \int_0^\infty d\tau\, e^{-s\tau} \hat{v}_1(\mathbf{x}, s, \tau) = \int_0^\infty d\tau\, e^{-s\tau} \int_0^\infty dt\, e^{-st} v_1(\mathbf{x}, t, \tau). \tag{5.5.26}$$

With this step, and after multiplication by s, the problem (5.5.25) becomes

$$s(s\hat{V}_1) - f - [\nabla^2 + c](s\hat{V}_1) = 0, \qquad a(s\hat{V}_1) + b\mathbf{n} \cdot \nabla(s\hat{V}_1) = \hat{B}(\mathbf{x}, \tau). \tag{5.5.27}$$

Upon comparing with (5.5.23), we see that $s\hat{V}_1$ satisfies exactly the same problem as $\hat{u}_1$ and the same conditions and, therefore, $s\hat{V}_1 = \hat{u}_1$.

To see the consequences of this observation, we need to establish a generalized convolution theorem for the iterated transform (5.5.26). For this purpose, let us calculate the Laplace transform of (Sneddon 1951, p. 46)

$$F_*(t) \equiv \int_0^t F(t - \tau, \tau)\, d\tau = \int_0^t F(\tau, t - \tau)\, d\tau. \tag{5.5.28}$$

After an interchange of the order of integration (p. 690), we find

$$\hat{F}_* = \int_0^\infty dt\, e^{-st} \int_0^t d\tau\, F(t - \tau, \tau) = \int_0^\infty d\tau\, e^{-s\tau} \int_\tau^\infty dt\, e^{-s(t-\tau)} F$$

$$= \int_0^\infty d\tau\, e^{-s\tau} \int_0^\infty dt\, e^{-st} F(t, \tau) \tag{5.5.29}$$

in which, in the last step, we have renamed the variable $t - \tau$ replacing it by t. In view of this relation and of the fact that $\hat{u}_1 = s\hat{V}_1$, we thus conclude that

$$u_1(\mathbf{x}, t) = \frac{\partial}{\partial t} \int_0^t v_1(\mathbf{x}, t - \tau, \tau) \, d\tau. \tag{5.5.30}$$

The problem for v_1 can be simplified further if we let $v_1 = v_{1a} + v_{1b}$ with

$$\nabla^2 v_{1a} + c(\mathbf{x}) v_{1a} = 0, \qquad a(\mathbf{x}) v_{1a} + b(\mathbf{x})\mathbf{n} \cdot \nabla v_{1a} = B(\mathbf{x}, \tau), \tag{5.5.31}$$

$$\frac{\partial v_{1b}}{\partial t} - [\nabla^2 + c(\mathbf{x})]v_{1b} = 0, \qquad a(\mathbf{x}) v_{1b} + b(\mathbf{x})\mathbf{n} \cdot \nabla v_{1b} = 0,$$
$$v_{1b}(\mathbf{x}, 0, \tau) = f(\mathbf{x}) - v_{1a}(\mathbf{x}, \tau) \tag{5.5.32}$$

because then

$$u_1(\mathbf{x}, t) = v_{1a}(\mathbf{x}, t) + \frac{\partial}{\partial t} \int_0^t v_{1b}(\mathbf{x}, t - \tau, \tau) \, d\tau. \tag{5.5.33}$$

In order to solve the second sub-problem (5.5.21), (5.5.22) we introduce another auxiliary function $v_2(\mathbf{x}, t, \tau)$ chosen to satisfy

$$\frac{\partial v_2}{\partial t} - \nabla^2 v_2 - c(\mathbf{x})v_2 = S(\mathbf{x}, \tau), \quad v_2(\mathbf{x}, 0, \tau) = 0. \tag{5.5.34}$$

Upon differentiating with respect to t, since S is independent of this variable, we have

$$\frac{\partial}{\partial t}\left(\frac{\partial v_2}{\partial t}\right) - [\nabla^2 + c(\mathbf{x})]\left(\frac{\partial v_2}{\partial t}\right) = 0 \tag{5.5.35}$$

with

$$a(\mathbf{x})\left(\frac{\partial v_2}{\partial t}\right) + b(\mathbf{x})\mathbf{n} \cdot \nabla\left(\frac{\partial v_2}{\partial t}\right) = 0. \tag{5.5.36}$$

If this problem is solved, then evidently

$$v_2(\mathbf{x}, t, \tau) = \int_0^t \frac{\partial}{\partial \theta} v_2(\mathbf{x}, \theta, \tau) \, d\theta \tag{5.5.37}$$

so that the initial condition appropriate for (5.5.35) is

$$\left.\frac{\partial v_2}{\partial t}\right|_{t=0} = [\nabla^2 + c(\mathbf{x})]v_2(\mathbf{x}, 0, \tau) + S(\mathbf{x}, \tau) = S(\mathbf{x}, \tau) \tag{5.5.38}$$

given that, from (5.5.34), $v_2(\mathbf{x}, 0, \tau) = 0$. The solution procedure is otherwise the same as before and we find at the end

$$u_2(\mathbf{x}, t) = \frac{\partial}{\partial t} \int_0^t v_2(\mathbf{x}, t - \tau, \tau) \, d\tau = \int_0^t \frac{\partial v_2}{\partial t}(\mathbf{x}, t - \tau, \tau) \, d\tau \tag{5.5.39}$$

where we have used $v_2(\mathbf{x}, 0, \tau) = 0$

The solution of the original problem (5.5.17), (5.5.18) is now found by adding u_1 given by (5.5.33) and u_2 given by (5.5.39). A theory similar to the one just described can also be developed for the wave equation.

5.6 Integral equations

Just as with the Fourier transform, the convolution theorem for the Laplace transform permits the solution of certain integral equations. Volterra equations of the *second kind* have the form

$$u(t) = g(t) + \int_0^t k(t - \tau)u(\tau)\,d\tau \qquad (5.6.1)$$

in which g and k are given. Upon taking the Laplace transform, using the convolution theorem and solving for $\hat{u}$ we have

$$\hat{u} = \frac{\hat{g}}{1 - \hat{k}}. \qquad (5.6.2)$$

If $k(t)$ admits an ordinary Laplace transform, $\hat{k}(s) \to 0$ at infinity and therefore $\mathscr{L}^{-1}\{(1 - \hat{k})^{-1}\}$ does not exist in the ordinary sense. Hence it may be preferable to re-write the solution in the form

$$\hat{u} = \hat{g} + \frac{\hat{k}}{1 - \hat{k}}\hat{g} \qquad (5.6.3)$$

from which

$$u(t) = g(t) + \int_0^t \mathscr{L}^{-1}\left\{\frac{\hat{k}}{1 - \hat{k}}\right\}(\tau)\,g(t - \tau)\,d\tau. \qquad (5.6.4)$$

The inverse transform indicated in this equation is the *resolvent kernel* of the integral equation.[3]

A Volterra equation of the *first kind* has the form

$$g(t) = \int_0^t k(t - \tau)u(\tau)\,d\tau \qquad (5.6.5)$$

with a formal Laplace transform $\hat{u} = \hat{g}/\hat{k}$. Since $\hat{k} \to 0$ at infinity, however, this expression cannot be inverted in general. A possible procedure is to differentiate (5.6.5) one or more times to transform it into an equation of the second kind. For example, if $k(0+)$ exists, we find

$$g'(t) = k(0+)u(t) + \int_0^t k'(t - \tau)\,u(\tau)\,d\tau. \qquad (5.6.6)$$

Another possibility is to proceed as in (5.1.5) writing

$$\hat{u}(s) = s^N \frac{\hat{g}(s)}{s^N \hat{k}(s)} \qquad (5.6.7)$$

with N chosen so that $(s^N\hat{k})^{-1}$ is well behaved at infinity. From rule 3 in Table 5.3 this is equivalent to regarding the N-th iterated integral of u as the unknown rather than u itself.

---

[3] By writing $\hat{k}/(1 - \hat{k}) = \hat{k} + \hat{k}^2 + \cdots$ and inverting term by term by repeated use of the convolution theorem, the solution takes the form of a Neumann series (§21.2.4 p. 633).

As an example of this latter procedure we may consider the problem mentioned on p. 116 of finding the heat flux $-q(t)$ which would result in a prescribed time-dependent temperature $f(t)$ at the surface of a semi-infinite solid. For this purpose we need to solve the equation (cf. (4.11.6) p. 116)

$$f(t) = \sqrt{\frac{D}{\pi}} \int_0^t \frac{q(\tau)}{\sqrt{t - \tau}} \, d\tau \qquad (5.6.8)$$

in which D is the constant diffusivity. By using an entry of Table 5.1 we immediately find

$$\hat{q} = \sqrt{\frac{s}{D}} \hat{f}. \qquad (5.6.9)$$

We multiply and divide by s and use the differentiation rule to find, if $f(t) \to 0$ sufficiently fast for $t \to 0$,

$$q(t) = \frac{d}{dt} \int_0^t \frac{f(\tau)}{\sqrt{\pi D(t - \tau)}} \, d\tau = -\frac{1}{2\sqrt{\pi D}} \int_0^t \frac{f(\tau)}{(t - \tau)^{3/2}} \qquad (5.6.10)$$

in agreement with the result (4.11.11) p. 117 found earlier by other means.

Equation (5.6.8) is a special case of Abel's equation

$$\int_0^t (t - \tau)^{-\alpha} \frac{du}{d\tau} \, d\tau = g(t), \qquad 0 < \alpha < 1, \qquad (5.6.11)$$

which can be solved in the same way to find

$$u(t) = u(0+) + \frac{\sin \alpha\pi}{\pi} \int_0^t (t - \tau)^{\alpha-1} g(\tau) \, d\tau. \qquad (5.6.12)$$

It may be of some interest to show how this equation may be solved by using the fractional derivative defined on p. 289. By comparing with (11.2.14) we see that (5.6.11) may be written as

$$\frac{d^{\alpha-1}}{dt^{\alpha-1}} \frac{du}{dt} = \frac{1}{\Gamma(1 - \alpha)} g(t). \qquad (5.6.13)$$

Apply to both sides the derivative of order $1 - \alpha$ taking $m = 1$ to find

$$\frac{du}{dt} = \frac{1}{\Gamma(1 - \alpha)} \frac{d^{1-\alpha}}{dt^{1-\alpha}} g(t) = \frac{1}{\Gamma(\alpha)\Gamma(1 - \alpha)} \frac{d}{dt} \int_0^t g(\tau)(t - \tau)^{\alpha-1} \, d\tau. \qquad (5.6.14)$$

Using one of the properties of the Gamma function from Table 17.2 p. 444 and integrating we recover (5.6.12).

5.7 Other applications

A few additional applications of the Laplace transform can be found in other chapters:

1. Transient diffusion in a cylindrical domain, see §6.5 p. 159;
2. Axially symmetric water waves, see §6.8 p. 166;
3. Slow viscous flow, see §7.18 p. 208.

Cylindrical Systems

6.1 Cylindrical coordinates

Cylindrical coordinates are generated by translating in the direction z normal to the (x, y) plane a system of plane polar coordinates (r, ϕ). Thus, the relation between the Cartesian coordinates (x, y, z) of a point $\mathbf{x}$ and its cylindrical coordinates (r, z, ϕ) are (see Figure 6.1)

$$x = r \cos \phi, \qquad y = r \sin \phi, \qquad z = z. \qquad (6.1.1)$$

The inverse relations are given in Table 6.1. The ranges of variation of the cylindrical coordinates are

$$0 \le r < \infty, \qquad -\infty < z < \infty, \qquad 0 \le \phi < 2\pi. \qquad (6.1.2)$$

The positions $\phi = 0$ and $\phi = 2\pi$ are indistinguishable and ϕ is undefined on the axis $r = 0$. Relations useful for converting derivatives between Cartesian and cylindrical coordinates are given in Table 6.2.

In some of the examples that follow we will use the δ-function. As explained on p. 38 and, in greater detail, in §20.7 p. 588, due to the appearance of the Jacobian upon transforming a volume integral from one coordinate system to another, the proper form of the multi-dimensional δ-function is coordinate-system dependent. The expressions to be used in cylindrical coordinates are given for various situations in Table 6.3.

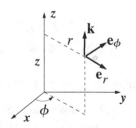

Fig. 6.1   Cylindrical coordinates and unit vectors in the coordinate directions at a generic point.

Table 6.1
Relations between Cartesian and cylindrical coordinates: $\mathbf{i}$, $\mathbf{j}$ and $\mathbf{k}$ are the unit vectors in the three Cartesian directions; $\mathbf{e}_r$, $\mathbf{e}_\phi$ and $\mathbf{k}$ are the unit vectors in the cylindrical coordinates directions (Figure 6.1).

Cartesian	Cylindrical
$\mathbf{x} = (x,\ y,\ z)$	$\mathbf{x} = (r\cos\phi,\ r\sin\phi,\ z)$
$\sqrt{x^2 + y^2}$	r
$(x,\ y)/\sqrt{x^2 + y^2}$	$(\cos\phi,\ \sin\phi)$
$ds^2 = dx^2 + dy^2 + dz^2$	$ds^2 = dr^2 + r^2 d\phi^2 + dz^2$
area element dS in planes $z = $ const.	$r\,d\phi\ dr$
volume element dV	$r\,d\phi\ dr\ dz$
$\mathbf{i} = \cos\phi\,\mathbf{e}_r - \sin\phi\,\mathbf{e}_\phi$	$\mathbf{e}_r = \cos\phi\,\mathbf{i} + \sin\phi\,\mathbf{j}$
$\mathbf{j} = \sin\phi\,\mathbf{e}_r + \cos\phi\,\mathbf{e}_\phi$	$\mathbf{e}_\phi = -\sin\phi\,\mathbf{i} + \cos\phi\,\mathbf{j}$
$\mathbf{k}$	$\mathbf{k}$

Table 6.2
Derivatives of the Cartesian with respect to the cylindrical coordinates and vice versa.

	x	y
$\dfrac{\partial}{\partial r}$	$\cos\phi = \dfrac{x}{r}$	$\sin\phi = \dfrac{y}{r}$
$\dfrac{\partial}{\partial\phi}$	$-r\sin\phi = -y$	$r\cos\phi = x$
	r	ϕ
$\dfrac{\partial}{\partial x}$	$\dfrac{x}{r} = \cos\phi$	$-\dfrac{y}{r^2} = -\dfrac{\sin\phi}{r}$
$\dfrac{\partial}{\partial y}$	$\dfrac{y}{r} = \sin\phi$	$\dfrac{x}{r^2} = \dfrac{\cos\phi}{r}$

6.2 Summary of useful relations

The form of the standard differential operators in cylindrical coordinates is summarized in Table 6.4. The Laplace operator takes the form

$$\nabla^2 u = \frac{1}{r}\frac{\partial}{\partial r}\left(r\frac{\partial u}{\partial r}\right) + \frac{1}{r^2}\frac{\partial^2 u}{\partial\phi^2} + \frac{\partial^2 u}{\partial z^2}. \tag{6.2.1}$$

Table 6.3	The form of $\delta(\mathbf{x} - \mathbf{x}')$ in cylindrical coordinates for several positions of the pole $\mathbf{x}'$; ϕ' is indeterminate when $r' = 0$.

	$\delta^{(3)}(\mathbf{x} - \mathbf{x}')$
$\mathbf{x}' = (r', \phi', z')$	$\delta(r - r')\,\delta(\phi - \phi')\,\delta(z - z')/r'$
$\mathbf{x}' = (0, -, z')$	$\delta(r)\,\delta(z - z')/(2\pi r)$
$\mathbf{x}' = (0, -, 0)$	$\delta(r)\,\delta(z)/(2\pi r)$

Table 6.4	Differential operators in cylindrical coordinates.

$$\nabla u = \frac{\partial u}{\partial r}\,\mathbf{e}_r + \frac{1}{r}\frac{\partial u}{\partial \phi}\,\mathbf{e}_\phi + \frac{\partial u}{\partial z}\,\mathbf{k}$$

$$\nabla \cdot \mathbf{A} = \frac{1}{r}\frac{\partial}{\partial r}(r A_r) + \frac{1}{r}\frac{\partial A_\phi}{\partial \phi} + \frac{\partial A_z}{\partial z}$$

$$\nabla \times \mathbf{A} = \left[\frac{1}{r}\frac{\partial A_z}{\partial \phi} - \frac{\partial A_\phi}{\partial z}\right]\mathbf{e}_r + \left[\frac{\partial A_r}{\partial z} - \frac{\partial A_z}{\partial r}\right]\mathbf{e}_\phi$$

$$+ \frac{1}{r}\left[\frac{\partial}{\partial r}(r A_\phi) - \frac{\partial A_r}{\partial \phi}\right]\mathbf{k} = \frac{1}{r}\begin{vmatrix} \mathbf{e}_r & r\mathbf{e}_\phi & \mathbf{k} \\ \partial/\partial r & \partial/\partial \phi & \partial/\partial z \\ A_r & r A_\phi & A_z \end{vmatrix}$$

$$\nabla^2 u = \frac{1}{r}\frac{\partial}{\partial r}\left(r\frac{\partial u}{\partial r}\right) + \frac{1}{r^2}\frac{\partial u}{\partial \phi^2} + \frac{\partial^2 u}{\partial z^2}$$

If, in the domain of interest, the angular coordinate ranges over $[0, 2\pi)$, it is usually convenient to expand u in a Fourier series in this variable:

$$u(r, z, \phi) = \sum_{m=-\infty}^{\infty} U_m(r, z)\frac{e^{im\phi}}{\sqrt{2\pi}} \tag{6.2.2}$$

with which[1]

$$\nabla^2 u = \sum_{m=-\infty}^{\infty} \frac{e^{im\phi}}{\sqrt{2\pi}}\left[\frac{1}{r}\frac{\partial}{\partial r}\left(r\frac{\partial U_m}{\partial r}\right) - \frac{m^2}{r^2}U_m + \frac{\partial^2 U_m}{\partial z^2}\right]. \tag{6.2.3}$$

For problems over the half range $[0, \pi]$ the real sine or cosine forms of the Fourier series may be more convenient.

---

[1] For the interchange of summation and differentiation see p. 227.

The next step is more problem-dependent: one may use expansions in eigenfunctions of the radial operator (i.e. a Fourier–Bessel series) or of the z-operator (i.e., a Fourier series) or integral transforms; or one may use a double expansion, one in r and one in z, and so forth. Homogeneous boundary conditions in one coordinate are usually an excellent reason to choose a series expansion in the eigenfunctions appropriate for that variable.

6.2.1 Bessel functions

The eigenfunction problem for the radial operator appearing in (6.2.3) leads to the equation

$$\frac{1}{r}\frac{d}{dr}\left(r\frac{dS}{dr}\right) - \frac{m^2}{r^2}S = -k^2 S, \tag{6.2.4}$$

where the eigenvalue has been written as $-k^2$ for later convenience. The change of variable $z = kr$ brings this equation into the standard Bessel form (12.2.1) p. 304. As explained in Chapter 12, the general solution can be expressed as a linear combination of any two Bessel functions

$$J_m(kr), \quad Y_m(kr), \quad H_m^{(1)}(kr), \quad H_m^{(2)}(kr) \tag{6.2.5}$$

with $m \geq 0$. When the index is a negative integer all four functions satisfy a relation of the type $J_{-m} = (-1)^m J_m$. The functions J_m and Y_m are Bessel functions of the first and second kind, respectively, while $H_m^{(1,2)} = J_m \pm i Y_m$ are Bessel functions of the third kind or Hankel functions. Any pair of Bessel functions can be used, although, depending on the problem, certain specific choices may lead to simpler algebra as will be seen in several examples. For the sake of generality, in the relations that follow we write z in place of kr and we do not restrict the index to be an integer and write ν in place of m.

Near the axis, the argument kr is small. For $z \to 0$ and ν not a negative integer one has

$$J_\nu(z) \simeq \frac{(z/2)^\nu}{\Gamma(\nu+1)}\left[1 - \frac{(z/2)^2}{\nu+1} + O(z^4)\right]. \tag{6.2.6}$$

For $z \to 0$ and ν not an integer

$$\begin{aligned}
Y_\nu(z) \simeq\ & -\frac{\Gamma(\nu)}{\pi}\left(\frac{2}{z}\right)^\nu\left[1 - \frac{(z/2)^2}{1-\nu} + O(z^4)\right] \\
& -\frac{\Gamma(-\nu)}{\pi}\cos\nu\pi\left(\frac{z}{2}\right)^\nu\left[1 - \frac{(z/2)^2}{1+\nu} + O(z^4)\right].
\end{aligned} \tag{6.2.7}$$

For $\nu = m > 0$ an integer, one has

$$\begin{aligned}
Y_m(z) \simeq\ & \frac{2}{\pi}J_m(z)\left[\log\frac{z}{2} + \gamma\right] - \frac{(z/2)^m}{\pi}\left[\frac{\psi(m+1)+\gamma}{m!} + O(z^2)\right] \\
& -\frac{(z/2)^{-m}}{\pi}\left[(m-1)! + O(z^2)\right]
\end{aligned} \tag{6.2.8}$$

with ψ the function defined on p. 445 and $\gamma = -\psi(1) \simeq 0.57721$ Euler's constant.[2] The $O(z^2)$ contribution in the last term is omitted if $m = 1$. The corresponding expression for $m = 0$ is

$$Y_0(z) \simeq \frac{2}{\pi} J_0(z) \left[\log \frac{z}{2} + \gamma \right] + \frac{2(z/2)^2}{\pi} \left[1 + O(z^2) \right]. \tag{6.2.9}$$

For large argument, $\mathrm{Re}\, z \to \infty$, with $\alpha_\nu = (\pi/2)(\nu + 1/2)$ and an error $O(z^{-2})$, one has

$$J_\nu(z) \simeq \sqrt{\frac{2}{\pi z}} \left[\cos (z - \alpha_\nu) - \left(\nu^2 - \frac{1}{4} \right) \frac{\sin (z - \alpha_\nu)}{2z} \right], \tag{6.2.10}$$

$$Y_\nu(z) \simeq \sqrt{\frac{2}{\pi z}} \left[\sin (z - \alpha_\nu) + \left(\nu^2 - \frac{1}{4} \right) \frac{\cos (z - \alpha_\nu)}{2z} \right], \tag{6.2.11}$$

$$H_\nu^{(1,2)}(z) \simeq \sqrt{\frac{2}{\pi z}} e^{\pm i(z - \alpha_\nu)} \left[1 \mp \left(\nu^2 - \frac{1}{4} \right) \frac{1}{2iz} \right]. \tag{6.2.12}$$

The leading order terms of these expressions are summarized in Table 6.5.

The only Bessel functions which reduce to elementary functions are those with index $\pm n + 1/2$ which are proportional to polynomials in $1/z$ multiplied by the circular functions. For these functions $Y_{n+1/2}(z) = (-1)^{n-1} J_{-n-1/2}(z)$ and, for example,

$$J_{1/2}(z) = Y_{-1/2}(z) = \sqrt{\frac{2}{\pi z}} \sin z, \quad J_{-1/2}(z) = -Y_{1/2}(z) = \sqrt{\frac{2}{\pi z}} \cos z, \tag{6.2.13}$$

$$H_{1/2}^{(1)}(z) = -i H_{-1/2}^{(1)}(z) = \sqrt{\frac{2}{\pi z}} \frac{e^{iz}}{i}, \quad H_{1/2}^{(2)}(z) = i H_{-1/2}^{(2)}(z) = \sqrt{\frac{2}{\pi z}} \frac{e^{-iz}}{-i}. \tag{6.2.14}$$

We encounter also the modified Bessel equation

$$\frac{1}{r} \frac{d}{dr} \left(r \frac{dS}{dr} \right) - \left(k^2 + \frac{m^2}{r^2} \right) S = 0 \tag{6.2.15}$$

which can be reduced to the form (6.2.4) by the substitution $k' = ik$. When kr is real, rather than dealing with the purely imaginary argument ikr it is convenient to introduce the *modified Bessel functions*, which are solutions of (6.2.15) and are real for real argument. The function corresponding to J_ν is denoted by I_ν and is defined as $I_\nu(z) = i^{-\nu} J_\nu(iz)$. The analog for the function of the second kind does not arise very frequently; the analog of the first Hankel function is denoted by $K_\nu(z)$. The asymptotic behavior of these modified functions can be obtained from that of the regular Bessel functions. For $z \to 0$ and $\mathrm{Re}\, \nu > 0$ one has

$$I_\nu(z) \simeq \frac{(z/2)^\nu}{\Gamma(\nu+1)} \left[1 + \frac{(z/2)^2}{\nu+1} + O(z^4) \right], \tag{6.2.16}$$

$$K_\nu(z) \simeq \frac{\Gamma(\nu)}{2} \left(\frac{2}{z} \right)^\nu, \quad K_0 \simeq - \left(\log \frac{z}{2} + \gamma \right). \tag{6.2.17}$$

[2] It follows from (17.5.36) p. 445 that $\psi(m+1) + \gamma = \psi(m+1) - \psi(1) = 1 + 1/2 + \cdots + 1/m$.

Table 6.5	Leading order terms of the asymptotic expressions of the regular and modified Bessel functions; here $\alpha_\nu = \frac{\pi}{2}\left(\nu + \frac{1}{2}\right)$ and γ is Euler's constant. See (6.2.8) for Y_m.	

	$z \to 0$	$z \to \infty$ $\quad(|\nu| \ll |z|)$
$J_\nu(z) \simeq$	$\dfrac{(z/2)^\nu}{\Gamma(\nu+1)}$ $\quad(\nu \neq -1, -2, \ldots)$	$\sqrt{\dfrac{2}{\pi z}}\,\cos\,(z - \alpha_\nu)$
$Y_\nu(z) \simeq$	$-\dfrac{\Gamma(\nu)}{\pi}\left(\dfrac{2}{z}\right)^\nu$ $\quad(\mathrm{Re}\,\nu > 0)$	$\sqrt{\dfrac{2}{\pi z}}\,\sin\,(z - \alpha_\nu)$
$Y_0(z) \simeq$	$\dfrac{2}{\pi}\left(\log\dfrac{z}{2} + \gamma\right)$	$\sqrt{\dfrac{2}{\pi z}}\,\sin\,\left(z - \dfrac{\pi}{4}\right)$
$H_\nu^{(1)}(z) \simeq$	$-\dfrac{i\Gamma(\nu)}{\pi}\left(\dfrac{2}{z}\right)^\nu$ $\quad(\mathrm{Re}\,\nu > 0)$	$\sqrt{\dfrac{2}{\pi z}}\,\exp\,[i(z - \alpha_\nu)]$
$H_\nu^{(2)}(z) \simeq$	$\dfrac{i\Gamma(\nu)}{\pi}\left(\dfrac{2}{z}\right)^\nu$ $\quad(\mathrm{Re}\,\nu > 0)$	$\sqrt{\dfrac{2}{\pi z}}\,\exp\,[-i(z - \alpha_\nu)]$
$I_\nu(z) \simeq$	$\dfrac{(z/2)^\nu}{\Gamma(\nu+1)}$ $\quad(\nu \neq -1, -2, \ldots)$	$\dfrac{e^z}{\sqrt{2\pi z}}$
$K_\nu(z) \simeq$	$\dfrac{\Gamma(\nu)}{2}\left(\dfrac{2}{z}\right)^\nu$ $\quad(\mathrm{Re}\,\nu > 0)$	$\sqrt{\dfrac{\pi}{2z}}\,e^{-z}$
$K_0(z) \simeq$	$-\left(\log\dfrac{z}{2} + \gamma\right)$	$\sqrt{\dfrac{\pi}{2z}}\,e^{-z}$

For large argument, $\mathrm{Re}\,z \to \infty$, one has

$$I_\nu(z) \simeq \frac{e^z}{\sqrt{2\pi z}}\left[1 - \frac{\nu^2 - 1/4}{2z} + O(z^{-2})\right], \tag{6.2.18}$$

$$K_\nu(z) \simeq \sqrt{\frac{\pi}{2z}}\,e^{-z}\left[1 + \frac{\nu^2 - 1/4}{2z} + O(z^{-2})\right]. \tag{6.2.19}$$

Similarly to (6.2.13) and (6.2.14) we have

$$I_{1/2} = \sqrt{\frac{2}{\pi z}}\,\sinh z, \qquad I_{-1/2} = \sqrt{\frac{2}{\pi z}}\,\cosh z, \qquad K_{\pm 1/2}(z) = \sqrt{\frac{\pi}{2z}}\,e^{-z}. \tag{6.2.20}$$

Working with Bessel functions is greatly simplified by the existence of several relations among them; some useful ones are summarized in Table 6.6.

Table 6.6	Some relations among contiguous Bessel functions and their derivatives; in the first group, Z_ν stands for any of J_ν, Y_ν or $H_\nu^{(1,2)}$.

$$zZ_{\nu-1} + zZ_{\nu+1} = 2\nu Z_\nu \qquad\qquad Z_{\nu-1} - Z_{\nu+1} = 2Z_\nu'$$

$$zZ_\nu' + \nu Z_\nu = zZ_{\nu-1} \qquad\qquad zZ_\nu' - \nu Z_\nu = -zZ_{\nu+1}$$

$$Z_1 = -Z_0' \qquad\qquad Z_2 = \frac{2}{z}Z_1 - Z_0$$

$$\left(\frac{1}{z}\frac{\mathrm{d}}{\mathrm{d}z}\right)^m \left(z^{\pm\nu} Z_\nu\right) = (\pm 1)^m z^{\pm\nu-m} Z_{\nu\mp m}$$

$$J_\nu(z)Y_{\nu-1}(z) - J_{\nu-1}(z)Y_\nu(z) = \frac{2}{\pi z}$$

$$J_{\nu-1}(z)H_\nu^{(1)}(z) - J_\nu(z)H_{\nu-1}^{(1)}(z) = \frac{2}{\pi i z}$$

$$zI_{\nu-1} - zI_{\nu+1} = 2\nu I_\nu \qquad\qquad I_{\nu-1} + I_{\nu+1} = 2I_\nu'$$

$$zI_\nu' + \nu I_\nu = zI_{\nu-1} \qquad\qquad zI_\nu' - \nu I_\nu = zI_{\nu+1}$$

$$I_1 = I_0' \qquad\qquad I_2 = -\frac{2}{z}I_1 + I_0$$

$$\left(\frac{1}{z}\frac{\mathrm{d}}{\mathrm{d}z}\right)^m \left(z^{\pm\nu} I_\nu\right) = z^{\pm\nu-m} I_{\nu\mp m}$$

$$zK_{\nu-1} - zK_{\nu+1} = -2\nu K_\nu \qquad\qquad K_{\nu-1} + K_{\nu+1} = -2K_\nu'$$

$$zK_\nu' + \nu K_\nu = -zK_{\nu-1} \qquad\qquad zK_\nu' - \nu K_\nu = -zK_{\nu+1}$$

$$K_1 = -K_0' \qquad\qquad K_2 = \frac{2}{z}K_1 + K_0$$

$$\left(\frac{1}{z}\frac{\mathrm{d}}{\mathrm{d}z}\right)^m \left(z^{\pm\nu} K_\nu\right) = (-1)^m z^{\pm\nu-m} K_{\nu\mp m}$$

$$I_\nu(z)K_{\nu+1}(z) + I_{\nu+1}(z)K_\nu(z) = \frac{1}{z}$$

6.2.2 Inhomogeneous Bessel equation

On occasion we need to solve the non-homogeneous form of the Bessel equation

$$\frac{1}{r}\frac{\mathrm{d}}{\mathrm{d}r}\left(r\frac{\mathrm{d}S}{\mathrm{d}r}\right) + \left(k^2 - \frac{\nu^2}{r^2}\right)S = g(r). \tag{6.2.21}$$

We do this by the usual method of variation of parameters (§2.2) using the following results for the Wronskian of two Bessel functions:

$$W\left(J_\nu, Y_\nu\right) = \frac{2}{\pi z}, \qquad W\left(H_\nu^{(1)}, H_\nu^{(2)}\right) = \frac{4}{i\pi z}, \tag{6.2.22}$$

$$W\left(J_\nu, H_\nu^{(1)}\right) = \frac{2i}{\pi z}, \qquad W\left(J_\nu, H_\nu^{(2)}\right) = -\frac{2i}{\pi z}. \tag{6.2.23}$$

With these expressions and adapting the general formula (2.2.13) p. 32, we may write the solution of (6.2.21) in the forms[3]

$$S(r) = \left[A - \frac{\pi}{2} \int^r s Y_\nu(ks) g(s)\, ds\right] J_\nu(kr)$$
$$+ \left[B + \frac{\pi}{2} \int^r s J_\nu(ks) g(s)\, ds\right] Y_\nu(kr), \tag{6.2.24}$$

$$S(r) = \left[C - \frac{i\pi}{4} \int^r s H_\nu^{(2)}(ks) g(s)\, ds\right] H_\nu^{(1)}(kr)$$
$$+ \left[D + \frac{i\pi}{4} \int^r s H_\nu^{(1)}(ks) g(s)\, ds\right] H_\nu^{(2)}(kr), \tag{6.2.25}$$

etc. Similarly, with

$$W\left(I_\nu, K_\nu\right) = -\frac{1}{z} \tag{6.2.26}$$

the solution of the modified non-homogeneous equation

$$\frac{1}{r} \frac{d}{dr}\left(r \frac{dS}{dr}\right) - \left(k^2 + \frac{\nu^2}{r^2}\right) S = g(r) \tag{6.2.27}$$

is

$$S(r) = \left[P + \int^r s K_\nu(ks) g(s)\, ds\right] I_\nu(kr)$$
$$+ \left[Q - \int^r s I_\nu(ks) g(s)\, ds\right] K_\nu(kr). \tag{6.2.28}$$

6.2.3 Fourier–Bessel series

The theory of the Fourier–Bessel series is described in §12.5 p. 312. A function $f(r)$ defined for $0 < r < R$ may be expanded as

$$f(r) = \sum_{n=1}^{\infty} a_{\nu n} J_\nu\left(j_{\nu,n} \frac{r}{R}\right), \qquad \nu \geq -\frac{1}{2}, \tag{6.2.29}$$

---

[3] As written, (6.2.24), (6.2.25) and (6.2.28) really contain four integration constants rather than two, namely the two unspecified integration limits and A, B, C, D or P, Q, respectively. As noted in §2.2, the flexibility inherent in this redundancy proves convenient in some cases.

where $j_{\nu,n}$ is the n-th (positive) zero of J_ν and the coefficients $a_{\nu n}$ are given by

$$a_{\nu n} = \frac{1}{N_{\nu n}^2} \int_0^R r\, J_\nu\left(j_{\nu,n}\frac{r}{R}\right) f(r)\, dr \,, \qquad N_{\nu n} = \frac{1}{\sqrt{2}} R J_{\nu+1}(j_{\nu,n}). \qquad (6.2.30)$$

For the application of the Fourier–Bessel expansion it is useful to note that, with $k = j_{\nu,n}/R$, (6.2.4) becomes

$$\left[\frac{1}{r}\frac{d}{dr}\left(r\frac{d}{dr}\right) - \frac{\nu^2}{r^2}\right] J_\nu\left(j_{\nu,n}\frac{r}{R}\right) = -\left(\frac{j_{\nu,n}}{R}\right)^2 J_\nu\left(j_{\nu,n}\frac{r}{R}\right). \qquad (6.2.31)$$

In several examples on the Fourier series we have seen how one can use, for example, the sine series even when the function does not vanish at the end points of the interval (see p. 57). A similar procedure is available here in spite of the fact that, for $\nu > 0$, $J_\nu\left(j_{\nu,n}r/R\right) = 0$ at $r = R$. With an eye to (6.2.3), rather than directly expanding U_m in (6.2.2), we expand the combination

$$\frac{1}{r}\frac{\partial}{\partial r}\left(r\frac{\partial U_m}{\partial r}\right) - \frac{m^2}{r^2}U_m = \sum_{n=1}^{\infty} A_{mn}(z) J_m\left(j_{m,n}\frac{r}{R}\right) \qquad (6.2.32)$$

where, for the time being, we have assumed $m \geq 0$ and, from (6.2.30),

$$A_{mn} = \frac{1}{N_{mn}^2} \int_0^R r\, J_m\left(j_{m,n}\frac{r}{R}\right)\left[\frac{1}{r}\frac{\partial}{\partial r}\left(r\frac{\partial U_m}{\partial r}\right) - \frac{m^2}{r^2}U_m\right] dr. \qquad (6.2.33)$$

Integrating by parts twice assuming that $r^m U_m \to 0$ for $r \to 0$ we find

$$N_{mn}^2 A_{mn}(z) = -j_{m,n} J_m'(j_{m,n}) U_m(R, z) - \left(\frac{j_{m,n}}{R}\right)^2 N_{mn}^2 a_{mn}(z) \qquad (6.2.34)$$

where

$$a_{mn}(z) = \frac{1}{N_{mn}^2} \int_0^R r\, J_m\left(j_{m,n}\frac{r}{R}\right) U_m(r, z)\, dr \qquad (6.2.35)$$

are the Fourier–Bessel coefficients of $U_m(r, z)$. Since no zero of J_m coincides with a zero of J_m', $J_m'(j_{m,n}) \neq 0$ and the first term of (6.2.34) will be non-zero if $U_m(R, z) \neq 0$. As noted on p. 149, $J_{-m} = (-1)^m J_m$ and, therefore, these relations remain valid for $m < 0$ provided that J_m and $j_{m,n}$ are interpreted as $J_{|m|}$ and $j_{|m|,n}$.

6.3 Laplace equation in a cylinder

Consider the Laplace equation in a cylinder of radius R and height L (Figure 6.2) with boundary conditions

$$u(r, \phi, 0) = 0, \qquad u(r, \phi, L) = f(r, \phi), \qquad u(R, \phi, z) = 0. \qquad (6.3.1)$$

To avoid discontinuous data we assume that $f(R, \phi) = 0$. We briefly consider other boundary conditions at the end of the section.

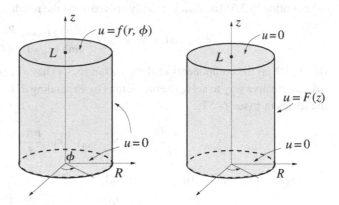

Fig. 6.2   Two of the problems considered in §6.3.

We start with the Fourier series expansion (6.2.2) and take scalar products to find an equation for the coefficients U_m:

$$\frac{1}{r}\frac{\partial}{\partial r}\left(r\frac{\partial U_m}{\partial r}\right) - \frac{m^2}{r^2}U_m + \frac{\partial^2 U_m}{\partial z^2} = 0 \tag{6.3.2}$$

with the boundary condition at $z = L$ following from (6.3.1)

$$U_m(r, L) = \int_{-\pi}^{\pi}\frac{e^{-im\phi}}{\sqrt{2\pi}}f(r, \phi)\,\mathrm{d}\phi \equiv f_m(r) \tag{6.3.3}$$

and $U_m = 0$ on the other boundaries.

For the next step, we demonstrate two different procedures. The first one is to solve (6.3.2) by a Fourier–Bessel expansion:

$$U_m(r, z) = \sum_{n=1}^{\infty} Z_{mn}(z)J_m\left(j_{m,n}\frac{r}{R}\right) \tag{6.3.4}$$

where, as before, the second index in $j_{m,n}$ numbers the zeros of the Bessel function of order m. For the reason mentioned at the end of the previous section we can focus on the case $m \geq 0$ and obtain results for $m < 0$ simply by interpreting J_m and $j_{m,n}$ as $J_{|m|}$ and $j_{|m|,n}$. Substitution into (6.3.2) and use of (6.2.31) gives

$$\frac{\mathrm{d}^2 Z_{mn}}{\mathrm{d}z^2} - \left(\frac{j_{m,n}}{R}\right)^2 Z_{mn} = 0. \tag{6.3.5}$$

The appropriate boundary condition is found from (6.3.3) and (6.3.4) written for $z = L$:

$$f_m(r) = \sum_{n=1}^{\infty} Z_{mn}(L)J_m\left(j_{m,n}\frac{r}{R}\right) \tag{6.3.6}$$

from which, by (6.2.30),

$$Z_{mn}(L) = \frac{1}{N_{mn}^2}\int_0^R r J_m\left(j_{m,n}\frac{r}{R}\right)f_m(r)\,\mathrm{d}r. \tag{6.3.7}$$

The equation (6.3.5) for Z_{mn} is readily solved with the result

$$Z_{mn}(z) = Z_{mn}(L) \frac{\sinh[j_{m,n}z/R]}{\sinh(j_{m,n}L/R)} \qquad (6.3.8)$$

with which all the components of the solution (6.3.4) have been determined.

An alternative way to solve the problem is by expanding $\partial^2 U_m/\partial z^2$ in (6.3.2) in a Fourier sine series in z (see p. 57):

$$\frac{\partial^2 U_m}{\partial z^2} = \sum_{n=1}^{\infty} b_{mn}(r) \sin \frac{n\pi z}{L}. \qquad (6.3.9)$$

By using the relation (3.1.16) p. 57 we find

$$b_{mn}(r) = (-1)^{n+1} \frac{2n\pi}{L^2} f_m(r) - \left(\frac{n\pi}{L}\right)^2 B_{mn}(r), \qquad (6.3.10)$$

in which the B_{mn} are the Fourier sine series coefficients of U_m:

$$U_m(r, z) = \sum_{n=1}^{\infty} B_{mn}(r) \sin \frac{n\pi z}{L}, \qquad B_{mn}(r) = \frac{2}{L} \int_0^L \sin \frac{n\pi z}{L} U_m(r, z) \, dz. \quad (6.3.11)$$

With this step the equation (6.3.2) for U_m leads to (see p. 62 for a similar calculation)

$$\frac{1}{r} \frac{d}{dr}\left(r \frac{dB_{mn}}{dr}\right) - \left[\left(\frac{n\pi}{L}\right)^2 + \frac{m^2}{r^2}\right] B_{mn} = (-1)^n \frac{2n\pi}{L^2} f_m(r) \qquad (6.3.12)$$

which has the form (6.2.27) of the non-homogeneous modified Bessel equation. Let us set $g_{mn}(r) = (-1)^n (2n\pi/L^2) f_m(r)$ and $k_n = n\pi/L$ for brevity. The general solution is given in (6.2.28). By taking in this formula $Q = 0$ and by extending the second integral between 0 and r we ensure that the solution is well behaved on the axis $r = 0$; we have then

$$B_{mn}(r) = \left[P_{mn} - \int_r^R s K_m(k_n s) g_{mn}(s) \, ds\right] I_m(k_n r)$$

$$-K_m(k_n r) \int_0^r s I_m(k_n s) g_{mn}(s) \, ds. \qquad (6.3.13)$$

To satisfy the remaining boundary condition $B_{mn}(R) = 0$ we have taken the other integral between r and R and choose

$$P_{mn} = \frac{K_m(k_n R)}{I_m(k_n R)} \int_0^R s I_m(k_n s) g_{mn}(s) \, ds. \qquad (6.3.14)$$

We have found other examples of the possibility of using different series expansions for the solution of the same problem, e.g. in §3.4. In principle, the different procedures are perfectly equivalent. In practice, differences arise in the rate of convergence of the series, which depends not only on the nature of the data (e.g., a non-smooth behavior causing a slow decrease of the coefficients), but also on the position of the point where the solution is evaluated.

If the boundary condition at $r = R$ had been one of vanishing gradient, $\partial u/\partial r = 0$, in applying the first method of solution it would have been necessary to choose the numbers $j_{m,n}$ as the zeros of J'_m rather than J_m. In this case we would have ended up with the Dini series explained on p. 315. The same type of series would enable us to deal with mixed boundary conditions of the form $au + b\partial u/\partial r = 0$ at $r = R$. For the second method of solution, these more general boundary conditions would only affect the constants P_{mn}.

A non-zero condition on $z = 0$ would modify the solution (6.3.8) with the first method and add a term similar to f_m in the right-hand side of (6.3.10) with the second one; similarly for the prescription of $\partial u/\partial z$ on $z = 0$ or L or both.

As demonstrated in other cases (see e.g. §3.2.4 p. 65), the linearity of the problem permits us to break up the solution into a number of terms each one of which satisfies the Laplace equation and the boundary condition on a single component of the boundary. Let us then consider such a sub-problem with a non-zero boundary condition at $r = R$, namely

$$u(r, \phi, 0) = u(r, \phi, L) = 0, \qquad u(R, \phi, z) = F(z). \tag{6.3.15}$$

A ϕ-dependence of F can be easily accommodated by means of a Fourier series in this variable; we just look at the axisymmetric case for simplicity. A Fourier sine series in z would naturally account for the homogeneous boundary conditions at the bottom and at the top of the cylinder, leading for the radial dependence of the coefficients to the homogeneous form of (6.3.12) with the non-homogeneous boundary condition arising from (6.3.15). Another possibility would be to use a Fourier–Bessel expansion of the radial operator as in (6.2.32):

$$\frac{1}{r}\frac{\partial}{\partial r}\left(r\frac{\partial u}{\partial r}\right) = \sum_{n=1}^{\infty} A_n(z) J_0\left(j_{0,n}\frac{r}{R}\right) \tag{6.3.16}$$

with, as in (6.2.34)

$$N_n^2 A_n(z) = -j_{0,n} J_0'(j_{0,n}) F(z) - \left(\frac{j_{0,n}}{R}\right)^2 N_n^2 a_n(z) \tag{6.3.17}$$

in which

$$a_n(z) = \frac{1}{N_n^2}\int_0^R r\, J_0\left(j_{0,n}\frac{r}{R}\right) u(r, z)\, dr \tag{6.3.18}$$

are the Fourier–Bessel coefficients of $u(r, z)$. Upon substitution of the u expansion into the Laplace equation, the linear independence of the J_0 leads to

$$\frac{d^2 a_n}{dz^2} - \left(\frac{j_{0,n}}{R}\right)^2 a_n = \frac{j_{0,n}}{N_n^2} J_0'(j_{0,n}) F(z) \tag{6.3.19}$$

to be solved subject to the boundary conditions $a_n(0) = a_n(L) = 0$.

6.4 Fundamental solution of the Poisson equation

The potential field generated in free space by a unit point source located at $\mathbf{X}$ is the solution of (see p. 402)

$$\nabla^2 u = -4\pi \delta^{(3)}(\mathbf{x} - \mathbf{X}) \tag{6.4.1}$$

and has the well-known form ("Coulomb's law")

$$u(\mathbf{x}) = \frac{1}{|\mathbf{x} - \mathbf{X}|}. \tag{6.4.2}$$

By using cylindrical coordinates to solve (6.4.1), we now find the expansion of this fundamental solution in this coordinate system. Thus, we consider

$$\frac{1}{r}\frac{\partial}{\partial r}\left(r\frac{\partial u}{\partial r}\right) + \frac{1}{r^2}\frac{\partial^2 u}{\partial \phi^2} + \frac{\partial^2 u}{\partial z^2} = -4\pi\delta^{(3)}(\mathbf{x} - \mathbf{X}). \tag{6.4.3}$$

If we write $\mathbf{x} = (r, \phi, z)$, $\mathbf{X} = (R, \Phi, Z)$ then (Table 6.3),

$$\delta^{(3)}(\mathbf{x} - \mathbf{X}) = \frac{\delta(r - R)}{R}\delta(\phi - \Phi)\delta(z - Z). \tag{6.4.4}$$

We require the solution to be regular everywhere except at $\mathbf{X}$.

We start with a Fourier expansion in the angular variable as in (6.2.2) to find, as in (6.2.3),

$$\sum_{m=-\infty}^{\infty} \frac{e^{im\phi}}{\sqrt{2\pi}}\left[\frac{1}{r}\frac{\partial}{\partial r}\left(r\frac{\partial U_m}{\partial r}\right) - \frac{m^2}{r^2}U_m + \frac{\partial^2 U_m}{\partial z^2}\right] = -4\pi\frac{\delta(r - R)}{R}\delta(\phi - \Phi)\delta(z - Z). \tag{6.4.5}$$

The scalar product with the generic Fourier basis function gives

$$\frac{1}{r}\frac{\partial}{\partial r}\left(r\frac{\partial U_m}{\partial r}\right) - \frac{m^2}{r^2}U_m + \frac{\partial^2 U_m}{\partial z^2} = -4\pi\frac{\delta(r - R)\delta(z - Z)e^{-im\Phi}}{R\sqrt{2\pi}}. \tag{6.4.6}$$

Since $-\infty < z < \infty$, we take a Fourier transform in z to solve this equation:

$$\frac{1}{r}\frac{\partial}{\partial r}\left(r\frac{\partial \tilde{U}_m}{\partial r}\right) - \left(k^2 + \frac{m^2}{r^2}\right)\tilde{U}_m = -\frac{2}{R}\delta(r - R)e^{ikZ}e^{-im\Phi} \tag{6.4.7}$$

in which $\tilde{U}_m(r, k)$ is the Fourier transform of $U_m(r, z)$ with respect to z. The solution of this non-homogeneous modified Bessel equation is given by (6.2.28); to cancel the divergent behavior of I_m at infinity we take $P = 0$ and extend the first integral from r to ∞ while to cancel the divergent behavior of K_m at $r = 0$ we take $Q = 0$ and extend the second integral from 0 to r. Hence, for $m \geq 0$,

$$\tilde{U}_m = \frac{2}{R}e^{ikZ}e^{-im\Phi}\left[I_m(|k|r)\int_r^{\infty} sK_m(|k|s)\delta(s - R)\,ds\right.$$
$$\left. + K_m(|k|r)\int_0^r sI_m(|k|s)\delta(s - R)\,ds\right] \tag{6.4.8}$$

where we write $|k|$ since the solution of the Bessel equation is only dependent on k^2. For a similar reason, as already noted, the index m on the Bessel functions should be interpreted as $|m|$. The integrations can be carried out by using the rules of Table 2.1 p. 37 to find

$$\tilde{U}_m = 2e^{ikZ}e^{-im\Phi}\left[H(R - r)I_m(|k|r)K_m(|k|R)\right.$$
$$\left. + K_m(|k|r)H(r - R)I_m(|k|R)\right] = 2e^{ikZ}e^{-im\Phi}I_m(|k|r_<)K_m(|k|r_>) \tag{6.4.9}$$

in which $r_{>,<} = \max, \min (r, R)$. Upon inverting the Fourier transform we have

$$
\begin{aligned}
U_m &= \sqrt{\frac{2}{\pi}} e^{-im\Phi} \int_{-\infty}^{\infty} e^{ik(Z-z)} I_m(|k|r_<) K_m(|k|r_>) \, dk \\
&= \sqrt{\frac{2}{\pi}} e^{-im\Phi} \int_0^{\infty} \left[e^{ik(Z-z)} + e^{-ik(Z-z)} \right] I_m(kr_<) K_m(kr_>) \, dk \\
&= 2\sqrt{\frac{2}{\pi}} e^{-im\Phi} \int_0^{\infty} \cos[k(Z-z)] \, I_m(kr_<) K_m(kr_>) \, dk.
\end{aligned}
\tag{6.4.10}
$$

The parity of the terms under the change of the sign of m can be exploited in a similar way and the final result is

$$
\frac{1}{|\mathbf{x} - \mathbf{X}|} = \frac{2}{\pi} \int_0^{\infty} \cos[k(z-Z)] I_0(kr_<) K_0(kr_>) \, dk
$$
$$
+ \frac{4}{\pi} \sum_{m=1}^{\infty} \cos[m(\phi - \Phi)] \int_0^{\infty} \cos[k(z-Z)] I_m(kr_<) K_m(kr_>) \, dk.
\tag{6.4.11}
$$

If we take $R = 0$ and $Z = 0$, $|\mathbf{x} - \mathbf{X}| = \sqrt{r^2 + z^2}$ and all the I_m vanish except I_0 which equals 1; then the previous result gives the interesting relation

$$
\frac{1}{\sqrt{r^2 + z^2}} = \frac{2}{\pi} \int_0^{\infty} \cos(kz) K_0(kr) \, dk.
\tag{6.4.12}
$$

If this is regarded as the cosine transform of $K_0(kr)$, upon inverting it (see (4.10.3) p. 113) we find an integral representation for K_0:

$$
K_0(kr) = \int_0^{\infty} \frac{\cos(kz)}{\sqrt{r^2 + z^2}} \, dz.
\tag{6.4.13}
$$

This is a special case of the integral representation (12.4.11) of K_ν given on p. 312.

The same problem in the space bounded by two parallel planes can be solved by using a Fourier series, rather than a transform, in z.

6.5 Transient diffusion in a cylinder

Let us now consider an example with the diffusion equation. The cases treated in the previous two sections demonstrate how to deal with the z-direction. Thus, for simplicity, we treat here the two-dimensional plane case:

$$
\frac{\partial u}{\partial t} = \frac{1}{r} \frac{\partial}{\partial r} \left(r \frac{\partial u}{\partial r} \right) + \frac{1}{r^2} \frac{\partial^2 u}{\partial \phi^2}
\tag{6.5.1}
$$

inside the disk (or infinite cylinder) of radius R. We study the initial-value problem assuming that $u(r, \phi, 0) = f(r, \phi)$ and that $u(R, \phi, t) = 0$; we are interested in the solution which is bounded at the origin $r = 0$.

After expanding in a Fourier series in ϕ as in (6.2.2) and separating the various angular components we find

$$\frac{\partial U_m}{\partial t} = \frac{1}{r}\frac{\partial}{\partial r}\left(r\frac{\partial U_m}{\partial r}\right) - \frac{m^2}{r^2}U_m. \tag{6.5.2}$$

In view of the boundary condition at $r = R$, we now expand U_m in a Fourier–Bessel series in r:

$$U_m(r,t) = \sum_{n=1}^{\infty} T_{mn}(t)J_m\left(j_{m,n}\frac{r}{R}\right) \tag{6.5.3}$$

in which, as already noted repeatedly, J_m and $j_{m,n}$ are to be interpreted as $J_{|m|}$ and $j_{|m|,n}$ for $m < 0$. We substitute into (6.5.2) to find

$$\frac{dT_{mn}}{dt} + \left(\frac{j_{m,n}}{R}\right)^2 T_{mn} = 0 \tag{6.5.4}$$

which has the solution

$$T_{mn}(t) = T_{mn}(0)\exp\left[-\left(\frac{j_{m,n}}{R}\right)^2 t\right] \tag{6.5.5}$$

so that

$$u(r,\phi,t) = \sum_{m=-\infty}^{\infty}\frac{e^{im\phi}}{\sqrt{2\pi}}\sum_{n=1}^{\infty} T_{mn}(0)J_m\left(j_{m,n}\frac{r}{R}\right)\exp\left[-\left(\frac{j_{m,n}}{R}\right)^2 t\right]. \tag{6.5.6}$$

The initial conditions $T_{mn}(0)$ are found by evaluating at $t = 0$

$$u(r,\phi,0) = f(r,\phi) = \sum_{m=-\infty}^{\infty}\frac{e^{im\phi}}{\sqrt{2\pi}}\sum_{n=1}^{\infty} T_{mn}(0)J_m\left(j_{m,n}\frac{r}{R}\right) \tag{6.5.7}$$

and taking scalar products:

$$T_{mn}(0) = \frac{1}{N_{mn}^2}\int_0^{2\pi}\frac{e^{-im\phi}}{\sqrt{2\pi}}\int_0^R sJ_m\left(j_{m,n}\frac{s}{R}\right)f(s,\phi)\,ds\,d\phi \tag{6.5.8}$$

with N_{mn} given in (6.2.30).

If the boundary condition on the lateral wall $r = R$ had been $\partial u/\partial r = 0$, we would have used a Dini series based on the zeros of J_m', but the procedure would have been the same. Similarly, with a mixed (or radiation) condition of the form

$$au(R,z,\phi) + b\left.\frac{\partial u}{\partial r}\right|_{r=R} = 0 \tag{6.5.9}$$

we would be led to choose the zeros of $aJ_m + bJ_{mn}'$ (p. 315).

An alternative method of solution is by means of the Laplace transform. To see the relation between the two methods it is sufficient to consider the axisymmetric case in which there is no ϕ-dependence. Application of the transform to (6.5.1) gives

$$\frac{1}{r}\frac{\partial}{\partial r}\left(r\frac{\partial \hat{u}}{\partial r}\right) - s\hat{u} = -f(r) \tag{6.5.10}$$

where the term in the right-hand side arises from the differentiation rule (see Table 5.3 p. 126). This is an inhomogeneous modified Bessel equation the solution of which is given by (6.2.28). By taking in this equation $Q = 0$ and extending the integral between 0 and r we avoid the singularity at $r = 0$; we also take the other limit in the first integral equal to R and find

$$\hat{u}(r) = \left[P + \int_r^R \xi K_0(\xi\sqrt{s}) f(\xi)\, d\xi \right] I_0(r\sqrt{s})$$
$$+ K_0(r\sqrt{s}) \int_0^r \xi I_0(\xi\sqrt{s}) f(\xi)\, d\xi. \tag{6.5.11}$$

The condition $\hat{u} = 0$ at $r = R$ requires that

$$P = -\frac{K_0(R\sqrt{s})}{I_0(R\sqrt{s})} \int_0^R \xi I_0(\xi\sqrt{s}) f(\xi)\, d\xi \tag{6.5.12}$$

so that

$$\hat{u}(r) = -\frac{K_0(R\sqrt{s}) I_0(r\sqrt{s})}{I_0(R\sqrt{s})} \int_0^R \xi I_0(\xi\sqrt{s}) f(\xi)\, d\xi$$
$$+ \int_0^R \xi I_0(r_<\sqrt{s}) K_0(r_>\sqrt{s}) f(\xi)\, d\xi \tag{6.5.13}$$

with $r_{<,>} = \min, \max(r, R)$. Keeping in mind (12.2.6) p. 305 we set

$$K_0(z) = -I_0(z) \log \frac{z}{2} + F(z) \tag{6.5.14}$$

where F is an entire function containing only even powers of z, to rewrite this expression as

$$\hat{u}(r) = \log \frac{R}{r_>} I_0(r\sqrt{s}) \int_0^R \xi I_0(\xi\sqrt{s}) f(\xi)\, d\xi$$
$$+ \int_0^R \xi I_0(r_<\sqrt{s}) F(r_>\sqrt{s}) f(\xi)\, d\xi$$
$$- F(R\sqrt{s}) \frac{I_0(r\sqrt{s})}{I_0(R\sqrt{s})} \int_0^R \xi I_0(\xi\sqrt{s}) f(\xi)\, d\xi. \tag{6.5.15}$$

Here we have used the facts that $I_0(r\sqrt{s}) I_0(\xi\sqrt{s}) = I_0(r_<\sqrt{s}) I_0(r_>\sqrt{s})$ and that $\log(R\sqrt{s}/2) - \log(r_>\sqrt{s}/2) = \log R/r_>$. The task is now to invert this transform, which can be done by contour integration. By Cauchy's theorem 17.3.1 p. 425 the first two terms contribute nothing because both I_0 and F only contain even powers of the argument so that they are entire functions and there are no cuts in the complex plane. The numerator of the last term is entire while the denominator has singularities at the zeros of $I_0(R\sqrt{s})$ or, which is the same thing, at the zeros of $J_0(iR\sqrt{s})$. Since J_0 only has real zeros at points $j_{0,n}$, the integrand has only simple poles at $iR\sqrt{s} = j_{0,n}$ i.e. $s = -(j_{0,n}/R)^2$ and, furthermore, $I_0(r\sqrt{s})/I_0(R\sqrt{s}) \to e^{-\sqrt{s}(R-r)}$ for $s \to \infty$ so that Jordan's lemma can be invoked to calculate the inversion integral by means of the residue theorem. By using (5.1.3) p. 124

we thus find

$$
\begin{aligned}
u(r,t) \;=\; & -\frac{2}{R^2} \sum_{n=1}^{\infty} \frac{(-i j_{0,n}) F(-i j_{0,n})}{J_0'(j_{0,n})} J_0\left(j_{0,n}\frac{r}{R}\right) \\
& \times \int_0^R \xi J_0\left(j_{0,n}\frac{\xi}{R}\right) f(\xi)\, d\xi \; \exp\left[-\left(\frac{j_{0,n}}{R}\right)^2 t\right].
\end{aligned}
\tag{6.5.16}
$$

To completely reconcile this result with the earlier one we return to (6.5.14), use the definition of $H_0^{(1)}$ and twice the fact that $I_0(-i j_{0,n}) = J_0(j_{0,n}) = 0$:

$$
F(-i j_{0,n}) \;=\; K_0(-i j_{0,n}) \;=\; \frac{i\pi}{2} H_0^{(1)}(j_{0,n}) \;=\; -\frac{\pi}{2} Y_0(j_{0,n}).
\tag{6.5.17}
$$

The last step uses the Jacobian (6.2.22), the relation $J_0'(z) = -J_1(z)$ and, again, $J_0(j_{0,n}) = 0$; in conclusion

$$
(-i j_{0,n}) F(-i j_{0,n}) \;=\; \frac{1}{J_1(j_{0,n})}
\tag{6.5.18}
$$

with which the earlier result is recovered.

The considerable amount of labor involved in the inversion of the Laplace transform leaves no doubt as to which of the two methods is more efficient.

6.6 Formulae for the Hankel transform

The Hankel transform is derived in §10.9 p. 282, where it is shown that a function $f(r)$ defined for $0 < r < \infty$ can be represented as

$$
f(r) \;=\; \mathcal{H}_\nu^{-1}\{\tilde{f}_\nu\} \;=\; \int_0^\infty k\, J_\nu(kr)\, \tilde{f}_\nu(k)\, dk
\tag{6.6.1}
$$

in which $\tilde{f}_\nu(k)$ is the *Hankel transform* of order $\nu \geq -1/2$ of f defined by[4]

$$
\tilde{f}_\nu(k) \;=\; \mathcal{H}_\nu\{f\} \;=\; \int_0^\infty r\, J_\nu(kr)\, f(r)\, dr.
\tag{6.6.2}
$$

These two relations are justified by the completeness relation

$$
\int_0^\infty k\, r\, J_\nu(kr)\, J_\nu(ks)\, dk \;=\; \delta(r - s)
\tag{6.6.3}
$$

valid for $\nu \geq -1/2$.

The most important operational rule satisfied by the Hankel transform is

$$
\mathcal{H}_\nu\left\{\frac{1}{r}\frac{d}{dr}\left(r\frac{df(r)}{dr}\right) - \frac{\nu^2}{r^2} f(r)\right\} \;=\; -k^2 \tilde{f}_\nu(k).
\tag{6.6.4}
$$

A short list of Hankel transforms of order 0 is given in Table 6.7. Note that, by the symmetry of the direct and inverse transform, the entries in the table can be used for both.

---

[4] Another definition used in the literature is the symmetric one

$$
\tilde{f}_\nu(k) = \int_0^\infty \sqrt{kx}\, J_\nu(kx) f(x)\, dx, \qquad f(x) = \int_0^\infty \sqrt{kx}\, J_\nu(kx) \tilde{f}(k)\, dk.
$$

Table 6.7

Hankel transforms of order 0; additional transforms can be obtained by differentiation with respect to the parameters. Interchanging k and r gives the inverse transforms.

$f(r) = \int_0^\infty k\,J_0(kr)\,\tilde{f}_0(k)\,\mathrm{d}k$	$\tilde{f}_0(k) = \int_0^\infty r\,J_0(kr)\,f(r)\,\mathrm{d}r$
$H(a-x)$	$(a/k)\,J_1(ka)$
$\delta(x-a)$	$(a/k^2)\,J_1(ka) - a\,J_1'(ka)$
r^{-1}	k^{-1}
$(r^2+a^2)^{-1/2}$ (Re $a>0$)	e^{-ak}/k
$r^{-1}(r^2+a^2)^{-1/2}$ (Re $a>0$)	$I_0(ak/2)\,K_0(ak/2)$
$(a^2-r^2)^{1/2}H(a-r)$	$(4a/k^3)\,J_1(ka) - (2a^2/k^2)\,J_0(ka)$
$(a^2-r^2)^{-1/2}H(a-r)$	$(\sin ak)/k$
$(r^2-a^2)^{-1/2}H(r-a)$	$(\cos ak)/k$
$r^{-1}(a^2-r^2)^{-1/2}H(a-r)$	$(\pi/2)[J_0(ak/2)]^2$
$r^{-1}(r^2-a^2)^{-1/2}H(r-a)$	$-(\pi/2)\,J_0(ak/2)\,Y_0(ak/2)$
e^{-ar}	$a/(a^2+k^2)^{3/2}$
e^{-ar^2}	$e^{-k^2/4a}/(2a)$
$(\sin ar)/r^2$	$(\pi/2)\,H(a-k) + H(k-a)\,\sin^{-1}(a/k)$
$(\sin ar)/[r(r^2+b^2)]$	$b^{-1}\sinh ab\,K_0(bk)$
$(\cos ar)/[r(r^2+b^2)]$	$(\pi/2)b^{-1}e^{-ab}\,I_0(bk)$
$J_1(ar)/r$	$H(k-a)/a$
$r^{-2}[1 - J_0(ar)]$	$H(a-k)\,\log(a/k)$
$J_0(ax)/(x^2+b^2)$	$I_0(bk_<)K_0(bk_>),\ \ k_{>,<} = \max, \min(a,k)$

6.7 Laplace equation in a half-space

Let us solve the axisymmetric Laplace equation

$$\frac{1}{r}\frac{\partial}{\partial r}\left(r\frac{\partial u}{\partial r}\right) + \frac{\partial^2 u}{\partial z^2} = 0 \tag{6.7.1}$$

in the half-space $0 < z < \infty$ subject to the boundary conditions

$$u(r,0) = f(r), \qquad u \to 0 \quad \text{for} \quad r, z \to \infty, \tag{6.7.2}$$

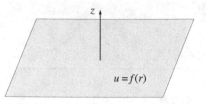

$$u = f(r)$$

Fig. 6.3   Laplace equation in a half-space, §6.7.

with u regular at $r = 0$. We approach this problem in two ways. The first one is a Hankel transform in r; from the operational formula (6.6.4) we find

$$\mathcal{H}_0 \left\{ \frac{1}{r} \frac{\partial}{\partial r} \left(r \frac{\partial u}{\partial r} \right) + \frac{\partial^2 u}{\partial z^2} \right\} = \frac{d^2 \tilde{u}_0}{dz^2} - k^2 \tilde{u}_0 = 0 \tag{6.7.3}$$

where

$$\tilde{u}_0(k, z) = \int_0^\infty r J_0(kr) u(r, z) \, dr \tag{6.7.4}$$

subject to the boundary condition $\tilde{u}_0(k, 0) = \tilde{f}_0(k)$ on the plane $z = 0$. The solution bounded at infinity is

$$\tilde{u}_0(k, z) = \tilde{f}_0(k) e^{-kz} \tag{6.7.5}$$

from which, using the inversion formula (6.6.1) and assuming that the nature of f permits the interchange of the order of the integrations (see p. 690),

$$\begin{aligned} u(r, z) &= \int_0^\infty k J_0(kr) \tilde{f}_0(k) e^{-kz} \, dk \\ &= \int_0^\infty s f(s) \, ds \int_0^\infty k J_0(kr) J_0(ks) e^{-kz} \, dk. \end{aligned} \tag{6.7.6}$$

From integral tables (e.g. Gradhsteyn et al. 2007) we have

$$\begin{aligned} \int_0^\infty k J_0(kr) J_0(ks) e^{-kz} \, dk &= -\frac{\partial}{\partial z} \int_0^\infty J_0(kr) J_0(ks) e^{-kz} \, dk \\ &= -\frac{1}{\pi \sqrt{rs}} \frac{\partial}{\partial z} Q_{-1/2} \left(\frac{r^2 + s^2 + z^2}{2rs} \right) \end{aligned} \tag{6.7.7}$$

where $Q_{-1/2}$ is a Legendre function (§13.5 p. 326).

In view of the complicated form of this result it is useful to consider some special cases. On the axis $r = 0$ (6.7.6) is (Table 6.7)

$$u(0, z) = \int_0^\infty s f(s) \, ds \int_0^\infty k J_0(ks) e^{-kz} \, dk = z \int_0^\infty \frac{s f(s)}{(s^2 + z^2)^{3/2}} \, ds. \tag{6.7.8}$$

For general r we can obtain an asymptotic expansion by noting that, formally, the inner integral in (6.7.6) is the Laplace transform of the function $k J_0(kr) J_0(ks)$ for which the asymptotic results provided by Watson's lemma (p. 296) are available. For large z, the most significant contribution to the integral comes from a neighborhood of $k = 0$. Using (6.2.6) we have

$$k\, J_0(kr)\, J_0(ks) \simeq k\left[1 - \frac{1}{4}k^2(r^2 + s^2)\right] \qquad (6.7.9)$$

with which, from (11.4.4) p. 296 (or, more simply, integrating by parts)

$$\int_0^\infty k\, J_0(kr)\, J_0(ks)e^{-kz}\, dk = \frac{1}{z^2} - \frac{3}{2}\frac{r^2 + s^2}{z^4} + \cdots \qquad (6.7.10)$$

With this approximation, (6.7.6) gives

$$u(r, z) = \frac{1}{z^2}\int_0^\infty s f(s)\, ds - \frac{3}{2z^4}\int_0^\infty s(r^2 + s^2)f(s)\, ds + \cdots \qquad z \to \infty \quad (6.7.11)$$

provided the indicated integrals exist.

A simple case which can be solved exactly is $f(r) = u_0/r$ for which, from Table 6.7, $\tilde{f}_0 = u_0/k$; (6.7.6) becomes

$$u(r, z) = u_0\int_0^\infty J_0(kr)\, e^{-kz}\, dk = \frac{u_0}{\sqrt{r^2 + z^2}}. \qquad (6.7.12)$$

The boundary condition is evidently satisfied, and so is the equation as $\sqrt{r^2 + z^2}$ is simply the distance from the origin. For $r = 0$ this is $u(0, z) = u_0/z$ in agreement with (6.7.8); (6.7.11) is not applicable as the integrals do not converge.

The problem can also be solved by using a Fourier sine transform in z. By taking the transform of (6.7.1) and using one of the operational rules in Table 4.6 p. 115 we find

$$\frac{1}{r}\frac{\partial}{\partial r}\left(r\frac{\partial \tilde{u}_s}{\partial r}\right) - h^2\tilde{u}_s = -\sqrt{\frac{2}{\pi}}\, hf(r). \qquad (6.7.13)$$

This is a modified inhomogeneous Bessel equation of order 0 the solution of which is given in (6.2.28). By requiring that $\tilde{u}_s$ be bounded as $r \to \infty$ we are led to take $P = 0$ and to extend the first integral between r and ∞. The requirement that $\tilde{u}_s$ be bounded at $r = 0$ is met by taking $Q = 0$ and extending the second integral between 0 and r. In conclusion

$$\tilde{u}_s = \sqrt{\frac{2}{\pi}}\, h\int_0^\infty s I_0(hr_<)\, K_0(hr_>)\, f(s)\, ds, \qquad \begin{matrix} r_> = \max(r, s) \\ r_< = \min(r, s) \end{matrix}. \qquad (6.7.14)$$

The solution $u(r, z)$ is found by inverting the Fourier sine transform:

$$u(r, z) = \frac{2}{\pi}\int_0^\infty s\, f(s)\, ds \int_0^\infty h\, I_0(hr_<)\, K_0(hr_>)\, \sin hz\, dh. \qquad (6.7.15)$$

By using the result

$$\int_0^\infty h\, K_0(hs)\, \sin hz\, dh = \frac{\pi}{2}\frac{z}{(s^2 + z^2)^{3/2}}, \qquad (6.7.16)$$

(6.7.15) is seen to be identical to (6.7.8) on the axis $r = 0$. Agreement with the previous result (6.7.6) for general r can be established by using the result (e.g. Gradshteyn *et al.*

2007)

$$\int_0^\infty I_0(hr_<) \, K_0(hr_>) \, \cos hz \, dh \;=\; \frac{1}{2\sqrt{rs}} Q_{-1/2}\left(\frac{r^2 + s^2 + z^2}{2rs}\right) \qquad (6.7.17)$$

and differentiating with respect to z. The logarithmic singularity of K_0 for small argument prevents the application of the asymptotic formulae for the sine transform given on p. 281.

The same problem with data $u = f(r, \phi)$ on $z = 0$ can be solved by first expanding in a Fourier series in ϕ and then using either one of the above methods for each Fourier component.

6.8 Axisymmetric waves on a liquid surface

We solve now the problem of the small-amplitude surface waves generated when an initial vertical velocity v is imparted at $t = 0$ to a horizontal liquid surface, e.g. by an instantaneous pressure pulse or by the impact of an object. As mentioned in §1.4, when viscous effects are negligible and the fluid density constant, an initially irrotational motion remains irrotational for all time and the fluid velocity $\mathbf{u}$ can be expressed as the gradient of a velocity potential φ, $\mathbf{u} = \boldsymbol{\nabla}\varphi$, which satisfies the Laplace equation. Let us assume for simplicity that the initial impulsive velocity is axially symmetric so that $v = v(r)$. The initial condition on φ is then

$$\frac{\partial}{\partial z} \varphi(r, z = 0, t = 0) \;=\; v(r). \qquad (6.8.1)$$

The pressure field is given by Bernoulli's equation (1.4.4) p. 13 which, identifying the force per unit mass $\mathbf{f}$ with the gravitational acceleration $\mathbf{g} = -g\mathbf{k}$ directed along the negative z-axis, we write as

$$\frac{p}{\rho} \;=\; -\frac{\partial\varphi}{\partial t} - gz. \qquad (6.8.2)$$

After the initial impulse, the pressure on the surface of the liquid must remain at its undisturbed reference level $p = 0.$[5] Therefore, if the instantaneous position of the surface is $z = \zeta(r, t)$, this equation implies that

$$\frac{\partial\varphi}{\partial t} \;=\; -g\zeta \qquad (6.8.3)$$

or, after taking a time derivative and noting that $\partial\zeta/\partial t$ is the vertical surface velocity and, therefore, it equals $\partial\varphi/\partial z,$[6]

$$\frac{\partial^2\varphi}{\partial t^2} + g\frac{\partial\varphi}{\partial z} \;=\; 0 \qquad \text{on} \qquad z = 0 \qquad (6.8.4)$$

---

[5] This condition has already been used to deduce that the function $F(t)$ appearing in the general form of the Bernoulli equation given in (1.4.4) vanishes assuming that, at large distance from the axis of symmetry, $\zeta = 0$ and $\varphi = 0$.

[6] This statement is valid only to first order in the disturbance. Also, the boundary condition (6.8.4) should be applied at $z = \zeta$ rather than at $z = 0$, but this refinement gives rise, through a Taylor series expansion, to a correction which is of second order in the disturbance; see §7.10 for a similar situation.

which has the form anticipated in (1.4.5) p. 13.

We assume that the fluid occupies the region $-\infty < z \le 0$ and, by proceeding as in the previous section, we find, in place of (6.7.5),

$$\tilde{\varphi}_0(k, z, t) = \tilde{\Phi}_0(k, t) e^{kz} \tag{6.8.5}$$

in which

$$\tilde{\varphi}_0(k, z, t) = \int_0^\infty r J_0(kr) \varphi(r, z, t) \, dr \tag{6.8.6}$$

is the zero-order Hankel transform of φ. The boundary condition (6.8.4) becomes

$$\frac{\partial^2 \tilde{\varphi}_0}{\partial t^2} + kg\tilde{\varphi}_0 = 0 \quad \text{on} \quad z = 0. \tag{6.8.7}$$

In order to proceed with the solution of the initial-value problem we take the Laplace transform of this relation noting that, at $t = 0$, $\partial\varphi/\partial t = 0$ from (6.8.3) as the surface is initially flat, while, from (6.8.1) and (6.8.5), $\hat{\Phi}_0(k, 0) = \tilde{v}_0(k)/k$; the result is

$$(s^2 + gk)\hat{\Phi}_0(k, s) - \frac{s}{k} \tilde{v}_0(k) = 0 \tag{6.8.8}$$

from which

$$\tilde{\Phi}_0(k, t) = \mathscr{L}^{-1}\left\{\frac{s}{s^2 + gk}\right\} \frac{\tilde{v}_0(k)}{k} = \frac{\tilde{v}_0(k)}{k} \cos \omega t, \quad \omega = \sqrt{gk}. \tag{6.8.9}$$

Upon inverting the Hankel transform, the velocity potential is found as

$$\varphi = \int_0^\infty J_0(kr)\tilde{v}_0(k) e^{kz} \cos \omega t \, dk. \tag{6.8.10}$$

The surface elevation follows from (6.8.3) as

$$\zeta = \frac{1}{\sqrt{g}} \int_0^\infty \sqrt{k} \, J_0(kr)\tilde{v}_0(k) \sin \omega t \, dk. \tag{6.8.11}$$

Rather than showing some special case in which the transform can be inverted, let us consider this expression for large values of r. We use the asymptotic form of J_0 for large argument from table 6.5 p. 151 and find[7]

$$\zeta(r, t) \simeq \sqrt{\frac{2}{\pi gr}} \int_0^\infty \cos\left(kr - \frac{\pi}{4}\right) \tilde{v}_0(k) \sin \omega t \, dk$$

$$\simeq \frac{1}{\sqrt{2\pi gr}} \text{Im} \, e^{-i\pi/4} \int_0^\infty \left[e^{i(\omega t - kr)} + e^{i(\omega t + kr)}\right] \tilde{v}_0(k) \, dk. \tag{6.8.12}$$

This integral is the sum of two terms each one of which has the form (10.7.2) p. 277 and lends itself therefore to an approximate evaluation by the method of stationary phase (p. 277). We set

$$\chi(k) = \sqrt{gk}\, t \mp kr. \tag{6.8.13}$$

---

[7] Strictly speaking, no matter how large r, there is a small interval around $k = 0$ where this approximation is invalid. The contribution of this small interval to the integral vanishes as $r \to \infty$.

The point(s) k_s of stationary phase are defined by the condition $\chi'(k_s) = 0$ and therefore $\sqrt{k_s} = \pm t\sqrt{g}/(2r)$. We see that the second term in the bracket does not have a point of stationary phase in the integration range and its contribution will therefore be subdominant. We then apply (10.7.9) p. 278 to the first term only (lower sign in (6.8.13)), for which we have $\chi(k_s) = gt^2/(4r^2)$ and $\chi''(k_s) = -2r^3/(gt^2)$; the result is

$$\zeta(r,t) \simeq \frac{t}{\sqrt{2}\,r^2} \sin\left(\frac{gt^2}{4r^2}\right) \tilde{v}_0\left(\frac{gt^2}{4r^2}\right). \tag{6.8.14}$$

6.9 Dual integral equations

In the application of the Hankel transform sometimes the problem is reduced to the solution of dual integral equations, i.e., a pair of simultaneous equations of the form

$$\int_0^\infty \xi^{-2\alpha} J_\nu(\xi x) B(\xi)\, d\xi = f(x) \qquad 0 < x < 1 \tag{6.9.1}$$

$$\int_0^\infty J_\mu(\xi x) B(\xi)\, d\xi = g(x) \qquad 1 < x. \tag{6.9.2}$$

By (6.2.13), these equations include as special cases the trigonometric kernels $\sin \xi x$ and and $\cos \xi x$ when $\nu = \pm 1/2$.

The solution of the problem takes different forms depending on the values of the parameters (see e.g. Sneddon 1951, p. 65 and 1966, chapter 4; Tranter 1966, chapter 8). For $\mu = \nu > -1$ the structure of the solution is

$$B(x) = 2^{-\alpha} x^{1+\alpha}\left[\int_0^1 \xi^{1+\alpha} J_{\nu-\alpha}(\xi x) h_1(\xi)\, d\xi + \int_1^\infty \xi^{1+\alpha} J_{\nu-\alpha}(\xi x) h_2(\xi)\, d\xi\right]. \tag{6.9.3}$$

The form of the two auxiliary functions $h_{1,2}(\xi)$ depends on α. For $0 < \alpha < 1$ one has

$$h_1(\xi) = \frac{2^{2\alpha}}{\Gamma(1-\alpha)\xi^{\nu+1}} \frac{d}{d\xi}\left[\xi^{2-2\alpha+\nu} \int_0^1 (1-\eta^2)^{-\alpha} \eta^{\nu+1} f(\xi\eta)\, d\eta\right], \tag{6.9.4}$$

$$h_2(\xi) = \frac{2}{\Gamma(\alpha)} \int_1^\infty (\eta^2-1)^{\alpha-1} \eta^{1-\nu} g(\xi\eta)\, d\eta, \tag{6.9.5}$$

while, for $-1 < \alpha < 0$,

$$h_1(\xi) = \frac{2^{1+2\alpha}}{\Gamma(-\alpha)\xi^{2\alpha}} \int_0^1 (1-\eta^2)^{-\alpha-1} \eta^{\nu+1} f(\xi\eta)\, d\eta, \tag{6.9.6}$$

$$h_2(\xi) = -\frac{\xi^{\nu-2\alpha-1}}{\Gamma(1+\alpha)} \frac{d}{d\xi}\left[\xi^{2+2\alpha-\nu} \int_1^\infty (\eta^2-1)^\alpha \eta^{1-\nu} g(\xi\eta)\, d\eta\right]. \tag{6.9.7}$$

As an example, let us consider the problem

$$\frac{1}{r}\frac{\partial}{\partial r}\left(r\frac{\partial u}{\partial r}\right) + \frac{\partial^2 u}{\partial z^2} = 0 \tag{6.9.8}$$

in the half-space $0 < r, z < \infty$ subject to

$$u(r, z = 0) = u_0(r) \quad 0 < r < R, \qquad \left.\frac{\partial u}{\partial z}\right|_{z=0} = v_0(r) \quad R < r \tag{6.9.9}$$

with u bounded at infinity. After taking the Hankel transform, as in the previous section the solution for $\tilde{u}$ must have the form $\tilde{u}(k, z) = \tilde{A}(k)e^{-kz}$ so that

$$u(r, z) = \int_0^\infty k J_0(kr) \, \tilde{A}(k) e^{-kz} \, dk \tag{6.9.10}$$

or, with $r = Rs, k = \xi/R$,

$$u(r, z) = \frac{1}{R^2} \int_0^\infty \xi J_0(\xi s) \, \tilde{A}(\xi/R) e^{-\xi z/R} \, d\xi. \tag{6.9.11}$$

The function $\tilde{A}$ must be such that

$$\frac{1}{R^2} \int_0^\infty \xi J_0(\xi s) \, \tilde{A}(\xi/R) \, d\xi = u_0(Rs), \qquad 0 < s < 1, \tag{6.9.12}$$

$$\frac{1}{R^3} \int_0^\infty \xi^2 J_0(\xi s) \, \tilde{A}(\xi/R) \, d\xi = v_0(Rs), \qquad 1 < s. \tag{6.9.13}$$

We can apply the formula (6.9.3) to the function $B = (\xi/R)^2 \tilde{A}$ with $\nu = 0, \alpha = 1/2$ with the result

$$\tilde{A}(x/R) = \frac{R^2}{x\sqrt{\pi}} \left[\int_0^1 \xi \cos(\xi x) \, h_1(\xi) \, d\xi + \int_1^\infty \xi \cos(\xi x) \, h_2(\xi) \, d\xi \right], \tag{6.9.14}$$

where we have used the relation (6.2.13) $J_{-1/2}(z) = (2/\pi z)^{1/2} \cos z$, and in which

$$h_1(\xi) = \frac{2}{\sqrt{\pi}\xi} \frac{d}{d\xi} \left[\xi \int_0^1 (1 - \eta^2)^{-1/2} \eta \, u_0(R\xi\eta) \, d\eta \right], \tag{6.9.15}$$

$$h_2(\xi) = -\frac{2}{\sqrt{\pi}} \int_1^\infty (\eta^2 - 1)^{-1/2} \eta \, v_0(R\xi\eta) \, d\eta. \tag{6.9.16}$$

In particular, with $u_0 = \text{const.}$, $v_0 = 0$ we find

$$\tilde{A}(x/R) = \frac{2u_0}{\pi} \frac{R^2}{x^2} \sin x \quad \text{or} \quad \tilde{A}(k) = \frac{2u_0}{\pi} \frac{\sin Rk}{k^2} \tag{6.9.17}$$

so that

$$u(r, z) = \frac{2u_0}{\pi} \int_0^\infty J_0(kr) \frac{\sin kr}{k} e^{-kz} \, dk = \sin^{-1} \frac{2r}{\sqrt{z^2 + 4r^2 + z}}. \tag{6.9.18}$$

In many ways the sphere is the prototypical three-dimensional body and the consideration of fields in the presence of spherical boundaries sheds light on several features of more general three-dimensional cases.

In all the examples of this chapter extensive use is made of expansions in series of Legendre polynomials, for axi-symmetric problems, or spherical harmonics, for the general three-dimensional case. After a review of the polar coordinate system, we begin with a summary of the properies of these functions which are dealt with in greater detail in Chapters 13 and 14, respectively. While the axi-symmetric situation is somewhat simpler, it is also contained as a special case in the general three-dimensional one and it is therefore expedient to treat it as a special case of the latter.

We start with the general solution of the Laplace and Poisson equations (§7.3) and apply it to several axisymmetric (§§7.4 and 7.5) and non-axisymmetric situations. In all these cases the radial part of the solution consists of powers of r. The examples in the second part of the chapter (§7.13 and §7.14) deal with the scalar Helmholtz equation, for which the radial dependence is expressed in terms of spherical Bessel functions, the fundamental properties of which are summarized in §7.12. The last four sections deal with problems involving vector fields and vector harmonics.

Expansions in terms of Legendre functions are also useful for some problems with conical boundaries, some examples of which are considered in §7.11.

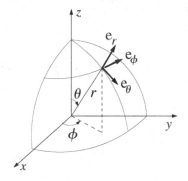

Fig. 7.1   Spherical polar coordinates and unit vectors in the coordinate directions at a generic point.

Table 7.1

Relations between Cartesian and polar coordinates; $\mathbf{i}$, $\mathbf{j}$ and $\mathbf{k}$ are the unit vectors in the three Cartesian directions; $\mathbf{e}_r$, $\mathbf{e}_\theta$ and $\mathbf{e}_\phi$ are the unit vectors in the spherical coordinates directions (see Figure 7.1).

Cartesian	Polar
$\mathbf{x} = (x, y, z)$	$\mathbf{x} = (r \sin\theta \cos\phi, r \sin\theta \sin\phi, r \cos\theta)$
$\sqrt{x^2 + y^2 + z^2}$	r
$(x, y)/\sqrt{x^2 + y^2}$	$(\cos\phi, \sin\phi)$
z/r	$\cos\theta$
$ds^2 = dx^2 + dy^2 + dz^2$	$ds^2 = dr^2 + r^2 d\theta^2 + r^2 \sin^2\theta \, d\phi^2$
area element dS	$r^2 \sin\theta \, d\theta \, d\phi$
volume element dV	$r^2 \sin\theta \, dr \, d\theta \, d\phi$

$\mathbf{i} = \sin\theta \cos\phi \, \mathbf{e}_r + \cos\theta \cos\phi \, \mathbf{e}_\theta$ $- \sin\phi \, \mathbf{e}_\phi$	$\mathbf{e}_r = \sin\theta \cos\phi \, \mathbf{i} + \sin\theta \sin\phi \, \mathbf{j}$ $+ \cos\theta \, \mathbf{k}$
$\mathbf{j} = \sin\theta \sin\phi \, \mathbf{e}_r + \cos\theta \sin\phi \, \mathbf{e}_\theta$ $+ \cos\phi \, \mathbf{e}_\phi$	$\mathbf{e}_\theta = \cos\theta \cos\phi \, \mathbf{i} + \cos\theta \sin\phi \, \mathbf{j}$ $- \sin\theta \, \mathbf{k}$
$\mathbf{k} = \cos\theta \, \mathbf{e}_r - \sin\theta \, \mathbf{e}_\theta$	$\mathbf{e}_\phi = -\sin\phi \, \mathbf{i} + \cos\phi \, \mathbf{j}$

7.1 Spherical polar coordinates

The analysis of all the examples of this chapter is most efficiently carried out in terms of spherical coordinates (r, θ, ϕ) (Figure 7.1) connected to Cartesian coordinates (x, y, z) by the relations

$$x = r \sin\theta \cos\phi, \quad y = r \sin\theta \sin\phi, \quad z = r \cos\theta. \qquad (7.1.1)$$

| Table 7.2 | Derivatives of the Cartesian with respect to the polar coordinates and vice versa. | | |

	x	y	z
$\dfrac{\partial}{\partial r}$	$\sin\theta\,\cos\phi = \dfrac{x}{r}$	$\sin\theta\,\sin\phi = \dfrac{y}{r}$	$\cos\theta = \dfrac{z}{r}$
$\dfrac{\partial}{\partial\theta}$	$r\cos\theta\,\cos\phi$ $= \dfrac{xz}{\sqrt{x^2+y^2}}$	$r\cos\theta\,\sin\phi$ $= \dfrac{yz}{\sqrt{x^2+y^2}}$	$-r\sin\theta$ $= -\sqrt{x^2+y^2}$
$\dfrac{\partial}{\partial\phi}$	$-r\sin\theta\,\sin\phi = -y$	$r\sin\theta\,\cos\phi = x$	0

	r	θ	ϕ
$\dfrac{\partial}{\partial x}$	$\dfrac{x}{r} = \sin\theta\,\cos\phi$	$\dfrac{xz}{r^2\sqrt{x^2+y^2}}$ $= \dfrac{\cos\theta\,\cos\phi}{r}$	$-\dfrac{y}{x^2+y^2}$ $= -\dfrac{\sin\phi}{r\sin\theta}$
$\dfrac{\partial}{\partial y}$	$\dfrac{y}{r} = \sin\theta\,\sin\phi$	$\dfrac{yz}{r^2\sqrt{x^2+y^2}}$ $= \dfrac{\cos\theta\,\sin\phi}{r}$	$\dfrac{x}{x^2+y^2}$ $= \dfrac{\cos\phi}{r\sin\theta}$
$\dfrac{\partial}{\partial z}$	$\dfrac{z}{r} = \cos\theta$	$-\dfrac{\sqrt{x^2+y^2}}{r^2} = -\dfrac{\sin\theta}{r}$	0

The inverse relations are given in Table 7.1. The range of variation of these coordinates are

$$0 \le r < \infty, \qquad 0 \le \theta \le \pi, \qquad 0 \le \phi < 2\pi. \qquad (7.1.2)$$

The line $\theta = 0$ and π – here the z-axis – is termed the *polar axis*; the positive and negative z-semi-axes correspond to $\theta = 0$ and $\theta = \pi$. The positions $\phi = 0$ and $\phi = 2\pi$ are indistinguishable; ϕ is undefined on the polar axis, where $\cos\theta = \pm 1$, and neither angle is defined at the origin $r = 0$.

It is sometimes useful to convert derivatives from Cartesian to spherical coordinates or vice versa by application of the chain rule. Some formulae useful for this purpose are given in Table 7.2.

Table 7.3 shows the form taken by several differential operators in spherical coordinates. Here $\mathbf{e}_r$, $\mathbf{e}_\theta$ and $\mathbf{e}_\phi$ denote the unit vectors in the direction of increasing r, θ and ϕ, respectively.

In several of the examples that follow we deal with the δ-function. For the reasons mentioned on p. 38 and explained in greater detail in §20.7 p. 588, it is necessary to use

Table 7.3   Differential operators in spherical coordinates.

$$\nabla u = \frac{\partial u}{\partial r}\, \mathbf{e}_r + \frac{1}{r}\frac{\partial u}{\partial \theta}\, \mathbf{e}_\theta + \frac{1}{r \sin \theta}\frac{\partial u}{\partial \phi}\, \mathbf{e}_\phi$$

$$\nabla \cdot \mathbf{A} = \frac{1}{r^2}\frac{\partial}{\partial r}\left(r^2 A_r\right) + \frac{1}{r \sin \theta}\frac{\partial}{\partial \theta}\left(\sin \theta\, A_\theta\right) + \frac{1}{r \sin \theta}\frac{\partial A_\phi}{\partial \phi}$$

$$\nabla \times \mathbf{A} = \frac{1}{r \sin \theta}\left[\frac{\partial}{\partial \theta}\left(\sin \theta\, A_\phi\right) - \frac{\partial A_\theta}{\partial \phi}\right] \mathbf{e}_r$$

$$+ \frac{1}{r}\left[\frac{1}{\sin \theta}\frac{\partial A_r}{\partial \phi} - \frac{\partial}{\partial r}\left(r A_\phi\right)\right] \mathbf{e}_\theta + \frac{1}{r}\left[\frac{\partial}{\partial r}\left(r A_\theta\right) - \frac{\partial A_r}{\partial \theta}\right] \mathbf{e}_\phi$$

$$= \frac{1}{r^2 \sin \theta}\begin{vmatrix} \mathbf{e}_r & r\,\mathbf{e}_\theta & r \sin \theta\, \mathbf{e}_\phi \\ \partial/\partial r & \partial/\partial \theta & \partial/\partial \phi \\ A_r & r A_\theta & r \sin \theta\, A_\phi \end{vmatrix}$$

$$\nabla^2 u = \frac{1}{r^2}\frac{\partial}{\partial r}\left(r^2 \frac{\partial u}{\partial r}\right) + \frac{1}{r^2 \sin \theta}\frac{\partial}{\partial \theta}\left(\sin \theta\, \frac{\partial u}{\partial \theta}\right) + \frac{1}{r^2 \sin^2 \theta}\frac{\partial^2 u}{\partial \phi^2}$$

Table 7.4   The δ-function in spherical coordinates.

		$\delta^{(3)}(\mathbf{x} - \mathbf{x}')$
1	$\mathbf{x}' = (r', \theta', \phi')$	$\delta(r - r')\,\delta(\theta - \theta')\,\delta(\phi - \phi')/\sin \theta'$
2	or	$= \delta(r - r')\,\delta(\cos \theta - \cos \theta')\,\delta(\phi - \phi')$
3	$\mathbf{x}' = (r', \theta' = 0, -)^*$	$\delta(r - r')\,\delta(\cos \theta - 1)/(2\pi r'^2)$
4	$\mathbf{x}' = (r', \theta' = \pi, -)$	$\delta(r - r')\,\delta(\cos \theta + 1)/(2\pi r'^2)$
5	$\mathbf{x}' = (0, -, -)^{**}$	$\delta(r)/4\pi r^2$

* The angle ϕ' is indeterminate for points on the polar axis.
** Both angles θ' and ϕ' are indeterminate when $\mathbf{x}'$ is at the origin.

the proper representations in spherical coordinates which are given for various cases in Table 7.4.

7.2 Spherical harmonics: useful relations

It is shown in Chapter 14 that a function $v(\theta, \phi)$ defined over the entire solid angle, i.e. for $0 \leq \theta \leq \pi$, $0 \leq \phi < 2\pi$, can be represented as a superposition of spherical surface harmonics $Y_\ell^m(\theta, \phi)$:

$$v(\theta, \phi) = \sum_{\ell=0}^{\infty} \sum_{m=-\ell}^{\ell} V_{\ell m} \, Y_\ell^m(\theta, \phi), \tag{7.2.1}$$

in which[1]

$$v_{\ell m} = \left(Y_\ell^m, v\right) = \int_0^{2\pi} d\phi \int_0^{\pi} \sin\theta d\theta \, \overline{Y}_\ell^m(\theta, \phi) \, v(\theta, \phi), \tag{7.2.2}$$

with the overline denoting the complex conjugate. Since the area element on a sphere of radius a is $a^2 \sin\theta \; d\theta \; d\phi$ (Table 7.1), the integral may be seen as the area integral over the surface of the unit sphere. When v depends on other variables, such as r or t, this dependence is transferred to the coefficients $V_{\ell m}$ of the expansion.

Some properties of these functions are summarized in Table 7.5. Writing (7.2.1) with the summation over m extending between $-\ell$ and ℓ implies the use of spherical harmonics expressed as the product of a Legendre function P_ℓ^m and a complex Fourier basis function, i.e.

$$Y_\ell^m(\theta, \phi) = N_{\ell m} P_\ell^m(\cos\theta) e^{im\phi}, \qquad N_{\ell m} = \sqrt{\frac{2\ell+1}{4\pi} \frac{(\ell-m)!}{(\ell+m)!}}, \tag{7.2.3}$$

where $N_{\ell m}$ is a normalization constant. These functions satisfy the orthonormality relation

$$(Y_{\ell'}^{m'}, Y_\ell^m) \equiv \int d\Omega \, \overline{Y}_{\ell'}^{m'} Y_\ell^m \equiv \int_0^{\pi} \sin\theta \, d\theta \int_0^{2\pi} d\phi \, \overline{Y}_{\ell'}^{m'} Y_\ell^m = \delta_{\ell\ell'}\delta_{mm'} \tag{7.2.4}$$

in which we indicate synthetically with $\int d\Omega \equiv \int_0^{\pi} \sin\theta \, d\theta \int_0^{2\pi} d\phi$ the integral over the solid angle.

Explicit expressions for the first few spherical harmonics with $m \geq 0$ are

$$\ell = 0: \qquad Y_0^0 = \frac{1}{\sqrt{4\pi}}; \tag{7.2.5}$$

$$\ell = 1: \qquad Y_1^1 = -\sqrt{\frac{3}{8\pi}} \sin\theta \, \exp i\phi, \qquad Y_1^0 = \sqrt{\frac{3}{4\pi}} \cos\theta; \tag{7.2.6}$$

$$\ell = 2: \qquad \begin{cases} Y_2^2 = \frac{1}{4}\sqrt{\frac{15}{2\pi}} \sin^2\theta \, \exp 2i\phi, \\[2mm] Y_2^1 = -\sqrt{\frac{15}{8\pi}} \sin\theta \, \cos\theta \, \exp i\phi, \\[2mm] Y_2^0 = \sqrt{\frac{5}{4\pi}} \left(\frac{3}{2} \cos^2\theta - \frac{1}{2}\right). \end{cases} \tag{7.2.7}$$

The corresponding formulae for negative m follow from the property

$$Y_\ell^{-m} = (-1)^m \, \overline{Y}_\ell^m. \tag{7.2.8}$$

Just as in the case of the Fourier series, it is also possible to use a purely real form writing

$$v(\theta, \phi) = \sum_{\ell=0}^{\infty} \left[\frac{1}{2} \hat{V}_\ell + \sum_{m=1}^{\ell} \left(\hat{V}_{\ell m} \hat{Y}_\ell^m(\theta, \phi) + \tilde{V}_{\ell m} \tilde{Y}_\ell^m(\theta, \phi) \right) \right] \tag{7.2.9}$$

---

[1] In principle the integrals appearing here and in the following are over a two-dimensional domain and should be written as $\int\int d\phi \, d\theta$. Writing them as in (7.2.2) implies their evaluation as iterated integrals (§A4.3).

Table 7.5	Some properties of the spherical harmonics.

1	$$Y_\ell^m(\theta, \phi) = N_{\ell m} P_\ell^m(\cos\theta)\, e^{im\phi}$$	definition
2	$$\left(Y_{\ell'}^{m'}, Y_\ell^m\right) = \int d\Omega\, \overline{Y}_{\ell'}^{m'}\, Y_\ell^m = \delta_{\ell\ell'}\delta_{mm'}$$	orthogonality
3	$$Y_\ell^{-m} = (-1)^m\, \overline{Y}_\ell^m$$	parity
4	$$\sum_{\ell=0}^{\infty} \sum_{m=-\ell}^{\ell} \overline{Y}_\ell^m(\theta', \phi')\, Y_\ell^m(\theta, \phi)$$ $$= \delta(\cos\theta - \cos\theta')\,\delta(\phi - \phi')$$	completeness
5	$$P_\ell(\cos\gamma) = \frac{4\pi}{2\ell+1} \sum_{m=-\ell}^{\ell} \overline{Y}_\ell^m(\theta', \phi')\, Y_\ell^m(\theta, \phi)$$ where: $\cos\gamma = \cos\theta\,\cos\theta' + \sin\theta\,\sin\theta'\,\cos(\phi' - \phi)$	addition theorem Fig. 7.2
6	$$L_\pm Y_\ell^m \equiv e^{\pm i\phi}\left(\pm\frac{\partial}{\partial\theta} + i\cot\theta\frac{\partial}{\partial\phi}\right) Y_\ell^m = \sqrt{(\ell\mp m)(\ell\pm m+1)}\, Y_\ell^{m\pm 1}$$	
7	$$L_z Y_\ell^m \equiv -i\frac{\partial}{\partial\phi} Y_\ell^m = m\, Y_\ell^m$$	
8	$$\mathscr{L}^2 Y_\ell^m \equiv \left[\frac{1}{2}(L_+ L_- + L_- L_+) + L_z^2\right] Y_\ell^m = \ell(\ell+1) Y_\ell^m$$	
9	$$\int (\nabla\overline{Y}_\ell^m)\cdot(\nabla Y_p^q)\, d\Omega = \ell(\ell+1)\delta_{\ell p}\delta_{mq}$$	

in which

$$\hat{Y}_\ell^m = M_{\ell m} P_\ell^m \cos\phi, \quad \tilde{Y}_\ell^m = M_{\ell m} P_\ell^m \sin\phi, \quad M_\ell^m = \sqrt{\frac{2\ell+1}{2\pi}\frac{(\ell-m)!}{(\ell+m)!}}. \quad (7.2.10)$$

In their real or complex form, the spherical harmonics are eigenfunctions of the angular part of the Laplacian operator

$$\mathscr{L}^2 Y_\ell^m \equiv -\frac{1}{\sin\theta}\frac{\partial}{\partial\theta}\left(\sin\theta\frac{\partial Y_\ell^m}{\partial\theta}\right) - \frac{1}{\sin^2\theta}\frac{\partial^2 Y_\ell^m}{\partial\phi^2} = \ell(\ell+1) Y_\ell^m \quad (7.2.11)$$

while

$$\frac{1}{r^2}\frac{\partial}{\partial r}\left(r^2\frac{\partial r^\ell}{\partial r}\right) = \ell(\ell+1) r^{\ell-2}. \quad (7.2.12)$$

Thus

$$\nabla^2\left(r^\ell Y_\ell^m\right) = \frac{1}{r^2}\left[\frac{\partial}{\partial r}\left(r^2\frac{\partial}{\partial r}\right) - \mathscr{L}^2\right]\left(r^\ell Y_\ell^m\right) = 0. \quad (7.2.13)$$

Table 7.6	Some properties of the solid harmonics of positive order; the corresponding relations for the solid harmonics of negative order $S^m_{-\ell-1} = r^{-\ell-1} Y^m_\ell$ are obtained by interchanging ℓ with $-\ell - 1$.

$$S^m_\ell = r^\ell Y^m_\ell$$

1	$\nabla^2 \left(S^m_\ell \mathbf{x} \right) = 2\nabla S^m_\ell$
2	$\nabla^2 \left(r^k S^m_\ell \right) = k(k + 2\ell + 1) r^{k-2} S^m_\ell$
3	$\mathbf{x} \cdot \nabla S^m_\ell = \ell \, S^m_\ell$
4	$r^2 \left[\nabla S^m_\ell - r^{2\ell+1} \nabla \left(r^{-2\ell-1} S^m_\ell \right) \right] = (2\ell + 1) \, \mathbf{x} \, S^m_\ell$

As mentioned on p. 341, the functions $r^\ell Y^m_\ell$ (or $r^\ell \hat{Y}^m_\ell$ and $r^\ell \tilde{Y}^m_\ell$) are *solid harmonics* of order ℓ and are just homogeneous polynomials in the Cartesian coordinates. The same relation (7.2.13) holds for the solid harmonics of negative order $r^{-\ell-1} Y^m_\ell$. Some properties of these functions are given in Table 7.6.

7.2.1 The axisymmetric case

Axisymmetric situations are of particular interest because they are simpler while, at the same time, retaining much of the structure of the more general case. In the absence of ϕ-dependence the two representations (7.2.1) and (7.2.9) coalesce into a single expression which is conveniently written using unnormalized functions as

$$v(\theta) = \sum_{\ell=0}^{\infty} V_\ell P_\ell(\mu) \qquad \mu = \cos\theta. \tag{7.2.14}$$

Here the P_ℓ are Legendre polynomials (Chapter 13); some useful properties of these polynomials are summarized in Table 7.7 and others, together with some derivations, can be found in Chapter 13. The first few Legendre polynomials are

$$P_0 = 1, \qquad P_1 = \mu, \qquad P_2 = \frac{1}{2}\left(3\mu^2 - 1\right), \qquad P_3 = \frac{1}{2}\left(5\mu^3 - 3\mu\right), \tag{7.2.15}$$

and the corresponding *zonal* (or axisymmetric) solid harmonics are

$$r P_1 = z, \qquad r^2 P_2 = z^2 - \frac{1}{2}(x^2 + y^2), \qquad r^3 P_3 = \frac{1}{2}z(5z^2 - 3r^2). \tag{7.2.16}$$

The orthogonality property

$$(P_k, P_\ell) \equiv \int_0^\pi \sin\theta \, P_k(\cos\theta) \, P_\ell(\cos\theta) \, d\theta = \int_{-1}^{1} P_k(\mu) \, P_\ell(\mu) \, d\mu = \frac{2}{2\ell + 1} \delta_{k\ell} \tag{7.2.17}$$

Table 7.7	Some properties of the Legendre polynomials; here $\mu = \cos\theta$.	
1	$\displaystyle\int_{-1}^{1} P_k(\mu)\,P_\ell(\mu)\,d\mu = \frac{2}{2\ell+1}\delta_{k\ell}$	orthogonality
2	$\displaystyle P_\ell(\mu) = \frac{1}{2^\ell \ell!}\frac{d^\ell}{d\mu^\ell}(\mu^2-1)^\ell$	Rodrigues formula
3	$\displaystyle (1-2t\mu+t^2)^{-1/2} = \sum_{0}^{\infty} t^\ell P_\ell(\mu)\ ,\ |t|<1$	generating function
4	$(\ell+1)P_{\ell+1}(\mu) - (2\ell+1)\mu P_\ell(\mu) + \ell P_{\ell-1}(\mu) = 0$	recurrence relation
5	$\displaystyle (\mu^2-1)\frac{dP_\ell}{d\mu} = \frac{\ell(\ell+1)}{2\ell+1}(P_{\ell+1}-P_{\ell-1})$	recurrence relation
6	$P_\ell(\mu) = P'_{\ell+1}(\mu) - 2\mu P'_\ell(\mu) + P'_{\ell-1}(\mu)$	recurrence relation
7	$(2\ell+1)P_\ell(\mu) = P'_{\ell+1}(\mu) - P'_{\ell-1}(\mu)$	recurrence relation
8	$\displaystyle P_{2\ell}(0) = (-1)^\ell\frac{(2\ell)!}{2^{2\ell}(\ell!)^2},\qquad P_{2\ell+1}(0)=0$	value at 0

coupled with the recurrence relations of Table 7.7 is useful, among others, to calculate scalar products; for example

$$
\begin{aligned}
\int_{-1}^{1} \mu\, P_\ell\, P_k\, d\mu &= \frac{1}{2\ell+1}\int_{-1}^{1}[(\ell+1)P_{\ell+1}+\ell P_{\ell-1}]\,P_k\,d\mu \\
&= \frac{2}{2\ell+1}\left[\frac{\ell+1}{2\ell+3}\delta_{k,\ell+1} + \frac{\ell}{2\ell-1}\delta_{k,\ell-1}\right] \\
&= \frac{2}{2k+1}\left[\frac{k}{2k-1}\delta_{\ell,k-1} + \frac{k+1}{2k+3}\delta_{\ell,k+1}\right].
\end{aligned}
\tag{7.2.18}
$$

The Legendre polynomials are eigenfunctions of the axisymmetric Laplacian:

$$
-\frac{1}{\sin\theta}\frac{d}{d\theta}\left(\sin\theta\frac{dP_\ell}{d\theta}\right) = -\frac{d}{d\mu}\left[(1-\mu^2)\frac{dP_\ell}{d\mu}\right] = \ell(\ell+1)P_\ell.
\tag{7.2.19}
$$

A useful deduction from the generating function of the Legendre polynomials (entry 3 in Table 7.7) is an expression for $|\mathbf{x}-\mathbf{x}'|^{-1} = (r^2 - 2rr'\cos\gamma + r'^2)^{-1/2}$, the inverse of the distance between two points $\mathbf{x} = (r,\theta,\phi)$ and $\mathbf{x}' = (r',\theta',\phi')$. Let $r_> = \max(r,r')$, $r_< = \min(r,r')$; then

$$
\frac{1}{|\mathbf{x}-\mathbf{x}'|} = \frac{1}{r_>}\frac{1}{\sqrt{1-2(r_</r_>)\cos\gamma + (r_</r_>)^2}} = \frac{1}{r_>}\sum_{\ell=0}^{\infty}\left(\frac{r_<}{r_>}\right)^\ell P_\ell(\cos\gamma)
\tag{7.2.20}
$$

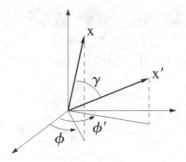

Fig. 7.2 The angle γ between position vectors $\mathbf{x}$ and $\mathbf{x}'$.

where γ is the angle between the two position vectors (figure 7.2) given by[2]

$$\cos \gamma = \frac{\mathbf{x} \cdot \mathbf{x}'}{|\mathbf{x}| \, |\mathbf{x}'|} = \cos \theta \, \cos \theta' + \sin \theta \, \sin \theta' \, \cos(\phi' - \phi). \tag{7.2.21}$$

7.3 General solution of the Laplace and Poisson equations

The Laplace operator in spherical polar coordinates is given in Table 7.3 and is

$$\nabla^2 u = \frac{1}{r^2} \frac{\partial}{\partial r} \left(r^2 \frac{\partial u}{\partial r} \right) + \frac{1}{r^2 \sin \theta} \frac{\partial}{\partial \theta} \left(\sin \theta \frac{\partial u}{\partial \theta} \right) + \frac{1}{r^2 \sin^2 \theta} \frac{\partial^2 u}{\partial \phi^2}. \tag{7.3.1}$$

By expanding u as in (7.2.1) or (7.2.14) we write[3]

$$u(r, \theta, \phi) = \sum_{\ell=0}^{\infty} \sum_{m=-\ell}^{\ell} R_{\ell m}(r) \begin{Bmatrix} Y_\ell^m \\ P_\ell, \end{Bmatrix} \tag{7.3.2}$$

so that, by (7.2.11)

$$\nabla^2 u = \sum_{\ell=0}^{\infty} \sum_{m=-\ell}^{\ell} \left[\frac{1}{r^2} \frac{d}{dr} \left(r^2 \frac{d R_{\ell m}}{dr} \right) - \frac{\ell(\ell+1)}{r^2} R_{\ell m}(r) \right] \begin{Bmatrix} Y_\ell^m \\ P_\ell \end{Bmatrix}. \tag{7.3.3}$$

On the basis of the remarks following (7.2.13) we can write the general solution of the Laplace equation $\nabla^2 u = 0$ as

$$u = \sum_{\ell=0}^{\infty} \sum_{m=-\ell}^{\ell} \left(A_{\ell m} r^\ell + B_{\ell m} r^{-\ell-1} \right) \begin{Bmatrix} Y_\ell^m \\ P_\ell \end{Bmatrix}, \tag{7.3.4}$$

---

[2] The most straightforward way to derive this result is to calculate the scalar product $\mathbf{x} \cdot \mathbf{x}'$ in Cartesian coordinates after expressing the Cartesian components of $\mathbf{x}$ and $\mathbf{x}'$ in terms of polar coordinates as in (7.1.1).

[3] For brevity we treat together the general (upper) and the axisymmetric cases. For the latter, all the summations over m should be ignored and $R_{\ell m}(r)$ understood as $R_l(r)$.

and we can base the solution of the Poisson equation

$$\nabla^2 u = -4\pi f(r, \theta, \phi) \tag{7.3.5}$$

on this homogeneous solution by the method of variation of parameters. Specifically, upon expanding u as in (7.3.2) and using (7.3.3), we have

$$\sum_{n=0}^{\infty} \sum_{k=-n}^{n} \frac{1}{r^2} \left[\frac{d}{dr} \left(r^2 \frac{dR_{\ell m}}{dr} \right) - n(n+1)R_{nk} \right] Y_n^k(\theta, \phi) = -4\pi f(r, \theta, \phi). \tag{7.3.6}$$

We take the scalar product of this equation by Y_ℓ^m (i.e., we multiply both sides by $\overline{Y}_\ell^m$ and integrate over the solid angle) and use the orthonormality relation (7.2.4) to find

$$\frac{1}{r^2} \frac{d}{dr} \left(r^2 \frac{dR_{\ell m}}{dr} \right) - \frac{\ell(\ell+1)}{r^2} R_{\ell m} = -4\pi \int d\Omega \, \overline{Y}_\ell^m \, f(r, \theta, \phi) \equiv -4\pi f_{\ell m}(r). \tag{7.3.7}$$

In the axisymmetric case the same procedure leads to

$$\frac{2}{2\ell+1} \left[\frac{1}{r^2} \frac{d}{dr} \left(r^2 \frac{dR_\ell}{dr} \right) - \frac{\ell(\ell+1)}{r^2} R_\ell \right] = -4\pi \int_0^\pi P_\ell(\cos\theta) \, f(r, \theta) \, \sin\theta \, d\theta \tag{7.3.8}$$

which can be reduced to the same form (7.3.7) by defining

$$f_\ell(r) = \left(\ell + \frac{1}{2} \right) (P_\ell, f) = \left(\ell + \frac{1}{2} \right) \int_0^\pi P_\ell(\cos\theta) \, f(r, \theta) \, \sin\theta \, d\theta. \tag{7.3.9}$$

The solution of (7.3.7) was given in Example 2.2.1 p. 32 as an application of the method of variation of parameters and is

$$R_{\ell m} = \left[A_{\ell m} - \frac{4\pi}{2\ell+1} \int^r f_{\ell m}(s) \, s^{-\ell+1} \, ds \right] r^\ell$$
$$+ \left[B_{\ell m} + \frac{4\pi}{2\ell+1} \int^r f_{\ell m}(s) \, s^{\ell+2} \, ds \right] r^{-\ell-1} \tag{7.3.10}$$

where the integration constants and lower limits of integration are to be chosen suitably depending on the specific problem to be solved.[4]

With these steps we have found the general solution of the Poisson equation in spherical coordinates. Much as in the case of linear ordinary differential equations, once this general solution has been found, all that needs to be done is to fit the integration constants to the problem at hand.

7.4 A sphere in a uniform field

As our first specific application let us consider an axisymmetric situation. We are interested in a vector field $\mathbf{U}$ which, in the absence of a sphere of radius a, has a spatially uniform

---

[4] As written, (7.3.10) really contains four integration constants rather than two, namely $A_{\ell m}$, $B_{\ell m}$ and the two unspecified integration limits. As already noted, the flexibility introduced by this redundancy proves convenient in some cases.

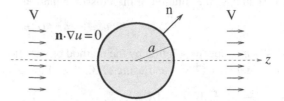

A sphere immersed in a uniform field, §7.4.

value $\mathbf{V}$ (Figure 7.3). We assume that the field is irrotational so that it admits a potential u, $\mathbf{U} = \nabla u$. We also assume that $\mathbf{U}$ is divergenceless so that $\nabla \cdot \mathbf{U} = \nabla^2 u = 0$. This would be the proper formulation e.g. for a steady heat conduction problem with a sphere immersed in a uniform heat flux, for the irrotational flow past a sphere, or for an electrostatic problem with a uniform field far from the sphere, in all cases in the absence of sources at finite distance from the sphere.

For the *Neumann problem* the normal derivative of the unknown function is taken to vanish on the boundary so that $\mathbf{n} \cdot \mathbf{U} = 0$. In heat conduction this would be the situation when the spherical boundary is non-conducting while, in electrostatics, $\mathbf{U}$ would be the electric field in the presence of a sphere with a very high permittivity; another situation described by this mathematical model is the potential flow past an impenetrable sphere.

Notice that the formulation of the problem only involves derivatives of u and, therefore, it will only be possible to determine the solution u up to an arbitrary constant devoid of physical meaning. This lack of uniqueness is a well-known feature of Neumann problems.

The formulation of the problem suggests to take the center of the sphere as the origin and the polar axis (i.e., the z-axis) in the direction of the constant vector $\mathbf{V}$ as, with these choices, the boundary conditions become

$$\left.\frac{\partial u}{\partial r}\right|_{r=a} = 0, \qquad \left(\frac{\partial u}{\partial r}, \frac{1}{r}\frac{\partial u}{\partial \theta}\right) \to V(\cos\theta, -\sin\theta) \quad \text{for} \quad r \to \infty \qquad (7.4.1)$$

with $V = |\mathbf{V}|$, so that the problem is axisymmetric.

The solution strategy consists in taking the general solution (7.3.4) of the Laplace equation and fitting the free constants A_ℓ and B_ℓ to satisfy the boundary conditions. To begin with, from the second of (7.4.1) we see that, of the constants A_ℓ, only A_0 and A_1 can be non-zero as all the other ones would introduce diverging r-dependent terms. Recalling that $P_1(\cos\theta) = \cos\theta$, for the latter we clearly have $A_1 = V$; A_0 is the arbitrary constant mentioned before and it can be taken equal to zero for convenience. With these steps, the general solution (7.3.4) is reduced to

$$u(r,\theta) = Vr P_1 + \sum_{\ell=0}^{\infty} B_\ell r^{-\ell-1} P_\ell(\cos\theta). \qquad (7.4.2)$$

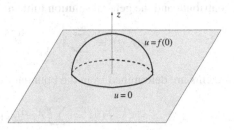

Fig. 7.4   A half-sphere on a plane, §7.5.

To determine the remaining constants we use the boundary condition on the sphere surface:

$$\left.\frac{\partial u}{\partial r}\right|_{r=a} = V P_1 - \sum_0^\infty (\ell+1) B_\ell\, a^{-\ell-2} P_\ell = 0. \tag{7.4.3}$$

By taking scalar products with the P_ℓ and using the orthogonality relation (7.2.17), we find $B_\ell = 0$ whenever $\ell \neq 1$ while, for $\ell = 1$,

$$V - 2B_1 a^{-3} = 0, \tag{7.4.4}$$

which determines B_1. We have therefore found the solution to be

$$u = Vr\, P_1(\cos\theta) + \frac{a^3 V}{2r^2} P_1(\cos\theta) = V\left(r + \frac{a^3}{2r^2}\right)\cos\theta. \tag{7.4.5}$$

This result may be written in a frame-independent form by noting that (Table 7.1) $V\cos\theta = Vz/r = \mathbf{V}\cdot\mathbf{x}/|\mathbf{x}|$ so that

$$u = \mathbf{V}\cdot\mathbf{x}\left(1 + \frac{a^3}{2|\mathbf{x}|^3}\right). \tag{7.4.6}$$

   The Dirichlet problem is treated in §7.8 for a case in which $u \to 0$ at infinity. Due to the linearity of the problem, the solution for the mixed problem with a Dirichlet condition on the sphere and a Neumann condition $\nabla u \to \mathbf{V}$ at infinity can be constructed by a suitable combination of the solution (7.8.8) of §7.8 and of the present solution (7.4.6).

7.5 Half-sphere on a plane

Let us now consider a Dirichlet problem in which the boundary consists of a half-sphere and the plane on which it rests (Figure 7.4). We look for a solution of the interior problem $0 < r < a$ satisfying $u(a,\theta) = f(\theta)$ for $0 \le \theta \le \frac{1}{2}\pi$ and $u(r, \theta = \pi/2) = 0$ for $0 \le r \le a$; for consistency, we assume that $f(\pi/2) = 0$.
   The problem is easily solved by extending it in an odd way to the entire sphere so that $u(r, \pi - \theta) = -u(r, \theta)$. Since the Legendre polynomials of even degree are even, they

will not contribute and the general solution finite at $r = 0$ will then have the form

$$u(r, \theta) = \sum_{\ell=0}^{\infty} A_\ell \left(\frac{r}{a}\right)^{2\ell+1} P_{2\ell+1}(\cos \theta). \tag{7.5.1}$$

The coefficients are determined from the boundary condition

$$f(\theta) = \sum_{\ell=0}^{\infty} A_\ell P_{2\ell+1}(\cos \theta) \tag{7.5.2}$$

from which we deduce

$$\begin{aligned}
A_\ell &= \left[\int_0^{\pi/2} P_{2\ell+1}^2(\cos \theta) \sin \theta \, d\theta\right]^{-1} \int_0^{\pi/2} P_{2\ell+1}(\cos \theta) f(\theta) \sin \theta \, d\theta \\
&= (4\ell + 3) \int_0^{\pi/2} P_{2\ell+1}(\cos \theta) f(\theta) \sin \theta \, d\theta.
\end{aligned} \tag{7.5.3}$$

the normalization constant is $1/2$ of the normal one as the integral is over the half-domain $0 \leq \theta \leq \frac{1}{2}\pi$.

This problem may be considered as a particular case of the conical boundary of §7.11 for the special value $\frac{1}{2}\pi$ of the cone semi-aperture.

7.6 The Poisson equation in free space

It is well known, e.g. from electrostatics, that the potential generated by a point charge q placed at a position $\mathbf{x}'$ in free space is the monopole potential

$$u(\mathbf{x}) = \frac{q}{|\mathbf{x} - \mathbf{x}'|} \tag{7.6.1}$$

which is a solution of

$$\nabla^2 u = -4\pi q \delta(\mathbf{x} - \mathbf{x}') \tag{7.6.2}$$

namely the fundamental solution of the Poisson equation (see p. 402). Because of linearity, the potential due to N charges is the sum of N terms of this type and a continuous distribution of charge, with a density $f(\mathbf{x}') \, d^3x'$ in each volume element d^3x', will give rise to the potential (see (16.1.16) p. 402)

$$u(\mathbf{x}) = \int \frac{f(\mathbf{x}')}{|\mathbf{x} - \mathbf{x}'|} \, d^3x'. \tag{7.6.3}$$

This expression represents the solution of the Poisson equation (7.3.5), namely $\nabla^2 u = -4\pi f$. The same expression gives the gravitational potential due to a prescribed mass distribution. We now recover this solution from the relations of §7.3. Among others, this simple application illustrates the use of the addition theorem for spherical harmonics (entry 5 in Table 7.5 p. 175) and of the relation (7.2.20), consequence of the generating function for the Legendre polynomials.

If u is required to be bounded at the origin and at infinity, the constants in the general solution (7.3.10) must be adjusted so that the term multiplying r^ℓ goes to zero as $r \to \infty$

and that multiplying $r^{-\ell-1}$ goes to zero as $r \to 0$. This objective can be achieved either by taking $A_{\ell m}$ and $B_{\ell m}$ equal to 0 and adjusting the limits of integration, or by arbitrarily fixing these limits and imposing the required conditions which will then determine $A_{\ell m}$ and $B_{\ell m}$. With either procedure the result is

$$
\begin{aligned}
R_{\ell m} &= \frac{4\pi}{2\ell+1} \left[r^\ell \int_r^\infty f_{\ell m}(s)\, s^{-\ell+1}\, ds + r^{-\ell-1} \int_0^r f_{\ell m}(s)\, s^{\ell+2}\, ds \right] \\
&= \frac{4\pi}{2\ell+1} \int_0^\infty (r_>)^{-\ell-1} (r_<)^\ell\, s^2\, f_{\ell m}(s)\, ds,
\end{aligned}
\tag{7.6.4}
$$

in which we have set $r_> = \max(r,s)$, $r_< = \min(r,s)$ so as to be able to combine the two integrals.

If f is such that integration and summation can be interchanged, upon substituting this result into (7.3.2) and effecting the sum over m first we have

$$
u = \sum_{\ell=0}^\infty \frac{1}{2\ell+1} \int_0^\infty \frac{r_<^\ell}{r_>^{\ell+1}} \left[\sum_{m=-\ell}^\ell f_{\ell m}(s)\, Y_{\ell m} \right] s^2\, ds.
\tag{7.6.5}
$$

In detail, after substituting the definition (7.3.7) of $f_{\ell m}$, the quantity in the square brackets is

$$
\sum_{m=-\ell}^\ell f_{\ell m}(s)\, Y_{\ell m}(\theta,\phi) = \int d\Omega'\, f(s,\theta',\phi') \sum_{m=-\ell}^\ell \overline{Y}_{\ell m}(\theta'\phi')\, Y_{\ell m}(\theta,\phi)
\tag{7.6.6}
$$

where the integral is due to the scalar product defining $f_{\ell m}$ and Ω' is the solid angle in the variables (θ',ϕ'). The summation can be effected by using the addition theorem (rule 5 in Table 7.5) with the result

$$
\sum_{m=-\ell}^\ell f_{\ell m}(s)\, Y_{\ell m}(\theta,\phi) = \frac{2\ell+1}{4\pi} \int d\Omega'\, P_\ell(\cos\gamma)\, f(s,\theta',\phi')
\tag{7.6.7}
$$

where the angle γ is defined in (7.2.21) and illustrated in Figure 7.2; upon substitution into (7.6.5) we have

$$
u = \int_0^\infty ds\, s^2 \int d\Omega'\, \frac{1}{r_>} \left[\sum_{\ell=0}^\infty \left(\frac{r_<}{r_>} \right)^\ell P_\ell(\cos\gamma) \right] f(s,\theta',\phi').
\tag{7.6.8}
$$

The summation was evaluated in (7.2.20) as an application of the generating function for the Legendre polynomials where it was seen to equal $\left(r^2 - 2rs\cos\gamma + s^2 \right)^{-1/2} = |\mathbf{x} - \mathbf{x}'|^{-1}$. In this way, we recover (7.6.3).

7.7 General solution of the biharmonic equation

As noted on p. 7, a convenient way to solve the biharmonic equation

$$
\nabla^4 u = 0
\tag{7.7.1}
$$

is to write it as the system

$$\nabla^2 u = v, \qquad \nabla^2 v = 0. \tag{7.7.2}$$

The solution of the second equation is given by (7.3.4) and, if we let (cf. footnote 3 p. 178)

$$u(r, \theta, \phi) = \sum_{\ell=0}^{\infty} \sum_{m=-\ell}^{\ell} S_{\ell m}(r) \begin{Bmatrix} Y_\ell^m \\ P_\ell \end{Bmatrix} \tag{7.7.3}$$

the developments of the previous section lead us directly to

$$\frac{1}{r^2} \frac{d}{dr} \left(r^2 \frac{dS_{\ell m}}{dr} \right) - \frac{\ell(\ell+1)}{r^2} S_{\ell m} = A_{\ell m} r^\ell + B_{\ell m} r^{-\ell-1} \tag{7.7.4}$$

where the right-hand side is the coefficient of the spherical harmonic expansion of v.

The general solution of this equation can be found from (7.3.10) or directly by postulating $S_{\ell m} \propto r^\alpha$. After renaming the arbitrary constants $A_{\ell m}$ and $B_{\ell m}$, we find

$$u = \sum_{\ell=0}^{\infty} \sum_{m=-\ell}^{\ell} \left(A_{\ell m} r^{\ell+2} + B_{\ell m} r^{-\ell+1} + C_{\ell m} r^\ell + D_{\ell m} r^{-\ell-1} \right) Y_\ell^m \tag{7.7.5}$$

in which the last two terms are the homogeneous solution of the first one of (7.7.2).

7.8 Exterior Poisson formula for the Dirichlet problem

We saw in §3.6 Poisson's formula for a disk. As a further example of a non-axisymmetric problem we derive now the analogous relation for a sphere. In this case we will consider the exterior problem, for which the Laplace equation is solved subject to the Dirichlet boundary condition

$$u(a, \theta, \phi) = f(\theta, \phi) \quad \text{for} \quad r = a, \quad u \to 0 \quad \text{for} \quad r \to \infty. \tag{7.8.1}$$

For the Laplace equation the quantities $f_{\ell m}$ in (7.3.10) all vanish and, in view of the condition at infinity, we can only have negative powers of r so that all the coefficients $A_{\ell m}$ also vanish; thus

$$u = \sum_{\ell=0}^{\infty} \sum_{m=-\ell}^{\ell} B_{\ell m} r^{-\ell-1} Y_\ell^m. \tag{7.8.2}$$

Upon evaluating at $r = a$ and taking the scalar product with the various spherical harmonics we find

$$B_{\ell m} a^{-\ell-1} = \left(Y_\ell^m, f \right) \equiv \int d\Omega' \, \overline{Y}_\ell^m (\theta', \phi') \, f(\theta', \phi') \tag{7.8.3}$$

where $d\Omega' = \sin \theta' \, d\theta' \, d\phi'$ is the element of solid angle.

If the problem asked for the solution inside the sphere, regular at the origin and satisfying the same boundary condition (7.8.1) at $r = a$, we would have to take all the $B_{\ell m} = 0$ and the same procedure would have led us to

$$A_{\ell m} a^\ell = \left(Y_\ell^m, f \right). \tag{7.8.4}$$

It is readily verified that this interior solution is related to the exterior one by *Kelvin's inversion* described in Theorem 7.9.1.

Let us return to the exterior solution; we substitute (7.8.3) into (7.8.2) and assume that f is such that summation and integration can be interchanged; we find

$$u = \int_0^\pi \sin\theta'\,d\theta' \int_0^{2\pi} d\phi'\, f(\theta', \phi') \sum_{\ell=0}^\infty \left(\frac{a}{r}\right)^{\ell+1} \sum_{m=-\ell}^\ell \overline{Y_\ell^m}(\theta', \phi')\, Y_\ell^m(\theta, \phi). \tag{7.8.5}$$

The inner summation can be effected with recourse to the addition theorem (rule 5 in Table 7.5) as already seen in §7.6 with the result

$$u = \frac{1}{2\pi}\frac{a}{r} \int_0^\pi \sin\theta'\,d\theta' \int_0^{2\pi} d\phi'\, f(\theta', \phi') \sum_{\ell=0}^\infty \left(\frac{a}{r}\right)^\ell \left(\ell + \frac{1}{2}\right) P_\ell(\cos\gamma). \tag{7.8.6}$$

Now, upon differentiating the relation giving the generating function for the Legendre polynomials (entry 3 in Table 7.7), we deduce that

$$\frac{\mu - t}{(1 - 2t\mu + t^2)^{3/2}} = \sum_{\ell=0}^\infty \ell t^{\ell-1} P_\ell(\mu). \tag{7.8.7}$$

With this result and the generating function, the summation in (7.8.6) can be expressed in closed form to find

$$u = \frac{a}{4\pi} \int \frac{r^2 - a^2}{(r^2 - 2ra\cos\gamma + a^2)^{3/2}}\, f(\theta', \phi')\, d\Omega', \tag{7.8.8}$$

which is the Poisson formula for the exterior Dirichlet problem. In a similar way one can work out the analogous formulae for the interior problem and for Neumann data.

7.9 Point source near a sphere

The reader is probably familiar with the elementary application of the so-called *method of images* for the problem of determining the electrostatic field due to a point charge of magnitude q placed at a distance d from an infinite conducting plane kept at zero potential. If the surface of the conductor is taken as the (x, y) plane and the point charge is at $(0, 0, z = d)$, the answer is obtained by noting that a charge $-q$ placed at the image point $(0, 0, -d)$ would result in a vanishing potential on the plane and a field with zero component parallel to the plane, as required. This is a limiting case (for $a \to \infty$) of a more general procedure known as *Kelvin inversion*:

Theorem 7.9.1 [Kelvin Inversion Theorem] *If $u(r, \theta, \phi)$ is a solution of Laplace's equation, then so is*

$$v(r, \theta, \phi) = \frac{a}{r} u\left(\frac{a^2}{r}, \theta, \phi\right) \tag{7.9.1}$$

and $u(a, \theta, \phi) = v(a, \theta, \phi)$.

The second part is obvious. The proof of the first part is immediate by using the chain rule with $s = a^2/r$ to find that $(1/r^2)\partial_r \left(r^2\partial_r v\right) = (a/r^3) \partial_s \left(s^2\partial_s u\right)$ from which the assertion readily follows by substituting into $\nabla^2 v = 0$ and using the fact that $\nabla^2 u = 0$ by hypothesis. Thus, a new solution of the Laplace equation can be generated from another given solution (see e.g. Panofsky & Phillips 1956, chapter 3; Dassios & Kleinman 1989).

Let us apply this result to the potential generated by a point charge (or source) q placed at $\mathbf{x}_b = (0, 0, b)$, namely on the z axis at a distance $b > a$ from the origin (Figure 7.5). This point charge generates the monopole potential (7.6.1), namely

$$U(r, \theta, \phi) = \frac{q}{\sqrt{r^2 - 2br \cos\theta + b^2}}. \tag{7.9.2}$$

From Kelvin's theorem we deduce that

$$V = \frac{a}{r} \frac{q}{\sqrt{(a^2/r)^2 - 2b(a^2/r)\cos\theta + b^2}} = \frac{(a/b)q}{\sqrt{r^2 - 2(a^2/b)\cos\theta + (a^2/b)^2}} \tag{7.9.3}$$

is also a solution of the Laplace equation, and the denominator of this expression will be recognized as the distance of the field point $\mathbf{x}$ from a point located at $\mathbf{x}_I = (0, 0, a^2/b)$, i.e. the image in the sphere of radius a of the point charge. It is also immediately verified that, as expected,

$$U(a, \theta, \phi) = V(a, \theta, \phi) = \frac{q}{\sqrt{a^2 - 2ab\cos\theta + b^2}}. \tag{7.9.4}$$

It is therefore evident that the function $u = U - V$ will be a solution of the Laplace equation vanishing on $r = a$ and will therefore represent the electric field generated by a point charge near a spherical conductor kept at zero potential.[5] We now recover this result by using the spherical harmonic expansion technique.

The mathematical statement of the problem is the following: To find the solution of the Poisson equation

$$\nabla^2 u = -4\pi q\, \delta\,(\mathbf{x} - \mathbf{x}_b), \qquad \mathbf{x}_b = (0, 0, b), \tag{7.9.5}$$

subject to the conditions

$$u(a, \theta, \phi) = 0, \qquad u \to 0 \quad \text{for} \quad r \to \infty. \tag{7.9.6}$$

By choosing the coordinate system shown in Figure 7.5 the problem is axisymmetric and we may represent u in the form of the expansion (7.2.14), with $R_\ell(a) = 0$, $R_\ell \to 0$ for $r \to \infty$.

---

[5] To recover the result for the plane case, it is necessary to set $b = a + d$ and then take the limit $a \to \infty$ with d fixed.

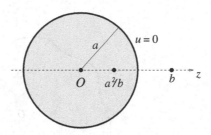

Fig. 7.5   Point charge near a spherical conductor, §7.9.

In order to apply the general result (7.3.10) we need to calculate the f_ℓ from (7.3.9); this task is readily effected since (see Table 7.4) $\delta\,(\mathbf{x}-\mathbf{x}_b) = \delta(r-b)\,\delta(\cos\theta-1)/(2\pi r^2)$ and $P_\ell(1) = 1$. Thus $f_\ell = (\ell+1/2)(q/2\pi r^2)\,\delta(r-b)$ and

$$R_\ell = \left[A_\ell - q\int_a^r s^{-\ell-1}\delta(s-b)\,\mathrm{d}s\right]r^\ell + \left[B_\ell + q\int_a^r s^\ell\delta(s-b)\,\mathrm{d}s\right]r^{-\ell-1}$$
$$= \left[A_\ell - qH(r-b)b^{-\ell-1}\right]r^\ell + \left[B_\ell + qH(r-b)b^\ell\right]r^{-\ell-1}. \qquad (7.9.7)$$

As remarked in the footnote on p. 179 the choice of the lower limit of integration is arbitrary and we have set them both equal to a. Since $H(r-b)=1$ for r large, the condition of vanishing at infinity evidently requires that $A_\ell = qb^{-\ell-1}$ while, since $H(a-b)=0$, vanishing at $r=a$ will be ensured provided that $B_\ell = -qa^{2\ell+1}b^{-\ell-1}$. Thus, using the obvious relation $1 - H(r-b) = H(b-r)$, we may cast the result in the compact form

$$R_\ell = q\left[r_<^\ell\, r_>^{-\ell-1} - a^{2\ell+1}\, r^{-\ell-1}\right]. \qquad (7.9.8)$$

where $r_> = \max(r,b), r_< = \min(r,b)$. We have thus found that

$$u(r,\theta) = \frac{q}{r_>}\sum_{\ell=0}^\infty \left(\frac{r_<}{r_>}\right)^\ell P_\ell(\cos\theta) - q\frac{a}{r}\sum_{\ell=0}^\infty \left(\frac{a^2}{r^2}\right)^\ell P_\ell(\cos\theta). \qquad (7.9.9)$$

Both summations can be calculated in closed form by using (7.2.20) with the expected result

$$u(r,\theta) = \frac{q}{\sqrt{r^2 - 2br\cos\theta + b^2}} - \frac{qa}{br}\frac{1}{\sqrt{(a^2/br)^2 - 2(a^2/br)\cos\theta + 1}}$$
$$= \frac{q}{|\mathbf{x}-\mathbf{x}_b|} - \frac{(a/b)q}{|\mathbf{x}-\mathbf{x}_I|}. \qquad (7.9.10)$$

7.10 A domain perturbation problem

In the example of §7.4 we solved Laplace's equation $\nabla^2 u = 0$ subject to $u \to Vr\cos\theta$ at infinity together with a homogeneous Neumann condition $\mathbf{n}\cdot\nabla u = 0$ on the

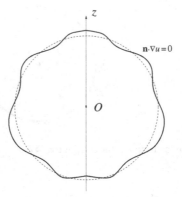

Fig. 7.6   A distorted sphere, §7.10.

sphere $r - a = 0$. We now consider the same problem with the only difference that the Neumann condition is imposed on a slightly distorted sphere (Figure 7.6). As long as any line drawn from the origin intersects the distorted surface in only one point, this surface can be represented in the form

$$S(r, \theta, \phi) \equiv r - a - \varepsilon a F(\theta, \phi) = 0, \qquad |\varepsilon| \ll 1, \tag{7.10.1}$$

and the function F can be expanded in a series of spherical harmonics as

$$F(\theta, \phi) = \sum_{\ell=0}^{\infty} \sum_{m=-\ell}^{\ell} F_{\ell m} Y_{\ell}^{m}(\theta, \phi). \tag{7.10.2}$$

It will be seen from the following developments that, as long as we are only interested in the first-order correction in ε, the effect of each spherical harmonic can be treated independently. With this justification, and adding the assumption of axial symmetry for simplicity, we will consider a distortion of the form

$$S(r, \theta, \phi) \equiv r - a \left[1 + \varepsilon P_n(\cos \theta)\right] = 0. \tag{7.10.3}$$

We look for an approximate solution based on the fact that the solution of the problem for $\varepsilon = 0$ is known and that, since $\varepsilon \ll 1$, the actual domain of the problem can be considered a perturbation of that for $\varepsilon = 0$. This justifies the denomination *domain perturbation* given to the procedure that will be demonstrated.

The exact solution $u(\mathbf{x}; \varepsilon)$ will depend not only on $\mathbf{x}$, but also on ε. Assuming that the dependence on ε is regular, we write

$$u(\mathbf{x}; \varepsilon) = u(\mathbf{x}; 0) + \varepsilon u_1(\mathbf{x}) + \cdots \tag{7.10.4}$$

In general it is not an easy task to decide a priori whether the solution of a problem does indeed admit a series expansion of this form. One usually adopts a heuristic stance based on the expectation that, if the dependence on ε is of a different form, difficulties would be encountered, either mathematical – for instance an equation with no solution – or physical – e.g. a meaningless solution.

The equations and boundary conditions for u_0, u_1, etc. are to be found by substituting this expansion into the Laplace equation and the boundary condition and expanding everything in powers of ε. It is important to realize that even though, in practice, we may be interested in a precise value of ε, the proper way in which the procedure is to be viewed mathematically is as the limit process $\varepsilon \to 0$.

Upon substitution into the Laplace equation we have

$$\nabla^2 (u_0 + \varepsilon u_1 + \cdots) = 0 \tag{7.10.5}$$

and, if this has to hold for every value of ε, we must require that

$$\nabla^2 u_k = 0 \quad \text{for} \quad k = 0, 1, \ldots \tag{7.10.6}$$

Upon substitution into the boundary condition at infinity we have

$$u_0 + \varepsilon u_1 + \cdots \to V r \cos \theta. \tag{7.10.7}$$

Since the right-hand side is independent of ε, we must require

$$u_0 \to V r \cos \theta, \quad \text{while} \quad u_k \to 0 \quad \text{for} \quad k = 1, \ldots. \tag{7.10.8}$$

Extricating the ε dependence of the boundary condition on $S = 0$ is more complicated. Let us start by calculating the unit normal $\mathbf{n}$ to the distorted sphere using

$$\mathbf{n} = \left. \frac{\nabla S}{|\nabla S|} \right|_{S=0} \tag{7.10.9}$$

with $S = 0$ the domain boundary given by (7.10.3).[6] We have

$$\nabla S = \left. \left(\frac{\partial S}{\partial r}, \frac{1}{r} \frac{\partial S}{\partial \theta} \right) \right|_{S=0} = \left. \left(1, \varepsilon \frac{a}{r} \sin \theta \, P_n' \right) \right|_{S=0} . \tag{7.10.10}$$

Since, on $S = 0$, $r = a + O(\varepsilon)$, to first order in ε this relation is

$$\nabla S = \left(1, \varepsilon \sin \theta \, P_n' \right) \tag{7.10.11}$$

while $|\nabla S| = \sqrt{1 + O(\varepsilon^2)} \simeq 1$. Here we see that, if terms of order ε^2 were retained, the dependence on the surface distortion would become non-linear and it would not be possible to consider the various Legendre modes separately. With this result, the homogeneous Neumann boundary condition on the surface is

$$0 = \left[n_r \frac{\partial u}{\partial r} + n_\theta \frac{1}{r} \frac{\partial u}{\partial \theta} \right]_{S=0} = \left. \frac{\partial u_0}{\partial r} \right|_{r=a(1+\varepsilon P_n)} + \varepsilon \left. \frac{\partial u_1}{\partial r} \right|_{r=a(1+\varepsilon P_n)} + \cdots$$

$$+ \left[\frac{\varepsilon \sin \theta \, P_n'}{r} \frac{\partial u_0}{\partial \theta} + \cdots \right]_{r=a(1+\varepsilon P_n)} . \tag{7.10.12}$$

---

[6] To justify this relation note that the domain boundary of interest here can be considered as a particular member of the family of surfaces $S = $ const. The vector field ∇S is normal to each member of this family and, evaluating it on $S = 0$, we have a vector field normal to our surface.

We approximate the first term by a truncated Taylor serias as

$$\left.\frac{\partial u_0}{\partial r}\right|_{r=a(1+\varepsilon P_n)} = \left.\frac{\partial u_0}{\partial r}\right|_{r=a} + \varepsilon a P_n \left.\frac{\partial}{\partial r}\frac{\partial u_0}{\partial r}\right|_{r=a} + O(\varepsilon^2) \tag{7.10.13}$$

and similarly for the other terms so that the boundary condition becomes

$$\frac{\partial u_0}{\partial r} + \varepsilon \frac{\partial u_1}{\partial r} + \varepsilon a P_n \frac{\partial^2 u_0}{\partial r^2} + \frac{\varepsilon \sin\theta P_n'}{a}\frac{\partial u_0}{\partial\theta} + O(\varepsilon^2) = 0 \tag{7.10.14}$$

where all the derivatives are now evaluated at $r = a$. Upon separating orders we thus find

$$\frac{\partial u_0}{\partial r} = 0, \tag{7.10.15}$$

$$\frac{\partial u_1}{\partial r} = -a P_n \frac{\partial^2 u_0}{\partial r^2} - \frac{\sin\theta P_n'}{a}\frac{\partial u_0}{\partial\theta}. \tag{7.10.16}$$

With these steps we have achieved a full specification of the problems for u_0 and u_1.

The problem for u_0 is the same as the one solved in §7.4 and therefore the solution is given by (7.4.5) from which we calculate

$$\left.\frac{\partial^2 u_0}{\partial r^2}\right|_a = 3\frac{V}{a}\cos\theta, \qquad \left.\frac{\partial u_0}{\partial\theta}\right|_a = -\frac{3}{2}\sin\theta\, a\, V, \tag{7.10.17}$$

so that the boundary condition (7.10.16) becomes

$$\left.\frac{\partial u_1}{\partial r}\right|_a = \frac{3}{2}V(\sin^2\theta P_n' - 2\cos\theta P_n). \tag{7.10.18}$$

By using the recurrence relations satisfied by the Legendre polynomials (Table 7.7 p. 177) we readily find, with $\mu = \cos\theta$,

$$(1-\mu^2)P_n' - 2\mu P_n = -\frac{(n+1)(n+2)}{2n+1}P_{n+1} + \frac{n(n-1)}{2n+1}P_{n-1} \tag{7.10.19}$$

so that

$$\left.\frac{\partial u_1}{\partial r}\right|_a = \frac{3}{2}V\left[\frac{n(n-1)}{2n+1}P_{n-1} - \frac{(n+1)(n+2)}{2n+1}P_{n+1}\right]. \tag{7.10.20}$$

Furthermore, at infinity, $u_1 \to 0$, so that it must have the form

$$u_1 = \sum_0^\infty \frac{B_\ell}{r^{\ell+1}} P_\ell \tag{7.10.21}$$

from which

$$\left.\frac{\partial u_1}{\partial r}\right|_a = -\sum_0^\infty (\ell+1)\frac{B_\ell}{a^{\ell+2}} P_\ell. \tag{7.10.22}$$

Upon equating to (7.10.20) and taking scalar products with all the P_k in turn we find that all the B_ℓ vanish except $B_{n\pm1}$ for which we have

$$n B_{n-1} = -\frac{3}{2}V\frac{n(n-1)}{2n+1}a^{n+1}, \qquad B_{n+1} = \frac{3}{2}V\frac{n+1}{2n+1}a^{n+3}. \tag{7.10.23}$$

Therefore

$$u_1 = \frac{3}{2} \frac{Va}{2n+1} \left[(n+1) \frac{a^{n+2}}{r^{n+2}} P_{n+1} - (n-1) \frac{a^n}{r^n} P_{n-1} \right]. \tag{7.10.24}$$

We have avoided simplifying the common factor n in the expression for B_{n-1} to point out that, if $n = 0$, this relation is automatically satisfied. Hence, in this special case, the previous result reduces to

$$u_1 = \frac{3}{2} V \frac{a^3}{r^2} P_1. \tag{7.10.25}$$

This expression allows us to test our approximate result for one case in which we know what the exact solution to the problem is. Indeed, if $n = 0$, since $P_0 = 1$, we see from (7.10.3) that we are dealing with a slightly larger sphere with radius $a(1 + \varepsilon)$ rather than a. The exact solution is then given by (7.4.5) with a replaced by $a(1 + \varepsilon)$:

$$u_0 = V \cos \theta \left(r + \frac{a^3(1+\varepsilon)^3}{2r^2} \right) \simeq V \cos \theta \left(r + \frac{a^3(1+3\varepsilon)}{2r^2} \right) + O(\varepsilon^2) \tag{7.10.26}$$

and indeed we recognize in the last expression the sum of u_0 given by (7.4.5) and εu_1, with u_1 given by (7.10.25). Another test of the solution can be effected by noting that the surface of a sphere centered at $(0, 0, \varepsilon a)$ can be represented, to first order in ε, as

$$S(r, \theta) \equiv r - a (1 + \varepsilon \cos \theta) = 0 \tag{7.10.27}$$

i.e. as in (7.10.3) with $n = 1$. The solution for this case is known exactly from (7.4.5) written with reference to the translated center of the sphere rather than $r = 0$. Upon retaining only terms of order ε, a simple calculation gives

$$u = V \left[\left(r + \frac{a^3}{2r^2} \right) \cos \theta + \varepsilon a \left(\frac{a^3}{r^3} P_2 - P_0 \right) \right] \tag{7.10.28}$$

the $O(\varepsilon)$ term of which coincides with (7.10.24) for $n = 1$ aside from the contribution $-\varepsilon a V P_0$ which, being a constant, has no effect on the physical field ∇u.

7.11 Conical boundaries

The half-space $z \geq 0$ considered in §7.5 is a particular example of a domain bounded by a cone, in that case with semi-aperture $\beta = \frac{1}{2}\pi$. We now consider more general geometries of this type, in which we are interested in solving the Laplace equation in a domain bounded by a cone of semi-aperture β and possibly spherical surfaces centered at the apex of the cone. The geometries for $0 < \beta < \frac{1}{2}\pi$ and $\frac{1}{2}\pi < \beta < \pi$ are shown in Figure 7.7; for simplicity we consider the axisymmetric case.

We have briefly mentioned in §13.7 the family of Legendre functions of the first kind P_{λ_n}, analogous to the Legendre polynomials, which can be used to expand $u(r, \theta)$ when

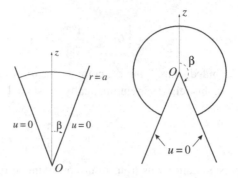

Conical boundaries with semi-aperture β smaller and larger than $\frac{1}{2}\pi$, §7.11.

$u(r, \beta) = 0$. Thus we set

$$u(r, \theta) = \sum_{n=0}^{\infty} R_n(r)\, P_{\lambda_n}(\cos \theta). \tag{7.11.1}$$

As mentioned in §13.7, for β small or close to π, respectively,

$$\lambda_n \simeq -\frac{1}{2} + \frac{j_{0,n}}{\beta}, \qquad \lambda_n \simeq n - 1 + \frac{1}{2}\left[\log \frac{2}{\pi - \beta}\right]^{-1} \tag{7.11.2}$$

with $j_{0,n}$ the n-th zero of the Bessel function J_0.

Now the P_{λ_n} are eigenfunctions of the angular part of the Laplacian, namely the operator $\mathcal{L}^2$ defined in (7.2.11), corresponding to the eigenvalue $\lambda_n(\lambda_n + 1)$ and, therefore, the only change necessary in the solution R_n of the radial equation that we used before is the use of exponents λ_n and $-\lambda_n - 1$ in place of ℓ and $-\ell - 1$:

$$R_n(r) = A_n r^{\lambda_n} + B_n r^{-\lambda_n - 1}. \tag{7.11.3}$$

As before, if the solution is required to be regular at the origin, we take $B_n = 0$. In this case, since λ_n increases with n, near $r = 0$ the leading term is the first one, $u(r, \theta) \simeq A_1 r^{\lambda_1} P_{\lambda_1}$. If β is small (a narrow "hole"), we see from the first one of (7.11.2) that λ_1 is large so that u and its gradient (i.e., the physical field of which u is the potential) are all very small near $r = 0$. Conversely, when β is close to π (a "sharp point"), λ_1 becomes less than 1 and, although u remains finite at $r = 0$, the field ∇u will blow up approaching the origin as is well known from electrostatics.

Another type of problem that can be posed is to require that u vanish on the portion of two spherical surfaces of radius a and b intercepted by the cone, while it is a prescribed function of r for $a \leq r \leq b, \theta = \beta$. In this case we have from (7.11.3)

$$A_n a^{\lambda_n} + B_n a^{-\lambda_n - 1} = 0, \qquad A_n b^{\lambda_n} + B_n b^{-\lambda_n - 1} = 0 \tag{7.11.4}$$

which can only be satisfied if $(b/a)^{2\lambda_n+1} = 1$, or

$$\lambda_n = -\frac{1}{2} + ink \qquad \text{with} \qquad k = \frac{\pi}{\log b/a}. \qquad (7.11.5)$$

This result leads to the appearance of Legendre functions of the type $P_{-1/2+ink}$, briefly treated on p. 332, which are known as *conical functions*. Since $\lambda_n(\lambda_n + 1) = -\frac{1}{4} - n^2k^2$, these functions can be taken real. Upon renaming the constant, the R_ℓ take the form

$$R_\ell(r) = \frac{A_\ell}{\sqrt{r}} \sin\left(\ell k \log\frac{r}{a}\right). \qquad (7.11.6)$$

By letting $x = k \log(r/a)$ we find

$$\int_a^b R_\ell(r)\, R_n(r)\, dr = \frac{A_\ell A_n}{k} \int_0^\pi \sin \ell x\, \sin nx\, dx = \frac{\pi}{2k} A_\ell^2 \delta_{\ell n}. \qquad (7.11.7)$$

The functions R_n are therefore orthogonal and can be used to represent the prescribed values of u on the conical surface $\theta = \beta$.

When the assigned boundary condition is over the entire cone wall $0 < r < \infty$, the series in r must be replaced by a suitable transform (see e.g. Hobson 1931, p. 448). As suggested by (7.11.6), let us make the change of dependent variable which converts between spherical and cylindrical Bessel functions, namely $u = r^{-1/2}v$ (p. 303). With this transformation, the axisymmetric Laplace equation becomes

$$r\frac{\partial}{\partial r}\left(r\frac{\partial v}{\partial r}\right) - \frac{1}{4}v + \frac{\partial}{\partial \mu}\left[(1-\mu^2)\frac{\partial v}{\partial \mu}\right] = 0. \qquad (7.11.8)$$

By inspection we see that, in terms of the new variable $s = \log(r/b)$, where b is an arbitrary length introduced to make the argument of the logarithm dimensionless, it becomes

$$\frac{\partial^2 v}{\partial s^2} - \frac{1}{4}v + \frac{\partial}{\partial \mu}\left[(1-\mu^2)\frac{\partial v}{\partial \mu}\right] = 0. \qquad (7.11.9)$$

As r ranges between 0 and ∞, s ranges between $-\infty$ and ∞ and we can therefore use the ordinary Fourier transform to find

$$\frac{\partial}{\partial \mu}\left[(1-\mu^2)\frac{\partial \tilde{v}}{\partial \mu}\right] - \left(k^2 + \frac{1}{4}\right)\tilde{v} = 0. \qquad (7.11.10)$$

This has the standard Legendre form with $\lambda = -1/2 + ik$ and we again encounter the conical functions $P_{-1/2+ik}(\mu)$; the solution regular on the polar axis may be written as

$$\tilde{v}(k, \theta) = \tilde{v}(k, \beta)\frac{P_{-1/2+ik}(\cos\theta)}{P_{-1/2+ik}(\cos\beta)}. \qquad (7.11.11)$$

If the boundary condition on the cone wall is $u(r, \beta) = U(r)$, the transformed boundary value $\tilde{v}(k, \beta)$ is given by

$$\tilde{v}(k, \beta) = \sqrt{\frac{b}{2\pi}} \int_{-\infty}^\infty e^{s/2}U(be^s)e^{iks}\, ds \qquad (7.11.12)$$

and the solution of (7.11.8) is

$$u(r, \theta) = \frac{1}{\sqrt{2\pi r}} \int_{-\infty}^{\infty} \frac{P_{-1/2+ik}(\cos\theta)}{P_{-1/2+ik}(\cos\beta)} \tilde{v}(k, \beta) \exp(-iks) \, dk. \tag{7.11.13}$$

As an example, let us consider a point charge on the axis of a cone with $0 < \beta < \pi$, the walls of which are kept at zero potential (Lebedev 1965, p. 210). We represent the potential in this system as

$$w(r, \theta) = \frac{q}{\sqrt{r^2 - 2br\cos\theta + b^2}} - u(r, \theta) \tag{7.11.14}$$

in which the first term is the potential due to a point charge in free space. The boundary condition $w = 0$ on the cone wall requires then that

$$u(r, \beta) = \frac{q}{\sqrt{r^2 - 2br\cos\beta + b^2}}. \tag{7.11.15}$$

Upon substitution into (7.11.12) we find

$$\begin{aligned}
\tilde{v}(k, \beta) &= \frac{q}{2\sqrt{\pi b}} \int_{-\infty}^{\infty} \frac{e^{iks}}{\sqrt{\cosh s - \cos\beta}} \, ds = \frac{q}{\sqrt{\pi b}} \int_{0}^{\infty} \frac{\cos ks}{\sqrt{\cosh s - \cos\beta}} \, ds \\
&= \sqrt{\frac{\pi}{2b}} \frac{q}{\cosh \pi k} P_{-1/2+ik}(-\cos\beta),
\end{aligned} \tag{7.11.16}$$

where the integral is calculated from (13.7.11) p. 332. As a consequence of the fact that $P_\lambda = P_{-\lambda-1}$ we have $P_{-1/2+ik} = P_{-1/2-ik}$ and $\tilde{v}(k, \beta)$ as given by (7.11.16) is an even function. In these conditions (7.11.13) gives

$$u(r, \theta) = \sqrt{\frac{2}{\pi r}} \int_{0}^{\infty} \tilde{v}(k, \beta) \frac{P_{-1/2+ik}(\cos\theta)}{P_{-1/2+ik}(\cos\beta)} \cos ks \, dk, \tag{7.11.17}$$

or, after substitution of (7.11.16),

$$u(r, \theta) = \frac{q}{\sqrt{br}} \int_{0}^{\infty} \frac{P_{-1/2+ik}(\cos\theta)}{P_{-1/2+ik}(\cos\beta)} P_{-1/2+ik}(-\cos\beta) \frac{\cos(k\log r/b)}{\cosh \pi k} \, dk. \tag{7.11.18}$$

7.12 Spherical Bessel functions: useful relations

In several important cases the angular dependence of the solution can be conveniently expanded on a basis of spherical harmonics, but the radial dependence involves Bessel functions. This is the situation e.g. with the Helmholtz equation and when Laplace transforms are used to solve diffusion or wave problems. Here we give a brief summary of the most useful relations to deal with the particular class of Bessel functions which arise in these contexts. The underlying theory is presented in §12.3 where it is shown that, in spherical coordinates, the solution of the Helmholtz equation $\nabla^2 u + k^2 u = 0$ may be represented in the form (7.3.2) in which the radial functions $R_{\ell m}(r)$ satisfy

$$\frac{1}{r^2} \frac{d}{dr}\left(r^2 \frac{dR_{\ell m}}{dr}\right) + \left(k^2 - \frac{\ell(\ell+1)}{r^2}\right) R_{\ell m} = 0. \tag{7.12.1}$$

Table 7.8	Some relations involving spherical Bessel functions valid for $n = 0, \pm 1, \ldots$; in the first two lines, ζ_n stands for any of j_n, y_n or $h_n^{(1,2)}$.

$$z\zeta_{n-1} + z\zeta_{n+1} = (2n+1)\,\zeta_n \qquad\qquad n\zeta_{n-1} - (n+1)\zeta_{n+1} = (2n+1)\zeta_n'$$

$$z\zeta_n' + (n+1)\zeta_n = z\zeta_{n-1} \qquad\qquad -z\zeta_n' + n\zeta_n = z\zeta_{n+1}$$

$$j_n(z) = z^n \left(-\frac{1}{z}\frac{d}{dz}\right)^n \frac{\sin z}{z} \qquad\qquad y_n(z) = -z^n \left(-\frac{1}{z}\frac{d}{dz}\right)^n \frac{\cos z}{z}$$

$$j_n(z) = (-1)^{n+1} y_{-n-1}(z) \qquad\qquad y_n(z) = (-1)^{n+1} j_{-n-1}(z)$$

$$h_n^{(1)}(z) = j_n(z) + iy_n(z) \qquad\qquad h_n^{(2)}(z) = j_n(z) - iy_n(z)$$

$$h_n^{(1)}(z) = z^n \left(-\frac{1}{z}\frac{d}{dz}\right)^n \frac{e^{iz}}{iz} \qquad\qquad h_n^{(2)}(z) = z^n \left(-\frac{1}{z}\frac{d}{dz}\right)^n \frac{e^{-iz}}{-iz}$$

$$j_n(z)y_{n-1}(z) - j_{n-1}(z)y_n(z) = \frac{1}{z^2} \qquad j_{n+1}(z)y_{n-1}(z) - j_{n-1}(z)y_{n+1}(z) = \frac{2n+1}{z^3}$$

The change of variable $R_{\ell m} = S_{\ell m}/\sqrt{r}$ reduces the equation to the standard Bessel form but, in view of the frequent occurrence of an equation of this type, it is useful to define so-called *spherical Bessel functions* which solve the equation directly avoiding the need for an explicit change of variable. The explicit relation between the regular and spherical Bessel function is given in §12.3; here we simply note that the general solution of (7.12.1) can be written as the sum of any two of the following functions:

$$j_\ell(kr), \qquad y_\ell(kr), \qquad h_\ell^{(1)}(kr), \qquad h_\ell^{(2)}(kr). \tag{7.12.2}$$

The spherical Bessel functions of the first and second kind, $j_\ell(z)$ and $y_\ell(z)$, are polynomials in $1/z$ multiplied by $\sin z$ and $\cos z$; for the Hankel functions $h_\ell^{(1)}(z)$ and $h_\ell^{(2)}(z)$ have a similar structure with $e^{\pm iz}$ in place of the sine and cosine. For example

$$j_0(z) = \frac{\sin z}{z}, \qquad y_0(z) = -\frac{\cos z}{z}, \qquad h_0^{(1)}(z) = \frac{e^{iz}}{iz}, \qquad h_0^{(2)}(z) = \frac{e^{-iz}}{-iz}. \tag{7.12.3}$$

If $\zeta_{1,\ell}$ and $\zeta_{2,\ell}$ denote any two of (7.12.2), their Wronskian has the form $W(\zeta_{1,n}, \zeta_{2,n}) = W_0/z^2$; specifically

$$W(j_n, y_n) = \frac{1}{z^2}, \qquad W(h_n^{(1)}, h_n^{(2)}) = -\frac{2i}{z^2}. \tag{7.12.4}$$

With this notation the solution of the non-homogeneous form of (7.12.1)

$$\frac{1}{r^2}\frac{d}{dr}\left(r^2 \frac{dR_{\ell m}}{dr}\right) + \left(k^2 - \frac{\ell(\ell+1)}{r^2}\right) R_{\ell m} = f_{\ell m}(r) \tag{7.12.5}$$

Table 7.9	Asymptotic behavior of the spherical Bessel functions; the semi-factorial is defined as $(2n+1)!! = (2n+1)(2n-1)(2n-3)\cdots 3\cdot 1$.

	$z \to 0$	$z \to \infty \quad (n \ll |z|)$
$j_n(z) \simeq$	$\dfrac{z^n}{(2n+1)!!}\left[1 - \dfrac{z^2}{2(2n+3)} + \cdots\right]$	$\dfrac{1}{z}\sin\left(z - \dfrac{n\pi}{2}\right)$
$y_n(z) \simeq$	$-\dfrac{(2n-1)!!}{z^{n+1}}\left[1 + \dfrac{z^2}{2(2n-1)} + \cdots\right]$	$-\dfrac{1}{z}\cos\left(z - \dfrac{n\pi}{2}\right)$
$h_n^{(1)}(z) \simeq$	$-i\,\dfrac{(2n-1)!!}{z^{n+1}}\,[1 + \cdots]$	$i^{-n-1}\dfrac{e^{iz}}{z}$
$h_n^{(2)}(z) \simeq$	$i\,\dfrac{(2n-1)!!}{z^{n+1}}\,[1 + \cdots]$	$i^{n+1}\dfrac{e^{-iz}}{z}$

may be expressed as

$$
\begin{aligned}
R_{\ell,m} =\ & \left[\alpha_{\ell m} - \frac{k}{W_0}\int^r s^2 \zeta_{2,\ell}(ks)\, f_{\ell m}(s)\, \mathrm{d}s\right]\zeta_{1,\ell}(kr) \\
& + \left[\beta_{\ell m} + \frac{k}{W_0}\int^r s^2 \zeta_{1,\ell}(ks)\, f_{\ell m}(s)\, \mathrm{d}s\right]\zeta_{2,\ell}(kr).
\end{aligned}
\tag{7.12.6}
$$

7.13 Fundamental solution of the Helmholtz equation

As a first application of the spherical Bessel functions let us consider the fundamental solution of the scalar three-dimensional Helmholtz equation

$$
\nabla^2 u + k^2 u = -4\pi Q\delta(\mathbf{x} - \mathbf{x}_s)
\tag{7.13.1}
$$

in which $\mathbf{x}_s$ is the position of the source point. We have seen in §1.2.3 p. 8 the connection between this equation and the wave equation in the case of monochromatic waves with frequency $\omega/2\pi = kc$, in which $k = 2\pi/\lambda$ is the wave number. We assume that the waves propagate in free space and that there is no source other than that at $\mathbf{x}_s$ (Figure 7.8).

The simplest procedure is to introduce the new variable $\mathbf{X} = \mathbf{x} - \mathbf{x}_s$, with which the equation becomes

$$
\frac{1}{R^2}\frac{\mathrm{d}}{\mathrm{d}R}\left(R^2 \frac{\mathrm{d}u}{\mathrm{d}R}\right) + k^2 u = -4\pi Q\delta(\mathbf{X}) = -Q\frac{\delta(R)}{R^2}
\tag{7.13.2}
$$

in which $R = |\mathbf{X}|$ and use has been made of the spherical symmetry of the situation and of the appropriate form of the δ-function given in Table 7.4. For $R > 0$, this equation is just

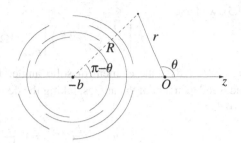

Fig. 7.8
Point source of scalar waves located at a distance b from the origin along the negative polar semi-axis, §7.13.

(7.12.1) with $\ell = 0$, but it is easier to recall the useful identity

$$\frac{1}{R^2}\frac{d}{dR}\left(R^2\frac{du}{dR}\right) = \frac{1}{R}\frac{d^2}{dR^2}(Ru) \tag{7.13.3}$$

to bring into the form

$$\frac{d^2}{dR^2}(Ru) + k^2(Ru) = -Q\frac{\delta(R)}{R}. \tag{7.13.4}$$

Away from $R = 0$ the right-hand side vanishes and the solution is[7]

$$u = \frac{1}{R}\left(Ae^{ikR} + Be^{-ikR}\right). \tag{7.13.5}$$

One piece of information which we have not yet used is that the strength of the source should be as specified in the right-hand side of (7.13.1) or (7.13.4), but we need two conditions to determine both A and B and it is clear that requiring $u \to 0$ at infinity is of no help. The situation here is similar to that encountered in §3.5 and the ambiguity is resolved in the same way (see the considerations on p. 73): if one restores the time dependence in the form of the factor $e^{-i\omega t}$, it is evident that the second term represents waves propagating inward from infinity. Since no such waves can be present in the problem as specified, we need to take $B = 0$. To determine A we return to the full equation and integrate over an arbitrary volume V containing $\mathbf{x}_s$; after applying the divergence theorem in the left-hand side we find

$$\oint_S \mathbf{n} \cdot \nabla u \, dS_\varepsilon = -4\pi Q \tag{7.13.6}$$

where S is the boundary of V. The value of the integral is independent of V and, for convenience, we can take it to be a very small sphere of radius $R = \varepsilon$ and consider the limit $\varepsilon \to 0$. If we rewrite the integral in terms of the transformed coordinate $\mathbf{X}$, after substitution

---

[7] It will be recognized from (7.12.3) that this solution is just a linear combination of $h_0^{(1)}(kR)$ and $h_0^{(2)}(kR)$ as expected.

of (7.13.5), we have

$$\int \left[\frac{d}{dR} \left(\frac{A}{R} e^{ikR} \right) \right]_{R=\varepsilon} d\Omega = -4\pi Q \tag{7.13.7}$$

since $dS_\varepsilon / R^2 = d\Omega$, the element of solid angle. The integral is actually trivial as the integrand is independent of the angles; taking the derivative and the limit $\varepsilon \to 0$ gives then $A = Q$ so that

$$u = \frac{Q}{R} e^{ikR} = Q \frac{e^{ik|\mathbf{x}-\mathbf{x}_s|}}{|\mathbf{x}-\mathbf{x}_s|}. \tag{7.13.8}$$

At distances R from the source much smaller than a wavelength, $k|\mathbf{x}-\mathbf{x}_s| \ll 1$, one expects that the term $k^2 u \sim u/\lambda^2$ in the Helmholtz equation would be small compared with $\nabla^2 u \sim u/R^2$ and indeed this result reduces to the monopole potential (7.6.1) $u \simeq Q/R$ which is the fundamental solution of the Poisson equation (§7.6).[8] It is intuitively clear that, if the source is very far from the observation point, the wave fronts would look approximately planar in the neightborhood of the z axis. We can verify this expectation as follows. Let us take a coordinate system in which the source position is on the negative polar axis at a distance b from the origin (Figure 7.8). Then

$$|\mathbf{x}-\mathbf{x}_s| = \sqrt{b^2 + r^2 + 2br \cos\theta} \simeq b\left(1 + \frac{r}{b}\cos\theta\right) = b + r\cos\theta \tag{7.13.9}$$

where we have used the fact that the angle between $\mathbf{x}$ and $\mathbf{x}_s$ is $\pi - \theta$ and the Taylor series approximation $\sqrt{1+2\eta} \simeq 1 + \eta$ valid for small η. Thus we find

$$u \simeq \frac{Qe^{ikb}}{b} e^{ikr\cos\theta} = \frac{Qe^{ikb}}{b} e^{ikz} \tag{7.13.10}$$

which is indeed a plane wave propagating in the positive direction of the polar axis as expected. Of course, as b increases, we must increase Q in proportion to maintain the amplitude of the wave.

We now find the same solution (7.13.8) retaining the original variables, both to demonstrate the method and to derive an important result that will be useful later. With the source placed at $z = -b$ the Helmholtz equation (7.13.1) is

$$\frac{1}{r^2}\frac{\partial}{\partial r}\left(r^2 \frac{\partial u}{\partial r}\right) + \frac{\partial}{\partial\mu}\left((1-\mu^2)\frac{\partial u}{\partial\mu}\right) + k^2 u = -2Q\frac{\delta(r-b)}{b^2}\delta(\mu+1) \tag{7.13.11}$$

in which, as usual, $\mu = \cos\theta$ and we have used the expression in line 3 of Table 7.4 p. 173 to express the δ-function. We expand u in a Legendre polynomial scrics as in (7.3.2) and we are led to

$$\frac{1}{r^2}\frac{d}{dr}\left(r^2 \frac{dR_\ell}{dr}\right) + \left(k^2 - \frac{\ell(\ell+1)}{r^2}\right)R_\ell = (-1)^{\ell+1}(2\ell+1)Q\frac{\delta(r-b)}{b^2} \tag{7.13.12}$$

where we have used the fact that $P_\ell(-1) = (-1)^\ell$. Since we expect the solution to be bounded at $r = 0$ and to have the form of outgoing waves away from the source, in applying

---

[8] Remembering how the Helmholtz equation is obtained from the wave equation, the argument can be restated by saying that, very near the source, the finite speed of propagation of the signal is immaterial and the time lag between the field and the source is negligible.

the method of variation of parameters we take the two solutions of the spherical Bessel equation to be $j_\ell(kr)$ and $h_\ell^{(1)}(kr)$. Their Wronksian is readily calculated from the definition $h_\ell^{(1)} = j_\ell + iy_\ell$ and (7.12.4):

$$W\left(j_\ell, h_\ell^{(1)}\right) = W(j_\ell, j_\ell) + iW(j_\ell, y_\ell) = \frac{i}{z^2}. \qquad (7.13.13)$$

We can now use the general solution (7.12.6) with $W_0 = i$ and f_ℓ given by the right-hand side of (7.13.12):

$$R_\ell = \left[\alpha_\ell + ik(-1)^{\ell+1}(2\ell+1)\frac{Q}{b^2}\int^r s^2 h_\ell^{(1)}(ks)\,\delta(s-b)\,ds\right] j_\ell(kr)$$

$$+ \left[\beta_\ell - ik(-1)^{\ell+1}(2\ell+1)\frac{Q}{b^2}\int^r s^2 j_\ell(ks)\,\delta(s-b)\,ds\right] h_\ell^{(1)}(kr). \qquad (7.13.14)$$

In order to have only outward propagating waves at infinity we take $\alpha_\ell = 0$ and the lower limit of the first integral as ∞ and, for regularity at 0, we take $\beta_\ell = 0$ and the lower limit of the second integral as 0. With these choices (see Table 2.1 p. 37),

$$\int_\infty^r s^2 h_\ell^{(1)}(ks)\,\delta(s-b)\,ds \;\;=\;\; -b^2 h_\ell^{(1)}(kb)\,H(b-r),$$

$$\int_0^r s^2 j_\ell(ks)\,\delta(s-b)\,ds \;\;=\;\; b^2 j_\ell(kb)H(r-b), \qquad (7.13.15)$$

in which H is the Heaviside function, so that

$$R_\ell = ik(-1)^\ell(2\ell+1)Q j_\ell(kr_<)h_\ell^{(1)}(kr_>) \qquad (7.13.16)$$

with $r_> = \max(r, b)$, $r_< = \min(r, b)$. In conclusion we have found that

$$u = iQk\sum_{\ell=0}^\infty (-1)^\ell(2\ell+1) j_\ell(kr_<)h_\ell^{(1)}(kr_>)\, P_\ell(\cos\theta). \qquad (7.13.17)$$

If, instead of the polar axis, the source had been placed at the generic angular position (θ', ϕ'), the term $(2\ell+1) P_\ell(\cos\theta)$ would be replaced by $(4\pi)^{-1}\sum_{m=-\ell}^\ell \overline{Y}_{\ell m}(\theta'\phi') Y_{\ell m}(\theta, \phi)$ which can be simplified with the aid of the addition theorem (rule 5 in Table 7.5) as already shown in a few examples.

By equating the two expressions of u given by (7.13.8) and (7.13.17) we have one instance of a class of results known as *addition theorems* for the Bessel functions (see e.g. Watson 1944, chapter 11). We will have use for a degenerate form of this theorem which is obtained by making on (7.13.17) the same approximation $r \ll b$ leading to (7.13.10). For large b, we have $r_> = b$, $r_< = r$ and we can use the asymptotic expression in Table 7.9 for $h_\ell^{(1)}(kb)$ with the result[9]

$$u \simeq \frac{Qe^{ikb}}{b}\sum_{\ell=0}^\infty (2\ell+1)i^\ell j_\ell(kr) P_\ell(\cos\theta). \qquad (7.13.18)$$

---

[9] The asymptotic form shown in the table holds when the argument of the function is larger than the index, a condition that will be satisfied only by a finite number of terms of the series for any given fixed b. Thus, while the conclusion is correct, the present derivation is not fully justified in a mathematical sense.

Upon equating to (7.13.10) and cancelling a common factor we find the important relation

$$e^{ikr\cos\theta} = \sum_{\ell=0}^{\infty}(2\ell+1)i^\ell j_\ell(kr)\,P_\ell(\cos\theta). \qquad (7.13.19)$$

As noted in §12.6, this is an instance of a series of the Neumann type associated with Bessel functions.

7.14 Scattering of scalar waves

In §1.2.3 we explain how the wave equation is reduced to the Helmholtz equation

$$\nabla^2 u + k^2 u = 0 \qquad (7.14.1)$$

in the case of monochromatic waves with frequency $\omega/2\pi = kc$, in which $k = 2\pi/\lambda$ is the wave number. We consider the solution of this equation when there is a scalar plane wave with wave number k, propagating from $-\infty$ along the z-axis, incident on a sphere of radius a centered at the origin (Figure 7.9). In the space surrounding the sphere the disturbance is composed of the incident wave

$$u_{in} = e^{ikz} = e^{ikr\cos\theta} \qquad (7.14.2)$$

which may be taken of unit amplitude in view of the linearity of the problem, and of a scattered wave u_{sc}. If we assume a condition of partial penetrability at the sphere surface, so that

$$[\alpha u + \beta \mathbf{n}\cdot\nabla u]_{r=a} = \left[\alpha u + \beta\frac{\partial u}{\partial r}\right]_{r=a} = 0 \qquad (7.14.3)$$

in which $|\alpha|+|\beta| \neq 0$, the problem is axisymmetric and the scattered wave u_{sc} will be independent of ϕ. Thus, with $u = u_{in} + u_{sc}$, we find that u_{sc} satisfies the same equation (7.14.1) with a boundary condition

$$\left[\alpha u_{sc} + \beta\frac{\partial u_{sc}}{\partial r}\right]_{r=a} = -\left[\alpha u_{in} + \beta\frac{\partial u_{in}}{\partial r}\right]_{r=a} = -\left[\left(\alpha+\beta\frac{\partial}{\partial r}\right)e^{ikr\cos\theta}\right]_{r=a} \tag{7.14.4}$$

at the sphere surface. Often, as in acoustics, ∇u has the physical meaning of a displacement or a velocity, which justifies the qualification of "hard" or "soft" often attributed to scatterers for which $\alpha = 0$ or $\beta = 0$, respectively.

We also expect of course that $u_{sc} \to 0$ at infinity, but we know from the previous section and §3.5 (see in particular the considerations on p. 73) that this condition is insufficient to specify the solution, and that a radiation condition will be needed. In this case, the radiation condition specifies that u_{sc} only contains waves propagating towards infinity rather than in the opposite direction.

We represent the scattered wave as a superposition of Legendre polynomials

$$u_{sc} = \sum_{\ell=0}^{\infty} R_\ell(r)\,P_\ell(\cos\theta) \qquad (7.14.5)$$

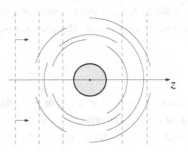

Fig. 7.9   Scattering of scalar plane waves by a sphere centered at the origin.

the coefficients R_ℓ of which satisfy (7.12.1). In view of the asymptotic form of $h_\ell^{(1)}$ and $h_\ell^{(2)}$ (Table 7.9), it is convenient to write

$$R_\ell = A_\ell h_\ell^{(1)}(kr) + B_\ell h_\ell^{(2)}(kr) \rightarrow \frac{1}{kr}\left[A_\ell\, i^{-\ell-1} e^{kr} + B_\ell\, i^{\ell+1} e^{-ikr}\right]. \qquad (7.14.6)$$

We see here that the second term corresponds to waves propagating inward from infinity and must therefore be discarded so that $B_\ell = 0$. We thus find

$$u_{sc} = \sum_{\ell=0}^{\infty} A_\ell\, h_\ell^{(1)}(kr)\, P_\ell(\cos\theta). \qquad (7.14.7)$$

The constants A_ℓ are determined from the boundary condition (7.14.4) by taking scalar products:

$$\frac{2}{2\ell+1}\left[\alpha h_\ell^{(1)}(ka) + k\beta \left(h_\ell^{(1)}\right)'(ka)\right] A_\ell = -\left[\left(\alpha + \beta \frac{\partial}{\partial r}\right)\left(P_\ell, e^{ikr\cos\theta}\right)\right]_{r=a}. \qquad (7.14.8)$$

The calculation of the scalar product in the right-hand side is easily carried out with the help of the relation (7.13.19) of the previous section and the final result is

$$A_\ell = -(2\ell+1)i^\ell \frac{\alpha j_\ell(ka) + k\beta j_\ell'(ka)}{\alpha h_\ell^{(1)}(ka) + k\beta \left(h_\ell^{(1)}\right)'(ka)}. \qquad (7.14.9)$$

By using the asymptotic form of $h^{(1)}$ from Table 7.9 we see that, at a large distance from the sphere, the scattered wave has the form[10]

$$u_{sc} \simeq \frac{e^{ikr}}{r}\frac{1}{k}\sum_{\ell=0}^{\infty} A_\ell\, i^{-\ell-1} P_\ell(\cos\theta). \qquad (7.14.10)$$

As we know from (7.13.8), the factor e^{ikr}/r is a spherically symmetric out-propagating solution of the wave equation while the summation depends on the scattering angle θ, which is also the direction of observation. We thus see that, far from the sphere as measured

---

[10] The same cautionary remark in the footnote on p. 199 applies here.

on the length scale k^{-1}, the wave consists of a spherical disturbance the amplitude of which depends on the scattering angle. The direction-dependent factor

$$f(\theta) = \frac{1}{k} \sum_{\ell=0}^{\infty} A_\ell \, i^{-\ell-1} P_\ell(\cos\theta) \qquad (7.14.11)$$

is known as the *scattering amplitude* or *form factor*.

In the case of a hard scatterer (i.e., $\alpha = 0$) and long wavelengths (i.e., $ka \ll 1$), using the results of Table 7.9, we have, for $\ell \geq 1$,

$$A_\ell = -(2\ell+1)i^\ell \frac{j_\ell'(ka)}{\left(h_\ell^{(1)}\right)'(ka)} \simeq -i \frac{\ell}{\ell+1} \frac{(ka)^{2\ell+1}\left[1 + O(ka)^2\right]}{(2\ell+1)!!(2\ell-1)!!} \qquad (7.14.12)$$

while, for $\ell = 0$,

$$A_0 \simeq \frac{i}{3}(ka)^3\left[1 + O(ka)^2\right]. \qquad (7.14.13)$$

It is evident from these expressions that, to leading order, it is sufficient to retain the first term of the monopole contribution A_0 and of the dipole contribution A_1 disregarding all higher terms and higher multipoles which would give contributions of higher order in ka; the result is then

$$f_{hard} \simeq \frac{1}{3}k^2 a^3 \left(\frac{3}{2}\cos\theta - 1\right). \qquad (7.14.14)$$

For a soft scatterer ($\beta = 0$) a similar argument shows that the dominant term comes from the monopole A_0 only and

$$f_{soft} \simeq -a. \qquad (7.14.15)$$

In the opposite limit of short waves, $ka \gg 1$, the series (7.14.11) and (7.14.18) converge slowly: this feature is the original motivation for the development of the Sommerfeld–Watson transformation described on p. 466.

The fraction of the incident wave energy per unit area scattered per unit time in the solid angle $d\Omega = \sin\theta \, d\theta \, d\phi$ comprised between θ and $\theta + d\theta$ and ϕ and $\phi + d\phi$ is known as the differential *scattering cross section* $d\sigma/d\Omega$ and is related to the scattering amplitude by

$$\frac{d\sigma}{d\Omega} = |f(\theta,\phi)|^2. \qquad (7.14.16)$$

The integral over the solid angle gives the total scattering cross section

$$\sigma = \int |f(\theta,\phi)|^2 \, d\Omega. \qquad (7.14.17)$$

Both quantities have the dimensions of an area and give a measure of the effectiveness of the scatterer in terms of what might be interpreted as an effective cross-sectional area. In view of the orthogonality of the Legendre polynomials (or, in a more general case, of the spherical harmonics), using (7.14.11) we find

$$\sigma = \frac{4\pi}{k^2} \sum_{\ell=0}^{\infty} \frac{|A_\ell|^2}{2\ell+1} \qquad (7.14.18)$$

Table 7.10	Some properties of the spherical vector harmonics; in the last group the prime denotes differentiation with respect to r.

$$\mathbf{B}_\ell^m = \frac{r\nabla Y_\ell^m}{\sqrt{\ell(\ell+1)}} = \frac{1}{\sqrt{\ell(\ell+1)}}\left[\frac{\partial Y_\ell^m}{\partial\theta}\mathbf{e}_\theta + \frac{1}{\sin\theta}\frac{\partial Y_\ell^m}{\partial\phi}\mathbf{e}_\phi\right]$$

$$\mathbf{C}_\ell^m = \frac{\nabla\times\left(Y_\ell^m\mathbf{x}\right)}{\sqrt{\ell(\ell+1)}} = \frac{\nabla Y_\ell^m\times\mathbf{x}}{\sqrt{\ell(\ell+1)}} = \frac{1}{\sqrt{\ell(\ell+1)}}\left[\frac{1}{\sin\theta}\frac{\partial Y_\ell^m}{\partial\phi}\mathbf{e}_\theta - \frac{\partial Y_\ell^m}{\partial\theta}\mathbf{e}_\phi\right]$$

$$\overline{\mathbf{P}}_\ell^m\cdot\mathbf{B}_\ell^m = 0 \qquad\qquad \overline{\mathbf{P}}_\ell^m\cdot\mathbf{C}_\ell^m = 0$$

$$\overline{\mathbf{B}}_\ell^m\cdot\mathbf{C}_\ell^m = 0 \qquad\qquad \overline{\mathbf{C}}_\ell^m\cdot\mathbf{C}_p^q = \overline{\mathbf{B}}_\ell^m\cdot\mathbf{B}_p^q$$

$$\int\overline{\mathbf{P}}_\ell^m\cdot\mathbf{P}_p^q\,d\Omega = \int\overline{\mathbf{B}}_\ell^m\cdot\mathbf{B}_p^q\,d\Omega = \int\overline{\mathbf{C}}_\ell^m\cdot\mathbf{C}_p^q\,d\Omega = \delta_{\ell p}\delta_{mq}$$

$$\mathbf{B}_\ell^m = \mathbf{e}_r\times\mathbf{C}_\ell^m \qquad\qquad \mathbf{C}_\ell^m = -\mathbf{e}_r\times\mathbf{B}_\ell^m$$

$$r\nabla\cdot\mathbf{P}_\ell^m = 2Y_\ell^m \qquad\qquad r\nabla\times\mathbf{P}_\ell^m = \sqrt{\ell(\ell+1)}\mathbf{C}_\ell^m$$

$$r\nabla\cdot\mathbf{B}_\ell^m = -\sqrt{\ell(\ell+1)}Y_\ell^m \qquad\qquad \nabla\cdot\mathbf{C}_\ell^m = 0$$

$$r\nabla\times\mathbf{B}_\ell^m = -\mathbf{C}_\ell^m \qquad\qquad r\nabla\times\mathbf{C}_\ell^m = \sqrt{\ell(\ell+1)}\mathbf{P}_\ell^m + \mathbf{B}_\ell^m$$

$$\nabla\times\nabla\times\left(f(r)\mathbf{P}_\ell^m\right) = \ell(\ell+1)(f/r^2)\mathbf{P}_\ell^m \qquad r^2\nabla[\nabla\cdot\left(f(r)\mathbf{P}_\ell^m\right)] = [(r^2 f')' - 2f]\mathbf{P}_\ell^m$$
$$+\sqrt{\ell(\ell+1)}(f'/r)\mathbf{B}_\ell^m \qquad\qquad +\sqrt{\ell(\ell+1)}[(r^2 f)'/r]\mathbf{B}_\ell^m$$

$$r^2\nabla\times\nabla\times\left(f(r)\mathbf{B}_\ell^m\right) = -(r^2 f')'\mathbf{B}_\ell^m \qquad \nabla[\nabla\cdot\left(f(r)\mathbf{B}_\ell^m\right)] = -\ell(\ell+1)(f/r^2)\mathbf{B}_\ell^m$$
$$-\sqrt{\ell(\ell+1)}(rf)'\mathbf{P}_\ell^m \qquad\qquad -\sqrt{\ell(\ell+1)}(f/r)'\mathbf{P}_\ell^m$$

$$r^2\nabla\times\nabla\times\left(f(r)\mathbf{C}_\ell^m\right) = [\ell(\ell+1)f - (r^2 f')']\mathbf{C}_\ell^m \qquad \nabla[\nabla\cdot\left(f(r)\mathbf{C}_\ell^m\right)] = 0$$

from which we see that the contribution of all the multipoles combine additively. For a hard sphere this expression gives, approximately,

$$\sigma = \frac{4}{9}\pi a^2\left(1 + \frac{3}{4}\right)(ka)^4 = \frac{7}{9}\pi a^2\,(ka)^4. \tag{7.14.19}$$

Among other physical effects, the proportionality to the fourth power of the wave number, and hence of the frequency, is a major cause of the blue color of the sky.

There is a considerable amount of physics embedded in these results, for which we must refer the reader to discipline-specific treatments.

7.15 Expansion of a plane vector wave in vector harmonics

We have introduced in §14.4 p. 345 the vector analogs of the scalar spherical harmonics. These are three families of vector fields defined by

$$\mathbf{P}_\ell^m = Y_\ell^m\mathbf{e}_r, \qquad \mathbf{B}_\ell^m = \frac{r\nabla Y_\ell^m}{\sqrt{\ell(\ell+1)}}, \qquad \mathbf{C}_\ell^m = \frac{\nabla\times\left(Y_\ell^m\mathbf{x}\right)}{\sqrt{\ell(\ell+1)}}. \tag{7.15.1}$$

Any vector field $\mathbf{V}$ can be expanded as

$$\mathbf{V}(r,\theta,\phi) = \sum_{\ell=0}^{\infty} \sum_{m=-\ell}^{\ell} \left[p_{\ell m}(r)\mathbf{P}_\ell^m + b_{\ell m}(r)\mathbf{B}_\ell^m + c_{\ell m}(r)\mathbf{C}_\ell^m \right]. \tag{7.15.2}$$

The more useful properties of these functions are summarized in Table 7.10.

In the next section we will need the expansion of a plane vector wave in vector harmonics analogous to (7.13.19) for the scalar case. To deduce it, we consider a circularly polarized wave and take the polar axis in the direction of propagation as before; thus

$$(\mathbf{i} \pm i\mathbf{j})\, e^{iX\cos\theta} = \sum_{\ell=0}^{\infty} \sum_{m=-\ell}^{\ell} \left[p_{\ell m}^\pm(X)\mathbf{P}_\ell^m + b_{\ell m}^\pm(X)\mathbf{B}_\ell^m + c_{\ell m}^\pm(X)\mathbf{C}_\ell^m \right] \tag{7.15.3}$$

where we have set $X = kr$ for brevity, with $k = \omega/c$ the wave number, and $\mathbf{i}$ and $\mathbf{j}$ are unit vectors in the x and y directions. Restoring the omitted time dependence proportional to $e^{-i\omega t}$, we see that the physically significant real part is

$$\mathrm{Re}\,(\mathbf{i} \pm i\mathbf{j})\, e^{-i\omega t} = \mathbf{i}\cos\omega t \pm \mathbf{j}\sin\omega t \tag{7.15.4}$$

so that the upper and lower signs corresponds to counter clockwise and clockwise rotation, respectively.

From the orthonormality relations of Table 7.10 we have

$$c_{\ell m}^\pm(X) = \int \left[\overline{\mathbf{C}}_\ell^m \cdot (\mathbf{i} \pm i\,\mathbf{j}) \right] e^{iX\cos\theta}\, d\Omega, \tag{7.15.5}$$

with similar relations for the other coefficients. Noting that, from Table 7.1 p. 171,

$$\mathbf{i} \pm i\,\mathbf{j} = e^{\pm i\phi}\left(\sin\theta\,\mathbf{e}_r + \cos\theta\,\mathbf{e}_\theta \pm i\mathbf{e}_\phi \right) \tag{7.15.6}$$

we have, by one of the properties of Table 7.5 p. 175,

$$\overline{(\mathbf{i} \pm i\,\mathbf{j}) \cdot \overline{\mathbf{C}}_\ell^m} = -\frac{ie^{\mp i\phi}}{\sqrt{\ell(\ell+1)}} \left[\mp \frac{\partial Y_\ell^m}{\partial\theta} - i\cot\theta\,\frac{\partial Y_\ell^m}{\partial\phi} \right]$$

$$= -\frac{ie^{\mp i\phi}}{\sqrt{\ell(\ell+1)}}\sqrt{(\ell \pm m)(\ell \mp m + 1)}\,Y_\ell^{m\mp1}. \tag{7.15.7}$$

Since the expansion of $e^{iX\cos\theta}$ only contains $P_\ell \propto Y_\ell^0$, the only terms surviving the integration will be for $m = \pm1$ and one finds that

$$c_{\ell,\pm1}^\pm(X) = \sqrt{4\pi(2\ell+1)}\,i^{\ell+1}j_\ell(X). \tag{7.15.8}$$

The calculation of the b_ℓ^m is facilitated by noting the identity[11]

$$\frac{1}{2\mu}\left[(1-\mu^2)P_\ell^2 + (1+\mu^2)\ell(\ell+1)P_\ell \right] = \frac{\ell(\ell+1)}{2\ell+1}\left[\ell P_{\ell+1} + (\ell+1)P_{\ell-1} \right] \tag{7.15.9}$$

---

[11] Here and in the following P_ℓ^2 does not denote the square of P_ℓ, but P_ℓ^m with $m = 2$.

which can be derived by using the Legendre equation to express $P_\ell^2 = (1 - \mu^2) \mathrm{d}^2 P_\ell / \mathrm{d}\mu^2$ and the recurrence relations. After some reduction with the aid of the relations among spherical Bessel functions in Table 7.8 p. 195, the result is

$$
\begin{aligned}
b_{\ell,\pm 1}^\pm(X) &= \int \left[\overline{\mathbf{B}}_\ell^m \cdot (\mathbf{i} \pm i \, \mathbf{j}) \right] e^{iX\cos\theta} \, \mathrm{d}\Omega \\
&= \pm\sqrt{4\pi(2\ell+1)} i^{\ell+1} \frac{1}{X} \frac{\mathrm{d}}{\mathrm{d}X} [X j_\ell(X)].
\end{aligned}
\tag{7.15.10}
$$

For $p_{\ell,m}^\pm$ we use the identity

$$
P_{\ell-1}^2 - P_{\ell+1}^2 = \ell(\ell+1)(P_{\ell+1} - P_{\ell-1})
\tag{7.15.11}
$$

obtained as before finding

$$
\begin{aligned}
p_{\ell,\pm 1}^\pm(X) &= \int \left[\overline{\mathbf{P}}_\ell^m \cdot (\mathbf{i} \pm i \, \mathbf{j}) \right] e^{iX\cos\theta} \, \mathrm{d}\Omega \\
&= \pm\sqrt{4\pi(2\ell+1)\ell(\ell+1)} i^{\ell+1} \frac{j_\ell(X)}{X}.
\end{aligned}
\tag{7.15.12}
$$

Collecting these results, we have the final form of the expansion as

$$
\begin{aligned}
(\mathbf{i} \pm i \, \mathbf{j}) e^{iX\cos\theta} = \sum_{\ell=1}^\infty i^{\ell+1} \sqrt{4\pi(2\ell+1)} &\left[\pm\sqrt{\ell(\ell+1)} \frac{j_\ell(X)}{X} \mathbf{P}_\ell^{\pm 1} \right. \\
&\left. \pm \frac{1}{X} \frac{\mathrm{d}}{\mathrm{d}X} (X j_\ell(X)) \, \mathbf{B}_\ell^{\pm 1} + j_\ell(X) \mathbf{C}_\ell^{\pm 1} \right].
\end{aligned}
\tag{7.15.13}
$$

A more compact form using the relation $\nabla \times \mathbf{C}_\ell^m = \sqrt{\ell(\ell+1)}\mathbf{P}_\ell^m + \mathbf{B}_\ell^m$ from Table 7.10 is (Jackson 1998, p. 769)

$$
(\mathbf{i} \pm i \, \mathbf{j}) e^{ikr\cos\theta} = \sum_{\ell=1}^\infty i^{\ell+1} \sqrt{4\pi(2\ell+1)} \left[j_\ell(kr)\mathbf{C}_\ell^{\pm 1} \pm \frac{1}{k}\nabla \times \left(j_\ell(kr)\mathbf{C}_\ell^m \right) \right].
\tag{7.15.14}
$$

7.16 Scattering of electromagnetic waves

In Cartesian coordinates the scattering of vector waves can be treated component-by-component, but this is not possible in any other coordinate systems because the unit vectors are not constant. With spherical coordinates one may have recourse to the vector harmonics described in §14.4 and summarized in the previous section. As an example we consider the scattering of a plane, linearly polarized monochromatic electromagnetic wave by a perfectly conducting sphere of radius a.

The analysis presented in §1.5 justifies the use of the Coulomb gauge in which the scalar potential vanishes and the electric and magnetic fields $\mathbf{E}$ and $\mathbf{H}$ are expressed in terms of a divergenceless vector potential $\mathbf{A}$ as

$$
\mathbf{E} = -\frac{1}{c}\frac{\partial \mathbf{A}}{\partial t} = ik\mathbf{A}, \qquad \mathbf{H} = \nabla \times \mathbf{A}
\tag{7.16.1}
$$

so that

$$\mathbf{H} = \frac{1}{ik}\mathbf{\nabla} \times \mathbf{E}. \tag{7.16.2}$$

We have omitted the explicit indication of the exponential time dependence proportional to $e^{-i\omega t}$ and $k = \omega/c$ is the wave number. By (7.16.2) it is sufficient to solve for $\mathbf{E}$ which is divergenceless and satisfies the vector Helmholtz equation

$$-\mathbf{\nabla} \times \mathbf{\nabla} \times \mathbf{E} + k^2\mathbf{E} = 0. \tag{7.16.3}$$

As in the scalar case, we take the direction of propagation as the polar axis and represent each field as the sum of incident and scattered components distinguished by indices *in* and *sc* respectively. We allow for a circular polarization as in the previous section so that

$$\mathbf{E}_{in} = (\mathbf{i} \pm i\mathbf{j})\ E_{in}e^{ikz}, \qquad \mathbf{H}_{in} = \mp i\mathbf{E}_{in}. \tag{7.16.4}$$

If the sphere is a perfect conductor, any tangential component of the electric field will immediately cause a current which tends to destroy it; the proper boundary condition is then the vanishing of the tangential component of $\mathbf{E}$, i.e.

$$\mathbf{x} \times \mathbf{E}_{sc} = -\mathbf{x} \times \mathbf{E}_{in} \qquad \text{on} \qquad |\mathbf{x}| = r = a. \tag{7.16.5}$$

The incident field already satisfies the vector Helmholtz equation. We represent the scattered field as in (14.4.11) p. 346

$$\mathbf{E}_{sc}^{\pm} = \sum_{\ell=0}^{\infty}\sum_{m=-\ell}^{\ell}\left[c_{\ell m}^{\pm}(r)\mathbf{C}_{\ell}^{m} + \mathbf{\nabla} \times \left(q_{\ell m}^{\pm}(r)\mathbf{C}_{\ell}^{m}\right)\right] \tag{7.16.6}$$

and substitute into (7.16.3) to find, with the help of Table 7.10, that $c_{\ell m}^{\pm}$ and $q_{\ell m}^{\pm}$ satisfy the radial Helmholtz equation (7.12.1) p. 194. As argued for the scalar case in §7.14, the proper solutions must be proportional to $h_{\ell}^{(1)}(kr)$ and therefore

$$\mathbf{E}_{sc}^{\pm} = E_{in}\sum_{\ell=0}^{\infty}\sum_{m=-\ell}^{\ell}\left[A_{\ell m}^{\pm}h_{\ell}^{(1)}(kr)\mathbf{C}_{\ell}^{m} \pm \frac{1}{k}\mathbf{\nabla} \times \left(B_{\ell m}^{\pm}h_{\ell}^{(1)}(kr)\mathbf{C}_{\ell}^{m}\right)\right]. \tag{7.16.7}$$

Forming $\mathbf{x} \times \mathbf{E}_{sc}$ and $\mathbf{x} \times \mathbf{E}_{in}$, the latter written in the form (7.15.14), evaluating at $r = a$ and substituting into the boundary condition (7.16.5) we have

$$\sum_{\ell=0}^{\infty}\sum_{m=-\ell}^{\ell}\left[A_{\ell m}^{\pm}h_{\ell}^{(1)}(ka)\mathbf{C}_{\ell}^{m} \pm \frac{1}{k}\mathbf{\nabla} \times \left(B_{\ell m}^{\pm}h_{\ell}^{(1)}(ka)\mathbf{C}_{\ell}^{m}\right)\right]$$
$$= -\sum_{\ell=1}^{\infty}i^{\ell+1}\sqrt{4\pi(2\ell+1)}\left[j_{\ell}(ka)\mathbf{C}_{\ell}^{\pm 1} \pm \frac{1}{k}\mathbf{\nabla} \times \left(j_{\ell}(ka)\mathbf{C}_{\ell}^{m}\right)\right]. \tag{7.16.8}$$

Since the fields $\mathbf{C}_{\ell}^{m}$ and $\mathbf{\nabla} \times \mathbf{C}_{\ell}^{m}$ are linearly independent, we conclude that

$$\left.\begin{array}{c}A_{\ell m}^{\pm}\\B_{\ell m}^{\pm}\end{array}\right\} = -i^{\ell+1}\sqrt{4\pi(2\ell+1)}\frac{j_{\ell}(ka)}{h_{\ell}^{(1)}(ka)} \tag{7.16.9}$$

and, therefore,

$$
\mathbf{E}_{sc}^{\pm} = -E_{in} \sum_{\ell=0}^{\infty} \sum_{m=-\ell}^{\ell} i^{\ell+1} \sqrt{4\pi(2\ell+1)} \, \frac{j_{\ell}(ka)}{h_{\ell}^{(1)}(ka)}
$$

$$
\times \left[h_{\ell}^{(1)}(kr)\mathbf{C}_{\ell}^{m} \pm \frac{1}{k}\nabla \times \left(h_{\ell}^{(1)}(kr)\mathbf{C}_{\ell}^{m} \right) \right]. \tag{7.16.10}
$$

7.17 The elastic sphere

In the equation of linear elasticity (1.6.3) p. 17

$$
\rho \frac{\partial^2 \mathbf{u}}{\partial t^2} = -\mu \nabla \times \nabla \times \mathbf{u} + (\lambda + 2\mu)\nabla(\nabla \cdot \mathbf{u}) + \mathbf{f} \tag{7.17.1}
$$

we substitute for the particle displacement $\mathbf{u}(\mathbf{x}, t)$ the expansion (7.15.2):

$$
\mathbf{u}(\mathbf{x}, t) = \sum_{\ell=0}^{\infty} \sum_{m=-\ell}^{\ell} \left[p_{\ell m}(r, t)\mathbf{P}_{\ell}^{m} + b_{\ell m}(r, t)\mathbf{B}_{\ell}^{m} + c_{\ell m}(r, t)\mathbf{C}_{\ell}^{m} \right]. \tag{7.17.2}
$$

With the aid of the relations of Table 7.10 we find, after taking scalar products with the vector families $\mathbf{P}_{\ell}^{m}$, $\mathbf{B}_{\ell}^{m}$ and $\mathbf{C}_{\ell}^{m}$,

$$
\rho \frac{\partial^2 p_{\ell m}}{\partial t^2} = \frac{\lambda}{r^2}\frac{\partial}{\partial r}\left[\frac{1}{r^2}\frac{\partial}{\partial r}\left(r^2 p_{\ell m}\right)\right] + \mu \left\{ 2\frac{\partial}{\partial r}\left[\frac{1}{r^2}\frac{\partial}{\partial r}\left(r^2 p_{\ell m}\right)\right] - \frac{\ell(\ell+1)}{r^2} p_{\ell m}\right\}
$$

$$
- \frac{\sqrt{\ell(\ell+1)}}{r^2}\left[\mu r^2 \frac{\partial}{\partial r}\left(\frac{b_{\ell m}}{r^3}\right) + \lambda \frac{\partial}{\partial r}\left(\frac{b_{\ell m}}{r}\right)\right] + (\mathbf{P}_{\ell}^{m}, \mathbf{f}), \tag{7.17.3}
$$

$$
\rho \frac{\partial^2 b_{\ell m}}{\partial t^2} = \frac{1}{r^2}\left[\mu \frac{\partial}{\partial r}\left(r^2 \frac{\partial b_{\ell m}}{\partial r}\right) - (\lambda + 2\mu)\ell(\ell+1)b_{\ell m}\right]
$$

$$
+ \frac{\sqrt{\ell(\ell+1)}}{r^3}\left[\lambda \frac{\partial}{\partial r}\left(r^2 p_{\ell m}\right) + \frac{\mu}{r^2}\frac{\partial}{\partial r}\left(r^4 p_{\ell m}\right)\right] + (\mathbf{B}_{\ell}^{m}, \mathbf{f}), \tag{7.17.4}
$$

$$
\rho \frac{\partial^2 c_{\ell m}}{\partial t^2} = \frac{\mu}{r^2}\left[\frac{\partial}{\partial r}\left(r^2 \frac{\partial c_{\ell m}}{\partial r}\right) - \ell(\ell+1)c_{\ell m}\right] + (\mathbf{C}_{\ell}^{m}, \mathbf{f}). \tag{7.17.5}
$$

Here the last terms are the scalar products of the body force with the base fields, e.g.

$$
(\mathbf{P}_{\ell}^{m}, \mathbf{f}) = \int \overline{\mathbf{P}}_{\ell}^{m} \cdot \mathbf{f}\, \mathrm{d}\Omega. \tag{7.17.6}
$$

In the static case and in the absence of body forces it is readily verified that the solutions of the first two equations regular at the origin are proportional to non-negative powers of r. By setting $p_{\ell m} = A_{\ell m} r^{\alpha}$ and $b_{\ell m} = B_{\ell m} r^{\alpha}$ we find a linear homogeneous system in the two unknowns $A_{\ell m}$ and $B_{\ell m}$. The condition that the determinant of this system vanish so that non-zero values of the unknowns are possible, gives two families of solutions (see e.g. Love 1927, p. 250). For the first family

$$p_{\ell m} \propto r^{\ell+1}, \qquad b_{\ell m} = \sqrt{\frac{\ell}{\ell+1}} \frac{(\ell+3)\lambda + (\ell+5)\mu}{\ell\lambda + (\ell-2)\mu} p_{\ell m} \qquad (7.17.7)$$

while, for the second one,

$$b_{\ell m} \propto r^{\ell-1}, \qquad p_{\ell m} = \sqrt{\frac{\ell+1}{\ell}} b_{\ell m}. \qquad (7.17.8)$$

The solutions of the last equation constitute a third family of the form

$$c_{\ell m} \propto r^{\ell}. \qquad (7.17.9)$$

By superposing solutions of these three types it is possible to fit arbitrary boundary conditions of imposed displacement or stress. The coefficients of the superposition are found in the usual way by taking scalar products thanks to the completeness and orthogonality of the base fields.

7.18 Toroidal–poloidal decomposition and viscous flow

In the case of divergenceless vector fields the toroidal–poloidal formalism is slightly more convenient than that of vector harmonics, to which however it is very closely related (§14.4.1 p. 346).

As noted on p. 6, the constraint of zero divergence limits the number of independent scalars which characterize any divergenceless vector field $\mathbf{u}$ to two, which may be chosen in such a way that

$$\mathbf{u} = \nabla \times (\psi \mathbf{x}) + \nabla \times \nabla \times (\chi \mathbf{x}) = \mathbf{T} + \mathbf{S}. \qquad (7.18.1)$$

The first term

$$\mathbf{T} \equiv \nabla \times (\psi \mathbf{x}) = \nabla\psi \times \mathbf{x} \qquad (7.18.2)$$

is the *toroidal component* of $\mathbf{u}$ with *defining scalar* ψ, while the second one,

$$\mathbf{S} = \nabla \times \nabla \times (\chi \mathbf{x}) = \nabla \times (\nabla\chi \times \mathbf{x}), \qquad (7.18.3)$$

is the *poloidal component* of $\mathbf{u}$ with defining scalar χ. If ψ is expanded in a series of scalar spherical harmonics we have

$$\mathbf{T} = \nabla \left[\sum_{\ell=0}^{\infty} \sum_{m=-\ell}^{\ell} \psi_{\ell m}(r) Y_{\ell}^{m} \right] \times \mathbf{x}$$

$$= \left[\sum_{\ell=0}^{\infty} \sum_{m=-\ell}^{\ell} \frac{d\psi_{\ell m}}{dr} Y_{\ell}^{m} \right] \mathbf{e}_r \times \mathbf{x} + \sum_{\ell=0}^{\infty} \sum_{m=-\ell}^{\ell} \psi_{\ell m}(r) \nabla Y_{\ell}^{m} \times \mathbf{x} \qquad (7.18.4)$$

and therefore, from Table 7.10,

$$\mathbf{T} = \sum_{\ell=0}^{\infty} \sum_{m=-\ell}^{\ell} \sqrt{\ell(\ell+1)}\, \psi_{\ell m}(r) \mathbf{C}_\ell^m, \tag{7.18.5}$$

where the $\mathbf{C}_\ell^m$ are the family of vector spherical harmonics defined in (7.15.1). A similar calculation gives

$$\mathbf{S} = \sum_{\ell=0}^{\infty} \sum_{m=-\ell}^{\ell} \nabla \times \left[\sqrt{\ell(\ell+1)}\, \chi_{\ell m}(r) \mathbf{C}_\ell^m \right]. \tag{7.18.6}$$

To connect the defining scalars to $\mathbf{u}$ we note that $\nabla \times [f(r,t)\mathbf{C}_\ell^m]$ is a linear combination of vector harmonics of the type $\mathbf{P}_\ell^m$ and $\mathbf{B}_\ell^m$ (see Table 7.10), which are both orthogonal to $\mathbf{C}_\ell^m$. Thus, a scalar product of (7.18.1) with $\mathbf{C}_\ell^m$ gives

$$\sqrt{\ell(\ell+1)}\, \psi_{\ell m} = (\mathbf{C}_{\ell m}, \mathbf{u}) = \int \overline{\mathbf{C}}_\ell^m \cdot \mathbf{u}\, d\Omega, \tag{7.18.7}$$

where the integration is over the solid angle. On the other hand, $\mathbf{x} \cdot \mathbf{C}_\ell^m = 0$, while

$$\mathbf{x} \cdot \nabla \times [f(r)\mathbf{C}_\ell^m] = \sqrt{\ell(\ell+1)}\, f(r) Y_\ell^m. \tag{7.18.8}$$

Thus, a scalar product of (7.18.1) with $Y_\ell^m \mathbf{x}$ gives

$$\ell(\ell+1)\, \chi_{\ell m} = \left(Y_\ell^m \mathbf{x}, \mathbf{u} \right) = \int \overline{Y}_\ell^m \mathbf{x} \cdot \mathbf{u}\, d\Omega. \tag{7.18.9}$$

As an example of the application of these relations we consider the momentum equation governing the slow viscous flow of an incompressible fluid given in (1.4.10) p. 14, which is

$$\rho \frac{\partial \mathbf{u}}{\partial t} = \mathbf{f} - \nabla p + \mu \nabla^2 \mathbf{u}, \tag{7.18.10}$$

in which ρ and μ are the density and the viscosity coefficient, both assumed constant. Due to the assumed constancy of ρ, the velocity field $\mathbf{u}$ satisfies (see §1.4)

$$\nabla \cdot \mathbf{u} = 0. \tag{7.18.11}$$

Furthermore, $\mathbf{f}$ is the external force per unit volume (rather than per unit fluid mass as in §1.4) and p is the pressure field. Upon taking the divergence of (7.18.10) and using (7.18.11) we find a Poisson equation for the pressure

$$\nabla^2 p = \nabla \cdot \mathbf{f} \tag{7.18.12}$$

the general solution of which we write as

$$p = \mu \sum_{\ell=0}^{\infty} \sum_{m=-\ell}^{\ell} q_{\ell m}(r) Y_\ell^m + p_p \tag{7.18.13}$$

where p_p is any particular solution, e.g. (see §7.6)

$$p_p(\mathbf{x}, t) = -\frac{1}{4\pi} \int \frac{\nabla' \cdot \mathbf{f}(\mathbf{x}', t)}{|\mathbf{x} - \mathbf{x}'|} d^3 x' \tag{7.18.14}$$

while the double sum, with

$$q_{\ell m} = P_{\ell m} r^{\ell} + Q_{\ell m} r^{-\ell-1} \tag{7.18.15}$$

is the general solution of the homogeneous equation (§7.3).

Upon noting that, because of (7.18.11), $\nabla^2 \mathbf{u} = -\nabla \times \nabla \times \mathbf{u}$ (cf. (1.1.10) p. 5) and using the results of Table 7.10, with the representation (7.18.1) for $\mathbf{u}$, the momentum equation (7.18.10) becomes

$$\sum_{k=0}^{\infty} \sum_{n=-k}^{k} \sqrt{k(k+1)} \left[\rho \frac{\partial \psi_{kn}}{\partial t} \mathbf{C}_k^n + \nabla \times \left(\rho \frac{\partial \chi_{kn}}{\partial t} \mathbf{C}_k^n \right) \right]$$

$$= \mathbf{f} - \nabla p_p - \mu \sum_{k=0}^{\infty} \sum_{n=-k}^{k} \nabla \left[q_{kn}(r) Y_k^n \right]$$

$$+ \mu \sum_{k=0}^{\infty} \sum_{n=-k}^{k} \frac{\sqrt{k(k+1)}}{r^2} \left[\frac{\partial}{\partial r} \left(r^2 \frac{\partial \psi_{kn}}{\partial r} \right) - k(k+1) \psi_{kn} \right] \mathbf{C}_k^n$$

$$+ \mu \sum_{k=0}^{\infty} \sum_{n=-k}^{k} \nabla \times \left\{ \frac{\sqrt{k(k+1)}}{r^2} \left[\frac{\partial}{\partial r} \left(r^2 \frac{\partial \chi_{kn}}{\partial r} \right) - k(k+1) \chi_{kn} \right] \mathbf{C}_k^n \right\}. \tag{7.18.16}$$

The scalar product of this equation with $\mathbf{C}_{\ell}^m$ gives, using (7.18.7) and noting that $\left(\mathbf{C}_{\ell}^m, \nabla(f(r) Y_k^n) \right) = 0$,

$$\rho \frac{\partial \psi_{\ell m}}{\partial t} = \frac{\mu}{r^2} \left[\frac{\partial}{\partial r} \left(r^2 \frac{\partial \psi_{\ell m}}{\partial r} \right) - \ell(\ell+1) \psi_{\ell m} \right] - \frac{(\mathbf{C}_{\ell}^m, \mathbf{f} - \nabla p_p)}{\sqrt{\ell(\ell+1)}}, \tag{7.18.17}$$

while the scalar product with $Y_{\ell}^m \mathbf{x}$ gives, by (7.18.8),

$$\rho \frac{\partial \chi_{\ell m}}{\partial t} = \frac{\mu}{r^2} \left[\frac{\partial}{\partial r} \left(r^2 \frac{\partial \chi_{\ell m}}{\partial r} \right) - \ell(\ell+1) \chi_{\ell m} \right]$$

$$+ \frac{1}{\ell(\ell+1)} \left[\left(Y_{\ell}^m, \mathbf{x} \cdot \mathbf{f} - \frac{\partial p_p}{\partial r} \right) - \mu r \frac{\partial q_{\ell m}}{\partial r} \right]. \tag{7.18.18}$$

7.18.1 Sphere in a time-dependent viscous flow

As an application, let us consider a sphere of radius a starting to move at $t = 0$ along a straight line with a velocity $\mathbf{v}_s(t)$ in a fluid initially at rest. It is more convenient to solve the problem in the frame in which the sphere is at rest at the origin. We neglect body forces so that $\mathbf{f} = 0$ and $p_p = 0$. If we take the line of motion as the polar axis, so that the undisturbed fluid velocity is $\mathbf{v}(t) = -\mathbf{v}_s(t)$, the problem is axisymmetric, $m = 0$ and we may drop the second index. Let us represent the total velocity field as $\mathbf{v} + \mathbf{u}$, with the incident flow given by

$$\mathbf{v} = -v(t)\mathbf{k} = v(t) \left(-\cos\theta\, \mathbf{e}_r + \sin\theta\, \mathbf{e}_\theta \right) = -\sqrt{\frac{2\pi}{3}} v(t) \nabla \times \left(r \mathbf{C}_1^0 \right) \tag{7.18.19}$$

where we have used (7.18.7) and (7.18.8) in the last step. In the flow (7.18.19) the pressure is given by $p_0 - \rho \mathbf{x} \cdot (d\mathbf{v}/dt)$ with p_0 an arbitrary constant.

The standard boundary condition for viscous flow over a solid body is the so-called *no slip* condition, according to which, at each point, there is no relative velocity between the fluid and the body surface. Thus, in our case, $\mathbf{u} = -\mathbf{v}$ at $r = a$ or

$$\psi_\ell(a) = 0, \qquad \sum_{\ell=0}^{\infty} \sqrt{\ell(\ell+1)} \nabla \times \left(\chi_\ell \mathbf{C}_\ell^0\right)\bigg|_{r=a} = \sqrt{\frac{2\pi}{3}}\, v\, \nabla \times \left(r\mathbf{C}_1^0\right)\bigg|_{r=a}.$$

$$(7.18.20)$$

We also require that the disturbance induced by the sphere vanish at infinity so that $\psi_\ell \to 0$ as $r \to \infty$. The two conditions on ψ_ℓ immediately show that these functions all vanish, so that there is no toroidal component. Upon taking the scalar product of the second condition (7.18.20) by $\mathbf{x}$ and using (7.18.8) we find

$$\sum_{\ell=0}^{\infty} \ell(\ell+1)\chi_\ell(a) = 2\sqrt{\frac{\pi}{3}}\, v(t)a,$$

$$(7.18.21)$$

Thus, all the χ_ℓ vanish except χ_1, which will satisfy

$$\chi_1(a) = \sqrt{\frac{\pi}{3}}\, av(t),$$

$$(7.18.22)$$

together with (7.18.18) which reduces to

$$\rho \frac{\partial \chi_1}{\partial t} = \frac{\mu}{r^2}\left[\frac{\partial}{\partial r}\left(r^2 \frac{\partial \chi_1}{\partial r}\right) - 2\chi_1\right] - \frac{1}{2}\mu r \frac{\partial q_1}{\partial r}.$$

$$(7.18.23)$$

Since only $\chi_1 \neq 0$, all the functions q_ℓ in the pressure expansion (7.18.15) vanish except q_1 in which, for the pressure disturbance to vanish at infinity, we must take $P_1 = 0$ so that $q_1 = Q_1(t)r^{-2}$.

We take the Laplace transform of (7.18.23) using the initial condition $\mathbf{u}(\mathbf{x}, t = 0) = 0$ and rearrange to find

$$\frac{1}{r^2}\frac{\partial}{\partial r}\left(r^2 \frac{\partial \hat{\chi}_1}{\partial r}\right) + \left(-\frac{s}{v} - \frac{2}{r^2}\right)\hat{\chi}_1 = -\frac{\hat{Q}_1}{r^2}$$

$$(7.18.24)$$

in which carets denote the Laplace-transformed functions and $v = \mu/\rho$ is the kinematic viscosity. This is an equation of the form (7.12.1) p. 194 modified in the sense that the term s/v is preceded by a minus sign. Since we have not introduced the seldom encountered modified spherical Bessel functions, we will use the regular functions giving them the argument ikr with $k = \sqrt{s/v}$ and the square root defined so that $\text{Re}\, k > 0$ when $\text{Re}\, s > 0$. Rather than applying the general formula (7.12.6) it is much simpler in this case to note that the equation can be rewritten as

$$\frac{d}{dr}\left[\frac{1}{r^2}\frac{d}{dr}(r^2\hat{\chi}_1)\right] - \frac{s}{v}\hat{\chi}_1 = -\frac{\hat{Q}_1}{r^2}.$$

$$(7.18.25)$$

It is obvious by inspection that a particular solution is given by $v\hat{Q}_1/(sr^2)$ so that the general solution of (7.18.24) may be written as

$$\hat{\chi}_1 = \hat{\alpha}\, j_1(ikr) + \hat{\beta}\, h_1^{(1)}(ikr) + \frac{v\hat{Q}_1}{sr^2}.$$

$$(7.18.26)$$

For $\mathrm{Re}\, kr > 0$, j_1 is unbounded as $r \to \infty$ (Table 7.9) and it is therefore necessary to take $\hat{\alpha}_1 = 0$. By imposing the boundary condition (7.18.22) and using two relations from Table 7.8 to eliminate derivaives of the spherical Bessel functions we find

$$\hat{\beta}_1(s) = \sqrt{3\pi}\, \frac{a\hat{v}(s)}{ikah_0^{(1)}(ika)}, \qquad \frac{\hat{Q}_1(s)}{k^2 a^2} = -\sqrt{\frac{\pi}{3}}\, \frac{h_2^{(1)}(ika)}{h_0^{(1)}(ika)}\, a\hat{v}(s) \qquad (7.18.27)$$

with which, after substituting explicit expressions for the $h_n^{(1)}$ functions,

$$\hat{\chi}_1(r, s) = \sqrt{\frac{\pi}{3}}\, a\hat{v}(s) \left[\left(1 + \frac{3}{ka} + \frac{3}{k^2 a^2}\right) \frac{a^2}{r^2} - \frac{3}{kr}\left(1 + \frac{1}{kr}\right) e^{-k(r-a)} \right] \qquad (7.18.28)$$

$$\hat{q}_1(r, s) = \sqrt{\frac{\pi}{3}}\, a\hat{v}(s) \left(1 + \frac{3}{ka} + \frac{3}{k^2 a^2}\right) \frac{k^2 a^2}{r^2}. \qquad (7.18.29)$$

An analysis of the flow and, in particular, of the force on the sphere can be built on these results (see e.g. Lamb 1932, p. 632; Landau & Lifshitz 1987a, section 24).

An interesting feature of the solution is that inversion of the transform by means of the convolution theorem (Rule 4 in Table 5.3 p. 126) shows that the flow at the current time has a slowly decaying memory of its previous history. This behavior, characteristic of diffusive processes, is due here to the diffusion of momentum by the action of viscosity.

Another aspect worth noting is the asymptotic behavior of the solution if $v(t) \to V = \mathrm{const.}$ for $t \to \infty$. For this purpose we note that the dependence on $k = \sqrt{s/\nu}$ gives (7.18.28) a branch point at $s = 0$ and, according to (5.1.18) p. 128, the behavior of χ_1 and q_1 for large t depends on the behavior of $\hat{\chi}_1$ and $\hat{q}_1$ near this branch point. Upon expanding (7.18.28) for small k, we readily find

$$\hat{\chi}_1 \simeq \sqrt{\frac{\pi}{12}}\, a\hat{v}(s) \left(3 - \frac{a^2}{r^2}\right), \qquad \hat{q}_1 \simeq \sqrt{3\pi}\, \frac{a\hat{v}(s)}{r^2}. \qquad (7.18.30)$$

In both expressions, terms of order kr and ka have been discarded. If $v(t) \to V$, the small-s behavior of $\hat{v}$ is $\hat{v} \simeq V/s$ (Theorem 11.4.3 p. 298). Thus, from the relation (5.1.18) p. 128, we conclude that

$$\chi_1 \to \sqrt{\frac{\pi}{12}}\, aV \left(3 - \frac{a^2}{r^2}\right), \qquad q_1 \to \sqrt{3\pi}\, \frac{aV}{r^2}, \qquad (7.18.31)$$

so that the disturbance velocity induced by the sphere at steady state is

$$\mathbf{u} = \frac{1}{2} V \frac{a}{r} \left[\left(3 - \frac{a^2}{r^2}\right) \cos\theta\, \mathbf{e}_r - \frac{1}{2}\left(3 + \frac{a^2}{r^2}\right) \sin\theta\, \mathbf{e}_\theta \right], \qquad (7.18.32)$$

$$p = \frac{3}{2} \frac{\mu a V}{r^2} \cos\theta. \qquad (7.18.33)$$

These expressions embody the well-known Stokes solution for the disturbance caused by a stationary sphere on a steady, slow viscous flow directed oppositely to the polar axis (see e.g. Landau & Lifshitz 1987a, section 20; Batchelor 1999, p. 230).

PART II

Essential Tools

8

Sequences and Series

The solution of many of the problems considered in Part I of this book takes the form of an infinite series. We have made occasional statements about the convergence properties of these series in connection with specific results. Here we begin to address in greater detail and in more general terms this issue, which is of fundamental importance for these and many other applications.

We start from the simplest case, numerical sequences and series, and we then proceed to consider sequences and series of functions. We conclude with some practical methods to generate closed-form expressions for the sum of an infinite series.

While some of the concepts introduced in this chapter may appear straightforward, their real importance is appreciated better later on when they are generalized to more abstract situations.

8.1 Numerical sequences and series

Let $\{c_n\} \equiv c_0, c_1, \ldots, c_n, \ldots$ be an unending, non-repeating *sequence* of real or complex numbers. If there is a real or complex number c such that $|c_n - c|$ can be made arbitrarily small for $n \geq N$, provided N is chosen large enough, we say that the sequence converges to c and write

$$\lim_{n \to \infty} c_n = c \quad \text{or} \quad c_n \to c \quad \text{as} \quad n \to \infty. \tag{8.1.1}$$

More formally, this concept is expressed as follows:

Definition 8.1.1 The sequence $\{c_n\}$ converges to the finite number c if and only if, for any arbitrarily small $\varepsilon > 0$, there is an $N = N(\varepsilon)$ such that, for all $n \geq N(\varepsilon)$, $|c_n - c| \leq \varepsilon$.

It is easily proven that the limit is unique, if it exists. Rational functions of convergent sequences converge to the same function of their limits, provided no denominator vanishes (see e.g. Knopp 1966, p. 71). Thus, for example, if $a_n \to a$ and $b_n \to b$, then $a_n \pm b_n \to a \pm b$, $a_n b_n \to ab$ and so on.

Every sequence which is not convergent according to this definition is termed *divergent*. Note that this characterization covers both sequences that "wander around" indefinitely on the real line (or in the complex plane), and sequences that tend to infinity, i.e. such that, given any arbitrarily large C, $|c_n| \geq C$ provided $n \geq M$, with $M = M(C)$ large enough.

Most often, in practice one knows the elements c_n and what is required is to find the limit c, if it exists. Thus, while the previous definition is pretty straightforward, it is seldom useful in practice. A more helpful result is

Theorem 8.1.1 [Cauchy] *Necessary and sufficient condition for the existence of a limiting value of a numerical sequence $\{c_n\}$ is that, for any positive ε, an integer $k = k(\varepsilon)$ can be found such that, for any $n > 0$,*

$$|c_{k+n} - c_k| < \varepsilon. \tag{8.1.2}$$

In other words, all points of the sequence beyond c_k lie within an interval of width 2ε around c_k or, in the complex case, within a circle of radius ε centered at c_k (see e.g. Bromwich 1926, p. 10; Knopp 1966, p. 84; Haaser & Sullivan 1991, p. 26). This is one of the most important theorems of analysis which is greatly generalized in Chapter 19. Sequences which satisfy the condition (8.1.2) are called *Cauchy sequences*. Another key result is

Theorem 8.1.2 [Bolzano–Weierstrass] *Every bounded sequence possesses at least one limiting point.*[1]

Bounded means that there is a finite positive number C such that $|c_n| < C$ for all n. A *limiting point* is a (real or complex) number such that every one of its neighborhoods, however small, contains an infinite number of terms of the sequence. A limiting point is not necessarily the limit. For instance, the sequence $0, 1, 0, 1, 0, 1, \ldots$ has limiting points 0 and 1 but no limit. For c to be the limit it is necessary that $|c_n - c| < \varepsilon$ *for all n* sufficiently large, not just for an infinity of n's.[2]

A consequence of the Bolzano–Weierstrass Theorem is a useful convergence criterion (see e.g. Knopp 1966, p. 80):

Theorem 8.1.3 *A bounded non-decreasing real sequence (i.e., such that $c_n \leq c_{n+1}$ for all n) is convergent.*

An analogous statement applies to non-increasing sequences.

8.1.1 Numerical series

Let $a_0, a_1, \ldots, a_k, \ldots$ be an unending sequence of real or complex numbers and define the partial sums

$$s_n = \sum_{k=0}^{n} a_k = a_0 + a_1 + \cdots + a_n. \tag{8.1.3}$$

[1] This is a special case of the theorem, which remains valid for sequences in arbitrary finite-dimensional spaces; see p. 636.
[2] However one can extract from the sequence one or more subsequences that converge to its limiting points; e.g., in the example just given, the sequences $0, 0, 0, \ldots$ and $1, 1, 1, \ldots$ (see §A.6).

The $\{s_n\}$ constitute a sequence which, if convergent, enables us to define the sum s of the series as the limit of the s_n. Applying Cauchy's Theorem to the sequence of partial sums, we deduce that a necessary and sufficient condition for the convergence of the series is that

$$|s_{k+n} - s_k| = |a_{k+1} + a_{k+2} + \cdots + a_{k+n}| < \varepsilon \qquad (8.1.4)$$

for any n and any positive ε, provided k is large enough. Since we can take $n = 1$, we deduce in particular that it must be possible to make $|a_{k+1}|$ as small as desired, which proves the obvious fact that a series cannot converge unless its generic term tends to zero. As is intuitively clear, this circumstance is necessary, but not sufficient (see §8.1.2). An illustration is provided by the following simple but powerful convergence criterion for a series with non-negative terms (see e.g. Bromwich 1926, p. 33; Widder 1971, p. 20):

Theorem 8.1.4 *Given $a_n \geq 0$, let $a(x)$ be a function such that $a_n = a(n)$; then*

$$\sum_{n>C}^{\infty} a_n \qquad \text{and} \qquad \int_C^{\infty} a(x)\, \mathrm{d}x \qquad (8.1.5)$$

where C is any positive constant, converge or diverge together.

When $a(x) > 0$, for the integral to converge at infinity it is necessary that $x\, a(x) \to 0$. Thus, it is not surprising that, for a series with positive monotonically decreasing terms, convergence requires that $n a_n \to 0$. A simple example is the *harmonic series*

$$1 + \frac{1}{2} + \frac{1}{3} + \cdots + \frac{1}{n} \simeq \log n + \gamma \qquad \text{for} \qquad n \to \infty, \qquad (8.1.6)$$

which diverges as $\int^{\infty} \mathrm{d}x/x$; here $\gamma \simeq 0.57721\ldots$ is Euler's constant (p. 445). As another application, the criterion (8.1.5) readily shows that the series $\sum [n\,(\log n)^p]^{-1}$ converges only for p strictly greater than 1. A connection between the two expressions in (8.1.5) (although not a proof of the statement), is given by the Poisson summation formula of §8.7.

8.1.2 Absolute convergence

Definition 8.1.2 A real or complex series $\sum a_k$ such that $\sum |a_k|$ converges, is called *absolutely convergent*.

It is obvious that $|a_1 + a_2| \leq |a_1| + |a_2|$ and, more generally, that $|s_n| \leq \sum_{k=0}^{n} |a_k|$. Thus, absolute convergence is sufficient to ensure convergence, but it is not necessary. For example, as we see in §8.7,

$$s \equiv 1 - \frac{1}{2} + \frac{1}{3} - \cdots + \frac{1}{n} - \cdots \to \log 2, \qquad (8.1.7)$$

while the series is not absolutely convergent as shown before in (8.1.6). The difference between ordinary and absolute convergence is illustrated by the following fairly obvious result (see e.g. Bromwich 1926, p. 55; Widder 1961, p. 293):

Theorem 8.1.5 [Leibniz] *If the terms of a real series alternate in sign, $a_0 - a_1 + a_2 - a_3 + \cdots$ and if $a_n \to 0+$[3] monotonically as $n \to \infty$, then the series converges and its sum is comprised between $a_0 - a_1$ and a_0.*

The series (8.1.7) illustrates this statement.

At first sight it may appear rather surprising that it is not always possible to *rearrange the terms* of a series without changing its sum. This fact may be made plausible by observing that the partial sums s_n of two series which differ in the order of terms will be different functions of n and, therefore, their limits for $n \to \infty$ may well be different. As a matter of fact, it has been proven by Riemann that the sum of a non-absolutely converging series can be made to have any arbitrary value by a suitable rearrangement of terms, and it can even be made oscillatory or divergent (see e.g. Bromwich 1926, p. 74). An example is the series (8.1.7) which, if rearranged so that each positive term is followed by two negative terms, sums to $\frac{1}{2} \log 2$. However (Bromwich 1926, p. 70; Whittaker & Watson 1927, p. 25; Knopp 1966, p. 138)

Theorem 8.1.6 *The sum of an absolutely convergent series is not affected by a change of the order in which the terms occur.*

An interesting consequence of these considerations for orthogonal bases in a Hilbert space is described on p. 559.

It will be seen in §8.3 that whether a numerical series does or does not converge absolutely has important implications for series of functions. It is therefore useful to state two important criteria for absolute convergence (see e.g. Bromwich 1926, p. 31; Whittaker & Watson 1927, p. 21; Knopp 1966, p. 116):

Theorem 8.1.7 [Cauchy's root test] *The series $\sum a_n$ converges absolutely or diverges according as*[4]

$$\varlimsup_{n \to \infty} |a_n|^{1/n} < 1 \quad \text{or} \quad \varlimsup_{n \to \infty} |a_n|^{1/n} > 1. \tag{8.1.8}$$

In the first case, $|a_n| \leq \rho^n < 1$ and the geometric series $\sum_{n=0}^{\infty} \rho^n$ converges if $\rho < 1$ while it diverges if $\rho > 1$ (see Example 8.3.1 p. 225). The first conclusion of the theorem then follows from the *comparison test*: if, term by term, the modulus is smaller than that of a convergent series, the series converges. In the second case $|a_n|$ does not tend to zero and the series cannot converge.

---

[3] Writing $a_0 - a_1 + a_2 - a_3 + \cdots$ implies that all a_n are non-negative; $a_n \to 0+$ means that the a_n tend to 0 from above.

[4] $\varlimsup$ and $\varliminf$ denote the upper and lower limits of the sequence, namely the largest, or smallest, of its limit values; see §A.6 for precise definitions. The condition $\varliminf |a_n|^{1/n} > 1$ is *not* necessary for divergence (Bromwich 1926, p. 31).

Theorem 8.1.8 [d'Alembert ratio test] *The series $\sum a_n$ converges absolutely or diverges according as*

$$\overline{\lim_{n\to\infty}} \left| \frac{a_{n+1}}{a_n} \right| < 1 \quad \text{or} \quad \underline{\lim}_{n\to\infty} \left| \frac{a_{n+1}}{a_n} \right| > 1. \tag{8.1.9}$$

No definite conclusion can be drawn if the limits equal 1.

As found e.g. in Chapter 13, the case in which the limit equals 1 is not uncommon. Sharper versions of the theorem are necessary in these cases. A useful fairly general criterion, essentially due to Gauss, is the following: If it is possible to express the quotient $|a_n/a_{n+1}|$ in the form

$$\left| \frac{a_n}{a_{n+1}} \right| = 1 + \frac{\theta}{n} + O\left(n^{-p}\right) \qquad \text{with } p > 1, \tag{8.1.10}$$

the series is divergent for $\theta \leq 1$ and convergent for $\theta > 1$ (see e.g. Bromwich 1926 p. 40 and p. 56 for a similar result on series with alternating signs).[5]

Usually d'Alembert's test is easier to apply than the root test, but it is less powerful because (Bromwich 1926, pp. 32 and 422):

$$\overline{\lim} \left| \frac{a_{n+1}}{a_n} \right| \geq \overline{\lim} |a_n|^{1/n} \geq \underline{\lim} |a_n|^{1/n} \geq \underline{\lim} \left| \frac{a_{n+1}}{a_n} \right|. \tag{8.1.11}$$

Therefore it is possible for Cauchy's test to be satisfied, while d'Alembert's is inconclusive.

8.2 Sequences of functions

Given a sequence of functions $\{F_n(x)\}$, $n = 0, 1, 2, \ldots$, by focusing on a particular value of x, we have an ordinary numerical sequence to which all the previous remarks apply. If, for $x = x_*$, the sequence is convergent in the sense of numerical sequences, its limit defines a number which we denote by $F(x_*)$. The points where the sequence converges in this sense constitute the domain of definition of a function $F(x)$ which is the *point-wise limit* of the sequence $\{F_n\}$.

Here and in the rest of this section we refer to real functions and intervals of the real line, but with relatively minor changes we might as well be referring to complex functions defined in the complex plane or to functions in spaces of higher dimension.

The previous definition is perhaps the most natural way to understand convergence of a sequence of functions, but it is actually rather restrictive and arguably not the most useful one, as will be seen later in this section and elsewhere in this book when other convergence modes are introduced.

---

[5] Given a series $\sum_n a_n$ and a divergent series with positive terms $\sum D_n^{-1}$, define $T_n = |a_n/a_{n+1}|D_n - D_{n+1}$. Then, if $\underline{\lim} T_n > 0$, the series $\sum_n a_n$ is convergent, if $\overline{\lim} T_n < 0$, it is divergent. Upon taking $D_n = 1$ we have d'Alembert's test. $D_n = n$ gives Raabe's test, namely convergence or divergence for $\underline{\lim} n(|a_n/a_{n+1}| - 1)$ greater and smaller than 1, respectively. When the limits given by these tests equal 1, more delicate tests are necessary. Some of these may be found by taking $D_n = n \log n$, $D_n = n \log n \log (\log n)$ and so forth (see Bromwich 1926, p. 37). On this basis it is possible to prove Gauss's criterion (8.1.10).

Whatever the type of convergence, new issues arise in the transition from numerical to functional sequences when we allow x to vary: If all the F_n are continuous, will F also be continuous? If the derivatives or integrals of the F_n exist, is F differentiable or integrable? Does the sequence of the derivatives or integrals converge to F' or $\int F \, dx$? We consider some of these issues in the context of series in §8.3 and, in a more general setting, in §A.4.6. Here we give a classic theorem on integration (see e.g. de La Vallée Poussin 1949, vol. 2 p. 41):

Theorem 8.2.1 [Osgood] *Let the sequence* $\{F_n(x)\} \to F(x)$ *point-wise in an interval* $a \leq x \leq b$ *and let the* F_n *and* F *be continuous in the same interval. If the sequence is uniformly bounded in* x,[6] *then*

$$\lim_{n \to \infty} \int_a^b F_n(x) \, dx = \int_a^b F(x) \, dx. \tag{8.2.1}$$

The result also holds if F *is bounded with a finite number of points of discontinuity.*

Example 8.2.1 As an illustration of the need for the hypothesis of uniform boundedness, consider for $0 \leq x \leq 1$ the sequence $F_n(x) = nx \, e^{-nx^2}$, which converges to $F(x) = 0$ in the point-wise sense. All the F_n, and F, are continuous, but the maximum of F_n occurs at $x = 1/\sqrt{2n}$ and is $\sqrt{n/2e}$ so that the sequence is not uniformly bounded; sure enough, we find $\int_0^1 F_n \, dx = 1/2$ which, being independent of n, is also the integral of the limit and is not 0.   □

For continuity of the limit to be a guaranteed consequence of that of the elements of the sequence we must require more than point-wise convergence and uniform boundedness.

8.2.1 Uniform convergence

Definition 8.2.1 The sequence $\{F_n(x)\}$ is said to be *uniformly convergent* in a finite or infinite interval I if, for any positive number ε, there exists a number $N(\varepsilon)$, *independent of x*, such that the inequality

$$|F(x) - F_n(x)| \leq \varepsilon \tag{8.2.2}$$

holds for all $n \geq N(\varepsilon)$ and *for all $x \in I$*.

If Cauchy's criterion (8.1.4) is satisfied at all points $x \in I$ with one and the *same* value of k for each value of ε, then convergence is uniform.

The crucial aspect of the definition is that the *same* value of N guarantees that $|F(x) - F_n(x)| \leq \varepsilon$ whatever x. Figure 8.1 illustrates the point: uniform convergence means that, for all sufficiently large values of n and for all $x \in I$, the graphs of all the $F_n(x)$ lie within a

---

[6] That is, for all n and all $x \in [a, b]$, there is a constant A independent of x such that $|F_n| < A$.

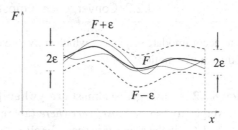

Fig. 8.1   Illustration of the notion of uniform convergence.

strip of width 2ε around the graph of $F(x)$. Thus, uniform convergence implies that the limit $F(x)$ can be approximated to within a certain fixed tolerance ε irrespective of the value of x. This is a very special type of convergence which endows the limit F with special properties. For example:

Theorem 8.2.2 *In a bounded interval $a \leq x \leq b$, a sufficient condition for the continuity of the limit $F(x)$ of a convergent sequence of functions $\{F_n(x)\}$, each continuous in $[a, b]$, is that convergence be uniform.*

Example 8.2.2 Consider the sequence $F_n = x^n$ for $n = 0, 1, 2, \ldots$ in the range $0 \leq x \leq 1$. It is obvious that, in the point-wise sense,

$$x^n \to F(x) = \begin{cases} 0 & 0 \leq x < 1 \\ 1 & x = 1 \end{cases}. \tag{8.2.3}$$

Thus, while all the F_n are continuous, the limit $F(x)$ is not. To examine whether the sequence converges uniformly, let us restrict our attention to the interval $0 \leq x \leq \rho < 1$ for some ρ. In this interval $|F_n(x) - F(x)| = x^n \leq \rho^n$, and this upper bound will be smaller than ε if and only if $n \geq N > (\log \varepsilon)/(\log \rho)$. The closest ρ is to 1, the largest N has to be and, in the limit $\rho \to 1$, the inequality cannot be satisfied with a finite N so that convergence is not uniform. $\square$

In this example non-uniform convergence is caused by a discontinuity of the limit, but this is not necessary. An example is the sequence

$$F_n(x) = \frac{nx}{1 + n^2 x^2} = \frac{1}{n} \frac{x}{x^2 + 1/n^2}. \tag{8.2.4}$$

Evidently $F_n \to 0$ for all $x \in [0, b]$ for any finite $b > 0$, but the maximum of F_n, occurring at $x = 1/n$, is $1/2$ independently of n and therefore $|F_n - F|$ cannot be made smaller than $1/2$. The sequence of Example 8.2.1 offers another instance of the same type.

As a consequence of Osgood's Theorem 8.2.1, integration and passage to the limit commute for uniformly converging sequences, and the sequence of indefinite integrals $\int_a^x F_n(x) \, dx$ is uniformly convergent to $\int_a^x F(x) dx$.

8.2.2 Convergence almost everywhere

In order to accommodate sequences which converge except in a "small" interval or domain we introduce a new notion:

Definition 8.2.2 [Convergence almost everywhere] In an interval $[a, b]$ a sequence $F_n(x)$ converges to $F(x)$ *almost everywhere* (abbreviated as a.e.), or *for almost all x*, if $\lim F_n(x) = F(x)$ in the point-wise sense for all $x \in [a, b]$ except for a set of total measure zero.[7]

The extension to regions of the complex plane or spaces of more than one dimension is obvious. The relation between convergence a.e. and uniform convergence is interesting (see e.g. Riesz & Sz.-Nagy 1955, p. 97; Kolmogorov & Fomin 1957, chapter 5):

Theorem 8.2.3 [Egorov] *Let $F_n(x)$ be a sequence of measurable functions[8] converging almost everywhere to a function $F(x)$ in a finite interval. Convergence can be made uniform by removing from the interval a set of arbitrarily small measure.*

As an illustration, the sequence of Example 8.2.2 converges to 0 almost everywhere on $[0, 1]$ and convergence can be made uniform by removing an arbitrarily short interval around $x = 1$; removal of an interval containing 0 accomplishes the same objective for the sequence (8.2.4).

Allowing for convergence almost everywhere, and understanding integration in the sense of Lebesgue rather than Riemann,[9] we have the following powerful extension of Osgood's Theorem (see e.g. Riesz & Sz.-Nagy 1955, p. 37; Kolmogorov & Fomin 1957, chapter 5; Haaser & Sullivan 1971, p. 136):

Theorem 8.2.4 [Lebesgue's dominated convergence theorem] *Let $\{F_n(x)\}$ be a sequence of integrable functions converging to $F(x)$ for almost all x in a finite or infinite interval I and such that, for almost all x and for all n, $|F_n(x)| < G(x)$ for some function $G(x)$ integrable in I. Then $F(x)$ is integrable in I and*

$$\lim_{n \to \infty} \int_I F_n(x)\, dx = \int_I F(x)\, dx. \tag{8.2.5}$$

A noteworthy aspect of this theorem is that integrability of the limit need not be assumed at the outset, only that of the bound G. However this theorem does not hold for the Riemann integral unless the limit $F(x)$ is known to be Riemann-integrable. Additional results on

---

[7] On the real line a set has *zero measure* if it can be covered by a collection of intervals with an arbitrarily small total length. For example, any finite or denumerably infinite set of points has zero measure. The definition generalizes readily to sets in the complex plane and in spaces of higher dimension; see §A.2 for further details.

[8] See §A.4.3 for the definition of this concept; functions absolutely integrable in the sense of Riemann are an example of measurable functions.

[9] See §A.4.4 for a summary of the differences between the two notions of integral.

passage to the limit under the integral sign are contained in the theorems of Levi and Fatou given in §A.4.6 p. 694.

8.2.3 Other modes of convergence

Point-wise convergence is not a sufficiently strong condition in several respects. For example, a sequence may converge point-wise, but fail to confer to its limit crucial properties such as continuity, as we have just seen. In other cases, what may be relevant for a certain application is not the point-wise behavior of the limit, but some sort of average value over a certain range. These and other considerations have led to several alternative notions of convergence besides the ones already mentioned. In contrast with point-wise convergence, these notions concern some global behavior of the elements of the sequence over an interval or, more generally, a domain.

In order to indicate that the (Lebesgue) integral of some power p of $|F|$ over a finite or infinite interval I is finite

$$\int_I |F(x)|^p \, dx \ < \infty \tag{8.2.6}$$

we write $F \in L^p(I)$. If $p = 2$, the function is said to be *square integrable* while, if $p = 1$, it is *absolutely integrable*.

A sequence of square-integrable functions $\{F_n(x)\}$ over an interval I is said to *converge in the mean* to a function $F(x)$, also square-integrable on the same interval, when

$$\lim_{n \to \infty} \int_I |F(x) - F_n(x)|^2 \, dx = 0. \tag{8.2.7}$$

This concept is also referred to as *mean-square convergence* or L^2-convergence. For absolutely integrable functions we may define another notion of convergence, *absolute* or L^1-*convergence*, by

$$\lim_{n \to \infty} \int_I |F(x) - F_n(x)| \, dx = 0. \tag{8.2.8}$$

From $0 \le |F - 1|^2$ we have $2|F| \le |F|^2 + 1$ and therefore, on a finite interval, mean-square convergence implies absolute convergence, but not vice versa and not if the interval is infinite.

Point-wise convergence and convergence in the L^1 and L^2 senses refer to genuinely different types of convergence and are not just alternative ways to say the same thing. A sequence may converge in the sense of one definition, but not according to another one, or the limits according to different definitions may be different. For example, in both the mean square and absolute senses, the sequence (8.2.3) converges to 0 rather than to its point-wise limit. For $0 \le x \le 1$, the sequence of Example 8.2.1 converges to 0 in the point-wise sense but, since the integral of F_n equals 1/2 independently of n, not in the L^1 sense. Since $\int_0^\infty (x + n)^{-2} \, dx = 1/n$, the sequence $(x + n)^{-1}$ converges to 0 in the sense of $L^2(0, \infty)$, but is not L^1-convergent.

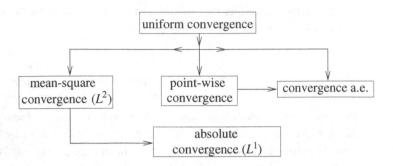

Relations among different types of convergence on a finite interval. The box labelled "point-wise convergence" indicates point-wise convergence at every point of the interval.

For a finite interval, the mutual relationship among the various modes of convergence is illustrated Figure 8.2 and can be summarized as follows:

1. Uniform convergence implies convergence almost everywhere, mean-square and absolute convergence.

2. Mean-square convergence implies absolute convergence.

3. Absolute convergence implies the existence of a subsequence converging almost everywhere.

4. Neither L^1-convergence nor mean-square convergence imply convergence almost everywhere. For example, one can construct sequences of functions converging to 0 both in the L^1 and L^2 sense, without the sequence converging to 0 *anywhere* in the point-wise sense.

5. If $F_n \to F$ in the L^1 sense, one can extract from the sequence a subsequence which converges almost everywhere.

It bears repeating that these results *do not hold* on an infinite interval, for which one can construct sequences which converge uniformly but neither absolutely nor in the mean-square sense.

When we consider abstract spaces in Chapter 19 we will introduce more general notions of convergence which, in some cases, can be exemplified by those defined in this section.

8.3 Series of functions

Just as in the case of numerical series, we can reduce the study of series of functions $\sum_{k=0}^{\infty} f_k(x)$ to that of sequences by focusing on the behavior of the partial sums:

$$s_n(x) = \sum_{k=0}^{n} f_k(x) \qquad n = 0, 1, 2, \dots \tag{8.3.1}$$

The material covered here is complemented by that presented in connection with the Fourier series in Chapter 9. As before, we consider explicitly the real case; extension to complex variables and functions is straightforward.

The series converges at a point x_* if the numerical sequence $s_n(x_*)$ is convergent; it converges point-wise in an interval if $s_n(x)$ is convergent for all points of that interval. The other modes of convergence described in §8.2.3 are defined similarly. For example, the series converges in the mean-square sense to $s(x)$ over the interval $[a, b]$ when

$$\lim_{n \to \infty} \int_a^b \left| s(x) - \sum_{k=0}^n f_k(x) \right|^2 dx = 0. \tag{8.3.2}$$

The implications shown in Figure 8.2 remain valid for a finite interval.

Definition 8.3.1 The series $\sum f_k(x)$ converges *absolutely* in an interval I if it converges absolutely for every point of the interval, i.e. $\sum |f_k(x)|$ converges for all $x \in I$.

Definition 8.3.2 The series $\sum f_k(x)$ converges *uniformly* in a closed interval[10] $a \leq x \leq b$ if, for any positive number ε, there exists a number $N(\varepsilon)$ *independent of* x such that the inequality

$$|s(x) - s_n(x)| \leq \varepsilon \tag{8.3.3}$$

holds for all $n \geq N(\varepsilon)$ and for all x in the interval.

Uniform convergence does not require absolute convergence at any point, nor conversely.

Example 8.3.1 [Geometric series] Consider, for $0 \leq x < 1$, the series $s(x) = \sum_{k=0}^{\infty} x^k$ and its partial sums

$$s_n = 1 + x + x^2 + \cdots + x^n. \tag{8.3.4}$$

By multiplying both sides by x we see that $x s_n = s_n - 1 + x^{n+1}$ so that

$$s_n = \frac{1 - x^{n+1}}{1 - x} \to \frac{1}{1 - x} = s(x) \tag{8.3.5}$$

provided $|x| < 1$. If we want to retain only as many terms in the series as are necessary to ensure that the error in estimating $s(x)$ is less than some small number ε, we must satisfy

$$s(x) - s_{n-1}(x) = \frac{x^n}{1 - x} < \varepsilon \quad \text{or} \quad n > \frac{\log[(1 - x)\varepsilon]}{\log x}. \tag{8.3.6}$$

We thus see that, as x approaches 1, n must be increased more and more to maintain a given accuracy ε. Thus, convergence is not uniform for $0 \leq x < 1$, but it would be uniform in any closed interval $0 \leq x \leq b < 1$ with $N(\varepsilon) \geq \log[(1 - b)\varepsilon]/\log b$. $\qquad\square$

---

[10] A closed interval $a \leq x \leq b$ includes the end points and is denoted by $[a, b]$. An open interval, denoted by (a, b), consists of all those x such that $a < x < b$. The symbols $[a, b)$ or $(a, b]$, denote $a \leq x < b$ or $a < x \leq b$ respectively. More generally, a closed domain includes all the points on its boundary; see §A.1 p. 678 for additional details.

Uniform convergence is important in view of three theorems due to Weierstrass which give *sufficient conditions* for continuity of the sum and term-by-term integration and differentiation; these theorems are direct consequences of the analogous ones given for sequences in §8.2.1 (see e.g. Bromwich 1926, chapter 7; Knopp 1966, chapter 11; Whittaker & Watson 1927, chapter 3). For continuity of the sum, Weierstrass's Theorem is:

Theorem 8.3.1 *If the functions $f_n(x)$ are continuous in $[a, b]$ and the series $\sum f_n(x) \to s(x)$ uniformly in $[a, b]$, then $s(x)$ is also continuous.*

If the terms of the series are positive, $f_n(x) > 0$, then uniform convergence is *also necessary* for continuity of the sum (see e.g. Knopp 1966, p. 344).

For term-by-term integration, Weierstrass's Theorem states:

Theorem 8.3.2 *If the functions $f_n(x)$ are continuous in a bounded closed interval $[a, b]$ and the series $\sum f_n(x) \to s(x)$ uniformly in $[a, b]$, then the series obtained by term-by-term integration of the given series converges uniformly to the integral of its sum:*

$$\int_\alpha^\chi \left[\sum_{n=0}^\infty f_n(x) \right] dx = \sum_{n=0}^\infty \int_\alpha^\chi f_n(x) \, dx \qquad a \le \alpha < \chi \le b \tag{8.3.7}$$

(see e.g. Bromwich 1926, p. 131; Whittaker & Watson 1927, p. 79). As already mentioned, uniform convergence is sufficient but not necessary. For example, the term-by-term integral between 0 and 1 of $1 - x + x^2 - x^3 + \cdots = (1 + x)^{-1}$ equals log 2, just as the integral of the sum, even though the series is not uniformly convergent over the entire interval, as we have seen.[11] Likewise, for the Fourier series and Sturm–Liouville expansions, term-by-term integration is legitimate under much less restrictive conditions (§9.8 and p. 371, respectively).

By relaxing the assumption of uniform convergence and only assuming L^2-convergence, we obtain that the series integrated term-by-term converges in the L^1 sense (8.2.8) (see e.g. Davis 1963, p. 131). This result can be extended to

$$\sum_{k=0}^n \int_a^b f_k(x) \, g(x) \, dx \to \int_a^b s(x) \, g(x) \, dx, \tag{8.3.8}$$

provided $g(x)$ is integrable in the L^2 sense. In the special case in which $\sum f_k$ is uniformly convergent and g absolutely integrable, this relation holds also if $b \to \infty$ (see e.g. Bromwich 1926 pp. 495, 499); other results, in particular concerning integration over infinite intervals, are given in §A.4.7 p. 694. We also have (see e.g. Bromwich 1926, p. 497):

---

[11] This can be justified by noting that the difference between the integral of $1/(1 + x)$ and that of the partial sums is smaller than $x^{n+1}/(n + 1)$ and therefore tends to 0 as $n \to \infty$ (see e.g. Bromwich 1926, p. 181).

Theorem 8.3.3 *Term-by-term integration over the finite range (a, b) is legitimate if either the integral*

$$\int_a^b |g(x)| \sum_{k=1}^\infty |f_k(x)| \, dx \quad \text{or the series} \quad \sum_{k=1}^\infty \int_a^b |g(x)| \, |f_n(x)| \, dx \quad (8.3.9)$$

with g integrable, converges.

As in the case of Weierstrass's theorem 8.3.1, these conditions are also necessary if g and $f_n(x)$ are positive. For differentiation, Weierstrass's Theorem is (see e.g. Bromwich 1926, p. 133; Knopp 1966, p. 342):

Theorem 8.3.4 *If the series $\sum f_n(x)$ converges at least at one point $x \in [a, b]$, its terms are differentiable in $[a, b]$, and the series $\sum_{n=1}^\infty f_n' \to \sigma(x)$ uniformly in $[a, b]$, then the given series is also uniformly convergent to a function $s(x)$ with $s' = \sigma$. In other words, the series can be differentiated term by term:*

$$\left(\sum_{n=1}^\infty f_n(x) \right)' = s'(x) = \sum_{n=1}^\infty f_n'(x). \quad (8.3.10)$$

Several examples given in connection with the Fourier series in Chapter 9 will demonstrate the failure of term-by-term differentiation for non-uniformly convergent series.[12]

If uniform convergence is replaced by a condition on the terms of the series we have (see e.g. Haaser & Sullivan 1971, p. 235; Kolmogorov & Fomin 1957, section 6.1):

Theorem 8.3.5 [Fubini's differentiation theorem] *If f_n is a sequence of non-decreasing functions on $[a, b]$ and $\sum f_n = s(x)$ for all $x \in [a, b]$, then, almost everywhere for $x \in [a, b]$,*

$$\sum f_n'(x) = s'(x). \quad (8.3.11)$$

We conclude with some criteria giving sufficient conditions for the uniform convergence of a series. One of the most useful ones is the following:

Theorem 8.3.6 [Weierstrass M-test] *If the series of positive numbers $M_1 + M_2 + \cdots + M_k + \cdots$ converges and if, for every $x \in [a, b]$, $|f_k(x)| \le M_k$, then the series $\sum f_k$ converges uniformly and absolutely in $[a, b]$.*

A typical application of the theorem is to series of the form

$$s_1(x) = \sum_{k=1}^\infty \frac{\cos kx}{k^q}, \quad s_2(x) = \sum_{k=1}^\infty \frac{\sin kx}{k^q} \quad (8.3.12)$$

---

[12] This theorem refers to differentiation in the classical sense. We will see in Chapter 20 that it is always possible to differentiate a series term-by-term in the sense of distributions.

for both of which $M_k = 1/k^q$. The series $\sum k^{-q}$ converges for $q > 1$ (see e.g. Theorem 8.1.4 p. 217) and therefore s_1 and s_2 are uniformly convergent under this condition. To examine the case $0 < q \le 1$ we need a more delicate test, such as the following one due to Dirichlet (see e.g. Knopp 1966, p. 347):

Theorem 8.3.7 *The series $\sum_k f_k(x)g_k(x)$ is uniformly convergent in a closed interval $[a, b]$ if the partial sums of the series $\sum_k f_k(x)$ are uniformly bounded in $[a, b]$[13] and if the functions $g_k(x)$ converge uniformly to 0 in $[a, b]$, the convergence being monotone (i.e., non-oscillatory) for every fixed x.*

As stated, the theorem is valid for the real case; versions applicable to complex series also exist (see e.g. Bromwich 1926, pp. 243, 246).

Example 8.3.2 Let us apply this test to the series (8.3.12); we take $f_k = \sin kx$ or $f_k = \cos kx$ and $g_k = k^{-q}$. Then, using (8.3.5) and recalling Euler's formula (p. 418)

$$e^{i\alpha} = \cos \alpha + i \sin \alpha, \qquad (8.3.13)$$

we find

$$\sum_{k=1}^{n-1} (\cos kx + i \sin kx) = \sum_{k=1}^{n-1} e^{ikx} = \frac{\sin [(n-1)x/2]}{\sin x/2} \exp \left(\frac{i}{2} nx \right). \qquad (8.3.14)$$

Thus, if $0 < \theta \le x \le 2\pi - \theta$ for some positive θ, the partial sums are bounded by $|\sin \theta/2|^{-1}$ and the series converge uniformly. □

Several other uniform convergence tests exist (see e.g. Bromwich 1926, p. 123; Knopp 1966, section 48).

8.4 Power series

A particularly important type of functional series are power series

$$s(z) = \sum_{n=0}^{\infty} c_n (z - z_0)^n. \qquad (8.4.1)$$

In view of the special importance of these series in the theory of analytic functions (see Chapter 17), in this section we switch from the real variable x to the complex variable z; in (8.4.1) z_0 and the c_n are fixed numbers, in general complex.

Let us consider points such that $|z - z_0| \le R$, with R a real positive number. Then, since $|c_n(z - z_0)^n| \le |c_n| R^n$, we can apply Weierstrass's M-test and examine the convergence

---

[13] That is, for all $x \in [a, b]$ there is a constant A independent of x such that the modulus of all the partial sums is less than A.

of the series with positive terms $\sum |c_n| R^n$. From Cauchy's root test we find that this series converges, and therefore *the power series converges absolutely and uniformly*, provided that $|c_n|^{1/n} R$ tends to a limit smaller than one, which implies that the power series has a radius of convergence given by

$$\frac{1}{R} = \overline{\lim}_{n\to\infty} |c_n|^{1/n} \tag{8.4.2}$$

which must be interpreted as zero if $|c_n|^{1/n}$ diverges.[14] Conversely, the power series diverges for $|z - z_0| > R$. On the circle of convergence $|z - z_0| = R$ the series may converge at some points but not at others (e.g., the geometric series at $z = 1$), converge everywhere, but not absolutely (see Bromwich 1926, p. 250), or diverge everywhere (see Example 17.5.7 p. 441). This is the essential content of the *Cauchy–Hadamard Theorem* for power series. Similarly, from d'Alembert's ratio test we deduce a radius of convergence

$$R = \underline{\lim}_{n\to\infty} \left| \frac{c_n}{c_{n+1}} \right|. \tag{8.4.3}$$

The radii of convergence deduced from the root and the ratio test can be shown to be equal.

Since *inside* its radius of convergence a power series converges uniformly, it follows from Theorems 8.3.1, 8.3.4 and 8.3.2 that

Theorem 8.4.1 *Inside its circle of convergence:*

(i) *the sum of a power series is a continuous function (in fact, analytic);*
(ii) *term-by-term differentiation and integration are valid, and the series so obtained have the same radius of convergence as the original series.*

The last statement is an easy consequence of the root test.

It is often the case that the coefficients of a power series are not constant, but depend on another variable:

$$s(z, t) = \sum_{n=0}^{\infty} c_n(t) (z - z_0)^n. \tag{8.4.4}$$

Continuity of all the coefficients with respect to the variable t in a closed interval $a \le t \le b$ is not sufficient to ensure continuity of s with respect to t. However if, in addition, for all n and $a \le t \le b$, a positive number Z can be found such that $|c_n(t)| Z^n < A n^p$, with A and p fixed positive constants, then s is continuous in t for $|z - z_0| < Z$ (see e.g. Bromwich 1926, p. 152). An example is the series

$$\sum_{n=1}^{\infty} \frac{\sin nt}{n} x^n = \tan^{-1} \frac{x \sin t}{1 - x \cos t}, \tag{8.4.5}$$

for which the constant Z of the theorem must be less than 1 for convergence. The explicit form of the sum is clearly continuous in t for $|x| < 1$ but, for $x = 1$, it is discontinuous at $t = 2k\pi$ even though the series is still convergent.

---

[14] See footnote on p. 218 and §A.6 p. 697 for the definitions of $\overline{\lim}$ and $\underline{\lim}$.

8.5 Other definitions of the sum of a series

While straightforward, the definition (8.1.3) of the sum of a series is not powerful enough for several applications. A simple example is the series $1 - 1 + 1 - 1 + \cdots$ which may be seen as the limit as $x \to 1-$ of $(1+x)^{-1}$ and which, therefore, "should be" summable to 1/2, rather than being divergent as it is according to the definition. In other situations – and for similar reasons – the numerical calculation of the sum according to the definition gives rise to unwelcome features of the result such as the Gibbs phenomenon (§9.10). These and other considerations have motivated the introduction of alternative definitions of the sum of a series. Of course, any such alternative definition must be *regular* in the sense that, if the given series is convergent in the ordinary sense, its sum according to the new definition must equal that according to the old one.

The oldest such extension, which is capable of dealing with the example just given, is the Abel (sometimes called Abel–Poisson) sum:

Definition 8.5.1 [Abel sum] Consider the power series

$$s(t) = \sum_{k=0}^{\infty} a_k t^k \tag{8.5.1}$$

associated to a series $\sum_{k=0}^{\infty} a_k$, and let it be convergent for $0 \leq t < 1$. If the limit

$$\lim_{t \to 1-} s(t) = s_A \tag{8.5.2}$$

exists, it defines the Abel sum s_A of the series $\sum_{k=0}^{\infty} a_k$.

Applications of the Abel sum notion are shown in some examples in §8.7, in §3.6 in connection with the Fourier series and in §13.4 and §14.3 in connection with expansions in series of Legendre polynomials and spherical harmonics. By noting that (see e.g. Hardy 1956, p. 108; Bary 1964, vol. 1 p. 13)

$$\sum_{n=0}^{\infty} a_n t^n = (1-t) \sum_{n=0}^{\infty} s_n t^n = \frac{\sum_{n=0}^{\infty} s_n t^n}{\sum_{n=0}^{\infty} t^n}, \tag{8.5.3}$$

where $s_n = \sum_0^n a_k$, Abel proved the regularity of this definition.

After this result, theorems concerning the limiting behavior of some weighted-average operation applied to a series (or sequence, or integral) from a hypothesis about its ordinary limit are called *Abelian theorems*. The fact that series exist which cannot be summed in a normal sense, although they can be summed in the Abel sense, shows that the converse of Abelian theorems is in general false. Theorems addressing such converse statements are called *Tauberian theorems* and involve additional hypotheses in addition to that of ordinary convergence.

The first result of this type, due to Tauber in 1897, states that the converse of Abel's theorem is true provided $ka_k \to 0$. The condition was weakened by Hardy and Littlewood who showed that boundedness from below of ka_k is sufficient (see e.g. Hardy 1956, p. 153; Korevaar 2004 p. 14).

Example 8.5.1 If we replace x by $iz = i(x + iy)$ with $y > 0$ in the geometric series of Example 8.3.1 we have $\sum_{n=0}^{\infty} e^{inz} = (1 - e^{iz})^{-1}$. The limit $y \to 0$ gives the Abel sum of the series on the left. Separating real and imaginary parts we find

$$\sum_{n=0}^{\infty} \cos nx = \frac{1}{2}, \quad \sum_{n=1}^{\infty} \sin nx = \frac{1}{2} \cot x.$$

$\square$

Another important generalization of the standard sum of a series is the following:

Definition 8.5.2 [Cesáro sum] Given the partial sums $s_n = \sum_{k=0}^{n} a_k$ of the series $\sum_{k=0}^{\infty} a_k$, form a sequence each term of which is the arithmetic mean of the first $n + 1$ partial sums:

$$\sigma_n = \frac{1}{n+1}(s_0 + s_1 + s_2 + \cdots + s_n) = \sum_{k=0}^{n}\left(1 - \frac{k}{n+1}\right)a_k. \qquad (8.5.4)$$

The limit of this sequence, when it exists, defines the Cesáro sum s_C of the series.

A famous application of the Cesáro sum is in Fejér's theory of the summability of the Fourier series (p. 264). The partial sums σ_n can be averaged in their turn and so on giving rise to a whole family of Cesáro summability methods of progressively higher order.

The Abelian theorem for the Cesáro sum states that, if a series converges, its Cesáro sum converges to the same sum while, if it diverges to $\pm\infty$, so does its Cesáro sum. For the example at the beginning of this section we have $s_{2k} = 1$, $s_{2k+1} = 0$, so that the limit of (8.5.4) also gives 1/2. When the Cesáro sum exists, the Abel sum also exists and the two are equal (Frobenius's theorem, see e.g. Bromwich 1926, p. 319; Bary 1964 vol. 1 p.13; Korevaar 2004, p. 4).

Similar definitions can be developed for divergent integrals. The Abel regularization of a divergent integral is motivated by replacing $\lim_{t \to 1-}$ by $\lim_{\varepsilon \to 0} e^{-\varepsilon x}$ in (8.5.2) or, equivalently, in (8.5.3) with s_n replaced by $\int_0^x f(\xi)\,d\xi$ with the result

$$\int_0^{\infty} f(x)\,dx \equiv \lim_{\varepsilon \to 0} \int_0^{\infty} e^{-\varepsilon x} f(x)\,dx. \qquad (8.5.5)$$

To develop the analog of the Cesáro sum we note that, when interchange of the integrations is justified,

$$\frac{1}{x}\int_0^x d\eta \int_0^{\eta} f(\xi)\,d\xi = \frac{1}{x}\int_0^x d\xi\, f(\xi) \int_{\xi}^x d\eta = \int_0^x \left(1 - \frac{\xi}{x}\right) f(\xi)\,d\xi. \qquad (8.5.6)$$

This result suggests to define the Cesáro regularization of a divergent integral by

$$\int_0^{\infty} f(\xi)\,d\xi \equiv \lim_{x \to \infty} \int_0^x \left(1 - \frac{\xi}{x}\right) f(\xi)\,d\xi. \qquad (8.5.7)$$

A broad class of summation methods relies on generalizations of (8.5.3) of the form

$$\lim_{t \to \infty} \frac{\sum_n p_n s_n t^n}{\sum_n p_n t^n} \tag{8.5.8}$$

with $p_n \geq 0$ and $\sum_{n=0}^{\infty} p_n t^n$ is an entire function of t (but not a polynomial).[15] By taking $p_n = 1/n!$, we find

$$\lim_{t \to \infty} \frac{\sum_{n=0}^{\infty} s_n t^n / n!}{\sum_{n=0}^{\infty} t^n / n!} = \lim_{t \to \infty} e^{-t} \sum_{n=0}^{\infty} s_n \frac{t^n}{n!}, \tag{8.5.9}$$

which defines the *Borel sum* (or B_1-sum) of the original series (see e.g. Hardy 1956, p. 79; Korevaar 2004, p. 18). For example, for the geometric series, $a_n = z^n$ and from (8.3.5) we have $s_n = (1 - z^{n+1})/(1 - z)$ so that

$$e^{-t} \sum_{n=0}^{\infty} \frac{1 - z^{n+1}}{1 - z} \frac{t^n}{n!} = \frac{e^{-t}}{1 - z} \left[\sum_{n=0}^{\infty} \frac{t^n}{n!} - z \sum_{n=0}^{\infty} \frac{(zt)^n}{n!} \right] = \frac{1 - z e^{-(1-z)t}}{1 - z}, \tag{8.5.10}$$

the limit of which is $1/(1 - z)$ provided z lies in the half-plane $\operatorname{Re} z < 1$. Thus, the domain of convergence $|z| < 1$ of the original geometric series is considerably enlarged by the Borel definition of sum. A variant and an application of the Borel summation method is described later on p. 239.

Some of the manipulations rendered possible by these general summation and integration methods (for a general theory of which see Hardy 1956; Korevaar 2004) are recovered more systematically and with greater generality in the theory of distributions (Chapter 20), but these ideas remain useful in numerical computation (e.g., to deal with the Gibbs phenomenon in spectral methods, see Canuto *et al.* 1988, p. 50; Jerry 1998), and in the theory of analytic continuation (§17.5 and §17.8). Furthermore, such methods may be applied to the summation of asymptotic series as shown on an example on p. 240.

8.6 Double series

In some problems (see e.g. §3.4) one may apply an eigenfunction expansion in two or more variables, which gives rise to double or multiple series. Imagine arranging the terms of a double series $\sum_{m,n} a_{mn}$ in an infinite matrix with rows $a_{11}, a_{12}, a_{13}, \ldots, a_{21}, a_{22}, a_{23}, \ldots$ and so on. Let $s_{M,N}$ denote the partial sums obtained by adding all the terms in the "rectangle" $1 \leq m \leq M$, $1 \leq n \leq N$, i.e. $a_{11} + (a_{21} + a_{22} + a_{12}) + \cdots$ The double series is said to converge to a finite s if $s_{M,N} \to s$ provided that $M, N \to \infty$ at the same time but *independently* of each other. The analog of Cauchy's criterion (8.1.4) now refers to the difference between the sums in two rectangles, as both rectangles tend to infinity.

---

[15] An entire function is a complex function analytic at all finite points of the complex plane.

The definition given is often not the most convenient one to calculate the sum of a double series. We might prefer to calculate the sum by row, i.e., by first forming, for each m, $b_m = \sum_{n=1}^{\infty} a_{mn}$ and then calculating $\sum_{m=1}^{\infty} b_m$ or, similarly, by column. In general the sum according to the definition does not equal either one of these iterated sums. However

Theorem 8.6.1 *If $s_{M,N}$ is convergent, and all the row and column series converge, then*

$$s = \sum_{m,n} a_{mn} = \sum_{m=1}^{\infty} \left(\sum_{n=1}^{\infty} a_{mn} \right) = \sum_{n=1}^{\infty} \left(\sum_{m=1}^{\infty} a_{mn} \right). \qquad (8.6.1)$$

Existence, or even equality, of the two iterated sums is no guarantee that the double series is convergent; likewise, existence of the sum of the double series does not imply the existence (let alone the equality) of the two iterated sums. The situation is similar to the calculation of a double integral which is not necessarily reducible to two successive one-dimensional integrals (see p. 690). The analogy is strengthened by the applicability of an analog of the Maclaurin convergence test (8.1.5): $\int^{\infty} dx \int^{\infty} dy\, f(x, y)$ and $\sum_m \sum_n f(m, n)$ converge or diverge together when $f > 0$ and steadily decreases to 0 as $x, y \to \infty$.

If the double series is absolutely convergent, existence of any member of (8.6.1) ensures the existence of the other two and the equality of the three series. Furthermore, in the case of an absolutely convergent double series, rather than by "rectangles," the sum can be calculated as the limit of a sequence of arbitrary "areas" progressively increasing in both directions, e.g. by "triangles," i.e. $a_{11} + (a_{21} + a_{12}) + (a_{31} + a_{22} + a_{13}) + \cdots$

A particular case of double series is found by multiplying two single series $A = \sum_m a_m$ and $B = \sum_n b_n$ to form $P = \sum_{mn} a_m b_n$ (see e.g. Bromwich 1926, p. 90; Whittaker & Watson 1927, p. 29; Knopp 1966, p. 146).

Theorem 8.6.2 [Cauchy] *Two absolutely convergent series can be multiplied term-by-term, with the products arranged in any order, to form a new absolutely convergent series the sum of which is the product of the sums of the original series.*

Since the product series is absolutely convergent, its terms can be summed in any order and, in particular, by "triangles":

$$\left(\sum_{n=0}^{\infty} a_n \right) \left(\sum_{k=0}^{\infty} b_k \right) = \sum_{n=0}^{\infty} \left(\sum_{k=0}^{n} a_k b_{n-k} \right). \qquad (8.6.2)$$

The result is still true, and (8.6.2) can be used, if one of the two series is absolutely convergent (*Merten's theorem*), and also if all three series converge (see e.g. Knopp 1966, p. 321; Boas 1987, p. 29), with the product series not necessarily absolutely convergent.[16]

---

[16] In this case, the words "in any order" in the statement of Cauchy's Theorem should be dropped; the proper order is that implicit in the summation by "triangles" in (8.6.2).

The previous considerations are in particular applicable to power series, which are absolutely convergent (p. 229). In this case the multiplication rule (8.6.2) takes the form

$$\left(\sum_{n=0}^{\infty} a_n z^n\right)\left(\sum_{k=0}^{\infty} b_k z^k\right) = \sum_{n=0}^{\infty}\left(\sum_{k=0}^{n} a_k b_{n-k}\right) z^n. \tag{8.6.3}$$

8.7 Practical summation methods

The calculation in closed form of the sum of a series is a task comparable to the evaluation of an integral. Here we review a few of the techniques available which, however, are not as powerful as those available for integration. Although here we focus on infinite series, it will be evident that some methods are applicable also to finite sums.

Summation by exact differences. Suppose that it be possible to find quantities A_k such that, for each term of the series $\sum_k a_k$,

$$a_k = A_{k+1} - A_k. \tag{8.7.1}$$

Then, for the partial sums (8.1.3) we have

$$s_n = (A_1 - A_0) + (A_2 - A_1) + \cdots + (A_{n+1} - A_n) = A_{n+1} - A_0. \tag{8.7.2}$$

Thus, if $A_n \to A_\infty$,

$$s = A_\infty - A_0. \tag{8.7.3}$$

A simple example is

$$a_k = \frac{1}{k(k+1)} = \left(-\frac{1}{k+1}\right) - \left(-\frac{1}{k}\right), \tag{8.7.4}$$

which can be summed by this method to find

$$\sum_{k=1}^{\infty} \frac{1}{k(k+1)} = 1. \tag{8.7.5}$$

This method is similar to the evaluation of a definite integral by taking the difference of the primitive evaluated at the two integration limits. Unfortunately it is seldom applicable as (8.7.1) is a difference equation for the A_k the solution of which is generally difficult, especially in the case of functional, rather than numerical, series.

Use of known series. Another simple, but more powerful, method is applicable when the series of interest has suitable convergence properties and can be related by differentiation or integration to a series the sum of which is known. An example is furnished by the series (8.1.7) encountered above. To evaluate it we consider

$$s(x) = \sum_{k=1}^{\infty} \frac{(-1)^{k-1}}{k} x^k \tag{8.7.6}$$

and note that the numerical series of interest is just $s(1)$, which exists as shown on p. 217. The ratio test (8.4.3) shows that this series converges for $|x| < 1$ and, by Theorem 8.4.1 (p. 229), it converges uniformly in this region. The series obtained by term-wise differentiation converges uniformly in any closed interval $|x| \le b < 1$ (Example 8.3.1 p. 225) and therefore, by Theorem 8.3.4 p. 227, s can be differentiated term by term to find

$$s'(x) = \sum_{k=1}^{\infty} (-x)^{k-1} = \sum_{n=0}^{\infty} (-x)^n = \frac{1}{1+x}, \tag{8.7.7}$$

where we have set $n = k - 1$ and we have used the known sum (8.3.5) of the geometric series. Since convergence is uniform for $|x| < 1$, we can integrate to find

$$s(x) = \log(1 + x), \tag{8.7.8}$$

with the integration constant vanishing because, from (8.7.6), $s(0) = 0$. We can now invoke Abel's regularity theorem to pass to the limit $x \to 1-$ and so recover the result shown in (8.1.7) (see also footnote p. 226). For $x \to -1$ this would be the series (8.1.6), the divergence of which is thus confirmed.

 A slight variation permits us to deal e.g. with the series

$$s(x) = \sum_{n=0}^{\infty} \frac{(-1)^n x^n}{2n + 1} = \frac{1}{2} x^{-1/2} \sum_{n=0}^{\infty} \frac{(-1)^n x^{n+1/2}}{n + 1/2}. \tag{8.7.9}$$

After multiplication by $x^{1/2}$, differentiation gives

$$\frac{d}{dx}\left(x^{1/2}s\right) = \frac{1}{2} x^{-1/2} \sum_{n=0}^{\infty} (-1)^n x^n = \frac{1}{2\sqrt{x}(1+x)}. \tag{8.7.10}$$

Upon integrating between 0 and x we find the sum of the original series

$$s(x) = \frac{\tan^{-1}\sqrt{x}}{\sqrt{x}}. \tag{8.7.11}$$

 In the previous examples, and in others shown later, we rely on the known sum of the geometric series. In the same way we can take advantage of other known sums such as

$$\exp z = \sum_{n=0}^{\infty} \frac{z^n}{n!} \tag{8.7.12}$$

and many others. For example, consider a series of the form

$$s(z; k) = \sum_{n=0}^{\infty} \frac{n^k z^n}{n!}, \tag{8.7.13}$$

the radius of convergence of which is infinite for any k. Upon introducing an auxiliary parameter λ, in view of its uniform convergence, the series can be re-written as

$$s(z; k) = \left[\frac{\partial^k}{\partial \lambda^k} \sum_{n=0}^{\infty} \frac{z^n e^{\lambda n}}{n!}\right]_{\lambda=0} = \left[\frac{\partial^k}{\partial \lambda^k} \sum_{n=0}^{\infty} \frac{(ze^\lambda)^n}{n!}\right]_{\lambda=0} \tag{8.7.14}$$

and summed using (8.7.12):

$$s(z; k) = \left[\frac{\partial^k}{\partial \lambda^k} \exp(ze^\lambda)\right]_{\lambda=0}. \tag{8.7.15}$$

Thus, for example, $s(z; 1) = \sum_{n=0}^{\infty} nz^n/n! = ze^z$. Similar series can be summed also by noting that, for example,

$$s(z; 2) = \sum_{n=0}^{\infty} \frac{n^2 z^n}{n!} = z\frac{d}{dz}\left(z\frac{d}{dz}\right)\sum_{n=0}^{\infty} \frac{z^n}{n!} = z\frac{d}{dz}\left(z\frac{d}{dz}e^z\right). \tag{8.7.16}$$

A series that can be used as a starting point to deal with series involving the semi-factorial $(2n + 1)!! = (2n + 1)(2n - 1)(2n - 3)\ldots 3 \cdot 1$ is the expansion of the error function:

$$\operatorname{erf} x = \frac{2}{\sqrt{\pi}} e^{-x^2} \sum_{n=0}^{\infty} \frac{2^n x^{2n+1}}{(2n + 1)!!}. \tag{8.7.17}$$

Additional examples of this general procedure can be found in connection with the the Legendre polynomials in §13.3.3 p. 322.

Trigonometric series. Since, as shown in Chapters 3 and 7, the solution of many problems leads to Fourier series, the summation of such series is a matter of great practical importance. The sum of some trigonometric series is given in Table 3.2 p. 56. Here we show how some of those results are derived.

If in the power series (8.4.1) we assume all the coefficients to be real, set $z - z_0 = re^{i\theta}$, use Euler's formula (8.3.13) for $e^{i\alpha}$ and separate real and imaginary parts, we find for $0 \leq r < 1$

$$\operatorname{Re} s(r, \theta) = \sum_{n=0}^{\infty} c_n r^n \cos n\theta, \qquad \operatorname{Im} s(r, \theta) = \sum_{n=0}^{\infty} c_n r^n \sin n\theta. \tag{8.7.18}$$

In this way, e.g. from the geometric series $\sum_{n=0}^{\infty} z^n = (1 - z)^{-1}$, we have

$$\sum_{n=0}^{\infty} r^n \cos n\theta = \frac{1 - r\cos\theta}{1 - 2r\cos\theta + r^2}, \qquad \sum_{n=1}^{\infty} r^n \sin n\theta = \frac{r\sin\theta}{1 - 2r\cos\theta + r^2}. \tag{8.7.19}$$

Similarly, by Euler's formula (8.3.13) and (8.7.12), we find for $0 \leq r < \infty$

$$\sum_{n=0}^{\infty} \frac{r^n}{n!} \cos n\theta = e^{r\cos\theta}\cos(r\sin\theta), \qquad \sum_{n=0}^{\infty} \frac{r^n}{n!} \sin n\theta = e^{r\cos\theta}\sin(r\sin\theta). \tag{8.7.20}$$

If the series (8.7.18) converge for $r < 1$, we can define their Abel sum (p. 230) by letting $r \rightarrow 1-$

$$\sum_{n=0}^{\infty} c_n \cos n\theta = \lim_{r \rightarrow 1-} \mathrm{Re}\, s(r, \theta), \qquad \sum_{n=0}^{\infty} c_n \sin n\theta = \lim_{r \rightarrow 1-} \mathrm{Im}\, s(r, \theta), \qquad (8.7.21)$$

even if the series in the left-hand sides were to diverge as written. Thus, allowing x to be complex in (8.7.6) and replacing it by $-re^{i\theta}$, we have

$$\sum_{n=1}^{\infty} \frac{\cos n\theta + i \sin n\theta}{n} = -\log(1 - e^{i\theta}). \qquad (8.7.22)$$

For $\theta \neq 0$ or 2π we find from (17.6.6) p. 449 $\log(1 - e^{i\theta}) = \log 2 \sin \theta/2 + i(\theta - \pi)/2$ so that, upon separating real and imaginary parts, we have for $0 < \theta < 2\pi$

$$\sum_{n=1}^{\infty} \frac{\cos n\theta}{n} = -\log\left(2 \sin \frac{\theta}{2}\right), \qquad \sum_{n=1}^{\infty} \frac{\sin n\theta}{n} = \frac{1}{2}(\pi - \theta). \qquad (8.7.23)$$

Changing θ to $\pi - \theta$ in these results we find

$$\sum_{n=1}^{\infty} (-1)^{n+1} \frac{\cos n\theta}{n} = \log\left(2 \cos \frac{\theta}{2}\right), \qquad \sum_{n=1}^{\infty} (-1)^{n+1} \frac{\sin n\theta}{n} = \frac{1}{2}\theta \qquad (8.7.24)$$

valid for $-\pi < \theta < \pi$. By adding (8.7.23) and (8.7.24) we have

$$\sum_{n=1}^{\infty} \frac{\cos(2n+1)\theta}{2n+1} = \frac{1}{2} \log\left(\cot \frac{\theta}{2}\right), \qquad \sum_{n=1}^{\infty} \frac{\sin(2n+1)\theta}{2n+1} = \frac{1}{4}\pi \qquad (8.7.25)$$

valid for $0 < \theta < \pi$. Another deduction from the second one of (8.7.23) obtained by integration and use of (8.7.52) is (Bernoulli)

$$2\sum_{n=1}^{\infty} \frac{\cos n\theta}{n^2} = \frac{1}{2}\theta^2 - \pi\theta + \frac{1}{3}\pi^2 \qquad \text{for} \qquad 0 \leq \theta \leq 2\pi. \qquad (8.7.26)$$

Other situations can be dealt with by a combination of the previous methods. For example consider, for $0 \leq r \leq 1$,

$$s(r, \theta) = \sum_{n=0}^{\infty} (-r)^n \frac{\sin(2n+1)\theta}{(2n+1)^2}. \qquad (8.7.27)$$

Differentiating twice we have

$$\frac{\partial^2 s}{\partial \theta^2} = -\sum_{n=0}^{\infty} (-r)^n \sin(2n+1)\theta = -\mathrm{Im}\, e^{i\theta} \sum_{n=0}^{\infty} (-r \exp 2i\theta)^n$$

$$= -\mathrm{Im}\, \frac{e^{i\theta}}{1 + r \exp 2i\theta} = \frac{(r-1)\sin \theta}{(r-1)^2 + 4r \cos^2 \theta}. \qquad (8.7.28)$$

Upon integrating once noting that $[\partial s/\partial \theta]_{\theta=0}$ is just the series (8.7.9) the sum of which is given in (8.7.11) we find

$$\frac{\partial s}{\partial \theta} = \frac{1}{2\sqrt{r}}\left[\tan^{-1}\left(\frac{2\sqrt{r}}{1-r}\cos\theta\right) - \tan^{-1}\left(\frac{2\sqrt{r}}{1-r}\right)\right] + \frac{\tan^{-1}\sqrt{r}}{\sqrt{r}}. \qquad (8.7.29)$$

A further integration and use of $s(r,0)=0$ gives the sum of the original series (8.7.27). For the more limited purpose of evaluating $s(1,\theta)$ we take the limit of (8.7.29) for $r \to 1-$ to find, for $0 < \theta < \frac{\pi}{2}$, $[\partial s/\partial \theta]_{r=1} = \frac{1}{4}\pi$. Another integration gives $s(1,\theta) = \pi\theta/4$ with the integration constant vanishing because $s(1,0) = 0$. The simplest way to deal with the range $\frac{\pi}{2} \leq \theta \leq \pi$ is to observe that, from the definition (8.7.27), $s(r,\theta) = s(r,\pi-\theta)$; in conclusion

$$\sum_{n=0}^{\infty}(-1)^n\frac{\sin(2n+1)\theta}{(2n+1)^2} = \begin{cases} \frac{1}{4}\pi\theta & 0 \leq \theta \leq \frac{1}{2}\pi \\ \frac{1}{4}\pi(\pi-\theta) & \frac{1}{2}\pi \leq \theta \leq \pi \end{cases}. \qquad (8.7.30)$$

Poisson summation formula. It is shown on p. 601[17] that, if $F(x)$ and $\tilde{F}(k)$ are cosine transforms of each another, so that (§10.5)

$$F(x) = \sqrt{\frac{2}{\pi}}\int_0^{\infty}\tilde{F}(k)\cos kx\, dk, \qquad \tilde{F}(k) = \sqrt{\frac{2}{\pi}}\int_0^{\infty}F(x)\cos kx\, dx, \qquad (8.7.31)$$

then the following relation holds:

$$\sqrt{\beta}\left[\frac{1}{2}F(0+) + \sum_{n=1}^{\infty}F(n\beta)\right] = \sqrt{\alpha}\left[\frac{1}{2}\tilde{F}(0+) + \sum_{n=1}^{\infty}\tilde{F}(n\alpha)\right], \qquad (8.7.32)$$

provided that the indicated integrals exist and the series are convergent. Here α and β are any two positive numbers such that $\alpha\beta = 2\pi$ and the notation $F(0+)$ indicates the limit value of $F(x)$ at $x=0$ is approached through positive values of x. In particular, with $\beta = 1$ and $\alpha = 2\pi$,

$$\frac{1}{2}F(0+) + \sum_{n=1}^{\infty}F(n) = \int_0^{\infty}F(x)\, dx + 2\sum_{n=1}^{\infty}\int_0^{\infty}F(x)\cos 2\pi nx\, dx. \qquad (8.7.33)$$

The reduction of the summation of one series to that of another one afforded by this formula is useful if the sum of the new series is known, or when the new series is more rapidly convergent than the original one. As an example of the first possibility, let us consider

$$s = \sum_{n=0}^{\infty}\frac{\cos bn}{n^2 + a^2} \qquad \text{for} \qquad 0 \leq b \leq 2\pi, \qquad (8.7.34)$$

so that the function $F(x)$ appearing in (8.7.32) is $F(x) = \cos bx/(x^2 + a^2)$. It is shown in Example 17.9.2 p. 463 that

$$\int_0^{\infty}\frac{\cos \lambda x}{x^2 + a^2}\, dx = \frac{\pi}{2a}\exp(-\lambda a) \qquad \text{for} \qquad a > 0, \qquad \lambda \geq 0 \qquad (8.7.35)$$

---

[17] This proof is based on the theory of distributions; for a proof along more traditional lines see e.g. Titchmarsh 1948, p. 60.

so that

$$\int_0^\infty \frac{\cos bx}{x^2 + a^2} \cos 2n\pi x \, dx = \frac{1}{2} \int_0^\infty \frac{\cos[(b - 2n\pi)x] + \cos[(b + 2n\pi)x]}{x^2 + a^2} \, dx$$

$$= \frac{\pi}{2a} e^{-2n\pi a} \cosh ab \tag{8.7.36}$$

where the condition $\lambda \geq 0$ for the validity of the formula (8.7.35) is ensured by having restricted the range of b as specified in (8.7.34). Application of (8.7.33) and of the known sum of the geometric series then shows that

$$s = \frac{\pi}{2a} \frac{\cosh a(\pi - b)}{\sinh \pi a} - \frac{1}{2a^2}. \tag{8.7.37}$$

As an example of the transformation of a slowly convergent series into a rapidly converging one we consider the series (related to a θ-function):

$$s(\mu) = \sum_{n=0}^\infty \exp(-\mu n^2) \tag{8.7.38}$$

with μ a positive constant,[18] and we again take $\beta = 1$ and $\alpha = 2\pi$. With the result

$$\int_0^\infty e^{-\mu x^2} \cos 2n\pi x \, dx = \frac{\sqrt{\pi}}{2\mu} \exp\left(-\pi^2 \frac{n^2}{\mu^2}\right), \tag{8.7.39}$$

application of (8.7.33) gives

$$s(\mu) = \frac{1}{2} + \frac{\sqrt{\pi}}{\mu} \left[\frac{1}{2} + \sum_{n=1}^\infty \exp\left(-\pi^2 \frac{n^2}{\mu^2}\right)\right]. \tag{8.7.40}$$

For small μ this series converges much faster than (8.7.38), while the latter one is preferable for μ large. This is but one example of several techniques available to speed up the convergence of series; a related example is considered in §3.2.

Borel summation formula. There is a variant of the Borel sum described in §8.5 which is useful for the calculation of the sum of series. For integer argument, $\Gamma(n + 1) = \int_0^\infty t^n e^{-t} \, dt = n!$ (p. 442) and therefore

$$\sum_{n=0}^\infty \frac{a_n}{n!} \int_0^\infty (zt)^n e^{-t} \, dt = \sum_{n=0}^\infty a_n z^n. \tag{8.7.41}$$

This relation shows that, whenever they exist, the functions

$$f(z) = \sum_{n=0}^\infty a_n z^n, \quad \text{and} \quad g(z) = \sum_{n=0}^\infty \frac{a_n z^n}{n!} \tag{8.7.42}$$

---

[18] Since, for example, $\exp(-\mu n^2) \leq \exp(-\mu_0 n)$, the series is uniformly convergent for $\mu \geq \mu_0 > 0$.

are related to each other by

$$f(z) = \int_0^\infty g(zt) e^{-t} dt, \tag{8.7.43}$$

provided that summation and integration can be interchanged.[19] When only the second series in (8.7.42) converges according to the standard definition, (8.7.43) *defines* the Borel sum of the divergent series $\sum_{n=0}^\infty a_n z^n$.[20] If d'Alembert's ratio test (8.4.3) is applied to the series defining g, the radius of convergence is found to be

$$R = \lim_{n\to\infty} (n+1) \left| \frac{a_n}{a_{n+1}} \right|, \tag{8.7.44}$$

which is obviously equal to, or greater than, the radius of convergence of the series defining f. For example, with $a_n = 1$, the series defining f is just the geometric series, which converges only for $|z| < 1$, while the series defining g sums to e^z and has an infinite radius of convergence. This is one way of effecting the *analytic continuation* of a function defined by a power series (see §17.5 and §17.8).

As an application to the summation of a divergent asymptotic series, consider the Euler series

$$f(x) = \sum_{n=0}^\infty (-1)^n n! x^n = 1 - 1!x + 2!x^2 - 3!x^3 + \cdots \tag{8.7.45}$$

Then $g(x) = \sum_{n=0}^\infty (-x)^n = (1+x)^{-1}$ so that[21]

$$f(x) = \int_0^\infty \frac{e^{-t}}{1+tx} dt = -\int_{-\infty}^{-1/x} \frac{e^{s+1/x}}{sx} ds = -\frac{e^{1/x}}{x} \mathrm{Ei}(-1/x), \tag{8.7.46}$$

where Ei is the exponential integral function defined by

$$\mathrm{Ei}(-x) = \int_{-\infty}^{-x} \frac{e^s}{s} ds \qquad \text{for} \qquad x > 0. \tag{8.7.47}$$

Expansion of the denominator and term-by-term integration shows that the series is an asymptotic series for the function defined by the integral (8.7.46) (see Bromwich 1926, p.323; Hardy 1956, p. 27; Knopp 1966, p. 549).

This example is related to *Stieltjes's problem of moments*, which consists in finding a function $f(x)$ given the asymptotic expansion

$$f(x) = a_0 + \frac{a_1}{x} + \frac{a_2}{x^2} + \cdots \tag{8.7.48}$$

---

[19] See pp. 226 and 694.

[20] A connection between the two variants of the Borel method is seen by observing that, if the function in the right-hand side of (8.5.9) is denoted by $y(t)$, we find $y'(t) = e^{-t} g'(t)$ so that $y = a_0 + \int_0^t e^{-s} g'(s) ds = \int_0^t e^{-s} g(s) ds$ by integration by parts, which equals $f(1)$ (see Knopp 1966, p. 551). These steps can only be taken under suitable hypotheses, the need for which renders the two variants not quite equivalent. Exact equivalence holds if and only if $e^{-x} \sum a_n x^n / n! \to 0$. There are series for which the right-hand side of (8.7.43) provides a finite sum, but for which the limit (8.5.9) diverges (see Hardy 1956, p. 182).

[21] This result was first found by Euler who showed by formal manipulation that $x(xf)' + f = 1$, the solution of which is given by (8.7.46).

Stieltjes showed that the solution is provided by

$$f(x) = \int_0^\infty \frac{\phi(t)}{x+t}\, dt \qquad (8.7.49)$$

where ϕ is such that $a_n = (-1)^{n-1} \int_0^\infty \phi(u)u^{n-1}\, du$.

Laplace transform. The Laplace transform of a function $f(t)$ is defined by (Chapter 5)

$$\hat{f}(s) = \int_0^\infty e^{-st} f(t)\, dt. \qquad (8.7.50)$$

Thus, if the order of summation and integration can be interchanged, we have

$$\sum_{n=0}^\infty \hat{f}(n) = \int_0^\infty \left(\sum_{n=0}^\infty e^{-nt} \right) f(t)\, dt = \int_0^\infty \frac{f(t)}{1-e^{-t}}\, dt. \qquad (8.7.51)$$

As an example, from the result $\int_0^\infty te^{-t(s+1)}dt = \int_0^\infty te^{-t}e^{-ts}dt = (s+1)^{-2}$, we find

$$\sum_1^\infty \frac{1}{n^2} = \sum_0^\infty \frac{1}{(n+1)^2} = \int_0^\infty \frac{te^{-t}}{1-e^{-t}}dt = \frac{\pi^2}{6} \qquad (8.7.52)$$

where we have taken care of the fact that the sum starts with from 1 rather than from 0 by subtracting the $n=0$ term of the geometric series; the integral can be evaluated by contour integration (§17.9).

Another particularly powerful method of calculating the sum of infinite series in closed form is the Sommerfeld–Watson transformation which relies on contour integration and is explained in §17.9.4 p. 466.

9

Fourier Series: Theory

A pragmatic view of the Fourier series together with numerous examples was presented in Chapter 3. Here we give more details on the underlying theory which is important not only in itself, but also because the point-wise convergence properties of many other infinite eigenfunction series expansions can be reduced essentially to those of the Fourier series for the reasons mentioned at the end of §15.2 p. 364.

The Fourier series may also be viewed as just one instance of the expansion of the elements, or vectors, of a Hilbert space onto an orthogonal basis. For this reason we defer the treatment of several aspects related to the global convergence of the series to the general exposition of the theory of Hilbert spaces in Chapter 19. Among others, in that chapter we discuss orthogonality, completeness, scalar products, strong and weak convergence, L^p-convergence and least-squares approximation. Completeness, eigenvalues and eigenvectors are treated in chapter 21.

9.1 The Fourier exponential basis functions

The method of eigenfunction expansion briefly summarized in §2.1 p. 27 is based on expressing a function as the superposition of eigenfunctions of a suitable operator. For the Fourier series this operator is $-i\,\mathrm{d}/\mathrm{d}x$ with periodicity boundary conditions. We consider the eigenvalue problem

$$-i\frac{\mathrm{d}e_n}{\mathrm{d}x} = \lambda_n e_n(x), \tag{9.1.1}$$

with the eigenfunction e_n subject to the periodicity condition

$$e_n(x + 2\pi) = e_n(x). \tag{9.1.2}$$

The solution of (9.1.1) is

$$e_n(x) = E_n \exp i\lambda_n x \tag{9.1.3}$$

and, since E_n cannot be zero for a non-trivial solution, the periodicity condition (9.1.2) requires that $\exp(i\lambda_n 2\pi) = 1$ from which[1]

$$\lambda_n = n, \qquad -\infty < n < \infty. \tag{9.1.4}$$

---

[1] Had we started by writing $\mathrm{d}e_n/\mathrm{d}x = \mu_n e_n$, we would have found, in place of (9.1.3), $e_n(x) = D_n \exp \mu_n x$. The periodicity condition would lead now to $\exp(\mu_n 2\pi) = 1$, from which $\mu_n = in$.

The integration constants E_n remain arbitrary as it is evident that any non-zero multiple of a given $e_n(x)$ would still satisfy the problem (9.1.1), (9.1.2).

The functions e_n are periodic with period 2π and therefore we can focus on any interval $a < x \leq a + 2\pi$ with a arbitrary.[2] We mostly consider $a = -\pi$, so that our fundamental interval is $-\pi < x \leq \pi$, although sometimes other choices, such as $0 \leq x < 2\pi$, may prove convenient. Of course, any interval can be mapped onto one of length 2π with an arbitrary starting point; this remark is utilized and amplified in §9.9.

Since we deal here with complex entities, for the scalar product between two functions $f(x)$ and $g(x)$ we use the definition (see §2.1 and §19.3)

$$(f, g) = \int_a^{a+2\pi} \overline{f}(x) \, g(x) \, \mathrm{d}x, \tag{9.1.5}$$

where the overline denotes the complex conjugate. The scalar product of an ordinary vector with itself gives the length of the vector. For a function the corresponding quantity is called *norm* (§19.2) and is indicated as

$$\|f\|^2 = (f, f) = \int_a^{a+2\pi} |f|^2 \, \mathrm{d}x. \tag{9.1.6}$$

The norm of f is only defined for functions such that the integral exists. Such functions are said to be *square integrable* and their class is denoted by L^2 or more precisely, where needed, $L^2(a, a + 2\pi)$. Two functions such that their scalar product vanishes are said to be *orthogonal* to each other; clearly, for the functions (9.1.3) with $\lambda_n = n$,

$$(e_m, e_n) = \int_a^{a+2\pi} \overline{e}_m(x) \, e_n(x) \, \mathrm{d}x = 0, \qquad m \neq n. \tag{9.1.7}$$

As in §2.1 p. 27 we choose the integration constants E_n so that the eigenfunctions have norm 1:

$$(e_n, e_n) = \int_a^{a+2\pi} \overline{e}_n(x) \, e_n(x) \, \mathrm{d}x = 2\pi |E_n|^2 = 1 \tag{9.1.8}$$

which can be satisfied by taking $E_n = 1/\sqrt{2\pi}$.[3]

The preceding arguments have led us to the family of functions

$$e_n = \frac{\exp(inx)}{\sqrt{2\pi}}, \qquad -\pi < x \leq \pi, \qquad -\infty < n < \infty, \tag{9.1.9}$$

which are orthogonal to each other and of "length" (i.e., norm) 1:

$$(e_m, e_n) = \delta_{nm}. \tag{9.1.10}$$

We refer to this property as *orthonormality*. The functions (9.1.9) constitute the set of basis functions on which the Fourier series is constructed.

---

[2] We could equally well write $a \leq x < a + 2\pi$; we want to take out one of the two extrema because $e_n(a + 2\pi) = e_n(a)$.

[3] There is a residual undetermined phase factor $e^{i\alpha}$ with modulus 1. This factor is inessential and can be chosen equal to 1 (i.e., $\alpha = 0$) for simplicity and without loss of generality; see footnote on p. 259.

9.2 Fourier series in exponential form

Given a periodic function $f(x)$ defined over the interval $-\pi < x < \pi$, we formally associate to it its *Fourier series*:

$$f(x) \sim \sum_{n=-\infty}^{\infty} f_n e_n(x), \qquad e_n(x) = \frac{\exp(inx)}{\sqrt{2\pi}}, \tag{9.2.1}$$

where the *Fourier coefficients* f_n are defined by

$$f_n = (e_n, f) = \int_{-\pi}^{\pi} \bar{e}_n(\xi)\, f(\xi)\, \mathrm{d}\xi = \int_{-\pi}^{\pi} \frac{\exp(-in\xi)}{\sqrt{2\pi}}\, f(\xi)\, \mathrm{d}\xi, \tag{9.2.2}$$

provided the integrals exist. Since, evidently,

$$\left| \int_{-\pi}^{\pi} f(\xi)\, \mathrm{d}\xi \right| \le \int_{-\pi}^{\pi} |f(\xi)|\, \mathrm{d}\xi, \tag{9.2.3}$$

we have

$$|f_n| \le \frac{1}{\sqrt{2\pi}} \int_{-\pi}^{\pi} |\exp(-in\xi)|\, |f(\xi)|\, \mathrm{d}\xi = \frac{1}{\sqrt{2\pi}} \int_{-\pi}^{\pi} |f(\xi)|\, \mathrm{d}\xi. \tag{9.2.4}$$

Hence, a *sufficient* (although not necessary) condition for the existence of the Fourier coefficients f_n is that $f(x)$ be absolutely integrable over interval $(-\pi, \pi)$, which we express by writing $f \in L^1(-\pi, \pi)$, or $f \in L^1(0, 2\pi)$ if the fundamental interval is $0 < x < 2\pi$. From $0 \le |f - 1|^2$ we have $2|f| \le |f|^2 + 1$ and therefore, if f is square integrable, it is also absolutely integrable and its Fourier coefficients exist. In either case, however, existence of the coefficients does not necessarily imply convergence of the series. Indeed, there are functions whose Fourier coefficients are well defined, but whose Fourier series diverges at single points or even almost everywhere (see e.g. Hardy & Rogosinski 1956, p. 70; Körner 1989, p.67). For this reason we have avoided the equal sign in (9.2.1) as, at this stage, we do not have any information as to the convergence of the infinite series, let alone the point-by-point equality of the two sides of (9.2.1). We consider these issues in the next section.

    Given a function, (9.2.1) associates to it a Fourier series. Conversely we may start with a series $\sum f_n e_n(x)$ and inquire whether there is function of which the f_n are the Fourier coefficients. The answer is not always in the affirmative as, for example, the series

$$\sum_{n=2}^{\infty} \frac{\sin nx}{\log n} \propto \sum_{n=2}^{\infty} \frac{e_n - e_{-n}}{\log n} \tag{9.2.5}$$

converges in any interval interior to $(0, 2\pi)$, but is not a Fourier series of any function if integration in (9.2.2) is understood in the Lebesgue sense (see e.g. Hardy & Rogosinski 1956, p. 4; Bary 1964, vol. 1 p. 117). However, if a series converges almost everywhere to a finite and integrable function f, then it is the Fourier series of f (see e.g. Hardy & Rogosinski 1956, p. 91).

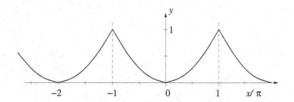

Fig. 9.1   Periodic repetition of the section of the function $y = (x/\pi)^2$ between $-\pi$ and π.

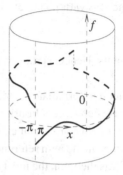

Fig. 9.2   The Cartesian plane wrapped around a circle of unit radius to form the cylinder $\mathscr{S}$ used in the discussion of the Fourier series.

9.3 Point-wise convergence of the Fourier series

While in applying the Fourier series a fundamental interval is often implicit in the very statement of the problem, one should not lose sight of the fact that, mathematically, the function f to be expanded is understood to be periodic. Thus, in principle, there is nothing "special" about the fundamental interval and any interval of length 2π is as good as any other. One way to visualize the situation is by imagining the entire x-axis filled with an infinite repetition of "copies" of f (Figure 9.1). Alternatively, we may imagine "wrapping" the fundamental interval on itself to form a closed cylinder of unit radius $\mathscr{S}$ on which the two points $x = -\pi$ and $x = \pi$ coincide (Figure 9.2). These considerations clarify the important point that a function for which $f(-\pi) \neq f(\pi)$ (or $f(0) \neq f(2\pi)$), is to be considered discontinuous even though it may be continuous at all other points of the interval $-\pi < x < \pi$ (or $0 < x < 2\pi$). Thus when, in the theory of the Fourier series, we refer to a "continuous function," we refer to a function whose graph extending over $-\infty < x < \infty$, or drawn over $\mathscr{S}$, does not exhibit any point of discontinuity.

Since the Fourier coefficients of an ordinary function f are given by integrals, they remain unchanged if the value of f at an isolated point is modified. In particular, at an isolated point of discontinuity x_*, $f(x_*)$ can be defined arbitrarily without affecting its

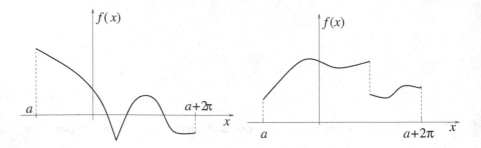

Examples of functions with points of discontinuity of the first kind.

Fourier coefficients. Hence, certainly, the two sides of (9.2.1) cannot be "equal" at a point of discontinuity.[4]

It is very difficult to formulate a complete characterization of the class of functions which admit a point-wise convergent Fourier series and the specialized treatises (e.g. Bary 1964) consider a large number of special cases. A fairly general statement which is sufficient for our purposes can be made with reference to the class of functions of *bounded variation*. For the purposes of this theorem, the most significant property of such functions is that they are either continuous or have a denumerable set of points of discontinuity of the first kind, i.e., such that at each point of discontinuity x_0, both $f(x_0 + 0)$ and $f(x_0 - 0)$ exist;[5] further characterization and properties of this class of functions are provided in §A.3 p. 684. For such functions Dirichlet in 1829 proved a famous convergence theorem later extended by Jordan. The theorem is stated in terms of the partial sums of the Fourier series for f

$$S_N(f, x) = \sum_{n=-N}^{N} f_n\, e_n(x) = \frac{1}{\pi} \int_{-\pi}^{\pi} \left[\frac{1}{2} + \sum_{n=1}^{N} \cos n(x - \xi) \right] f(\xi)\, d\xi. \qquad (9.3.1)$$

The sum in the integrand can be expressed in closed form as[6]

$$\frac{1}{2} + \sum_{n=1}^{N} \cos nx = \frac{\sin\left(N + \frac{1}{2}\right)x}{2 \sin \frac{1}{2}x}. \qquad (9.3.2)$$

This is the *Dirichlet kernel*, which plays a key role in the analysis of the convergence of the Fourier series.

---

[4] On the basis of the previous considerations it is clear that this remark applies also at the end points of the fundamental interval, namely $\pm\pi$, (or a and $a + 2\pi$).

[5] $f(x_0 + 0)$ and $f(x_0 - 0)$ denote the limits of $f(x)$ as $x \to x_0$ from the right and from the left, respectively.

[6] The last step in (9.3.1) is proven by combining the terms e_k and e_{-k} in the sum to find a multiple of the cosine. The equality in (9.3.2) follows by writing $1 + \sum_{n=1}^{N} \cos nx = \text{Re}\left(\sum_{n=0}^{N} e^{inx} \right)$ and using (8.3.5) p. 225 with e^{inx} in place of x.

Theorem 9.3.1 [Dirichlet–Jordan] *If the function f has bounded variation in some interval [a, b] of the cylinder 𝒮 then, for each x_0 interior to [a, b],*

$$\lim_{N \to \infty} S_N(f, x_0) = \frac{1}{2} \left[\lim_{x \to x_0-} f(x) + \lim_{x \to x_0+} f(x) \right]. \qquad (9.3.3)$$

In particular, if f is continuous at x_0, the two limits are equal and

$$S_N(f, x_0) \to f(x_0). \qquad (9.3.4)$$

Convergence is uniform *inside any closed interval in which f is continuous.*

If the function has bounded variation and is continuous over the *whole* 𝒮, then convergence is uniform everywhere in 𝒮. In particular, the convergence relation (9.3.3) holds true:

1. If $f(x)$ is *piecewise smooth* in 𝒮, i.e. piecewise continuous with a piecewise continuous first derivative.[7] This is the class of functions originally considered by Dirichlet; some examples of such functions are shown in Figure 9.3.
2. At points where $f(x)$ possesses a finite derivative; if $f(x)$ is differentiable everywhere in 𝒮, then its Fourier series converges everywhere in 𝒮.
3. If the function satisfies a Lipschitz condition of order α, i.e., for $0 \le h < \delta$,

$$|f(x + h) - f(x)| \le K |h|^\alpha, \qquad (9.3.5)$$

with K and α independent of h, for some positive α and δ (see p. 684).

The proof of some version of this theorem can be found in a very large number of books (see e.g. Hardy & Rogosinski 1956; Bary 1964; Champeney 1987, p. 156; Haaser & Sullivan 1991, p. 315 and many others) and will not be given here.

In spite of the appearance of an integral sign in (9.3.1), the convergence or divergence of a Fourier series at a point x depends only on the behavior of the function in an arbitrarily small neighborhood of the point x (*Riemann's principle of localization*). Thus, for example, the function may have an absolutely integrable singularity at a point, and its Fourier series can still converge in any subinterval not containing that point. An example is (see (8.7.24) p. 237)

$$\log \left| 2 \cos \frac{1}{2} x \right| = \sum_{n=1}^{\infty} (-1)^{n+1} \frac{\cos nx}{n}. \qquad (9.3.6)$$

The series in the right-hand side converges uniformly in every interval completely contained *inside* $[-\pi, 0]$ or $[0, \pi]$, i.e., not including $x = 0$.

The generalization of Theorem 9.3.1 has occupied mathematicians for a long time. Dirichlet himself thought that continuity of f would be enough to ensure convergence. In 1876, however, du Bois-Reymond was able to explicitly construct a continuous function whose

---

[7] A function is said to be piecewise continuous in a domain if it has only a finite number of points of discontinuity at which it possesses finite one-sided limits.

Fourier series diverges at a point (se e.g. Körner 1989, p. 66). The relation between continuity and convergence was clarified surprisingly recently in a theorem published only in 1966:

Theorem 9.3.2 [Carleson] *If the (real or complex) function f defined in $\mathcal{S}$ is square-integrable,[8] then*

$$S_N(f, x) \to f(x) \qquad (9.3.7)$$

as $N \to \infty$, except possibly for x in a set of zero measure.

Upon substituting the coefficients (9.2.2) into the expansion (9.2.1), for those functions for which summation and integration can be interchanged,[9] we have

$$f(x) = \int_{-\pi}^{\pi} \sum_{n=-\infty}^{\infty} \frac{\exp[in(x - \xi)]}{2\pi} f(\xi)\, d\xi. \qquad (9.3.8)$$

This relation suggests that (cf. (9.3.2))

$$\delta(x - \xi) = \frac{1}{2\pi} \sum_{n=-\infty}^{\infty} \exp[in(x - \xi)] = \frac{1}{2\pi} + \frac{1}{\pi} \sum_{n=1}^{\infty} \cos[n(x - \xi)], \quad -\pi < x, \xi < \pi. \qquad (9.3.9)$$

Without a basis in the theory of generalized functions developed in Chapter 20, this may be regarded as a useful short-hand for

$$\lim_{N \to \infty} \frac{1}{2\pi} \sum_{n=-N}^{N} e^{inx} \int_{-\pi}^{\pi} e^{-in\xi} f(\xi)\, d\xi = f(x) \qquad (9.3.10)$$

at any point of continuity of $f(x)$, which is essentially the content of the Dirichlet–Jordan theorem.

Formally, (9.3.9) can be obtained from the definition (9.2.2) of the Fourier coefficients by using the sifting property (2.3.2) of the δ-function; a more satisfactory derivation is given on p. 603.

9.4 Uniform and absolute convergence

As stated in Theorem 9.3.1, convergence of the Fourier series is uniform in the interior of any interval in which the function is continuous and of bounded variation (see e.g. Champeney 1987, p. 157). This result can be strengthened if the function f satisfies further conditions (see e.g. Whittaker & Watson 1927, chapter 9; Kolmogorov & Fomin 1957, chapter 8; Sansone 1959; Bary 1964; Körner 1989).

---

[8] The theorem has been extended by Hunt to apply to functions belonging to L^p with $p > 1$.
[9] On this issue see pp. 226 and 694.

Theorem 9.4.1 [Uniform convergence I] *If f is continuous with bounded variation in the cylinder $\mathscr{S}$ and its derivative is square-integrable in $\mathscr{S}$, the Fourier series of f converges absolutely and uniformly in $\mathscr{S}$.*

Uniform convergence inside any subinterval where the stated properties hold is also ensured. In particular, the theorem holds true if both f and f' are continuous in $\mathscr{S}$.

As in the case of general series of functions, there are Fourier series which converge uniformly but not absolutely. However it follows from Weierstrass's M-test (p. 227) that

Theorem 9.4.2 [Uniform convergence II] *If f is continuous in $\mathscr{S}$, and if*

$$\sum_{-\infty}^{\infty} |(e_n, f)| < \infty, \tag{9.4.1}$$

then $S_N(f, x) \to f(x)$ in $\mathscr{S}$ not only absolutely, but also uniformly.

As regards absolute convergence, one may state the following sufficiency conditions:

Theorem 9.4.3 [Absolute convergence.] *The Fourier series of f converges absolutely if:*

(1) $f(x)$ *satisfies the Lipschitz condition (9.3.5) with $\alpha > 1/2$; or*
(2) $f(x)$ *is of bounded variation and satisfies the Lipschitz condition with $\alpha > 0$; or*
(3) $f(x)$ *is absolutely continuous[10] and $f' \in L^p$ with $p > 1$.*

In the case of the Fourier series absolute convergence is not a local property: absolute convergence at one point implies it everywhere.

9.5 Behavior of the coefficients

As pointed out in §8.1, a series can only converge if the modulus of the n-th term tends to zero as $n \to \infty$. In the case of the Fourier series this condition is

$$\lim_{n \to \infty} \int_{-\pi}^{\pi} \exp(inx) f(x) \, dx = 0, \tag{9.5.1}$$

which is the essential content of the *Riemann–Lebesgue lemma* for $f \in L^1(-\pi, \pi)$ (see e.g. Titchmarsh 1948, p. 11; Champeney 1987, p. 23; Rudin 1987, p.103). By separating real and imaginary parts we also deduce that

$$\lim_{n \to \infty} \int_{-\pi}^{\pi} \sin nx \, f(x) \, dx = 0, \qquad \lim_{n \to \infty} \int_{-\pi}^{\pi} \cos nx \, f(x) \, dx = 0. \tag{9.5.2}$$

---

[10] For a definition of absolute continuity see p. 685. Absolutely continuous functions are continuous, but not necessarily vice versa. Absolutely continuous functions in a closed interval have bounded variation in that interval.

These relations actually hold for any interval, finite or infinite, over which $|f|$ is integrable. The integer n can be replaced by a continuous parameter also tending to infinity. For a differentiable function, integration by parts is sufficient to establish the validity of these results. What is non-trivial is that f does not even have to be continuous for their validity.

The Riemann–Lebesgue lemma in itself does not say anything about the rate of decrease of the coefficients f_n as $|n|$ increases, which is of course a matter of considerable practical importance. We now show that this rate is tied to the smoothness properties of $f(x)$.

Example 9.5.1 Consider, for $-\pi < x < \pi$, $f(x) = 1/\sqrt{|x|}$. Then $f_0 = 2\sqrt{2}$ while, for $n \neq 0$,

$$f_n = \int_{-\pi}^{\pi} \frac{e^{-inx} dx}{\sqrt{2\pi |x|}} = \frac{2}{\sqrt{|n|}} C\left(\sqrt{\pi |n|}\right), \tag{9.5.3}$$

where

$$C(x) = \sqrt{\frac{2}{\pi}} \int_0^x \cos y^2 \, dy \tag{9.5.4}$$

is one of the two Fresnel integrals. Since $C(x) \to \frac{1}{2}$ for $x \to \infty$, it is seen that the Fourier coefficients in this case decay as $|n|^{-1/2}$. The Fourier series for f is

$$\frac{1}{\sqrt{|x|}} = \frac{2}{\sqrt{\pi}} + \frac{4}{\sqrt{2\pi}} \sum_{n=1}^{\infty} C\left(\sqrt{\pi |n|}\right) \frac{\cos nx}{\sqrt{n}} \tag{9.5.5}$$

and it converges uniformly in any closed interval not containing $x = 0$. □

Suppose now that f is not singular but has a point of discontinuity of the first kind (p. 246) at x_0. Then, upon integration by parts,

$$\sqrt{2\pi} f_n = \frac{e^{-inx_0}}{in}[f(x_0 + 0) - f(x_0 - 0)] + \frac{1}{in} \int_{-\pi}^{\pi} e^{-inx} f'(x) \, dx, \tag{9.5.6}$$

where the last integral is understood as the sum of the separate integrals over $[-\pi, x_0)$ and $(x_0, \pi]$ where f' exists. We thus see that if f has a point of discontinuity, f_n is of order $1/n$.

Example 9.5.2 Consider the sign (or signum) function $\text{sgn } x$ defined in (2.3.13) p. 36. This function is discontinuous at $x = 0$ and, if extended periodically beyond the interval $-\pi < x < \pi$, also at all multiples of $\pm\pi$. We have

$$\frac{f_n}{\sqrt{2\pi}} = \frac{1}{2\pi} \int_{-\pi}^{\pi} e^{-inx} \, \text{sgn } x \, dx = \frac{i}{\pi} \int_0^{\pi} \sin nx \cdot 1 \, dx = \frac{i}{\pi n}[(-1)^n - 1], \tag{9.5.7}$$

so that

$$\text{sgn } x = \sum_{-\infty}^{\infty} f_n \frac{\exp(inx)}{\sqrt{2\pi}} = \frac{4}{\pi} \sum_0^{\infty} \frac{\sin(2k+1)x}{2k+1}, \tag{9.5.8}$$

as anticipated. □

If f is continuous but f' has a discontinuity, a double integration by parts shows that f_n is of order $1/n^2$.

Example 9.5.3 Let $f = |x|$ in $-\pi < x < \pi$, with $f(x + 2\pi) = f(x)$. Then while f is continuous, $f' = \operatorname{sgn} x$ is not, although it has bounded variation; we find

$$f_n = \frac{1}{\sqrt{2\pi}} \int_{-\pi}^{\pi} e^{-inx} |x| \, dx = \frac{1}{\sqrt{2\pi}} \frac{2}{n^2} \cos n\pi$$

which shows that $|n^2 f_n| \le M$ with $M = 2$. □

More generally we have the following result:

Theorem 9.5.1 *If f is continuous in $\mathscr{S}$ with its first $p-1$ derivatives and its pth derivative $f^{(p)}$ has bounded variation, then*

$$\left| f_n \, n^{p+1} \right| \le M \qquad\qquad (9.5.9)$$

for all n or, in other words, f_n tends to zero no faster than $1/n^{(p+1)}$.

If f is only integrable, but not of bounded variation, the rate of decrease can be slower than n^{-1}, as seen in Example 9.5.1. If $f \notin C^0$ but is of bounded variation, its Fourier coefficients decay at least as fast as n^{-1} as in (9.5.6). A converse result has been proven by Wiener who gave the following simple criterion for the continuity of $f(x)$ (see e.g. Hardy & Rogosinski 1956, p. 28):

Theorem 9.5.2 *Necessary and sufficient condition for the continuity of the sum of a Fourier series is that $\sum_{-n}^{n} |k f_k| \to 0$.*

In particular, if the coefficients decay proportionally to $1/n$, the function cannot be continuous. If $p \ge 1$, it follows from Weierstrass's M-test (p. 227) and the estimate (9.5.9) that the function defined by the series is continuous and its Fourier series uniformly convergent.

A very important practical consequence of the previous theorems is:

Theorem 9.5.3 [Spectral convergence] *If f is infinitely differentiable, $f \in C^\infty$, then $f_n \to 0$ faster than any power of n.*

This property of *spectral convergence* is obviously extremely desirable in applications, but it should be remembered that, for it to hold, f should be continuous and infinitely differentiable also at the endpoints of the fundamental interval.

In practice, a slow rate of convergence of the series can be a problem. The rate of convergence can be improved by extracting the slowly converging part and summing it separately. The remaining series will then converge faster than the original one. We demonstrate the procedure in §9.7.

9.6 Trigonometric form

In view of Euler's formula (17.12) p. 415 $\exp i\alpha = \cos\alpha + i\sin\alpha$, the exponential Fourier series may be written

$$f(x) = \frac{1}{\sqrt{2\pi}} \sum_{n=-\infty}^{\infty} f_n (\cos nx + i\sin nx). \qquad (9.6.1)$$

If the series converges absolutely, the order of the terms can be rearranged (Theorem 8.1.6, §8.1) so that, observing the parity of the sine and cosine functions, we may rewrite this as

$$f(x) = \frac{f_0}{\sqrt{2\pi}} + \sum_{n=1}^{\infty} \left[\frac{f_n + f_{-n}}{\sqrt{2}} \frac{\cos nx}{\sqrt{\pi}} + i \frac{f_n - f_{-n}}{\sqrt{2}} \frac{\sin nx}{\sqrt{\pi}} \right]. \qquad (9.6.2)$$

Now, from the definition (9.2.2) of the Fourier coefficients,

$$f_0 = \frac{1}{\sqrt{2\pi}} \int_{-\pi}^{\pi} f(\xi)\, d\xi \equiv a_0, \qquad \frac{f_n + f_{-n}}{\sqrt{2}} = \int_{-\pi}^{\pi} \frac{\cos n\xi}{\sqrt{\pi}} f(\xi)\, d\xi \equiv a_n, \qquad (9.6.3)$$

$$i \frac{f_n - f_{-n}}{\sqrt{2}} = \int_{-\pi}^{\pi} \frac{\sin n\xi}{\sqrt{\pi}} f(\xi)\, d\xi \equiv b_n. \qquad (9.6.4)$$

We thus have the Fourier series in the trigonometric form:

$$f(x) = \frac{a_0}{\sqrt{2}} \frac{1}{\sqrt{2\pi}} + \sum_{n=1}^{\infty} \left[a_n \frac{\cos nx}{\sqrt{\pi}} + b_n \frac{\sin nx}{\sqrt{\pi}} \right], \qquad (9.6.5)$$

in which the first term has been written so that a_0 is given by the second one of (9.6.3) with $n = 0$. With the same definition (9.1.5) of scalar product and (9.1.6) of norm the functions

$$\frac{1}{\sqrt{2\pi}}, \qquad \frac{1}{\sqrt{\pi}} \cos(nx), \qquad \frac{1}{\sqrt{\pi}} \sin(nx), \qquad (9.6.6)$$

are pairwise orthogonal and have norm ("length") 1 on any interval of length 2π.

Frequently the series is written without bothering to insist on eigenfunctions of norm 1 and the various $\sqrt{\pi}$ are combined together. In this case the expansion has the perhaps more familiar form

$$f(x) = \frac{A_0}{2} + \sum_{n=1}^{\infty} (A_n \cos nx + B_n \sin nx), \qquad (9.6.7)$$

where

$$A_n = \frac{1}{\pi} \int_{-\pi}^{\pi} \cos(n\xi) f(\xi)\, d\xi, \qquad B_n = \frac{1}{\pi} \int_{-\pi}^{\pi} \sin(n\xi) f(\xi)\, d\xi. \qquad (9.6.8)$$

9.7 Sine and cosine series

Any function can be identically decomposed into an even and an odd part:

$$f(x) = \frac{1}{2} [f(x) + f(-x)] + \frac{1}{2} [f(x) - f(-x)]. \tag{9.7.1}$$

Upon substituting this decomposition into the definition (9.6.8) of A_n, by parity, only the first term gives a non-zero contribution while, for the same reason, only the second one gives a contribution to the coefficient B_n. In addition, when the integrand is even, by using the parity property the integral from $-\pi$ to π may be replaced by twice the integral between 0 and π. Therefore we may rewrite (9.6.8) as

$$A_n = \frac{1}{\pi} \int_0^\pi [f(\xi) + f(-\xi)] \cos(n\xi) \, d\xi, \tag{9.7.2}$$

$$B_n = \frac{1}{\pi} \int_0^\pi [f(\xi) - f(-\xi)] \sin(n\xi) \, d\xi. \tag{9.7.3}$$

If the function f is odd to start with, then all A_n's vanish and its Fourier series reduces to

$$f(x) = \sum_{n=1}^\infty B_n \sin nx, \qquad B_n = \frac{2}{\pi} \int_0^\pi f(\xi) \sin(n\xi) \, d\xi. \tag{9.7.4}$$

This is the *Fourier sine series*. Conversely, if f is even, all the B_n's vanish and we have the Fourier *cosine series*:

$$f(x) = \frac{A_0}{2} + \sum_{n=1}^\infty A_n \cos nx, \qquad A_n = \frac{2}{\pi} \int_0^\pi f(\xi) \cos(n\xi) \, d\xi. \tag{9.7.5}$$

Assuming that summation and differentiation can be interchanged (see footnote p. 248), we can easily see how the correct $f(x)$ is reconstructed with these series. Consider for example the sine series. Then we may write

$$f(x) = \int_0^\pi \left[\frac{2}{\pi} \sum_{n=1}^\infty \sin n\xi \, \sin nx \right] f(\xi) \, d\xi. \tag{9.7.6}$$

But, using (9.3.9),

$$\frac{2}{\pi} \sum_{n=1}^\infty \sin n\xi \, \sin nx = \frac{1}{\pi} \sum_{n=1}^\infty [\cos n(x - \xi) - \cos n(x + \xi)]$$

$$= \left[\delta(x - \xi) - \frac{1}{2\pi} \right] - \left[\delta(x + \xi) - \frac{1}{2\pi} \right] = \delta(x - \xi) - \delta(x + \xi), \tag{9.7.7}$$

which is the odd extension of $\delta(x - \xi)$ and is equivalent to $\delta(x - \xi)$ over the range $[0, \pi]$ since $x + \xi \neq 0$ over this range so that $\delta(x + \xi) = 0$.

Consider now a function f defined only for $0 < x < \pi$ and absolutely integrable in this interval. The definitions (9.7.4) and (9.7.5) of B_n and A_n both make sense and we conclude

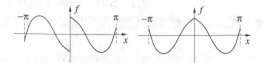

In the diagram on the left a function f defined for $0 < x < \pi$ is extended to $(-\pi, 0)$ in an odd way, $f(-x) = -f(x)$. In the diagram on the right the same function is extended evenly, $f(-x) = f(x)$. In reality the Fourier sine and cosine expansions are expansions of these odd and even extensions, respectively. The even extension maintains continuity of the function, but not of the derivative, and conversely for the odd extension.

that $f(x)$ can be represented either by a sine or a cosine series.[11] An important point to note, however, is that the sine and cosine series of the *same* function $f(x)$ in general behave differently at 0 and π. Indeed, expanding f in a sine series is equivalent to first prolonging it to $[-\pi, 0]$ as an odd function, $f(-x) = -f(x)$, and then using a regular two-sided Fourier series for this prolonged function (Figure 9.4 left). If $f(0) \neq 0$, the prolonged function will have a discontinuity at 0 and its regular Fourier series, and therefore the sine series as well, will not converge to $f(0)$ but to the average of $f(0)$ and $-f(0)$, i.e. to 0. For example, consider the function $\cos x$ in $0 < x < \pi$. According to the definition (9.7.4) of B_n, this can be expanded in a Fourier sine series as follows:

$$\cos x = \frac{8}{\pi} \sum_{n=1}^{\infty} \frac{n}{4n^2 - 1} \sin 2nx, \qquad 0 < x < \pi, \qquad (9.7.8)$$

which implies that the function $\cos x$ has been extended to the interval $-\pi < x < 0$ in an odd way. As another example

$$\sin x = \frac{2}{\pi} - \frac{4}{\pi} \sum_{n=1}^{\infty} \frac{\cos 2nx}{4n^2 - 1}, \qquad 0 < x < \pi. \qquad (9.7.9)$$

The method of summation by exact differences (p. 234) shows that the sum of the series at $x = 0$ and π vanishes.

Over a different fundamental interval, e.g. $0 < x < 2\pi$, use of the cosine series for a function defined for $0 < x < \pi$ amounts to extending the function according to $f(2\pi - x) = f(x)$, while the sine series would converge to a function extended by the rule $f(2\pi - x) = -f(x)$.

At the end of §9.5 we have referred to the idea of improving the rate of convergence by dealing separately with the slowest converging part of the series. To demonstrate the

---

[11] At first sight it may appear paradoxical that, while one needs *both* sines and cosines over the range $-\pi < x < \pi$, one can use *either* a sine or a cosine series over the half interval $[0, \pi]$ and therefore, apparently, "fewer" coefficients. We return on this point in §9.9.

approach let us consider as an example the series

$$f(x) = \sum_{n=1}^{\infty} \frac{n}{n^2 + a^2} \sin nx \qquad \text{for} \qquad 0 < x < \pi. \qquad (9.7.10)$$

For large n the coefficient is $O(1/n)$ which, as noted before on p. 251, implies that the sum is discontinuous. In general, if a function and its first derivative have discontinuities Δ and Δ' at a single point x_0 internal to $[0, \pi]$, we can calculate the coefficient B_n of its sine series proceeding as in the derivation of (9.5.6): we break up the integral from 0 to π into two integrals over $[0, x_0)$ and $(x_0, \pi]$ and integrate by parts twice to find

$$B_n = \frac{2}{\pi} \frac{1}{n} \left[f(0) - (-1)^n f(\pi) + \Delta \cos nx_0 - \frac{\Delta'}{n} \sin nx_0 \right] - \frac{1}{n^2} b_n, \qquad (9.7.11)$$

where b_n is the Fourier coefficient of the second derivative, calculated in the same way as the sum of the contributions of $[0, x_0)$ and $(x_0, \pi]$. Upon writing

$$\frac{n}{n^2 + a^2} = \frac{1}{n} - \frac{a^2}{n(n^2 + a^2)}, \qquad (9.7.12)$$

and comparing with (9.7.11), from the absence of an $O(1/n^2)$ term we conclude that $\Delta' = 0$, so that the derivative has no discontinuity *inside* $[0, \pi]$. From the fact that the coefficient of the $O(1/n)$ term is independent of n, we conclude that $\Delta = 0$ and $f(\pi) = 0$, so that $f(x)$ must be continuous in $0 < x < \pi$, with discontinuities possible only at $x = 0$ or at $x = \pi$. And indeed, since (9.7.11) and (9.7.12) can only be equal if $(2/\pi) f(0) = 1$, we see that we must have $f(0) = \pi/2 \neq f(\pi)$. The discontinuous part is (see (8.7.23) p. 237)

$$\sum_{n=1}^{\infty} \frac{\sin nx}{n} = \frac{1}{2}(\pi - x), \qquad (9.7.13)$$

and the remaining series, having coefficients $O(1/n^3)$, converges much faster. Several powerful variations on this basic idea can be found in Bromwich (1926), p. 364.

We have shown this argument as an example of how the slowly convergent part of a series can be separated out but, in this particular case, the original series (9.7.10) can be summed exactly. To this end we note that, with the previous deductions, (9.7.11) implies that

$$B_n = \frac{1}{n} - \frac{a^2}{n(n^2 + a^2)} = \frac{1}{n} - \frac{1}{n^2} b_n, \qquad (9.7.14)$$

so that $b_n = [na^2/(n^2 + a^2)] = a^2 B_n$, from which $f'' = a^2 f$. Upon integrating and imposing $f(0) = \pi/2$ and $f(\pi) = 0$ we have the sum of the series in closed form as

$$f(x) = \frac{\pi}{2} \frac{\sinh a(\pi - x)}{\sinh a\pi}. \qquad (9.7.15)$$

The transformation $x \to 2\pi - x$ changes the sign of this function as well as that of the series (9.7.10), and we therefore conclude that this representation is valid for $0 \leq x \leq 2\pi$.

9.8 Term-by-term integration and differentiation

In applications we are often concerned with the operations of term-by-term integration and differentiation of the Fourier series. As explained in §8.3, in general there are restrictions on the legitimacy of these operations, which we now consider.

Term-by-term integration of a Fourier series is actually freer than for general series of functions (cf. Theorem 8.3.2), a fact that has already been pointed out and used to great advantage in Chapter 3 (see e.g. p. 57).

Theorem 9.8.1 [Term-by-term integration] *Let f be integrable over the cylinder $\mathscr{S}$ and let*

$$f(x) \sim \frac{A_0}{2} + \sum_{n=1}^{\infty} (A_n \cos nx + B_n \sin nx). \tag{9.8.1}$$

Then, for any x,

$$\int_0^x f(\xi)\,d\xi = \frac{A_0}{2}x + \sum_{n=1}^{\infty} \left[\frac{A_n}{n} \sin nx - \frac{B_n}{n} (\cos nx - 1) \right] \tag{9.8.2}$$

and this series is uniformly convergent whether the original series be convergent or not.

In other words, the Fourier series of f, whether convergent or not, can be integrated between any limits resulting in a uniformly convergent series. An analogous result holds for the Fourier series in exponential form. This theorem rests on the fact that, if $f(x)$ is integrable, its primitive $F(x)$ exists, is absolutely continuous and has bounded variation. Furthermore, it is easily seen that $F_0 = F - f_0 x$ is periodic. Thus, by the Dirichlet–Jordan theorem 9.3.1, the Fourier series for F_0 converges uniformly. Upon writing the Fourier coefficients of F_0 and integrating by parts, one finds the coefficients of the term-by-term integrated series for f (see e.g. Sansone 1959, p. 78; Jeffreys & Jeffreys 1999, p. 440).

Example 9.8.1 To illustrate the theorem, let us return to Example 9.5.1 on p. 250. Upon integrating term by term the Fourier series (9.5.5) of f we have

$$\frac{2}{\sqrt{\pi}}x + \frac{4}{\sqrt{2\pi}} \sum_{n=1}^{\infty} C\left(\sqrt{\pi|n|}\right) \frac{\sin nx}{n^{3/2}}, \tag{9.8.3}$$

while

$$\int_0^x \frac{dx}{\sqrt{|x|}} = 2\,\mathrm{sgn}(x)\,\sqrt{|x|}. \tag{9.8.4}$$

The direct Fourier expansion of this function has coefficients

$$\frac{1}{\sqrt{2\pi}} \int_{-\pi}^{\pi} 2\,\mathrm{sgn}(\xi)\,\sqrt{|\xi|}\,e^{-in\xi}\,d\xi = -\frac{4i}{n\sqrt{2\pi}} \left[(-1)^{n+1}\sqrt{\pi} + \frac{C\left(\sqrt{\pi|n|}\right)}{\sqrt{|n|}} \right], \tag{9.8.5}$$

which is easily seen to reproduce (9.8.3) in view of (8.7.24) p. 237.   □

The previous theorem can be recast in the form

$$\int_0^x \left[f(x) - \frac{1}{2} A_0 \right] dx = \frac{1}{2}\alpha_0 + \sum_{n=1}^{\infty} \frac{1}{n} (-B_n \cos nx + A_n \sin nx), \qquad (9.8.6)$$

where α_0 is the mean value of $f(x) - \frac{1}{2}A_0$. Since we are guaranteed that the series converges, if we take $x = 0$ we deduce that $\frac{1}{2}\alpha_0 = \sum_{n=1}^{\infty} B_n/n$ so that this series is always convergent. However, it is not generally true that $\sum_{n=1}^{\infty} A_n/n$ converges.

An immediate consequence of Theorem 9.8.1 is that the series (9.2.1), with coefficients given by (9.2.2), is indeed the Fourier series of f if f is integrable in $\mathscr{S}$. Indeed, multiply (9.2.1) by $\bar{e}_m(x)$ (which obviously does not change the convergence properties of the series) and integrate term by term. By using the orthonormality property (9.1.10) of the e_n's, we find (9.2.2).

Another result which is easily derived using term-by-term integration is the *Parseval equality* for square integrable functions. In the calculation of the norm of f according to the definition (9.1.6), substitute for f in the integral its Fourier series. By using the orthonormality property of the e_n's we find

$$\| f \|^2 = \int_{-\pi}^{\pi} \left(\sum_{m=-\infty}^{\infty} \bar{f}_m \bar{e}_m \right) \left(\sum_{n=-\infty}^{\infty} f_n e_n \right) dx$$

$$= \left(\sum_{m=-\infty}^{\infty} \bar{f}_m \right) \left(\sum_{n=-\infty}^{\infty} f_n \right) \int_{pi}^{\pi} \bar{e}_m \, e_n \, dx$$

$$= \left(\sum_{m=-\infty}^{\infty} \bar{f}_m \right) \left(\sum_{n=-\infty}^{\infty} f_n \right) \delta_{nm} = \sum_{n=-\infty}^{\infty} |f_n|^2. \qquad (9.8.7)$$

This relation states that the square of the "length" of the "vector" f equals the sum of the squares of its "projections" on the unit vectors and, therefore, may be viewed as a generalization of Pythagoras's theorem. By Theorem 8.6.2, p. 233, the term-by-term multiplication used to derive the previous result is certainly legitimate when the series is absolutely convergent. The general theory of Hilbert spaces developed in §19.3 shows that the Parseval equality is a general property valid under less restrictive assumptions. In the same way we readily prove that the scalar product of two functions f and g can be calculated by multiplying their Fourier coefficients f_n and g_n, respectively, and adding:

$$(g, f) \equiv \int_{-\pi}^{\pi} \bar{g}(x) f(x) \, dx = \sum_{n=\infty}^{\infty} \bar{g}_n f_n. \qquad (9.8.8)$$

As noted on p. 559, the series in the right-hand side is absolutely convergent.

In many applications of the Fourier series we are concerned with term-by-term differentiation, for which uniform convergence is important as we have seen in §8.3. Two useful theorems are the following:

Theorem 9.8.2 [Term-by-term differentiation I] *Let f be continuous and differentiable in $\mathscr{S}$ with a derivative f' piecewise continuous in $\mathscr{S}$. Then the Fourier series for f' can be obtained by term-wise differentiation of that for f. The differentiated series converges* *point-wise to f' at points of continuity and to $\frac{1}{2}\left[f'(x+0)+f'(x-0)\right]$ at points of discontinuity.*

This is a consequence of Theorem 9.8.1 on term-by-term integration: under the hypotheses stated, f' is integrable term by term to reproduce f.

Theorem 9.8.3 [Term-by-term differentiation II] *If f is continuous in $\mathcal{S}$, and if*

$$\sum_{-\infty}^{\infty} |k\,(e_k, f)| < \infty, \tag{9.8.9}$$

then it follows that f is once continuously differentiable and

$$\sum_{k=-n}^{n} i\,k\,(e_k, f)\,e_k \;\to\; f'(x) \tag{9.8.10}$$

uniformly in $\mathcal{S}$ as $n \to \infty$.

The result follows by showing that, as a convergence of the stated hypotheses, the conditions of Theorem 8.3.4 (p. 227) are met.

By the method demonstrated in §9.7 it may be possible to extract explicitly the non-smooth part of the series; the remaining series will then converge rapidly and term-by-term differentiation will be legitimate.

It will be seen on p. 587 that, in a distributional sense, any Fourier series can be differentiated term by term. However, as mentioned at the end of §9.11, this freedom comes at a price as point-wise convergence will then be lost in general.

9.9 Change of scale

So far we have focused on intervals of length 2π. To see how to write the Fourier series over a general interval $[a, b]$, we map it onto $[-\pi, \pi]$ by the change of variable:

$$x' = -\pi + 2\pi\frac{x-a}{b-a}, \qquad x = \frac{b-a}{2\pi}(x'+\pi)+a. \tag{9.9.1}$$

Define now the new function

$$f_*(x') = f\left(\frac{b-a}{2\pi}(x'+\pi)+a\right), \tag{9.9.2}$$

and expand $f_*(x')$ in Fourier series:

$$f_*(x') = \sum_{-\infty}^{\infty}\left(\frac{\exp(ikx')}{\sqrt{2\pi}}, f_*(x')\right)\frac{\exp(ikx')}{\sqrt{2\pi}}. \tag{9.9.3}$$

Upon changing the integration variable back to x to find

$$\left(\frac{\exp{(ikx')}}{\sqrt{2\pi}}, f_*(x')\right) = \frac{1}{\sqrt{2\pi}}\int_{-\pi}^{\pi}\exp{(-ikx')}f_*(x')\,dx'$$

$$= \frac{1}{\sqrt{2\pi}}\int_a^b\exp\left[-ik\pi\left(2\frac{x-a}{b-a}-1\right)\right]f(x)\,\frac{2\pi}{b-a}\,dx$$

$$= \frac{\sqrt{2\pi}}{b-a}\exp\left(ik\pi\frac{b+a}{b-a}\right)\int_a^b\exp\left(-\frac{2ik\pi x}{b-a}\right)f(x)\,dx \quad (9.9.4)$$

so that

$$\left(\frac{\exp{(ikx')}}{\sqrt{2\pi}}, f_*(x')\right)\frac{\exp{(ikx')}}{\sqrt{2\pi}} = \int_a^b\frac{1}{\sqrt{b-a}}\exp\left(-ik\frac{2\pi\xi}{b-a}\right)f(\xi)\,d\xi.$$

$$\times\left[\frac{1}{\sqrt{b-a}}\exp\left(ik\frac{2\pi x}{b-a}\right)\right]. \quad (9.9.5)$$

The factor $\exp{[ik\pi(b+a)/(b-a)]}$ of (9.9.4) is cancelled by its complex conjugate coming from $\exp{(ikx')}$. Since this always happens in writing down the Fourier series, these factors of modulus 1 will be dropped in the following.[12]

In this way we have proven that, over the interval $[a, b]$, the appropriate family of orthonormal eigenfunctions is given by

$$\hat{e}_k(x) = \frac{1}{\sqrt{b-a}}\exp\left(ik\frac{2\pi x}{b-a}\right), \quad (9.9.6)$$

and the Fourier coefficients by

$$f_k = \frac{1}{\sqrt{b-a}}\int_a^b\exp\left(-ik\frac{2\pi x}{b-a}\right)f(x)\,dx. \quad (9.9.7)$$

In terms of these eigenfunctions and coefficients we can write a series expansion formally identical to (9.2.1):

$$f(x) \sim \sum_{n=-\infty}^{\infty}f_n\,\hat{e}_n(x), \qquad a \le x < b. \quad (9.9.8)$$

A similar procedure may be used for the trigonometric form. After some reduction, the final formulae are:

$$f(x) = \frac{A_0}{2} + \sum_{n=1}^{\infty}\left[A_n\cos\frac{2\pi nx}{b-a} + B_n\sin\frac{2\pi nx}{b-a}\right] \quad (9.9.9)$$

$$A_n = \frac{2}{b-a}\int_a^b\cos\frac{2\pi n\xi}{b-a}f(\xi)\,d\xi \quad (9.9.10)$$

$$B_n = \frac{2}{b-a}\int_a^b\sin\frac{2\pi n\xi}{b-a}f(\xi)\,d\xi. \quad (9.9.11)$$

---

[12] For the same reason the phase factor $e^{i\alpha}$ mentioned in the footnote on p. 243 is immaterial.

The relevant formulae for the sine and cosine series are readily found following the argument of §9.7. In particular, the relations applicable to the cosine and sine series over the range $0 \le x < L$ are given in (3.1.11) and (3.1.12) p. 55, respectively.

Example 9.9.1 As mentioned in the footnote on p. 253, a function defined for $0 < x < \pi$ can be expanded in a Fourier sine or cosine series, but it can also be expanded in a complete Fourier series by making the change of variable $x' = 2x - \pi$ which causes x' to range between $-\pi$ and π as x ranges between 0 and π. Upon so doing, we have from (9.9.9):

$$f(x) = \frac{A_0}{2} + \sum_{n=1}^{\infty} [A_n \cos 2nx + B_n \sin 2nx], \qquad (9.9.12)$$

with

$$A_n = \frac{2}{\pi} \int_0^\pi \cos 2n\xi \, f(\xi) \, d\xi, \qquad B_n = \frac{2}{\pi} \int_0^\pi \sin 2n\xi \, f(\xi) \, d\xi. \qquad (9.9.13)$$

If we had used a sine or a cosine series we would have had only the B_n or A_n coefficients for all n, even and odd. Here we see that we have *both* types of coefficients, but only for even integers. The two expansions are therefore inherently different, and (9.9.12) cannot be directly reduced to either (9.7.4) or (9.7.5).

Nevertheless, (9.9.12) is a valid expansion. To show it we assume that integration and summation can be interchanged and find

$$f(x) = \int_0^\pi \left[\frac{1}{\pi} + \frac{2}{\pi} \sum_{k=1}^{\infty} \cos 2n(x - \xi) \right] f(\xi) \, d\xi, \qquad (9.9.14)$$

and the quantity in brackets must equal $\delta(x - \xi)$ for any value of x and ξ in the range $(0, \pi)$. Using (9.3.9) we have

$$\frac{1}{\pi} + \frac{2}{\pi} \sum_{k=1}^{\infty} \cos 2n(x - \xi) = 2\delta \left(2(x - \xi) \right) = \delta(x - \xi), \quad |x - \xi| < \frac{\pi}{2}, \qquad (9.9.15)$$

where we have used the property $\delta(\lambda x - \beta) = \delta(x - \beta/\lambda)/|\lambda|$ of the δ-function (Table 2.1 p. 37). If $0 < x, \xi < \pi$, then $|x - \xi| < \pi$ and the range of validity of this relation is insufficient. However, since the series in the left-hand side has periodicity π in the variable $x - \xi$, for $|x - \xi| < (3/2)\pi$ we can write

$$\frac{1}{\pi} + \frac{2}{\pi} \sum_{k=1}^{\infty} \cos 2n(x - \xi) = \delta(x - \xi) + \delta(x - \xi - \pi) + \delta(x - \xi + \pi). \quad (9.9.16)$$

Upon substitution into (9.9.14) if, for example, $\pi/2 < x - \xi < \pi$ we have

$$\int_0^\pi \left[\frac{1}{\pi} + \frac{2}{\pi} \sum_{k=1}^{\infty} \cos 2n(x - \xi) \right] f(\xi) \, d\xi = f(x - \pi). \qquad (9.9.17)$$

But use of the series (9.9.12) presupposes that the function has been extended periodically with a period of π and, therefore, the correct $f(x)$ is recovered.

As a concrete example we use $f(x) = x$ for $0 < x < \pi$. Since this is an odd function over $-\pi < x < \pi$, an odd extension to $-\pi < x < 0$ is natural, which makes a Fourier sine series as in (9.7.4) appropriate; we find

$$x = 2 \sum_{k=1}^{\infty} (-1)^k \frac{\sin kx}{k}, \tag{9.9.18}$$

while the expansion (9.9.12) is

$$x = \frac{\pi}{2} - \sum_{k=1}^{\infty} \frac{\sin 2kx}{k}. \tag{9.9.19}$$

As periodically extended, at $x = 0$ the function jumps from 0 to π and the complete Fourier series converges to $\pi/2$ as expected. $\square$

9.10 The Gibbs phenomenon

If f is discontinuous, the Fourier series cannot converge uniformly in any interval containing the discontinuity because a uniformly convergent series of continuous functions converges to a continuous function (Theorem 8.2.2 §8.2). To see how uniform convergence is spoiled by the presence of a discontinuity, let us examine the convergence of the Fourier series (9.5.8) for the sign function (2.3.13) p. 36.[13]

Let us consider therefore the partial sums of the series (9.5.8). We have[14]

$$
\begin{aligned}
S_N &= \frac{4}{\pi} \sum_{0}^{N} \frac{\sin (2k+1)x}{2k+1} = \frac{4}{\pi} \sum_{0}^{N} \int_{0}^{x} \cos (2k+1)y \, dy \\
&= \frac{2}{\pi} \int_{0}^{x} \frac{\sin 2(N+1)y}{\sin y} \, dy.
\end{aligned} \tag{9.10.1}
$$

Since we are interested in the neighborhood of the discontinuity at the origin, we may limit our analysis to small x. Then, in the integration range $0 \le y \le x$, y will be small and $\sin y \simeq y$ so that

$$S_N \simeq \frac{2}{\pi} \int_{0}^{x} \frac{\sin 2(N+1)y}{y} dy = \frac{2}{\pi} \int_{0}^{2(N+1)x} \frac{\sin u}{u} = \frac{2}{\pi} \text{Si}\,[2(N+1)x], \tag{9.10.2}$$

where Si is a special function, the *sine integral*, defined by the last equality. A graph of the function $(2/\pi)\text{Si}(x)$ is shown in Figure 9.5; since $\text{Si} \to \frac{1}{2}\pi$ as the argument goes to infinity, $(2/\pi)\text{Si} \to 1$.

---

[13] This is sufficient because, in the neighborhood of a discontinuity of the first kind (definition on p. 246), e.g. at 0, a function can be represented as $f(x) = \frac{1}{2} [f(0+) - f(0-)] \operatorname{sgn} x + f_c(x)$ with f_c continuous as is readily verified.

[14] The summation is carried out by writing it as $\operatorname{Re} \sum_{0}^{N} \exp[i\,(2k+1)y]$ and using the result (8.3.5) for the geometric sum.

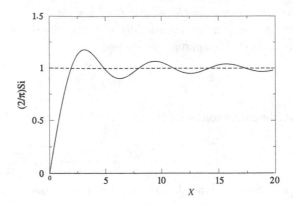

Fig. 9.5 Graph of the function $(2/\pi)\mathrm{Si}(x)$ defined in (9.10.2).

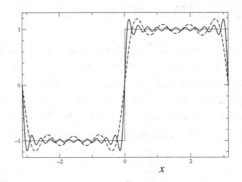

Fig. 9.6   The Fourier representation (9.10.1) of the function sgn x truncated at $N = 3$ (dashed line) and $N = 10$ (solid line).

The first positive maximum of Si occurs at π where $(2/\pi)\mathrm{Si}(\pi) \simeq 1.18$. Hence, at $x_N = \pi/(2N + 2)$, $S_N(x_N) \simeq 1.18$ no matter how large N is. On the other hand, at every fixed point, $S_N \to \mathrm{sgn}\,(x)$, although convergence is not uniform in any interval including 0. In the neighborhood of $x = 0$, therefore the graph of the truncated Fourier series of sgn x exhibits oscillations with a fixed maximum amplitude that bunch up closer and closer to $x = 0$ as the order of truncation N is increased. This behavior is known as the "Gibbs phenomenon" after the first scientist who remarked on it. Figure 9.6 illustrates this behavior by comparing the sum (9.10.1) truncated at $N = 3$ and $N = 10$ with the exact graph of sgn x. Since the region of overshooting is of vanishing small width as $N \to \infty$, an integral of the series would not be affected by the Gibbs phenomenon in agreement with Theorem 9.8.1 p. 256 according to which the series obtained by term-by-term integration is uniformly convergent.

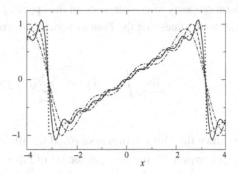

Fig. 9.7 Graph of the N-term partial sums of the Fourier sine series for the periodic extension of the function x/π (dotted line) for $N = 10$ (solid line), 5 (dashed) and 3 (dash-dots).

Another illustration of the Gibbs phenomenon is provided by the Fourier series expansion of the function x/π given, for $-\pi < x < \pi$, by

$$\frac{x}{\pi} = \frac{2}{\pi}\sum_{k=1}^{\infty}(-1)^k\frac{\sin kx}{k}, \tag{9.10.3}$$

and extended periodically outside of this interval. The partial sums of the Fourier series for this function are shown in Figure 9.7 for $N = 10$ (solid line), 5 (dashed) and 3 (dash–dots). Discontinuities are present at $\pm\pi$, $\pm 3\pi$ etc and it is evident from the figure that, as the number of terms increases, the error to the left and right of $x = \pm\pi$ gets pushed closer and closer to $x = \pm\pi$ without decreasing in magnitude. On the other hand, if $f(x)$ is reconstructed by Fejér's method, namely calculating the Cesáro sum of the series as explained on p. 264, the Gibbs phenomenon is avoided (see e.g. Canuto *et al.* 1988, p. 50; Sansone 1959, p. 147).[15]

9.11 Other modes of convergence

As we have seen in §8.2.3, convergence of a series can be understood in several different ways of which the point-wise one of §9.3 is not necessarily the most significant in spite of its intuitive appeal.[16]

---

[15] According to the *Fejér–Lebesgue theorem*, the Cesáro sum of the Fourier series converges at all Lebesgue points of f, namely at all points at which the function equals the derivative of its indefinite integral (see e.g. Natanson 1961, vol. 2 p. 25); all points of continuity are Lebesgue points.

[16] MacCluer (2004) refers to "the ill-behaved classical notion of point-wise convergence" (p. 111) and Hardy & Rogosinski (1956) warn the reader that the point-wise convergence problem "has lost a good deal of its importance." From a practical (rather than mathematical) viewpoint, the behavior of quantities at single points cannot have physical significance – what matters is always the behavior over finite (if possibly small) intervals.

A general property of convergence in the spaces L^2 (see §19.4.3 p. 558), specialized to the case at hand, states that the Fourier series converges in the mean-square sense:

$$\lim_{n\to\infty} \int_{-\pi}^{\pi} \left| f(x) - \sum_{k=-n}^{n} (e_k, f)\, e_k \right|^2 dx = 0. \qquad (9.11.1)$$

It can be shown that this relation is satisfied for all functions $f \in L^2(-\pi, \pi)$ which, in the language of Chapter 19, implies that the set of basis functions $\{e_n(x)\}$ is *complete*.

As mentioned in §8.2.3, over a finite interval mean-square convergence implies convergence in the mean so that, if (9.11.1) is satisfied, then also

$$\lim_{n\to\infty} \int_{-\pi}^{\pi} \left| f(x) - \sum_{k=-n}^{n} (e_k, f)\, e_k \right| dx = 0. \qquad (9.11.2)$$

The converse, however, is not true: satisfaction of (9.11.2) does not imply (9.11.1).

These and many other aspects of the Fourier series are really special cases of expansions on an orthonormal basis and it is more efficient to bring them all together in the general treatment of these issues to be found in Chapter 19.

Another form of convergence is distributional convergence (§20.6 p. 584). In particular, when convergence is understood in this sense, *any* Fourier series may be differentiated term by term any number of times but then, when we write that the generalized function (or distribution) "equals" its Fourier series, the equality in general holds only distributionally and cannot be understood in the point-wise sense.

In §8.5 we have introduced two different definitions of the sum of a series, the Abel–Poisson sum and the Cesáro sum. Some considerations on the convergence of the Abel–Poisson sum of the Fourier series are given on p. 76. The Cesáro sum of the Fourier series was first considered by Fejér who showed that, in the case of a Riemann-integrable function f, the Cesáro sum of the Fourier series converges at every point of continuity of f with no need for additional hypotheses on f (see e.g. Hardy & Rogosinski 1956, p. 53; Bary 1964, vol. 1, p. 135; Körner 1989, p. 6). In this case the Dirichlet kernel in the partial sums (9.3.1) is replaced by the *Fejér kernel*

$$\frac{1}{2(N+1)} \left[\frac{\sin (N+1)x/2}{\sin x/2} \right]^2. \qquad (9.11.3)$$

Thus *any* continuous function can be reconstructed from its Fourier coefficients by using the Cesáro summation procedure in place of the ordinary one. The crucial difference here is that the Fejér kernel is non-negative. If $f(x)$ is continuous in $\mathscr{S}$, the convergence of the partial Cesáro sums is uniform.

9.12 The conjugate series

In the complex power series

$$F(z) = \frac{1}{2}A_0 + \sum_{n=1}^{\infty} C_n z^n, \qquad C_n = A_n - i B_n, \tag{9.12.1}$$

assumed convergent inside the unit circle, let $z = re^{i\theta}$. Then

$$u(r, \theta) \equiv \operatorname{Re} F(z) = \frac{1}{2}A_0 + \sum_{n=1}^{\infty} r^n \left(A_n \cos n\theta + B_n \sin n\theta \right) \tag{9.12.2}$$

is a harmonic function given, for any fixed r, by a Fourier series in θ. It is natural to associate to this series the series

$$v(r, \theta) \equiv \operatorname{Im} F(z) = \sum_{n=1}^{\infty} r^n \left(-B_n \cos n\theta + A_n \sin n\theta \right), \tag{9.12.3}$$

so that $F(z) = u + iv$; u and v are conjugate harmonic functions (p. 423) and this second series is the *conjugate series* to (9.12.2).

The series conjugate to a Fourier series is not necessarily a Fourier series. For example, the series (9.2.5) is conjugate to $\sum_{n=2}^{\infty} \cos nx / \log n$, which is a Fourier series, while (9.2.5) is not (see e.g. Hardy & Rogosinski 1956, p. 34; Bary 1964, vol. 1 pp. 94 and 117). The Abel–Poisson sum of the conjugate series can be calculated as on p. 76 and is found to be

$$\frac{1}{\pi} \lim_{r \to 1-} \int_{-\pi}^{\pi} \frac{r \sin (x - \xi)}{1 - 2r \cos (x - \xi) + r^2} \, f(\xi) \, d\xi. \tag{9.12.4}$$

Upon setting $r = 1$, this expression becomes

$$\frac{1}{2\pi} \int_{-\pi}^{\pi} \cot \frac{1}{2}(x - \xi) \; f(\xi) \, d\xi, \tag{9.12.5}$$

and, understood as a Cauchy principal value at $\xi = x$, it furnishes the sum of the conjugate series almost everywhere when f is absolutely integrable (Hardy & Rogosinski 1956, p. 79); this same reference gives other results for the convergence of the conjugate series on the basis of properties of $f(x)$ (see especially pp. 45 and 64).

10   The Fourier and Hankel Transforms

In this chapter the Fourier and Hankel transforms are presented in a rather formal way without attempting to provide rigorous justifications of the various statements. The reason is that stringent hypotheses make proofs relatively straightforward, and these can be found in a large number of texts. The Fourier transform, however, is a much more powerful tool than these simplified results imply, and the general proofs are more difficult; a useful resource is the book by Champeney (1987). The Fourier transform of generalized functions is treated in §20.10 p. 595.

10.1 Heuristic motivation

According to §9.9 of the previous chapter, a function f having bounded variation in the interval $-L < x \leq L$ can be expanded in a Fourier series of the form[1]

$$f(x) = \sum_{n=-\infty}^{\infty} f_n \frac{\exp(-in\pi x/L)}{\sqrt{2L}}, \tag{10.1.1}$$

where

$$f_n = \frac{1}{\sqrt{2L}} \int_{-L}^{L} \exp(in\pi \xi/L) \, f(\xi) \, d\xi. \tag{10.1.2}$$

Take now L very large. On a k-axis consider all the points

$$k_n = \frac{\pi n}{L}, \qquad -\infty < n < \infty, \tag{10.1.3}$$

and rewrite (10.1.1) as

$$f(x) = \sum_{\text{all } k_n's} \frac{\exp(-ik_n x)}{\sqrt{2L}} f_n. \tag{10.1.4}$$

A given segment of the k-axis with length Δk will contain, approximately, $\Delta n = (L/\pi)\Delta k$ points and, for L large, all the points k_n in this segment are very close to each other. The

---

[1] In §9.9 we wrote $\exp ikx$ in place of $\exp -ikx$. Since the summation is between $-\infty$ and ∞ the present expression is equivalent and it will lead us to the more customary definition of the Fourier transform. Definitions with the opposite signs in the exponentials are also in use however; see footnote on p. 267.

corresponding values of f_n will likewise be close. If we pick one representative value $\bar{n}$ for each Δk, we can therefore write, approximately,

$$f(x) \simeq \sum_{\text{all } \Delta k's} f_{\bar{n}} \frac{\exp(-ik_{\bar{n}} x)}{\sqrt{2L}} \frac{L}{\pi} \Delta k$$

$$\simeq \sum_{\text{all } \Delta k's} \left[\frac{1}{\sqrt{2\pi}} \int_{-L}^{L} \exp(ik_{\bar{n}} \xi) f(\xi) \, d\xi \right] \frac{\exp(-ik_{\bar{n}} x)}{\sqrt{2\pi}} \Delta k. \quad (10.1.5)$$

The reader will recognize here the Riemann sums (cf. p. 687) which, in the limit $\Delta k \to 0$, give rise to an integral.

On the basis of this heuristic argument, we are led to *define* the *Fourier transform* of the function f by

$$\mathscr{F}\{f\} \equiv \tilde{f}(k) = \frac{1}{\sqrt{2\pi}} \int_{-\infty}^{\infty} e^{ikx} f(x) \, dx, \quad (10.1.6)$$

with a candidate inverse

$$\mathscr{F}^{-1}\{\tilde{f}\} \equiv f(x) = \frac{1}{\sqrt{2\pi}} \int_{-\infty}^{\infty} e^{-ikx} \tilde{f}(k) \, dk. \quad (10.1.7)$$

Equation (10.1.6) defines the *exponential Fourier transform*, so denominated to distinguish it from the sine and cosine transforms of §10.5 and the one-sided transforms of §10.6. While purely formal, the previous derivation has the merit of illustrating how an integral transform arises from an eigenfunction expansion when the eigenvalues coalesce into a continuum; this point is pursued further in §15.4.[2]

It is evident that, whenever the indicated transforms exist,

$$\mathscr{F}^2\{f\} \equiv \mathscr{F}\{\mathscr{F}\{f\}\} = f(-x), \qquad \mathscr{F}^4\{f\} = f(x), \quad (10.1.8)$$

and, for a real function,

$$\overline{\tilde{f}}(k) = \tilde{f}(-k), \qu(10.1.9)$$

where the overline denotes the complex conjugate.

Before continuing, let us note that several slightly different definitions of the Fourier transform can be found in the literature. In some cases the signs plus and minus in the exponential are switched between f and $\tilde{f}$. Sometimes one encounters the less symmetric definition

$$\tilde{f}(k) = \int_{-\infty}^{\infty} e^{ikx} f(x) \, dx, \quad \text{with} \quad f(x) = \frac{1}{2\pi} \int_{-\infty}^{\infty} e^{-ikx} \tilde{f}(k) \, dk. \quad (10.1.10)$$

In other definitions the exponentials are written as $\exp(\pm 2\pi ikx)$, which eliminates the 2π factors in front of the integrals, but leads to somewhat more cumbersome operational formulae.[3]

---

[2] For a rigorous version of this argument see Titchmarsh 1948, p. 70.

[3] If the transform is written as $\tilde{f} = A \int_{-\infty}^{\infty} f(x) \exp(-ibkx) \, dx$, then the inversion formula is $f = [|b|/(2\pi A)] \int_{-\infty}^{\infty} \tilde{f} \exp(ibkx) \, dk$ and the convolution theorem (10.3.8) is $f * g = [|b|/(2\pi A^2)] \int \tilde{f}(k)\tilde{g}(k) \exp(ibkx) \, dk$, with $f * g$ as defined in (10.3.6) and (10.3.7).

10.2 The exponential Fourier transform

Since

$$\left| \int_{-\infty}^{\infty} e^{ikx} f(x)\, dx \right| \le \int_{-\infty}^{\infty} |f(x)|\; dx, \tag{10.2.1}$$

a *sufficient condition* for the existence of the integral defining the Fourier transform is that f be absolutely integrable, which we express by writing $f \in L(-\infty, \infty)$ or $f \in L^1(-\infty, \infty)$. In the case of square-integrable functions $f \in L^2(-\infty, \infty)$, the Fourier transform is defined as the function $\tilde{f}(k)$ such that

$$\lim_{L \to \infty} \int_{-\infty}^{\infty} \left| \tilde{f}(k) - \frac{1}{\sqrt{2\pi}} \int_{-L}^{L} e^{ikx} f(x)\, dx \right|^2 dk = 0, \tag{10.2.2}$$

if such a function exists.[4] An alternative approach to the definition of the Fourier transform for square-integrable functions is given on p. 644.

Both $f \in L^1$ and $f \in L^2$ are sufficient conditions only: no simple necessary and sufficient criterion is known for the existence of the Fourier transform of ordinary functions. It is only with the introduction of generalized functions that such criteria can be formulated (see e.g. Lighthill 1958; Champeney 1987, p. 47).

The inversion formula (10.1.7) for the transform is properly interpreted as the Cauchy principal value[5]

$$f(x) = \lim_{K \to \infty} \frac{1}{\sqrt{2\pi}} \int_{-K}^{K} e^{-ikx}\, \tilde{f}(k)\, dk. \tag{10.2.3}$$

If we substitute in this integral the definition of $\tilde{f}$, we have

$$\begin{aligned}
\frac{1}{\sqrt{2\pi}} \int_{-K}^{K} e^{-ikx}\, \tilde{f}(k)\, dk &= \frac{1}{2\pi} \int_{-K}^{K} dk\, e^{-ikx} \int_{-\infty}^{\infty} e^{ik\xi} f(\xi)\, d\xi \\
&= \int_{-\infty}^{\infty} d\xi\; f(\xi) \frac{\sin K(\xi - x)}{\pi(\xi - x)} d\xi,
\end{aligned} \tag{10.2.4}$$

where the interchange of the order of integration is legitimate because both the individual integrals exist (p. 690). The inversion formula (10.2.3) then implies that, in the sense of distributions (§20.6 p. 584)[6]

$$\lim_{K \to \infty} \frac{1}{2\pi} \int_{-K}^{K} \exp[ik(\xi - x)]dx = \lim_{K \to \infty} \frac{\sin K(\xi - x)}{\pi(\xi - x)} = \delta(\xi - x). \tag{10.2.5}$$

---

[4] As pointed out on p. 244, on a finite interval $f \in L^2$ implies $f \in L^1$. No such implication exists for an infinite interval. For example, $1/\sqrt{a^2 + x^2}$ belongs to $L^2(-\infty, \infty)$ but not to $L^1(-\infty, \infty)$, while the converse is true for $|x|^{-1/2} \exp(-|x|)$.

[5] If $\tilde{f}$ is locally and absolutely integrable over $-\infty < x < \infty$, the inverse (10.1.7) exists as an ordinary improper integral (see p. 688).

[6] For a direct proof under fairly broad hypotheses see e.g. Titchmarsh 1948, p. 22 or Rudin 1987, p. 182.

There are several sufficient criteria which ensure the point-wise convergence of the inversion integral almost everywhere (see e.g. Champeney 1987, chapters 8 and 9). If $f(x) \in L^2(-\infty, \infty)$ and x_0 is a point of discontinuity of the first kind (p. 246) then, just as in the case of the Fourier series (see e.g. Titchmarsh 1948, p. 83),

$$\lim_{K \to \infty} \frac{1}{\sqrt{2\pi}} \int_{-K}^{K} e^{-ikx} \, \tilde{f}(k) \, dk = \frac{1}{2}[f(x+0) + f(x-0)]. \tag{10.2.6}$$

Whenever $f(x)$ is sufficiently smooth, the inversion relation (10.2.3) holds point-wise. This is the case most commonly encountered in practice (at least when dealing with ordinary functions rather than distributions) and, accordingly, we omit the explicit consideration of discontinuities and other pathologies using rather liberally the relation

$$\frac{1}{2\pi} \int_{-\infty}^{\infty} e^{ikx} \, dx = \frac{1}{2\pi} \int_{-\infty}^{\infty} e^{-ikx} \, dx = \delta(k), \tag{10.2.7}$$

as well as the analogous one

$$\frac{1}{\pi} \int_{0}^{\infty} \cos kx \, dx = \delta(k). \tag{10.2.8}$$

For a square integrable function, a consequence of these relations is the *Plancherel–Parseval equality*:[7]

$$\| f \|^2 = \int_{-\infty}^{\infty} |f(x)|^2 \, dx = \int_{-\infty}^{\infty} |\tilde{f}(k)|^2 \, dk = \| \tilde{f} \|^2, \tag{10.2.9}$$

which can be generalized to

$$(f, g) = \int_{-\infty}^{\infty} \overline{f}(x) \, g(x) \, dx = \int_{-\infty}^{\infty} \overline{\tilde{f}}(k) \, \tilde{g}(k) \, dk = (\tilde{f}, \tilde{g}), \tag{10.2.10}$$

as readily follows by applying (10.2.9) to $f + g$ and $f - g$ and subtracting.

As explained in §19.2 p. 541 and Example 19.2.5 p. 543, the integrals $\| f \|^2$ and $\| \tilde{f} \|^2$ in (10.2.9) are the L^2-norms of the "vectors" f and $\tilde{f}$ and can be interpreted, in an abstract sense, as their "length." Parseval's relation then shows that the Fourier transform, considered as a mapping from the space to which f belongs to the space to which $\tilde{f}$ belongs, is an *isometry*, i.e., it preserves lengths. Similarly, (10.2.10) states that the transformation preserves scalar products.[8] A brief presentation of the Fourier transform in the framework of the theory of unitary operators is given in §21.4 p. 644.

When $f(x)$ and $\tilde{f}(k)$ can be continued into the complex plane, by Jordan's lemma (p. 461), in many cases the integration path along the real line defining the direct and inverse Fourier transforms can be closed by a large semi-circle in the upper or lower half-planes. Contour integration (§17.9.1 p. 461) is then a particularly powerful tool for the evaluation of the transforms. The need for the actual evaluation of transforms is much reduced by the existence of extensive integral tables, mathematical software and websites

---

[7] A superficially simple proof is immediate: replace f by its expression in terms of $\tilde{f}$, interchange the order of the integrations and use (10.2.5). The difficulty is that each one of these steps should be justified.

[8] The isometric nature of the transformation would be lost with the alternative definition (10.1.10) of the transform pair.

which directly supply many of the transform relations that arise in practice; a short table of Fourier transforms is given on p. 86.

A function of two variables x and y defined for $-\infty < x,\ y < \infty$ can be transformed in both variables by iterating the Fourier transform:[9]

$$\tilde{f}(k_x, k_y) = \frac{1}{2\pi} \int_{-\infty}^{\infty} dx \int_{-\infty}^{\infty} dy \, \exp\left[i(k_x x + k_y y)\right] f(x, y), \tag{10.2.11}$$

and, more generally,

$$\tilde{f}(\mathbf{k}) = \frac{1}{(2\pi)^{d/2}} \int \exp\left(i\mathbf{k}\cdot\mathbf{x}\right) f(\mathbf{x}) d^d x, \tag{10.2.12}$$

where $\mathbf{x}$ and $\mathbf{k}$ are vectors in d-dimensional spaces and the integral is extended to the entire space. The inversion formula is

$$f(\mathbf{x}) = \frac{1}{(2\pi)^{d/2}} \int \exp\left(-i\mathbf{k}\cdot\mathbf{x}\right) \tilde{f}(\mathbf{k}) d^d k. \tag{10.2.13}$$

Parseval's equalities (10.2.9) and (10.2.10) hold for this case as well with

$$\|f\|^2 \equiv \int |f(\mathbf{x})|^2 d^d x = \int |\tilde{f}(\mathbf{k})|^2 d^d k \equiv \|\tilde{f}\|^2. \tag{10.2.14}$$

10.3 Operational formulae

The examples of Chapter 4 show the great value of the Fourier transform in simplifying many operations which frequently arise in practice. These properties are collectively known as *operational rules* or *operational formulae*. Here we present a few; additional ones are given in Table 4.1 on p. 85. The derivation of these rules often requires manipulations, such the interchange of the order of integration, or differentiation under the integral sign, which are legitimate under very broad conditions, at least in the sense of distributions; for a statement of precise conditions of validity the reader is referred, e.g., to Champeney 1987 (see also p. 690).

Integration by parts directly shows that, for $f \to 0$ at infinity,

$$\mathscr{F}\left\{\frac{df}{dx}\right\} = \frac{1}{\sqrt{2\pi}} \int_{-\infty}^{\infty} e^{ikx} \frac{df}{dx} dx = -\frac{ik}{\sqrt{2\pi}} \int_{-\infty}^{\infty} e^{ikx} f(x) \, dx = -ik\tilde{f}, \tag{10.3.1}$$

and, similarly, that

$$\mathscr{F}^{-1}\{\frac{d\tilde{f}}{dk}\} = ix \, f(x), \qquad \mathscr{F}\{xf\} = -i\frac{d\tilde{f}}{dk}, \tag{10.3.2}$$

which can be generalized to

$$\mathscr{F}\left\{f^{(n)}\right\} = (-ik)^n \tilde{f}, \qquad \mathscr{F}^{-1}\{\tilde{f}^{(n)}\} = (ix)^n f(x), \tag{10.3.3}$$

---

[9] More properly (10.2.11) should be regarded as an integral over the (x, y) plane, which can be reduced to two iterated integrals such as the form shown under suitable hypotheses, e.g. $f \in L^2(-\infty, \infty; -\infty, \infty)$; see Fubini's and Tonelli's theorems in §A.4.3.

An analogous relation for integration ("derivative of order -1") is

$$\mathscr{F}\left\{ \int_{-\infty}^{x} f(\xi)\, d\xi \right\} = -\frac{1}{ik}\tilde{f}. \tag{10.3.4}$$

We also have the translation theorems

$$\mathscr{F}\{f(ax - b)\} = \frac{\exp(ibk/a)}{|a|}\,\tilde{f}\left(\frac{y}{a}\right), \qquad \mathscr{F}^{-1}\left\{\tilde{f}(k + c)\right\} = \exp(icx)\, f(x),$$

$$\tag{10.3.5}$$

both of which are simple consequences of a change of variable in the definition of the transform and its inverse.

A remarkably useful property is the *convolution theorem*. The *convolution product* or simply *convolution* of two integrable functions $f(x)$ and $g(x)$, denoted by $f * g$, is defined by

$$h(x) = (f * g)(x) = \int_{-\infty}^{\infty} f(x - \xi)g(\xi)\, d\xi, \tag{10.3.6}$$

when the integral exists. By the change of variable $\eta = x - \xi$ we find that also

$$h(x) = \int_{-\infty}^{\infty} f(\eta)g(x - \eta)\, d\eta = (g * f)(x), \tag{10.3.7}$$

so that the convolution product is commutative. By substituting for f and g their expressions (10.1.7) in terms of the inverse Fourier transformation and using the relation (10.2.7) for the δ-function, one proves that

$$\mathscr{F}\{f * g\} = \sqrt{2\pi}\,\tilde{f}\,\tilde{g}, \qquad \mathscr{F}^{-1}\left\{\tilde{f}\,\tilde{g}\right\} = \frac{1}{\sqrt{2\pi}}(f * g)(x). \tag{10.3.8}$$

Several illustrations of the great usefulness of this property are given in Chapter 4.

10.4 Uncertainty relation

An interesting application of some of the previous relations is to the derivation of "uncertainty relations" for the Fourier transform. These relations express the fact that the transform of a function which is appreciably non-zero only over a narrow x-range is significantly non-zero over a broad k-range, and vice versa.[10]

For a given $f(\mathbf{x})$, we define the mean value of the i-th component of $\mathbf{x}$ as

$$X_i = \frac{1}{\|f\|^2} \int x_i\, |f(\mathbf{x})|^2 d^d x, \tag{10.4.1}$$

---

[10] By imagining that a time factor $e^{i\omega t}$ is understood, the Fourier inversion theorem may be interpreted as stating that the function f is represented by a superposition of waves $e^{-ikx}\tilde{f}(k)$. In this picture the uncertainty relation states that "many" waves are necessary to represent a narrow pulse f, while a very broad pulse can be represented by "few" waves. The statement is trivially true for an infinite wave consisting of a single sinusoid.

when the integral exists. The "spread" of f around this mean value can be characterized by the variance, defined by

$$(\Delta X_i)^2 = \frac{1}{\|f\|^2} \int |x_i - X_i|^2 |f(\mathbf{x})|^2 \mathrm{d}^d x = \frac{\|(x_i - X_i)f\|^2}{\|f\|^2}, \tag{10.4.2}$$

provided it is finite. Similar definitions can be used for K_j, the mean value of k_j, and the variance of $\tilde{f}$:

$$K_j = \frac{1}{\|\tilde{f}\|^2} \int k_j |\tilde{f}(\mathbf{k})|^2 \mathrm{d}^d k, \quad (\Delta K_j)^2 = \frac{1}{\|\tilde{f}\|^2} \int |k_j - K_j|^2 |\tilde{f}(\mathbf{k})|^2 \mathrm{d}^d k. \tag{10.4.3}$$

By using the differentiation rule (10.3.1) we have

$$(\Delta K_j)^2 = \frac{1}{\|f\|^2} \int \left| -i\frac{\partial f}{\partial x_j} - K_j f \right|^2 \mathrm{d}^d x = \frac{1}{\|f\|^2} \int \left| \frac{\partial f}{\partial x_j} - iK_j f \right|^2 \mathrm{d}^d x, \tag{10.4.4}$$

where we have used the fact that $\|f\| = \|\tilde{f}\|$ by Parseval's equality. Thus

$$\|f\|^4 (\Delta X_i)^2 (\Delta K_j)^2 = \|(x_i - X_i)f\|^2 \left\| \frac{\partial f}{\partial x_j} - iK_j f \right\|^2. \tag{10.4.5}$$

At this point we use the Schwarz inequality $\|g_1\| \|g_2\| \geq |(g_1, g_2)|$ (p. 549), where the scalar product is defined in (10.2.10), and note that $|(g_1, g_2)| \geq |\mathrm{Re}\,(g_1, g_2)| = \frac{1}{2}|(g_1, g_2) + (g_2, g_1)|$, to write

$$\|f\|^4 (\Delta X_i)^2 (\Delta K_j)^2 \geq \frac{1}{4} \left| \int \left[(x_i - X_i)\overline{f} \left(\frac{\partial f}{\partial x_j} - iK_j f \right) + (x_i - X_i)f \right. \right.$$

$$\left. \left. \times \left(\frac{\partial \overline{f}}{\partial x_j} + iK_j \overline{f} \right) \right] \mathrm{d}^d x \right|^2 = \frac{1}{4} \left| \int (x_i - X_i) \left[\overline{f}\frac{\partial f}{\partial x_j} + f\frac{\partial \overline{f}}{\partial x_j} \right] \mathrm{d}^d x \right|^2$$

$$= \frac{1}{4} \left| \int \frac{\partial}{\partial x_j} \left[(x_i - X_i)|f|^2 \right] \mathrm{d}^d x - \delta_{ij} \int |f|^2 \mathrm{d}^d x \right|^2. \tag{10.4.6}$$

Upon applying the generalized divergence theorem (p. 397) with the assumption that $(x_i - X_i)|f|^2$ tends to zero at infinity sufficiently fast, the first integral vanishes and we find

$$\Delta X_i \, \Delta K_j \geq \frac{1}{2}\delta_{ij}. \tag{10.4.7}$$

This lower bound cannot be improved as it is actually attained when f is a Gaussian. In quantum mechanics $|f(\mathbf{x})|^2$ represents the probability distribution of the particle position $\mathbf{x}$ and $\mathbf{k} = -\mathbf{p}/\hbar$ with $\mathbf{p}$ the particle momentum. In this case then (10.4.7) becomes Heisenberg's famous uncertainty relation $\Delta X_i \, \Delta p_j \geq (\hbar/2)\delta_{ij}$.

10.5 Sine and cosine transforms

Just as the sine and cosine Fourier series are derived from the complete Fourier series, one can derive the sine and cosine transforms from the exponential Fourier transform.

If $f(x) = f(-x)$, i.e., f is an even function, after a change of the integration variable we have

$$\mathscr{F}\{f\} = \frac{1}{\sqrt{2\pi}} \left[\int_{-\infty}^{0} e^{ikx} f(x) \, dx + \int_{0}^{\infty} e^{ikx} f(x) \, dx \right]$$

$$= \frac{1}{\sqrt{2\pi}} \int_{0}^{\infty} \left(e^{ikx} + e^{-ikx} \right) f(x) \, dx = \sqrt{\frac{2}{\pi}} \int_{0}^{\infty} \cos kx \, f(x) \, dx. \quad (10.5.1)$$

This relation defines the *Fourier cosine transform* of f:

$$\mathscr{F}_c\{f\} = \sqrt{\frac{2}{\pi}} \int_{0}^{\infty} \cos kx \, f(x) \, dx = \tilde{f}_c(k), \quad (10.5.2)$$

while $\mathrm{Im}\, \mathscr{F}\{f\} = 0$. Since, evidently, $\tilde{f}(-k) = \tilde{f}(k)$ if f is even, the Fourier inversion formula shows that

$$\mathscr{F}_c^{-1}\{\tilde{f}_c\} = \sqrt{\frac{2}{\pi}} \int_{0}^{\infty} \cos kx \, \tilde{f}_c(k) \, dk = \frac{1}{2}[f(x+0) + f(x-0)]. \quad (10.5.3)$$

In a similar way, for an odd function, $f(-x) = -f(x)$, we find

$$\mathscr{F}\{f\} = i\sqrt{\frac{2}{\pi}} \int_{0}^{\infty} \sin kx \, f(x) \, dx, \quad (10.5.4)$$

and we define the *Fourier sine transform* as

$$\mathscr{F}_s\{f\} = \sqrt{\frac{2}{\pi}} \int_{0}^{\infty} \sin kx \, f(x) \, dx = \tilde{f}_s(k), \quad (10.5.5)$$

while $\mathrm{Re}\, \mathscr{F}\{f\} = 0$; the inverse is

$$\mathscr{F}_s^{-1}\{\tilde{f}_s\} = \sqrt{\frac{2}{\pi}} \int_{0}^{\infty} \sin kx \, \tilde{f}_s(k) \, dk = \frac{1}{2}[f(x+0) + f(x-0)]. \quad (10.5.6)$$

While these transforms have been defined starting from a function defined for $-\infty < x < \infty$, they make sense even if the function is only defined for $0 < x < \infty$. In this case, however, if we evaluate (10.5.6) at $x = 0$, we find 0 as, whenever $f(0) \neq 0$, use of the Fourier sine transform is equivalent to an odd extension of the definition of $f(x)$ over the entire real line (cf. (10.2.6) and the analogous phenomenon with the Fourier series on p. 55).

The operational relations for the sine and cosine transforms are readily derived by integration by parts and differentiation under the integral sign. The most important ones are the following (see Table 4.5 p. 114 and Table 4.6, p. 115 for additional relations):

$$\mathscr{F}_s\{f'\} = -k\,\mathscr{F}_c\{f\}, \qquad \mathscr{F}_s\{f''\} = -k^2\,\mathscr{F}_s\{f\} + \sqrt{\frac{2}{\pi}}\, k\, f(0), \quad (10.5.7)$$

$$\mathscr{F}_c\{f'\} = k\,\mathscr{F}_s\{f\} - \sqrt{\frac{2}{\pi}}\, f(0), \qquad \mathscr{F}_c\{f''\} = -k^2\,\mathscr{F}_c\{f\} - \sqrt{\frac{2}{\pi}}\, f'(0). \quad (10.5.8)$$

These relations show that these transforms are not generally useful for an equation containing both even and odd derivatives. As shown in many examples of Chapter 4, the explicit appearance of $f(0)$ and $f'(0)$ permits the solution of boundary value problems when f or

f' are assigned on the boundaries; in this respect these relations are similar to (3.1.14) and (3.1.16) p. 57 for the Fourier series.

It is not possible to define a symmetric convolution on the half range $0 < x < \infty$ and, therefore, the convolution theorem of the exponential Fourier transform cannot be applied directly to the sine or cosine transforms. Nevertheless we have seen in several examples in Chapter 4 that, by the use of parity considerations, it is often possible to turn the inversion of a sine or cosine transform into that of an exponential transform. After this step, by using the convolution theorem and again parity considerations, the final result can be written as an integral over the proper half-infinite range of the variable.

The Fourier sine and cosine transforms are particular examples of a general relation of the type

$$f(x) = \int_0^\infty K(kx)\, dk \int_0^\infty K(ky) f(y)\, dy, \qquad (10.5.9)$$

for some kernel function K. They are by no means the only ones for which a relation of this type holds. There is a great variety of kernels which satisfy (10.5.9) for suitable classes of functions, and each one of these defines a transform having certain operational properties and, therefore, useful in particular situations. For example, if $K(x) = \sqrt{x}\, J_\nu(x)$ we have essentially the Hankel transform of §10.9. In other cases, in place of (10.5.9), we have a more general relation of the type

$$f(x) = \int_0^\infty H(kx)\, dk \int_0^\infty K(ky) f(y)\, dy, \qquad (10.5.10)$$

which defines an unsymmetrical transform. We return to these general transforms in §15.4.3 p. 385 and §21.4.1 p. 646. A different approach to the general theory can be found in Titchmarsh 1948, chapter 8.

10.6 One-sided and complex Fourier transform

Even if the behavior of f at infinity is such that $\tilde{f}$ does not exist, the function $\exp(-hx)\, f(x)$ may be integrable at infinity for sufficiently large h, $h \geq a$ say. In this case, the function

$$\tilde{f}_+(k + ih) = \frac{1}{\sqrt{2\pi}} \int_0^\infty f(x) \exp[i(k + ih)x]\, dx \qquad (10.6.1)$$

may be well defined for sufficiently large h and it may be interpreted as the Fourier transform of a function equal to 0 for $x < 0$ and equal to $\exp(-hx)\, f(x)$ for $x > 0$. The exponential convergence of the integral permits us to differentiate under the integral sign an arbitrary number of times, which shows that $\tilde{f}_+$ is an entire function (i.e., free of finite singularities) in the half-plane $a < h < \infty$.

Application of the inversion theorem then gives

$$\frac{1}{\sqrt{2\pi}} \int_{-\infty}^\infty \tilde{f}_+(k + ih) \exp(-ikx)\, dk = \begin{cases} \exp(-hx)\, f(x) & \text{for } x > 0 \\ 0 & \text{for } x < 0 \end{cases} \qquad (10.6.2)$$

or

$$\frac{1}{\sqrt{2\pi}} \int_{-\infty}^{\infty} \tilde{f}_+(k+ih) \exp\left[-i(k+ih)x\right] dk = \begin{cases} f(x) & \text{for } x > 0 \\ 0 & \text{for } x < 0 \end{cases} \quad (10.6.3)$$

which, with $\kappa = k + ih$ complex, may also be written as

$$\frac{1}{\sqrt{2\pi}} \int_{ia-\infty}^{ia+\infty} \tilde{f}_+(\kappa) \exp(-i\kappa x) d\kappa = \begin{cases} f(x) & \text{for } x > 0 \\ 0 & \text{for } x < 0 \end{cases} \quad (10.6.4)$$

where a is a sufficiently large positive number.

Proceeding similarly, if the function $\exp(hx) f(x)$ is integrable at minus infinity for $h \geq b$, with b sufficiently large and positive, we can define

$$\tilde{f}_-(k+ih) = \frac{1}{\sqrt{2\pi}} \int_{-\infty}^{0} f(x) \exp\left[i(k+ih)x\right] dx, \quad (10.6.5)$$

with relations similar to the previous ones for the inverse. $\tilde{f}_-$ is an entire function in the half-plane $h < b$. Adding the two transforms we have

$$f(x) = \frac{1}{\sqrt{2\pi}} \int_{ia-\infty}^{ia+\infty} \tilde{f}_+(\kappa) e^{-i\kappa x} d\kappa + \frac{1}{\sqrt{2\pi}} \int_{-ib-\infty}^{-ib+\infty} \tilde{f}_-(\kappa) e^{-i\kappa x} d\kappa . \quad (10.6.6)$$

As an example, if $f(x) = e^x$, we have $\tilde{f}_\pm = \mp(1+i\kappa)^{-1}/\sqrt{2\pi}$, with $\operatorname{Im} \kappa$ greater and smaller than 1, respectively.

Example 10.6.1 [Connection with the Fourier series] The one-sided transforms can be applied to a periodic function of period 2π. For $\operatorname{Im} \kappa > 0$, we have

$$\begin{aligned} \tilde{f}_+(\kappa) &= \frac{1}{\sqrt{2\pi}} \sum_{n=0}^{\infty} \int_{2n\pi}^{2(n+1)\pi} f(x) e^{i\kappa x} dx = \frac{1}{\sqrt{2\pi}} \sum_{n=0}^{\infty} \int_{0}^{2\pi} f(y) e^{i\kappa(y+2n\pi)} dy \\ &= \frac{1}{\sqrt{2\pi}} \int_{0}^{2\pi} f(y) \frac{\exp(i\kappa y)}{1 - \exp(2\pi i\kappa)} dy = \frac{\hat{f}(\kappa)}{1 - \exp(2\pi i\kappa)}, \end{aligned} \quad (10.6.7)$$

where

$$\hat{f}(\kappa) = \frac{1}{\sqrt{2\pi}} \int_{0}^{2\pi} f(y) \exp(i\kappa y) dy, \quad (10.6.8)$$

and we have interchanged the order of integration and summation and used the known sum (8.3.5) p. 225 of the geometric series. A similar calculation shows that

$$\hat{f}_-(\kappa) = -\frac{\hat{f}(\kappa)}{1 - \exp(2\pi i\kappa)} \qquad \text{for} \quad \operatorname{Re} \kappa < 0, \quad (10.6.9)$$

so that (10.6.6) gives

$$f(x) = \frac{1}{\sqrt{2\pi}} \int_{ia-\infty}^{ia+\infty} \frac{\exp(-i\kappa x)}{1 - \exp(2\pi i\kappa)} \hat{f}(\kappa) d\kappa - \frac{1}{\sqrt{2\pi}} \int_{-ib-\infty}^{-ib+\infty} \frac{\exp(-i\kappa x)}{1 - \exp(2\pi i\kappa)} \hat{f}(\kappa) d\kappa . \quad (10.6.10)$$

If the behavior of $\tilde{f}$ at infinity is such that the integrals can be calculated by the method of residues in the usual way, we recover the Fourier series representation of f:[11]

$$f(x) = \frac{1}{\sqrt{2\pi}} \sum_{-\infty}^{\infty} \tilde{f}(n) \, \exp(-inx).$$

□

So far we have dealt with real functions, but the Fourier transform may also be defined for complex-valued functions $g(z)$ defined and regular in a strip $-a_- < \operatorname{Im} z < a_+$ (with $a_\pm > 0$) including the real line:

$$\tilde{g}(\kappa) = \frac{1}{\sqrt{2\pi}} \int_{-\infty}^{\infty} e^{i\kappa x} g(x) \, dx. \tag{10.6.11}$$

It is found that $\tilde{g}$ is an analytic function of the complex variable κ (see e.g. Titchmarsh 1948, p. 44; Champeney 1987, p. 95):

Theorem 10.6.1 *Let $g(z)$ be an analytic and regular function of the complex variable z in any strip interior to the interval $-a_- < \operatorname{Im} z < a_+$, with $a_\pm > 0$, and assume that*

$$g(z) = O\left(e^{\mp(\lambda_\pm - \varepsilon)x}\right) \qquad \text{for} \qquad x = \operatorname{Re} z \to \pm\infty \tag{10.6.12}$$

for any $\varepsilon > 0$, with $\lambda_\pm > 0$. Then $\tilde{f}(\kappa)$ is analytic in the strip $-\lambda_- < \operatorname{Im} \kappa < \lambda_+$, satisfies a relation similar to (10.6.12) with $\lambda_\pm$ replaced by $a_\mp$ and

$$f(z) = \frac{1}{\sqrt{2\pi}} \int_{-\infty}^{\infty} \tilde{f}(\kappa) \, e^{-i\kappa z} d\kappa \tag{10.6.13}$$

for every z in the strip $-a_- < \operatorname{Im} z < a_+$.

10.7 Integral asymptotics

It is a consequence of the Riemann–Lebesgue lemma (§9.5) that, if f is absolutely integrable (see e.g. Titchmarsh 1948, p. 11; Champeney 1987, p. 23; Rudin 1987, p. 103),

$$\lim_{k \to \pm\infty} \tilde{f}(k) = \lim_{k \to \pm\infty} \frac{1}{\sqrt{2\pi}} \int_{-\infty}^{\infty} e^{ikx} f(x) \, dx = 0. \tag{10.7.1}$$

Intuitively this result derives from the fact that, for large k, the exponential oscillates faster than any length scale present in f. Thus, for k large enough, f is essentially constant over each interval $2n\pi \leq kx \leq 2(n+1)\pi$ and the integral vanishes.

More detailed information on the behavior of $\tilde{f}$ for large k can be obtained by applying methods of the so-called saddle-point type to the integral defining the transform. We will

---

[11] For the difference in the sign of the exponential see footnote p. 266.

not deal with these methods in general (see e.g. Erdélyi 1956; de Bruijn 1958; Jeffreys 1962; Bleistein & Handelsman 1975); we will only focus on some special cases that arise in the problems of concern in this book (see e.g. Hinch 1991; Ablowitz & Fokas 1997). While asymptotic series can be obtained by carrying these methods to higher order, we will content ourselves with the leading-order terms.

We start by considering an integral of the general form

$$I_1(k) = \int_a^b f(x)e^{ik\varphi(x)}\, dx, \tag{10.7.2}$$

with $\varphi(x)$ real. We are interested in the behavior of I_1 for k real with $k \to \infty$. The intuitive argument just given to justify the Riemann–Lebesgue lemma indicates that the biggest contribution to the integral will derive from the neighborhood of those points X where φ changes as slowly as possible. Provided $\varphi''(X) \neq 0$, these points are such that $\varphi'(X) = 0$, whence the denomination *stationary phase* by which this method is known. In the neighborhood of such points

$$\varphi(x) \simeq \varphi(X) + \frac{1}{2}(x - X)^2\varphi''(X) + \cdots \tag{10.7.3}$$

and there will always be a short interval around X where $k(x - X)^2$ is not large, irrespective of the magnitude of k. Suppose that there is only one such point and write

$$I_1(k) = \left(\int_a^{X-\delta} + \int_{X-\delta}^{X+\delta} + \int_{X+\delta}^b\right) f(x)e^{ik\varphi(x)}\, dx \tag{10.7.4}$$

for some small δ. Let us focus on the middle term. We have, approximately,

$$\int_{X-\delta}^{X+\delta} f(x)e^{ik\varphi(x)}\, dx \simeq f(X)e^{ik\varphi(X)} \int_{X-\delta}^{X+\delta} \exp\left[\frac{i}{2}k(x - X)^2\varphi''(X)\right] dx. \tag{10.7.5}$$

In the integral we set $\xi = (x - X)\sqrt{k|\varphi''(X)|/2}$ with which $X \pm \delta$ becomes $\pm\delta\sqrt{k|\varphi''(X)|/2}$. For k large, these integration limits can be replaced by $\pm\infty$ so that

$$\int_{X-\delta}^{X+\delta} f(x)e^{ik\varphi(x)}\, dx \simeq \frac{e^{ik\varphi(X)} f(X)}{\sqrt{k|\varphi''(X)|/2}} \int_{-\infty}^{\infty} e^{i[\operatorname{sgn}\phi''(X)]\xi^2}\, d\xi. \tag{10.7.6}$$

Since the remaining integral exists[12] we can evaluate it by taking any appropriate limit. Let therefore

$$J = \int_{-\infty}^{\infty} e^{i(\pm 1+i\varepsilon)\xi^2}\, d\xi \tag{10.7.7}$$

and consider J^2, which can be seen as an area integral over a plane (ξ, η) (Fubini's theorem, p. 690). Upon passing to polar coordinates in this plane we find

$$J^2 = 2\pi \int_0^{\infty} r e^{i(\pm 1+i\varepsilon)r^2}\, dr = \pi \int_0^{\infty} e^{i(\pm 1+i\varepsilon)\rho}\, d\rho = -\frac{\pi}{\pm i - \varepsilon} \to \pm i\pi \tag{10.7.8}$$

[12] To see this one can, for example, write it is a sum of integrals from $-L$ to L and integrals between $\pm L$ and $\pm\infty$. The change of variable $\xi = x/L$ in the latter gives integrands of the form $\exp(iL^2\xi^2)$ which vanish by the Riemann–Lebesgue lemma for $L \to \infty$. See e.g. Ablowitz & Fokas 1997, p. 233 for a direct evaluation by contour methods.

so that $J = \sqrt{\pi}\,e^{\pm i\pi/4}$. Thus, in conclusion,

$$I_1(k) = \sqrt{\frac{2\pi}{k|\varphi''(X)|}}\,\exp\left[ik\varphi(X) + \frac{i}{4}\pi\,\mathrm{sgn}\,\phi''(X)\right]f(X) + O\left(\frac{1}{k}\right), \qquad (10.7.9)$$

where the error estimate follows by keeping track of the various approximations used to find the result. If the point of stationary phase coincides with one of the end points of the interval, the integral in (10.7.6) is over a half line and this result should be divided by 2. If there is more than one point where $\varphi' = 0$, to leading order the result is the sum of as many terms of the form (10.7.9). It may happen that some of these terms are subdominant with respect to others and, in this case, they can be dropped. When they are of the same order of the error affecting the leading order terms it may actually be necessary to drop them for consistency. The method can be extended to cases in which the first non-zero term after $\varphi(X)$ in (10.7.3) is proportional to $\varphi^{(n)}(X)$, with $n > 2$; we refer the reader to the references for this situation.

The order of magnitude of the remaining two integrals in (10.7.4) can be estimated by integration by parts. For example, if $\varphi(x) = x$, writing $c = X - \delta$ we have

$$\int_a^c f(x)e^{ikx}\,\mathrm{d}x = \frac{1}{ik}\left[f(c)e^{ikb} - f(a)e^{ika}\right] - \frac{1}{ik}\int_a^c f'(x)e^{ikx}\,\mathrm{d}x. \qquad (10.7.10)$$

Thus, the leading order behavior is proportional to $1/k$ and therefore subdominant in comparison with (10.7.9). The integration by parts can be continued to generate an asymptotic series. Its application to the second term gives a contribution of order k^{-2} and so on. The Taylor series approximation (10.7.3) can be avoided by proceeding as before after effecting the change of variable $\xi = \varphi(x)$, if necessary breaking up the interval of integration into sub-intervals in each one of which $\varphi(x)$ is monotonic.

Example 10.7.1 Let us apply the previous development to the integral

$$\int_0^\infty \frac{\cos a\sqrt{x}}{\sqrt{x}}\cos kx\,\mathrm{d}x = \frac{1}{2}\int_{-\infty}^\infty \frac{\cos a\sqrt{|x|}}{\sqrt{|x|}}e^{ikx}\,\mathrm{d}x \qquad (10.7.11)$$

$$= \frac{1}{4}\int_{-\infty}^\infty \left[e^{i(kx-a\sqrt{|x|})} + e^{i(kx+a\sqrt{|x|})}\right]\frac{\mathrm{d}x}{\sqrt{|x|}} = I_- + I_+,$$

with I_- and I_+ designating the first and second integral, respectively; we assume $a > 0$. With

$$\varphi_-(x) = x - \frac{a}{k}\sqrt{|x|}, \qquad (10.7.12)$$

we find only one point where $\varphi'_-(X) = 0$, namely $X = (a/2k)^2$. At this point $\varphi_-(X) = -a^2/4k^2$, $\varphi''_-(X) = 2k^2/a^2$ and (10.7.9) gives

$$I_- \simeq \frac{1}{2}\sqrt{\frac{\pi}{k}}\,\exp\left(-\frac{ia^2}{4k} + i\frac{\pi}{4}\right) + O\left(\frac{1}{k}\right). \qquad (10.7.13)$$

In principle it would be necessary to drop the first term in the exponential because otherwise, for large k, after a Taylor series expansion it would lead to a correction of order $k^{-3/2}$ which is of the same order as other terms discarded in arriving to the asymptotic estimate; we retain it however due to a fortuitous circumstance mentioned shortly. In a similar way we find

$$I_+ \simeq \frac{1}{2}\sqrt{\frac{\pi}{k}} \exp\left(\frac{ia^2}{4k} - i\frac{\pi}{4}\right) + O\left(\frac{1}{k}\right), \tag{10.7.14}$$

so that, combining the two integrals,

$$I = \sqrt{\frac{\pi}{k}} \cos\left(\frac{\pi}{4} - \frac{a^2}{4k}\right). \tag{10.7.15}$$

This actually happens to be the exact value of the integral but, without this knowledge, the only legitimate conclusion that we could draw would be

$$I = \sqrt{\frac{\pi}{k}} \left(\cos\frac{\pi}{4}\right)\left[1 + O\left(\frac{1}{k}\right)\right] = \sqrt{\frac{\pi}{2k}}\left[1 + O\left(\frac{1}{k}\right)\right]. \tag{10.7.16}$$

□

An application to the derivation of the asymptotic behavior of Bessel functions is shown on p. 309 and to an integral arising in a model of waves at a liquid surface can be found in §6.8 p. 166.

Let us now turn to the Laplace-type integral

$$I_2(h) = \int_a^b f(x)e^{-h\varphi(x)}\,dx \qquad h \to \infty. \tag{10.7.17}$$

With both h and φ real the oscillation argument is inapplicable, but it is clear that, for large h, the greatest contribution to the integral will come from the neighborhood of the minima of φ, where $\varphi' = 0$, $\varphi'' > 0$. Similarly to (10.7.4), the integral can be broken up into a sum of integrals over small intervals centered at such points, plus a remainder. As before, the latter can be estimated by integration by parts and is found to be subdominant. For a point X where φ is a minimum we proceed as before and find

$$I_2(h) \simeq \sqrt{\frac{2\pi}{h\varphi''(X)}}\, e^{-h\varphi(X)}\, f(X). \tag{10.7.18}$$

As in the previous case, the result should be divided by 2 when the point X coincides with one of the end points of the interval.

Example 10.7.2 Applying (10.7.18) to the integral

$$I = \int_0^\infty \frac{e^{-hx^2}}{x^2 + b^2}\,dx \tag{10.7.19}$$

we find, after dividing by 2 because φ' vanishes at the integration limit $X = 0$, with $\varphi''(X) = 2$,

$$I \simeq \frac{1}{2b^2}\sqrt{\frac{\pi}{h}} \tag{10.7.20}$$

which is the first term in the asymptotic expansion of the exact result $I = (\pi/2b)\mathrm{erfc}\,(b\sqrt{h})\,\exp(hb^2)$ with erfc the complementary error function defined in (5.5.15) p. 140.   □

Another application to the derivation of the Stirling formula for the Gamma function is shown in §17.5.1 p. 442.

Integrals of the form I_1 and I_2 discussed before are special cases of the integral

$$I_3(k) = \int_C f(z) e^{k\chi(z)} \, dz, \qquad (10.7.21)$$

with χ in general complex, taken along a path C in the complex plane; we can take k real and positive, attributing its phase, if any, to χ. By writing $\chi = \mathrm{Re}\,\chi\,(z) + i\,\mathrm{Im}\,\chi\,(z)$, on the basis of the considerations developed before, for $k \to \infty$ we expect the dominant contribution to come from the neighborhood of point(s) Z where $\chi'(Z) = 0$. Even if the original integration path does not go through these points, it can be deformed so that it does provided no singularity of f or χ is traversed.[13] Since, by the maximum modulus principle (p. 432), any contour surrounding Z must contain points z where $|\chi\,(z)| > |\chi\,(Z)|$ and points where $|\chi\,(z)| < |\chi\,(Z)|$, the point Z will not be a maximum or a minimum but a *saddle point*, from which the method takes its name.

It will clearly be advantageous to choose the direction of traversal of the point Z in such a way as to maximize the contribution of the immediate neighborhood of this point to the integral. The argument used for integrals of the type (10.7.17) suggests that this direction must be such that $\mathrm{Re}\,\chi\,(z)$ decreases as fast as possible moving away from Z; since

$$\mathrm{Re}\,[\chi\,(z) - \chi\,(Z)] \simeq \frac{1}{2}|z - Z|^2 |\chi''(Z)| \cos\,[2\arg\,(z - Z) + \arg\,\chi''(Z)], \quad (10.7.22)$$

the maximum decrease will take place for $2\arg\,(z - Z) + \arg\,\chi''(Z) = \pi$, which is seen to be a path of constant $\mathrm{Im}\,\chi$. With this choice, the auxiliary variable ξ is now properly defined as

$$z = Z + \sqrt{k|\chi''(X)|/2}\ \xi\ e^{(i/2)[\pi - \arg\,\chi''(z)]}. \qquad (10.7.23)$$

With this substitution, the integral can be evaluated as before with the result

$$I_3 \simeq \sqrt{\frac{2\pi}{k|\chi''(Z)|}}\ f(Z)\,\exp\left[k\chi(Z) + \frac{i}{2}\,(\pi - \arg\,\chi''(Z))\right]. \qquad (10.7.24)$$

For integrals of the type I_1, $\arg\chi'' = \arg\varphi'' + \pi/2$ and $\pi - \arg\,\chi''(Z) = \pi/2 - \arg\phi''(Z) = \pm\pi/2$ according as $\phi''(Z) > 0$ or $\phi''(Z) < 0$, which agrees with (10.7.9). For integrals of the type I_2, $\arg\chi'' = \arg\varphi'' + \pi$ and $\pi - \arg\,\chi''(Z) = -\arg\phi''(Z)$, which agrees with (10.7.18). Again, the result needs to be halved if the saddle point is one of the end points of the integration path.

---

[13] If singularities are present, the integration path can still be deformed provided their contribution is properly taken into consideration by accounting for their residues; see Chapter 17.

In all these cases the arguments can be extended to generate asymptotic series for the integrals by what is essentially a generalization of Watson's lemma given on p. 296 (see e.g. Erdélyi 1956, chapter 2; Jeffreys 1962, chapter 2; Bleistein & Handelsman 1975).

10.8 Asymptotic behavior of the sine and cosine transforms

In order to deduce the asymptotic behavior of the Fourier cosine transform we may proceed as follows. Let $f(x) = x^{-\alpha}\varphi(x)$, where $0 < \alpha < 1$, with $\varphi(0+)$ finite, and write

$$\tilde{f}_c(k) = \sqrt{\frac{2}{\pi}} \int_0^\infty x^{-\alpha} \{\varphi(0+) + [\varphi(x) - \varphi(0+)]\} \cos kx \, dx. \tag{10.8.1}$$

The first term can be integrated directly using

$$\int_0^\infty x^{\mu-1} \left\{ \begin{array}{c} \cos kx \\ \sin kx \end{array} \right\} dx = \Gamma(\mu) k^{-\mu} \left\{ \begin{array}{c} \cos \frac{\mu\pi}{2} \\ \sin \frac{\mu\pi}{2} \end{array} \right\} \tag{10.8.2}$$

valid for $0 < \mu < 1$. The second term can be estimated by means of an integration by parts and the Riemann–Lebesgue lemma, which shows that it is $o(1/k)$ for large k if φ is of bounded variation in $(0, \infty)$ (see Titchmarsh 1948, p. 172). In conclusion

$$\tilde{f}_c(k) \simeq \varphi(0+)\sqrt{\frac{2}{\pi}} k^{\alpha-1} \Gamma(1-\alpha) \sin \frac{\pi\alpha}{2} \qquad \text{for} \qquad k \to \infty. \tag{10.8.3}$$

We can estimate $\tilde{f}_s(k)$ similarly:

$$\tilde{f}_s(k) \simeq \varphi(0+)\sqrt{\frac{2}{\pi}} k^{\alpha-1} \Gamma(1-\alpha) \cos \frac{\pi\alpha}{2} \qquad \text{for} \qquad k \to \infty. \tag{10.8.4}$$

If $\alpha = 0$ the integrals can be estimated more simply by an integration by parts. In this case the first estimate becomes

$$\tilde{f}_c(k) \simeq \sqrt{\frac{2}{\pi}} k^{-2} \left[-f'(0+) + k^{-2} f'''(0+) + \cdots \right] \qquad k \to \infty, \tag{10.8.5}$$

and the second one

$$\tilde{f}_s(k) \simeq \sqrt{\frac{2}{\pi}} k^{-1} \left[f(0+) - k^{-2} f''(0+) + \cdots \right] \qquad k \to \infty. \tag{10.8.6}$$

If $\alpha = 1$ (10.8.4) becomes

$$\tilde{f}_s(k) \simeq \sqrt{\frac{\pi}{2}} \varphi(0+) \qquad \text{for} \qquad k \to \infty, \tag{10.8.7}$$

while the cosine transform does not exist.

When $f(x) = x^{-\alpha}\varphi(x)$ for $x \to \infty$, with $0 < \alpha < 1$ and $\varphi(\infty)$ finite, similar results hold for $k \to 0$ by replacing $\varphi(0+)$ by $\varphi(\infty)$:[14]

$$\tilde{f}_c(k) \simeq \varphi(\infty)\sqrt{\frac{2}{\pi}}\, k^{\alpha-1}\, \Gamma(1-\alpha)\, \sin\frac{\pi\alpha}{2} \qquad \text{for} \qquad k \to 0, \qquad (10.8.8)$$

$$\tilde{f}_s(k) \simeq \varphi(\infty)\sqrt{\frac{2}{\pi}}\, k^{\alpha-1}\, \Gamma(1-\alpha)\, \cos\frac{\pi\alpha}{2} \qquad \text{for} \qquad k \to 0. \qquad (10.8.9)$$

These are justified by setting $f(x) = x^{-\alpha}[\varphi(\infty) + \varphi_1(x)]$. By (10.8.2) the first term gives (10.8.8) and (10.8.9), while the second one can be proven to be negligible with respect to the first one by use of the mean value theorem. An intuitive argument is that, by setting $y = x/k$, we have

$$\int_0^\infty x^{-\alpha}\varphi_1(x)\, \sin kx\, dx = k^{\alpha-1}\int_0^\infty y^{-\alpha}\varphi_1(y/k)\, \sin y\, dy, \qquad (10.8.10)$$

and similarly for the cosine transform. By hypothesis φ_1 vanishes for large argument and, therefore, $\varphi_1(y/k) \to 0$ as $k \to 0$. Thus this contribution will be subdominant with respect to the leading term proportional to $k^{\alpha-1}$.

10.9 The Hankel transform

Let us consider the double Fourier transform (10.2.11) of a function $f(x, y)$. Upon converting to polar coordinates in the (x, y)-plane we have

$$\tilde{f}(k_x, k_y) = \frac{1}{2\pi}\int_0^\infty r\, dr \int_0^{2\pi} d\theta\, \exp\left[ir\left(k_x \cos\theta + k_y \sin\theta\right)\right] f(r, \theta). \qquad (10.9.1)$$

Let us expand f in a Fourier series in θ:

$$f(r, \theta) \sum_{n=-\infty}^{\infty} f_n(r)\frac{\exp in\theta}{\sqrt{2\pi}}. \qquad (10.9.2)$$

Then, upon adopting polar coordinates also in the plane (k_x, k_y) of the transformed variable, we may write

$$k_x = k \cos\eta, \qquad k_y = k \sin\eta, \qquad (10.9.3)$$

and, assuming that the interchange of summation and integration is permissible (see pp. 226 and 694), we find

$$\tilde{f}(k, \eta) = \frac{1}{(2\pi)^{3/2}}\sum_{n=-\infty}^{\infty}\int_0^\infty r\, f_n(r)\, dr \int_0^{2\pi} d\theta\, e^{ikr\,\cos(\theta-\eta)+in\theta}. \qquad (10.9.4)$$

---

[14] See Titchmarsh 1948 p. 172 for additional results of this type.

Because of the periodicity of the integrand, with $\alpha = \theta - \eta$, we may write

$$\int_0^{2\pi} d\theta \, e^{ikr\cos(\theta-\eta)+in\theta} = e^{in\eta} \int_0^{2\pi} d\alpha \, e^{ikr\cos\alpha+in\alpha} = e^{in\eta} \, 2\pi \, J_n(kr), \qquad (10.9.5)$$

where use has been made of the integral representation (12.2.25) p. 308 of the Bessel function J_n. Thus, we find

$$\tilde{f}(k,\eta) = \sum_{n=-\infty}^{\infty} \frac{\exp in\eta}{\sqrt{2\pi}} \int_0^{\infty} r \, J_n(kr) \, f_n(r) \, dr, \qquad (10.9.6)$$

which is the Fourier series expansion of the transformed function $\tilde{f}$. We define the *Hankel transform* of order n as

$$\mathcal{H}_n\{f_n\} = \tilde{f}_n(k) = \int_0^{\infty} r \, J_n(kr) \, f_n(r) \, dr. \qquad (10.9.7)$$

The relation (10.9.6) can be inverted using the inversion theorem for the Fourier transform. Executing steps that parallel those leading to (10.9.6), an easy calculation shows that

$$f(r,\theta) = \sum_{n=-\infty}^{\infty} \frac{\exp in\theta}{\sqrt{2\pi}} \int_0^{\infty} k \, J_n(kr) \, \tilde{f}_n(k) \, dk, \qquad (10.9.8)$$

so that the inversion formula for the Hankel transform (10.9.7) is

$$\mathcal{H}_n^{-1}\{\tilde{f}_n\} = f_n(r) = \int_0^{\infty} k \, J_n(kr) \, \tilde{f}_n(k) \, dk. \qquad (10.9.9)$$

We thus conclude that

$$\int_0^{\infty} k \, r \, J_n(kr) \, J_n(ks) \, dk = \delta(r-s). \qquad (10.9.10)$$

Another definition of the direct and inverse transforms encountered in the literature is the symmetric one (e.g. Erdélyi *et al.* 1954)[15]

$$\tilde{f}_\nu(k) = \int_0^{\infty} \sqrt{kr}\, J_\nu(kr) f(r)\, dr, \qquad f(r) = \int_0^{\infty} \sqrt{kx}\, J_\nu(kr) \tilde{f}(k)\, dk. \qquad (10.9.11)$$

As shown in several examples of Chapter 6, the Hankel transform is very useful in dealing with one piece of the Laplace operator expressed in plane polar coordinates:

$$\nabla_2^2 f = \frac{1}{r}\frac{\partial}{\partial r}\left(r\frac{\partial f}{\partial r}\right) + \frac{1}{r^2}\frac{\partial^2 f}{\partial\theta^2}, \qquad (10.9.12)$$

which becomes

$$\mathcal{H}\{\nabla_2^2 f\} = \sum_{n=-\infty}^{\infty} \frac{\exp in\theta}{\sqrt{2\pi}} \int_0^{\infty} r \, J_n(kr) \left[\frac{1}{r}\frac{d}{dr}\left(r\frac{df_n}{dr}\right) - \frac{n^2}{r^2} f_n\right] dr. \qquad (10.9.13)$$

---

[15] In the sense of ordinary functions (as opposed to distributions), these formulae hold for $\nu \geq -1/2$ provided that $\int_0^{\infty} \sqrt{r}|f(r)|dr < \infty$ (see Erdélyi *et al.* 1953, volume 2 p. 73).

A double integration by parts and use of the differential equation satisfied by J_n readily shows that

$$\mathcal{H}_n \left\{ \frac{1}{r} \frac{d}{dr} \left(r \frac{df_n}{dr} \right) - \frac{n^2}{r^2} f_n \right\} = -k^2 \tilde{f}_n(k), \qquad (10.9.14)$$

which is the most important operational formula concerning the Hankel transform. Using this result, (10.9.13) becomes

$$\mathcal{H} \left\{ \frac{1}{r} \frac{\partial}{\partial r} \left(r \frac{\partial f}{\partial r} \right) + \frac{1}{r^2} \frac{\partial^2 f}{\partial \theta^2} \right\} = -k^2 \sum_{n=-\infty}^{\infty} \frac{\exp in\theta}{\sqrt{2\pi}} \tilde{f}_n(k) = -k^2 \tilde{f}(k, \theta), \qquad (10.9.15)$$

as it must, given that $k^2 = k_x^2 + k_y^2$ and that we have taken a double Fourier transform.

More generally, one can define the Hankel transform of order ν as

$$\mathcal{H}_\nu \{g\} = \tilde{g}_\nu(k) = \int_0^\infty r J_\nu(kr) g(r) \, dr, \qquad (10.9.16)$$

which, for $\nu = \pm 1/2$ reduce to Fourier sine and cosine transforms respectively.

A general theorem on the direct and inverse Hankel transform is the following (see e.g. Watson 1944, p. 456; Titchmarsh 1948, p. 240):

Theorem 10.9.1 *If $f(x) \in L(0, \infty)$ and is of bounded variation near the point x, then, for $\nu \geq 1/2$,*

$$\int_0^\infty J_\nu(kx) \sqrt{kx} \, dk \int_0^\infty J_\nu(ky) \sqrt{ky} \, f(y) \, dy = \frac{1}{2} [f(x+0) + f(x-0)]. \qquad (10.9.17)$$

As always, these hypotheses can be considerably weakened and the transform studied in the context of the theory of distributions (see e.g. Gelfand & Shilov 1964, vol. 2; Champeney 1987).

We can obtain asymptotic estimates of the transform by proceeding as in the previous section. If we assume that $g(r) = r^{-\alpha} \varphi(r)$ for small r, with $1/2 < \alpha < \nu + 2,$[16] then

$$\tilde{g}_\nu(k) = \int_0^\infty r^{1-\alpha} J_\nu(kr) \{\varphi(0+) + [\varphi(r) - \varphi(0+)]\} \, dr, \qquad (10.9.18)$$

and we find

$$\tilde{g}_\nu(k) \simeq 2^{-\alpha}(\nu - \alpha) \frac{\Gamma\left(\frac{1}{2}(\nu - \alpha)\right)}{\Gamma\left(\frac{1}{2}(\nu + \alpha)\right)} k^{\alpha-2} \varphi(0+) \qquad k \to \infty. \qquad (10.9.19)$$

In particular, for $\nu = 0$,

$$\tilde{g}_0(k) \simeq -2^{-\alpha} \alpha \frac{\Gamma(-\alpha/2)}{\Gamma(\alpha/2)} k^{\alpha-2} \varphi(0+) \qquad k \to \infty. \qquad (10.9.20)$$

---

[16] The first condition ensures convergence of the integral at infinity, the second one at 0.

The Laplace Transform

11.1 Direct and inverse Laplace transform

Consider a function $f(t)$ such that $f = 0$ for $-\infty < t < 0$, and let $e^{-\sigma t} f(t)$, with σ a real number, be absolutely integrable. This condition ensures the existence of the Fourier transform of $e^{-\sigma t} f(t)$ given by

$$\mathscr{F}\{e^{-\sigma t} f(t)\} = \frac{1}{\sqrt{2\pi}} \int_{-\infty}^{\infty} e^{-\sigma t} f(t) e^{i\omega t} \, dt = \frac{1}{\sqrt{2\pi}} \int_{0}^{\infty} e^{i(\omega+i\sigma)t} f(t) \, dt. \quad (11.1.1)$$

As we have seen in discussing the one-sided Fourier transform in §10.6, the integral can be considered as the Fourier transform of $f(t)$ evaluated for the complex argument $\omega + i\sigma$, $\tilde{f}_+(\omega + i\sigma) = \mathscr{F}\{e^{-\sigma t} f(t)\}$. The *Laplace transform* $\hat{f}$ of f is defined as $\tilde{f}_+$ considered as a function of the variable $s = -i(\omega + i\sigma)$:[1]

$$\mathscr{L}\{f\}(s) = \hat{f}(s) \equiv \sqrt{2\pi} \; \tilde{f}(is) = \int_{0}^{\infty} e^{-st} f(t) \, dt. \quad (11.1.2)$$

In the following we denote the real and imaginary parts of s by σ and τ respectively, so that $s = \sigma + i\tau$; from the definition of s we then have $\tau = -\omega$.

Just as $\tilde{f}_+$ is an analytic function of its argument, $\hat{f}$ is an analytic function of s. Clearly, if $\hat{f}$ exists for some s_1, it will also exist for any s with $\operatorname{Re} s > \operatorname{Re} s_1$. In this region, convergence of the integral is uniform (see e.g. Doetsch 1974, p. 13). The smallest $\operatorname{Re} s$ such that $\hat{f}$ exists is called the *abscissa of convergence* σ_c (see e.g. Hille & Phillips 1965, p. 215).

Theorem 11.1.1 [Landau–Pincherle] *The abscissa of convergence* σ_c *of the Laplace transform of* $f(t)$ *is given by*[2]

$$\sigma_c = \lim_{t \to \infty} \frac{1}{t} \log \left| \int_{t}^{\infty} f(t) \, dt \right| \quad or \quad \sigma_c = \lim_{t \to \infty} \frac{1}{t} \log \left| \int_{0}^{t} f(t) \, dt \right| \quad (11.1.3)$$

according as the integral does or does not converge as $t \to \infty$.

---

[1] With regard to the upper limit, the integral defining the Laplace transform is to be understood as an improper one, i.e. as the limit as $T \to \infty$ of the integral between 0 and T. A similar interpretation is necessary at all other singular points of the integrand, e.g. the function equal to $1/\sqrt{t(1-t)}$ for $0 < t < 1$ and vanishing everywhere else.

[2] A more precise statement is that σ_c equals the supremum of the indicated limits.

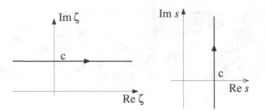

Fig. 11.1 Inversion paths for $\tilde{f}_+$ (left) and for the Laplace transform $\hat{f}$ (right).

If $\sigma_c = -\infty$ (e.g., $f = \exp e^{-t}$ for $t > 0$), $\hat{f}$ is an entire function while, if $\sigma_c = \infty$ (e.g. $f = \exp e^t$ for $t > 0$) $f(t)$ does not admit a Laplace transform. It is intuitively clear that $\hat{f}(s) \to 0$ as $\operatorname{Re} s - \sigma_c \to \infty$.

The domain of convergence of the Laplace transform may or may not contain point(s) on the boundary $\operatorname{Re} s = \sigma_c$ of the region of convergence. For example, for the functions $(1 + t^n)^{-1}$, the domain of convergence is the open half plane $\operatorname{Re} s > 0$ for $n = 0$, while it is $\operatorname{Re} s \geq 0$, $s \neq 0$ for $n = 1$, and $\operatorname{Re} s \geq 0$ for $n > 1$. In any event, in most cases of interest, $\hat{f}$ can be continued analytically to the left of the line $\operatorname{Re} s = \sigma_c$ once the nature and position of the singularities has been elucidated.[3] In a similar way one can define the *abscissa of absolute convergence* σ_a and the *abscissa of uniform convergence* σ_u. In general $-\infty \leq \sigma_c \leq \sigma_u \leq \sigma_a \leq \infty$ (see e.g. Hille & Phillips 1965, p. 215). As an example, for $f(t) = e^t \sin e^t$, $\sigma_c = 0$, while $\sigma_a = 1$ as can be proven with the change of variable $\tau = e^t$.

An important *uniqueness theorem* states that two functions having the same Laplace transform for $\operatorname{Re} s$ greater than some finite value are equal almost everywhere.

To recover $f(t)$ from a knowledge of $\hat{f}$, we may apply the Fourier inversion theorem to $\mathscr{F}\{e^{-\sigma t} f(t)\}$ finding, as in (10.6.4),

$$f(t) = \frac{1}{\sqrt{2\pi}} \int_{ic-\infty}^{ic+\infty} \tilde{f}_+(\zeta) e^{-i\zeta t} d\zeta \qquad \text{for} \qquad t > 0. \qquad (11.1.4)$$

Since $\tilde{f}_+$ is analytic for $\operatorname{Im} \zeta > \sigma_c$, the path of integration $\operatorname{Im} \zeta = c$ is arbitrary provided it is chosen above all the singularities of $\tilde{f}_+(\zeta)$ (Figure 11.1, left). Upon setting $s = -i\zeta$ and noting that, from the definition (11.1.2), $\hat{f}(\zeta) = \sqrt{2\pi}\, \tilde{f}(i\zeta)$, the previous integral becomes

$$\mathscr{L}^{-1}\left\{\hat{f}\right\} \equiv H(t) f(t) = \frac{1}{2\pi i} \int_{c-i\infty}^{c+i\infty} \hat{f}(s) e^{st} ds, \qquad (11.1.5)$$

where now the path of integration $\operatorname{Re} s = c$ is to the right of all the singularities of $\hat{f}(s)$ (Figure 11.1, right); here $H(t)$ is Heaviside's step function

$$H(t) = \begin{cases} 0 & t < 0 \\ 1 & t > 0 \end{cases}, \qquad (11.1.6)$$

---

[3] The smallest value of $\operatorname{Re} s$ for which this is possible is the *abscissa of holomorphy* σ_h of $\hat{f}$.

introduced to stress the fact that the integral equals 0 for $t < 0$; (11.1.5) is the *Mellin inversion theorem* for the Laplace transform.

To account for the possibility of discontinuities of $f(t)$, a more precise statement of this inversion theorem is

$$\frac{1}{2\pi i} \int_{c-i\infty}^{c+i\infty} \hat{f}(s)\, e^{st}\, ds = \begin{cases} \frac{1}{2}[f(t+0) + f(t-0)] & t > 0 \\ \frac{1}{2} f(0+) & t = 0 \\ 0 & t < 0 \end{cases} . \qquad (11.1.7)$$

It is very frequently the case that, at $t = 0$, $f(t)$ is continuous from the right so that $\lim_{t\to 0+} f(t) = f(0+) \neq 0$. Since $f(t)$ is assumed to vanish for $t < 0$, in this case, $f(t)$ has a discontinuity at $t = 0$ and, from the second line in this relation,

$$\lim_{t\to 0+} \frac{1}{2\pi i} \int_{c-i\infty}^{c+i\infty} \hat{f}(s)\, e^{st}\, ds = f(0+) \quad \text{while} \quad \frac{1}{2\pi i} \int_{c-i\infty}^{c+i\infty} \hat{f}(s)\, ds = \frac{1}{2} f(0+).$$

$$(11.1.8)$$

A simple example is $f(t) = \cos t$, for which $\hat{f} = s/(s^2 + 1)$. For this function $\int [s/(s^2 + 1)]\,ds = (1/2) \int dz/(z+1) \to \log(-1) = i\pi$ so that $\mathscr{L}^{-1}\{s/(s^2+1)\}(0) = 1/2$.

11.2 Operational formulae

As with the Fourier transforms, many operational formulae valid for the Laplace transform greatly enhance its usefulness. Here we provide a brief look at some of the most important ones; others are given in Table 5.3 p. 126 and in several references (see e.g. Erdélyi *et al.* 1954 vol. 1). The great convenience of formulae of this type is illustrated in the applications described in Chapter 5.

1. Step function. The Laplace transform of 1 – or, since the function must vanish for $t < 0$, more properly of the Heaviside step function $H(t)$ – can be calculated directly and is

$$\mathscr{L}\{H(t)\} = 1/s. \qquad (11.2.1)$$

2. Derivative. Upon integrating by parts, the Laplace transform of the first derivative of f is found to be[4]

$$\mathscr{L}\{f'\} = s\,\mathscr{L}\{f\} - f(0+) = s\,\hat{f} - f(0+). \qquad (11.2.2)$$

By a repeated application of this result we have

$$\mathscr{L}\{f''\} = s\,\mathscr{L}\{f'\} - f'(0+) = = s^2\hat{f} - sf(0+) - f'(0+). \qquad (11.2.3)$$

---

[4] If f is discontinuous at $t_0 > 0$, then $\mathscr{L}\{df/dt\} = s\hat{f} - f(0+) + \exp(-st_0)[f(t_0-) - f(t_0+)]$ as can be easily seen by breaking the integral from 0 to ∞ into two parts at t_0 and integrating by parts in each one of them, or by expressing the derivative with the aid of the δ function.

We will justify in §20.12 p. 605 the relation

$$\mathscr{L}\{\delta(t)\} = 1. \tag{11.2.4}$$

With this result, we may reconsider (11.2.2) by noting that, as a function defined over the entire real line, $f(t)$ should be interpreted as $H(t)f(t)$. For this function $(d/dt)(Hf) = Hf' + f(t)\delta(t) = Hf' + f(0)\delta(t)$ so that $\mathscr{L}\{Hf'\} = \mathscr{L}\{(d/dt)(Hf)\} - f(0).$[5]

Since the integral defining the Laplace transform converges uniformly, differentiation can be carried out under the integral sign (p. 691) which shows that, for $\mathrm{Re}\, s > \sigma_a$,

$$\mathscr{L}\{tf(t)\} = -\frac{d}{ds}\hat{f}. \tag{11.2.5}$$

4. Similarity and shift rules. By a simple change of variable we immediately prove the similarity rule:

$$\mathscr{L}\{f(bt)\}(s) = (1/b)\,\hat{f}(s/b) \qquad \text{for} \qquad b > 0, \tag{11.2.6}$$

where b must positive since f vanishes for negative argument. In a similar way we prove the shift rule

$$\mathscr{L}\{f(t-a)\}(s) = e^{-as}\,\hat{f}(s) \qquad \text{for} \qquad a > 0, \tag{11.2.7}$$

where we have used the fact that f vanishes for negative argument so that $f(t-a) = 0$ for $t < a$. By combining with (11.2.6) we have

$$\mathscr{L}\{f(bt-a)\}(s) = \frac{1}{b}e^{-as/b}\,\hat{f}\left(\frac{s}{b}\right) \qquad \text{for} \qquad a, b > 0. \tag{11.2.8}$$

5. Attenuation rule. Directly from the definition of the transform we have

$$\mathscr{L}\{e^{-at}f(t)\}(s) = \hat{f}(s+a). \tag{11.2.9}$$

6. Integration. By interchanging the order of integrations (p. 690), we have

$$\mathscr{L}\left\{\int_0^t f(\tau)\,d\tau\right\} = \hat{f}(s)/s. \tag{11.2.10}$$

7. Periodic functions. For a periodic function, $f(t+T) = f(t)$, the integral can be broken up into a series of integrals from nT to $(n+1)T$ which can be summed with the result

$$\mathscr{L}\{f\} = \left(1 - e^{sT}\right)^{-1}\int_0^T f(t)\,e^{-st}\,dt. \tag{11.2.11}$$

---

[5] The ambiguity of an expression like $\int_0 \delta(t)f(t)\,dt$ has been noted in §2.3 and p. 577. In applications involving the Laplace transform the simplest and most consistent option is to interpret it as $f(0+)$.

8. Convolution theorem. Just as in the case of the Fourier transform, extremely useful is the convolution theorem for the Laplace transform:

$$\mathscr{L}^{-1}\left\{\hat{f}\,\hat{g}\right\} = \int_0^t f(t-\tau)\,g(\tau)\,d\tau = \int_0^t f(\tau)\,g(t-\tau)\,d\tau. \qquad (11.2.12)$$

This result follows directly from the corresponding theorem for the Fourier transform due to the vanishing of $f(t)$ and $g(t)$ for $t < 0$. As a simple application, we may derive the inverse of the formula (11.2.10) by noting that

$$\mathscr{L}^{-1}\{\hat{f}/s\} = \int_0^t H(t-\tau)\,f(\tau)\,d\tau = \int_0^t f(\tau)\,d\tau. \qquad (11.2.13)$$

9. Fractional derivative. The relations (11.2.2) and (11.2.3) suggest the possibility of defining the fractional derivative of order β of a function $f(t)$. A straightforward definition would be $d^\beta f/dt^\beta = \mathscr{L}^{-1}\{s^\beta\,\hat{f}\}$ with which, upon using the convolution theorem and the fact that, for $\beta > 0$, $\mathscr{L}^{-1}\{s^\beta\} = t^{-\beta-1}/\Gamma(-\beta)$ (see Table 5.1 p. 124), we would have

$$\frac{d^\beta f}{dt^\beta} = \frac{1}{\Gamma(-\beta)}\int_0^t f(t-\tau)\,\tau^{-\beta-1}\,d\tau = \frac{1}{\Gamma(-\beta)}\int_0^t f(\tau)\,(t-\tau)^{-\beta-1}\,d\tau. \quad (11.2.14)$$

The problem with this definition is that, if $\beta > 0$, in general the integrand is singular. To circumvent this difficulty we define the *Riemann–Liouville fractional derivative* of order β by

$$\frac{d^\beta f}{dt^\beta} = \frac{d^m}{dt^m}\mathscr{L}^{-1}\{s^{\beta-m}\,\hat{f}\} = \frac{1}{\Gamma(m-\beta)}\frac{d^m}{dt^m}\int_0^t f(\tau)\,(t-\tau)^{m-\beta-1}\,d\tau, \quad (11.2.15)$$

where m is the smallest integer greater than β. In this way, for example, by using the relation between the beta function and the Gamma functions (p. 445), we find[6]

$$\frac{d^\beta t^a}{dt^\beta} = \frac{\Gamma(a+1)}{\Gamma(a-\beta+1)}\,t^{a-\beta}. \qquad (11.2.16)$$

An example of the application of the fractional derivative to Abel's equation is shown in §5.6 p. 143.

The derivative defined in this way has some strikingly different properties from the ordinary derivative. For example, from (11.2.16), $d^{1/2}1/dt^{1/2} = (\pi t)^{-1/2}$ although $(d^{1/2}/dt^{1/2})(d^{1/2}/dt^{1/2})1 = d1/dt = 0$. Furthermore, while the ordinary derivative depends only on the behavior of the function in the neighborhood of a point, the fractional derivative has a non-local dependence on the variable. For this reason, it is encountered most frequently in the description of processes with memory, e.g. in rheology, heat conduction, control theory and others. The most common fractional derivative is that of order 1/2 which is also called the half-derivative. By taking $\beta < 0$, the same rule (11.2.14) defines

---

[6] It is interesting to note that the same result would be found by proceeding formally directly from (11.2.14). The reason is that the expression for the B function in terms of the Γ function gives an analytic continuation of the former in the region where the integral fails to converge.

the *fractional integral* of f. An extensive treatment of fractional calculus can be found in Samko *et al.* (1993).

A corresponding set of formulae may be derived by applying the inverse transform operator to the previous relations. For example, from (11.2.6),

$$\mathscr{L}^{-1}\left\{\hat{f}(s/b)\right\} = b\,f(bt). \tag{11.2.17}$$

These and other operational rules are summarized in Table 5.3 on p. 126.

Several other rules can be proven. An example, useful for instance when solving the diffusion equation as the balance of $\partial^2/\partial x^2$ and $\partial/\partial t$ then results in $\sqrt{s}$ (see §5.5 p. 138), is

$$\mathscr{L}^{-1}\{\hat{f}(\sqrt{s})\} = \frac{1}{2\sqrt{\pi}\,t^{3/2}} \int_0^\infty \tau \exp\left(-\frac{\tau^2}{4t}\right) f(\tau)\,d\tau. \tag{11.2.18}$$

In words: to invert $\hat{f}(\sqrt{s})$, invert first $\hat{f}(s)$ and then calculate the integral (11.2.18). As an application, consider $\hat{f} = \exp(-\alpha\sqrt{s})/\sqrt{s}$. We invert first $\exp(-as)/s$ using (11.2.8) and (11.2.1) to find $\mathscr{L}^{-1}\{\exp(-as)/s\} = H(t-a)$. Upon substitution into (11.2.18), we then have

$$f(t) = \frac{1}{2\sqrt{\pi}\,t^{3/2}} \int_a^\infty \tau \exp\left(-\frac{\tau^2}{4t}\right) d\tau = \frac{\exp(-a^2/4t)}{\sqrt{\pi t}}. \tag{11.2.19}$$

The result (11.2.18) is proven by writing

$$\hat{f}(\sqrt{s}) = \int_0^\infty e^{-\tau\sqrt{s}} f(\tau)\,d\tau, \tag{11.2.20}$$

so that

$$\mathscr{L}^{-1}\{\hat{f}(\sqrt{s})\}(t) = \int_0^\infty \mathscr{L}^{-1}\left\{e^{-\tau\sqrt{s}}\right\}(t)\,f(\tau)\,d\tau, \tag{11.2.21}$$

in which the notation implies that, in taking the transform of the exponential, the variable conjugate to s should be taken as t, regarding τ as a parameter. Since

$$\mathscr{L}^{-1}\left\{e^{-\tau\sqrt{s}}\right\} = \frac{\tau}{2\sqrt{\pi}\,t^{3/2}} \exp\left(-\frac{\tau^2}{4t}\right), \tag{11.2.22}$$

the expression (11.2.18) follows. Several other relations of this type may be proven similarly, and so can relations for the direct transform, e.g.

$$\mathscr{L}\{f(t^2)\} = \frac{1}{\sqrt{\pi}} \int_0^\infty \exp\left(-s^2/4u^2\right) \hat{f}(u^2)\,du. \tag{11.2.23}$$

A partial list is given in Table 11.1.

The Laplace transform of the product of two functions f and g is not simple:

$$\mathscr{L}\{f(t)\,g(t)\}(s) = \frac{1}{2\pi i} \int_{c-i\infty}^{c+i\infty} \hat{f}(s')\,\hat{g}(s-s')\,ds', \tag{11.2.24}$$

where c is to the right of all the singularities of both $\hat{f}$ and $\hat{g}$.

Table 11.1	Additional operational rules for the Laplace transform; see also Table 5.3 p. 126.	
	$f(t)$	$\hat{f}(s) = \int_0^\infty e^{-st} f(t)\, dt$
1	$t^{-3/2} \int_0^\infty \tau \exp[-(\tau^2/4t)] f(\tau)\, d\tau$	$2\sqrt{\pi}\, \hat{f}(\sqrt{s})$
2	$(\pi t)^{-1/2} \int_0^\infty \exp[-(\tau^2/4t)] f(\tau)\, d\tau$	$\hat{f}(\sqrt{s})/\sqrt{s}$
3	$(\pi t)^{-1/2} \int_0^\infty \cos(2\sqrt{\tau\tau}) f(\tau)\, d\tau$	$s^{-1/2}\hat{f}(1/s)$
4	$(\pi t)^{-1/2} \int_0^\infty \cosh(2\sqrt{\tau\tau}) f(\tau)\, d\tau$	$s^{-1/2}\hat{f}(-1/s)$
5	$(\pi t)^{-1/2} \int_0^\infty \sin(2\sqrt{\tau\tau}) f(\tau)\, d\tau$	$s^{-3/2}\hat{f}(1/s)$
6	$(\pi t)^{-1/2} \int_0^\infty \sinh(2\sqrt{\tau\tau}) f(\tau)\, d\tau$	$s^{-3/2}\hat{f}(-1/s)$
7	$\int_0^\infty J_0(2\sqrt{\tau\tau}) f(\tau)\, d\tau$	$\hat{f}(1/s)/s$
8	$\int_0^t J_0(a\sqrt{t^2-\tau^2}) f(\tau)\, d\tau$	$\tilde{f}(\sqrt{s^2+a^2})/\sqrt{s^2+a^2}$
9	$\int_0^t I_0(a\sqrt{t^2-\tau^2}) f(\tau)\, d\tau$	$\tilde{f}(\sqrt{s^2-a^2})/\sqrt{s^2-a^2}$
10	$\int_0^t J_0(2\sqrt{(t-\tau)\tau}) f(\tau)\, d\tau$	$\tilde{f}(s+s^{-1})/s$
11	$f(t^2)$	$\pi^{-1/2} \int_0^\infty \exp\left(-s^2/4u^2\right) \hat{f}(u^2)\, du$
12	$t^{\nu-1}f(t^{-1})$ (Re $\nu > -1$)	$s^{-\nu/2} \int_0^\infty u^{\nu/2} J_\nu(2\sqrt{us})\, \hat{f}(u)\, du$

11.3 Inversion of the Laplace transform

Much of the usefulness of the Laplace transform arises from the possibility of using the powerful methods of complex contour integration (§17.9.1 p. 461) for the evaluation of the inverse. A key result for this purpose is *Jordan's lemma* which is given in full in §17.9.1 p. 461; for the present purposes it may be stated as follows:

Theorem 11.3.1 [Jordan's lemma] *Let there be a sequence of arcs of circle C_n of radius $R_n \to \infty$ such that, when $|s| = R_n$, with Re s less than some fixed constant c, the function $\hat{f}$ tends to zero uniformly in* arg *s (Figure 11.2). Then, for any positive t,*

$$\lim_{R_n \to \infty} \int_{C_n} e^{st}\, \hat{f}(s)\, ds = 0. \tag{11.3.1}$$

A similar relation holds for arcs to the right of the line $z = c$ when $t < 0$.

On the basis of this result, the integration path in (11.1.7) can be closed by using a sequence of large semi-circles which contribute nothing as their radius becomes infinite. When $t < 0$, the semi-circles must be closed to the right of the line $z = c$. Since $\hat{f}$ is

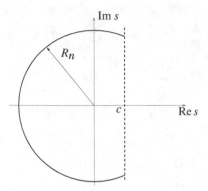

Fig. 11.2 Integration path for Jordan's lemma 11.3.1.

analytic for $\mathrm{Re}\, s > c$, the result is zero. When $t > 0$, on the other hand, the path is closed as in Figure 11.2 and the integral receives contributions from the singularities and branch cuts of $\hat{f}$. Let us consider a few specific cases.

Rational functions. Rational functions are often encountered when solving systems of ordinary differential equations with constant coefficients (§5.2.1). If $\hat{f}(s) = p_m(z)/q_n(s)$, where p_m and q_n are polynomials of degree m and n respectively (with $m < n$ for convergence at infinity),[7] the integration path must be chosen to the right of the zero of q_n with the largest real part. If q_n has only simple zeros $s_1, s_2, \ldots, s_n$, the residue theorem gives immediately

$$f(t) = \sum_{k=1}^{n} \frac{p_m(s_k)}{q_n'(s_k)} \exp(s_k t). \tag{11.3.2}$$

The behavior of f as $t \to \infty$ will be dominated by the zero with largest real part. More generally, from §17.9 p. 460, a pole of order m at s_k will contribute a term

$$\frac{1}{(m-1)!} \lim_{s \to s_k} \frac{d^{m-1}}{ds^{m-1}} \left[(s - s_k)^m \hat{f}(s) e^{st} \right]. \tag{11.3.3}$$

A result with the same formal structure, known as the *Heaviside expansion theorem*, holds when $\hat{f} = p(s)/q(s)$, where $p(s)$ is an entire function (i.e., having only singularities at infinity) and $q(s)$ has a finite or infinite number of simple poles (in the latter case, the sum in (11.3.2) becomes an infinite series):

$$f(t) = \sum_{k} \frac{p(s_k)}{q'(s_k)} \exp(s_k t). \tag{11.3.4}$$

As an example, consider

$$\hat{f}(s) = 1 / \left[s \cosh(\alpha \sqrt{s}) \right], \tag{11.3.5}$$

---

[7] See (5.1.5) p. 125 for the situation $m \geq n$.

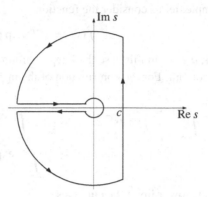

Integration path for a transform with a branch point at the origin.

which arises in certain problems governed by the diffusion equation. Note that, since the hyperbolic cosine is even, this function does not have a branch point in spite of the presence of $\sqrt{s}$. The denominator $q(s)$ has simple poles at $s = 0$, where $q'(0) = 1$, and at $\alpha \sqrt{s_k} = i(k - 1/2)\pi$ where $q'(s_k) = (-1)^k (\pi/2)(k - 1/2)$; since cosh is even, we recover all distinct zeros by considering only positive values of k. Heaviside's formula (11.3.4) then gives

$$f(t) = 1 + \frac{2}{\pi} \sum_{k=1}^{\infty} (-1)^k \frac{\exp\left[-(k-1/2)^2 \pi^2 t/\alpha\right]}{k - 1/2}. \tag{11.3.6}$$

It should be remembered that the result (11.3.4) is only valid when Jordan's lemma applies. For example, e^{-as}/s has a pole at $s = 0$ but, due to the exponential, Jordan's lemma is not applicable and, indeed, (11.3.4) would give $f(t) = 1$, rather than the correct result $H(t - a)$.

Branch point. In the solution of problems involving diffusion or wave motion one often encounters transforms with branch points. Suppose that $\hat{f}(s)$ has a branch point at the origin, and possibly poles. In this case we use the integration contour C shown in Figure 11.3 and note that

$$\oint_C \mathrm{d}s = \left(\int_{c-iR}^{c+iR} + \int_{C_R^+} + \int_{-R+i\varepsilon}^{(-1+i)\varepsilon} + \int_{C_\varepsilon} + \int_{(-1-i)\varepsilon}^{-R+i\varepsilon} + \int_{C_R^-} \right) \mathrm{d}s, \tag{11.3.7}$$

where the limits $R \to \infty$ and $\varepsilon \to 0$ are understood. Here $C_R^{\pm}$ denote the large arcs above and below the real axis, and C_ε is the smalle circle centered at the origin to be considered in the limit $\varepsilon \to 0$. For suitable $\hat{f}$, Jordan's lemma can be invoked to show that the integrals over the large arcs vanish and, therefore,

$$f(t) = \frac{1}{2\pi i} \left(\oint_C - \int_{-\infty+i\varepsilon}^{(-1+i)\varepsilon} - \int_{(-1-i)\varepsilon}^{-\infty+i\varepsilon} - \int_{C_\varepsilon} \right) \hat{f}(s)\, e^{st}\, \mathrm{d}s. \tag{11.3.8}$$

As an example, let us consider the function

$$\hat{f}(s) = s^{-\beta} \exp(-a\sqrt{s}), \tag{11.3.9}$$

with $\beta \le 1$, $a > 0$. In this case the integral around C vanishes as the integrand is analytic inside the contour. For the contribution of the upper lip of the cut, upon setting $s = e^{i\pi}\sigma$, we have

$$\int_{-\infty+i\varepsilon}^{(-1+i)\varepsilon} e^{st-\alpha\sqrt{s}} \frac{ds}{s^{\beta}} = -\int_{\infty}^{\varepsilon} \frac{\exp(e^{i\pi}\sigma t - ae^{i\pi/2}\sqrt{\sigma})}{\sigma^{\beta}\exp(i\beta\pi)} d\sigma$$

$$= \int_{\varepsilon}^{\infty} \frac{\exp(-\sigma t - ia\sqrt{\sigma} - i\beta\pi)}{\sigma^{\beta}} d\sigma. \tag{11.3.10}$$

Similarly, the lower lip of the cut gives

$$\int_{(-1-i)\varepsilon}^{-\infty+i\varepsilon} e^{st-\alpha\sqrt{s}} \frac{ds}{s^{\beta}} = -\int_{\varepsilon}^{\infty} \frac{\exp(-\sigma t + ia\sqrt{\sigma} + i\beta\pi)}{\sigma^{\beta}} d\sigma. \tag{11.3.11}$$

When the two terms are combined, since $\beta \le 1$, it is seen that there is no singularity at $\sigma = 0$ so that the limit $\varepsilon \to 0$ can be taken with the result

$$\left(\int_{-\infty+i\varepsilon}^{(-1+i)\varepsilon} + \int_{(-1-i)\varepsilon}^{-\infty+i\varepsilon}\right) e^{st-\alpha\sqrt{s}} \frac{ds}{s^{\beta}} = -2i\int_{0}^{\infty} \frac{\sin(a\sqrt{\sigma} + \beta\pi)}{\sigma^{\beta}} e^{-\sigma t} d\sigma. \tag{11.3.12}$$

To carry out the integration around the small circle, we set $s = \varepsilon e^{i\theta}$ to find

$$\int_{C_\varepsilon} e^{st-\alpha\sqrt{s}} \frac{ds}{s^{\beta}} = i\varepsilon^{1-\beta}\int_{\pi}^{-\pi} e^{i(1-\beta)\theta} \exp\left(\varepsilon e^{i\theta}t - a\sqrt{\varepsilon}e^{i\theta/2}\right) d\theta. \tag{11.3.13}$$

If $\beta < 1$, the limit of this integral as $\varepsilon \to 0$ vanishes while, if $\beta = 1$, the limit value is $-2\pi i$. In conclusion, if $0 < \beta < 1$, we find

$$f(t) = \frac{1}{\pi}\int_{0}^{\infty} \frac{\sin(a\sqrt{\sigma} + \beta\pi)}{\sigma^{\beta}} e^{-\sigma t} d\sigma. \tag{11.3.14}$$

With $\beta = 1/2$ this gives

$$f(t) = \frac{1}{\pi}\int_{0}^{\infty} \frac{\cos(\alpha\sqrt{\sigma})}{\sqrt{\sigma}} e^{-\sigma t} d\sigma = \frac{\exp(-\alpha^2/4t)}{\sqrt{\pi t}}, \tag{11.3.15}$$

and we recover the result (11.2.19). For $\beta = 1$ we have instead

$$f(t) = -\frac{1}{\pi}\int_{0}^{\infty} \frac{\sin(\alpha\sqrt{\sigma})}{\sigma} e^{-\sigma t} d\sigma + 1 = \operatorname{erfc}\frac{\alpha}{2\sqrt{t}}. \tag{11.3.16}$$

For $\beta > 1$, the integral can be calculated using the inverse of the operational formula (11.2.10). For example, for $\beta = 3/2$, we may write

$$\mathscr{L}^{-1}\left\{\frac{\exp(-a\sqrt{s})}{s^{3/2}}\right\} = \mathscr{L}^{-1}\left\{\frac{1}{s}\left[\frac{\exp(-a\sqrt{s})}{\sqrt{s}}\right]\right\} = \int_{0}^{t} \frac{\exp(-\alpha^2/4t')}{\sqrt{\pi t'}} dt', \tag{11.3.17}$$

which is reduced to an incomplete Gamma function (p. 445) by the substitution $\tau = 1/t$. In this case, while separate limits for $\varepsilon \to 0$ of (11.3.12) and (11.3.13) do not exist, it is readily verified that the limit of the combination of the two terms is well defined.

Theorem 11.3.2 [Series expansion] *If, for* $\text{Re } s \geq \sigma$, $\hat{f}$ *can be expanded in an absolutely convergent series of the form*

$$\hat{f} = \sum_{n=0}^{\infty} a_n s^{-\lambda_n}, \qquad 0 < \lambda_0 < \lambda_1 < \lambda_2 \ldots \to \infty, \qquad (11.3.18)$$

then it is the Laplace transform of the series

$$f(t) = \sum_{n=0}^{\infty} \frac{a_n}{\Gamma(\lambda_n)} t^{\lambda_n - 1} \qquad (11.3.19)$$

obtained by term-wise inversion; this latter series converges absolutely for all $t \neq 0$.

This result rests on Lebesgue's dominated convergence theorem (p. 222) to establish the convergence of the partial sums $e^{-\sigma t} \sum_{n=0}^{N} |a_n| |t^{\lambda_n - 1}| / \Gamma(\lambda_n)$ (see e.g. Doetsch 1974, p. 195). As will be seen in the next section, the converse of this result is only true to the extent that absolute convergence of (11.3.19) guarantees the asymptotic nature of (11.3.18), not necessarily the convergence of the series so obtained.

While the Mellin inversion theorem recovers the original function $f(t)$ in terms of $\hat{f}(s)$ evaluated for complex values of s, inversion formulae also exist which involve $\hat{f}$ for real values of the variable; the reader is referred e.g. to van der Pol & Bremmer 1959, p. 148; Widder 1971, chapter 6; Sneddon 1955, p. 238. Generally speaking, however, these formulae have a rather limited usefulness.

11.4 Behavior for small and large t

The inversion of a Laplace transform in closed form is often not possible and, even when it is, it may give rise to complicated functions. For these reasons, it is useful to be able to establish the asymptotic behavior of the original function $f(t)$ directly from that of its image $\hat{f}(s)$. The appearance of the combination st in the definition of the transform and its inverse suggests that, qualitatively, small t should correspond to large s and vice versa. This is generally true, although some care is needed in practice.

By analogy with Abelian and Tauberian theorems for series (§8.5), theorems which deduce features of $\hat{f}(s)$ from known properties of $f(t)$ are called *Abelian theorems*, while deductions in the opposite direction are the province of Laplace transform *Tauberian theorems*.

11.4.1 Small t

A typical *Abelian theorem* is the following (see e.g. van der Pol & Bremmer 1959, p. 124; Doetsch 1974, p. 231; Davies 2002, p. 33):

Theorem 11.4.1 *Let $\hat{f}$ exist for sufficiently large* $\mathrm{Re}\, s$ *and let* $\lim_{t\to 0} t^{1-\lambda} f(t) = A$ *with* $\mathrm{Re}\,\lambda > 0$. *Then* $\lim_{s\to\infty} \hat{f}(s) = A\Gamma(\lambda)s^{-\lambda}$ *uniformly in* $|\arg s| < \frac{1}{2}\pi$. *In particular,* $f(0+) = \lim_{s\to\infty} s\hat{f}(s)$ *provided that* $f(0+)$ *exists.*[8]

Proof The proof uses the identity

$$s^\lambda \hat{f} - \Gamma(\lambda)A = s^{\lambda+1} \int_0^\infty \left[F(t) - \frac{A}{\lambda} t^\lambda \right] e^{-st}\, dt \qquad (11.4.1)$$

in which $F(t) = \int_0^t f(\tau)\, d\tau$. The integral is broken up into two parts, from 0 to T and from T to ∞. The latter part is found to tend to zero exponentially and uniformly as $\mathrm{Re}\, s \to \infty$. For the integral between 0 and T we have

$$\left| \frac{s^{\lambda+1}}{\lambda} \int_0^T e^{-st} t^\lambda \left[\lambda t^{-\lambda} F(t) - A \right]\, dt \right| \leq \sup_{0\leq t \leq T} \left| \lambda t^{-\lambda} F(t) - A \right| \Gamma(\lambda), \qquad (11.4.2)$$

which tends to 0 in the limit $T \to 0$ by hypothesis.[9] $\square$

As with all Abelian theorems, in general it is not possible to interchange hypothesis and conclusion: the fact that $\lim_{s\to\infty} \hat{f}(s) = \Gamma(\lambda)As^{-\lambda}$ does not necessarily imply that $\lim_{t\to 0} t^{1-\lambda} f(t)$ exists. For example, the Laplace transform of $f(t) = (\pi t)^{-1/2} \cos(1/t)$ is $\hat{f} = s^{-1/2} e^{-\sqrt{s}} \cos\sqrt{s}$ which tends to 0 as $\mathrm{Re}\, s \to \infty$, while $f(t)$ does not tend to any limit for $t \to 0$. We return on this point in §11.4.3.

Repeated application of the previous theorem leads to the following very useful conclusion:

Theorem 11.4.2 [Watson's lemma] *Suppose that, for* $t \to 0+$, $f(t)$ *has the asymptotic expansion*[10]

$$f(t) \sim \sum_{n=0}^\infty a_n t^{\lambda_n - 1} \qquad \text{with} \qquad 0 < \mathrm{Re}\,\lambda_0 < \mathrm{Re}\,\lambda_1 < \cdots \qquad (11.4.3)$$

Then, as $s \to \infty$, $\hat{f}$ *has the asymptotic expansion*

$$\hat{f}(s) \sim \sum_{n=0}^\infty \Gamma(\lambda_n) a_n s^{-\lambda_n} \qquad (11.4.4)$$

uniformly in $0 \leq |\arg s| < \frac{1}{2}\pi$.

---

[8] For $\lambda = 1$ a correspondence with Abel's theorem for series asserting that, if $\sum_{n=0}^\infty a_n = S$ exists and is finite, then $\lim_{x\to 1} \sum_{n=0}^\infty a_n x^n = S$ (§8.5 p. 230), is readily established by setting $x = e^{-ns}$ and taking the limit $s \to 0+$. By using Hardy and Littlewood's Tauberian theorem for series (p. 231) we get a corresponding Tauberian theorem for the Laplace transform.

[9] In the proof we have integrated an asymptotic relation, which is permissible for asymptotic relations of the power type (see e.g. Erdélyi 1956 p. 21; Hinch 1991 p. 23; Holmes 1995 p. 15; Bender & Orszag 2005 p. 126).

[10] A more precise way to indicate an asymptotic expansion would be as a finite sum plus a remainder, with the remainder going to zero faster than the last term in the finite sum. For simplicity, here and in the following, we write the expansion as an infinite sum using the symbol "$\sim$" to signify the asymptotic nature of the series which is not necessarily assumed to converge. Since a convergent series can be considered as a special case of an asymptotic one, all the statements made for asymptotic series are true *a fortiori* for convergent series.

It should be noted that only the asymptotic nature of the series is required: (11.4.3) need not converge. However, if f can be expanded in a convergent Taylor series around 0, we have a representation as in (11.4.3) in which $\lambda_n = n + 1$, $a_n = f^{(n)}(0+)/n!$. If this Taylor series converges uniformly for $0 \le t < \infty$ (i.e., if f is an entire function), the integral defining $\hat{f}$ can be calculated term by term (Theorem 8.3.2 p. 226; see also §A.4.7 p. 694) and we find the convergent series expansion

$$\hat{f}(s) = \sum_{n=0}^{\infty} f^{(n)}(0+)s^{-n-1} \qquad \text{as} \qquad s \to \infty, \tag{11.4.5}$$

which is the Laurent series of $\hat{f}$ around the point at infinity. In this case we recover a converse of Theorem 11.3.2. If (11.4.3) converges but not uniformly, (11.4.5) in general will only be an asymptotic approximation to $\hat{f}$. An example is the function $f = (1+t)^{-1}$, whose Laplace transform is $-e^s \, \mathrm{Ei}\,(-s)$, with Ei the exponential integral (p. 240). Term-by-term integration of the Taylor series for $(1+t)^{-1}$ gives the non-convergent series $-\sum_{k=1}^{\infty}(k-1)!s^{-k}$ which, however, truncated to N terms, is a valid asymptotic approximation to $-e^s \, \mathrm{Ei}\,(-s)$.

The power of the estimate given by Watson's lemma may be reduced by the presence of exponentially small terms in $\hat{f}(s)$. For example, for $f(t) = H(a-t)$ (with $a > 0$), (11.4.3) reduces to the first term $f(t) \simeq 1$. The exact Laplace transform is $\hat{f}(s) = s^{-1}\left(1 - e^{-as}\right)$ and only the leading order behavior $\hat{f} \simeq s^{-1}$ is recovered as, for large s, an expansion of the type (11.4.4) does not exist.

Another point to note is that a result similar to Watson's lemma is true for asymptotic approximations of $f(t)$ in terms of functions other than powers of t provided a bound such as (11.4.2) can be established. In this way, for example, using the fact that $\mathscr{L}\{t^{\lambda-1} \log t\} = \Gamma(\lambda)\,[\psi(\lambda) - \log s]\, s^{-\lambda}$, in which $\psi = \Gamma'/\Gamma$ is the logarithmic derivative of the Gamma function (p. 445), we can establish that if, for $t \to 0$,

$$f(t) \sim -\log t \sum_{n=1}^{\infty} a_n t^{\lambda_n - 1}, \qquad 0 < \operatorname{Re}\lambda_1 < \operatorname{Re}\lambda_2 < \cdots \to \infty, \tag{11.4.6}$$

then, as $s \to \infty$ (see e.g. Doetsch 1974 p. 227),

$$\hat{f}(s) \sim \sum_{n=1}^{\infty} \Gamma(\lambda_n) a_n s^{-\lambda_n} [\log s - \psi(\lambda_n)]. \tag{11.4.7}$$

11.4.2 Large t

For most image functions satisfying Jordan's lemma the inversion can be carried out by closing the integration path with a large semi-circle. In this way, the integral to be calculated is reduced to the contribution of poles, loops around cuts connecting finite branch points, and loops extending from a finite branch point to infinity as sketched in Figure 11.4. The proviso about the applicability of Jordan's lemma should be kept in mind as the example given in §11.3 after (11.3.4) shows.

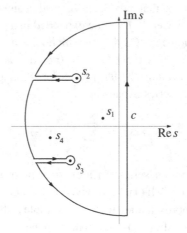

Fig. 11.4   Inversion of a transform satisfying the conditions of Jordan's lemma and having poles and branch points in the image plane.

Let s_1 be the singularity with the largest real part and let it be a simple pole; with the notation of (11.3.2) we have

$$f(t) \simeq \frac{p_m(s_1)}{q'_n(s_1)} \exp(s_1 t), \qquad t \to \infty. \tag{11.4.8}$$

More generally if, near s_1, $\hat{f}$ has the expansion

$$\hat{f} \simeq \frac{a_1}{s - s_1} + \frac{a_2}{(s - s_1)^2} + \cdots + \frac{a_n}{(s - s_1)^n}, \tag{11.4.9}$$

then, as $t \to \infty$,

$$f(t) \simeq \left(a_1 + a_2 t + \cdots + \frac{a_n}{(n-1)!} t^{n-1} \right) e^{s_1 t}. \tag{11.4.10}$$

Several poles of the same order and with the same real part contribute all equally as $t \to \infty$. Other poles to the left of s_1 can be treated similarly but their contribution is exponentially smaller than (11.4.8) or (11.4.10) as $t \to \infty$ (see e.g. Doetsch 1974, p. 238).

An adaptation of the method followed to prove Theorem 11.4.1 gives the result (see e.g. Widder 1971, p. 195):

Theorem 11.4.3 *Let $\hat{f}$ exist for* $\mathrm{Re}\, s > 0$ *and let* $\lim_{t\to\infty} t^{1-\lambda} f(t) = A$ *with* $\mathrm{Re}\, \lambda > 0$. *Then* $\lim_{s\to 0+} s^\lambda \hat{f}(s) = A\,\Gamma(\lambda)$.

In this case, after breaking up the integral in (11.4.1) as before, the first integral goes to zero as $s \to 0$ as it is multiplied by $s^{\lambda+1}$, while the second integral vanishes by the assumed behavior of f for large t. One can base on this result an expansion analogous to Watson's lemma.

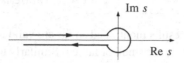

Fig. 11.5 Integration path for (11.4.12).

When $\hat{f}$ has a branch point at $s = 0$ and Jordan's lemma is applicable, the converse result is also true as the inversion integral can be reduced to the integral along a path beginning and ending at $-\infty$ and encircling the origin once in the positive direction as shown in Figure 11.5. In this case, we can prove an analog of Watson's lemma (see e.g. Davies 2002, p. 50):

Theorem 11.4.4 [Watson's lemma for loop integrals] *Let $\hat{f}(s) \to 0$ for $|s| \to \infty$ with* Re *s less than some positive number and let $\hat{f}$ have the asymptotic expansion*

$$\hat{f}(s) \sim \sum_{n=0}^{\infty} a_n s^{\lambda_n}, \qquad \text{Re } \lambda_0 < \text{Re } \lambda_1 < \cdots \to \infty \tag{11.4.11}$$

for $s \to 0$ with $|\arg s| \leq \pi$. Then

$$f(t) = \frac{1}{2\pi i} \int_{-\infty}^{0+} \hat{f}(s) \, e^{st} \, \mathrm{d}s, \tag{11.4.12}$$

where the integration path is as shown in Figure 11.5, has the asymptotic expansion

$$f(t) \sim \sum_{n=0}^{\infty} \frac{a_n}{\Gamma(-\lambda_n)} t^{-\lambda_n - 1}, \qquad t \to \infty, \tag{11.4.13}$$

where all terms with λ_n equal to a non-negative integer are omitted.

A branch point at s_2 can be transferred to the origin simply by defining $s' = s - s_2$, after which one can apply the rule (11.2.9) to recover the result in terms of the original variable. In this way, we can extend the previous theorem to conclude that, if

$$\hat{f}(s) \sim \sum_{n=0}^{\infty} a_n (s - s_2)^{\lambda_n}, \qquad s \to s_2, \qquad \text{Re } \lambda_0 < \text{Re } \lambda_1 < \ldots \to \infty, \tag{11.4.14}$$

then

$$f(t) \sim e^{s_2 t} \sum_{n=0}^{\infty} \frac{a_n}{\Gamma(-\lambda_n)} t^{-\lambda_n - 1}, \qquad t \to \infty. \tag{11.4.15}$$

A consideration of (11.4.8) and (11.4.15) illustrates the basic point that, for $t \to \infty$, the dominant contributions arise from the singularities of $\hat{f}$ with the largest real part.

11.4.3 Tauberian theorems

Many of the results mentioned so far are valid under certain hypotheses on the asymptotic behavior of $f(t)$ for small or large t. Similarly to the case of series (§8.5 p. 230), Tauberian theorems make assertions on the asymptotic behavior of $f(t)$ on the basis of that of $\hat{f}(s)$, although some conditions on $f(t)$ are usually required. An important result of this type is the following one (see e.g. Widder 1946, p. 192; Sneddon 1955, p. 242):

Theorem 11.4.5 [Karamata] *Let* $F(t) = \int_0^t f(\tau)\,d\tau$ *be non-decreasing and* $\hat{f}(s)$ *exist for* $\mathrm{Re}\,s > 0$. *If, for some* $\lambda \geq 0$,
(i) $\lim_{s\to\infty} s^\lambda \hat{f}(s) = A$, *then* $\lim_{t\to 0+} \Gamma(\lambda+1)t^{-\lambda}F(t) = A$;
(ii) *If* $\lim_{s\to 0+} s^\lambda \hat{f} = A$ *then* $\lim_{t\to\infty} \Gamma(\lambda+1)t^{-\lambda}F(t) = A$.

If f is absolutely continuous, $F' = f$. The requirement that F be non-decreasing is then that $f \geq 0$, which is also a sufficient condition to deduce from $F(t) \sim At^\lambda/\Gamma(\lambda+1)$ that $f(t) = F'(t) \sim At^{\lambda-1}/\Gamma(\lambda)$ (see e.g. Doetsch 1943 p. 209).[11] Thus, we conclude that $\hat{f} \to As^{-\lambda}$ as $s \to \infty$ or 0 implies $f(t) \sim At^{\lambda-1}/\Gamma(\lambda)$ as $t \to 0$ or ∞ provided $f(t) \geq 0$. This is then the Tauberian condition sufficient for the validity of the converse of Theorems 11.4.1 and 11.4.3. Another useful result is the following:

Theorem 11.4.6 *Necessary and sufficient condition for* $f(t) \to A$ *for* $t \to \infty$, *given that* $\lim_{s\to 0+} s\hat{f} = A$, *is that* $t^{-1}\int_0^t \tau f'(\tau)\,d\tau \to 0$ *for* $t \to \infty$.

If f is absolutely continuous integration by parts is legitimate (p. 686) and, provided $\lim_{t\to 0} tf(t) = 0$, the condition of the theorem is equivalent to requiring that $f(t) - t^{-1}\int_0^t f(\tau)\,d\tau \to 0$. It is not surprising that, if $f \to$ const., the importance of the non-constant part must ultimately vanish for the result to hold.

A nice example that illustrates the theorem is $f(t) = (\sin t)/t$, for which $\hat{f} = \pi/2 - \tan^{-1} s$. The condition of the theorem is evidently satisfied and, since $\hat{f} \to \pi/2$ for $s \to 0$, we conclude that $\int_0^t (\sin \tau/\tau)\,d\tau \to \pi/2$ for $t \to \infty$, a result which can also be proved by other means. A simple counterexample is $f(t) = \cos t$, for which $\lim_{s\to 0} \hat{f} = \lim_{s\to 0}[s/(s^2+1)] = 0$, while $f(t)$ itself has no limit as $t \to \infty$. In this case, $t^{-1}\int_0^t \tau f(\tau)\,d\tau$ does not tend to 0 for large t.

Watson's lemma for loop integrals 11.4.4 has the following analog, which is the converse of the result stated in (11.4.6) and (11.4.7): If, for $s \to \infty$ with $|\arg s| \leq \pi$, $\hat{f}$ has the asymptotic expansion

$$\hat{f}(s) \sim \log s \sum_{n=1}^\infty a_n s^{-\lambda_n}, \qquad \mathrm{Re}\,\lambda_1 < \mathrm{Re}\,\lambda_2 < \cdots \to \infty, \qquad (11.4.16)$$

---

[11] The differentiation of asymptotic relations is a delicate matter with its own Tauberian theorems; see e.g. Widder 1946, p. 193–194; Hinch 1991, p. 23; Bender & Orszag 2005, p. 127 and Korevaar 2004, p. 34.

then the loop integral (11.4.12) has the asymptotic expansion

$$f(t) \sim \sum_{n=1}^{\infty} \frac{a_n}{\Gamma(-\lambda_n)} \frac{\psi(-\lambda_n) - \log t}{t^{\lambda_n+1}}. \tag{11.4.17}$$

    Tauberian theorems are more complicated than the corresponding Abelian results and we will not pursue this topic further referring the reader to several available references (e.g. Doetsch 1943, 1950, 1974; van der Pol & Bremmer 1959; Widder 1946, 1971; Arendt *et al.* 2001; Korevaar 2004).

12

The Bessel Equation

It would be difficult to overestimate the importance of the Bessel equation and its solutions, the Bessel functions, which arise in a huge variety of problems. Their properties have been studied in minute detail by generations of mathematicians and are recorded in many treatises such as the classic book by Watson (1944). A large collection of formulae and properties can be found especially in Erdélyi *et al.* (1954), and in Abramowitz & Stegun (1964), Gradshteyn *et al.* (2007), with many other books carrying at least a selection of the most important relations. Our purpose here is to present a concise framework to illuminate the origin of these relations and their mutual dependencies rather than giving detailed derivations.

12.1 Introduction

In view of the central role played by the Laplace operator in the problems of concern in this book, we introduce the Bessel functions by considering the form of this operator in cylindrical coordinates:

$$\nabla^2 u = \frac{1}{r}\frac{\partial}{\partial r}\left(r\frac{\partial u}{\partial r}\right) + \frac{\partial^2 u}{\partial z^2} + \frac{1}{r^2}\frac{\partial^2 u}{\partial \phi^2}. \tag{12.1.1}$$

For problems in which the coordinate ϕ ranges over $[0, 2\pi)$, it is natural to expand the function u in a Fourier series in this coordinate:

$$u(r, z, \phi) = \sum_{m=-\infty}^{\infty} U_m(r, z)\frac{e^{im\phi}}{\sqrt{2\pi}}, \tag{12.1.2}$$

so that[1]

$$\nabla^2 u = \sum_{-\infty}^{\infty}\left[\frac{1}{r}\frac{\partial}{\partial r}\left(r\frac{\partial U_m}{\partial r}\right) - \frac{m^2}{r^2}U_m + \frac{\partial^2 U_m}{\partial z^2}\right]\frac{e^{im\phi}}{\sqrt{2\pi}}. \tag{12.1.3}$$

In some problems, in order to deal with the dependence on the coordinate r, one may try an expansion in terms of eigenfunctions $R_{mn}(r)$ of the radial operator:

$$\frac{1}{r}\frac{d}{dr}\left(r\frac{dR_{mn}}{dr}\right) - \frac{m^2}{r^2}R_{mn} = -k^2 R_{mn}. \tag{12.1.4}$$

---

[1] See the considerations in §3.1 and §9.8 for the legitimacy of term-by-term differentiation.

With this approach one would write

$$U_m = \sum_n R_{mn}(r)\, Z_{mn}(z), \tag{12.1.5}$$

so that (see previous footnote)

$$\nabla^2 u = \sum_{-\infty}^{\infty} \frac{e^{im\phi}}{\sqrt{2\pi}} \sum_n R_{mn}(r) \left(\frac{d^2 Z_{mn}}{dz^2} - k^2 Z_{mn} \right). \tag{12.1.6}$$

The change of variable $x = kr$, brings (12.1.4) to the standard form of the *Bessel equation*, namely

$$\frac{1}{x}\frac{d}{dx}\left(x \frac{dR_{mn}}{dx} \right) + \left(1 - \frac{m^2}{x^2} \right) R_{mn} = 0. \tag{12.1.7}$$

In other problems the z-dependence of the solution is conveniently expressed in terms of a Fourier series or integral, and therefore we would set

$$U_m(r, z) = \int S_m(r, k)\, e^{-ikz}\, dk \quad \text{or} \quad U_m(r, z) = \sum_n S_{mn}(r)\, e^{in\pi z/L}. \tag{12.1.8}$$

Substitution into (12.1.3) would then lead to

$$\nabla^2 u = \sum_{-\infty}^{\infty} \frac{e^{im\phi}}{\sqrt{2\pi}} \int dk\, e^{-ikz} \left[\frac{1}{r}\frac{d}{dr}\left(r \frac{dS_m}{dr} \right) - \left(k^2 + \frac{m^2}{r^2} \right) S_m \right] \tag{12.1.9}$$

or a similar expression with the integral over k replaced by a summation over $k_n = n\pi/L$. If we were solving for example the Laplace equation $\nabla^2 u = 0$, in view of the linear independence of the exponentials in ϕ and z, we would need to require that

$$\frac{1}{r}\frac{d}{dr}\left(r \frac{dS_{mn}}{dr} \right) - \left(k^2 + \frac{m^2}{r^2} \right) S_{mn} = 0, \tag{12.1.10}$$

which, with the substitution $k \to ik$, becomes identical to (12.1.4). For the Poisson equation, we would find a non-homogeneous form of the same equation.

Another way in which the Bessel equation arises is in the process of solving the Helmholtz equation $\nabla^2 u + k^2 u = 0$ in spherical coordinates. We would start by expanding u in spherical harmonics as e.g. in (7.3.2) p. 178 and we would then be led to

$$\sum_{\ell=0}^{\infty} \sum_{m=-\ell}^{\ell} \left[\frac{1}{r^2}\frac{d}{dr}\left(r^2 \frac{dR_{\ell m}}{dr} \right) + \left(k^2 - \frac{\ell(\ell+1)}{r^2} \right) R_{\ell m} \right] Y_\ell^m = 0. \tag{12.1.11}$$

In view of the linear independence of the spherical harmonics, satisfaction of this equation requires that

$$\frac{1}{r^2}\frac{d}{dr}\left(r^2 \frac{dR_{\ell m}}{dr} \right) + \left(k^2 - \frac{\ell(\ell+1)}{r^2} \right) R_{\ell m} = 0. \tag{12.1.12}$$

The change of variables $x = kr$ and $R_{\ell m} = S_{\ell m}/\sqrt{x}$ gives then

$$\frac{1}{x}\frac{d}{dx}\left(x \frac{dS_{\ell m}}{dx} \right) + \left[1 - \frac{(\ell+1/2)^2}{x^2} \right] S_{\ell m} = 0, \tag{12.1.13}$$

which is again of the Bessel form (12.1.7).

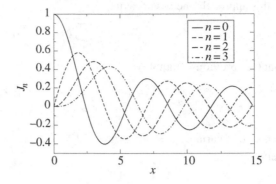

Fig. 12.1 Bessel functions of the first kind $J_n(x)$ for $n = 0, 1, 2,$ and 3

We have described just a few of the very many ways in which the Bessel equation is encountered in applied mathematics. There are many more, but these examples suffice for our purposes.

12.2 The Bessel functions

The previous examples point to the importance of an equation the general form of which we write as

$$\frac{1}{z}\frac{d}{dz}\left(z\frac{du}{dz}\right) + \left(1 - \frac{v^2}{z^2}\right)u = 0 \qquad (12.2.1)$$

or

$$\frac{d^2u}{dz^2} + \frac{1}{z}\frac{du}{dz} + \left(1 - \frac{v^2}{z^2}\right)u = 0. \qquad (12.2.2)$$

This is the general form of the *Bessel equation*. It is advantageous to consider the general case of a complex independent variable, which we have denoted by z. The order of the equation is indicated by v which, in general, can be real or complex.

Since the equation is linear, the possible singularities of its solutions coincide with those of the coefficients. Thus, we recognize immediately that $z = 0$ is a singular point of the equation and so is infinity after the substitution $\zeta = 1/z$ (see §2.5 p. 42). These are the only points where solutions can be singular. Following the method of §2.5, we set

$$u = z^\alpha \sum_{n=0}^{\infty} a_n z^n \qquad (12.2.3)$$

and substitute into (12.2.1). As shown on p. 45, with the standard normalization, one of the solutions generated in this way is

$$J_\nu(z) = \left(\frac{z}{2}\right)^\nu \sum_{k=0}^\infty \frac{(-1)^k}{k!\Gamma(k+\nu+1)} \left(\frac{z}{2}\right)^{2k}, \qquad (12.2.4)$$

with the second one obtained by writing $-\nu$ in place of ν. Here Γ is the Gamma function defined in §17.5.1 p. 442. This expression defines the *Bessel function of the first kind*. The ratio test (p. 218) readily shows that the radius of convergence of the series is infinite. The function $z^{-\nu} J_\nu(z)$ is therefore an entire function and, since it is not a constant, it must have a singularity at infinity (Liouville's Theorem, p. 431). Graphs of the first few Bessel functions for integer ν and real argument are shown in Figure 12.1.

As noted in §2.5, when $\nu = N$ is a non-negative integer, the formula (12.2.4) gives only one solution and, in fact, it can be shown that $J_{-N}(Z) = (-1)^N J_N(Z)$. The general structure of a second solution in this case is shown in (2.5.29) p. 46. In practice it can be more readily determined with the observation that, for non-integer ν, a second solution of the Bessel equation may be taken in the form[2]

$$Y_\nu = \frac{1}{\sin \pi \nu} [\cos \pi \nu \, J_\nu(z) - J_{-\nu}(z)]. \qquad (12.2.5)$$

The limit of this function for ν approaching an integer N exists and gives a second solution of the Bessel equation linearly independent of J_N. The relation (12.2.5) defines the *Bessel function of the second kind*. For $N = 0$ one finds

$$Y_0(z) = \frac{2}{\pi} \sum_{k=0}^\infty \frac{(-1)^k}{(k!)^2} \left[\log \frac{z}{2} - \psi(k+1)\right] \left(\frac{z}{2}\right)^{2k}, \qquad (12.2.6)$$

which may also be written as

$$Y_0(z) = \frac{2}{\pi} \left[\log \frac{z}{2} + \gamma\right] J_0(z) - \frac{2}{\pi} \sum_{k=0}^\infty \frac{(-1)^k}{(k!)^2} [\psi(k+1) + \gamma] \left(\frac{z}{2}\right)^{2k}. \qquad (12.2.7)$$

Here ψ is the logarithmic derivative of the Gamma function (p. 445) and $\gamma = -\psi(1) \simeq 0.57721$ is Euler's constant.[3] For $N \geq 1$ we have

$$\pi Y_N(z) = \left(\frac{z}{2}\right)^N \sum_{k=0}^\infty \frac{(-1)^k}{k!(N+k)!} \left[2\log \frac{z}{2} - \psi(k+1) - \psi(N+k+1)\right] \left(\frac{z}{2}\right)^{2k}$$
$$- \left(\frac{2}{z}\right)^N \sum_{k=0}^{N-1} \frac{(N-k-1)!}{k!} \left(\frac{z}{2}\right)^{2k}, \qquad (12.2.8)$$

---

[2] Sometimes this function is indicated as $N_\nu(z)$.

[3] It follows from (17.5.36) p. 445 that $\psi(m+1) + \gamma = \psi(m+1) - \psi(1) = 1 + 1/2 + \cdots + 1/m$.

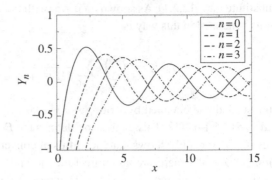

Fig. 12.2 Bessel functions of the second kind $Y_n(x)$ for $n = 0, 1, 2$, and 3

or, equivalently,

$$
\begin{aligned}
\pi Y_N(z) &= 2\left(\log\frac{z}{2}+\gamma\right)J_N(z) - \left(\frac{2}{z}\right)^N\sum_{k=0}^{N-1}\frac{(N-k-1)!}{k!}\left(\frac{z}{2}\right)^{2k} \\
&\quad - \left(\frac{z}{2}\right)^N\sum_{k=0}^{\infty}(-1)^k\frac{\psi(N+k+1)+\psi(k+1)-2\psi(1)}{k!(k+N)!}\left(\frac{z}{2}\right)^{2k}
\end{aligned}
$$

(12.2.9)

These functions exhibit a logarithmic singularity at $z = 0$ and poles at $z = 0$ if $N > 0$; graphs of the first few are shown in Figure 12.2.

It will be seen later that the Bessel functions of the first and second kind behave in some sense similarly to the ordinary sine and cosine. Just as the combinations $\cos z \pm i \sin z = e^{\pm iz}$ often arise, the combinations

$$
H_\nu^{(1)}(z) = J_\nu(z) + iY_\nu(z) = \frac{i}{\sin\pi\nu}\left[e^{-i\pi\nu}J_\nu(z) - J_{-\nu}(z)\right]
$$

(12.2.10)

and

$$
H_\nu^{(2)}(z) = J_\nu(z) - iY_\nu(z) = -\frac{i}{\sin\pi\nu}\left[e^{i\pi\nu}J_\nu(z) - J_{-\nu}(z)\right]
$$

(12.2.11)

of the Bessel functions prove useful. These are known as *Hankel functions* or *Bessel functions of the third kind.* Since they involve the function $J_{-\nu}$, both are singular at the origin. It follows from these definitions that

$$
\begin{aligned}
H_\nu^{(1)}(e^{i\pi}z) &= -H_{-\nu}^{(2)}(z) = -e^{-i\pi\nu}H_\nu^{(2)}(z), \\
H_\nu^{(2)}(e^{-i\pi}z) &= -H_{-\nu}^{(1)}(z) = -e^{i\pi\nu}H_\nu^{(1)}(z),
\end{aligned}
$$

(12.2.12)

and, furthermore, $\overline{H^{(2)}}_\nu(z) = H_{\bar\nu}^{(1)}(\bar z)$. An example of the use of these functions in the solution of the wave equation is given in §7.13.

Unless ν is an integer, the origin is a branch point for all the Bessel functions. The definition is made unique by cutting the complex plane along the negative real axis between

$-\infty$ and 0. The transition to different branches across the cut is made using the relations

$$J_\nu(e^{im\pi} z) = e^{im\pi\nu} J_\nu(z), \qquad m = \pm 1, \pm 2, \dots \tag{12.2.13}$$

$$Y_\nu(e^{im\pi} z) = e^{-im\pi\nu} Y_\nu(z) + 2i \sin(m\nu\pi) \cot(\nu\pi) J_\nu(z), \tag{12.2.14}$$

and the corresponding ones for $H_\nu^{(1)}$ and $H_\nu^{(2)}$ that follow from the definitions (12.2.10) and (12.2.11).

12.2.1 Wronskians

Adapted to the Bessel equation, the general relation (2.2.17) p. 32 for the Wronskian of the solutions of a second-order equation gives

$$W(u_1, u_2) = W_0 \exp\left(-\int (1/z)\, dz\right) = W_0/z. \tag{12.2.15}$$

The constant W_0 depends on the particular two solutions u_1 and u_2 of the Bessel equation the Wronskian of which is desired. This constant can be determined by evaluating $W(u_1, u_2) = u_1 u_2' - u_1' u_2$ at any convenient z. Since we have the power series (12.2.4) valid near the origin, $z = 0$ is particularly suitable for this purpose; we have

$$u_1 = J_{-\nu} \simeq \frac{(z/2)^{-\nu}}{\Gamma(-\nu+1)}, \qquad u_2 = J_\nu \simeq \frac{(z/2)^\nu}{\Gamma(\nu+1)}, \tag{12.2.16}$$

from which we find

$$W(J_{-\nu}, J_\nu) = \frac{2\sin\pi\nu}{\pi z}, \tag{12.2.17}$$

which confirms the linear independence of the two functions as long as ν is not an integer. Proceeding similarly we also find

$$W(J_\nu, Y_\nu) = \frac{2}{\pi z}, \qquad W\left(H_\nu^{(1)}, H_\nu^{(2)}\right) = \frac{4}{i\pi z}. \tag{12.2.18}$$

As demonstrated in many examples of Chapter 6, these relations are of great help in the solution of non-homogeneous problems by the method of variation of parameters.

12.2.2 Recurrence relations

The following relations can be proven directly from the power series definition (12.2.4) after some reordering of terms, which is legitimate in view of the absolute convergence of the series:

$$J_{\nu-1} + J_{\nu+1} = \frac{2\nu}{z} J_\nu, \qquad J_{\nu-1} - J_{\nu+1} = 2J_\nu'. \tag{12.2.19}$$

As a consequence,

$$\frac{\nu}{z} J_\nu + J_\nu' = J_{\nu-1}, \qquad \frac{\nu}{z} J_\nu - J_\nu' = J_{\nu+1}, \tag{12.2.20}$$

and, in particular, $J_0' = -J_1$. Proceeding by induction, these relations can be generalized to

$$\left(\frac{1}{z}\frac{d}{dz}\right)^m \left(z^\nu J_\nu\right) = z^{\nu-m} J_{\nu-m}, \qquad \left(\frac{1}{z}\frac{d}{dz}\right)^m \left(z^{-\nu} J_\nu\right) = (-1)^m z^{-\nu-m} J_{\nu+m}, \quad (12.2.21)$$

valid for a general integer $m \geq 0$. Since the Bessel functions of the second and third kind are linear combinations of those of the first kind, these relations are also valid for Y_ν and $H_\nu^{(1,2)}$.

12.2.3 Integral representations

By taking $z = 0$, $w(z) = e^z$ in Equation (17.3.29) p. 430 giving the derivative of an analytic function, we have

$$\frac{1}{n!} = \frac{1}{2\pi i} \oint_C e^\zeta \zeta^{-n-1} \, d\zeta, \qquad (12.2.22)$$

where C is any finite closed path encircling the origin. Substitution into the power series representation (12.2.4) of J_n leads to the exponential series which can be summed with the result

$$J_n(z) = \frac{1}{2\pi i} \left(\frac{z}{2}\right)^n \oint_C \zeta^{-n-1} e^{\zeta - z^2/4\zeta} \, d\zeta. \qquad (12.2.23)$$

In particular, take C to be a circle of radius $(1/2)|z|$ centered at the origin, and let $\zeta = (z/2)e^{i\theta}$ on this circle. Then

$$J_n(z) = \frac{1}{2\pi} \int_\alpha^{\alpha+2\pi} e^{iz\sin\theta - in\theta} \, d\theta = \frac{i^{-n}}{2\pi} \int_\beta^{\beta+2\pi} e^{iz\cos\eta + in\eta} \, d\eta, \qquad (12.2.24)$$

with α, β arbitrary; the second form is obtained by setting $\theta = \pi/2 - \eta$. In particular, upon taking $\alpha = -\pi$ and writing the integral as the sum of the contributions from $[-\pi, 0]$ and $[0, \pi]$, after a change of variable in the first contribution, we find

$$J_n(z) = \frac{1}{\pi} \int_0^\pi \cos\left(z\sin\theta - n\theta\right) \, d\theta. \qquad (12.2.25)$$

An analogous representation can be found for general ν (with $\mathrm{Re}\,\nu > -1/2$) by using Hankel's contour integral for $1/\Gamma(\nu)$ (see p. 444) in place of (12.2.22). Many other integral representations of the various Bessel functions exist.

12.2.4 Asymptotic behavior

Since infinity is an irregular singular point for the Bessel equation, to study the behavior of the solutions in its vicinity it is appropriate to set $u(z) = e^{v(z)}$ (see e.g. Bender & Orszag 2005, p. 484) to find

$$v'' + v'^2 + \frac{1}{z}v' + 1 - \frac{\nu^2}{z^2} = 0. \qquad (12.2.26)$$

If we set $v \simeq a + b/z + \cdots$ and assume that $|z| \gg |v|$ we find

$$v' \simeq \pm i - \frac{1}{2z} + O(z^{-2}) \tag{12.2.27}$$

from which, upon integration,

$$u \simeq A_+ \frac{e^{iz}}{\sqrt{z}} + A_- \frac{e^{-iz}}{\sqrt{z}} \tag{12.2.28}$$

which gives the general form of the asymptotic behavior of *any* solution of the Bessel equation.

To determine the integration constants $A_\pm$ for J_n we apply the method of stationary phase (p. 277) to the integral representation (12.2.24) with $\alpha = -\pi$; we set $\varphi(\theta) = \sin\theta - n\theta/x$ as taking the argument real cannot affect the values of the constants. The function φ is stationary at points $\cos\theta_\pm = n/x$, or $\theta_\pm = \pm\cos^{-1} n/x$. For x large, $\theta_\pm$ will be close to $\pm\pi/2$ and, indeed, a simple calculation gives $\theta_\pm = \pm(\pi/2 - n/x)$ to leading order. Thus $\varphi(\theta_\pm) \simeq \pm(1 - \pi n/2x)$, $\varphi''(\theta_\pm) \simeq \mp 1$. Application of formula (10.7.9) to each one of $\theta_\pm$ and addition of the results then gives, in terms of the original variable z,

$$J_n(z) \simeq \frac{e^{-i\pi(n+1/2)/2} e^{iz} + e^{i\pi(n+1/2)/2} e^{-iz}}{\sqrt{2\pi z}} = \sqrt{\frac{2}{\pi z}} \cos\left[z - \frac{\pi}{2}\left(n + \frac{1}{2}\right)\right], \tag{12.2.29}$$

so that $A_\pm = e^{\mp i\pi(n+1/2)/2}/\sqrt{2\pi}$; the same relation with n replaced by v is valid.

From (12.2.10) and (12.2.11) we have $J_v(x) = (1/2)[H_v^{(1)}(x) + H_v^{(2)}(x)]$ which, by the first one of (12.2.12), becomes $J_v(x) = (1/2)[H_v^{(1)}(x) - e^{i\pi v} H_v^{(1)}(-x)]$. Upon substituting the general asymptotic form (12.2.28) for $H_v^{(1)}(x)$ and equating to (12.2.29) one readily determines the appropriate integration constants with the result

$$H_v^{(1,2)}(z) \simeq \sqrt{\frac{2}{\pi z}} \exp\left[\pm iz \mp \frac{\pi}{2} i\left(v + \frac{1}{2}\right)\right], \tag{12.2.30}$$

from which

$$Y_v(z) \simeq \sqrt{\frac{2}{\pi z}} \sin\left[z - \frac{\pi}{2}\left(n + \frac{1}{2}\right)\right]. \tag{12.2.31}$$

The next term in these asymptotic formulae is shown on p. 150.

In both methods used to derive these results we have explicitly assumed that $|z| \gg v$. If this condition is not satisfied the above formulae are invalid and one needs to use different approximations which can be found e.g. in Watson 1944, chapter 8, Gradshteyn *et al.* (2007), section 8.45 and others.

12.3 Spherical Bessel functions

We have seen in §7.12 and §12.1 that Bessel functions of half-integer order arise in certain problems for which spherical polar coordinates are appropriate. It just so happens that this is also the only case in which Bessel functions can be reduced to elementary functions, specifically powers and trigonometric functions. A direct manipulation of the series (12.2.4) shows that

$$J_{-1/2}(z) = \left(\frac{2}{\pi z}\right)^{1/2} \cos z, \qquad J_{1/2}(z) = \left(\frac{2}{\pi z}\right)^{1/2} \sin z, \tag{12.3.1}$$

$$H_{1/2}^{(1)}(z) = \sqrt{\frac{2}{\pi z}} \frac{e^{iz}}{i}, \qquad H_{1/2}^{(2)}(z) = \sqrt{\frac{2}{\pi z}} \frac{e^{-iz}}{-i}, \tag{12.3.2}$$

and, more generally,

$$J_{n+1/2}(z) = \sqrt{\frac{2}{\pi}} z^{n+1/2} \left(-\frac{1}{z}\frac{d}{dz}\right)^n \frac{\sin z}{z}, \tag{12.3.3}$$

which is just a special case of (12.2.21).

The equation (12.1.12) is encountered sufficiently frequently that it is convenient to avoid carrying out the change of variables leading to (12.1.13) and to deal with solutions of (12.1.12) directly. This objective is achieved by defining the *spherical Bessel functions*:

$$j_\ell(z) = \left(\frac{\pi}{2z}\right)^{1/2} J_{\ell+1/2}(z), \qquad y_\ell(z) = \left(\frac{\pi}{2z}\right)^{1/2} Y_{\ell+1/2}(z), \tag{12.3.4}$$

$$h_\ell^{(1)}(z) = \left(\frac{\pi}{2z}\right)^{1/2} H_{\ell+1/2}^{(1)}(z), \qquad h_\ell^{(2)}(z) = \left(\frac{\pi}{2z}\right)^{1/2} H_{\ell+1/2}^{(2)}(z). \tag{12.3.5}$$

With these definitions we have, for example,

$$j_0(z) = \frac{\sin z}{z}, \qquad y_0(z) = -\frac{\cos z}{z}. \tag{12.3.6}$$

The Wronskians of these functions are

$$W(j_n, y_n) = \frac{1}{z^2}, \qquad W(h_n^{(1)}, h_n^{(2)}) = \frac{2}{iz^2}, \tag{12.3.7}$$

and the recurrence relations analogous to those of §12.2.2

$$j_{n-1} + j_{n+1} = \frac{2n+1}{z} j_n, \qquad n j_{n-1} - (n+1) j_{n+1} = (2n+1) j_n', \tag{12.3.8}$$

$$\frac{n+1}{z} j_n + j_n' = j_{n-1}, \qquad \frac{n}{z} j_n - j_n' = j_{n+1}, \tag{12.3.9}$$

valid for y_n, $h_n^{(1)}$ and $h_n^{(2)}$ as well.

12.4 Modified Bessel functions

We have seen in §12.1 an equation similar to Bessel's, except for a sign in the last term, namely (12.1.10). The substitution $z = kr$ gives

$$\frac{d^2u}{dz^2} + \frac{1}{z}\frac{du}{dz} - \left(1 + \frac{\nu^2}{z^2}\right) u = 0. \tag{12.4.1}$$

The further substitution $z' = iz$ brings the equation to the standard Bessel form (12.2.2), which implies that the general solution of (12.4.1) may be written e.g. as $u = A J_\nu(iz) + B Y_\nu(iz)$. We see from the series expansion (12.2.4) that, aside from the phase factor i^ν, J_ν is real if its argument is purely imaginary. Since this situation is common, for $-\pi < \arg z < \frac{1}{2}\pi$ it is convenient to define a *modified Bessel function* of the first kind by

$$I_\nu(z) = e^{-i\pi\nu/2} J_\nu(e^{i\pi/2} z) = \left(\frac{z}{2}\right)^\nu \sum_{k=0}^\infty \frac{1}{k!\Gamma(k+\nu+1)} \left(\frac{z}{2}\right)^{2k}. \tag{12.4.2}$$

The functions of the second kind do not have a definite character when their variable is purely imaginary, but those of the third kind do and we define the modified Bessel function of the third kind by[4]

$$K_\nu(z) = \frac{i\pi}{2} e^{i\nu\pi/2} H_\nu^{(1)}(iz), \qquad -\pi < \arg z < \frac{\pi}{2}. \tag{12.4.3}$$

By expressing $J_\nu(iz)$ in terms of the Hankel functions (cf. the procedure followed in §12.2.4), it can be shown that

$$K_\nu(z) = \frac{\pi}{2} \frac{I_{-\nu}(z) - I_\nu(z)}{\sin \pi\nu}, \tag{12.4.4}$$

from which, in particular, we see that K_ν is real when its argument is real and that $K_{-\nu} = K_\nu$; furthermore

$$\lim_{\nu \to m} K_\nu(z) = K_m(z). \tag{12.4.5}$$

Graphs of the functions $e^{-x} I_n$ and $e^x K_n$ are shown in Figure 12.3 for $n = 0$ and 1. As before, these functions reduce to simpler forms when $\nu = \pm n + 1/2$; examples are given in (6.2.13) and (6.2.14) p. 150.

By expressing Y_ν in terms of J_ν and $H_\nu^{(1)}$ one finds

$$Y_\nu(iz) = i e^{i\pi\nu/2} I_\nu(z) - \frac{2}{\pi} e^{-i\pi\nu/2} K_\nu(z), \qquad -\pi < \arg z < \frac{\pi}{2}. \tag{12.4.6}$$

Since I_ν and K_ν diverge at infinity and at 0, respectively (p. 151), this relation explains the limited usefulness of the functions $Y_\nu(iz)$.

The recurrence relations satisfied by these functions are readily obtained from those of §12.2.2 by the change of variable $z' = iz$; in this way one finds, for example,

$$I_{\nu-1}(z) - I_{\nu+1}(z) = \frac{2\nu}{z} I_\nu(z), \qquad I_{\nu-1}(z) + I_{\nu+1}(z) = 2 I_\nu'(z) \tag{12.4.7}$$

$$K_{\nu-1}(z) - K_{\nu+1}(z) = -\frac{2\nu}{z} K_\nu(z), \qquad K_{\nu-1}(z) + K_{\nu+1}(z) = -2 K_\nu'(z) \tag{12.4.8}$$

---

[4] The extra factor $\pi/2$ spoils the parallelism with the relation between I_ν and J_ν and also introduces other differences, e.g. in the prefactors of the asymptotic expressions (second part of Table 6.5 p. 151), in the expressions of the functions for half-integer indices (see (6.2.20) p. 151) and others. The origin of this anomaly is due to the original definition of the function K_ν given by Macdonald in 1899. He included this factor and the numerical tables published in the following years retained it. By the time it was realized that it would have made more sense to avoid the extra factor, the effort of recalculating and publishing new tables seemed unwarranted and the factor remained (see Watson 1944, p. 79).

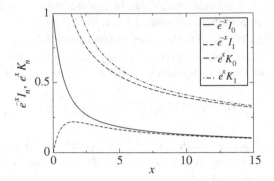

Fig. 12.3 Modified Bessel functions $e^{-x} I_n$ and $e^x K_n$ with $n = 0$ and 1

etc.; see table 6.6 p. 152 for others. A point to note is that the recurrence relations satisfied by I_ν and K_ν are slightly different from those of the ordinary Bessel functions as these examples show.

Several integral representations of the modified Bessel functions are available. Examples are

$$I_\nu(z) = \frac{(z/2)^\nu}{\Gamma(\nu + 1/2)\,\Gamma(1/2)} \int_0^\pi \cosh(z \cos\theta)\, \sin^{2\nu}\theta\, d\theta\,, \qquad \mathrm{Re}\,\nu > -\frac{1}{2}, \quad (12.4.9)$$

$$K_\nu(z) = \int_0^\infty e^{-z\cosh t}\, \cosh \nu t\, dt\,, \qquad |\arg z| < \frac{\pi}{2}, \qquad (12.4.10)$$

$$K_\nu(xz) = \frac{\Gamma(\nu + 1/2)}{\Gamma(1/2)} \left(\frac{2z}{x}\right)^\nu \int_0^\infty \frac{\cos kx}{(k^2 + z^2)^{\nu+1/2}}\, dk,\ x > 0,\ |\arg z| < \frac{\pi}{2}, \mathrm{Re}\,\nu > -\frac{1}{2}.$$
$$(12.4.11)$$

A derivation of this last relation for the special case $\nu = 0$ is found on p. 159 as a byproduct of the calculations for the example in §6.4.

12.5 The Fourier–Bessel and Dini series

By exploiting the infinity of zeros of the sine function $\sin \pi\xi$, we construct the orthogonal family $\sin \lambda_n \pi x$, with $\lambda_n = n$, on which the ordinary Fourier sine series is based. The oscillating nature of the Bessel functions evident e.g. in Figure 12.1 suggests that we might similarly construct an orthogonal family by using their zeros. The first step in this process is to investigate the nature of these zeros; we will only summarize the results necessary for our purposes. As for any other aspect of Bessel function theory, a great deal is known on this topic; see in particular Watson 1944, chapter 15; Hochstadt 1971, p. 249, among others, gives an abbreviated treatment.

12.5.1 The zeros of Bessel functions

The asymptotic expansion (12.2.29) and the explicit expression (12.3.3) for $v = n + 1/2$ both suggest that $J_v(z)$ has an infinity of real zeros. This is indeed true for any real value of $v > -1$: there is an infinite number of zeros of $J_v(z)$, all of which are real and simple with the possible exception of $z = 0$. For large k, $j_{v,k}$, the k-th zero of $z^{-v}J_v$ (i.e., the k-th zero of J_v not counting $z = 0$) is given approximately by

$$ j_{v,k} \simeq k\pi + \left(v - \frac{1}{2} \right) \frac{\pi}{2}. \qquad (12.5.1) $$

Similar properties are enjoyed by the zeros of $J_v'(z)$: they are all real and simple, except possibly $z = 0$. Furthermore, the zeros of J_{v+1} and J_v, and of J_v' and J_v interlace. Since $J_0' = -J_1$, the zeros of J_0' coincide with those of J_1.

These results can be extended to the linear combinations $A J_v(z) + B Y_v(z)$ and $C J_v(z) + D z J_v'(z)$ for $v > -1$ and real A, B, C, D, both of which have an infinity of simple real zeros.[5] The large-order zeros $\zeta_{v,k}$ of the function $J_v(z) \cos \alpha - Y_v(z) \sin \alpha$, with $0 \leq \alpha < \pi$, are given by (see e.g. Watson 1944, p. 490)

$$ \zeta_{v,k} \simeq k\pi + \left(v - \frac{1}{2} \right) \frac{\pi}{2} - \alpha - \frac{1}{8} \frac{4v^2 - 1}{k\pi + \left(v - \frac{1}{2} \right) \frac{\pi}{2} - \alpha} - \cdots \qquad (12.5.2) $$

which extends (12.5.1) for $d = D$.

12.5.2 Eigenfunction expansion

With an understanding of the main properties of the zeros of Bessel functions, we can now proceed to build a family of orthogonal functions which can be used to expand a general function of r. The developments described here are a particular case of the Sturm–Liouville theory of §15.2 p. 359 and the method of investigation is the same.

The function $Z_v(kr) = J_v(kr) \cos \alpha - Y_v(kr) \sin \alpha$, with $0 \leq \alpha < \pi$, is a general solution of the equation

$$ -\frac{\mathrm{d}}{\mathrm{d}r} \left(r \frac{\mathrm{d}Z_v}{\mathrm{d}r} \right) + \frac{v^2}{r} Z_v = k^2 r Z_v, \qquad 0 \leq R < r < S. \qquad (12.5.3) $$

Write the equations satisfied by $Z_v^a = Z_v(k_a r)$ and $Z_v^b = Z_v(k_b r)$ with $k_a \neq k_b$, multiply the first one by Z_v^b, the second one by Z_v^a and subtract; the result may be written as

$$ \frac{\mathrm{d}}{\mathrm{d}r} \left[r \left(Z_v^a \frac{\mathrm{d}Z_v^b}{\mathrm{d}r} - Z_v^b \frac{\mathrm{d}Z_v^a}{\mathrm{d}r} \right) \right] = \left(k_b^2 - k_a^2 \right) r Z_v^a Z_v^b. \qquad (12.5.4) $$

Integrate between R and S to find

$$ \left[r \left(Z_v^a \frac{\mathrm{d}Z_v^b}{\mathrm{d}r} - Z_v^b \frac{\mathrm{d}Z_v^a}{\mathrm{d}r} \right) \right]_R^S = \left(k_b^2 - k_a^2 \right) \int_R^S r Z_v(k_a r) Z_v(k_b r) \, \mathrm{d}r. \qquad (12.5.5) $$

[5] The function $C J_v(z) + D z J_v'(z)$ has a pair of purely imaginary zeros when $C/D + v < 0$.

Suppose that R and S are such that both Z_ν^a and Z_ν^b vanish for $r = R$ and $r = S$. This relation then shows that the product in the right-hand side must vanish so that $Z_\nu(k_a r)$ and $Z_\nu(k_b r)$ are orthogonal with respect to the scalar product[6]

$$(f, g) = \int_R^S r \, \bar{f}(r) g(r) \, dr. \tag{12.5.6}$$

The same result is found if one Z_ν and the first derivative of the other one vanish, or both first derivatives vanish.

Let us specialize the discussion to the case $\alpha = 0$, for which $Z_\nu = J_\nu$, and let us take $R = 0$, $S = 1$ and $k_a = j_{\nu,m} > 0$, $k_b = j_{\nu,n} > 0$. Then the relation (12.5.5) gives

$$\int_0^1 r J_\nu(j_{\nu,n} r) J_\nu(j_{\nu,m} r) \, dr = 0, \qquad n \neq m. \tag{12.5.7}$$

The value of the integral when $m = n$ can be calculated explicitly with the result

$$\int_0^1 r J_\nu(j_{\nu,n} r) J_\nu(j_{\nu,m} r) \, dr = N_{\nu n}^2 \delta_{mn}, \qquad N_{\nu n} = \frac{1}{\sqrt{2}} J_{\nu+1}(j_{\nu,n}). \tag{12.5.8}$$

We thus conclude that, for $\nu > -1$, the functions $e_n = J_\nu(j_{\nu,n} r)/N_{\nu n}$ constitute an orthonormal set in the range $(0, 1)$ with the scalar product (12.5.6). This set is complete in $L^2(0, 1)$ and we may write

$$f(r) = \sum_{n=1}^{\infty} a_{\nu n} J_\nu(j_{\nu,n} r), \qquad a_{\nu n} = \frac{1}{N_{\nu n}^2} \int_0^1 r J_\nu(j_{\nu,n} r) f(r) \, dr, \tag{12.5.9}$$

where the $j_{\nu,n}$ are the positive zeros of J_ν arranged in ascending order of magnitude. This is the *Fourier–Bessel series expansion* of f. An example is

$$x^\nu = 2 \sum_{m=1}^{\infty} \frac{J_\nu(j_{\nu,m} x)}{j_{\nu,m} J_{\nu+1}(j_{\nu,m})}, \qquad 0 \leq x < 1. \tag{12.5.10}$$

Concerning the convergence properties of the Fourier–Bessel series we have the following result (see e.g. Watson 1944, pp. 591, 593):

Theorem 12.5.1 *Let $f(r)$ be defined for $0 < r < 1$ and let $\int_0^1 \sqrt{r} f(r) \, dr$ exist (and be absolutely convergent if improper).[7] Let f have bounded variation in the interval $[a, b]$, with $0 < a < b < 1$, and let $r \in (a, b)$. Then the series (12.5.9) with $\nu \geq -1/2$ is convergent with sum $\frac{1}{2}[f(r + 0) + f(r - 0)]$. Convergence is uniform in any interval interior to (a, b) if f is continuous in $[a, b]$.*

Since all the terms of the series vanish at $r = 1$, for convergence of the series at this point it is necessary that $f(1) = 0$; continuity of f in $a \leq r \leq 1$ is then a sufficient condition for uniform convergence in $[a, 1]$. Concerning convergence of the series at $r = 0$, it may be stated that the series $\sum_{n=1}^{\infty} a_n \sqrt{r} J_\nu(j_{\nu,m} r)$ converges uniformly to $\sqrt{r} f(r)$ in $0 \leq r < b$ if

---

[6] The derivation is trivially adaptable to the complex case.
[7] See (15.2.28) p. 364 for the origin and significance of the factor $\sqrt{r}$.

$\sqrt{r}\, f(r)$ has bounded variation in $[0, b]$ and $r^{-\nu} f(r)$ is continuous in that interval (Watson 1944, p. 615).

Just as the ordinary Fourier series can be used to deal with problems in which the function does not vanish at the end points of the interval (see p. 57), so can the Fourier–Bessel series after a suitable integration by parts, as shown on p. 154.

The previous developments can be repeated for functions of the form $c J_\nu(z) + z J_\nu'(z)$ with c a real constant and $\nu \geq -1/2$. If $\sigma_{\nu,n}$ are the zeros of this function, one has the *Dini series expansion*

$$f(r) = \sum_{n=1}^{\infty} b_{\nu n} J_\nu(\sigma_{\nu,n} r), \qquad b_n = 2\frac{\sigma_{\nu,n}^2}{M_{\nu n}^2} \int_0^1 r\, J_\nu(\sigma_{\nu,n} r)\, f(r)\, dr, \qquad (12.5.11)$$

$$M_{\nu n}^2 = \sigma_{\nu,n}^2 [J_\nu'(\sigma_{\nu,n})]^2 + (\sigma_{\nu,n}^2 - \nu^2)[J_\nu(\sigma_{\nu,n})]^2. \qquad (12.5.12)$$

For convergence results on the Dini series see Watson (1944), p. 600.

Both the Fourier–Bessel and the Dini expansions are unique in the sense that a function such that all the coefficients $a_{\nu n}$ in (12.5.9) or $b_{\nu n}$ in (12.5.11), vanish must vanish almost everywhere provided $\int_0^1 \sqrt{r}\, |f(r)|\, dr < \infty$ (Watson 1944, p. 616).

12.6 Other series expansions

Several other series expansions are associated to the Bessel functions. The *Neuman series* is based on the relation, valid for $\nu > -1$, (see e.g. Watson 1944, chapter 16)

$$\int_0^\infty J_{\nu+2n+1}(x)\, J_{\nu+2m+1}(x)\frac{dx}{x} = \frac{\delta_{mn}}{4n + 2\nu + 2}, \qquad (12.6.1)$$

from which one derives the expansion

$$f(x) = \sum_{n=0}^{\infty} (4n + 2\nu + 2) J_{\nu+2n+1}(x) \int_0^\infty f(x)\, J_{\nu+2n+1}(x)\frac{dx}{x}. \qquad (12.6.2)$$

An important instance is the expansion of $e^{i\mu x}$ given in (7.13.19) p. 199.

The *Kapteyn series* has the form (see e.g. Watson 1944, chapter 17)

$$f(x) = \sum_{n=0}^{\infty} a_{\nu n} J_{\nu+n}[(\nu + n)x]. \qquad (12.6.3)$$

For example

$$\frac{1}{1 \pm x} = 1 \mp 2\sum_{n=1}^{\infty} (\mp 1)^{n-1} J_n(nx), \qquad 0 \leq x < 1. \qquad (12.6.4)$$

The *Schlömilch series* has the form (see e.g. Watson 1944, chapter 19)

$$f(x) = \frac{1}{2}a_0 + \sum_{n=1}^{\infty} a_n J_0(nx). \qquad (12.6.5)$$

A remarkable example is

$$0 = \frac{1}{2} + \sum_{n=0}^{\infty} (-1)^n J_0(nx), \qquad 0 < x < 1. \tag{12.6.6}$$

If, in the Fourier expansion,

$$\frac{\pi}{8}(\pi - 2|x|) = \sum_{n=1}^{\infty} \frac{\cos(2n-1)x}{(2n-1)^2}, \qquad |x| < \pi, \tag{12.6.7}$$

we replace x by $x \sin \theta$ and integrate over $0 \le \theta \le \pi/2$, we find the Schlömilch expansion

$$|x| = \frac{\pi^2}{4} - 2 \sum_{n=1}^{\infty} \frac{J_0[(2n-1)x]}{(2n-1)^2}, \qquad |x| < \pi. \tag{12.6.8}$$

13 The Legendre Equation

13.1 Introduction

In spherical polar coordinates the Laplacian operator has the form

$$\nabla^2 u = \frac{1}{r^2} \frac{\partial}{\partial r} \left(r^2 \frac{\partial u}{\partial r} \right) + \frac{1}{r^2 \sin \theta} \frac{\partial}{\partial \theta} \left(\sin \theta \frac{\partial u}{\partial \theta} \right) + \frac{1}{r^2 \sin^2 \theta} \frac{\partial^2 u}{\partial \phi^2}. \qquad (13.1.1)$$

Our objective here is to simplify the action of this operator by expanding u on a suitable set of basis of functions. Since, in polar coordinates, the range of the coordinate ϕ is $0 \leq \phi < 2\pi$ (p. 171), it makes sense to start by expanding u in a Fourier series in ϕ:

$$u(r, \theta, \phi) = \sum_{-\infty}^{\infty} c_m(r, \theta) \, e^{im\phi}. \qquad (13.1.2)$$

Upon substituting this expansion into $\nabla^2 u$ and interchanging differentiation and summation,[1] we have

$$\nabla^2 u = \sum_{-\infty}^{\infty} \frac{1}{r^2} \left[\frac{\partial}{\partial r} \left(r^2 \frac{\partial c_m}{\partial r} \right) + \frac{1}{\sin \theta} \frac{\partial}{\partial \theta} \left(\sin \theta \frac{\partial c_m}{\partial \theta} \right) - \frac{m^2}{\sin^2 \theta} c_m \right] e^{im\phi}. \qquad (13.1.3)$$

A consideration of this equation suggests to expand the c_m in the form

$$c_m = \sum_{\ell} R_{\ell m}(r) w_{\ell}^{(m)}, \qquad (13.1.4)$$

where the $w_{\ell}^{(m)}$ are eigenfunctions of the θ operator, i.e. such that

$$\frac{1}{\sin \theta} \frac{d}{d\theta} \left(\sin \theta \frac{d w_{\ell}^{(m)}}{d\theta} \right) - \frac{m^2}{\sin^2 \theta} w_{\ell}^{(m)} = -\kappa_{\ell}^{(m)} w_{\ell}^{(m)} \qquad (13.1.5)$$

because in so doing we would find (see footnote 1)

$$\nabla^2 u = \sum_{m=-\infty}^{\infty} \sum_{\ell} \frac{1}{r^2} \left[\frac{d}{dr} \left(r^2 \frac{d R_{\ell m}}{dr} \right) - \kappa_{\ell}^{(m)} R_{\ell m} \right] w_{\ell}^{(m)} e^{im\phi} \qquad (13.1.6)$$

in which only differentiation in the radial direction remains. Thus, we are led to study the equation (13.1.5).

---

[1] For this operation see p. 227 and §9.8 p. 256.

13.2 The Legendre equation

In terms of the new variable

$$\mu = \cos\theta, \tag{13.2.1}$$

equation (13.1.5) becomes

$$\frac{d}{d\mu}\left[(1-\mu^2)\frac{dw^{(m)}}{d\mu}\right] + \left(\kappa - \frac{m^2}{1-\mu^2}\right)w^{(m)} = 0, \tag{13.2.2}$$

where we write κ in place of κ_ℓ^m and have dropped the subscript ℓ from w as, for now, we are interested in the differential equation rather than eigenfunctions and eigenvalues. This is the *associated Legendre equation*.[2] Since only m^2 enters the equation, without loss of generality, we may consider $m \geq 0$.

The solution procedure is greatly facilitated by the observation that, once the solutions $w^{(0)}$ of this equation for $m = 0$ are known, those for $m > 0$ can be constructed simply as[3]

$$w^{(m)} = (1-\mu^2)^{m/2}\,\frac{d^m w^{(0)}}{d\mu^m}. \tag{13.2.3}$$

Note also that the solutions for $m = 0$ are useful in themselves as they are applicable to axisymmetric problems in which there is no ϕ dependence. Hence let us focus on the equation

$$\frac{d}{d\mu}\left[(1-\mu^2)\frac{dw}{d\mu}\right] + \kappa w = 0 \tag{13.2.4}$$

or

$$\frac{d^2 w}{d\mu^2} - \frac{2\mu}{1-\mu^2}\frac{dw}{d\mu} + \frac{\kappa}{1-\mu^2}w = 0 \tag{13.2.5}$$

which is known as the *Legendre equation*.

Since the solutions of a linear differential equation in standard form (13.2.5) may be singular only at the points of singularity of the coefficients, we see that the possible finite singularities of the solutions of (13.2.2) and (13.2.4) can only be at $\mu = \pm 1$. In spherical polar coordinates $\mu = 1$ and $\mu = -1$ correspond to $\theta = 0$ and $\theta = \pi$, namely the positive and negative z-semi-axes. In many cases we are concerned with situations in which portions of both semi-axes are in the domain of interest and, in these cases, we will need solutions of the Legendre equation which are regular at $\mu = \pm 1$. We begin by focusing on this case. For a problem in which one or both of the semi-axes are outside the domain of interest (e.g., a domain bounded by the cone $\theta = \beta < \pi$), this condition can be relaxed and other solutions, considered in the later sections of this chapter, become relevant.

---

[2] More properly, in the general form of the associated Legendre equation, the constant m^2 is an arbitrary real or complex number.

[3] By differentiating equation (13.2.4) m times one finds that $d^m w^{(0)}/d\mu^m$ satisfies the same equation satisfied by $(1-\mu^2)^{-m/2}w^{(m)}$; see e.g. Hobson (1931), p. 89.

To generate a solution of (13.2.4) we expand w in a power series around $\mu = 0$, which is a regular point, by writing $w = \sum_{n=0}^{\infty} a_n \mu^n$. We substitute the expansion into (13.2.4) and multiply by $1 - \mu^2$ to find the recurrence relation (§2.5)

$$a_{n+2} = \frac{n(n+1) - \kappa}{(n+1)(n+2)} a_n \qquad n = 0, 1, 2, \ldots . \tag{13.2.6}$$

Starting with $a_0 \neq 0$, and $a_1 = 0$, all the coefficients of the odd powers vanish and the solution only contains even powers of μ; likewise, by choosing $a_1 \neq 0$ and $a_0 = 0$, we are left with odd powers only. To investigate the convergence of these series we use consider the ratio $a_{n+2}\mu^{n+2}/(a_n\mu^n)$ to find, for large n,

$$\left| \frac{a_{n+2}\mu^{n+2}}{a_n\mu^n} \right| \simeq \left(1 - \frac{2}{n} + \frac{4 - \kappa}{n^2} \right) \mu^2. \tag{13.2.7}$$

If $|\mu| < 1$, the series converges, while $|\mu| = 1$ is critical. Both the simple d'Alembert ratio test (p. 218) and the more refined Raabe test fail in this case, but Gauss's criterion (p. 219) shows that the series is divergent at $\mu = \pm 1$ unless it terminates. It is evident from (13.2.6) that the only way in which this may occur is if, for some integer ℓ,

$$\kappa = \kappa_\ell = \ell(\ell + 1). \tag{13.2.8}$$

The value of κ_ℓ remains the same if ℓ is changed into $-\ell - 1$ and, since only the value of κ_ℓ is of interest, rather than ℓ per se, it is sufficient to consider $\ell = 0, 1, 2, \ldots$ With this choice of κ, the recurrence relation (13.2.6) will give $a_{\ell+2} = 0$, so that all subsequent a_n will also vanish. A value κ_ℓ which causes the even series to terminate would not terminate the odd series, and vice versa. Hence, in order to have a solution regular at $\mu = \pm 1$, if ℓ is even we must take $a_1 = 0$ so that all the odd coefficients vanish and only $a_0, a_2, \ldots, a_\ell$ are non-zero, and conversely if ℓ is odd.

13.3 Legendre polynomials

We have concluded that the solutions of the Legendre equation

$$\frac{d}{d\mu}\left[(1 - \mu^2)\frac{dw}{d\mu} \right] + \ell(\ell + 1)w = 0 \tag{13.3.1}$$

bounded at $\mu = \pm 1$ are polynomials, even or odd according to the parity of ℓ; with the conventional normalization $P_\ell(1) = 1$ these are the *Legendre polynomials*.[4] In view of their definite parity, $P_\ell(-1) = (-1)^\ell$ and $P_{2k+1}(0) = 0$. It can be shown that $P_{2\ell}(0) = (-1)^\ell(2\ell)!/[2^{2\ell}(\ell!)^2] = (-1)^\ell(2\ell - 1)!!/(2^\ell \ell!)$.

---

[4] We will be led to these polynomials by other considerations in Example 19.4.1 p. 556. An infinite number of polynomial families exist which share similar properties; some of the most important ones are introduced in the final section of this chapter.

By manipulating the recurrence relations to be described later, it is possible to show that the Legendre polynomials can be expressed as

$$P_\ell(\mu) = \frac{1}{2^\ell \ell!} \frac{d^\ell}{d\mu^\ell} (\mu^2 - 1)^\ell, \tag{13.3.2}$$

which is *Rodrigues's formula* for the Legendre polynomials. From the fact that $(\mu^2 - 1)^\ell$ has only two zeros, each one of order ℓ, at $\mu = \pm 1$, it follows that all the zeros of P_ℓ are real and lie between -1 and 1. Rodrigues's formula is a key result that enables us to derive many properties.

13.3.1 Integral representations

The mathematical structure of the solutions of ordinary differential equations with analytic coefficients is usually best revealed by considering them in the complex plane and we therefore write z in place of μ. By using Cauchy's formula (17.3.29) (p. 430) for the ℓ-th derivative of an analytic function in conjunction with Rodrigues's formula we can express the Legendre polynomial $P_\ell(z)$ as

$$P_\ell(z) = \frac{1}{2\pi i} \int_C \frac{1}{2^\ell} \frac{(\zeta^2 - 1)^\ell}{(\zeta - z)^{\ell+1}} \, d\zeta, \tag{13.3.3}$$

which is also known as *Schläfli's integral* for the Legendre polynomials. Since the numerator of the integrand is an entire function, here C is any closed contour containing z in its interior and encircling it once counterclockwise. We find a useful integral representation for P_ℓ by taking the integration contour in a special way. Let $z^2 - 1 = R e^{i\alpha}$ and choose C as a circle of radius R centered at z, so that $\zeta - z = R e^{i\beta}$. This step transforms the integral into an integration over β in any 2π range, $\beta_0 \le \beta \le \beta_0 + 2\pi$ with β_0 arbitrary. A simple calculation then shows that Schläfli's integral becomes

$$P_\ell(z) = \frac{1}{2\pi} \int_{\beta_0}^{\beta_0+2\pi} [z + \sqrt{z^2 - 1} \, \cos(\beta - \alpha)]^\ell \, d\theta. \tag{13.3.4}$$

To bring this to standard form, let $\eta = \beta - \alpha$, choose $\beta_0 = -\pi + \alpha$ and use the fact that the integrand is even in the angular variable to find

$$P_\ell(z) = \frac{1}{\pi} \int_0^\pi (z \pm \cos \eta \, \sqrt{z^2 - 1})^\ell \, d\eta. \tag{13.3.5}$$

Since only even powers of $\cos \eta$ give a non-zero contribution to the integral, the choice between the two signs is immaterial. This is *Laplace's first integral* for P_ℓ from which we immediately deduce, for example, that

$$P_0 = 1, \qquad P_1 = z. \tag{13.3.6}$$

The introduction of a new variable η' by the relation $(z \pm \cos \eta \sqrt{z^2 - 1})(z \mp \cos \eta' \sqrt{z^2 - 1}) = 1$ gives *Laplace's second integral*

$$P_\ell(z) = \pm \frac{1}{\pi} \int_0^\pi \frac{d\eta'}{(z \mp \sqrt{z^2 - 1} \, \cos \eta')^{\ell+1}}, \tag{13.3.7}$$

where the upper or lower sign in front of the integral is to be taken according as Re z is positive or negative while the one inside the integral remains arbitrary. The two forms of the Laplace integral reflect the invariance of the Legendre equation (13.3.1) under the substitution $\ell \to -\ell - 1$. Two other integral representations which can be obtained starting from the generating function (see next section) are due to Mehler (see e.g. Hobson 1931 p. 26):

$$P_\ell(\cos\theta) = \frac{\sqrt{2}}{\pi} \int_0^\theta \frac{\cos\left(\ell + \frac{1}{2}\right)\eta}{\sqrt{\cos\eta - \cos\theta}} \, d\eta, \tag{13.3.8}$$

$$P_\ell(\cos\theta) = \frac{\sqrt{2}}{\pi} \int_\theta^\pi \frac{\sin\left(\ell + \frac{1}{2}\right)\eta}{\sqrt{\cos\theta - \cos\eta}} \, d\eta. \tag{13.3.9}$$

13.3.2 Generating function

If in a circle $|t| < R$ we have the expansion

$$F(t, z) = \sum_{n=0}^\infty f_n(z)\, t^n, \tag{13.3.10}$$

where $\{f_n(z)\}$ is a family of functions, then the function $F(t, z)$ is called the generating function of the family $\{f_n\}$. Using Laplace's first integral, we can deduce the generating function for the $P_n(z)$. To this end, consider

$$F(t, z) = \sum_0^\infty t^n P_n(z) = \frac{1}{\pi} \sum_0^\infty \int_0^\pi t^n \left(z + \sqrt{z^2 - 1}\, \cos\eta\right)^n \, d\eta. \tag{13.3.11}$$

The series converges uniformly for all η provided $|t|$ is smaller than the smaller one of $|z \pm \sqrt{z^2 - 1}|^{-1} = |z \mp \sqrt{z^2 - 1}|$ which, for $z = \mu = \cos\theta$, implies $|t| < 1$. Thus, we can interchange summation and integration[5] to find the integral of a geometric series which can be readily summed with the result

$$F(t, z) = \frac{1}{\pi} \int_0^\pi \frac{d\eta}{1 - tz - t\sqrt{z^2 - 1}\, \cos\eta}. \tag{13.3.12}$$

The substitution $\zeta = (tz - 1)/\sqrt{1 - 2tz + z^2}$ brings this expression to the form

$$F(t, z) = -\frac{1}{\pi\sqrt{1 - 2tz + t^2}} \int_0^\pi \frac{d\eta}{\zeta + \sqrt{\zeta^2 - 1}\, \cos\eta} = \frac{1}{\sqrt{1 - 2tz + t^2}} \tag{13.3.13}$$

where the last step follows from the observation that the integral is just Laplace's second integral (13.3.7) for $P_0 = 1$. We have shown therefore that

$$\frac{1}{\sqrt{1 - 2tz + t^2}} = \sum_{\ell=0}^\infty t^\ell P_\ell(z). \tag{13.3.14}$$

---

[5] For the interchange of summation and integration see pp. 226 and 694.

A useful alternative representation for the P_ℓ can be found by setting, in this expression, $t = h/r$ and $z = \cos\theta$ so that

$$\frac{1}{\sqrt{r^2 - 2hr\cos\theta + h^2}} = \frac{1}{r}\sum_{\ell=0}^{\infty}\frac{h^\ell}{r^\ell}P_\ell(\cos\theta). \tag{13.3.15}$$

If $r = \sqrt{x^2 + y^2 + z^2}$ is the distance of a point $\mathbf{x} = (x, y, z)$ from the origin and $\mathbf{h} = (0, 0, h)$ is the position vector of a point $\mathbf{h}$ on the z-axis, the square root in the left-hand side is just $\sqrt{x^2 + y^2 + (z - h)^2} = |\mathbf{x} - \mathbf{h}|$, namely the distance between $\mathbf{x}$ and $\mathbf{h}$. Upon expanding in a Taylor series in z we have

$$\frac{1}{\sqrt{x^2 + y^2 + (z - h)^2}} = \sum_{\ell=0}^{\infty}\frac{(-h)^\ell}{\ell!}\frac{\partial^\ell}{\partial z^\ell}\frac{1}{r}, \tag{13.3.16}$$

so that comparison with (13.3.15) gives

$$P_\ell(\mu) = \frac{(-1)^\ell}{\ell!}r^{\ell+1}\frac{\partial^\ell}{\partial z^\ell}\frac{1}{r}. \tag{13.3.17}$$

Readers with some familiarity e.g. with electrostatics will recognize here a connection with the theory of multipoles, which will be pursued further in Chapter 14.

13.3.3 Recurrence relations

From the generating function we can prove a number of very useful *recurrence relations* among the P_ℓ, valid for real or complex argument:

$$(\ell + 1)P_{\ell+1}(z) - (2\ell + 1)zP_\ell(z) + \ell P_{\ell-1}(z) = 0, \tag{13.3.18}$$

$$P_\ell(z) = P_{\ell+1}'(z) - 2zP_\ell'(z) + P_{\ell-1}'(z), \tag{13.3.19}$$

$$(2\ell + 1)P_\ell(z) = P_{\ell+1}'(z) - P_{\ell-1}'(z). \tag{13.3.20}$$

For example, to prove (13.3.18), we differentiate the generating function (13.3.14) with respect to t, clear the denominator, and combine the resulting series in the two sides of the result into a single series which equals zero. For this result to be valid for all t it is necessary that all the coefficients of the different powers of t equal zero, which gives the desired result. To prove (13.3.19) we differentiate (13.3.14) with respect to z and proceed similarly. To prove (13.3.20), we differentiate (13.3.18), solve for P_ℓ' and substitute into (13.3.19).

From a manipulation of the preceding formulae others easily follow:[6]

$$zP_\ell' - P_{\ell-1}' = \ell P_\ell, \qquad P_\ell' - zP_{\ell-1}' = \ell P_{\ell-1}, \tag{13.3.21}$$

---

[6] These four recurrence relations plus (13.3.18) and (13.3.20) are special cases of the formulae of Gauss connecting contiguous hypergeometric functions.

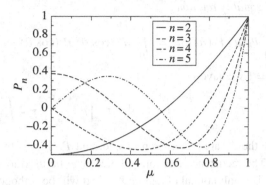

Fig. 13.1 The Legendre polynomials P_2, P_3, P_4 and P_5; $P_0 = 1$ and $P_1 = \mu$. Since $P_n(-\mu) = (-1)^n P_n(\mu)$, only the range $0 \le \mu \le 1$ is shown.

$$(z^2 - 1) P'_\ell = \ell z P_\ell - \ell P_{\ell-1} = \frac{\ell(\ell+1)}{2\ell+1} (P_{\ell+1} - P_{\ell-1}), \tag{13.3.22}$$

$$(z^2 - 1) P'_\ell = (\ell+1)(P_{\ell+1} - z P_\ell). \tag{13.3.23}$$

We can now use these relations to construct the polynomials sequentially. We have already seen that $P_0 = 1$, $P_1 = z$. Then, using e.g. (13.3.18), we have

$$P_2 = \frac{1}{2}\left(3z^2 - 1\right), \qquad P_3 = \frac{1}{2}\left(5z^3 - 3z\right), \tag{13.3.24}$$

$$P_4 = \frac{1}{8}\left(35z^4 - 30z^2 + 3\right), \qquad P_5 = \frac{1}{8}\left(63z^5 - 70z^3 + 15z\right). \tag{13.3.25}$$

Graphs of a few P_ℓ for real argument are shown in Figure 13.1.

13.3.4 Orthogonality

In applications, a crucial property of the Legendre polynomials is their mutual orthogonality relationship. This can be established by following the general procedure described on p. 361, namely multiplying the equation satisfied by P_k by P_ℓ and, conversely, the equation satisfied by P_ℓ by P_k. Upon subtracting, integrating between any two limits μ_1 and μ_2 and integrating by parts one readily finds

$$(\ell - k)(\ell + k + 1) \int_{\mu_1}^{\mu_2} P_k(\mu) P_\ell(\mu) \, d\mu = (\mu^2 - 1) \left[P_k \frac{dP_\ell}{d\mu} - P_\ell \frac{dP_k}{d\mu} \right]_{\mu_1}^{\mu_2}. \tag{13.3.26}$$

This result is actually quite general: k and ℓ can be complex numbers and the P any two solutions of the Legendre equation such that the integral exists. If P_k and P_ℓ are Legendre polynomials and we take $\mu_1 = -1$, $\mu_2 = 1$, the right-hand side vanishes and we find the *orthogonality relation*

$$\int_{-1}^{1} P_k(\mu)\, P_\ell(\mu)\, d\mu = \int_0^\pi P_k(\cos\theta)\, P_\ell(\cos\theta)\, \sin\theta\, d\theta = 0, \qquad k \neq \ell. \quad (13.3.27)$$

If $k = \ell$ we have

$$\int_{-1}^{1} P_\ell^2\, d\mu = \left[\mu P_\ell^2\right]_{-1}^{1} - 2\int_{-1}^{1} \mu\, P_\ell' P_\ell\, d\mu. \quad (13.3.28)$$

Since the P_ℓ are either even or odd, and $P_\ell(1) = 1$ by definition, it follows that $P_\ell^2(\pm 1) = 1$. The recurrence relation (13.3.21) can be used to set $\mu\, P_\ell' = P_{\ell-1}' + \ell P_\ell$. But since $P_{\ell-1}'$ is a polynomial of degree $\ell - 2$, it will be orthogonal to P_ℓ, while the second term is proportional to the left-hand side. Thus $(2\ell + 1)\int_{-1}^{1} P_\ell^2\, d\mu = 2$ and we have shown that

$$\int_{-1}^{1} P_k(\mu)\, P_\ell(\mu)\, d\mu = \int_0^\pi P_k(\cos\theta)\, P_\ell(\cos\theta)\, \sin\theta\, d\theta = \frac{2}{2\ell + 1}\, \delta_{k\ell}. \quad (13.3.29)$$

Changing $\mu = \cos\theta$ to the new variable $(2\ell + 1)\sin(\theta/2)$ brings the Legendre equation to a form which reduces to the Bessel equation of order 0 for $|1 - \mu|$ small (see e.g. Hobson 1931, p. 406; Forsyth 1956, p. 186; Erdélyi *et al.* 1953, vol. 1 p. 147); in this way one finds, for small values of θ,[7]

$$P_\ell(\cos\theta) \simeq J_0\left((2\ell + 1)\sin\frac{\theta}{2}\right), \qquad \ell \gg 1 \quad (13.3.30)$$

which, by its derivation, is valid also for non-integer ℓ.

By the method of §15.2 p. 363 one finds that, for θ away from 0 or π (see e.g. Hobson 1931, p. 303),[8]

$$P_\ell(\cos\theta) \simeq \left(\frac{2}{\pi\ell\sin\theta}\right)^{1/2} \cos\left[\left(\ell + \frac{1}{2}\right)\theta - \frac{\pi}{4}\right] + O(\ell^{-3/2}), \qquad \ell \gg 1, \quad (13.3.31)$$

which, again, holds also for non-integer ℓ.

13.4 Expansion in series of Legendre polynomials

The orthogonality of the Legendre polynomials suggests to associate to a function $f(\mu)$ the series

$$f(\mu) \sim \sum_{\ell=0}^{\infty} a_\ell P_\ell(\mu) \quad (13.4.1)$$

[7] This approximation may be interpreted as the result of replacing a sphere by its tangent plane in the neighborhood of its "north pole."
[8] More precisely, this estimate is valid for $0 < \varepsilon \leq \theta \leq \pi - \varepsilon < \pi$ and $\ell \gg 1/\varepsilon$, with ε arbitrarily small.

with

$$a_\ell = \left(\ell + \frac{1}{2}\right) \int_{-1}^{1} f(\mu) \, P_\ell(\mu) \, d\mu \qquad (13.4.2)$$

provided the integrals exist. The expression (13.4.1) is purely formal. The important questions are under what conditions the symbol "$\sim$" can be replaced by an equal sign, in what sense the equality should be interpreted, the mode of convergence, and so on. The answer to these questions hinges on the fact that, with an important proviso, the analysis of the convergence of the Legendre expansion (13.4.1) can be reduced to that of the convergence of the Fourier series for the function $f(\cos \theta) \sin^{1/2} \theta$ (Hobson 1931; see also the general results of p. 363 and §15.3).

Theorem 13.4.1 [Expansion Theorem] *Let $f(\mu)/(1 - \mu^2)^{1/4}$ be integrable between -1 and 1 or, equivalently, $f(\cos \theta) \sin^{1/2} \theta$ be integrable between 0 and π. Then the series (13.4.1), with a_n as defined in (13.4.2), is convergent and*

$$\sum_{0}^{\infty} a_n P_n(\mu) = \frac{1}{2} [f(\mu + 0) + f(\mu - 0)] \qquad (13.4.3)$$

at any μ interior to the interval $(-1, 1)$ such that any one of the convergence criteria for the Fourier series corresponding to the function $f(\cos \theta) \sin^{1/2} \theta$ is satisfied.

From §9.3 we thus deduce that (13.4.3) will hold if, for example, $f(\cos \theta) \sin^{1/2} \theta$ has bounded variation in some neighborhood of μ, or it has bounded derivatives at μ, or satisfies a Lipschitz condition.

Integrability of f over $-1 \leq \mu \leq 1$ is not sufficient to ensure convergence at the end points ± 1. Somewhat more stringent conditions are needed, such as being of bounded variation over the entire interval, or having, inside the interval, at most a finite number of singularities of the type $|\mu - \mu_0|^{-\alpha}$, with $0 < \alpha < 1/2$. If $\alpha \geq 1/2$ at one such point, the series will not converge to $f(\pm 1)$. By an argument similar to that given in §14.3 and resting on the result (3.6.11) p. 75, it can be shown that the Abel–Poisson sum (see §8.5) of the Legendre series satisfies (13.4.3) and that, at the points ± 1, it equals $f(\pm 1 \mp 0)$ whenever these limits exist (see e.g. Hobson 1931, p. 339).

In the theorem, the hypothesis on the integrability of $f(\mu)/(1 - \mu^2)^{1/4}$ is crucial. Alternatively stated, it is equivalent to the assumptions that $f(\mu)$ be integrable in the open interval $(-1, 1)$ and that, in the neighborhood of the points $\mu = \pm 1$, $|f(\mu)| < A_\mp/(1 \mp \mu)^k$, with $A_\mp$ constants and $k < 3/4$. It can be shown (see e.g. Hobson 1932 p. 329) that, if $k \geq 3/4$, the series *does not converge at any point* μ interior to $(-1,1)$. It is remarkable that the convergence of the series is disrupted *throughout the interval* by the behavior of f near the end points. This fact, which has no counterpart for the Fourier series, is a consequence of the singular nature of the points ± 1 for the Legendre equation.

Sufficient conditions for the uniform convergence of the Legendre series in any interval *interior* to $(-1, 1)$ can also be reduced to sufficient conditions for the uniform convergence of the Fourier series for $f(\cos \theta) \sin^{1/2} \theta$ in the corresponding interval (see §9.3). For example, the series will converge uniformly inside a sub-interval of an interval in which $f(\cos\theta)\sin^{1/2}\theta$ is of bounded variation and continuous.

When f is such that summation and integration in the Legendre series can be interchanged (in particular, when convergence is uniform; see also p. 227 and §9.8 p. 256), we have

$$f(\mu) = \int_{-1}^{1}\left[\sum_{\ell=0}^{\infty}\left(\ell+\frac{1}{2}\right)P_{\ell}(\mu)\,P_{\ell}(\mu')\right]f(\mu')\,\mathrm{d}\mu' \tag{13.4.4}$$

from which we deduce the relation

$$\sum_{\ell=0}^{\infty}\left(\ell+\frac{1}{2}\right)P_{\ell}(\mu)\,P_{\ell}(\mu') = \delta(\mu-\mu'), \tag{13.4.5}$$

which can also be obtained formally by expanding $\delta(\mu-\mu')$ as in (13.4.1) with coefficients given by (13.4.2).

As defined in Chapter 19, a (generally infinite) set of orthogonal functions v_n is *complete* over the interval (a, b) if the relation $\int_a^b v_n(x)\,f(x)\,\mathrm{d}x = 0$ for all n implies $f(x) = 0$ almost everywhere. The recurrence relation (13.3.18) affords an easy way to prove the completness of the Legendre polynomials because, on its basis, if $\int_{-1}^{1}P_{\ell}(\mu)\,f(\mu)\,\mathrm{d}\mu = 0$ for all ℓ, it follows from (13.3.18) that $\int_{-1}^{1}\mu\,P_{\ell}(\mu)\,f(\mu)\,\mathrm{d}\mu = 0$. Repeating the argument we conclude that $\int_{-1}^{1}\mu^k\,P_{\ell}(\mu)f(\mu)\,\mathrm{d}\mu = 0$ for any ℓ and any non-negative integer k. Now, the Taylor series for $\cos n\pi\mu$ and $\sin n\pi\mu$ converge uniformly and, therefore, they can be multiplied by $f(\mu)\,P_{\ell}(\mu)$ and integrated term by term with the result

$$\int_{-1}^{1}P_{\ell}(\mu)\,f(\mu)\sin n\pi\mu\,\mathrm{d}\mu = 0, \qquad \int_{-1}^{1}P_{\ell}(\mu)\,f(\mu)\cos n\pi\mu\,\mathrm{d}\mu = 0. \tag{13.4.6}$$

But these are just the Fourier coefficients of the function $P_{\ell}(\mu)\,f(\mu)$ which are thus seen to be all equal to 0 for any ℓ. From the completeness of the Fourier series we thus conclude that f vanishes almost everywhere. The relation (13.4.5) is another way to express the completeness of the Legendre polynomials.

13.5 Legendre functions

We now briefly consider the second solution of the Legendre equation for integer ℓ. A straightforward application of (2.2.33) p. 35 shows that, for real $|\mu| < 1$, another linearly independent solution of the equation is given by

$$Q_{\ell}(\mu) = A_{\ell}\,P_{\ell}(\mu)\int\frac{\mathrm{d}\xi}{(1-\xi^2)P_{\ell}^2(\xi)}, \tag{13.5.1}$$

with A_{ℓ} an arbitrary normalization constant. Since all the zeros of the denominator are real, the integrand can be decomposed in partial fractions as the sum of terms $(1\pm\xi)^{-1}$ and $(\xi-\mu_k)^{-2}$, where $\mu_k, k = 1, 2, \ldots, \ell$ are the roots of P_{ℓ}. Upon integration the first pair of terms gives logarithms while all the remaining ones are proportional to $(\mu-\mu_k)^{-1}$. Each

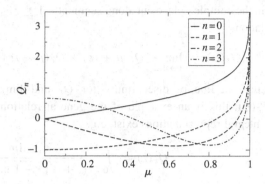

Fig. 13.2 The Legendre functions Q_0, Q_1, Q_2 and Q_3; since $Q_n(-\mu) = (-1)^n Q_n(\mu)$, only the range $0 \le \mu < 1$ is shown.

one of these terms divides P_ℓ exactly so that, when the combination (13.5.1) is formed, the result is a polynomial of degree $\ell - 1$. With a suitable normalization constant, the result of the integration may be written as

$$Q_\ell(\mu) = \frac{1}{2} P_\ell(\mu) \log \frac{1+\mu}{1-\mu} - W_{\ell-1}(\mu) \tag{13.5.2}$$

with the polynomial $W_{\ell-1}$ given by

$$W_{\ell-1}(\mu) = \sum_{k=1}^{\ell} \frac{1}{k} P_{k-1}(\mu)\, P_{\ell-k}(\mu). \tag{13.5.3}$$

The functions Q_ℓ thus defined are the *Legendre functions of the second kind*. The analog of Rodrigues's formula is

$$Q_\ell(\mu) = \frac{1}{2^\ell \ell!} \frac{d^\ell}{d\mu^\ell} \left[(\mu^2 - 1)^\ell \log \frac{1+\mu}{1-\mu} \right] - \frac{1}{2} P_\ell \log \frac{1+\mu}{1-\mu} \tag{13.5.4}$$

from which

$$Q_0 = \frac{1}{2} \log \frac{1+\mu}{1-\mu} = \tanh^{-1} \mu, \qquad Q_1 = \frac{1}{2}\mu \log \frac{1+\mu}{1-\mu} - 1, \tag{13.5.5}$$

$$Q_2 = \frac{1}{4}(3\mu^2 - 1) \log \frac{1+\mu}{1-\mu} - \frac{3}{2}\mu. \tag{13.5.6}$$

Graphs of the first few of these functions are shown in Figure 13.2 for real argument.

Functions of higher degree can be found by using recurrence relations such as (13.3.18), (13.3.21), (13.3.22) which are readily established using Neumann's result

$$Q_\ell(z) = \frac{1}{2} \int_{-1}^{1} \frac{P_\ell(\mu)}{z-\mu}\, d\mu \tag{13.5.7}$$

valid in the complex plane cut along segment $-1 \leq z \leq 1$ provided z is not on the cut, and the relation

$$Q_\ell(\mu) = \lim_{\varepsilon \to 0} \frac{1}{2} [Q_\ell(\mu + i\varepsilon) + Q_\ell(\mu - i\varepsilon)], \qquad |\mu| < 1. \tag{13.5.8}$$

The cut is a line of discontinuty for Q_ℓ and $\lim_{\varepsilon \to 0} [Q_\ell(\mu + i\varepsilon) - Q_\ell(\mu - i\varepsilon)] = -\pi i \, P_\ell(\mu)$; this is an example of the general relation (17.5.1). In addition to (13.5.7), other integral representations exist, e.g.

$$Q_\ell(z) = \int_0^\infty \frac{d\eta}{(z + \sqrt{z^2 - 1} \, \cosh \eta)^{\ell+1}}, \tag{13.5.9}$$

which is similar to Laplace's second integral (13.3.7) for the Legendre polynomials.

Solutions of Legendre's differential equation (13.2.5) also exist for general $\kappa = \lambda(\lambda + 1)$ and they are denoted by P_λ and Q_λ; the P_λ are the *Legendre functions of the first kind*, and they reduce to the Legendre polynomials for integer λ. We return to these functions in §13.7.

13.6 Associated Legendre functions

Let us now go back to the full associated Legendre equation (13.2.2), which we consider for $-1 \leq \mu \leq 1$. The symbol m will be taken to refer to a non-negative integer; we will explicitly write $-m$ when we refer to negative order.

By using the property (13.2.3), we define the associated Legendre functions of the first and second kind of degree ℓ and order m by[9]

$$P_\ell^m = (-1)^m (1 - \mu^2)^{m/2} \frac{d^m P_\ell}{d\mu^m}, \qquad Q_\ell^m = (-1)^m (1 - \mu^2)^{m/2} \frac{d^m Q_\ell}{d\mu^m}. \tag{13.6.1}$$

From Rodrigues's formula we see that

$$P_\ell^m = \frac{(-1)^m}{2^\ell \ell!} (1 - \mu^2)^{m/2} \frac{d^{\ell+m}}{d\mu^{\ell+m}} (\mu^2 - 1)^\ell, \tag{13.6.2}$$

from which it immediately follows that $P_\ell^m = 0$ for $m > \ell$. This of course does not mean that the differential equation ceases to possess solutions, but only that they cannot be given in this form. If we interpret a derivative of negative order as an integration, formally (13.2.3) suggests to consider as a possible solution

$$P_\ell^{-m}(\mu) = (1 - \mu^2)^{-m/2} \underbrace{\int_\mu^1 d\mu' \int_{\mu'}^1 d\mu'' \cdots P_\ell}_{m\text{-fold}}, \qquad |\mu| \leq 1. \tag{13.6.3}$$

---

[9] When the argument is not a real number between -1 and 1, the factor $(-1)^m (1 - \mu^2)^{m/2}$ is usually replaced by $(z^2 - 1)^{m/2}$ and the complex plane is cut between $-\infty$ and 1 to avoid multivaluedness.

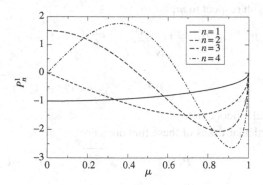

Fig. 13.3 Graphs of $P_n^1(\mu)$ for $n = 1, 2, 3$ and 4

These functions are indeed solutions of the equation which, for $m \leq \ell$ and $|\mu| \leq 1$, satisfy

$$P_\ell^{-m} = (-1)^m \frac{(\ell - m)!}{(\ell + m)!} P_\ell^m. \tag{13.6.4}$$

The existence of a proportionality relation between P_ℓ^m and P_ℓ^{-m} was to be expected from the dependence of the equation on m^2. However, for $m > \ell$, $P_\ell^m = 0$ whereas P_ℓ^{-m} does not vanish. Thus, for $\ell \geq m$, two linearly independent solutions may be taken as P_ℓ^m, Q_ℓ^m while, for $m > \ell$, we may use P_ℓ^{-m} and Q_ℓ^m. The reason the functions P_ℓ^{-m} with $m > \ell$ are not frequently encountered is that they are singular at ± 1. For example

$$P_0^{-1} = \sqrt{\frac{1 - \mu}{1 + \mu}}, \qquad P_1^{-2} = \frac{1}{6} \frac{1 - \mu}{1 + \mu} (2 + \mu). \tag{13.6.5}$$

Accordingly, we do not consider them further.

The orthogonality relation for unequal degree is

$$\int_{-1}^{1} P_k^m(\mu) P_\ell^m(\mu) \, \mathrm{d}\mu = \frac{2}{2\ell + 1} \frac{(\ell + m)!}{(\ell - m)!} \delta_{k\ell}, \tag{13.6.6}$$

while, for unequal order,

$$\int_{-1}^{1} P_\ell^m(\mu) P_\ell^{m'}(\mu) \frac{\mathrm{d}\mu}{1 - \mu^2} = \frac{(\ell + m)!}{m(\ell - m)!} \delta_{mm'}. \tag{13.6.7}$$

The associated Legendre functions satisfy relations similar to those of the regular functions. For example, for $-1 \leq \mu \leq 1$, two recurrence relations with respect to ℓ are

$$(2\ell + 1)\mu P_\ell^m = (\ell - m + 1) P_{\ell+1}^m + (\ell + m) P_{\ell-1}^m, \tag{13.6.8}$$

$$(1 - \mu^2) \frac{\mathrm{d} P_\ell^m}{\mathrm{d}\mu} = (\ell + 1)\mu \, P_\ell^m - (\ell - m + 1) P_{\ell+1}^m, \tag{13.6.9}$$

and, with respect to m,

$$P_\ell^{m+2} + 2(m+1)\frac{\mu}{\sqrt{1-\mu^2}} P_\ell^{m+1} + (\ell - m)(\ell + m + 1) P_\ell^m = 0, \qquad (13.6.10)$$

$$P_{\ell-1}^m - P_{\ell+1}^m = (2\ell + 1)\sqrt{1-\mu^2} P_\ell^{m-1}. \qquad (13.6.11)$$

Several others exist.

Particular cases of these functions are

$$P_1^1(\mu) = -\sqrt{1-\mu^2}, \qquad P_1^{-1}(\mu) = \frac{1}{2}\sqrt{1-\mu^2}, \qquad (13.6.12)$$

$$P_2^1(\mu) = -3\mu\sqrt{1-\mu^2}, \qquad P_2^2(\mu) = 3(1-\mu^2), \qquad (13.6.13)$$

$$P_2^{-1}(\mu) = \frac{1}{2}\mu\sqrt{1-\mu^2}, \qquad P_2^{-2} = \frac{1}{8}(1-\mu^2). \qquad (13.6.14)$$

Graphs of the first few P_n^1 are shown in Figure 13.3.

A useful relation in applications is *Legendre's addition theorem* for the associated functions of the first kind:

$$P_\ell(\cos\gamma) = P_\ell(\cos\theta) P_\ell(\cos\theta') \qquad (13.6.15)$$

$$+2\sum_1^\ell \frac{(\ell-m)!}{(\ell+m)!} P_\ell^m(\cos\theta) P_\ell^m(\cos\theta') \cos m(\phi - \phi'),$$

where

$$\cos\gamma = \cos\theta \cos\theta' + \sin\theta \sin\theta' \cos(\phi - \phi'), \qquad 0 \le \theta_1, \theta_2 \le \pi. \quad (13.6.16)$$

As shown in Figure 7.2 p. 178, the angle γ thus defined is the angle between the two directions given, in polar coordinates, by (θ, ϕ) and (θ', ϕ'). For a proof, see e.g. Hobson 1931 p. 143.

Bounds and asymptotic estimates generalizing (13.3.30) and (13.3.31) also exist for these functions. For example, for small θ,

$$\left[\left(\ell + \frac{1}{2}\right)\cos\frac{\theta}{2}\right]^m P_\ell^{-m}(\cos\theta) \simeq J_m\left((2\ell+1)\sin\frac{\theta}{2}\right), \qquad \ell \gg 1. \quad (13.6.17)$$

13.7 Conical boundaries

We were led to the Legendre polynomials by requiring solutions of the Legendre equations to be finite for $\cos\theta = \pm 1$, i.e., on both z-semi-axes. With a conical boundary, one (or both) of the z-semi-axes are outside the domain of interest and more general solutions of the equation become relevant. Consider for example a domain with a conical boundary of semi-aperture β (see Figure 7.7 p. 192), and suppose we are interested in solutions which vanish on the surface of the cone. In this case, a suitable set of eigenfunctions is constituted by Legendre functions of the first kind P_λ, which are solutions of

$$\frac{d}{d\mu}\left[(1 - \mu^2)\frac{dP_\lambda}{d\mu}\right] + \lambda(\lambda + 1)P_\lambda = 0 \qquad (13.7.1)$$

regular for $\mu = 1$ and vanishing when $\mu = \cos\beta$. In the case $\beta = \frac{1}{2}\pi$ the cone is the plane $z = 0$ and the example of §7.5 shows that $P_{\lambda_1} = P_1$, $P_{\lambda_2} = P_3$ and so on. Except for this case, the functions P_λ are not polynomials, but the general theory of Sturm–Liouville problems of §15.2 shows that there is an infinite number of them, which we may label P_{λ_1}, P_{λ_2} and so on, and that the corresponding eigenvalues $\lambda_n(\lambda_n + 1)$ are real.

We can gain some insight into the situation if the cone semi-aperture β is small so that the asymptotic expression (13.3.30) applies:

$$P_{\lambda_n}(\cos\theta) \simeq J_0\left((2\lambda_n + 1)\sin\frac{\theta}{2}\right). \qquad (13.7.2)$$

By using $2\sin(\theta/2) \simeq \theta$, valid for small angles, we thus see that the zeros of $P_{\lambda_n}(\cos\beta)$ occur approximately for

$$\lambda_n + \frac{1}{2} = \frac{j_{0,n}}{\beta}\left[1 + O\left(\frac{\beta}{2}\right)^2\right], \qquad (13.7.3)$$

where the $j_{0,n}$ are non-vanishing zeros of the Bessel function J_0. In particular, $j_{0,1} \simeq 2.405$ and we thus find

$$\lambda_1 \simeq \frac{2.405}{\beta} - \frac{1}{2}, \qquad (13.7.4)$$

which increases with decreasing β. For β close to π, one has instead (Hobson 1931, p. 408)

$$\lambda_{n+1} \simeq n + \frac{1}{2}\left[\log\frac{2}{\pi - \beta}\right]^{-1} \qquad n = 0, 1, 2, \ldots \qquad (13.7.5)$$

Thus, as $\beta \to \pi$, the λ_n approach integer values as expected, although very slowly.

For general β, the solution of the Legendre equation regular on the positive z-semi-axis $\mu = 1$ can be expressed in terms of the hypergeometric function (see e.g. Erdélyi et al. 1953, vol. 1 chapter 2 and §2.5):

$$F(a, b; c; z) = 1 + \frac{ab}{1!c}z + \frac{a(a + 1)b(b + 1)}{2!c(c + 1)}z^2 + \cdots \qquad (13.7.6)$$

as

$$P_\lambda(\mu) = F\left(-\lambda, \lambda + 1; 1; \frac{1 - \mu}{2}\right) = F\left(-\lambda, \lambda + 1; 1; \sin^2\frac{1}{2}\theta\right)$$

$$= 1 - \frac{\lambda(\lambda + 1)}{(1!)^2}\sin^2\frac{\theta}{2} + \frac{(\lambda - 1)\lambda(\lambda + 1)(\lambda + 2)}{(2!)^2}\sin^4\frac{\theta}{2} + \cdots \qquad (13.7.7)$$

If $0 \leq \mathrm{Re}\,(a + b - c) < 1$ the hypergeometric series converges throughout the unit circle except at $z = 1$, while it diverges on $|z| = 1$ if $1 \leq \mathrm{Re}\,(a + b - c)$. Thus $P_\lambda(\cos\theta) = 1$ for $\theta = 0$ while it is singular for $\theta = \pi$ as expected from the analysis of §13.2. Conversely, upon changing θ to $\pi - \theta$, we see that $P_\lambda(-\cos\theta)$ diverges for $\theta = 0$ and equals 1 for

$\theta = \pi$. These latter functions are therefore the relevant regular solutions when the negative, rather than the positive, z-semi-axis is in the physical domain of interest.

The relation (13.3.26) holds for this family of functions and proves that different members of the family are orthogonal. In place of the expansion (13.4.1) we now have

$$f(\mu) = \sum_{n=0}^{\infty} a_n P_{\lambda_n}(\mu) \tag{13.7.8}$$

with

$$a_n = \left[\int_{\cos \beta}^{1} P_{\lambda_n}^2(\mu) \, d\mu \right]^{-1} \int_{\cos \beta}^{1} P_{\lambda_n}(\mu) f(\mu) \, d\mu. \tag{13.7.9}$$

The normalization constant in this case will be a function of $\cos \beta$.

In §7.11 we saw the need for Legendre functions of the type $P_{-1/2+ik}$, with k real, the so-called *conical functions* (see e.g. Hobson 1931). Their expression in terms of the hypergeometric function (13.7.6) is

$$P_{-1/2+ik}(\cos \theta) = F\left(\frac{1}{2} - ik, \frac{1}{2} + ik; 1; \sin^2 \frac{\theta}{2} \right)$$

$$= 1 + \frac{\frac{1}{4} + k^2}{(1!)^2} \sin^2 \frac{\theta}{2} + \frac{(\frac{1}{4} + k^2)(\frac{9}{4} + k^2)}{(2!)^2} \sin^4 \frac{\theta}{2} + \cdots \tag{13.7.10}$$

We see that $P_{-1/2+ik} = P_{-1/2-ik}$, which is also evident from the invariance of the differential equation (13.7.1) upon the transformation $\lambda \to -\lambda - 1$. As noted before $P_{-1/2+ik}(\cos \theta)$ is singular for $\theta = \pi$.

The conical functions have an infinite number of zeros which are all real and greater than 1. An integral representation is (see e.g. Hobson 1931, p. 272; Gradshteyn *et al.* 2007, p. 1020)

$$P_{-1/2+ik}(\cos \theta) = \frac{\sqrt{2}}{\pi} \cosh(\pi k) \int_0^{\infty} \frac{\cos kx}{\sqrt{\cosh x + \cos \theta}} \, dx. \tag{13.7.11}$$

13.8 Extensions

In the previous section and in §7.11 we have seen how the need for Legendre functions of the type $P_{-1/2+ip}(\mu)$, with $|\mu| < 1$ arises with conical boundaries. The use of other coordinate systems leads to the appearance of an equation of the general Legendre form (13.7.1) but with the independent variable outside the range $|\mu| \leq 1$. For example, prolate spheroidal coordinates lead to Legendre polynomials of the form $P_\ell(\cosh \eta)$ and to the Legendre functions $Q_\ell(\cosh \eta)$, with integer ℓ and $0 \leq \eta < \infty$. The $P_\ell(\cosh \eta)$ diverge at infinity and, therefore, can only be used in bounded domains while, conversely, the $Q_\ell(\cosh \eta)$ diverge for $\eta = 1$, which corresponds to the origin, and therefore can only be used for domains not containing this point.

Oblate spheroidal coordinates lead to $P_\ell(i \sinh \eta)$ and $Q_\ell(i \sinh \eta)$, which are real, and can be used inside or outside the spheroid $\eta = \eta_0$ respectively.

In toroidal coordinates the functions $P_{\ell-1/2}(\cosh \eta)$ and $Q_{\ell-1/2}(\cosh \eta)$ with integer ℓ arise, also known as toroidal functions. The functions regular inside the torus $\eta = \eta_0$ are the $Q_{\ell-1/2}$, and those outside the $P_{\ell-1/2}$.

Conical functions with argument $\cosh \eta \geq 1$ arise for boundaries having the form of two intersecting spheres or of a hyperboloid. In this case we have an integral representation reminiscent of (13.3.9):

$$P_{-1/2+ip}(\cosh \eta) = \frac{2}{\pi} \cosh (\pi p) \int_\eta^\infty \frac{\sin (\theta p)}{\sqrt{2 (\cosh \theta - \cosh \eta)}} \, d\theta. \tag{13.8.1}$$

It may also be of some interest to mention the Mehler–Fock transform, which is useful e.g. for boundaries in the shape of hyperboloids (see e.g. Lebedev 1965, p. 221)

$$\tilde{f}(p) = \int_1^\infty f(x) \, P_{-1/2+ip}(x) \, dx, \tag{13.8.2}$$

and its inverse

$$f(x) = \int_0^\infty \tilde{f}(p) \, \tanh(\pi p) \, P_{-1/2+ip}(x) \, p \, dp. \tag{13.8.3}$$

This last relation is valid at any point of continuity of a function f defined in $1 < x < \infty$, piecewise continuous and of bounded variation in any finite sub-interval provided, for any $a > 1$, the two integrals

$$\int_1^a |f(x)| (x - 1)^{-3/4} \, dx \, , \qquad \int_a^\infty |f(x)| x^{-1/2} \log x \, dx \tag{13.8.4}$$

are finite.

When the problem does not have axial symmetry, the corresponding associated functions arise. Most of the properties of the Legendre functions described above admit generalizations to these more general situations. In particular, expansion theorems similar to Theorem 13.4.1 hold. Several examples of the application of these coordinate systems can be found e.g. in Hobson 1931 and Lebedev 1965. A point to keep in mind in dealing with general Legendre functions is that, to avoid multi-valuedness, as interest before, they are defined on the complex plane cut between $-\infty$ and 1. For this reason some relations between functions with real and complex argument involve phase factor that should not be overlooked (see e.g. Copson 1955 chapter 11; Endély, et al. 1953, vol. 1 chapter 3; Gradshteyn et al. 2007).

Name	$X(x)$	$a < x < b$	$s(x)$	N_n
Legendre $P_n(x)$	$1 - x^2$	$-1 < x < 1$	1	$2/(2n+1)$
Hermite $H_n(x)$	1	$-\infty < x < \infty$	$\exp(-x^2)$	$2^n \sqrt{\pi}\, n!$
Laguerre $L_n(x)$ ($\alpha = 0$)	x	$0 < x < \infty$	$\exp(-x)$	1
Chebyshev $T_n(x)$	$1 - x^2$	$-1 < x < 1$	$(1-x^2)^{-1/2}$	$\frac{1}{2}(1 + \delta_{n,0})\pi$
Chebyshev $U_n(x)$	$1 - x^2$	$-1 < x < 1$	$(1-x^2)^{1/2}$	$\frac{1}{2}\pi$

Table 13.1 Some classical families of polynomials; $T_n(x)$ and $U_n(x)$ are the Chebyshev polynomials of the first and second kind.

13.9 Orthogonal polynomials

As already mentioned, there is an infinity of families of polynomials, each one orthogonal according to a different scalar product. A general way to develop such families is to apply the the Gram–Schmidt orthogonalization to the appropriate scalar product; we demonstrate this procedure in §19.4.1 p. 555. Here we take an alternative approach starting from a generalized Rodrigues formula which is sufficient to generate the so-called classical polynomial families (see e.g. Erdélyi *et al.* 1953, vol. 2 p. 164).

We consider the family of functions $\{p_n(x)\}$ given by the relation

$$p_n = \frac{1}{k_n s(x)} \frac{d^n}{dx^n} \left[s(x) X^n(x) \right] \tag{13.9.1}$$

in which k_n is a constant, $s(x)$ is a positive weight function and $X(x)$ a fixed polynomial having degree $0 \le \kappa \le 2$; both $s(x)$ and $X(x)$ are independent of n. We assume that $s(x)X(x)$ vanishes at the end points of the interval of interest $[a, b]$.[10] The Legendre polynomials P_n are found by taking $k_n = (-1)^n 2^n n!$, $s =$const., $X = 1 - x^2$ and the interval as $[-1, 1]$.

For $n = 1$ (13.9.1) is

$$k_1 p_1(x) = X'(x) + \frac{s'(x)}{s(x)} X(x), \tag{13.9.2}$$

which can be solved for $s(x)$ with the result

$$s(x) \propto \frac{1}{X(x)} \exp\left(k_1 \int \frac{p_1(\xi)}{X(\xi)} \, d\xi \right). \tag{13.9.3}$$

---

[10] There is an infinity of families of orthogonal polynomials which violate one or more of these hypotheses. The importance of the classical families that can be constructed in this way resides in the eigenvalue equation (13.9.7) that they satisfy, which is frequently encountered and which leads to eigenfunction expansions useful in various contexts.

The requirement that p_1 be a polynomial of the first degree limits the possible functions $s(x)$. If $\kappa = 0$ so that $X =$ const. we have

$$s(x) \propto \exp\left(\frac{k_1}{X}(ax^2 + b)\right).$$ (13.9.4)

By taking the interval as $(-\infty, \infty)$ and after a suitable change of variable we may take $s(x) = e^{-x^2}$; the polynomials so generated are the Hermite polynomials H_n. With $\kappa = 1$ and $[0, \infty)$ we find $s(x) = x^\alpha e^{-x}$ and the generalized Laguerre polynomials L_n^α, while $\kappa = 2$ gives $s(x) = (1-x)^\alpha(1+x)^\beta$ and the Jacobi polynomials $J_n^{\alpha,\beta}$ in the interval $[-1, 1]$ (see Table 13.1). The standard Laguerre polynomials have $\alpha = 0$. The Jacobi polynomials with $\alpha = \beta = -1/2$ and $\alpha = \beta = 1/2$ are the Chebyshev polynomials of the first and second kind, respectively.

By taking the scalar product of (13.9.1) with a polynomial $q(x)$ and integrating by parts n times we have

$$(q, p_n)_s \equiv \int_a^b q(x)\, p_n(x)\, s(x)\, dx \;=\; \frac{(-1)^n}{k_n} \int_a^b q^{(n)}(x)\, sX^n(x)\, s(x)\, dx$$ (13.9.5)

where we have used the postulated vanishing of sX at the end points. The second integral vanishes if q has a degree lower than n. In this case this relation shows that the polynomials $\{p_n\}$ form an orthogonal system over $[a, b]$ with respect to the weight $s(x)$ so that

$$\int_a^b p_k(x)\, p_n(x)\, s(x)\, dx \;=\; N_n \delta_{k,n}$$ (13.9.6)

with the normalization constants N_n depending on the particular conventional normalizations used in each case.

By differentiating (13.9.1) twice and using (13.9.2) we find

$$\frac{d}{dx}\left[X(x)\, s(x)\, \frac{dp_n}{dx}\right] = n\left[p_1' k_1 + \frac{1}{2}(n-1)X''\right] p_n$$ (13.9.7)

so that the polynomials are eigenfunctions of the (Sturm–Liouville) operator on the left-hand side of the equation. Thus, one can construct series expansions of suitably restricted classes of functions similar to the Legendre series (13.4.1) in terms of these polynomials (see Theorem 15.3.3 p. 371).

If we write

$$p_n(x) = a_n x^n + b_n x^{n-1} + \cdots,$$ (13.9.8)

then one can show that

$$p_{n+1} = (A_n x + B_n)\, p_n - C_n p_{n-1}$$ (13.9.9)

in which

$$A_n = \frac{a_{n+1}}{a_n}, \quad B_n = A_n\left(\frac{b_{n+1}}{a_{n+1}} - \frac{b_n}{a_n}\right), \quad C_n = \frac{N_n\, A_n}{N_{n-1}\, A_{n-1}},$$ (13.9.10)

with

$$N_n = (p_n, p_n)_s = (-1)^n \frac{a_n n!}{k_n} \int_a^b X^n(x)\, s(x)\, dx$$ (13.9.11)

	Constants for the relations satisfied by the classical polynomial families of Table 13.1; here $g_n = 2^{-2n}(2n)!/(n!)^2$.				
Table 13.2					

Family	k_n	a_n	b_n	α_n	β_n
$P_n(x)$	$(-1)^n 2^n n!$	$2^n g_n$	0	0	n
$H_n(x)$	$(-1)^n$	2^n	0	0	$2n$
$L_n(x)$	$n!$	$(-1)^n/n!$	$(-1)^{n-1}n/(n-1)!$	n	$-n$
$T_n(x)$	$(-1)^n 2^n n! g_n$	$2^{n-1}\ (n \geq 1)$	0	0	n
$U_n(x)$	$(-1)^n 2^{n+1} n! g_{n+1}$	2^n	0	0	$n+1$

and

$$X p_n' = \left(\alpha_n + \frac{1}{2}nxX''\right) p_n + \beta_n p_{n-1} \tag{13.9.12}$$

with

$$\alpha_n = nX'(0) - \frac{b_n}{2a_n}X'', \qquad \beta_n = -\left[p_1'k_1 + \left(n - \frac{1}{2}\right)X''\right]\frac{C_n}{A_n}. \tag{13.9.13}$$

The relevant numerical constants for the families of polynomials of Table 13.1 are shown in Table 13.2.

These orthogonal families satisfy many relations of a nature similar to those enjoyed by the Legendre polynomials, such as the existence of a generating function, integral representations, addition theorems and others (see e.g. Erdélyi *et al.* 1953, vol. 2 chapter 10; Abramowitz & Stegun 1964, chapter 22; Gradshteyn *et al.* 2007, section 8.9).

Spherical Harmonics

14.1 Introduction

In the previous chapter we were led to a consideration of the associated Legendre equation by attempting to represent, in spherical coordinates, the angular dependence of a function $u(r, \theta, \phi)$ defined over the entire range of the angular variables, $0 \leq \theta \leq \pi$, $0 \leq \phi \leq 2\pi$. We started by expressing u in a Fourier series in ϕ

$$u(r, \theta, \phi) = \sum_{-\infty}^{\infty} S_m(r, \theta) \, e^{im\phi} \tag{14.1.1}$$

and then we looked for suitable functions of θ useful to expand S_m. In this way, we were led to the associated Legendre functions of the first kind $P_\ell^m(\cos \theta)$. We found that, in order for S_m to be finite on the polar axis $\cos \theta = \pm 1$, it was necessary to restrict the index ℓ to be greater than or equal to m so that

$$S_m = \sum_{\ell=|m|}^{\infty} R_{\ell m}(r) \, P_\ell^m(\cos \theta) N_\ell^m, \tag{14.1.2}$$

where the N_ℓ^m are normalization constants to be chosen later for convenience. Upon substitution into (14.1.1) we thus have

$$u(r, \theta, \phi) = \sum_{m=-\infty}^{\infty} \sum_{\ell=|m|}^{\infty} R_{\ell m}(r) \, N_\ell^m \, P_\ell^m \, e^{im\phi}. \tag{14.1.3}$$

With reference to Figure 14.1, one may say that the double summation indicated here is carried out "vertically": for each fixed m, all the terms with $\ell \geq m$ are added, and then the series thus generated are summed over m. A more convenient expression, which avoids dealing with doubly infinite series, can be found by carrying out the summation "horizontally": for each fixed ℓ, we sum over $-\ell \leq m \leq \ell$, and then add up the finite sums corresponding to

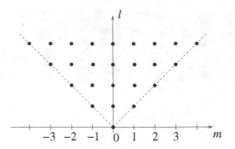

Fig. 14.1 The summations indicated in (14.1.3) are done "vertically" i.e., for each m, lettering ℓ range between $|m|$ and infinity. The summations indicated in (14.1.4) are done "horizontally", allowing m to range between $-\ell$ and ℓ for each ℓ.

each ℓ.[1] In this way we find

$$u(r, \theta, \phi) = \sum_{\ell=0}^{\infty} \sum_{m=-\ell}^{\ell} R_{\ell m}(r) \, N_{\ell}^{m} \, P_{\ell}^{m} \, e^{im\varphi}. \tag{14.1.4}$$

14.2 Spherical harmonics

For each ℓ, the $2\ell + 1$ functions

$$Y_{\ell}^{m} = N_{\ell}^{m} \, P_{\ell}^{m}(\cos\theta) \, e^{im\phi}, \qquad -\ell \leq m \leq \ell, \tag{14.2.1}$$

are called *spherical* (surface) *harmonics* of degree ℓ and order m. They are eigenfunctions of the Laplace operator on the unit sphere:

$$-\nabla^2 Y_{\ell}^{m}\big|_{r=1} \equiv -\frac{1}{\sin\theta} \frac{\partial}{\partial\theta}\left(\sin\theta \frac{\partial Y_{\ell}^{m}}{\partial\theta}\right) - \frac{1}{\sin^2\theta} \frac{\partial^2 Y_{\ell}^{m}}{\partial\phi^2} = \ell(\ell+1)Y_{\ell}^{m}. \tag{14.2.2}$$

We select the normalization constants N_{ℓ}^{m} so as to have the orthonormality relation

$$\left(Y_{\ell}^{m}, Y_{k}^{n}\right) = \int_{\Omega} \overline{Y}_{\ell}^{m} Y_{k}^{n} \, d\Omega = \delta_{\ell k} \, \delta_{mn}, \tag{14.2.3}$$

where the integration is over the solid angle $\Omega = (0 \leq \theta \leq \pi, \, 0 \leq \phi \leq 2\pi)$ or, alternatively put, the surface of the unit sphere. Since the surface element on the unit sphere is given by

---

[1] As mentioned in §8.1, rearranging the order of summation in general is legitimate only for absolutely convergent series. Since we have imposed no restrictions on $R_{\ell m}$, this step is not necessarily legitimate in all cases. One may take a pragmatic view: derive the solution under some relatively stringent hypotheses and then check whether the validity of the result can be extended to other cases. Doetsch (1974, p. 74) calls this procedure the *principle of extension*.

$\sin\theta\, d\theta\, d\phi$ (Table 7.1 p. 171), explicitly, we have

$$\left(Y_\ell^m, Y_k^n\right) = \int_0^\pi \sin\theta\, d\theta \int_{-\pi}^\pi d\phi\, \overline{Y}_\ell^m\, Y_k^n \tag{14.2.4}$$

or, upon setting $\mu = \cos\theta$,

$$
\begin{aligned}
\left(Y_\ell^m, Y_k^n\right) &= \int_{-1}^1 d\mu \int_{-\pi}^\pi d\phi \left(\overline{N}_\ell^m\, P_\ell^m(\mu)\, e^{-im\phi}\right)\left(N_k^n\, P_k^n(\mu)\, e^{in\phi}\right) \\
&= 2\pi \overline{N}_\ell^m\, N_k^m\, \delta_{mn} \int_{-1}^1 d\mu\, P_\ell^m(\mu)\, P_k^m(\mu) \\
&= \delta_{mn}\, \delta_{\ell k}\, \left|N_\ell^m\right|^2 \frac{4\pi}{2\ell+1}\frac{(\ell+m)!}{(\ell-m)!},
\end{aligned}
\tag{14.2.5}
$$

where use has been made of (13.6.6) to express the last integral. Thus, in order to satisfy (14.2.3), we must take

$$N_\ell^m = \sqrt{\frac{2\ell+1}{4\pi}\frac{(\ell-m)!}{(\ell+m)!}} \tag{14.2.6}$$

so that (14.2.1) is

$$Y_\ell^m = \sqrt{\frac{2\ell+1}{4\pi}\frac{(\ell-m)!}{(\ell+m)!}}\; P_\ell^m(\cos\theta)\, e^{im\phi}. \tag{14.2.7}$$

Sometimes it is not useful to carry the normalization constants but, if they are dropped, one must be careful to remember that there is a multiplicative factor in the right-hand side of (14.2.3). Furthermore, as with other special functions, different normalizations are also encountered.

A relation which will useful in the following is

$$\int \left(\nabla\overline{Y}_\ell^m\right)\cdot\left(\nabla Y_p^q\right) d\Omega = \ell(\ell+1)\delta_{\ell p}\delta_{mq}, \tag{14.2.8}$$

where the gradient is evaluated on the unit sphere. This result follows from the identity

$$\left(r\nabla\overline{Y}_\ell^m\right)\cdot\left(r\nabla Y_p^q\right) = r\nabla\cdot\left(r\overline{Y}_\ell^m \nabla Y_p^q\right) - r^2 \overline{Y}_\ell^m \nabla^2 Y_p^q \tag{14.2.9}$$

evaluated at $r = 1$ after taking the derivatives. To prove it, it is sufficient to use (14.2.2) observing that, for any vector $\mathbf{y}(\theta, \phi)$, the integral of $\nabla\cdot\mathbf{y}$ on the unit sphere vanishes as is readily checked. Explicit expressions for the first few spherical harmonics with $m \geq 0$ are given on p. 174; expressions for negative m follow from the relation

$$Y_\ell^{-m} = (-1)^m\, \overline{Y}_\ell^m \tag{14.2.10}$$

which is a consequence of property (13.6.4) p. 329 of the P_ℓ^m.

We have seen in (13.3.17) a connection between the Legendre polynomials and multipoles along the z-axis. Not surprisingly, a similar connection exists between the spherical harmonics and multipoles with different axes; for $0 \leq m \leq \ell$ we have

$$\frac{\partial^{\ell-m}}{\partial z^{\ell-m}}\left(\frac{\partial}{\partial x} \pm i\frac{\partial}{\partial y}\right)^m \frac{1}{r} = \frac{(-1)^{\ell-m}(\ell-m)!}{r^{\ell+1}} P_\ell^m(\mu)\, e^{\pm im\phi}. \tag{14.2.11}$$

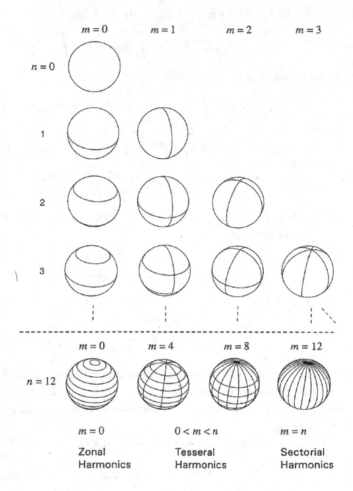

Fig. 14.2 Nodal lines, across which the function changes sign, for some low-degree (real) spherical harmonics (from Fornberg 1996, with permission).

When it is inconvenient to deal with complex quantities, one may use the equivalent set of functions

$$\hat{Y}_{\ell m} = M_\ell^m P_\ell^m (\cos\theta)\cos m\phi, \quad \tilde{Y}_{\ell m} = M_\ell^m P_\ell^m (\cos\theta)\sin m\phi \qquad (14.2.12)$$

with

$$M_\ell^0 = N_\ell^0, \qquad M_\ell^m = \sqrt{\frac{2\ell+1}{2\pi}\frac{(\ell-m)!}{(\ell+m)!}} \quad (m \neq 0). \qquad (14.2.13)$$

Clearly $\hat{Y}_{\ell 0} = Y_\ell^0$ while $\tilde{Y}_{\ell 0} = 0$.

The spherical harmonics can be represented in polar form, i.e., as a three-dimensional rendering of the functions

$$r = 1 + \varepsilon\, \hat{Y}_{\ell m} \qquad \text{and} \qquad r = 1 + \varepsilon\, \tilde{Y}_{\ell m} \tag{14.2.14}$$

for some relatively small value of ε. The nodal lines of such plots (i.e., the lines along which the $Y_{\ell m}$ vanish) are shown on the unit sphere for some low-degree (real) spherical harmonics in Figure 14.2. These diagrams justify the following terminology:

- *Zonal harmonics*, for $m = 0$. In this case there are ℓ nodal lines corresponding to the ℓ zeros of P_{ℓ}. These are therefore lines $\theta = $ constant, i.e. "parallels" on the unit sphere which divide it into "belts" or zones. The lines are symmetrically placed about the "equator," with the equator itself being one of them for odd ℓ.
- *Sectorial harmonics*, for $\ell = |m|$. In this case the P_{ℓ}^m are constants and the nodal lines are the "meridians" along which $\cos m\phi$ or $\sin m\phi$ vanish. The pattern is similar to the wedges of an orange.
- *Tesseral harmonics*, for all other values of m. In this case there are $\ell - m$ nodal lines, symmetrically placed about the equator, along which P_{ℓ}^m vanishes, and m great circles through the poles, i.e. meridians on which $\cos m\phi$ or $\sin m\phi$ vanish. The unit sphere is thus divided into curved "rectangles" like the tesserae of a mosaic.

The nodal lines and Cartesian plots of the functions $N_{\ell}^m P_{\ell}^m(\cos\theta) \cos m\phi$ for $\ell = 0, 4, 8$ and 12 are shown in Figure 14.3.

14.2.1 Solid harmonics

The functions $r^{\ell} Y_{\ell}^m$, or their real analogs $r^{\ell}\hat{Y}_{\ell m}$ and $r^{\ell}\tilde{Y}_{\ell m}$, are called *solid harmonics* of degree ℓ. Similarly, the functions $r^{-\ell-1} Y_{\ell}^m$ and their real analogs are solid harmonics of degree $-\ell - 1$. As shown in Chapter 7, these functions are solutions of the Laplace equation regular, respectively, at the origin and at infinity.

When expressed in terms of Cartesian coordinates, the $r^{\ell} Y_{\ell}^m$ are homogeneous polynomials of degree ℓ in x, y, z. For example, recalling the relation between Cartesian and polar coordinates (Table 7.1, p. 171) we find

$$r\, Y_1^0 = \sqrt{\frac{3}{4\pi}}\, z, \qquad r\, Y_1^{\pm 1} = -\sqrt{\frac{3}{8\pi}}\, (x \pm iy) \tag{14.2.15}$$

which are homogeneous polynomials of degree 1, and

$$r^2\, Y_2^0 = \sqrt{\frac{5}{4\pi}} \left[z^2 - \frac{1}{2}(x^2 + y^2) \right], \qquad r^2\, Y_2^{\pm 1} = -\sqrt{\frac{15}{8\pi}}\, z\,(x \pm iy), \tag{14.2.16}$$

$$r^2\, Y_2^{\pm 2} = \frac{1}{4}\sqrt{\frac{15}{2\pi}}\, (x \pm iy)^2 \tag{14.2.17}$$

which are homogeneous polynomials of degree 2. For each ℓ, there are only $2\ell + 1$ distinct such polynomials which satisfy the Laplace equation.

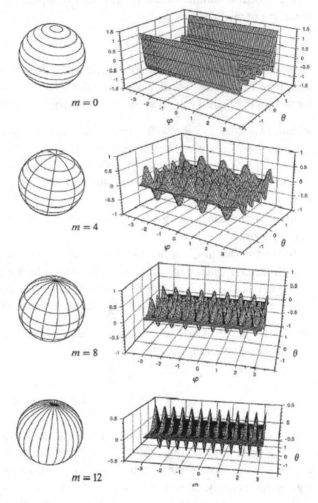

Fig. 14.3 Spherical harmonics $N_\ell^m P_\ell^m (\cos\theta) \cos m\phi$ for $\ell = 0$, 4, 8 and 12. The diagrams on the left show the nodal lines, across which the function changes sign, and the diagrams on the right are the numerical values in the $(\cos\theta, \phi)$ plane (from Fornberg 1996, with permission).

14.2.2 Addition theorem

We saw in §13.6 the addition theorem for the associated Legendre functions of the first kind. That relation can be written compactly in terms of the spherical harmonics as

$$P_\ell(\cos\gamma) = \frac{4\pi}{2\ell + 1} \sum_{m=-\ell}^{\ell} \overline{Y}_\ell^m(\theta', \phi') Y_\ell^m(\theta, \phi), \qquad (14.2.18)$$

where the angle γ, illustrated in Figure 7.2 p. 178, is defined by

$$\cos \gamma = \frac{\mathbf{x} \cdot \mathbf{x}'}{|\mathbf{x}| \, |\mathbf{x}'|} = \cos \theta \, \cos \theta' + \sin \theta \, \sin \theta' \cos(\phi' - \phi). \tag{14.2.19}$$

In particular, if $\theta = \theta'$ and $\phi = \phi'$, $\gamma = 0$ and, since $P_\ell(1) = 1$, we have the "sum rule"

$$\sum_{m=-\ell}^{\ell} |Y_\ell^m(\theta, \phi)|^2 = \frac{2\ell + 1}{4\pi} \tag{14.2.20}$$

valid irrespective of the values of θ and ϕ.

14.3 Expansion in series of spherical harmonics

As was done in §13.4, by virtue of the orthogonality of the spherical harmonics, we can formally associate to a function $f(\theta, \phi)$ defined for $0 \leq \theta \leq \pi$ and 2π-periodic in ϕ the series

$$f(\theta, \phi) \sim \sum_{\ell=0}^{\infty} \sum_{m=-\ell}^{\ell} f_{\ell m} Y_\ell^m \tag{14.3.1}$$

the coefficients of which are given by

$$f_{\ell m} = (Y_\ell^m, f) = \int_\Omega \overline{Y}_\ell^m f \, d\Omega = \int_0^\pi d\theta \, \sin \theta \int_{-\pi}^\pi d\phi \, \overline{Y}_\ell^m(\theta, \phi) \, f(\theta, \phi). \tag{14.3.2}$$

The study of the convergence of the *Laplace series* (14.3.1) can be reduced to that of the Legendre series considered in §13.4 by making a change of variables in which the point (θ, ϕ) on the unit sphere is taken as the new origin (see e.g. Hobson 1931 p. 342). This point and the one diametrically opposite to it play a role analogous to that of the points $\mu = \pm 1$ in the theory of the Legendre series, with corresponding restrictions on the behavior of f. Without entering into details, it can be stated that, under suitable conditions on f similar to those mentioned in connection with the Legendre series (and usually met in practice), the series converges to $f(\theta, \phi)$ at points of continuity, or to $\frac{1}{2}[f_1(\theta, \phi) + f_2(\theta, \phi)]$ if the point (θ, ϕ) is such that, through it, passes a line of discontinuity with a continually turning tangent such that f_1 and f_2 are the limits of f at (θ, ϕ) taken from the two sides of the line. The conditions for the applicability of this result are satisfied, in particular, if $f(\theta, \phi)$, expressed as a function of the angle γ defined in (14.2.19) and of $\overline{\phi} = \phi - \phi'$, is of bounded variation in the interval $0 \leq \gamma \leq \pi$ for each value of $\overline{\phi}$, and such that the total variation in such interval is bounded for all values of $\overline{\phi}$. Under conditions which extend those mentioned in §13.4 for the Legendre series (in particular, continuity and bounded variation), convergence of the Laplace series is uniform.

When summation and integration can be interchanged in the Laplace series, we have

$$f(\theta, \phi) = \int_0^\pi \sin \theta \, d\theta' \int_{-\pi}^\pi d\phi' \left[\sum_{\ell=0}^{\infty} \sum_{m=-\ell}^{\ell} \overline{Y}_\ell^m(\theta', \phi') \, Y_\ell^m(\theta, \phi) \right] f(\theta', \phi') \tag{14.3.3}$$

from which

$$\sin\theta \sum_{\ell=0}^{\infty} \sum_{m=-\ell}^{\ell} \overline{Y}_\ell^m(\theta',\phi') Y_\ell^m(\theta,\phi) = \delta(\theta-\theta')\delta(\phi-\phi') \tag{14.3.4}$$

or, since $\delta(\cos\theta - \cos\theta') = \delta(\theta-\theta')/\sin\theta = \delta(\theta-\theta')/\sin\theta'$ (see the table 2.1 p. 37 of properties of the δ, next to the last line)

$$\sum_{\ell=0}^{\infty} \sum_{m=-\ell}^{\ell} \overline{Y}_\ell^m(\theta',\phi') Y_\ell^m(\theta,\phi) = \delta(\cos\theta - \cos\theta')\delta(\phi-\phi'). \tag{14.3.5}$$

These relations express the completeness of the system of spherical harmonics.

By using (13.3.14) for the generating function of the Legendre polynomials and proceeding similarly to the derivation of (7.8.7) on p. 185, we can show that

$$\frac{1-t^2}{(1-2t\cos\gamma+t^2)^{3/2}} = \sum_{\ell=0}^{\infty} (2\ell+1) t^\ell P_\ell(\cos\gamma), \tag{14.3.6}$$

where convergence is uniform for $|t| < 1$. With $\cos\gamma$ defined as in (14.2.19), the series can therefore be multiplied by $f(\theta',\phi')$ and integrated term by term over the surface of the unit sphere to give, after using the addition theorem (14.2.20),

$$\sum_{\ell=0}^{\infty} t^\ell \sum_{m=-\ell}^{\ell} f_{\ell m} Y_\ell^m(\theta,\phi) = \frac{1}{4\pi} \int \frac{f(\theta',\phi')}{\sqrt{1-2t\cos\gamma+t^2}} d\Omega' \tag{14.3.7}$$

with the $f_{\ell m}$ as given in (14.3.2). By definition, the limit of the left-hand side as $t \to 1$ is the Abel–Poisson sum (§8.5) of the Laplace series (14.3.1). It can be proven that the Laplace series summed in this way converges to f at any point of continuity, while the behavior is as mentioned before at points of discontinuity. Over any closed set of points of continuity convergence of the Abel-Poisson sum is uniform (see e.g. Hobson 1931, p. 347). We have thus found the relation (see (7.8.8) p. 185)

$$\lim_{t\to1-} \frac{1}{4\pi} \int \frac{1-t^2}{(1-2t\cos\gamma+t^2)^{3/2}} d\Omega' = \delta(\cos\theta - \cos\theta')\delta(\phi-\phi'). \tag{14.3.8}$$

We can now return to the expansion (14.1.4), which we rewrite as

$$u(r,\theta,\phi) = \sum_{\ell=0}^{\infty} \sum_{m=-\ell}^{\ell} R_{\ell m}(r) Y_\ell^m(\theta,\phi). \tag{14.3.9}$$

The functions $R_{\ell m}(r)$ are related to u by (14.3.2) which becomes

$$R_{\ell m}(r) = \left(Y_\ell^m, u\right) = \int_0^\pi d\theta \, \sin\theta \int_{-\pi}^\pi d\phi \, \overline{Y}_\ell^m(\theta,\phi) u(r,\theta,\phi). \tag{14.3.10}$$

In terms of the differential operator

$$\mathscr{L}^2 = -\frac{1}{\sin\theta} \frac{\partial}{\partial\theta}\left(\sin\theta \frac{\partial}{\partial\theta}\right) - \frac{1}{\sin^2\theta} \frac{\partial^2}{\partial\varphi^2}, \tag{14.3.11}$$

the Laplacian operator in spherical coordinates can be written as

$$\nabla^2 = \frac{1}{r^2}\frac{\partial}{\partial r}\left(r^2\frac{\partial}{\partial r}\right) - \frac{\mathscr{L}^2}{r^2},$$ (14.3.12)

Since, by construction, the Y_ℓ^m are eigenfunctions of $\mathscr{L}^2$ (see (14.2.2))

$$\mathscr{L}^2\,Y_\ell^m \;=\; \ell\,(\ell+1)\,Y_\ell^m,$$ (14.3.13)

upon substituting the expansion (14.3.9) into $\nabla^2 u$ and assuming that summation and differentiation can be interchanged, we have

$$\nabla^2 u = \sum_{\ell=0}^{\infty}\sum_{m=-\ell}^{\ell}\left[\frac{1}{r^2}\frac{\mathrm{d}}{\mathrm{d}r}\left(r^2\frac{\mathrm{d}R_{\ell m}}{\mathrm{d}r}\right) - \frac{\ell(\ell+1)}{r^2}R_{\ell m}\right]Y_\ell^m\,(\theta,\phi).$$ (14.3.14)

14.4 Vector harmonics

Just as scalar spherical harmonics allow us to represent the angular dependence of scalar fields, *vector harmonics* permit a representation of vector fields. Details on the underlying theory can be found e.g. in Morse & Feshbach 1953, p. 1898, Jackson 1998, chapter 16 and Chandrasekhar 1961, pp. 225 and 622. Here we present a summary of the main results.

A complete set of vectors on the unit sphere is given by linear combinations of the three families

$$\mathbf{P}_\ell^m\,(\theta,\phi) \;=\; Y_\ell^m(\theta,\phi)\,\mathbf{e}_r,$$ (14.4.1)

$$\begin{aligned}\mathbf{B}_\ell^m\,(\theta,\phi) &= \frac{r}{\sqrt{\ell(\ell+1)}}\nabla Y_\ell^m(\theta,\phi) = \mathbf{e}_r \times \mathbf{C}_\ell^m(\theta,\phi)\\[2mm] &= \frac{1}{\sqrt{\ell(\ell+1)}}\left[\frac{\partial Y_\ell^m}{\partial\theta}\mathbf{e}_\theta + \frac{1}{\sin\theta}\frac{\partial Y_\ell^m}{\partial\phi}\mathbf{e}_\phi\right],\end{aligned}$$ (14.4.2)

$$\begin{aligned}\mathbf{C}_\ell^m\,(\theta,\phi) &= \frac{1}{\sqrt{\ell(\ell+1)}}\nabla\times\left(Y_\ell^m\mathbf{x}\right) = \frac{1}{\sqrt{\ell(\ell+1)}}\left(\nabla Y_\ell^m\right)\times\mathbf{x} = -\mathbf{e}_r\times\mathbf{B}_\ell^m\\[2mm] &= \frac{1}{\sqrt{\ell(\ell+1)}}\left[\frac{1}{\sin\theta}\frac{\partial Y_\ell^m}{\partial\phi}\mathbf{e}_\theta - \frac{\partial Y_\ell^m}{\partial\theta}\mathbf{e}_\phi\right].\end{aligned}$$ (14.4.3)

Thus, any vector field $\mathbf{V}$ can be expanded in the form

$$\mathbf{V}(r,\theta,\phi) = \sum_{\ell=0}^{\infty}\sum_{m=-\ell}^{\ell}\left[p_{\ell m}(r)\mathbf{P}_\ell^m + b_{\ell m}(r)\mathbf{B}_\ell^m + c_{\ell m}(r)\mathbf{C}_\ell^m\right].$$ (14.4.4)

It is evident that

$$\overline{\mathbf{P}}_\ell^m\cdot\mathbf{B}_\ell^m \;=\; \overline{\mathbf{P}}_\ell^m\cdot\mathbf{C}_\ell^m \;=\; \overline{\mathbf{B}}_\ell^m\cdot\mathbf{C}_\ell^m \;=\; 0,$$ (14.4.5)

while

$$\overline{\mathbf{C}}_\ell^m\cdot\mathbf{C}_p^q \;=\; \overline{\mathbf{B}}_\ell^m\cdot\mathbf{B}_p^q.$$ (14.4.6)

Integral orthogonality relations readily follow from (14.2.3) and (14.2.8)

$$\int \overline{\mathbf{P}}_\ell^m \cdot \mathbf{P}_p^q \, d\Omega = \int \overline{\mathbf{B}}_\ell^m \cdot \mathbf{B}_p^q \, d\Omega = \int \overline{\mathbf{C}}_\ell^m \cdot \mathbf{C}_p^q \, d\Omega = \delta_{\ell p} \delta_{mq}. \qquad (14.4.7)$$

These and several other properties of the vector harmonics are summarized in Table 7.10 p. 203. By using (14.4.1) to (14.4.3), the expansion (14.4.4) can be written in terms of $\mathbf{e}_r$ and the vectors $\mathbf{B}_\ell^m$ only as

$$\mathbf{V}(r, \theta, \phi) = \sum_{\ell=0}^{\infty} \sum_{m=-\ell}^{\ell} \left[\frac{p_{\ell m}(r)}{r} \mathbf{e}_r + b_{\ell m}(r) \mathbf{B}_\ell^m - c_{\ell m}(r) \mathbf{e}_r \times \mathbf{B}_\ell^m \right] \qquad (14.4.8)$$

or of the vectors $\mathbf{C}_\ell^m$ only as

$$\mathbf{V}(r, \theta, \phi) = \sum_{\ell=0}^{\infty} \sum_{m=-\ell}^{\ell} \left[c_{\ell m} \mathbf{C}_\ell^m + \nabla \times \left(\frac{r p_{\ell m}}{\sqrt{\ell(\ell+1)}} \mathbf{C}_\ell^m \right) \right.$$
$$\left. - \left(\frac{1}{r\sqrt{\ell(\ell+1)}} \frac{d}{dr} \left(r^2 p_{\ell m} \right) - b_{\ell m} \right) \mathbf{e}_r \times \mathbf{C}_\ell^m \right]. \qquad (14.4.9)$$

14.4.1 Solenoidal and irrotational vector fields

The divergence of $\mathbf{V}$ is

$$\nabla \cdot \mathbf{V} = \sum_{\ell=0}^{\infty} \sum_{m=-\ell}^{\ell} \frac{\sqrt{\ell(\ell+1)}}{r} \left[\frac{1}{r\sqrt{\ell(\ell+1)}} \frac{d}{dr} \left(r^2 p_{\ell m} \right) - b_{\ell m} \right] Y_\ell^m \qquad (14.4.10)$$

and, therefore, comparing with (14.4.9), we see that a divergenceless field $\mathbf{V}_D$ can be represented as

$$\mathbf{V}_D(r, \theta, \phi) = \sum_{\ell=0}^{\infty} \sum_{m=-\ell}^{\ell} \left[c_{\ell m} \mathbf{C}_\ell^m + \nabla \times \left(\frac{r p_{\ell m}}{\sqrt{\ell(\ell+1)}} \mathbf{C}_\ell^m \right) \right], \qquad (14.4.11)$$

i.e., in the toroidal–poloidal form $\mathbf{V}_D = \nabla \times (\psi \mathbf{x}) + \nabla \times \nabla \times (\chi \mathbf{x})$ of (1.1.15) p. 6 with defining scalars given by

$$\psi = \sum_{\ell=0}^{\infty} \sum_{m=-\ell}^{\ell} \frac{c_{\ell m}(r) Y_\ell^m}{\sqrt{\ell(\ell+1)}}, \qquad \chi = \sum_{\ell=0}^{\infty} \sum_{m=-\ell}^{\ell} \frac{r p_{\ell m}(r) Y_\ell^m}{\ell(\ell+1)}. \qquad (14.4.12)$$

The curl of $\mathbf{V}$ may be written as

$$\nabla \times \mathbf{V} = \sum_{\ell=0}^{\infty} \sum_{m=-\ell}^{\ell} \left[\frac{1}{r} \left(\frac{p_{\ell m}(r)}{\sqrt{\ell(\ell+1)}} - \frac{d}{dr} (r b_{\ell m}(r)) \right) \mathbf{C}_\ell^m + \nabla \times \left(c_{\ell m}(r) \mathbf{C}_\ell^m \right) \right], \qquad (14.4.13)$$

so that the conditions of irrotationality are

$$\frac{p_{\ell m}(r)}{\sqrt{\ell(\ell+1)}} - \frac{d}{dr} (r b_{\ell m}(r)) = 0, \qquad c_{\ell m}(r) = 0. \qquad (14.4.14)$$

When these relations are satisfied, an irrotational field $\mathbf{V}_I$ admits a potential $\tilde{\varphi}$, $\mathbf{V}_I = \boldsymbol{\nabla}\tilde{\varphi}$, with

$$\tilde{\varphi} = \sum_{\ell=0}^{\infty} \sum_{m=-\ell}^{\ell} \frac{r b_{\ell m}(r) Y_\ell^m}{\sqrt{\ell(\ell+1)}}. \tag{14.4.15}$$

For a general vector field, however, this is not the full longitudinal component in the sense of the decomposition $\mathbf{V} = \boldsymbol{\nabla}\varphi + \boldsymbol{\nabla}\times\mathbf{A}$ mentioned on p. 5. If we write $\varphi = \hat{\varphi} + \tilde{\varphi}$, the missing part is a particular solution of

$$\nabla^2\hat{\varphi} = \sum_{\ell=0}^{\infty} \sum_{m=-\ell}^{\ell} \frac{1}{r^2} \frac{d}{dr} \left(r^2 p_{\ell m} - \frac{r^2}{\ell(\ell+1)} \frac{d}{dr}(r b_{\ell m}) \right) Y_\ell^m \tag{14.4.16}$$

which vanishes when $\boldsymbol{\nabla}\cdot\mathbf{V} = 0$ as expected.

15 Green's Functions: Ordinary Differential Equations

Green's functions permit us to express the solution of a non-homogeneous linear problem in terms of an integral operator of which they are the kernel. We have already presented in simple terms this idea in §2.4. We now give a more detailed theory with applications mainly to ordinary differential equations. The next chapter deals with Green's functions for partial differential equations.

The determination of a Green's function requires the solution of a problem similar to (although somewhat simpler than) the original one, but the effort required is balanced by several advantages. In the first place, and at the most superficial level, once the Green's function G is known, it is unnecessary to solve the problem *ex novo* for every new set of data: it is sufficient to allow G to act on the new data to have the solution directly. Secondly, and most importantly for our purposes, Green's function theory provides a foundation for the various eigenfunction expansion and integral transform methods used in Part I of this book. Thirdly, even if the Green's function cannot be determined explicitly, one can base on it the powerful boundary integral numerical method outlined in §16.1.3 of the next chapter. Furthermore, once an expression for the solution of a problem – even if not fully explicit – is available, it becomes possible to deduce several important features of it, including bounds existence, uniqueness and others.

Most of this chapter is devoted to Green's function theory for two-point boundary value problems of the so-called Sturm–Liouville type. We show the intimate relation of this case with eigenfunction expansions and integral transforms. In this sense, the material in §15.3 and §15.4 lies at the core of much of our work. The same topics can also be presented in the context of abstract operator theory, which we do to some extent in Chapter 21. The distance between theory and applications is however made somewhat smaller by adopting the less general view presented here.

In §15.5 we describe Green's function theory for the initial-value problem of ordinary differential equations and, in the final two sections, present a more general view of the subject.

15.1 Two-point boundary value problems

On an interval $a < x < b$ we consider the second-order equation

$$\frac{d^2 u}{dx^2} + m(x)\frac{du}{dx} + n(x)u(x) = F(x) \tag{15.1.1}$$

with two-point boundary conditions which will be specified later. The transformation

$$p(x) \propto \exp\left(\int m(x)\,dx\right), \quad q(x) = -p(x)n(x), \quad f(x) = -p(x)F(x), \quad (15.1.2)$$

brings the equation to a form in which the operator is of the standard *Sturm–Liouville* type

$$\mathsf{L}u \equiv -\frac{\mathrm{d}}{\mathrm{d}x}\left[p(x)\,\frac{\mathrm{d}u}{\mathrm{d}x}\right] + q(x)\,u(x) = f(x). \quad (15.1.3)$$

We will develop the theory of the Green's function for two-point boundary value problems of this form rather than the original (15.1.1) because then, as is readily checked,

$$\bar{v}(\mathsf{L}u) - (\overline{\mathsf{L}v})u = \frac{\mathrm{d}}{\mathrm{d}x}\left\{p(x)\left[\bar{v}'u - \bar{v}u'\right]\right\} \quad (15.1.4)$$

provided p and q are real; here and in the following primes denote differentiation. The developments that follow will show the significance of this relation, known as the *Lagrange identity*. By a simple integration of both sides, it leads to the important relation[1]

$$(v, \mathsf{L}u) \equiv \int_a^b \bar{v}\,\mathsf{L}u\,dx = \int_a^b (\overline{\mathsf{L}v})\,u\,dx + J[\bar{v}, u] = (\mathsf{L}v, u) + J[\bar{v}, u]. \quad (15.1.5)$$

The last term

$$J[\bar{v}, u] = \left[p(\bar{v}'u - \bar{v}u')\right]_a^b \quad (15.1.6)$$

is the *concomitant* of u and v and the notation $[\ldots]_a^b$ indicates the difference between the values of the enclosed quantity at $x = b$ and at $x = a$. As will be seen in §15.6.1 and §21.6.1, the fact that the same operator appears in both sides of (15.1.5) makes L symmetric.

We assume that the given functions $p(x)$, $p'(x)$ and $q(x)$ are real and continuous. A point of $[a, b]$ where $p = 0$ would be a singular point for the equation.[2] We assume for the time being that no such point exists. Thus, p does not change sign in $[a, b]$ and therefore, by changing the sign of q and f if necessary, we can assume that $p > 0$. If $[a, b]$ is finite and the assumptions on the coefficients hold in the closed interval (i.e., including the end points), the problem is said to be *regular*, otherwise it is *singular*. In this section we treat the regular case; §15.4 is devoted to singular problems. In order to make sure that the operator L is well defined, we will also assume that u is twice continuously differentiable.[3]

Alongside the original problem (15.1.3) we consider the auxiliary problem

$$-\frac{\partial}{\partial x}\left[p(x)\,\frac{\partial G}{\partial x}\right] + q(x)\,G = \delta(x - y), \quad (15.1.7)$$

---

[1] The same result can also be found by multiplying (15.1.3) by $\bar{v}$ and integrating by parts twice.

[2] As noted in §2.5 p. 42, the solutions of a linear ordinary differential equation written so that the highest derivative has coefficient 1 can have singularities only at the singularities of the coefficients (see e.g. Coddington & Levinson 1955, p. 111; Bender & Orszag 2005, p. 146).

[3] These assumptions can be relaxed somewhat. For example, one could take q measurable and bounded, p absolutely continuous and u such that both u and $p'u$ are absolutely continuous, with $(p'u)' \in L^2(a, b)$.

where $G = G(x, y)$ is the Green's function for the problem (15.1.3) and $a < y < b$. By replacing v by G in (15.1.5) and using the property of the δ to effect the integration in the right-hand side we find

$$u(y) = \int_a^b G(x, y) f(x)\,dx - J[G, u], \tag{15.1.8}$$

where we have used the relation $Lu = f$ and have dispensed with the complex conjugation as everything is real with the hypotheses made. In spite of appearances, this is not yet a solution of the original problem as knowledge of the concomitant

$$J[G, u] = \left[p \left(\frac{\partial G}{\partial x} u - Gu' \right) \right]_{x=a}^{x=b} \tag{15.1.9}$$

would require specification of u and u' at both $x = a$ and $x = b$, namely four pieces of data rather than the two allowed for second-order equations. The way out of this difficulty is to impose on G boundary conditions which make the unknown data unnecessary. This is a general procedure consistently following in this and the next chapter. To see how to proceed, we need to solve the equation (15.1.7) for G first.

15.1.1 Solution for the Green's function

The presence of the δ in the right-hand side of (15.1.7) effectively breaks the interval into two parts and therefore, while G satisfies the same homogeneous equation for $x \neq y$, its form for $x < y$ is not necessarily equal to that for $x > y$. Thus, we write[4]

$$G(x, y) = \lambda_b(y) w_-(x) H(x - y) + \lambda_a(y) w_+(x) H(y - x), \tag{15.1.10}$$

where $w_\pm$ are two *linearly independent general solutions* of the homogeneous equation to be determined in the course of the analysis and H is Heaviside's step function; $\lambda_{a,b}(y)$ are constants of integration with respect to x which may, however, depend on y. Upon substitution of (15.1.10) into (15.1.7), since $w_\pm$ satisfy the homogeneous equation, what is left is

$$p(y) \left[\lambda_a w_+(y) - \lambda_b w_-(y) \right] \delta'(x - y)$$
$$+ p(y) \left[\lambda_a w'_+(y) - \lambda_b w'_-(y) \right] \delta(x - y) = \delta(x - y), \tag{15.1.11}$$

where we have used the property that $f(x)\delta(x - c) = f(c)\delta(x - c)$. Since there is no δ' in the right-hand side to balance that in the left-hand side (see (20.13.11) p. 610), we must require that

$$\lambda_a w_+(y) - \lambda_b w_-(y) = 0 \tag{15.1.12}$$

i.e., that the Green's function be continuous at $x = y$,

$$G(x = y - 0, y) = G(x = y + 0, y), \tag{15.1.13}$$

where $x = y \pm 0$ denote the limits $x \to y$ from the right or left, respectively.

---

[4] Since $H(x - y) + H(y - x) = 1$, one could equivalently start by writing $G(x, y) = \mu_1(y) w_1(x) + \mu_2(y) H(y - x) w_2(x)$, with $w_{1,2}$ suitable solutions of the homogeneous equation; the form (15.1.10) however is preferable due to its symmetry.

Balancing the other terms then leads us to

$$\lambda_a w'_+(y) - \lambda_b w'_-(y) = \frac{1}{p(y)}. \tag{15.1.14}$$

Once it is known that G is continuous, this result can be obtained directly by integrating (15.1.7) between $x = y - 0$ and $x = y + 0$:

$$-p(y) \left[\frac{\partial G}{\partial x} \right]_{x=y-0}^{x=y+0} = 1 \tag{15.1.15}$$

which, by (15.1.10), is (15.1.14). Equations (15.1.12) and (15.1.14) can be solved for $\lambda_{a,b}$ to find

$$\lambda_a = \frac{w_-(y)}{p(y)\,W(y; w_-, w_+)}, \qquad \lambda_b = \frac{w_+(y)}{p(y)\,W(y; w_-, w_+)}, \tag{15.1.16}$$

in which $W(y; w_-, w_+) = w_-(y)w'_+(y) - w_+(y)w'_-(y)$ is the Wronskian of the two solutions $w_\pm$ evaluated at y. For this Wronskian we use Abel's formula (2.2.17) p. 32 and find

$$W(y; w_-, w_+) = W_0 \exp\left(-\int \frac{p'}{p}\, dy \right) = \frac{W_0}{p(y)}, \tag{15.1.17}$$

where the constant W_0 is the only quantity depending on the specific solutions $w_\pm$ selected for the homogeneous equation. Thus (15.1.16) become

$$\lambda_a = \frac{w_-(y)}{W_0}, \qquad \lambda_b = \frac{w_+(y)}{W_0} \tag{15.1.18}$$

and we find

$$G(x, y) = \frac{1}{W_0} \left[w_+(y)w_-(x)H(x - y) + w_-(y)w_+(x)H(y - x) \right]. \tag{15.1.19}$$

This result may be written in a more compact form by noting that, in both terms, the function w_- has as its argument the greater one of x and y, while the function w_+ has the smaller one. Therefore, with

$$x_> = \max(x, y), \qquad x_< = \min(x, y), \tag{15.1.20}$$

we may write

$$G(x, y) = \frac{w_-(x_>)w_+(x_<)}{W_0} = G(y, x), \tag{15.1.21}$$

where the last step follows from the fact that max and min are symmetric functions of their arguments. Thus the Green's function is symmetric in its two variables; we will see in §15.6 that this feature is a direct consequence of the fact that that the operator L is symmetric.

Upon substituting this result into (15.1.8), we see that the solution of the original problem may be written as the sum of two terms:

$$u(y) = u_p(y) + u_h(y). \tag{15.1.22}$$

The first term is the particular solution

$$
\begin{aligned}
u_p(y) &= \int_a^b G(x, y) f(x) \, dx \\
&= \frac{w_-(y)}{W_0} \int_a^y w_+(x) f(x) \, dx + \frac{w_+(y)}{W_0} \int_y^b w_-(x) f(x) \, dx, \quad (15.1.23)
\end{aligned}
$$

which will be seen to satisfy homogeneous boundary conditions. The second term is a solution of the homogeneous equation:

$$
\begin{aligned}
u_h(y) = \; & p(a) \left[u(a) \left. \frac{\partial G}{\partial x} \right|_{x=a} - u'(a) G(a, y) \right] \\
& - p(b) \left[u(b) \left. \frac{\partial G}{\partial x} \right|_{x=b} - u'(b) G(b, y) \right] \quad (15.1.24)
\end{aligned}
$$

and carries the boundary conditions.

A full specification of G requires the imposition of boundary conditions on the two functions $w_\pm$. As mentioned before, these conditions will be chosen in such a way that the concomitant $J[G, u]$ given by (15.1.9) does not include unknown boundary information on u.

15.1.2 Separated boundary conditions

We consider boundary conditions of the form

$$
a_{11} u(a) + a_{12} u'(a) = U_a, \qquad a_{21} u(b) + a_{22} u'(b) = U_b, \quad (15.1.25)
$$

where the constants a_{ij} satisfy $a_{11}^2 + a_{12}^2 \neq 0$ and $a_{21}^2 + a_{22}^2 \neq 0$ so that both members of the same pair do not vanish at the same time. While fairly general, these conditions are of the separated (or unmixed) type as they do not involve simultaneously the values of u at *both* end points. We will consider mixed boundary conditions in §15.1.4.

The contribution of the point a to the concomitant (15.1.9) is proportional to

$$
u(a) \left. \frac{\partial G}{\partial x} \right|_{x=a} - u'(a) G(a, y). \quad (15.1.26)
$$

Suppose that $a_{12} \neq 0$, and multiply this relation by a_{12}. Upon using the first one of (15.1.25) we then find

$$
\begin{aligned}
a_{12} \left(u(a) \left. \frac{\partial G}{\partial x} \right|_{x=a} - u'(a) G(a, y) \right) = \; & -U_a G(a, y) \\
& + \left(a_{11} G(a, y) + a_{12} \left. \frac{\partial G}{\partial x} \right|_{x=a} \right) u(a). \quad (15.1.27)
\end{aligned}
$$

Here $u(a)$ is unknown, but we can cause its coefficient to vanish by imposing that

$$
a_{11} G(a, y) + a_{12} \left. \frac{\partial G}{\partial x} \right|_{x=a} = 0, \quad (15.1.28)
$$

which accomplishes the desired objective also when $a_{12} = 0$ as is readily seen. Similarly, at $x = b$ we impose

$$a_{21} G(b, y) + a_{22} \left. \frac{\partial G}{\partial x} \right|_{x=b} = 0, \tag{15.1.29}$$

so that G satisfies the homogeneous form of the boundary conditions (15.1.25) on u. In view of the symmetry of G, the same boundary conditions (15.1.28) and (15.1.29) will be satisfied by $G(x, a)$ and $G(x, b)$, so that the particular solution (15.1.23) satisfies homogeneous boundary conditions at the end points as already anticipated. We can verify this statement directly by using the fact that, from the expression (15.1.21) of G, the boundary conditions (15.1.28) and (15.1.29) require

$$a_{11} w_+(a) + a_{12} w'_+(a) = 0, \qquad a_{21} w_-(b) + a_{22} w'_-(b) = 0. \tag{15.1.30}$$

A convenient way to satisfy these conditions is to choose[5]

$$w_+(a) = -a_{12}, \qquad w'_+(a) = a_{11}, \tag{15.1.31}$$

as then the first bracket in (15.1.24) is proportional to $a_{11} u(a) + a_{12} u'(a) = U_a$. Similarly, upon choosing

$$w_-(b) = a_{22}, \qquad w'_-(b) = -a_{21}, \tag{15.1.32}$$

the second bracket in (15.1.24) becomes proportional to $a_{21} u(b) + a_{22} u'(b) = -U_b$ and

$$u_h(y) = \frac{p(a)}{W_0} U_a \, w_-(y) + \frac{p(b)}{W_0} U_b \, w_+(y). \tag{15.1.33}$$

This expression satisfies the boundary conditions (15.1.25). For example, at $x = a$ the contribution of the second term vanishes due to (15.1.32) and the first one reduces to U_a due to (15.1.31). In this way, the solution of the problem (15.1.3) is completely determined. The procedure is summarized in Table 15.1. A point worthy of note is that, unlike (15.1.25), both pairs of conditions (15.1.31) and (15.1.32) apply at the *same* point.

Example 15.1.1 Consider the one-dimensional Helmholtz equation

$$-u'' - k^2 u = f(x), \qquad 0 < x < b, \tag{15.1.34}$$

subject to homogeneous boundary conditions $u(0) = 0$, $u(b) = 0$, so that $a_{11} = a_{21} = 1$, $a_{12} = a_{22} = 0$. By (15.1.31) and (15.1.32), the solution w_+ of the homogeneous equation should vanish at $x = 0$ and the solution w_- at $x = b$ while the derivatives at these points should equal 1 and -1 respectively; thus

$$w_+ = \frac{1}{k} \sin kx, \qquad w_- = \frac{1}{k} \sin k(b - x), \tag{15.1.35}$$

---

[5] The general theory for the existence of solutions of the initial-value problem for ordinary differential equations guarantees that such a solution of the homogeneous equation exists when $p(a) \neq 0$ as assumed here (see e.g. Coddington & Levinson 1955, p. 15).

Table 15.1

Construction of the Green's function for a regular two-point boundary value problem; a general differential equation can be reduced to the form in the first line by effecting the transformation (15.1.2).

$$-\frac{d}{dx}\left[p(x)\frac{du}{dx}\right]+q(x)u(x) = f(x) \qquad -\frac{d}{dx}\left[p(x)\frac{dw_{\pm}}{dx}\right]+q(x)w_{\pm} = 0$$

$$a_{11}u(a)+a_{12}u'(a) = U_a \qquad\qquad w_+(a) = -a_{12}, \quad w'_+(a) = a_{11}$$

$$a_{21}u(b)+a_{22}u'(b) = U_b \qquad\qquad w_-(b) = a_{22}, \quad w'_-(b) = -a_{21}$$

$$u(y) = u_p(y)+u_h(y) \qquad\qquad G(x,y) = \frac{w_-(x_>)w_+(x_<)}{W_0}$$

$$u_p(y) = \int_a^b G(x,y)f(x)\,dx \qquad x_> = \max(x,y), \quad x_< = \min(x,y)$$

$$u_h = \frac{p(a)}{W_0}U_a\,w_-(y)+\frac{p(b)}{W_0}U_b\,w_+(y) \qquad W_0 = p\left[w_-w'_+ - w_+w'_-\right]$$

with which $W = (\sin kb)/k$ is a constant (and therefore equal to W_0) as expected, since $p = 1$. We thus find

$$G(x,y) = \frac{\sin k(b-x_>)\sin kx_<}{k\sin kb}. \qquad (15.1.36)$$

Note for future reference (§15.7) that this expression breaks down if $kb = n\pi$ as then $\sin kb = 0$. In this case, if the equation (15.1.7) for G is multiplied by $\sin kx$ and integrated between 0 and b one finds

$$G(0,y) - (-1)^n G(b,y) = \frac{1}{k}\sin ky, \qquad (15.1.37)$$

so that G cannot vanish at both end points. It will be shown in §15.7 that, in this case, there is no solution unless

$$\int_0^b f(x)\sin kx\,dx = 0. \qquad (15.1.38)$$

To deal with this situation we need to use a modified Green's function as described in §15.7.

□

Example 15.1.2 Consider now

$$-\frac{d^2u}{dx^2} - \frac{1}{x}\frac{du}{dx} = F(x) \qquad (15.1.39)$$

in $a < x < b$, subject to $u(a) = U_a$, $u'(b) = U_b$. With the transformation (15.1.2) we can rewrite this equation as

$$-\frac{d}{dx}\left(x\frac{du}{dx}\right) = xF(x) = f(x). \qquad (15.1.40)$$

Comparing with (15.1.25) we see that $a_{11} = a_{22} = 1$, $a_{12} = a_{21} = 0$. The functions $w_\pm$ satisfying (15.1.31) and (15.1.32) are

$$w_- = 1, \qquad w_+ = a\log(x/a), \qquad (15.1.41)$$

with which $G(x, y) = \log(x_</a)$ and

$$u(y) = U_a + bU_b\log(y/a) + \int_a^y \log(x/a)f(x)\,dx + \log(y/a)\int_y^b f(x)\,dx. \qquad (15.1.42)$$

Note for future reference that, if we try to solve the problem with $u'(a) = U_a$, $u'(b) = U_b$ a difficulty arises. Indeed, upon integrating (15.1.40) between a and b we find

$$aU_a - bU_b = \int_a^b f(x)\,dx \qquad (15.1.43)$$

and, unless this condition is satisfied, no solution can be found. But even if this condition is satisfied, the Green's function cannot be constructed according to the previous rules, as now we would require $\partial G/\partial x = 0$ at both end points, while integration of (15.1.7) would require that

$$a\frac{\partial G}{\partial x}\bigg|_{x=a} - b\frac{\partial G}{\partial x}\bigg|_{x=b} = 1. \qquad (15.1.44)$$

We address this problem in §15.7 dealing with modified Green's functions.   □

15.1.3 Properties of the Green's function for separated boundary conditions

The solution of the problem (15.1.3) with boundary conditions (15.1.25) constructed in the course of the previous developments is given by

$$u(x) = \int_a^b G(x, y)f(y)\,dy + \frac{p(a)}{W_0}U_a\, w_-(x) + \frac{p(b)}{W_0}U_b\, w_+(x), \qquad (15.1.45)$$

where we have interchanged the symbols x and y and used the symmetry of G in the first term. A direct calculation will show that this is precisely the solution that one would find by the method of variation of parameters of §2.2. The integral in (15.1.45) is readily interpreted by noting that, if $f(y) = f(y_0)\delta(y - y_0)$, its contribution to u would be $f(y_0)G(x, y_0)$. Thus, as already pointed out in §2.4, $G(x, y)$ represents the effect at x of a unit "load" at y and the integral is the superposition of loads each with a magnitude $f(y)$. Let us recapitulate the properties of the Green's function that emerged in the course of the analysis:

1. Since *max* and *min* are symmetric functions of their arguments, G given by (15.1.21) is symmetric:

$$G(x, y) = \frac{w_-(x_>)w_+(x_<)}{W_0} = G(y, x).$$ (15.1.46)

2. G is continuous at $x = y$.
3. Since

$$\frac{\partial G}{\partial x_>} = \frac{w'_-(x_>)w_+(x_<)}{W_0}, \qquad \frac{\partial G}{\partial x_<} = \frac{w_-(x_>)w'_+(x_<)}{W_0},$$ (15.1.47)

we have

$$\left[\frac{\partial G}{\partial x}\right]_{x=y-}^{x=y+} = -\frac{1}{p(y)}, \qquad \left[\frac{\partial G}{\partial y}\right]_{y=x-}^{y=x+} = -\frac{1}{p(x)},$$ (15.1.48)

in which the brackets are the jumps of the derivative across y and x respectively;
4. If we express the second derivative of G similarly to (15.1.47) we see that, since $w_\pm$ satisfy the homogeneous form of (15.1.3), so does G in both its variables away from the point $x = y$.
5. G satisfies homogeneous boundary conditions of the same form as those satisfied by u:[6]

$$a_{11}G(x, a) + a_{12}\left.\frac{\partial G}{\partial y}\right|_{x=a} = 0, \qquad a_{21}G(x, b) + a_{22}\left.\frac{\partial G}{\partial y}\right|_{x=b} = 0.$$ (15.1.49)

In view of the symmetry of G, the same boundary conditions are satisfied in the variable y, so that the particular solution (15.1.23) satisfies homogeneous conditions.

The construction of the Green's function requires a non-zero Wronskian of the two solutions $w_\pm$, i.e., a pair of linearly independent solutions of the homogeneous equation with the non-homogeneous boundary conditions (15.1.31) and (15.1.32):

$$w_+(a) = -a_{12}, \quad w'_+(a) = a_{11}, \quad w_-(b) = a_{22}, \quad w'_-(b) = -a_{21}.$$ (15.1.50)

Such a pair does not exist when the completely homogeneous problem

$$-\frac{d}{dx}\left(p\frac{dv}{dx}\right) + qv = 0,$$ (15.1.51)

$$a_{11}v(a) + a_{12}v'(a) = 0, \qquad a_{21}v(b) + a_{22}v'(b) = 0,$$ (15.1.52)

has a non-zero solution. For the present case of unmixed boundary conditions only one such solution can exist. We will prove these statements presently; for the time being, we note that we have reached the following conclusion:

Theorem 15.1.1 *If the completely homogeneous boundary value problem (15.1.51), (15.1.52) has only the trivial solution, then the inhomogeneous problem (15.1.3), (15.1.25) with f, U_a, U_b given, admits the solution (15.1.45), which is unique.*

---

[6] This fact is a consequence of the self-adjointness of the present operator L when operating on functions which satisfy the homogeneous form of the boundary conditions (15.1.25); see §21.6.1.

Uniqueness is a direct consequence of the fact that the difference between two solutions would satisfy the completely homogeneous problem and must therefore vanish by hypothesis.

Let us now show the difficulty that arises when the homogeneous problem (15.1.51), (15.1.52) has a non-trivial solution. We first show that only one such solution can exist. Indeed, any three solutions of a second-order equation must be connected by a linear relation. Since 0 is obviously a solution of the homogeneous problem, if two other solutions v_1 and v_2 existed it would be necessary that $c_1 v_1 + c_2 v_2 = 0$ for some non-zero c_1 and c_2, which would make the two solutions proportional to each other and therefore not independent.

So let us assume that there is one solution $v \neq 0$ of the completely homogeneous problem, and let w be another linearly independent solution satisfying some general non-homogeneous boundary conditions. Then we may write

$$w_+ = \alpha\, v + \beta\, w, \qquad w_- = \gamma\, v + \delta\, w. \tag{15.1.53}$$

Upon imposing the boundary condition (15.1.31) we then have

$$a_{11}\left[\alpha\, v(a) + \beta\, w(a)\right] + a_{12}\left[\alpha v'(a) + \beta\, w'(a)\right] = 0 \tag{15.1.54}$$

from which, since v satisfies the homogeneous condition,

$$a_{11}\, w(a) + a_{12}\, w'(a) = 0 \tag{15.1.55}$$

and similarly, at b,

$$a_{21}\, w(b) + a_{22}\, w'(b) = 0. \tag{15.1.56}$$

But these relations show that w also satisfies the completely homogeneous problem and, by the previous argument, must then be proportional to v. We thus conclude that both w_+ and w_- are proportional to v and therefore to each other. In this case, as is well known, their Wronskian vanishes and the construction of the Green's function fails. This is the situation encountered with Neumann conditions in Example 15.1.1 when $\sin kb = 0$, and in Example 15.1.2, where $u = \sin kx$ and $u = 1$, respectively, are solutions of the completely homogeneous problem. These cases require a modification of the Green's function construction as shown later in §15.7.

15.1.4 Mixed boundary conditions

Let us now briefly consider more general boundary conditions of the "mixed" type, namely involving the values of u at both end points; we write the required two conditions as

$$\begin{vmatrix} a_{11} & a_{12} \\ a_{21} & a_{22} \end{vmatrix} \begin{vmatrix} u(a) \\ u'(a) \end{vmatrix} + \begin{vmatrix} b_{11} & b_{12} \\ b_{21} & b_{22} \end{vmatrix} \begin{vmatrix} u(b) \\ u'(b) \end{vmatrix} = \begin{vmatrix} U_a \\ U_b \end{vmatrix} \tag{15.1.57}$$

and denote the two 2×2 matrices by A and B respectively.[7] The concomitant may be written in a similar matrix notation as

$$J[G, u] = p(b) \left| \begin{matrix} \partial G/\partial x|_b \\ -G(b, y) \end{matrix} \right|^{\mathsf{T}} \left| \begin{matrix} u(b) \\ u'(b) \end{matrix} \right| - p(a) \left| \begin{matrix} \partial G/\partial x|_a \\ -G(a, y) \end{matrix} \right|^{\mathsf{T}} \left| \begin{matrix} u(a) \\ u'(a) \end{matrix} \right| \qquad (15.1.58)$$

in which the superscript T denotes the transpose.

To identify the proper boundary conditions on G, let us suppose for the moment that $\det \mathsf{A} \neq 0$. Then, upon eliminating $|u(a) \ u'(a)|^{\mathsf{T}}$ between (15.1.57) and (15.1.58), we find

$$J[G, u] = \left\{ p(b) \left| \begin{matrix} \partial G/\partial x|_b \\ -G(b, y) \end{matrix} \right|^{\mathsf{T}} + p(a) \left| \begin{matrix} \partial G/\partial x|_a \\ -G(a, y) \end{matrix} \right|^{\mathsf{T}} \mathsf{A}^{-1} \mathsf{B} \right\} \left| \begin{matrix} u(b) \\ u'(b) \end{matrix} \right|$$

$$- p(a) \left| \begin{matrix} \partial G/\partial x|_a \\ -G(a, y) \end{matrix} \right|^{\mathsf{T}} \mathsf{A}^{-1} \left| \begin{matrix} U_a \\ U_b \end{matrix} \right|. \qquad (15.1.59)$$

By the usual strategy for the construction of the Green's function, we need to eliminate the unknown quantities $u(b)$ and $u'(b)$, which requires that the term in braces be 0; if $\det \mathsf{B} \neq 0$, after multiplying from the right by B^{-1} and taking the transpose the result may be written as

$$\frac{p(b)}{\det \mathsf{B}} \left| \begin{matrix} b_{22} & -b_{21} \\ -b_{12} & b_{11} \end{matrix} \right| \left| \begin{matrix} \partial G/\partial x|_b \\ -G(b, y) \end{matrix} \right|$$

$$+ \frac{p(a)}{\det \mathsf{A}} \left| \begin{matrix} a_{22} & -a_{21} \\ -a_{12} & a_{11} \end{matrix} \right| \left| \begin{matrix} \partial G/\partial x|_a \\ -G(a, y) \end{matrix} \right| = 0. \qquad (15.1.60)$$

Cases where the determinants vanish may be treated directly or recovered from this expression by taking suitable limits. For example, by setting $b_{11} = b_{12} = a_{21} = a_{22} = \varepsilon$ and taking the limit $\varepsilon \to 0$, we obtain the conditions (15.1.28) and (15.1.29) of the unmixed case (in the present slightly different notation).

Of particular interest, as we will see later, is the case in which the boundary conditions have the form

$$u(a) \mp u(b) = 0, \qquad u'(a) \mp u'(b) = 0, \qquad (15.1.61)$$

i.e., they impose periodicity (upper sign) or semi-periodicity (lower sign). In this case (15.1.60) gives

$$p(b) \left. \frac{\partial G}{\partial x} \right|_{x=b} \mp p(a) \left. \frac{\partial G}{\partial x} \right|_{x=a} = 0, \qquad p(b)G(b, y) \mp p(a)G(a, y) = 0. \quad (15.1.62)$$

The differential equation satisfied by the Green's function is (15.1.7) as before, so that we have the same conditions (15.1.13) and (15.1.15) of continuity of G and discontinuity of $\partial G/\partial x$ at $x = y$. To construct the Green's function it is convenient in this case to write

$$G(x, y) = \frac{w_-(x_>)w_+(x_<)}{W_0} + C_-(y)w_-(x) + C_+(y)w_+(x), \qquad (15.1.63)$$

---

[7] In order to have two truly independent conditions the elements of the matrices must be such that at least two of the four ratios $a_{1k}/a_{2k}, b_{1k}/b_{2k}$ are not equal.

in which the first term takes care of the jump condition (15.1.48) and $C_{\pm}(y)$ will be adjusted to satisfy (15.1.62). In this way it will be possible to choose the solutions $w_{\pm}$ of the homogeneous solution to satisfy some simple conditions at $x = a$, namely

$$w_+(a) = w_-'(a) = 0, \qquad w_-(a) = w_+'(a) = p(b)/p(a). \qquad (15.1.64)$$

As a consequence $W_0 = p(a)W(a; w_-, w_+) = p^2(b)/p(a) = p(b)W(b; w_-, w_+)$, so that $W(b; w_-, w_+) = p(b)/p(a)$. Upon substitution into (15.1.62) we find

$$[w_-(b) \mp 1]C_- + w_+(b)C_+ = -\frac{w_-(b)w_+(y)}{W_0}, \qquad (15.1.65)$$

$$w_-'(b)C_- + [w_+'(b) \mp 1]C_+ = -\frac{w_-'(b)w_+(y) \pm w_-(y)}{W_0}, \qquad (15.1.66)$$

with solution

$$C_-(y) = \frac{\pm[w_-(b)w_+(y) - w_+(b)w_-(y)] - [p(b)/p(a)]w_+(y)}{W_0\{[1 + p(b)/p(a) \mp [w_-(b) + w_+'(b)]\}}, \qquad (15.1.67)$$

$$C_+(y) = \frac{\pm[w_-(b)w_-(y) + w_-'(b)w_+(y)] - w_-(y)}{W_0\{1 + p(b)/p(a) \mp [w_-(b) + w_+'(b)]\}}. \qquad (15.1.68)$$

The upper and lower signs are to be chosen in correspondence with those of the boundary conditions (15.1.61).

It is readily verified that G is symmetric in x and y only if $p(b) = p(a)$, which causes the operator to be self-adjoint see (§21.6.1).

Example 15.1.3 A simple case is that of Example 15.1.1, p. 353 to which we return imposing the conditions (15.1.61). The solutions of the homogeneous equation satisfying (15.1.64) are

$$w_+(x) = \frac{\sin kx}{k}, \qquad w_-(x) = \cos kx, \qquad (15.1.69)$$

and (15.1.67) and (15.1.68) give

$$C_-(y) = \frac{\pm \sin k(y - b) - \sin ky}{2k(1 \mp \cos kb)}, \qquad C_+(y) = \frac{\pm \cos k(y + b) - \cos ky}{2(1 \mp \cos kb)}.$$
$$(15.1.70)$$

Just as in Example 15.1.1, these expressions become indeterminate when $\cos kb = \pm 1$. $\square$

15.2 The regular eigenvalue Sturm–Liouville problem

The families of eigenfunctions used in many of the examples in the first part of this book were based on eigenvalue problems governed by an equation with the general structure

$$\mathsf{L}u = -\frac{d}{dx}\left[p(x)\frac{du}{dx}\right] + q(x)u = \lambda s(x)u(x), \qquad a < x < b, \qquad (15.2.1)$$

with λ an eigenvalue. For example, for the Legendre functions we considered

$$-\frac{d}{dx}\left[(1-x^2)\frac{du}{dx}\right]+\frac{m^2}{1-x^2}u = \lambda u, \qquad -1 < x < 1, \tag{15.2.2}$$

and, for the functions used in the Fourier–Bessel expansion,

$$-\frac{d}{dx}\left(x\frac{dw}{dx}\right)+\frac{m^2}{x}w = \lambda x w, \qquad 0 \le R_1 < x < R_2. \tag{15.2.3}$$

The Fourier series can be based on the solutions of

$$-\frac{d^2u}{dx^2} = \lambda u, \qquad -\pi < x < \pi. \tag{15.2.4}$$

The differential operators in all these problems have the standard (regular or singular) Sturm–Liouville form (15.2.1).

Problems of this type are classified as regular or singular in the same way specified earlier after Eq. (15.1.3). Thus, the Legendre case (15.2.2) is singular because $p(x) = 1 - x^2$ vanishes at the end points and $q = m^2/(1 - x^2)$ is singular there. The Bessel case (15.2.3) is regular if $R_1 > 0$, but singular if $R_1 = 0$ as $p(x) = x$ vanishes at 0. The Fourier case is regular. All these problems share some general properties that we now consider for the regular case. The singular case will be taken up in §15.4. For reasons which will be evident in the following, we assume that $s(x) > 0$ in $a < x < b$, which is the case in the examples of the regular problem that we have just given.

We start with the case of unmixed boundary conditions which, in the regular case, after an appropriate normalization, may be written as

$$\cos\alpha\, u(a) - \sin\alpha\, u'(a) = 0, \qquad \cos\beta\, u(b) + \sin\beta\, u'(b) = 0, \tag{15.2.5}$$

with $0 \le \alpha, \beta < \pi$ given real parameters. If two functions u and v both satisfy these conditions, then their concomitant $J[\bar{v}, u] = \left[p(\bar{v}'u - \bar{v}u')\right]_a^b$ vanishes and (15.1.5) simply becomes

$$(v, \mathsf{L}u) = (\mathsf{L}v, u). \tag{15.2.6}$$

When this happens, the operator is *self-adjoint* rather than simply symmetric (see §21.6.1).

A simple heuristic argument shows that one way to find the eigenvalues is to look for values of λ such that the non-homogeneous equation associated to (15.2.1)

$$\mathsf{L}\Phi - \lambda s\Phi = sf \tag{15.2.7}$$

does not have an ordinary Green's function and therefore *is not solvable for all f*. Indeed, if u is an eigenfunction associated to an eigenvalue λ so that $\mathsf{L}u = \lambda su$, we would have

$$\int_a^b u\mathsf{L}\Phi\, dx - \lambda\int_a^b u\Phi s\, dx = \int_a^b (\mathsf{L}u)\Phi\, dx - \lambda\int_a^b u\Phi s\, dx = 0 \tag{15.2.8}$$

which would be inconsistent with (15.2.7) unless $\int_a^b ufs\, dx = 0$. This condition, of which we have seen an instance in (15.1.38) of Example 15.1.1, restricts the class of functions f for which (15.2.7) is solvable. We will see several other instances of this feature in §15.4.

15.2.1 Properties of Sturm–Liouville eigenvalues and eigenfunctions

There are many results on the solutions of the Sturm–Liouville eigenvalue problem, some of which we now briefly summarize; for complete proofs and additional properties see e.g. Ince 1926, Morse & Feshbach 1953, Coddington & Levinson 1955, Levitan & Sargsjan 1991 and many others.

Multiplication of the eigenvalue equation (15.2.1) by $\bar{u}$ and integration by parts gives

$$\lambda \int_a^b s|u|^2 \, \mathrm{d}x = [p\,\bar{u}u']_a - [p\,\bar{u}u']_b + \int_a^b \left(p|u'|^2 + q|u|^2 \right) \mathrm{d}x. \qquad (15.2.9)$$

By using the boundary conditions (15.2.5), for real α and β, the integrated term can be re-expressed to find

$$\lambda \int_a^b s|u|^2 \, \mathrm{d}x = \tan\beta\,[p\,|u'|^2]_b + \tan\alpha\,[p\,|u'|^2]_a + \int_a^b \left(p|u'|^2 + q|u|^2 \right) \mathrm{d}x, \qquad (15.2.10)$$

from which we see that the eigenvalues are certainly positive if $q \geq 0$ and $0 \leq \alpha, \beta < \pi/2$.

The argument of p. 357 can be adapted to conclude that, up to a multiplicative constant, for any given λ at most one single solution of the homogeneous problem (15.2.1) and (15.2.5) can exist: this is the defining property of *simple eigenvalues* (see pp. 630 and 664).

To make further progress, together with (15.2.1) we consider the equation

$$\tilde{L}v \equiv -\frac{\mathrm{d}}{\mathrm{d}x}\left[\tilde{p}(x)\frac{\mathrm{d}v}{\mathrm{d}x} \right] + \tilde{q}(x)\,v = \tilde{\lambda}s(x)v(x), \qquad a < x < b, \qquad (15.2.11)$$

subject to the same boundary conditions (15.2.5). Thus, u is an eigenfunction of the problem (p, q, λ) while v is an eigenfunction of the problem $(\tilde{p}, \tilde{q}, \tilde{\lambda})$. Upon multiplying (15.2.1) by $\bar{v}$, the complex conjugate of (15.2.11) by u, subtracting and integrating over $a \leq x_1 < x_2 \leq b$ we have

$$\left[p\bar{v}u' - \tilde{p}\bar{v}'u \right]_{x_1}^{x_2} = \int_{x_1}^{x_2} \left[(p - \tilde{p})\bar{v}'u' + (q - \tilde{q})\bar{v}u + s(\tilde{\bar{\lambda}} - \lambda)\bar{v}u \right] \mathrm{d}x. \qquad (15.2.12)$$

Several deductions can be made from this relation:

1. If $\tilde{p} = p, \tilde{q} = q$, then u and v satisfy the same equation and are therefore both eigenfunctions of the same Sturm–Liouville operator. Upon taking $x_1 = a$, $x_2 = b$, the left-hand side of (15.2.12) vanishes by the boundary conditions and we deduce that

$$(\tilde{\bar{\lambda}} - \lambda) \int_a^b s(x)\bar{v}(x)u(x) \, \mathrm{d}x = 0. \qquad (15.2.13)$$

If $v = u$, so that $\tilde{\lambda} = \lambda$, we deduce that $\bar{\lambda} = \lambda$ and all eigenvalues are real, so that the eigenfunctions can be taken real without loss of generality. If $v \neq u, \tilde{\lambda} \neq \lambda$ and this relation implies that eigenfunctions corresponding to different eigenvalues are orthogonal with respect to the scalar product (15.2.13). These are special cases of general results on the eigenvalues and eigenfunctions of self-adjoint operators given in §21.9.2.

2. Let us now assume $\tilde{p} = p$ and take v real and x_1 and x_2 two consecutive zeros of u:

$$\left[pu'v \right]_{x_1}^{x_2} = \int_{x_1}^{x_2} \left[(q - \lambda s) - (\tilde{q} - \tilde{\lambda}s) \right] u(x)v(x)\, dx. \qquad (15.2.14)$$

Assume without loss of generality that $u < 0$ between x_1 and x_2, so that $u'(x_2) > 0$, $u'(x_1) < 0$; if v does not change sign in the interval, the left-hand side has the sign of v. Suppose that $q - \lambda s > \tilde{q} - \tilde{\lambda}s$ for $x_1 < x < x_2$. Then, since $u < 0$, the right-hand side would have the opposite sign unless v had at least one zero between x_1 and x_2. Following this line of reasoning, we conclude that decreasing q throughout the interval must increase the number of zeros of the eigenfunction.

3. Likewise, the number of zeros of the eigenfunction increases with λ. If λ_n and λ_{n+1} are consecutive eigenvalues with $\lambda_n < \lambda_{n+1}$, the zeros of u_n and u_{n+1} interlace and u_{n+1} must have one more zero than u_n.

4. An extension of the argument based on integrating $(d/dx)[(u/v)(pu'v - \tilde{p}uv')]$ rather than the Lagrange identity (15.1.4) shows that the same result holds if p is decreased. These considerations prove

Theorem 15.2.1 [Sturm's oscillation theorem] *The solution of (15.2.1) oscillates more rapidly when p or q are decreased, or λ or s are increased.*

For example, If p, q and s are equal to constants p_0, q_0 and s_0, (15.2.1) may be recast in the form

$$-\frac{d^2 u}{dx^2} = \frac{\lambda s_0 - q_0}{p_0} u(x), \qquad (15.2.15)$$

from which the statement of the theorem is evident. The previous results may be strengthened as follows (see e.g. Ince 1926):

1. Let $p \geq p_m > 0$ and $q \geq q_m$ in $[a, b]$, with p_m and q_m constants. Then the solutions of (15.2.1) do not oscillate more rapidly than those of the same equation with p and q replaced by p_m and q_m. The solutions of (15.2.1) are non-oscillatory if $[\min (\lambda s - q)]/[\min p] < \pi^2/(b - a)^2$.

2. Let now $0 < p \leq p_M$ and $q \leq q_M$ in $[a, b]$, with p_M and q_M constants. Then the solutions of (15.2.1) oscillate at least as rapidly as those of the same equation with p and q replaced by p_M and q_M. A sufficient condition for a solution of (15.2.1) to have at least k zeros in (a, b) is that $[\max (\lambda s - q)]/[\max p] \geq k^2\pi^2/(b - a)^2$.

3. If λ_0 is the smallest eigenvalue and λ_n the n-th one, u_0 has no zeros in the open interval (a, b) while u_n has n. As λ increases, the existing zeros move toward a and new zeros appear near $x = b$.

4. The eigenvalues are decreasing functions of the parameters α and β introduced in (15.2.5).

5. If $q \geq 0$, $0 \leq \alpha$, $\beta \leq \pi/2$ and $s(x)$ changes sign in the interval (a, b), the previous arguments can be applied to each subinterval where s has the same sign and one ends up with two families of eigenvalues $\lambda_n^+ \to \infty$ and $\lambda_m^- \to -\infty$ with corresponding eigenfunctions.

Example 15.2.1 In §3.2.4 we were led to the eigenvalue problem

$$-\frac{d^2 u_n}{dx^2} = \lambda_n u_n \quad \text{in} \quad 0 < x < b, \tag{15.2.16}$$

subject to the conditions (15.2.5) with $\alpha = 0$, namely

$$u_n(0) = 0, \qquad \cos \beta\, u_n(b) + \sin \beta\, u'_n(b) = 0. \tag{15.2.17}$$

The eigenfunctions are

$$u_n \propto \sin \sqrt{\lambda_n} x \tag{15.2.18}$$

with the eigenvalues determined by the characteristic equation which is conveniently written in the form

$$\frac{\tan \sqrt{\lambda_n} b}{\sqrt{\lambda_n}\, b} = -\frac{\tan \beta}{b}. \tag{15.2.19}$$

A simple graphical argument shows all the roots $\sqrt{\lambda_n}$ are real if $\tan \beta > 0$ or $\tan \beta < -b$, which illustrates the fact that the condition $0 \le \alpha, \beta \le \pi/2$ is sufficient but not necessary for positivity of the eigenvalues. There are two equal and opposite imaginary roots (corresponding to a negative eigenvalue) if $0 < -\tan \beta < b$. If $\tan \beta = -b$, the smallest eigenvalue is $\lambda_0 = 0$ and the corresponding eigenfunction $u_0 \propto x$. By taking the derivative of (15.2.19) with respect to β we find

$$\frac{d\lambda_n}{d\beta} = -\frac{2\lambda_n(1 + \tan^2 \beta)}{b(1 + \lambda_n \tan^2 \beta) + \tan \beta} \tag{15.2.20}$$

which illustrates property 4 according to which λ_n is a decreasing function of β. □

The *Liouville transformation*

$$\xi = \int_a^x \sqrt{\frac{s}{p}}\, dx, \qquad u = (ps)^{-1/4} w(\xi) \tag{15.2.21}$$

leaves the boundary conditions (15.2.5) unchanged and brings the equation (15.2.1) to the form

$$\frac{d^2 w}{d\xi^2} + [\lambda - Q(\xi)]\, w = 0, \tag{15.2.22}$$

with

$$Q = \frac{q}{s} - \left(\frac{p}{s^3}\right)^{1/4} \frac{d}{dx}\left[p \frac{d}{dx}(ps)^{-1/4}\right] = \frac{q}{s} + (ps)^{-1/4} \frac{d^2}{d\xi^2}(ps)^{1/4} \tag{15.2.23}$$

For the purpose of using eigenfunctions for expansions in infinite series we are interested in the behavior of the eigenfunctions for large λ; in this case the equation may be approximated by

$$\frac{d^2 w}{d\xi^2} + \lambda w \simeq 0. \tag{15.2.24}$$

Substitution of the general solution of this equation into the boundary conditions (15.2.5) produces approximately the equation determining the eigenvalues, which may be written as

$$\tan \left(\sqrt{\lambda_n}\,\Xi\right) \simeq \frac{\tan \alpha + \tan \beta}{\lambda_n \tan \alpha \, \tan \beta - 1}\sqrt{\lambda_n} \quad \text{with} \quad \Xi = \int_a^b \sqrt{\frac{s}{p}}\,\mathrm{d}x. \quad (15.2.25)$$

We recognize Ξ as the length of the ξ-domain as x ranges over the original domain $[a, b]$. For large λ_n and general α and β this equation can be approximated further as

$$\tan \left(\sqrt{\lambda_n}\,\Xi\right) \simeq \frac{\cot \alpha + \cot \beta}{\sqrt{\lambda_n}}, \quad (15.2.26)$$

which shows that, as λ_n increases,

$$\sqrt{\lambda_n}\,\Xi \simeq n\pi + O(1/n). \quad (15.2.27)$$

For either $\alpha = 0$ or $\beta = 0$, (15.2.25) reduces to (15.2.19) and the estimate becomes $\sqrt{\lambda_n}\,\Xi \simeq (n + \frac{1}{2})\pi + O(1/n)$ while, for $\alpha = \beta = 0$, $\sqrt{\lambda_n}\,\Xi \simeq (n + 1)\pi + O(1/n)$. In any case, for the eigenfunctions we have to leading order

$$u_n \propto (ps)^{-1/4} \cos \left[\sqrt{\lambda_n} \int_a^x \sqrt{\frac{s}{p}}\,\mathrm{d}x \right]. \quad (15.2.28)$$

These results can be refined by a systematic use of the WKB asymptotic method (see e.g. Bender & Orszag 2005, p. 490; Fedoriouk 1987, chapter 2; Murdock 1991, chapter 9; see also Morse & Feshbach 1953, p. 740) or otherwise (see e.g. Ince 1926, p. 270). In any event, they show why the Fourier cosine series plays such an important role in the point-wise convergence of eigenfunction expansions, as will be seen in Theorem 15.3.3 p. 371.

15.2.2 Mixed boundary conditions

For the sake of brevity we will only consider the cases (15.1.61) of periodic or semi-periodic boundary conditions, which are the most important ones in applications. For a more general treatment the reader may consult e.g. Coddington & Levinson 1955, chapter 8, or Ince 1926, chapters 9–11.

According to the argument on p. 361, the concomitant must vanish for the operator to be self-adjoint and the eigenvalues real; with the boundary conditions (15.2.29) this condition is verified provided $p(a) = p(b)$, which we assume.

For the identification of the eigenvalues we use the shortcut suggested by the argument associated with (15.2.7), which remains valid in the present case. So we try to construct a Green's function for the non-homogeneous equation (15.2.7) with boundary conditions

$$\Phi(a; \lambda) = \pm\Phi(b; \lambda), \qquad \Phi'(a; \lambda) = \pm\Phi'(b; \lambda). \quad (15.2.29)$$

As in §15.1.4 we seek the Green's function in the form

$$G(x, y; \lambda) = \frac{\phi_-(x_>)\phi_+(x_<)}{W_0(\lambda)} + C_-(y)\phi_-(x) + C_+(y)\phi_+(x), \quad (15.2.30)$$

the dependence of $\phi_\pm$ on λ being understood. It is convenient to choose $\phi_\pm$ to satisfy

$$\phi_+(a) = \phi_-'(a) = 0, \qquad \phi_-(a) = \phi_+'(a) = 1, \qquad (15.2.31)$$

so that $W_0 = p(a)W(a) = p(a) = p(b)W(b)$ and $W(b) = 1$. The expressions (15.1.67) and (15.1.68) for $C_\pm$ are readily adapted to the present case. We will not write them down, however, as here we are only interested in determining the values of λ such that the analog of the linear system (15.1.65) and (15.1.66) has no solution. This happens when the determinant of the system vanishes:

$$\begin{vmatrix} \phi_-(b; \lambda) \mp 1 & \phi_+(b; \lambda) \\ \phi_-'(b; \lambda) & \phi_+'(b; \lambda) \mp 1 \end{vmatrix} = 2 \mp [\phi_-(b; \lambda) + \phi_+'(b; \lambda)] = 0, \qquad (15.2.32)$$

where use has been made of the fact that $W(b) = 1$. This is then the equation determining the eigenvalues.

If (15.2.32) is satisfied, and at least some of the elements of the determinant are non-zero, the algebraic system (15.1.65) and (15.1.66) has a simple infinity of solutions, which corresponds to the existence of one eigenfunction determined up to the usual multiplicative constant. However, a look at the determinant (15.2.32) reveals that another possibility exists here, unlike the case of separated boundary conditions. Indeed, it may happen that

$$\phi_-'(b; \lambda_n) = 0 \qquad \text{and} \qquad \phi_+(b; \lambda_n) = 0 \qquad (15.2.33)$$

$$\phi_-(b; \lambda_n) = \pm 1 \qquad \text{and} \qquad \phi_+'(b; \lambda_n) = \pm 1 \qquad (15.2.34)$$

which clearly satisfy the eigenvalue equation but set to 0 all the elements of the determinant. In this case the system has a *double infinity* of solutions, corresponding to the fact that the eigenvalue is degenerate. As we have seen, this cannot happen when the boundary conditions are unmixed.

Example 15.2.2 To illustrate this point we can go back to Example 15.1.3:

$$-\frac{d^2\Phi}{dx^2} = \lambda\Phi + f, \qquad 0 < x < b, \qquad (15.2.35)$$

with periodicity boundary conditions, i.e., the upper sign in (15.2.29). Now, provided $\lambda \neq 0$,

$$\phi_+ = \frac{\sin \sqrt{\lambda}x}{\sqrt{\lambda}}, \qquad \phi_- = \cos \sqrt{\lambda}x. \qquad (15.2.36)$$

The eigenvalue equation (15.2.32) is

$$2 \cos \sqrt{\lambda_n}b = 2 \qquad \text{from which} \qquad \lambda_n = \left(\frac{2n\pi}{b}\right)^2. \qquad (15.2.37)$$

In this case the conditions (15.2.33) and (15.2.34) are satisfied and the eigenvalues are degenerate with eigenfunctions

$$u_n^{(1)} \propto \cos 2n\pi \frac{x}{b}, \qquad u_n^{(2)} \propto \sin 2n\pi \frac{x}{b}. \qquad (15.2.38)$$

There is one non-degenerate eigenvalue, namely $\lambda_0 = 0$, the corresponding eigenfunction being a constant. $\qquad \square$

15.3 The eigenfunction expansion

Eigenfunction expansions have been one of the main tools used to solve the problems of Part I of this book. We have already repeatedly remarked on the formal similarity between different expansions, which suggests the existence of an underlying general framework. It is this framework that we now describe in the regular Sturm–Liouville case; the singular case is deferred to §15.4.

We present an argument based on a general result of the theory of compact operators, covered in §21.3, which leads to the mean-square convergence of the expansion, and a more direct proof which concerns point-wise convergence. We consider the eigenfunctions of the differential operator (15.2.1) subject to the unmixed boundary conditions (15.2.5); a brief comment on the case of mixed boundary conditions is provided in §15.3.4.

15.3.1 The Hilbert–Schmidt theorem

We start by noting that we can assume without loss of generality that $\lambda = 0$ is not an eigenvalue as, if it were, we could redefine q to be $\tilde{q} = q + cs$ for any non-zero constant c, which would redefine the eigenvalues as $\tilde{\lambda} = \lambda + c \neq 0$. With this assumption, the problem $Lu = 0$ with the boundary conditions (15.2.5) does not have a non-zero solution and we can therefore construct a Green's function for the problem $L\Phi = f$, with the aid of which the eigenvalue problem for the differential equation (15.2.1) becomes an eigenvalue problem for the integral equation

$$\int_a^b G(\xi, x) u(\xi) s(\xi) \, d\xi \;=\; \frac{1}{\lambda} u(x). \tag{15.3.1}$$

In §21.3 dealing with compact operators we prove the general Hilbert–Schmidt theorem (Theorem 21.3.4, p. 639) which, adapted to the present case, states the following:

Theorem 15.3.1 *The eigenvalue problem (15.3.1) admits an infinity of solutions $\{u_n\}$, each one corresponding to a real eigenvalue λ_n with $\lambda_n \to \infty$. These solutions are mutually orthogonal in the sense that*

$$(u_n, u_m)_s \;\equiv\; \int_a^b \bar{u}_n(x) u_m(x) \, s(x) \, dx \;\propto\; \delta_{nm}. \tag{15.3.2}$$

Every function $f(x)$ such that

$$\|f\|_s^2 \;\equiv\; \int_a^b |f(x)|^2 s(x) \, dx \;<\; \infty \tag{15.3.3}$$

can be uniquely represented as

$$f(x) \;=\; \sum_{k=0}^{\infty} c_k \, u_k(x) \qquad where \qquad c_k = \frac{(u_k, f)_s}{(u_k, u_k)_s} \tag{15.3.4}$$

in the mean-square sense, namely

$$\lim_{N \to \infty} \int_a^b \left| f(x) - \sum_{k=0}^N c_k u_k(x) \right|^2 s(x)\, dx = 0 \tag{15.3.5}$$

and

$$Lf = s(x) \sum_{k=0}^\infty \lambda_k c_k u_k(x). \tag{15.3.6}$$

This theorem unifies all the expansions that we have extensively used in this book, at least in the regular case.

In stating the theorem we have used the symbol $\| f \|_s$ for the norm of f defined in (15.3.3) and the expression $(u_n, f)_s$ to denote the scalar product of u_n and f defined by

$$(u_n, f)_s = \int_a^b \overline{u_n}(x) f(x) s(x)\, dx. \tag{15.3.7}$$

In our general treatment of these notions in §19.2 and §19.3 we point out that these are good definitions provided $s > 0$ and also, with suitable hypotheses, if s vanishes at isolated points. It may be noted that, with this scalar product, the operator $L/s(x)$ is self-adjoint for real λ as, using (15.1.5) and the vanishing of the concomitant,

$$\left(v, s^{-1}Lu \right)_s = \int_a^b \overline{v}(Lu)\, dx = \int_a^b \overline{Lv}\, u\, dx = \left(s^{-1}Lv, u \right)_s. \tag{15.3.8}$$

Note also that (15.3.6) does not justify term-by-term differentiation of the series but only the term-by-term application of L to an expansion on a basis of its own eigenfunctions.

By multiplying (15.3.4) by $s\,\overline{f}$ and integrating term by term (p. 227) we find

$$\| f \|_s^2 = \sum_{k=0}^\infty \frac{|(u_k, f)_s|^2}{(u_k, u_k)_s}, \tag{15.3.9}$$

which is a general form of Parseval's equality to be considered in yet greater generality in §19.3. This or (15.3.5) only guarantee the equality (15.3.4) in the mean-square sense. We consider point-wise convergence next.

15.3.2 Direct proof of the expansion theorem

The results summarized in Theorem 15.3.1 rely on operator theory. It is interesting to give a direct proof of the expansion (15.3.4) also because a rather straightforward modification of it applies to the singular case (§15.4.3).

Let λ_n be the set of eigenvalues of the Sturm–Liouville problem (15.2.1) with boundary conditions (15.2.5) and e_n the corresponding orthogonal eigenfunctions, assumed real and normalized according to (15.3.3) so that

$$\| e_n \|_s^2 = \int_a^b e_n^2(x) s(x)\, dx = 1. \tag{15.3.10}$$

Consider the function

$$\Phi(x; \lambda) = \sum_{n=0}^{\infty} \frac{c_n e_n(x)}{\lambda_n - \lambda},$$ (15.3.11)

where λ is a real or complex fixed parameter not equal to any of the eigenvalues and $\{c_n\}$ is a sequence of real or complex numbers tending to zero fast enough that the series can be differentiated term by term. As a function of λ, Φ has simple poles at $\lambda = \lambda_n$ with residues (p. 435) $-c_n e_n(x)$, and it will be remembered from (15.2.27) that, asymptotically, $\lambda_n \sim n^2$.

If we apply the operator L to Φ we find[8]

$$\frac{1}{s} L\Phi = \sum_{n=0}^{\infty} \frac{\lambda_n c_n e_n}{\lambda_n - \lambda} = \lambda \sum_{n=0}^{\infty} \frac{c_n e_n}{\lambda_n - \lambda} + \sum_{n=0}^{\infty} c_n e_n = \lambda\Phi + f$$ (15.3.12)

where

$$f(x) = \sum_{n=0}^{\infty} c_n e_n(x) = -\sum_{n=0}^{\infty} \text{Res}\,[\Phi, \lambda_n].$$ (15.3.13)

We thus see that, if $\Phi(x; \lambda)$ is the solution of the differential equation

$$\frac{1}{s(x)} L\Phi(x) - \lambda\Phi(x) = f(x)$$ (15.3.14)

the sum of its λ-residues equals $-f$. By calculating explicitly these residues, we will determine the coefficients of the expansion (15.3.13). This expansion will be valid in a point-wise sense for all those f which satisfy the boundary conditions of the eigenvalue problem and such that (15.3.14) can be solved. These functions belong to the *range* of the operator (definition p. 621) and are square-integrable in the sense of (15.3.3) i.e., they are *compatible* with the operator.

To achieve our objective, we solve (15.3.14) by means of the appropriate Green's function which we write, as in (15.1.21), in the form

$$G(x, y; \lambda) = \frac{\phi_-(x_>; \lambda)\phi_+(x_<; \lambda)}{W_0(\lambda)},$$ (15.3.15)

where $\phi_\pm(x; \lambda)$ are solutions of the homogeneous equation (15.2.1) subject to the conditions (15.1.31) and (15.1.32) with the a_{ij} given as in (15.2.5), namely

$$\phi_+(a; \lambda) = \sin\alpha, \qquad \phi_+'(a; \lambda) = \cos\alpha,$$
$$\phi_-(b; \lambda) = \sin\beta, \qquad \phi_-'(b; \lambda) = -\cos\beta.$$ (15.3.16)

A point to note here is the dependence of $\phi_\pm$ on the parameter λ, which also causes a λ-dependence of $W_0 = pW(\phi_-, \phi_+)$. With the Green's function (15.3.15), the solution of

---

[8] This step requires a justification of the term-by-term application of L that we do not give. The proof is facilitated by the fact that the series to which L is applied here has a very special form insofar as its terms are proportional to the eigenfunctions of L.

(15.3.14) for Φ reduces to the particular solution (15.1.23) since the completely homogeneous problem has only the zero solution given that λ is not an eigenvalue by hypothesis. Thus, explicitly,[9]

$$\Phi(x; \lambda) = \int_a^b G(\xi, x; \lambda) f(\xi) s(\xi) \, d\xi \qquad (15.3.17)$$

$$= \frac{\phi_-(x)}{W_0(\lambda)} \int_a^x \phi_+(\xi; \lambda) f(\xi) s(\xi) \, d\xi + \frac{\phi_+(x)}{W_0(\lambda)} \int_x^b \phi_-(\xi; \lambda) f(\xi) s(\xi) \, d\xi \, .$$

By the hypotheses made on p, q and s, the solutions of the differential equation are entire functions of λ (see e.g. Stakgold 1997, p. 417). Thus, the poles of Φ can only arise from the vanishing of $W_0(\lambda)$, which will occur for those λ where the functions $\phi_\pm$ are proportional to each other. But, when this happens, each one of them satisfies homogeneous boundary conditions at *both* end points and, therefore, they are both eigenfunctions of L. Since, as shown in §15.1.3, up to a multiplicative constant there is only one eigenfunction for each λ_n, we conclude that $\phi_\pm(x; \lambda_n) = k_n^\pm e_n(x)$, with $k_n^\pm$ independent of x. Furthermore, again because eigenvalues are not degenerate, W_0 must have simple zeros for $\lambda = \lambda_n$ so that (see (17.9.6) p. 460)

$$\text{Res}[\Phi, \lambda_n] = \frac{k_n^+ k_n^-}{W_0'(\lambda_n)} e_n(x) \int_a^b e_n(\xi) f(\xi) s(\xi) \, d\xi \, , \qquad (15.3.18)$$

where $W_0'(\lambda) = \partial W_0/\partial \lambda$. It is not difficult to calculate the derivative $W_0'(\lambda_n)$. For this purpose we need the derivatives with respect to λ of $\phi_\pm$. If we take $\partial/\partial\lambda$ of the equation (15.2.1) satisfied by these functions we find

$$\mathsf{L}\frac{\partial\phi_\pm}{\partial\lambda} - \lambda s\frac{\partial\phi_\pm}{\partial\lambda} = s\phi_\pm \qquad (15.3.19)$$

with homogeneous boundary conditions at $x = a$ and $x = b$ for both ϕ_+ and ϕ_-. This problem can be solved by using the same Green's function as before to find

$$\frac{\partial\phi_+}{\partial\lambda} = -\frac{\phi_-(x)}{W_0(\lambda)} \int_a^x \phi_+^2 s \, d\xi + \frac{\phi_+(x)}{W_0(\lambda)} \int_a^x \phi_+\phi_- s \, d\xi \, , \qquad (15.3.20)$$

$$\frac{\partial\phi_-}{\partial\lambda} = -\frac{\phi_-(x)}{W_0(\lambda)} \int_x^b \phi_+\phi_- s \, d\xi + \frac{\phi_+(x)}{W_0(\lambda)} \int_x^b \phi_-^2 s \, d\xi \, , \qquad (15.3.21)$$

in which the modification of the integration limits ensures the satisfaction of the homogeneous boundary conditions.[10] If we now substitute this expression into

$$W_0' = p(x)\frac{\partial}{\partial\lambda}\left(\phi_-\phi_+' - \phi_+\phi_-'\right) \qquad (15.3.22)$$

---

[9] This is the explicit representation of the resolvent of the operator $(1/s(x))\mathsf{L}$ in the Hilbert space with a scalar product defined in (15.3.7); see §21.9, Chapter 21.

[10] Satisfaction of the boundary conditions requires the addition of a suitable solution of the homogeneous equation to the particular solution expressed in terms of the Green's function as in (15.3.17). When combined with the particular solution, the net result is a change of the integration limits as shown.

we find

$$W_0' = -\int_a^b \phi_-\phi_+s\,\mathrm{d}\xi \;\to\; -k_n^+ k_n^- \qquad \text{as} \quad \lambda \to \lambda_n \tag{15.3.23}$$

where use has been made of the normalization (15.3.10). We thus conclude that $k_n^+ k_n^- / W_0' = -1$ so that

$$c_n = -\mathrm{Res}[\Phi, \lambda_n] = \int_a^b e_n(\xi)f(\xi)s(\xi)\,\mathrm{d}\xi = (e_n, f)_s \tag{15.3.24}$$

and therefore

$$f(x) = \sum_{n=0}^\infty \left(\int_a^b e_n(\xi)f(\xi)s(\xi)\,\mathrm{d}\xi \right) e_n(x). \tag{15.3.25}$$

These developments lead to the following expansion theorem, which complements Theorem 15.3.1 (see e.g. Levitan & Sargsjan 1991, p. 28; Titchmarsh 1962, p. 12; Ince 1926, p. 277; Courant & Hilbert 1953, vol. 1 p. 293):

Theorem 15.3.2 [Expansion theorem] *If $f(x)$ is continuous with piece-wise continuous first and second derivative and satisfies the boundary conditions (15.2.5), then it can be expanded in an absolutely and uniformly converging series of eigenfunctions of the boundary-value problem (15.2.1), (15.2.5):*

$$f(x) = \sum_{n=0}^\infty c_n e_n(x), \qquad c_n = \int_a^b f(\xi)e_n(\xi)s(\xi)\,\mathrm{d}\xi . \tag{15.3.26}$$

If $f(x)$ has a continuous second derivative, the series may be differentiated once term by term retaining the properties of absolute and uniform convergence.

Example 15.3.1 Let us see how the previous general theory works for the operator of Example 15.2.1. We normalize the eigenfunctions (15.2.18) according to (15.3.10) writing

$$e_n = a_n \sin \sqrt{\lambda_n}\, x \tag{15.3.27}$$

and find

$$a_n = \left[\frac{1}{2} \left(b + \frac{\tan \beta}{1 + \lambda_n \tan^2 \beta} \right) \right]^{-1/2}. \tag{15.3.28}$$

Proceeding as in the proof of the expansion, we consider

$$-\frac{\mathrm{d}^2 \Phi}{\mathrm{d}x^2} - \lambda \Phi = f \tag{15.3.29}$$

with $\Phi(x; \lambda)$ subject to the boundary conditions (15.2.17). The functions $\phi_\pm$ necessary to construct the Green's function are found to be

$$\phi_+(x, \lambda) = \frac{1}{\sqrt{\lambda}} \sin \sqrt{\lambda}\, x, \tag{15.3.30}$$

$$\phi_-(x, \lambda) = \frac{\cos \beta}{\sqrt{\lambda}} \sin \sqrt{\lambda} (b - x) + \sin \beta \cos \sqrt{\lambda} (b - x), \qquad (15.3.31)$$

and, when $\lambda \to \lambda_n$, they become

$$\phi_+(x, \lambda_n) = \frac{e_n(x)}{a_n \sqrt{\lambda_n}}, \qquad \phi_-(x, \lambda_n) = -\frac{\cos \beta}{a_n \sqrt{\lambda_n} \cos \sqrt{\lambda_n} b} e_n(x), \qquad (15.3.32)$$

so that

$$k_n^+ = \frac{1}{a_n \sqrt{\lambda_n}}, \qquad k_n^- = -\frac{\cos \beta}{a_n \sqrt{\lambda_n} \cos \sqrt{\lambda_n} b}. \qquad (15.3.33)$$

The Wronskian (which equals W_0 in this case since $p = 1$) is

$$W(\phi_-, \phi_+) = W_0 = \frac{\cos \beta}{\sqrt{\lambda}} \sin \sqrt{\lambda} b + \sin \beta \cos \sqrt{\lambda} b \qquad (15.3.34)$$

and its derivative evaluated at $\lambda = \lambda_n$ is

$$W_0' = \frac{\cos \beta}{2\lambda_n \sqrt{1 + \lambda_n \tan^2 \beta}} \left[b(1 + \lambda_n \tan^2 \beta) + \tan \beta \right]; \qquad (15.3.35)$$

we thus see that, as expected, $k_n^+ k_n^- / W_0' = -1$. The expansion (15.3.25), written out in full, is then

$$f(x) = \frac{2}{b} \sum_{n=0}^{\infty} \frac{1 + \lambda_n \tan^2 \beta}{1 + \lambda_n \tan^2 \beta + (\tan \beta)/b} \sin \sqrt{\lambda_n} x \int_0^b f(\xi) \sin \sqrt{\lambda_n} \xi \, d\xi . \qquad (15.3.36)$$

This expansion can be used for the solution of the problems mentioned in §3.11 on applications of the Fourier series. Other examples can be found in Titchmarsh 1962, vol. 1 chapter 1. □

A further result on point-wise convergence is:

Theorem 15.3.3 *Let* $(ps)^{-1/4} f(x)$ *be absolutely Lebesgue-integrable over* $[a, b]$. *Then the Sturm–Liouville expansion* (15.3.4) *behaves as regards convergence in the same way as an ordinary Fourier series. In particular, it converges to* $\frac{1}{2}[f(x + 0) + f(x - 0)]$ *if* f *is of bounded variation in the neighborhood of* x.

Without presenting a proof (for which see e.g. Titchmarsh 1962, vol. 1, p. 12; Morse & Feshbach 1953, p. 743), we note that this result is a consequence of the estimate of the eigenfunctions and eigenvalues given earlier at the end of §15.2.1. Indeed, from (15.2.28), we have, for large λ_n,

$$c_n \sim \frac{1}{(u_n, u_n)_s} \int_a^b \cos \left[\sqrt{\lambda_n} \xi(x) \right] f(x)(ps)^{-1/4} dx . \qquad (15.3.37)$$

In view of the close correspondence between Sturm–Liouville and Fourier series, it is not surprising that the Sturm–Liouville series can be integrated term-by-term between any two points $a \le x_1 < x_2 \le b$:

$$\int_{x_1}^{x_2} f(x)\,dx = \sum_{k=0}^{\infty} c_k \int_{x_1}^{x_2} u_k(x)\,dx, \qquad (15.3.38)$$

the series on the right being uniformly convergent (see e.g. Smirnov 1964, vol. 2, p. 454).

15.3.3 Remarks

Results similar to those to be discussed here also hold for other self-adjoint operators with real eigenvalues but complex eigenfunctions. With an eye to this extension, we will therefore write $\bar{e}_n$ when the eigenfunction is the first element of a scalar product.

The expansion (15.3.26) may be rewritten as

$$f(x) = \sum_{n=0}^{\infty} e_n(x) \int_a^b f(\xi)\bar{e}_n(\xi)s(\xi)\,d\xi \qquad (15.3.39)$$

which shows that, in the sense of distributions,[11]

$$\sum_{n=0}^{\infty} \bar{e}_n(\xi)e_n(x) = \frac{\delta(x-\xi)}{s(\xi)} = \frac{\delta(x-\xi)}{s(x)}. \qquad (15.3.40)$$

Furthermore, from (15.3.11) and (15.3.17), we may write

$$\Phi(x;\lambda) = \int_a^b G(x,\xi)f(\xi)s(\xi)\,d\xi = \sum_{n=0}^{\infty} \frac{e_n(x)}{\lambda_n - \lambda} \int_a^b f(\xi)\bar{e}_n(\xi)s(\xi)\,d\xi \qquad (15.3.41)$$

and, since the series converges uniformly due to the fact that, from (15.2.27), $\lambda_n \sim n^2$, we can interchange summation and integration (see p. 227) obtaining the *bilinear expansion* of the Green's function

$$G(x,\xi;\lambda) = \sum_{n=0}^{\infty} \frac{\bar{e}_n(\xi)e_n(x)}{\lambda_n - \lambda}. \qquad (15.3.42)$$

This series also converges uniformly by Weierstrass's M-test (p. 227) due to the behavior of eigenfunctions and eigenvalues for large n. From this expression we see that the residues of G at the poles λ_n give a convenient way to determine the normalized eigenfunctions:

$$\text{Res}[G;\lambda_n] = -\bar{e}_n(\xi)e_n(x). \qquad (15.3.43)$$

An application of this remark is shown in Example 15.4.8 p. 378.

By rewriting (15.3.14) identically as

$$L\Phi - \mu s(x)\Phi = s(x)[(\lambda - \mu)\Phi + f(x)] \qquad (15.3.44)$$

---

[11] The reader will realize that this step is not entirely warranted without further analysis as (15.3.39) has been established only for continuous functions which satisfy the boundary conditions of the eigenfunctions, rather than for all test functions.

we see that

$$\Phi(x;\lambda) = (\lambda - \mu)\int_a^b G(\xi,x;\mu)\Phi(\xi;\lambda)s(\xi)\,\mathrm{d}\xi + \int_a^b G(\xi,x;\mu)f(\xi)\,s(\xi)\,\mathrm{d}\xi.$$

(15.3.45)

An alternative expression for $\Phi(x;\lambda)$ is provided by (15.3.17) which, upon substitution here, gives

$$(\lambda - \mu)\int_a^b \mathrm{d}\xi\,s(\xi)\,G(\xi,x;\mu)\int_a^b G(y,\xi;\lambda)\,f(y)\,s(y)\,\mathrm{d}y$$

$$+ \int_a^b G(\xi,x;\mu)f(\xi)\,s(\xi)\,\mathrm{d}\xi = \int_a^b G(\xi,x;\lambda)\,f(\xi)\,s(\xi)\,\mathrm{d}\xi.$$
(15.3.46)

By Fubini's Theorem (§A.4.3), this relation will hold for all f if and only if

$$G(x,y;\lambda) - G(x,y;\mu) = (\lambda - \mu)\int_a^b G(x,\xi;\lambda)G(\xi,y;\mu)\,\mathrm{d}\xi.$$
(15.3.47)

This is a special case of the general resolvent equation of §21.10, p. 669. If, in (15.3.44), we take $f = 0$, $\lambda = \lambda_n$ and $\Phi = \bar{e}_n$, we also immediately have

$$\bar{e}_n(x) = (\lambda_n - \mu)\int_a^b G(\xi,x;\mu)\bar{e}_n(\xi)s(\xi)\,\mathrm{d}\xi,$$
(15.3.48)

which gives an expression for the coefficient of the expansion of G on the eigenfunctions e_n and agrees with the bilinear expansion (15.3.42) and with (15.3.43).

It is useful to recast the previous relations in terms of contour integrals in the complex λ plane to better illustrate the transition to a continuous spectrum in §15.4.3. For this purpose we use the residue theorem (p. 459) which, applied to (15.3.13), permits us to write

$$f(x) = -\frac{1}{2\pi i}\oint_{C_\infty}\Phi(x;\lambda)\,\mathrm{d}\lambda = -\frac{1}{2\pi i}\oint_{C_\infty}\mathrm{d}\lambda\int_a^b G(\xi,x;\lambda)\,f(\xi)\,s(\xi)\,\mathrm{d}\xi,$$
(15.3.49)

where use has been made of the Green's function representation (15.3.17) of Φ; C_∞ is a large path encircling the poles of Φ, or G, in the counterclockwise sense. In the sense of distributions we can then write

$$-\frac{1}{2\pi i}\oint_{C_\infty}\mathrm{d}\lambda\,G(\xi,x;\lambda) = \frac{\delta(x-\xi)}{s(x)}$$
(15.3.50)

or, from (15.3.40),

$$\frac{1}{2\pi i}\oint_{C_\infty}\mathrm{d}\lambda\,G(\xi,x;\lambda)\,\mathrm{d}\xi = -\sum_{n=0}^{\infty}\bar{e}_n(\xi)e_n(x)$$
(15.3.51)

in agreement with the residue theorem applied to the bilinear expansion (15.3.42).

Another way to look at these results is to use the bilinear expansion (15.3.42) of G to rewrite (15.3.13) as

$$f(x) = \frac{1}{2\pi i} \oint_{C_\infty} d\lambda \sum_{n=0}^{\infty} e_n(x) \int_a^b s(\xi) \frac{e_n(\xi)}{\lambda - \lambda_n} f(\xi) \, d\xi \,. \tag{15.3.52}$$

If we define the finite transform of f by

$$\tilde{f}_n(\lambda) = \int_a^b s(\xi) \frac{e_n(\xi)}{\lambda - \lambda_n} f(\xi) \, d\xi \,, \tag{15.3.53}$$

(15.3.52) provides the inversion formula

$$f(x) = \frac{1}{2\pi i} \oint_{C_\infty} d\lambda \sum_{n=0}^{\infty} e_n(x) \tilde{f}_n(\lambda). \tag{15.3.54}$$

The results described here can be regarded as special cases of a more general framework concerning the eigenfunction of a continuous square-integrable real symmetric integral operator

$$\int_a^b K(x, \xi) e_n(\xi) \, d\xi = \frac{1}{\lambda_n} e_n(x). \tag{15.3.55}$$

If the operator is not degenerate (i.e., its kernel K cannot be expressed as the finite sum of products of functions of x by functions of ξ), then it has an infinity of orthonormal eigenfunctions and real eigenvalues (see e.g. Riesz & Sz.-Nagy 1955, p. 242; Courant & Hilbert 1953, vol. 1 p. 122) and the expansion

$$K(x, \xi) = \sum_{n=0}^{\infty} \frac{1}{\lambda_n} \bar{e}_n(\xi) e_n(x) \tag{15.3.56}$$

converges in the mean square sense. *Mercer's theorem* strengthens this result: convergence is uniform if all the eigenvalues, except possibly a finite number, have the same sign.

15.3.4 Mixed boundary conditions

Other than for the possible presence of degenerate eigenvalues, the situation for periodic boundary conditions is similar to that described above: there is an infinite number of eigenvalues which tend to infinity and the eigenfunctions oscillate with more and more zeros as the corresponding eigenvalue grows (see e.g. Ince 1926, p. 242; Levitan & Sargsjan 1975, p. 39; Levitan & Sargsjan 1991, p. 33) An expansion theorem similar to the previous one can also be proven; in fact, the expansion for the case of Example 15.2.2 is just the ordinary Fourier series.

15.4 Singular Sturm–Liouville problems

§15.2 was devoted to the Sturm–Liouville eigenvalue problem and, in particular, to the use of its eigenfunctions as a basis for series expansions. The regular case was treated in some detail, but it was pointed out that there are important practical cases, such as the Legendre and Bessel equations, in which the problem is singular. In this section we present a brief description of the new features encountered in this situation with the purpose of giving some understanding of why, in some cases, singular problems give rise to series expansions while, in others, the expansion takes the form of an integral transform, and while, in others still, one has a mixture of the two.

As defined in §15.1, a Sturm–Liouville problem is termed singular when $p^{-1}(x)$ or $q(x)$ have singularities at one or both end points of a finite interval $[a, b]$ or p vanishes at an interior point. We can limit ourselves to the case in which one of the end points is regular and the other one singular as, if there is more than one singular point, or the singularity is not at an end point, we can divide the interval into parts, each one bounded by a singular point and an arbitrarily chosen regular point. The problem is also singular when the interval extends to infinity in one or both directions. In all these cases the argument leading to the eigenfunction expansion needs to be modified. Since a change of variable can be used to map a singular problem with singularities of p or q at the end points of a finite interval to a problem on an infinite interval with p and q regular (except, possibly, at infinity), the semi-infinite interval case contains, in essence, all singular problems. Before turning to general results, it is interesting to consider an example which gives an indication of the type of behavior that may be expected.

Example 15.4.1 For clarity we consider a simple first-order problem (Stakgold 1997, p. 430) which was already used in §2.4 p. 40 as an introduction to the idea of the Green's function, namely

$$L\Phi = -i\frac{d\Phi}{dx} = \lambda s(x)\Phi + f(x), \qquad \Phi(0) = \Phi(b). \tag{15.4.1}$$

We assume that $s(x) > 0$. With the periodicity boundary conditions indicated, this operator is self-adjoint (Example 21.6.5, p. 655) in the function space with the scalar product (15.3.7); because of this important feature, this example mimics the more complex Sturm–Liouville case. The homogeneous equation associated to (15.4.1) is

$$-i\frac{du}{dx} = \lambda s(x)u, \qquad u(0) = u(b), \tag{15.4.2}$$

with solutions

$$u(x) = Ae^{i\lambda S(x)} \quad \text{where} \quad S(x) = \int_0^x s(\xi)\,d\xi. \tag{15.4.3}$$

The Green's function found in §2.4 is given in (2.4.19) p. 41 and is

$$G(x, y) = ie^{i\lambda[S(y)-S(x)]}\left[H(y-x) - \frac{1}{1 - e^{-i\lambda S(b)}}\right]. \tag{15.4.4}$$

From its poles we see immediately that the eigenvalues λ_n are

$$\lambda_n = \frac{2n\pi}{S(b)}, \qquad -\infty < n < \infty, \tag{15.4.5}$$

which, of course, can also be determined directly from the general solution (15.4.3) of the homogeneous equation by imposing the boundary conditions. Upon normalizing as in (15.3.10), the eigenfunctions are found to be

$$e_n(x) = [S(b)]^{-1/2} e^{i\lambda_n S(x)} \tag{15.4.6}$$

and they are readily seen to be orthogonal with the scalar product (15.3.7). We can expand the Green's function on the basis of the e_n as

$$G(x, y) = \sum_{-\infty}^{\infty} (e_n(x), G(x, y))_s \, e_n(y) = \sum_{-\infty}^{\infty} (e_n(y), G(x, y))_s \, e_n(x). \tag{15.4.7}$$

The coefficients are readily calculated using the fact that $\exp[\pm i\lambda_n S(b)] = 1$ with the result

$$\sum_{-\infty}^{\infty} \frac{\bar{e}_n(x)e_n(y)}{\lambda_n - \lambda} = G(x, y) \tag{15.4.8}$$

which is the bilinear expansion in this case. Upon substituting this expansion into $u(y) = \int_0^b G(x, y) f(x) \, dx$ we find

$$u(y) = \sum_{-\infty}^{\infty} \frac{c_n \, e_n(y)}{\lambda_n - \lambda}, \qquad \text{with} \qquad c_n = \int_0^b \bar{e}_n(\xi) f(\xi) s(\xi) \, d\xi \tag{15.4.9}$$

which, other than for the extended summation range, is the same as (15.3.11) p. 368.

Since $S(x)$ is the integral of a positive function, it is monotonically increasing and therefore either it tends to infinity as $b \to \infty$, or it tends to a finite limit. In the latter case the spectrum remains discrete and the situation is not different from the one that prevails on a finite interval. The eigenvalues and eigenfunctions are simply given by (15.4.5) and (15.4.6) with $S(b)$ replaced by $S(\infty)$. What we have here is essentially the Fourier series in exponential form in the variable $\sigma = S(x)$. Furthermore the homogeneous solutions (15.4.3) have a finite norm for any λ as

$$\|u\|_s^2 \propto \int_0^\infty e^{i(\lambda - \bar{\lambda})S(x)} s(x) \, dx = \int_0^{S(\infty)} e^{-2(\mathrm{Im}\,\lambda)S} dS < \infty. \tag{15.4.10}$$

If $S(b) \to \infty$, on the other hand, the situation is quite different: (15.4.5) shows that the eigenvalues coalesce on the real line and the spectrum becomes continuous (§21.9, p. 662). Heuristically, we may get an insight on what might happen in this case by repeating the argument used at the beginning of §10.1 to introduce the Fourier transform. Suppose $S(b)$ is very large. Then the eigenvalues are very closely spaced and one may expect that the coefficients c_n of the expansion $\sum c_n e_n$ of f do not change much from one value of n to a neighboring one. In this case, we can carry out the summation by grouping terms with values of n between N and $N + \Delta N$:

$$f(x) \simeq \sum_N c_N e_N(x) \Delta N = \frac{S(b)}{2\pi} \sum_{\lambda_N} c_N e_N(x) \Delta \lambda$$

$$= \frac{1}{2\pi} \sum_{\lambda_N} \Delta \lambda \, e^{i\lambda_N S(x)} \int_0^b f(\xi) e^{-i\lambda_N S(\xi)} s(\xi) \, d\xi , \qquad (15.4.11)$$

where (15.4.5) has been used to replace ΔN by $\Delta \lambda$. As $b \to \infty$, one would expect then that

$$f(x) \simeq \sum_N c_N e_N(x) \Delta n \to \frac{1}{\sqrt{2\pi}} \int_{-\infty}^{\infty} e^{i\lambda S(x)} \tilde{f}(\lambda) d\lambda \qquad (15.4.12)$$

with

$$\tilde{f}(\lambda) = \frac{1}{\sqrt{2\pi}} \int_0^{\infty} e^{-i\lambda S(\xi)} f(\xi) s(\xi) \, d\xi \qquad (15.4.13)$$

which are, respectively, the inverse and direct Fourier transforms in the variable $\sigma = S(x)$ of a function which vanishes for $S < 0$. It is also evident from (15.4.10) that in this case, for the solutions of the homogeneous equation to have a finite norm, it is necessary that $\text{Im}\,\lambda > 0$. We will see that Sturm–Liouville problems exhibit a behavior similar to the two types encountered in this example.   □

Let us return to the Sturm–Liouville case

$$\mathsf{L}u = -\frac{d}{dx}\left[p(x) \frac{du}{dx} \right] + q(x)\,u = \lambda s(x) u(x), \qquad a < x < b, \qquad (15.4.14)$$

with one regular and one singular end point. We have the important

Theorem 15.4.1 [Weyl]

(i) *Whenever* $\text{Im}\,\lambda \neq 0$, *(15.4.14) has one solution with a finite norm:*

$$\|u\|_s^2 \equiv \int_a^b |u(x)|^2 s(x)\,dx \; < \infty \qquad (15.4.15)$$

where the integral is over the (finite or infinite) interval of interest.

(ii) *If, for any one particular value of* λ, *real or complex, all solutions of (15.4.14) have a finite norm, then the same is true for all other real or complex values of* λ.

The second situation is referred to as the *limit circle* case and, similarly to what happens for $S(\infty) < \infty$ in Example 15.4.1, it leads to infinite-series expansions as in the regular case. Problems such that two linearly independent finite-norm solutions do not exist are of the *limit-point* type and may result in a discrete spectrum (see Example 15.4.8), a continuous spectrum (cf. the situation $S(b) \to \infty$ in Example 15.4.1), or a combination of the two. In this second case, for λ real, there may or may not be one finite-norm solution. A point worth stressing is that, to decide which case applies, it is sufficient to consider one single value of λ. A proof of this theorem, and a justification of its terminology, can be found e.g.

in Titchmarsh 1962, vol. 1 chapter 2; Levitan & Sargsjan 1975, chapter 2; or Levitan & Sargsjan 1991, chapter 2. We illustrate the theorem with a few examples.

Example 15.4.2 Consider

$$-\frac{d^2 u}{dx^2} = \lambda u. \tag{15.4.16}$$

If the interval of interest is $a < x < \infty$, the right point is singular. For $\lambda \neq 0$, two linearly independent solutions are $u_\pm(x; \lambda) = \exp(\pm i\sqrt{\lambda}x)$, with the square root defined so that $\sqrt{\lambda}$ is real and positive when λ is real and positive. For Im $\lambda > 0$, only u_+ has a finite norm and we are therefore in the limit-point case. For λ real and negative again only one solution has finite norm, while for real $\lambda \geq 0$ no solution with finite norm exists. The situation is similar if the interval of interest is $-\infty < x < b$, and also if $-\infty < x < \infty$, in which case both end points are singular and belong to the limit-point case.   □

Example 15.4.3 Consider now Bessel's equation

$$-\frac{d}{dx}\left(x\frac{du}{dx}\right) + \frac{v^2}{x}u = \lambda x u \tag{15.4.17}$$

for $v \geq 0$, with the associated norm and scalar product[12]

$$\|u\|^2 = \int_a^b x|u|^2 \, dx, \qquad (u, v) = \int_a^b x\overline{u}(x)v(x) \, dx. \tag{15.4.18}$$

If the interval is $0 < a < x < b < \infty$, the problem is regular. On $0 < x < b < \infty$, the left end point is singular. For $\lambda = 0$ the equation has solutions $u_1 = x^v$ and $u_2 = x^{-v}$, both of which have finite norm provided $0 < v < 1$. For $v = 0$ the solutions are $u_1 = 1$ and $u_2 = \log x$, both of which also have a finite norm. Thus, for $0 \leq v < 1$, we have the limit-circle case. For $v \geq 1$, on the other hand, only one solution has finite norm and we are in the limit-point case. We can confirm Weyl's result by considering, for general λ, the two linearly independent solutions

$$u_1 = J_v(\sqrt{\lambda}x) \qquad \text{and} \qquad u_2 = Y_v(\sqrt{\lambda}x) \tag{15.4.19}$$

which, for small x and $v \neq 0$, behave like

$$u_1 \propto \left(\frac{\sqrt{\lambda}x}{2}\right)^v, \qquad u_2 \propto \left(\frac{\sqrt{\lambda}x}{2}\right)^{-v}. \tag{15.4.20}$$

The first one is integrable near 0 while the second one is not if $v \geq 1$ as $x|u_2|^2 \propto x^{1-2v}$.

□

---

[12] When $a = 0$, the weight function x vanishes at 0. To avoid dealing with this problem directly, one could effect the Liouville transformation $u = w/\sqrt{x}$ (15.2.21), which brings the equation to the form (15.4.21). The norm would now be the integral of $|w|^2 = x|u|^2$ and no issue would then arise. For simplicity, we deal with (15.4.17) directly.

Example 15.4.4 We now consider the same equation, but in the interval $a < x < \infty$. With Liouville's transformation (15.2.21), $w = u/\sqrt{x}$, the equation becomes

$$\frac{d^2w}{dx^2} + \left(\lambda - \frac{\nu^2 - 1/4}{x^2}\right) w = 0, \tag{15.4.21}$$

from which we see that the behavior at the right singular point is similar to that of Example 15.4.2 so that this is a limit-point case for all ν. Over the interval $0 < x < \infty$, we have a combination of the results of this example and the previous one, except for $\nu = \frac{1}{2}$ in which case there is no singularity at the left end point. □

Example 15.4.5 For the Legendre equation in $-1 < x < 1$

$$-\frac{d}{dx}\left[(1 - x^2)\frac{du}{dx}\right] + \left(\lambda - \frac{m^2}{1 - x^2}\right) u = 0 \tag{15.4.22}$$

both end points $x = \pm 1$ are singular due to the simultaneous vanishing of p and divergence of q. For $\lambda = 0$, the transformation $\xi = (1/2)\log(1 + x)/(1 - x)$ brings the equation to the form

$$-\frac{d^2u}{d\xi^2} - m^2 u = 0 \tag{15.4.23}$$

with independent solutions $u_\pm = \exp(\pm im\xi)$. Both solutions are square integrable and the end points are therefore of the limit-circle type. For $m = 0$ two solutions are $u_1 = 1$ and $u_2 = x$, both of which are square integrable over $[-1, 1]$ confirming the limit-circle nature of this problem for all real m. □

15.4.1 Limit-point case

Let us consider the problem

$$\frac{1}{s}L\Phi - \lambda\Phi = f, \qquad a < x < b, \tag{15.4.24}$$

when one of the end points is singular while the other one is regular with an unmixed boundary condition. Proceeding to construct the Green's function as in §15.1 we are led to

$$L_x G(x, y) = \delta(x - y), \tag{15.4.25}$$

while the concomitant is

$$J[G, \Phi] = \left[p\left(\frac{\partial G}{\partial x}\Phi - \Phi'G\right)\right]_{x=a}^{x=b} \tag{15.4.26}$$

to be interpreted in a limit sense at the singular end point. The relevant solution of the homogeneous analog of (15.4.24) to be used for the construction of G in the interval bounded by the regular end point and $x = y$ is selected as before in §15.1.2, choosing boundary conditions ensuring that unknown boundary values of Φ and Φ' do not appear in the concomitant.

Examples 15.4.2 and 15.4.4 show that the single square-integrable solution of the homogeneous equation available in this case decays exponentially at the singular end point, so that the contribution of this point to the concomitant vanishes; this vanishing of the concomitant is a general result (see e.g. Naimark 1967, vol. 2 p. 78). Thus, the solution to be chosen to construct the Green's function between y and the singular point is simply a multiple of the only square-integrable solution.

Example 15.4.6 Let us consider the case of Example 15.4.2 on $0 < x < \infty$, with $\Phi(0) = 0$. Then, for $x < y$, $G(x, y; \lambda) = A(y) \sin \sqrt{\lambda} x$ while, for $x > y$, $G(x, y; \lambda) = B(y) \exp(i\sqrt{\lambda} x)$. The derivative jump condition and continuity at $x = y$ determine A and B with the final result

$$G(x, y; \lambda) = \frac{\sin \sqrt{\lambda} x_< \exp(i\sqrt{\lambda} x_>)}{\sqrt{\lambda}}, \qquad \lambda \notin [0, \infty) \tag{15.4.27}$$

with which the concomitant vanishes at infinity if Φ is bounded. We see that, in its dependence on λ, the Green's function has no poles, but it has a branch cut along the positive real axis across which it jumps by an amount

$$G\left(x, y; |\lambda| e^{i0+}\right) - G\left(x, y; |\lambda| e^{i2\pi-}\right) = \frac{2i}{\sqrt{|\lambda|}} \sin \sqrt{|\lambda|} x \, \sin \sqrt{|\lambda|} y. \tag{15.4.28}$$

We can see how this cut arises in the limit $b \to \infty$ by adapting (15.1.36), which gives us an expression for the Green's function for finite b:

$$G_b(x, y) = \frac{\sin \sqrt{\lambda} x_< \sin \sqrt{\lambda}(b - x_>)}{\sqrt{\lambda} \sin \sqrt{\lambda} b}. \tag{15.4.29}$$

This Green's function has simple poles at $\sqrt{\lambda_n} = n\pi/b$, which get closer and closer as b increases. It is easy to see that the limit of G_b for $b \to \infty$ is precisely (15.4.27) provided $\lambda \notin [0, \infty)$, which indicates that the branch cut arises from the coalescence of the poles. If λ is real and positive, (15.4.27) has no limit as $b \to \infty$. $\qquad\square$

Example 15.4.7 Let us generalize the previous problem by imposing, at $x = 0$,

$$\cos \alpha \, \Phi(0) + \sin \alpha \, \Phi'(0) = 0, \qquad 0 < x < \infty. \tag{15.4.30}$$

The Green's function is readily found to be

$$G(x, y; \lambda) = \frac{i e^{i\sqrt{\lambda} x_>}}{\sqrt{\lambda} - i \cot \alpha} \left[\cos \sqrt{\lambda} x_< - \cot \alpha \frac{\sin \sqrt{\lambda} x_<}{\sqrt{\lambda}} \right] \tag{15.4.31}$$

which reduces to (15.4.27) as $\alpha \to 0$. Here again we encounter a cut along the real positive axis of the λ plane but the new feature, which will play a role later in Example 15.4.12, is the presence of a pole at $\sqrt{\lambda} = i \cot \alpha$.[13] $\square$

Example 15.4.8 Let us now consider the Bessel function case of Example 15.4.3 on $0 < x < b < \infty$, with $\Phi(b) = 0$. If $\nu \geq 1$, we are in the limit-point case at $x = 0$ while the right end point is regular. In constructing the Green's function, therefore, the relevant solution for $y < x < b$ is $G(x, y; \lambda) = A(y)J_\nu(\sqrt{\lambda}x) + B(y)Y_\nu(\sqrt{\lambda}x)$, with $A(y)J_\nu(\sqrt{\lambda}b) + B(y)Y_\nu(\sqrt{\lambda}b) = 0$. For $0 < x < y$, we need to choose a square-integrable solution at $0+$, which requires $G(x, y; \lambda) = C(y)J_\nu(\sqrt{\lambda}x)$. The usual conditions at $x = y$ determine A, B and C:

$$G(x, y; \lambda) = \frac{\pi}{2}\left[Y_\nu(\sqrt{\lambda}x_>) - \frac{Y_\nu(\sqrt{\lambda}b)}{J_\nu(\sqrt{\lambda}b)}J_\nu(\sqrt{\lambda}x_>)\right]J_\nu(\sqrt{\lambda}x_<). \tag{15.4.32}$$

It might seem from this expression that, as a function of λ, G has a branch cut along the positive semi-axis as found before for (15.4.27). However, the argument of a $\lambda = \lambda_0 + i\varepsilon$ just above the positive real axis is small, δ say, so that $\sqrt{\lambda} \to \sqrt{\lambda_0}$. Just below the axis, the argument of $\lambda_0 - i\varepsilon$ is $2\pi - \delta$, and therefore $\sqrt{\lambda} \to -\sqrt{\lambda_0}$. By using the analytic continuation relations expressing $J_\nu(-z)$ and $Y_\nu(-z)$ in terms of $J_\nu(z)$ and $Y_\nu(z)$ (p. 307), it is readily shown that G is a continuous function of λ across the real axis so that no cut, in fact, exists. There are however poles at

$$\sqrt{\lambda_n} = \frac{j_{\nu,n}}{b} \qquad \text{with} \qquad J_\nu(j_{\nu,n}) = 0. \tag{15.4.33}$$

The zeros $j_{\nu,n}$ of J_ν are all real (p. 313). Since in this case $p(x) = x$, near $x = 0$ the concomitant is a linear combination of terms proportional to $\Phi(x)x^\nu$ and $\Phi'(x)x^{\nu+1}$, both of which tend to 0 if Φ is bounded.

This example illustrates the fact that a limit-point case is not necessarily associated with a continuous spectrum.

We can determine the normalized eigenfunctions corresponding to the eigenvalues (15.4.33) by calculating the residue of G at the poles as in (15.3.43). From the Wronskian relation $W(J_\nu, Y_\nu; z) = 2/(\pi z)$ with $z = j_{\nu,n}$ we have $Y_\nu(j_{\nu,n}) = 2[\pi j_{\nu,n} J_\nu'(j_{\nu,n})]^{-1}$ so that

$$\lim_{\lambda \to \lambda_n}(\lambda - \lambda_n)G(x, y; \lambda) = \left[\frac{\sqrt{2}}{bJ_\nu'(j_{\nu,n})}J_\nu\left(j_{\nu,n}\frac{x}{b}\right)\right]\left[\frac{\sqrt{2}}{bJ_\nu'(j_{\nu,n})}J_\nu\left(j_{\nu,n}\frac{y}{b}\right)\right] \tag{15.4.34}$$

[13] In writing the boundary condition (15.4.30) we have changed the sign of α with respect to similar conditions used before, e.g. (15.2.5), so that the pole arises for α in the convenient range $0 < \alpha < \pi/2$. Indeed, due to the presence of the cut, the argument of of $\sqrt{\lambda}$ ranges between 0 and π and therefore $\sqrt{\lambda}$ can equal $i \cot \alpha$ only as long as $\cot \alpha > 0$.

which has been written in the form suggested by (15.3.43) and which permits us to identify the eigenfunctions:

$$e_n(x) = \frac{\sqrt{2}}{bJ'_\nu(j_{\nu,n})} J_\nu\left(j_{\nu,n}\frac{x}{b}\right) = \frac{\sqrt{2}}{bJ_{\nu+1}(j_{\nu,n})} J_\nu\left(j_{\nu,n}\frac{x}{b}\right) \tag{15.4.35}$$

normalized so that

$$\int_0^b \bar{e}_k(x)e_n(x)x\,\mathrm{d}x = \delta_{kn}. \tag{15.4.36}$$

The sign difference between (15.4.34) and (15.3.43) is due to the fact that the singular point is the left end point of the interval. According to (15.3.26), a function f can then be expanded as

$$f(x) = \sum_{n=1}^\infty c_n e_n(x) \qquad \text{where} \qquad c_n = \frac{\sqrt{2}}{bJ'_\nu(j_{\nu,n})} \int_0^b J_\nu\left(j_{\nu,n}\frac{x}{b}\right) f(x)x\,\mathrm{d}x.$$

$$\tag{15.4.37}$$

It will be recognized that this is nothing other than the Fourier–Bessel series of §12.5. If, in place of $\Phi(b) = 0$, we had imposed a boundary condition of the more general form $a_{21}\Phi(b) + a_{22}\Phi'(b) = U_b$, we would have been led to an expansion similar to (15.4.37) with eigenfunctions (Erdélyi *et al.* 1953, vol. 2 p. 71)

$$e_n(x) = \frac{k_{\nu,n}}{b}\left[\frac{2}{(k_{\nu,n})^2 J'^2_\nu(k_{\nu,n}) + [k^2_{\nu,n} - \nu^2]J^2_\nu(k_{\nu,n})}\right]^{1/2} J_\nu\left(k_{\nu,n}\frac{x}{b}\right) \tag{15.4.38}$$

with positive real numbers $k_{\nu,n}$ such that $a_{21}J_\nu(k_{\nu,n}) + a_{22}k_{\nu,n}J'_\nu(k_{\nu,n}) = 0$ (see §12.5).
□

Example 15.4.9 If we now consider Example 15.4.8, but with the interval of interest unbounded on the right, we would still pick $G(x, y; \lambda) = C(y)J_\nu(\sqrt{\lambda}x)$ as the appropriate solution for $0 < x < y$, while, for $y < x$, the solution bounded at infinity is $G(x, y; \lambda) = B(y)H^{(1)}_\nu(\sqrt{\lambda}x)$. The final result is

$$G(x, y; \lambda) = \frac{\pi}{2}iJ_\nu(\sqrt{\lambda}x_<)H^{(1)}_\nu(\sqrt{\lambda}x_>). \tag{15.4.39}$$

As in Example 15.4.6, provided λ is not real and positive, this is also the limit of (15.4.32) for $b \to \infty$ since $Y_\nu(\sqrt{\lambda}b)/J_\nu(\sqrt{\lambda}b) \to i$ in this limit. On the other hand, (15.4.32) has no limit as $b \to \infty$ for λ real and positive.
□

15.4.2 Limit-circle case

In the limit-point case, up to a multiplicative constant, there was only one square-integrable solution of the homogeneous equation in the interval between y and the singular end point, and no ambiguity arose in the construction of the Green's function. In the limit-circle case, however, two linearly independent solutions exist, which can be combined in different ways to give rise to different Green's functions. Although two linearly independent solutions also exist for the regular problem, there is no ambiguity there as the condition that no unknown boundary information be involved in the final solution gives an unambiguous prescription for the choice of the correct linear combination, as we have seen. In the singular case, however, this procedure is not available in general as the functions we are dealing with may not have a definite value at the singular point. Thus, the conventional notion of boundary condition fails and a generalization is necessary. The basic idea is to "balance" the behavior of the solutions w of the homogeneous equation by using functions $f_1(x)$ and $f_2(x)$ such that, as x approaches the singular point, the limits

$$\lim \, p(x)\,(f_n w' - f_n' w) = \lim \, p(x)\,W(f_n, w), \qquad n = 1, 2, \tag{15.4.40}$$

exist for all w. For given f_n, the values of these limits then distinguish different solutions of the homogeneous equation (see e.g. Dunford & Schwartz 1963, vol. 2, chapter 13).

The general theory of the limit-circle case is not simple. Heuristic arguments and examples can be found e.g. in Stakgold 1967, vol. 1, p. 310 and 1997, p. 444. The complete theory from the point of view of self-adjoint linear operators is given in detail in Dunford & Schwartz 1963, vol. 2, chapter 13; Titchmarsh 1962, vol. 1, chapters 2 and 4; Levitan & Sargsjan 1975, chapters 2 and 5. Levitan & Sargsjan 1991, chapter 2, follow a more direct approach that does not appeal to operator theory and Hajmirzaahmad & Krall 1992 give a self-contained presentation of the subject with several examples. A different line of attack is briefly presented in Richtmyer 1978, vol. 1, p. 209. The brief treatment that follows aims to give only a broad outline of the subject.

Let the singular point be at $x = a$, and let $\phi_\pm$ be two linearly independent solutions of the homogeneous equation (15.4.14). By variation of parameters the solution of the non-homogeneous equation (15.4.24) will have the form

$$
\begin{aligned}
\Phi(x) \;=\;& \frac{1}{W_0}\left(c_- + \int_a^x \phi_+(\xi)s(\xi)f(\xi)\,\mathrm{d}\xi \right)\phi_-(x) \\
& + \frac{1}{W_0}\left(c_+ - \int_a^x \phi_-(\xi)s(\xi)f(\xi)\,\mathrm{d}\xi \right)\phi_+(x),
\end{aligned}
\tag{15.4.41}
$$

where W_0 is the constant

$$W_0 \;=\; p(x)\,W(\phi_-, \phi_+) \;=\; p(x)\left[\phi_-(x)\phi_+'(x) - \phi_-'(x)\phi_+(x) \right], \tag{15.4.42}$$

which is finite in the present limit-circle case if both $\phi_\pm$ and $L\phi_\pm$ are square integrable in the sense of (15.4.15) (see e.g. Stakgold 1997, p. 444). For given $\phi_\pm$, the constants $c_\pm$ identify different solutions of (15.4.24) much as ordinary boundary conditions specify different solutions. In other words, for given $\phi_\pm$, the constants $c_\pm$ select a unique solution Φ. A simple direct calculation shows that

$$\lim_{x \to a} p(x)W(\phi_+, \Phi) = c_+, \qquad \lim_{x \to a} p(x)W(\phi_+, \Phi) = -c_-, \tag{15.4.43}$$

so that $\phi_\pm$ can play the role of the f_1 and f_2 in (15.4.40). If we were to prescribe, therefore, a condition of the type

$$\cos\alpha\,[p(x)W(\phi_-, \Phi)]_a - \sin\alpha\,[p(x)W(\phi_+, \Phi)]_a = 0, \tag{15.4.44}$$

coupled with a standard condition at the regular point $x = b$,

$$\cos \beta \, \Phi(b) + \sin \beta \Phi'(b) = \Phi_b, \tag{15.4.45}$$

we would have two meaningful equations determining the constants $c_\pm$. It may be noted that (15.4.44) does not involve unknown information on Φ. Depending on the specific problem, there may be alternative ways to express the condition (15.4.44), for example by requiring that Φ be regular, but the condition is meaningful as stated in (15.4.44) at least when $\|f\|_s < \infty$, with the norm defined as in (15.4.15). This follows from the identity $\|f\|_s^2 = \|\mathsf{L}\Phi - \lambda\Phi\|_s^2$ and the Schwarz inequality, which shows that $\|\mathsf{L}\Phi\|_s$ is also finite (see e.g. Stakgold 1997, p. 444). The parameter α in (15.4.44) is arbitrary and it is necessary to rely on problem-specific external information in order to make a definite choice.

To investigate the eigenvalues of the operator L we may rewrite the eigenvalue equation in the form

$$\frac{1}{s}\mathsf{L}u - \mu u = (\lambda - \mu)u, \quad a - x < b. \tag{15.4.46}$$

where μ is an arbitrary constant. The equation has now the form (15.4.24) and, if $(\lambda - \mu)u$ is substituted into (15.4.41) in place of f requiring that u satisfy (15.4.44) and the homogeneous form of (15.4.45) one finds an integral equation of the Hilbert–Schmidt type (15.3.1) p. 366 with λ replaced by $\lambda - \mu$, which guarantees the existence of a discrete spectrum with a countable infinity of real eigenvalues and oscillatory eigenfunctions. Thus, in the limit-circle case, the situation is similar to the regular case.

Example 15.4.10 Let us consider once again the Bessel equation for $0 < x < 1$ in the limit-circle case $\nu = 0$ (Example 15.4.3 p. 378). With an eye toward the solution of the eigenvalue problem, as (15.4.46) shows, it does not matter which value of λ we consider and it proves convenient to take $\lambda = 0$ so that the problem becomes

$$-\frac{1}{x}\frac{d}{dx}\left(x\frac{d\Phi}{dx}\right) = f. \tag{15.4.47}$$

We also require that $\beta = 0$ in the boundary condition (15.4.45) and $\Phi_b = 0$ as well. Two solutions of the homogeneous equations are $\phi_- = 1$ and $\phi_+ = \log x$, so that $W_0 = 1$. The solution given by (15.4.41) is then

$$\Phi(x) = c_- + \int_0^x \xi \log(\xi) f(\xi) \, d\xi + \left(c_+ - \int_0^x \xi f(\xi) \, d\xi\right) \log x. \tag{15.4.48}$$

Upon imposing the boundary conditions we readily find

$$\Phi = -\int_x^1 \xi \log \xi \, f(\xi) \, d\xi + \left(\tan \alpha \int_0^1 \xi \log \xi \, f(\xi) \, d\xi + \int_0^x \xi f(\xi) \, d\xi\right) \log x$$

$$= \int_0^1 \xi \left[\tan \alpha \log x \log \xi - \log x_>\right] f(\xi) \, d\xi \tag{15.4.49}$$

with $x_> = \max(x, \xi)$. We therefore see that the Green's function is given by

$$G(\xi, x) = \tan \alpha \log x \log \xi - \log x_> ; \tag{15.4.50}$$

G is symmetric in x and ξ, as expected.

According to (15.4.46), for the eigenvalue problem we replace Φ by the eigenfunction u and f by λu to find

$$\frac{1}{\lambda} u = \int_0^1 \xi \left[\tan \alpha \, \log \xi \, \log x - \log x_> \right] u(\xi) \, d\xi . \tag{15.4.51}$$

The kernel of the integral operator is clearly square integrable over the unit square so that the Hilbert–Schmidt theorem (p. 366 and p. 639) applies with the consequence that a denumerable infinity of eigenvalues and eigenfunctions exists. To investigate the eigenvalue problem directly we solve

$$-\frac{1}{x} \frac{d}{dx} \left(x \frac{du}{dx} \right) = \lambda u \tag{15.4.52}$$

subject to (15.4.44) at $x = 0$ and $u(1) = 0$. Upon imposing this latter condition, the general solution of the equation is

$$u(x) \propto \frac{J_0(\sqrt{\lambda} \, x)}{J_0(\sqrt{\lambda})} - \frac{Y_0(\sqrt{\lambda} \, x)}{Y_0(\sqrt{\lambda})}. \tag{15.4.53}$$

The condition (15.4.44) at $x = 0$ becomes

$$\lim_{x \to 0} \left[\cos \alpha \, x \frac{du}{dx} - \sin \alpha \left(x \log x \frac{du}{dx} - u \right) \right] = 0, \tag{15.4.54}$$

which, using

$$J_0(z) \simeq 1 - \frac{z^2}{4}, \qquad Y_0(z) \simeq \frac{2}{\pi} (\log z + \gamma) + O(x^2 \log x), \tag{15.4.55}$$

with γ Euler's constant, reduces to

$$-\frac{2 \cos \alpha}{\pi Y_0(\sqrt{\lambda})} - \sin \alpha \left[\frac{2}{\pi Y_0(\sqrt{\lambda})} \left(\log \frac{\sqrt{\lambda}}{2} + \gamma \right) - \frac{1}{J_0(\sqrt{\lambda})} \right] = 0 \tag{15.4.56}$$

or

$$\left[1 + \tan \alpha \left(\log \frac{\sqrt{\lambda}}{2} + \gamma \right) \right] J_0(\sqrt{\lambda}) - \frac{\pi}{2} \tan \alpha \, Y_0(\sqrt{\lambda}) = 0. \tag{15.4.57}$$

By picking different values of α one can generate an infinity of different eigenfunction sets and corresponding orthogonal expansions. In particular, for $\alpha = 0$, $\sqrt{\lambda} = j_{0,n}$, the zeros of J_0, and the normalized eigenfunctions give the Fourier–Bessel series for $\nu = 0$ (§12.5 p. 312).

While all solutions (15.4.53) are square integrable, only $\alpha = 0$ gives rise to eigenfunctions bounded at 0, which would often be a requirement stemming from the specific application one is interested in. This remark illustrates the comment made earlier that often one can rely on problem-specific information rather than imposing a condition of the form (15.4.44) in order to resolve the ambiguity which affects the choice of the relevant homogeneous solution in the limit-circle situation.   □

15.4.3 "Eigenfunction expansion" with a continuous spectrum

We have seen in Example 15.4.6 how a continuous spectrum can arise from the coalescence of the eigenvalues as one of the end points of the interval of interest becomes singular. It is natural to enquire what happens to the eigenfunction expansion (15.3.13) in such a situation.

For the case of a discrete spectrum we have derived in §15.3.3 the relation (15.3.49):

$$f(x) = -\frac{1}{2\pi i} \oint_{C_\infty} d\lambda \int_a^b G(\xi, x; \lambda)\, f(\xi)\, s(\xi)\, d\xi \,. \tag{15.4.58}$$

Let us accept that this relation is also valid when the poles of G coalesce into a branch cut along (part of) the real λ-axis, as would happen in the limit-point case with a self-adjoint operator. If there are no discrete eigenvalues the integration contour can be deformed into a path coming from $+\infty$ just above the cut and returning to $+\infty$ just below the cut. This part of the spectrum would then give a contribution

$$\int_{\lambda_0}^\infty \frac{[G(\xi, x; \lambda)]}{2\pi i}\, d\lambda, \tag{15.4.59}$$

in which λ_0 is the leftmost point of the cut (which could be at $-\infty$) and $[G]$ denotes the jump in G in traversing the cut:[14]

$$[G(x, \xi; \lambda)] = \lim_{\varepsilon \to 0} [G(\xi, x; \lambda + i\varepsilon) - G(\xi, x; \lambda - i\varepsilon)] \,. \tag{15.4.60}$$

Allowing for the possible presence of discrete eigenvalues, (15.4.58) then becomes

$$f(x) = \sum_n \left(\int_a^b \bar{e}_n(\xi) f(\xi) s(\xi)\, d\xi \right) e_n(x) + \int_{\lambda_0}^\infty d\lambda \int_a^b \frac{[G]}{2\pi i} f(\xi) s(\xi)\, d\xi \,, \tag{15.4.61}$$

in which the first summation is the contribution of the discrete spectrum, with e_k the corresponding normalized eigenfunctions and the coefficients given as usual by (15.3.24). In particular, if the discrete spectrum is empty and $[G]$ can be split as

$$\frac{[G](x, \xi; \lambda)}{2\pi i} = \Lambda(\lambda)\, g_1(x, \lambda)\, g_2(\xi, \lambda) \tag{15.4.62}$$

we have the transform pair

$$\tilde{f}(\lambda) = \int_a^b g_2(\xi, \lambda)\, f(\xi)\, s(\xi)\, d\xi, \qquad f(x) = \int_{\lambda_0}^\infty \Lambda(\lambda) g_1(x, \lambda) \tilde{f}(\lambda)\, d\lambda. \tag{15.4.63}$$

The content of (15.4.61) can be equivalently expressed as

$$\frac{\delta(x - \xi)}{s(\xi)} = \sum_n \bar{e}_n(\xi) e_n(x) + \int_{\lambda_0}^\infty d\lambda\, \frac{[G](\xi, x; \lambda)}{2\pi i} \tag{15.4.64}$$

---

[14] When λ is just above the cut, the integral runs from $+\infty$ in the negative direction; this fact explains the sign change between (15.4.58) and (15.4.61).

and, similarly, that of (15.4.63) as

$$\frac{\delta(x-\xi)}{s(\xi)} = \int_{\lambda_0}^{\infty} \Lambda(\lambda) g_1(x, \lambda) g_2(\xi, \lambda) \, d\lambda. \tag{15.4.65}$$

This is the common structure underlying the well known and most of the lesser known transforms of applied mathematics.

Example 15.4.11 [Fourier sine transform] Consider the case of Example 15.4.6. We had found the Green's function to be

$$G(x, y; \lambda) = \frac{\sin \sqrt{\lambda} x_< \exp(i \sqrt{\lambda} x_>)}{\sqrt{\lambda}}, \qquad \lambda \notin [0, \infty) \tag{15.4.66}$$

which has a branch cut extending from 0 to ∞. Above the cut the argument of $\lambda + i\varepsilon$ is small, δ say, so that $\sqrt{\lambda + i\varepsilon} \to \sqrt{\lambda}$. Below the cut, the argument of $\lambda - i\varepsilon$ is $2\pi - \delta$, and therefore $\sqrt{\lambda - i\varepsilon} \to -\sqrt{\lambda}$. Thus, the jump across the cut is

$$\frac{[G]}{2i\pi} = \frac{\sin \sqrt{\lambda} x \, \sin \sqrt{\lambda} \xi}{\pi \sqrt{\lambda}}, \tag{15.4.67}$$

which has the factorized form (15.4.62) with $\Lambda = 1/2\sqrt{\lambda}$ and

$$g_1 = g_2 = \sqrt{\frac{2}{\pi}} \sin \sqrt{\lambda} x. \tag{15.4.68}$$

Upon setting $\sqrt{\lambda} = k^2$, (15.4.63) become

$$f(x) = \sqrt{\frac{2}{\pi}} \int_0^{\infty} \tilde{f} \sin kx \, dk, \qquad \tilde{f}(k) = \sqrt{\frac{2}{\pi}} \int_0^{\infty} f(\xi) \sin kx \, d\xi, \tag{15.4.69}$$

which are the defining relations for the Fourier sine transform (§10.5 p. 272). A similar development starting from Example 15.4.9 p. 382 leads to the Hankel transform.   □

Example 15.4.12 [Generalized Fourier transform] Consider now the case of Example 15.4.7, the Green's function of which was given in (15.4.31):

$$G(x, y; \lambda) = i \frac{e^{i\sqrt{\lambda} x_>}}{\sqrt{\lambda} - i \cot \alpha} \left[\cos \sqrt{\lambda} x_< - \cot \alpha \frac{\sin \sqrt{\lambda} x_<}{\sqrt{\lambda}} \right]. \tag{15.4.70}$$

The residue at the pole $\sqrt{\lambda} = i \cot \alpha$ is

$$\text{Res}\,[G(x, y; \lambda); i \cot \alpha] = -2 \cot \alpha \, \exp(-x \cot \alpha) \, \exp(-y \cot \alpha) \tag{15.4.71}$$

while the jump across the branch cut is

$$\frac{[G]}{2\pi i} = \frac{\sqrt{\lambda} g(x, \sqrt{\lambda}) g(y, \sqrt{\lambda})}{2(\lambda + \cot^2 \alpha)}, \qquad g(x, \sqrt{\lambda}) = \sqrt{\frac{2}{\pi}} \left[\cos \sqrt{\lambda} x - \cot \alpha \frac{\sin \sqrt{\lambda} x}{\sqrt{\lambda}} \right] \tag{15.4.72}$$

which has the form (15.4.62). Thus, on setting $\lambda = k^2$, we have

$$f(x) = 2\cot\alpha\, e^{-x\cot\alpha} \int_0^\infty f(y)e^{-y\cot\alpha}\,dy + \int_0^\infty \frac{k^2}{k^2 + \cot^2\alpha} g(x, k)\tilde{f}(k)\,dk$$

$$(15.4.73)$$

with

$$\tilde{f}(k) = \int_0^\infty g(y, k) f(y)\,dy. \tag{15.4.74}$$

In particular, for $\cot\alpha = 0$, we have

$$\tilde{f}(k) = \sqrt{\frac{2}{\pi}} \int_0^\infty f(y)\cos ky\,dy, \qquad f(x) = \sqrt{\frac{2}{\pi}} \int_0^\infty \tilde{f}(k)\cos kx\,dk, \quad (15.4.75)$$

i.e., the ordinary Fourier cosine transform while, for $\alpha \to 0$, we recover the sine transform of the previous example. This can be seen by taking the limit $\cot\alpha \to \infty$ in the expression (15.4.72) of $[G]$ and by noting that the first term of (15.4.73) between 0 in the limit. □

15.5 Initial-value problem for ordinary differential equations

In preparation for the development of Green's function theory for the diffusion and wave equations in the next chapter, let us consider the initial-value problem for an ordinary differential equation:

$$\mathsf{L}_t u(t) \equiv \frac{d^2 u}{dt^2} + a(t)\frac{du}{dt} + b(t)u(t) = f(t), \qquad 0 < t, \tag{15.5.1}$$

in which we interpret the variable as time for definiteness. We assume general initial conditions of the form

$$u(0) = u_0, \qquad \frac{du}{dt}\bigg|_{t=0} = u_0', \tag{15.5.2}$$

and that a, b, f and the initial conditions are such that a unique solution exists and, for simplicity, that all quantities are real.

To construct the Green's function we proceed as in the simple-minded approach used in the introductory example 2.4.2 p. 40 deferring a better justification to the next section. Let $\tau > 0$ be the generic future time at which the solution u is desired. Then, we multiply (15.5.1) by a function $G(t, \tau)$ and integrate by parts to find

$$\int_0^\tau G(t, \tau)\mathsf{L}_t u(t)\,dt = \int_0^\tau \left[\frac{\partial^2 G}{\partial t^2} - \frac{\partial}{\partial t}[a(t)G(t, \tau)] + b(t)G(t, \tau) \right] u(t)\,dt$$

$$+ \left[G(t, \tau)\left(au + \frac{du}{dt} \right) - \frac{\partial G}{\partial t}u \right]_0^\tau = \int_0^\tau G(t, \tau)f(t)\,dt. \tag{15.5.3}$$

If we impose that G satisfy

$$\mathsf{L}_t^* G(t, \tau) \equiv \frac{\partial^2 G}{\partial t^2} - \frac{\partial}{\partial t}[a(t)G(t, \tau)] + b(t)G(t, \tau) = \delta(t - \tau) \tag{15.5.4}$$

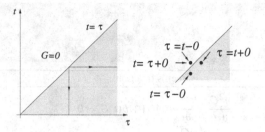

Fig. 15.1   Two ways to construct the Green's function for an initial-value problem.

and use the initial conditions (15.5.2), (15.5.3) becomes

$$u(\tau) = \int_0^{\tau+0} G(t, \tau) f(t)\, dt - G(\tau+0, \tau)\left[au + \frac{du}{d\tau}\right]_{t=\tau+0}$$

$$+ \left.\frac{\partial G}{\partial t}\right|_{t=\tau+0} u(\tau) + G(0, t)\left[a(0)u_0 + u_0'\right] - \left.\frac{\partial G}{\partial t}\right|_{t=0} u_0. \qquad (15.5.5)$$

Here we have extended the integration to $\tau+0$ so as to have the full benefit of the δ in calculating $\int_0^\tau \delta(t - \tau)u(t)\, dt = u(\tau).$[15] We also note that, because of causality, the solution up to a time τ can only depend on the values of f between 0 and τ so that $G(t, \tau)$ must vanish identically for $t > \tau$. Thus, in fact, integration up to $t = \tau$ does not entail any particular limitation.

Future information on u at $t = \tau$ is of course unknown but we can remove it from (15.5.5) by imposing that G satisfy[16]

$$G(t = \tau + 0, \tau) = 0, \qquad \left.\frac{\partial G}{\partial t}\right|_{t=\tau+0} = 0. \qquad (15.5.6)$$

In this way we find an explicit expression for $u(\tau)$, namely

$$u(\tau) = \int_0^\tau G(t, \tau) f(t)\, dt + \left[u_0' + a(0)u_0\right] G(0, \tau) - \left.\frac{\partial G}{\partial t}\right|_{t=0} u_0. \qquad (15.5.7)$$

As for the conditions necessary to specify a unique solution of (15.5.4) for $0 < t < \tau$ we use the same argument used before in connection with (15.1.11) p. 350 to conclude that G will be continuous at $t = \tau$, while its derivative undergoes a jump. Upon integrating (15.5.4) from $t = \tau - 0$ to $t = \tau + 0$, by the second of (15.5.6), we find

$$\left.\frac{\partial G}{\partial t}\right|_{t=\tau-0} = -1 \qquad \text{while} \qquad G(t = \tau - 0, \tau) = 0. \qquad (15.5.8)$$

---

[15] As remarked on p. 38, provided it is adhered to consistently, this choice is arbitrary and it is convenient here for obvious reasons.

[16] Note that $G = 0$ for $t > \tau$ is the proper solution of (15.5.4) subject to (15.5.6) at $t = \tau + 0$.

	Green's function	Impulse response

Table 15.2 For τ fixed, the Green's function can be constructed according to the left column as a function of t. For t fixed, it can be constructed according to the right column as a function of τ and it is then known as the impulse response.

Green's function	Impulse response
$\dfrac{\partial^2 G}{\partial t^2} - \dfrac{\partial}{\partial t}[a(t)G] + b(t)G = 0$	$\dfrac{\partial^2 G}{\partial \tau^2} + a(\tau)\dfrac{\partial G}{\partial \tau} + b(\tau)G = 0$
$\dfrac{\partial G}{\partial t}\bigg|_{t=\tau-0} = -1$	$\dfrac{\partial G}{\partial \tau}\bigg|_{\tau=t+0} = 1$
$G(t = \tau \pm 0, \tau) = 0$	$G(t, \tau = t \pm 0) = 0$
$G(t, \tau) = 0$ for $t > \tau$	$G(t, \tau) = 0$ for $\tau < t$

$$u(\tau) = \int_0^\tau G(t, \tau) f(t)\, dt + \left[u_0' + a(0)u_0\right] G(0, \tau) - \dfrac{\partial G}{\partial t}\bigg|_{t=0} u_0.$$

These are "final" conditions on G which, together with the differential equation (15.5.4), specify it completely. From this point of view, G describes the propagation "backward in time" of a unit δ-pulse occurring at time τ.

We can get a "forward" interpretation of G by drawing some further conclusions from the previous construction. Since, by the way it has been obtained, (15.5.7) satisfies the original equation (15.5.1), it must be that[17]

$$L_\tau u(\tau) = \int_0^\tau L_\tau G(t, \tau) f(t)\, dt + L_\tau \left[\left(u_0' + a(0)u_0\right) G(0, \tau) - \dfrac{\partial G}{\partial t}\bigg|_{t=0} u_0 \right] = f(\tau). \tag{15.5.9}$$

For this relation to hold it is necessary that

$$L_\tau G(t, \tau) = \dfrac{\partial^2 G}{\partial \tau^2} + a(t)\dfrac{\partial G}{\partial \tau} + b(\tau)G = \delta(\tau - t), \tag{15.5.10}$$

which also causes the second term of (15.5.9) to vanish as $L_\tau G(0, \tau) = 0$ for $\tau > 0$. Integration of (15.5.10) across $t = \tau$, in view of the continuity of G, gives

$$\dfrac{\partial G}{\partial \tau}\bigg|_{\tau=t+0} = 1, \quad \text{and} \quad G(t, \tau = t + 0) = 0. \tag{15.5.11}$$

Thus, in its second variable, G is a solution the original equation with a δ-forcing and it therefore represents the response of the system to a unit impulse occurring at $\tau = t$, which justifies the denomination *impulse response*, or *causal fundamental solution*, by which this function is often designated (Figure 15.1). As noted before, G vanishes for $t > \tau$, i.e., $\tau < t$ which is the time at which the impulse is applied. At $\tau = t + 0$ it satisfies the initial

---

[17] When L_τ is brought inside the integral, a term $a(\tau)G(\tau, \tau) + \partial_\tau[G(\tau, \tau)f(\tau)]$ also results, which vanishes by (15.5.6).

conditions (15.5.11) with this second interpretation of G, the integral in (15.5.7) may be seen as the summation of the effects of f at time t "propagated" forward by G to the desired time τ.

As summarized in Table 15.2, the Green's function can be constructed either by solving (15.5.4) "backward" in t, subject to the "final" conditions (15.5.8), or by solving (15.5.10) forward in τ subject to the initial conditions (15.5.11). The result will be the same with either procedure. In the first line of the table we omit the δ's in the right-hand sides as their effect is incorporated in the jump conditions.

Example 15.5.1 As an example, let us consider a forced oscillator with negative damping:

$$\frac{d^2 u}{dt^2} - 2c\frac{du}{dt} + (1 + c^2)u(t) = f(t) \tag{15.5.12}$$

with the general initial conditions (15.5.2); c is a constant. Proceeding according to the left column of Table 15.2, we start from

$$\frac{\partial^2 G}{\partial t^2} + 2c\frac{\partial G}{\partial t} + (1 + c^2)G = \delta(t - \tau), \tag{15.5.13}$$

which has the general solution

$$G(t, \tau) = [A(\tau)\sin t + B(\tau)\cos t]\, e^{-ct}. \tag{15.5.14}$$

By imposing the conditions (15.5.8) we find

$$A\sin\tau + B\cos\tau = 0, \qquad [A\cos\tau - B\sin\tau]e^{-c\tau} = -1, \tag{15.5.15}$$

with solution $A = -e^{c\tau}\cos\tau$, $B = e^{c\tau}\sin\tau$. Thus

$$G(t, \tau) = e^{c(\tau-t)}\sin(\tau - t)\, H(\tau - t) \tag{15.5.16}$$

where the last factor ensures that $G = 0$ for $t \geq \tau + 0$. According to the "impulse response" procedure summarized in the right column of Table 15.2, we start from

$$\frac{\partial^2 G}{\partial\tau^2} - 2c\frac{\partial G}{\partial\tau} + (1 + c^2)G = \delta(t - \tau) \tag{15.5.17}$$

finding the general solution

$$G(t, \tau) = [C(t)\sin\tau + B(t)\cos\tau]\, e^{c\tau}. \tag{15.5.18}$$

The initial conditions at $\tau = t$ are

$$[C(t)\sin t + D(t)\cos t]\, e^{ct} = 0, \qquad [C(t)\cos t - D(t)\sin t]\, e^{ct} = 1, \tag{15.5.19}$$

with solution $C(t) = e^{-ct}\cos t$, $D(t) = -e^{-ct}\sin t$, with which (15.5.16) is recovered. $\square$

We have not yet verified that (15.5.7) satisfies the initial conditions (15.5.2). For this purpose we note that the limit $\tau \to 0$ implies evaluation at time $\tau = 0+$. By the second of (15.5.8), we have $G(0, 0+) = 0$ while, by the first, $\partial G/\partial t|_{t=\tau-0} = -1$, so that $u(\tau =$

$0+) = u_0$ as required. Verification of the second initial conditions is slightly more involved and hinges on the fact that

$$\left.\frac{\partial^2 G}{\partial \tau \partial t}\right|_{t=\tau-0} = a(\tau), \tag{15.5.20}$$

which can be proven by applying $\partial/\partial\tau$ to the equation (15.5.4) for G and integrating over t between $\tau - 0$ and $\tau + 0$.

15.6 A broader perspective on Green's functions

We now present a more general view of Green's functions which is useful to better appreciate the developments of the previous §15.1 and §15.5 and to set the stage for the next chapter.

We wish to solve a linear non-homogeneous ordinary or partial differential equation of the general form

$$\mathsf{L}_x u(x) = f(x), \qquad x \in \Omega. \tag{15.6.1}$$

The domain of interest Ω may in general be multi-dimensional and the subscript x is appended to the linear operator L to specify the variable, or variables, on which it acts. We associate to (15.6.1) some appropriate boundary conditions which we write generically as[18]

$$\mathsf{B}_x u = U(x), \qquad x \in \partial\Omega, \tag{15.6.2}$$

in which B_x is a suitable linear operator which evaluates u on the boundary $\partial\Omega$ of the domain Ω and $U(x)$ indicates some general boundary data, not necessarily of the Dirichlet type. We may suppose that the function u belongs to a function space with a scalar product (v, u) between u and v defined by

$$(v, u) = \int_\Omega \bar{v}(x)\, u(x)\mathrm{d}\Omega. \tag{15.6.3}$$

Due to linearity, the difference $w = u_2 - u_1$ between two solutions of (15.6.1) and (15.6.2) would satisfy the completely homogeneous problem

$$\mathsf{L}_x w(x) = 0, \qquad \mathsf{B}_x w = 0. \tag{15.6.4}$$

Thus, for the given problem to have a unique solution, it is necessary that the homogeneous problem (15.6.4) only have the solution $w = 0$, which we assume here. Since the existence of an ordinary Green's function leads to a well-defined unique solution of the problem, we may well expect difficulties in this case. Indeed, the procedure will have to be modified when non-zero solutions exist as will be shown in the next section.

Let L^* be the formal adjoint operator of L defined so that[19]

$$(v, \mathsf{L}u) = (\mathsf{L}^* v, u) + J[\bar{v}, u], \tag{15.6.5}$$

---

[18] These might be, at least in part, initial conditions when the set of variables x inlcudes time.

[19] The adjective "formal" is introduced to distinguish this operator from the usual adjoint operator, which requires the specification of suitable boundary conditions; this notion is treated in some detail in §21.2.1 and §21.6.1.

in which J is the concomitant of u and v and involves only boundary values of the two functions. The complex conjugation shown explicitly in $J[\bar{v}, u]$ is a consequence of the definition (15.6.3) of the scalar product. In practice the relation (15.6.5) is found by an integration by parts, as a result of which one gets explicit expressions for both L^* and J as exemplified by the derivation of (15.1.5) and (15.5.3); other examples will be found in the next chapter. An operator such that $L = L^*$ is termed *symmetric* or *formally self-adjoint*.

The concomitant is a bilinear form in its arguments and in particular, for fixed v, it depends linearly on the boundary data on u. In general, $J[\bar{v}, u]$ involves integration over a manifold with one fewer dimension than the space where L operates, e.g. a surface in three-dimensional space.[20] In the special case of one variable, the concomitant simply requires the evaluation of its arguments at the end points as we have seen.

With (15.6.1) we consider the equation

$$L_x^* G(x, y) = \delta(x - y). \tag{15.6.6}$$

Here the variable is x; G depends on y through the appearance of this variable in the argument of the δ-function. Now we take the scalar product (on the variable x) of (15.6.1) by G and use (15.6.5) to rewrite the result identically as

$$(L_x^* G, u) + J[\bar{G}, u] = (G, f). \tag{15.6.7}$$

The first term is simply $(\delta(x - y), u) = u(y)$ by (15.6.6) and, therefore,

$$u(y) = (G(x, y), f(x)) - J\left[\bar{G}(x, y), u(x)\right], \tag{15.6.8}$$

where, it will be noted, the variable x in the concomitant is on the boundary where information on $u(x)$ is available from the boundary condition (15.6.2); for both terms, the dependence on y is through the dependence of G on this variable.

As in §15.1 and §15.5, and in spite of appearances, (15.6.8) is not an explicit result for u as the concomitant involves pieces of information which are known on the boundary from the conditions imposed on u, but also pieces which are not. We thus are led to impose on G boundary conditions which remove from (15.6.8) the unknown boundary information. A simple way to identify these *adjoint conditions* is to require $J\left[\bar{G}(x, y), u(x)\right]$ to vanish when the available data on u are homogeneous.[21] The adjoint conditions will therefore be homogeneous themselves and have the general form

$$B_x^* G(x, y) = 0. \tag{15.6.9}$$

We assume that the problem (15.6.6), (15.6.9) for G has a unique solution.[22] The expression (15.6.8) embodies therefore the inverse operator L^{-1} of L. When $L^* = L$ and $B_x^* = B_x$ the

[20] More precisely, $J[\bar{v}, u]$ is a sesquilinear form as it is linear in the second argument and antilinear in the first one. One can localize J on $\partial\Omega$ by introducing suitable δ-functions. We do not pursue this avenue here; the interested reader may consult e.g. Friedman 1956, chapter 3.

[21] For a fixed v satisfying the adjoint conditions, $J[\bar{v}, u]$ becomes in fact a linear functional of the data on u, see §19.6 p. 563.

[22] The usual differential operators are closed (definition p. 650) and therefore, for them, the difference between the number of linearly independent solutions of the homogeneous direct problem (15.6.4) and of the homogeneous adjoint problem $L_x^* v = 0$, $B_x^* v = 0$ is the *index* of the operator L (see pp. 621 and 656). In most situations of practical interest the index vanishes, which we assume to be the case here. Thus, since we have stipulated that the

problem is termed *self-adjoint*. This is the case, for example, of the Sturm–Liouville problem of §15.1.

As mentioned in §2.4, physically G may be interpreted as the response of the system to a point source or a unit impulse. For this reason, depending on the physical context, it is sometimes referred to as the *fundamental solution* (see e.g. §20.13.4 p. 613) or the *impulse response* (see e.g. §15.5 p. 388).

15.6.1 Symmetry properties of Green's functions

Accepting that, by construction, $u(y)$ as given by (15.6.8) is a solution of the original differential equation (15.6.1), we must have

$$\mathsf{L}_y u(y) = \mathsf{L}_y \left(G(x, y), f(x) \right) - \mathsf{L}_y J \left[\overline{G}(x, y), u(x) \right] = f(y), \qquad (15.6.10)$$

which can only hold if, in its second variable, G satisfies

$$\mathsf{L}_y \overline{G}(x, y) = \delta(x - y). \qquad (15.6.11)$$

Since the concomitant is linear in each one of its arguments, we have $\mathsf{L}_y J[\overline{G}, u] = J[\mathsf{L}_y \overline{G}, u]$ and therefore, if (15.6.11) is satisfied, then also $\mathsf{L}_y J[\overline{G}, u] = J[\delta(x - y), u] = 0$ as long as y is not on the boundary, given that the variable x in the concomitant is on the boundary.[23] As for the boundary condition (15.6.2) (written in terms of the variable y rather than x) we have

$$\mathsf{B}_y u = \left(\mathsf{B}_y G(x, y), f(x) \right) - J \left[\mathsf{B}_y \overline{G}(x, y), u(x) \right]. \qquad (15.6.12)$$

In order for the boundary condition on u to be independent of f we must have

$$\mathsf{B}_y \overline{G}(x, y) = 0, \qquad x \in \Omega, \qquad (15.6.13)$$

while

$$J \left[\mathsf{B}_y \overline{G}(x, y), u(x) \right] = -U(x), \qquad x \in \partial\Omega, \qquad (15.6.14)$$

so that $\mathsf{B}_y \overline{G}$ behaves in some sense as a sort of δ function when x is on the boundary. We have verified this property for the Sturm–Liouville case on p. 352 and we will see other instances in the next chapter (see e.g. p. 406).

We thus conclude that $G(x, y)$ *satisfies the adjoint problem* (15.6.6), (15.6.9) *in the first variable* while $\overline{G}(x, y)$ *satisfies the direct problem* (15.6.11) *with the homogeneous conditions* (15.6.13) *in the second one*. As a consequence, when the operator L is symmetric so that $\mathsf{L} = \mathsf{L}^*$ and, in addition, $\mathsf{B} = \mathsf{B}^*$,[24] $G(x, y)$ satisfies the same problem as $\overline{G}(y, x)$

---

homogeneous direct problem (15.6.4) only has the trivial solution, the same will be true for the homogeneous adjoint problem, which ensures the solvability of the problem for G.

[23] The situation in which both x and y are on the boundary is a special one and needs to be addressed by a suitable limit process; see e.g. §§16.1.1. and 16.1.2.

[24] A symmetric operator acting on functions satisfying the homogeneous form of these conditions is in fact *self-adjoint*; see p. 653.

and the two may therefore be expected to the equal. A direct proof of this fact may be found by wirting (15.6.6) twice, for $G(x, y)$ and $G(x, z)$:

$$L_x^* G(x, y) = \delta(x - y), \qquad L_x^* G(x, z) = \delta(x - z). \tag{15.6.15}$$

We take the scalar product of the first one by $G(x, z)$ on the right, of the second one by $G(x, y)$ on the left and subtract to find

$$G(y, z) - \overline{G}(z, y) = \left(L_x^* G(x, y), G(x, z) \right) - \left(G(x, y), L_x^* G(x, z) \right)$$
$$= \left(G(x, y), \left(L_x - L_x^* \right) G(x, z) \right) - J \left[\overline{G}(x, y), G(x, z) \right]. \tag{15.6.16}$$

When L is symmetric, $L = L^*$ and the first term vanishes. As noted before, $J \left[\overline{G}(x, y), u(x) \right]$ is a linear functional of the boundary conditions on the second term, expressed by $B_x u$, and vanishes when these conditions are homogeneous. But by (15.6.9) this is precisely the case when $B_x^* = B_x$, which implies that $J \left[\overline{G}(x, y), G(x, z) \right] = 0$. Thus, we conclude that G satisfies the symmetry relation

$$G(y, z) = \overline{G}(z, y). \tag{15.6.17}$$

For example, this relation is satisfied by the bilinear expansion (15.3.42) of the Green's function for Sturm–Liouville problems when λ is real, as is necessary for the operator to be symmetric. Often this symmetry relation (15.6.17) has an interesting physical meaning, as the two arguments of the Green's functions may be interpreted as the location of a source and of a field point where the effect of the source is felt. In this context, the symmetry of G means that there is a *reciprocity relation* between the two: the effect at y of a point source at x is the same as the effect at x of a point source at y.

Example 15.6.1 One of the simplest symmetric differential operator is $L = -i d/dx$ on the interval $(0, 1)$ (see examples on pp. 654 and following). Indeed

$$\int_0^1 \overline{G}(x, y) \left(-i \frac{du}{dx} \right) dx = \int_0^1 \overline{\left(-i \frac{\partial G}{\partial x} \right)} u(x) dx - i \left[\overline{G}, u \right]_{x=0}^{x=1}$$
$$= \left(-i \frac{\partial G}{\partial x}, u \right) + i \overline{G}(0, y) u(0) - i \overline{G}(1, y) u(1). \tag{15.6.18}$$

With the boundary condition $Bu(x) = u(0) = U_0$, the Green's function satisfies

$$-i \frac{\partial G}{\partial x} = \delta(x - y), \qquad B_x^* G = G(1, y) = 0, \tag{15.6.19}$$

and is readily calculated to be

$$G(x, y) = i \left[H(x - y) - H(1 - y) \right] = -i H(y - x), \tag{15.6.20}$$

so that

$$u(y) = i \overline{G}(0, y) U_0 + \int_0^1 \overline{G}(x, y) f(x) dx = U_0 + i \int_0^y f(x) dx \tag{15.6.21}$$

as can readily be found by elementary means. It is evident that

$$L_y \overline{G} = \delta(x - y) \tag{15.6.22}$$

and

$$B_y \overline{G}(x, y) = \overline{G}(x, 0) = iH(0 - x) = 0, \tag{15.6.23}$$

as expected from (15.6.11) and (15.6.13). Even though, in this case, L = L*, we find

$$G(y, z) - \overline{G}(z, y) = -iH(z - y) - iH(y - z) = -i \tag{15.6.24}$$

which, as expected from (15.6.17), is the same as $-J\left[\overline{G}(x, y), G(x, z)\right]$ given that, as shown by (15.6.18),

$$J\left[\overline{G}(x, y), G(x, z)\right] = i\overline{G}(0, y)G(0, z) - i\overline{G}(1, y)G(1, z)$$
$$i\,[iH(y)]\,[-iH(z)] - i\,[iH(y - 1)]\,[-iH(z - 1)] = i. \tag{15.6.25}$$

Thus we see that, even if the first term of (15.6.16) vanishes, the symmetry of the operator is not sufficient to ensure that of the Green's function for which we must also have B = B*.

As a check we can return to the introductory example 2.4.2 p. 40 given in §2.4. With the periodicity boundary condition imposed in that case the problem is self-adjoint. The Green's function was[25]

$$G(x, y) = -ie^{-i\lambda[S(y)-S(x)]}\left[H(y - x) - \frac{1}{1 - e^{i\lambda S(b)}}\right], \tag{15.6.26}$$

and, using $H(x - y) = 1 - H(y - x)$, we have

$$\overline{G}(y, x) = ie^{i\lambda[S(x)-S(y)]}\left[1 - H(y - x) - \frac{1}{1 - e^{-i\lambda S(b)}}\right]$$
$$= ie^{-i\lambda[S(y)-S(x)]}\left[-H(y - x) - \frac{e^{-i\lambda S(b)}}{1 - e^{-i\lambda S(b)}}\right]$$
$$= ie^{-i\lambda[S(y)-S(x)]}\left[-H(y - x) + \frac{1}{1 - e^{i\lambda S(b)}}\right] = G(x, y) \tag{15.6.27}$$

as expected from (15.6.17). □

15.7 Modified Green's function

If u satisfies $Lu = f$, the adjoint relation (15.6.5) shows that

$$(v, f) = (L^*v, u) + J[\overline{v}, u]. \tag{15.7.1}$$

If v is a solution of the the completely homogeneous adjoint problem

$$L^*v = 0, \qquad B^*v = 0, \tag{15.7.2}$$

[25] We give here the complex conjugate of the expression given in (2.4.19) in keeping with the present use of a scalar product defined by (15.6.3).

where $B^*v = 0$ denotes homogeneous adjoint boundary conditions, then (15.7.1) leads to the conclusion that, unless

$$(v, f) = J[\overline{v}, u], \tag{15.7.3}$$

the original problem has no solution. In particular, when u satisfies homogeneous conditions as well, $J[\overline{v}, u] = 0$ and the solvability condition (15.7.3) becomes

$$(v, f) = 0. \tag{15.7.4}$$

This is exactly the condition (15.1.38) of Example 15.1.1 where $\sin kx$ is evidently a solution of the homogeneous problem when $\sin kb = 0$. This is also the solvability condition we would get for Example 15.1.43 if $u'(a) = u'(b) = 0$ as, in that case, any constant solves the homogeneous problem. Of course, if the homogeneous adjoint problem has only the trivial solution, no restriction on f is imposed by (15.7.3) or (15.7.4), but not so when a non-zero v exists. In §21.7 we will study the solvability condition (15.7.3) in greater generality. Here we consider its implications for the construction of the Green's function.

In most situations of interest (and certainly when $L^* = L$), the number of non-zero solutions for (15.7.2) equals that of non-zero solutions of the homogeneous direct problem (15.6.4).[26] Therefore, a solvability condition analogous to (15.7.3) will be encountered when we try to solve (15.6.6) for the Green's function itself. Indeed, a scalar product of $L_x^* G = \delta(x - y)$ by a solution $w(x)$ of the homogeneous problem $Lw = 0$ gives

$$(L_x^* G, w) = w(y) \tag{15.7.5}$$

but also

$$(L_x^* G, w) = (G, L_x w) - J[\overline{G}, w] = -J[\overline{G}, w] = 0 \tag{15.7.6}$$

as w satisfies homogeneous boundary conditions which cause the concomitant to vanish if G is constructed according to the procedure of the previous section (cf. comment just before (15.6.9) p. 393). The contradiction between (15.7.5) and (15.7.6) shows that the solvability condition is not satisfied and no ordinary G can be constructed. This is the situation encountered in Example 15.1.1 when $\sin kb = 0$. In cases such as these we have to have recourse to a *modified Green's function*, which can be constructed in several ways.

A first approach consists in modifying the right-hand side of the equation (15.6.6) for G in such a way that the solvability condition is satisfied. An obvious choice is

$$L_x^* \tilde{G} = \delta(x - y) - \frac{\overline{w}(x)w(y)}{(w, w)}, \tag{15.7.7}$$

as now

$$(L_x^* \tilde{G}, w) = w(y) - w(y) = 0. \tag{15.7.8}$$

With this $\tilde{G}$ the solution (15.6.8) for u becomes

$$u(y) = (\tilde{G}, f) - J[\overline{\tilde{G}}, u] + Cw(y), \qquad C = \frac{(w, u)}{(w, w)}. \tag{15.7.9}$$

---

[26] See footnote p. 393. If the operator is closed and (15.6.4) has non-zero solutions, then 0 is an eigenvalue for L. According to §21.9 (see in particular pp. 666 and 668) this implies that 0 also belongs to the point or residual spectrum of L^*. Thus, in either case, (15.7.2) has a non-zero solution.

This will be a solution of the problem whatever the value of the constant C, which cannot be determined without further specifications.

A different modified Green's function can be found by relinquishing part of the boundary conditions on G so that the last step in (15.7.6) is no longer true and (15.7.5) can be satisfied without modification of the equation for G. In this case, again, the solution will contain an arbitrary multiple of w.

We illustrate these ideas for the previous Example 15.1.1. Further examples will be shown in §16.1.2 for the Poisson equation.

Example 15.7.1 Let us return to Example 15.1.1 p. 353 when $kb = n\pi$. Here $w = \sin kx$ and, since the concomitant vanishes, the solvability condition (15.7.3) reduces to

$$\int_0^b \sin kx \; f(x)\,dx \; = \; 0. \tag{15.7.10}$$

Equation (15.7.7) for G is

$$\frac{\partial^2 G}{\partial x^2} + k^2 G = \delta(x - y) - \frac{2}{b}\sin kx \; \sin ky. \tag{15.7.11}$$

For $x < y$, the solution satisfying $G(0, y) = 0$ is

$$G = \beta_-(y)\sin kx + \frac{x}{kb}\sin ky \; \cos kx, \tag{15.7.12}$$

while, for $x < y$, the solution satisfying $G(b, y) = 0$ is

$$G = \beta_+(y)\sin kx + \left(\frac{x}{b} - 1\right)\frac{\sin kx}{k}\cos kx. \tag{15.7.13}$$

Continuity at $x = y$ requires $\beta_+(y) - \beta_-(y) = (\cos ky)/k$, and the jump condition gives the same result. Thus one of β_+ or β_- remains undetermined and

$$G(x, y) = \beta_-(y)\sin kx + \left(\frac{x}{b} - 1\right)\frac{\sin ky \; \cos kx}{k} + H(y - x)\frac{\sin k(y - x)}{k}. \tag{15.7.14}$$

The solution u is therefore

$$u(y) \; = \; \frac{1}{k}\int_0^y \sin k(y - x) \; f(x)\,dx \; + \beta_-(y)\int_0^b \sin kx \; f(x)\,dx$$
$$+ \left[\int_0^b \left(\frac{x}{b} - 1\right)\cos kx \; f(x)\,dx \; + C\right]\frac{\sin ky}{k}. \tag{15.7.15}$$

Because of the solvability condition (15.7.10) the second term vanishes irrespective of the unknown value of β_-.

If we were to take the second approach mentioned before, we would be solving the original equation for G which, in this case, is

$$\frac{\partial^2 G}{\partial x^2} + k^2 G = \delta(x - y) \tag{15.7.16}$$

imposing e.g. $G(0, y) = 0$, but leaving $G(b, y)$ free to satisfy (15.7.5). Thus, for $x < y$ and $x > y$, respectively

$$G = \beta_-(y)\sin kx, \qquad G = \beta_+(y)\sin kx + \alpha(y)\cos kx. \qquad (15.7.17)$$

Continuity at $x = y$ and the jump conditions lead to

$$(\beta_+ - \beta_-)\sin ky + \alpha\cos ky = 0, \qquad (\beta_+ - \beta_-)\cos ky - \alpha\sin ky = 1/k \quad (15.7.18)$$

so that

$$G = \beta_-\sin kx + \frac{\sin k(x-y)}{k} + H(y-x)\frac{\sin k(y-x)}{k}. \qquad (15.7.19)$$

After discarding the same term multiplied by β_- in (15.7.15), the solution may be written as

$$u(y) = \frac{1}{k}\int_0^y \sin k(y-x)\, f(x)\,dx$$
$$+ \left(ku'(b)\cos kb - \int_0^b \cos kx\, f(x)\,dx\right)\frac{\sin ky}{k} \qquad (15.7.20)$$

with $u'(b)$ unknown. The problem can also be solved by variation of parameters with the result

$$u(y) = \frac{1}{k}\int_0^y \sin k(y-x)\, f(x)\,dx + B\sin ky \qquad (15.7.21)$$

with the constant B undetermined. The equivalence between all three expressions (15.7.15), (15.7.20) and (15.7.21) for u is evident. We return on this problem in the context of linear operators in Example 21.7.1 p. 659.

The role of the concomitant term in (15.7.3) can be illustrated if the boundary condition at $x = b$ is modified to $u(b) = U_b$, still with $\sin kb = 0$. The solution satisfying $u(0) = 0$ is

$$u(x) = -\frac{\cos kx}{k}\int_0^x f(\xi)\sin k\xi\,d\xi + \frac{\sin kx}{k}\left(C + \int_0^x f(\xi)\cos k\xi\,d\xi\right) \quad (15.7.22)$$

with C an arbitrary constant. In order for $u(b)$ to equal U, we must evidently have

$$-\frac{\cos kb}{k}\int_0^b f(\xi)\sin k\xi\,d\xi = U_b \qquad (15.7.23)$$

which is (15.7.3).   □

16

Green's Functions: Partial Differential Equations

We have introduced the general idea of the Green's function in §2.4 and we have presented some of the underlying theory in §15.6 and §15.7 of the previous chapter, which was mostly devoted to ordinary differential equations. Here we construct Green's functions for partial differential equations. We will not repeat the theoretical considerations given earlier, but we will provide appropriate reference to that material.

16.1 Poisson equation

In all the cases treated in this chapter an essential ingredient for the construction of the adjoint operator and the identification of the concomitant (p. 391) is the *divergence* or *Gauss theorem* in d-dimensional space (see e.g. Kemmer 1977; Stroock 1999, p. 105):[1]

Theorem 16.1.1 *Let $A(\mathbf{x})$ denote any sufficiently smooth function. Then*

$$\int_\Omega \frac{\partial A}{\partial x_j} \, \mathrm{d}^d x \; = \; \oint_S A \, n_j \, \mathrm{d}S \qquad (16.1.1)$$

where $S = \partial\Omega$ is the boundary of Ω, which may consist of several disjoint pieces and include a surface at infinity, and $\mathbf{n}$ is the unit normal directed out of Ω.

Here $\mathrm{d}^d x = \mathrm{d}x_1 \, \mathrm{d}x_2 \cdots \mathrm{d}x_d$ is the volume element in d-dimensional space and $\mathrm{d}S$ the element of a $(d-1)$-dimensional (hyper)surface. If Ω is unbounded, S in this formula includes a surface at infinity and, in this case, the result is valid in a limit sense.[2] This is actually a generalization of the conventional divergence theorem, which is recovered by applying (16.1.1) separately to the d components $\partial A_j/\partial x_j$ of a vector field $\mathbf{A}$ and adding the results to find (p. 4)

$$\int_\Omega \boldsymbol{\nabla} \cdot \mathbf{A} \, \mathrm{d}^d x \; = \; \oint_S \mathbf{A} \cdot \mathbf{n} \, \mathrm{d}S. \qquad (16.1.2)$$

Another well-known theorem is *Stokes's Theorem* which we state in a slightly generalized form (see e.g. Kemmer 1977):

---

[1] A generalized form of this theorem is given on p. 590.
[2] If the integral is only conditionally convergent at infinity, its value will be dependent on how the surface is made to approach infinity in the limit.

Theorem 16.1.2 [Stokes] *Let $A(\mathbf{x})$ denote any sufficiently smooth function; then*[3]

$$\oint_L A\, t_k\, d\ell = \int_S n_i \varepsilon_{ijk} \frac{\partial A}{\partial x_j}\, dS, \qquad (16.1.3)$$

*where the integral in the left-hand side is along a closed line L in the direction specified by the unit tangent **t** and the integral in the right-hand side is over any surface attached to L, with the direction of the unit normal **n** determined by the right-hand rule.*

The usual form of the theorem is recovered by taking the function A equal in turn to the three components of a vector $\mathbf{V}$ and adding to find

$$\oint_L \mathbf{V} \cdot \mathbf{t}\, d\ell = \int_S \mathbf{n} \cdot (\nabla \times \mathbf{V})\, dS, \qquad (16.1.4)$$

We start with the Poisson equation in a spatial domain Ω. Following convention, we write the equation as

$$Lu \equiv \nabla^2 u(\mathbf{x}) = -S_d f(\mathbf{x}) \qquad \mathbf{x} \in \Omega, \qquad (16.1.5)$$

where S_d is the area of the unit sphere in d dimensions; for ordinary three-dimensional space $S_3 = 4\pi$ while, in the plane, $S_2 = 2\pi$.[4] Without loss of generality we take all the functions we deal with to be real as the same developments can be carried out separately for the real and imaginary parts of a problem with complex data.

Let v be another sufficiently smooth function defined in Ω and note that

$$v\nabla^2 u - u\nabla^2 v = \nabla \cdot [v\nabla u - u\nabla v]. \qquad (16.1.6)$$

Integrate over Ω and apply the divergence theorem (16.1.2) to find

$$(v, \nabla^2 u) - (\nabla^2 v, u) \equiv \int_\Omega \left(v\,\nabla^2 u - u\,\nabla^2 v \right) d^d x = \oint_S (v\nabla u - u\nabla v) \cdot \mathbf{n}\, dS,$$
$$(16.1.7)$$

where it has been assumed that the surface integral exists. This is sometimes called *Green's second identity* or *Green's theorem*. We can now read directly from (16.1.7) L^* and J (see p. 391):

$$L_x^* v = \nabla_x^2 v, \qquad J[v, u] = \oint_S (v\nabla u - u\nabla v) \cdot \mathbf{n}\, dS. \qquad (16.1.8)$$

---

[3] The convention of summation over repeated indices is adopted in (16.1.3). ε_{ijk}, with the indices i, j, k ranging over over 1, 2 and 3, is the alternating tensor; $\varepsilon_{ijk} = 0$ if any two indices are equal, $\varepsilon_{ijk} = 1$ if (i, j, k) is an even permutation of (1, 2, 3), and $\varepsilon_{ijk} = -1$ if (i, j, k) is an odd permutation of (1, 2, 3).

[4] An interesting way of calculating S_d is the following. Consider the integral $\int_{-\infty}^{\infty} \exp(-x^2)\, dx = \sqrt{\pi}$ (cf. footnote p. 117) and its d-fold product:

$$I^d = \int_{-\infty}^{\infty} \exp(-x_1^2)\, dx_1 \cdots \int_{-\infty}^{\infty} \exp(-x_d^2)\, dx_d = \int \exp[-(x_1^2 + \cdots + x_d^2)]\, d^d x = \pi^{d/2}.$$

This integral may also be considered as taken over the whole d-dimensional space and calculated alternatively by using polar coordinates with $x_1^2 + \cdots + x_d^2 = r^2$:

$$I^d = \int_0^{\infty} S_d r^{d-1} \exp(-r^2)\, dr = \tfrac{1}{2} S_d \Gamma(d/2).$$

Equating we find $S_d = 2\pi^{d/2}/\Gamma(d/2)$. For $d = 1$ this formula gives $S_1 = 2$, but in this case it is customary not to include this factor in (16.1.5) as we have seen in the previous chapter.

In this case, motivated by (16.1.5), we write the equation for the Green's function (15.6.6) p. 393 slightly differently by inserting the factor S_d in the right-hand side:

$$\nabla_x^2 G(\mathbf{x}, \mathbf{y}) = -S_d \delta(\mathbf{y} - \mathbf{x}) \tag{16.1.9}$$

and find the analog of (15.6.8) p. 393 valid for points $\mathbf{y} \in \Omega$:

$$u(\mathbf{y}) = \int_\Omega f(\mathbf{x}) \, G(\mathbf{x}, \mathbf{y}) \, d^d x + \frac{1}{S_d} \oint_S (G \nabla_x u - u \nabla_x G) \cdot \mathbf{n}_x \, dS_x. \tag{16.1.10}$$

When $\mathbf{y}$ is neither inside Ω nor on the boundary, the domain of integration does not include the pole of $\delta(\mathbf{y} - \mathbf{x})$ and

$$0 = \int_\Omega f(\mathbf{y}) \, G(\mathbf{x}, \mathbf{y}) \, d^d x + \frac{1}{S_d} \oint_S (G \nabla_x u - u \nabla_x G) \cdot \mathbf{n}_x \, dS_x. \tag{16.1.11}$$

The case when $\mathbf{y} \in \partial\Omega$ is a special one and will be discussed later in §16.1.3.

If the mathematical nature of the problem (16.1.5) required the specification of both u and $\mathbf{n} \cdot \nabla u$ on the boundary, the formula (16.1.10) would be a closed-form solution of the problem expressed in terms of the data. This, however, would be an over-specification of boundary data for an elliptic equation and result in an ill-posed problem (see §1.2.2 p. 7 and the example in §4.7 p. 108). Proper boundary conditions are of the *Dirichlet type*, when u is prescribed on the boundary and $\mathbf{n} \cdot \nabla u$ is unknown, or of the *Neumann type*, when $\mathbf{n} \cdot \nabla u$ is prescribed and u is unknown. Dirichlet conditions may also be prescribed on part of the boundary and Neumann on the remainder, or a linear combination of u and $\mathbf{n} \cdot \nabla u$ may be prescribed. We will treat these different possibilities separately in the following. For the time being, let us make some general comments.

The equation for G is linear and non-homogeneous and, therefore, its general solution is the sum of any one particular solution G_p and the general solution G_h of the homogeneous equation:

$$G(\mathbf{x}, \mathbf{y}) = G_p(\mathbf{x}, \mathbf{y}) + G_h(\mathbf{x}, \mathbf{y}). \tag{16.1.12}$$

If the number of dimensions is greater than two we have (p. 615)

$$\nabla^2 \frac{1}{(d-2)|\mathbf{x} - \mathbf{y}|^{d-2}} = -S_d \delta^{(d)}(\mathbf{x} - \mathbf{y}), \tag{16.1.13}$$

while, for $d = 2$,

$$\nabla^2 (-\log |\mathbf{x} - \mathbf{y}|) = -2\pi \, \delta^{(2)}(\mathbf{x} - \mathbf{y}), \tag{16.1.14}$$

both valid whether ∇^2 is ∇_x^2 or ∇_y^2. Thus, for $d = 3$ and $d = 2$, respectively,

$$G_p = \frac{1}{|\mathbf{x} - \mathbf{y}|} \quad \text{or} \quad G_p = -\log |\mathbf{x} - \mathbf{y}|. \tag{16.1.15}$$

In particular, in free space with $u \to 0$ at infinity and $d > 2$, the solution of (16.1.5) is[5]

$$u(\mathbf{y}) = \frac{1}{d-2} \int \frac{f(\mathbf{x})}{|\mathbf{x} - \mathbf{y}|^{d-2}} \, d^d x. \tag{16.1.16}$$

---

[5] This result is valid provided u decreases faster than any negative power of $|\mathbf{x} - \mathbf{y}|$ at infinity.

In electrostatics, with $d = 3$, this is the familiar expression for the potential generated by a charge distributed with density $f(\mathbf{x})$ (§7.6 p. 182).

G_p or, more generally, G itself, can be interpreted as the potential generated by a point source of unit strength located at $\mathbf{x}$, commonly referred to as *monopole potential*. Writing

$$\mathbf{n} \cdot \nabla_x G = \lim_{\varepsilon \to 0} \frac{G(\mathbf{x} + \varepsilon\mathbf{n}, \mathbf{y}) - G(\mathbf{x}, \mathbf{y})}{\varepsilon} \tag{16.1.17}$$

then justifies reference to $\mathbf{n} \cdot \nabla_x G$ as the potential of a unit dipole. For this reason the two terms in the surface integrals of (16.1.10) and (16.1.11) are referred to as the *single layer* and *double layer* potentials, respectively.

16.1.1 Dirichlet problem

In the Dirichlet problem u is prescribed on the boundary S, $u = U(\mathbf{x})$, but $\mathbf{n} \cdot \nabla u$ is unknown. According to the general procedure outlined in §15.6, in order to eliminate the need for information on this latter quantity in (16.1.10), we prescribe that

$$G(\mathbf{x}, \mathbf{y}) = 0 \quad \text{for} \quad \mathbf{x} \in S, \tag{16.1.18}$$

so that (16.1.10) becomes an explicit solution to the problem:

$$u(\mathbf{y}) = \int_\Omega f(\mathbf{x}) \, G(\mathbf{x}, \mathbf{y}) \, \mathrm{d}^d x - \frac{1}{S_d} \oint_S U(\mathbf{x}) \, \mathbf{n}_x \cdot \nabla_x G \, \mathrm{d}S_x. \tag{16.1.19}$$

This expression represents u as the combined effect of point sources of strength $f(\mathbf{x})$ distributed throughout the volume and of dipoles of strength $U(\mathbf{x})$ distributed over the bounding surface.

The existence of a solution in this form hinges on the solvability of the problem for G, consisting of the differential equation (16.1.9) and of the Dirichlet boundary condition (16.1.18). The existence and uniqueness of the solution to this problem is guaranteed under very general conditions on the nature of the boundary (see e.g. Kress 1989, chapter 6; Renardy & Rogers 1993, chapter 4), which is the only feature distinguishing the Green's function for one Dirichlet problem from that for another one. Once G has been determined for a particular geometry, the Dirichlet problem is solved by (16.1.19) for any source function $f(\mathbf{x})$ and any boundary data such that the integrals in (16.1.19) are meaningful.

The relation (15.6.16) p. 394 readily shows that the Green's function for the Poisson equation with Dirichlet boundary conditions is symmetric. Indeed, in this case $L = L^* = \nabla^2$ and G has been so constructed as to vanish when its first argument is on the boundary. As a consequence it is not necessary to treat differently the variables x and y and, in particular, G satisfies the same equation in $\mathbf{y}$ as it does in $\mathbf{x}$:

$$\nabla_y^2 G(\mathbf{x}, \mathbf{y}) = -S_d \delta(\mathbf{x} - \mathbf{y}) \quad \mathbf{y} \notin S. \tag{16.1.20}$$

The symmetry of G reflects the reciprocity between source point and field point: the potential at a field point $\mathbf{y}$ generated by a unit point source at $\mathbf{x}$ is the same as that at $\mathbf{x}$ of a unit source at $\mathbf{y}$.

Example 16.1.1 As an application let us consider an exterior problem in three-dimensional space in which the boundary is a sphere of radius a on which $u = U(\mathbf{x})$ is given; in order to be able to use a more compact notation we write $\mathbf{x}'$ in place of $\mathbf{y}$. To calculate the Green's function we need to solve

$$\nabla_x^2 G = -4\pi \delta^{(3)}(\mathbf{x} - \mathbf{x}'), \qquad G(\mathbf{x}, \mathbf{x}') = 0, \qquad \mathbf{x} \in S. \qquad (16.1.21)$$

In the example of §7.9 we have already solved this problem when the boundary condition on the sphere surface is a constant. By setting this constant to 0, that solution, (7.9.10) p. 187, becomes

$$G(\mathbf{x}, \mathbf{x}') = \frac{1}{|\mathbf{x} - \mathbf{x}'|} - \frac{a/|\mathbf{x}'|}{|\mathbf{x} - \mathbf{x}'_I|} \qquad \text{with} \qquad \mathbf{x}'_I = \frac{a^2}{|\mathbf{x}'|} \frac{\mathbf{x}'}{|\mathbf{x}'|} \qquad (16.1.22)$$

or

$$G = \frac{1}{\sqrt{r^2 + r'^2 - 2rr' \cos \gamma}} - \frac{1}{\sqrt{r^2 r'^2/a^2 + a^2 - 2rr' \cos \gamma}} \qquad (16.1.23)$$

where $\cos \gamma = \mathbf{x} \cdot \mathbf{x}'/|\mathbf{x}||\mathbf{x}'| = \cos\theta \cos\theta' + \sin\theta \sin\theta' \cos(\phi - \phi')$ (Figure 7.2 p. 178). Clearly $G(\mathbf{x}, \mathbf{x}') = G(\mathbf{x}', \mathbf{x})$, and $G = 0$ on the sphere. To apply Green's formula, we need $\partial G/\partial n'$:

$$\frac{\partial G}{\partial n'} = -\frac{\partial G}{\partial r} = \frac{r' - r \cos \gamma}{[\sqrt{r^2 + r'^2 - 2rr' \cos \gamma}]^3} - \frac{(r^2/a^2)r' - r \cos \gamma}{[\sqrt{r^2 r'^2/a^2 + a^2 - 2rr' \cos \gamma}]^3}, \qquad (16.1.24)$$

where the minus sign arises from the fact that the normal is directed out of the domain of interest and, therefore, into the sphere. Upon evaluating this expression on the sphere, we find

$$\left. \frac{\partial G}{\partial n'} \right|_{r=a} = \frac{a^2 - r^2}{a \, [\sqrt{r^2 + a^2 - 2ra \cos \gamma}]^3} \qquad (16.1.25)$$

so that

$$u(\mathbf{x}) = \int d^3 x' \, f(\mathbf{x}') \, G(\mathbf{x}, \mathbf{x}') + \frac{a(r^2 - a^2)}{4\pi} \int_\omega \frac{U(\theta', \phi')}{[\sqrt{r^2 + a^2 - 2ar \cos \gamma}]^3} \, d\omega', \qquad (16.1.26)$$

where the second integral is over the total solid angle $\omega = 4\pi$. For the Laplace equation, $f = 0$ and this reduces to Poisson's exterior formula which is derived by other means in §7.8 p. 184. For the interior problem $\partial/\partial n' = \partial/\partial r$ and the normal derivative of G changes sign. □

Since G satisfies (16.1.17) and differentiation under the integral sign is legitimate for (16.1.17), u as given by this formula satisfies the Poisson equation. We now sketch a proof that $u(\mathbf{y}) = U(\mathbf{y})$ for $\mathbf{y} \in S$. We have constructed G so that $G(\mathbf{x}, \mathbf{y}) = 0$ for $\mathbf{x} \in S$, but $G(\mathbf{x}, \mathbf{y}) = G(\mathbf{y}, \mathbf{x})$ and, therefore, $G(\mathbf{x}, \mathbf{y}) = 0$ also for $\mathbf{y} \in S$ so that the volume integral vanishes for $\mathbf{y} \in S$. Thus, we need only consider the second term as $\mathbf{y} \to \mathbf{x}_0 \in S$. As long as $\mathbf{y} \notin S$ we can write

$$\oint_S U(\mathbf{x}) \, \mathbf{n}_x \cdot \nabla_x G(\mathbf{x}, \mathbf{y}) \, dS_x = \oint_S [U(\mathbf{x}) - U(\mathbf{x}_0)] \, \mathbf{n}_x \cdot \nabla_x G \, dS_x$$

$$+ U(\mathbf{x}_0) \oint_S \mathbf{n}_x \cdot \nabla_x G \, dS_x. \qquad (16.1.27)$$

Integration of the equation (16.1.9) for G over the domain Ω and use of the divergence theorem (16.1.2) with $\mathbf{A} = \nabla G$ gives

$$-\frac{1}{S_d} \oint_S \mathbf{n}_x \cdot \nabla G \, dS_x = 1 \qquad (16.1.28)$$

so that

$$\lim_{\mathbf{y} \to \mathbf{x}_0 \in S} u(\mathbf{y}) = U(\mathbf{x}_0) - \frac{1}{S_d} \lim_{\mathbf{y} \to \mathbf{x}_0} \oint_S [U(\mathbf{x}) - U(\mathbf{x}_0)] \, \mathbf{n}_x \cdot \nabla_x G(\mathbf{x}, \mathbf{y}) \, dS_x. \qquad (16.1.29)$$

The first term is the one desired; the remaining task, which we omit, is then to prove that the second one vanishes as $\mathbf{y} \to \mathbf{x}_0$. Detailed and elaborate proofs of this fact which minimize the regularity requirements on S are available (see e.g. Kress 1989 p. 68 and references therein).

16.1.2 Neumann problem

For the Neumann problem u is unknown on the boundary while

$$\mathbf{n}_x \cdot \nabla_x u = V(\mathbf{x}), \qquad \mathbf{x} \in S, \qquad (16.1.30)$$

is prescribed. Integration of the Poisson equation (16.1.5) over the domain Ω and use of the divergence theorem shows that a solution exists only if the following compatibility condition is satisfied:

$$\oint_S V(\mathbf{x}) \, dS_x = -S_d \int_\Omega f(\mathbf{x}) \, d^d x. \qquad (16.1.31)$$

From (16.1.10) the natural choice for the boundary condition on G would seem to be $\mathbf{n}_x \cdot \nabla G = 0$, but this choice would be evidently incompatible with (16.1.28). We encounter here the same situation as in §15.7 which requires the introduction of a modified Green's function. Indeed, in this case, it is evident that a constant solves the completely homogeneous problem the statement of which only involves derivatives of u.

The same two approaches described earlier in §15.7 are applicable here. For the first one (p. 397), we modify the right-hand side of the equation (16.1.9) for G writing

$$\nabla_x^2 G(\mathbf{x}, \mathbf{y}) = -S_d \delta(\mathbf{y} - \mathbf{x}) + \frac{S_d}{V_\Omega}, \qquad (16.1.32)$$

where V_Ω is the volume of the domain Ω of interest. With this modification (16.1.28) is satisfied and we can impose $\mathbf{n}_x \cdot \nabla_x G = 0$ for $\mathbf{x}$ on the boundary. As given by (16.1.10) the solution is then

$$u(\mathbf{y}) = \frac{1}{V_\Omega} \int_\Omega u \, d^d x + \int_\Omega f(\mathbf{x}) \, G(\mathbf{x}, \mathbf{y}) \, d^d x + \frac{1}{S_d} \oint_S G(\mathbf{x}, \mathbf{y}) \, V(\mathbf{x}) \, dS_x. \qquad (16.1.33)$$

The first term in the right-hand side is the volume average of u and is undetermined.

If the domain has an infinite volume, the last term in (16.1.32) vanishes. For example, for a half-space bounded by a plane in three dimensions, the Neumann Green's function satisfying $\mathbf{n} \cdot \nabla G = 0$ on the plane is readily determined by the method of images as

$$G(\mathbf{x}, \mathbf{y}) = \frac{1}{|\mathbf{x} - \mathbf{y}|} + \frac{1}{|\mathbf{x} - \mathbf{y}_I|}, \tag{16.1.34}$$

where $\mathbf{y}_I$ is the image point of $\mathbf{y}$ in the plane.

The second approach (p. 397) is to set $\mathbf{n}_x \cdot \nabla_x G = -S_d b(\mathbf{x})$ on the boundary, with the function b chosen so that (16.1.28) is satisfied. The result is

$$u(\mathbf{y}) = \oint b(\mathbf{x}) u(\mathbf{x}) \, \mathrm{d}S_x + \int_\Omega f(\mathbf{x}) \, G(\mathbf{x}, \mathbf{y}) \, \mathrm{d}^d x + \frac{1}{S_d} \oint_S G \, V(\mathbf{x}) \, \mathrm{d}S_x. \tag{16.1.35}$$

In particular, if one were to take $1/b$ equal to the area S of the boundary,

$$u(\mathbf{y}) = \frac{1}{S} \oint u(\mathbf{x}) \, \mathrm{d}S_x + \int_\Omega f(\mathbf{x}) \, G(\mathbf{x}, \mathbf{y}) \, \mathrm{d}^d x + \frac{1}{S_d} \oint_S G \, V(\mathbf{x}) \, \mathrm{d}S_x, \tag{16.1.36}$$

and the first term in the right-hand side would then be the average of u over the boundary. Also in this case u is determined up to an arbitrary constant as expected. For the exterior problem, if it is required that $u \to 0$ at infinity (which is really a Dirichlet condition), the undetermined constants vanish.

With the first approach the Green's function is symmetric as, due to the vanishing normal derivative, the concomitant in (15.6.16) p. 394 vanishes. With the second approach, symmetry is not guaranteed.

An interpretation of either procedure followed to construct the modified Green's function may be given if we regard the δ as a source of a field ∇G. In a bounded domain this source needs to be balanced by an equal sink, either distributed throughout the domain (first method), or on the boundary (second method). In an unbounded domain, the sink is at infinity and, in most cases, needs not be shown explicitly.

Application of the operator ∇_y^2 to (16.1.33) or (16.1.35) shows that the Poisson equation is satisfied when the compatibility relation (16.1.31) holds. To prove that the boundary condition (16.1.30) is also met, let us consider the form (16.1.33), for which G is symmetric, and calculate, for $\mathbf{y} \to \mathbf{x}_0 \in S$, the limit of

$$\mathbf{n}_{x_0} \cdot \nabla_y u(\mathbf{y}) = \int_\Omega (\mathbf{n}_{x_0} \cdot \nabla_y G) \, f(\mathbf{x}) \, \mathrm{d}^3 x + \frac{1}{S_d} \oint_S (\mathbf{n}_{x_0} \cdot \nabla_y G) \, V(\mathbf{x}) \, \mathrm{d}S_x. \tag{16.1.37}$$

The first integral is regular at $\mathbf{x}_0$ due to the volume differential and vanishes since the normal component of $\nabla_x G$ vanishes for $\mathbf{x}$ on the boundary which, by the symmetry of G, implies that the normal component of $\nabla_y G$ vanishes for $\mathbf{y}$ on the boundary. The same conclusion does not follow for the second term because the singularity of the integrand is not cancelled by the surface differential. By writing $V(\mathbf{x}) = [V(\mathbf{x}) - V(\mathbf{x}_0)] + V(\mathbf{x}_0)$, if V is regular near $\mathbf{x}_0$, the integral of the first term vanishes because the approximation $V(\mathbf{x}) - V(\mathbf{x}_0) \simeq (\mathbf{x} - \mathbf{x}_0) \cdot \nabla V$ shows that this part of the integrand is regular. What remains to show is then that

$$\frac{1}{S_d} \oint_S (\mathbf{n}_{x_0} \cdot \nabla_y G) \, \mathrm{d}S_x = 1. \tag{16.1.38}$$

An argument suggesting that this relation is valid, e.g. for the three-dimensional case, is the following. Let us assume for simplicity that at the point $\mathbf{x}_0$ the surface S is sufficiently regular to have a tangent plane. At distances from $\mathbf{x}_0$ much smaller than the smallest radius of curvature of S, the Green's function is dominated by its singular part (16.1.34) and, since what remains is regular at $\mathbf{x}_0$, it gives a vanishing contribution as argued before. As $\mathbf{y}$ approaches the plane from above, the integral in (16.1.38) approaches the integral over an infinite plane for which G is given by (16.1.34). With a local coordinate system in which $\mathbf{x} = (\rho, \theta, 0)$, $\mathbf{y} = (0, 0, \zeta)$ and the normal directed out of the domain of interest and, therefore, in the negative direction, we have

$$\frac{1}{4\pi} \int_0^\infty \left(-\frac{\partial}{\partial \zeta} \frac{2}{[\rho^2 + \zeta^2]^{1/2}} \right) 2\pi\rho \, d\rho = -\int_0^\infty \zeta \left(\frac{\partial}{\partial \rho} \frac{1}{[\rho^2 + \zeta^2]^{1/2}} \right) d\rho = 1.$$
(16.1.39)

In order to turn this argument into a proof one needs to provide careful stimates of the neglected terms. This is a highly technical matter dealt with in many references (see e.g. Kress 1989, p. 68; Kellogg 1967).

16.1.3 The boundary integral method

While, as noted before, (16.1.10) is not an explicit solution for u, by taking the point $\mathbf{y}$ on the boundary we can obtain from it an integral equation for the unknown boundary values, namely $\mathbf{n}_x \cdot \nabla_x u$ for the Dirichlet problem or u for the Neumann problem. Once this missing boundary information is known, (16.1.10) itself can be used to calculate u at all points of the domain. This idea is at the basis of the so-called *boundary integral method* (see e.g. Pozrikidis 1992; Pomp 1998; Constanda 2000).

We re-write (16.1.10) as

$$S_d u(\mathbf{y}) = S_d \int_\Omega f(\mathbf{x}) \, G(\mathbf{x}, \mathbf{y}) d^d x + \oint_S (G \nabla_x u - u \nabla_x G) \cdot \mathbf{n}_x \, dS_x.$$
(16.1.40)

Very near a boundary point $\mathbf{x}_0$, G is dominated by its singular part (16.1.15) and this mild singularity is cancelled by the area differential dS_x. Thus, the monopole surface term, namely the first surface integral, is regular and continuous as $\mathbf{y}$ approaches the surface and, in fact, $u(\mathbf{x}_0+) = u(\mathbf{x}_0-)$, if $u(\mathbf{x}_0\pm)$ denote the limit of u as the point $\mathbf{x}_0$ is approached from the positive (external to Ω) or negative side of the surface. The same argument shows that the volume integral is also regular and continuous.

The second surface integral, namely the dipole contribution, is singular and needs to be regularized. We write

$$\lim_{\mathbf{y}\to\mathbf{x}_0} \oint_S u \, \mathbf{n}_x \cdot \nabla_x G \, dS_x = \lim_{\mathbf{y}\to\mathbf{x}_0} \left[\int_{S-C_\varepsilon+S_\varepsilon} u \, \mathbf{n}_x \cdot \nabla_x G \, dS_x + u(\mathbf{x}_0) \int_{S_\varepsilon-C_\varepsilon} \mathbf{n}_x \cdot \nabla_x G \, dS_\varepsilon \right]$$
(16.1.41)

in which the domains C_ε and S_ε are as shown in Figure 16.1. In words, we replace the small area C_ε by a small hemisphere S_ε and then add the second term to correct the result. The limit of the first integral defines the Cauchy principal value. Upon reversing the direction

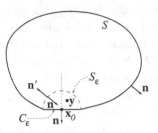

Fig. 16.1 Illustration of the nomenclature used in (16.1.41).

of the normal on S_ε from $\mathbf{n}$ to $\mathbf{n}'$, the second term is the integral of $\mathbf{n}_x \cdot \nabla_x G$ over the closed surface $S_\varepsilon + C_\varepsilon$, but this does not equal S_d as before in (16.1.28) if the point $\mathbf{y} \to \mathbf{x}_0 \in S$. In fact, one has

$$\oint_S \mathbf{n}_x \cdot \nabla G \, \mathrm{d}S_x = -\omega, \qquad (16.1.42)$$

where ω is the angle, in two dimensions, or the solid angle, in three, under which the domain of interest Ω is "viewed" from the point $\mathbf{x}_0$. This relation can be proven by noting that, near $\mathbf{x}_0$, G is dominated by its singular part $|\mathbf{x} - \mathbf{y}|^{-1}$ so that $\mathbf{n}_x \cdot \nabla G \, \mathrm{d}S_x = -\mathrm{d}\omega$ by definition of (solid) angle. Thus (16.1.40) becomes

$$(S_d - \omega)u(\mathbf{y}) = S_d \int_\Omega f(\mathbf{x}) \, G(\mathbf{x}, \mathbf{y}) \mathrm{d}^d x + \oint_S (G \nabla_x u - u \nabla_x G) \cdot \mathbf{n}_x \, \mathrm{d}S_x, \quad (16.1.43)$$

in which the second surface integral is to be interpreted in the sense of a Cauchy principal value as defined before.

In two dimensions, if the boundary is smooth at $\mathbf{x}_0$, it admits a tangent and $\omega = \pi$. Similarly, in three dimensions, a smooth domain admits a tangent plane and $\omega = 2\pi$. Thus, in this latter case, for example, we have

$$u(\mathbf{y}) = 2 \int_\Omega f(\mathbf{x}) \, G(\mathbf{x}, \mathbf{y}) \, \mathrm{d}^3 x + \frac{1}{2\pi} \oint_S (G \nabla_x u - u \nabla_x G) \cdot \mathbf{n}_x \, \mathrm{d}S_x. \qquad (16.1.44)$$

It should be noted that any convenient form for G can be used here, as the starting point for this relation is (16.1.10) which is valid for any G satisfying (16.1.9). In three dimensions the simplest choice might be, for example, $G = |\mathbf{x} - \mathbf{y}|^{-1}$ with which (16.1.44) takes the commonly encountered form

$$u(\mathbf{y}) = 2 \int_\Omega \frac{f(\mathbf{x})}{|\mathbf{x} - \mathbf{y}|} \, \mathrm{d}^3 x + \frac{1}{2\pi} \oint_S \left(\frac{\nabla_x u}{|\mathbf{x} - \mathbf{y}|} - u \nabla_x \frac{1}{|\mathbf{x} - \mathbf{y}|} \right) \cdot \mathbf{n}_x \, \mathrm{d}S_x. \qquad (16.1.45)$$

16.1.4 Eigenvalue problem

We can parallel here for the Laplace operator the developments of the previous chapter for Sturm–Liouville operators. We consider eigenfunctions and eigenvalues satisfying

$$-\nabla^2 u = \lambda u, \qquad \mathbf{x} \in \Omega, \tag{16.1.46}$$

subject, for example, to homogeneous Dirichlet conditions $u = 0$ for $\mathbf{x} \in S$. With these boundary conditions, if λ were 0, we would conclude from (16.1.19) that $u = 0$ as well. Thus, the eigenvalues cannot vanish.

If (λ, u) and (μ, v) are two eigenvalue–eigenfunction pairs, from Green's identity (16.1.7) we have

$$(\overline{\mu} - \lambda) \int_\Omega \overline{v} u \, \mathrm{d}^d x = 0, \tag{16.1.47}$$

since the surface integral vanishes due to the assumed homogeneous Dirichlet conditions. If we take $v = u$, the volume integral is positive definite and this relation implies $\overline{\mu} - \lambda = 0$, so that all eigenvalues are real. If $\mu \neq \lambda$, (16.1.47) implies that the scalar product $(v, u) = 0$ so that eigenfunctions corresponding to different eigenvalues are orthogonal. The same conclusions follow for similar reasons for an equation of the type

$$-\nabla \cdot [p(\mathbf{x}) \nabla u] + q(\mathbf{x})u = \lambda u, \tag{16.1.48}$$

provided the given functions $p(\mathbf{x})$ and $q(\mathbf{x})$ are positive in Ω.

Since $\lambda \neq 0$, use of the Dirichlet Green's function permits us to rewrite (16.1.46) in the form (16.1.19), which may be regarded as an eigenvalue problem for an integral operator G:

$$\mathsf{G}u \equiv \frac{1}{S_d} \int_\Omega G(\mathbf{x}, \mathbf{y}) u(\mathbf{x}) \, \mathrm{d}^d x = \frac{1}{\lambda} u(\mathbf{y}). \tag{16.1.49}$$

In a bounded domain the Green's function is square integrable and therefore the kernel of the integral operator is of the Hilbert–Schmidt type and thus compact (see p. 637). The Hilbert–Schmidt Theorem 21.3.4 p. 639 therefore applies and we may draw a conclusion analogous to that described in Theorem 15.3.1 p. 366 as to the existence of a complete set of eigenfunctions and eigenvalues.

This result can be applied to solving problems by the same method of eigenfunction expansion used throughout the first part of this book. For example, to attack the problem

$$\frac{\partial w}{\partial t} - \nabla^2 w = f(\mathbf{x}, t), \qquad \mathbf{x} \in V, \tag{16.1.50}$$

subject to

$$w(\mathbf{x}, t) = 0 \quad \text{for} \quad \mathbf{x} \in S, \qquad w(\mathbf{x}, 0) = w_0(\mathbf{x}), \tag{16.1.51}$$

we would set $w(\mathbf{x}, t) = \sum_{n=0}^{\infty} c_n(t) e_n(\mathbf{x})$, where the e_n's are the normalized eigenfunctions of the Laplacian (16.1.46), substitute into the equation and take scalar products to find

$$\frac{\mathrm{d}c_n}{\mathrm{d}t} + \lambda_n c_n = (e_n, f), \qquad c_n(0) = (e_n, w_0). \tag{16.1.52}$$

Two ways of calculating the Green's function for the diffusion equation. For τ fixed, as a function of t, G is constructed according to the left column by integrating over $0 < t < \tau$ with a "final" condition at $t = \tau + 0$. For t fixed, as a function of τ, it is constructed according to the right column integrating over $\tau > t$ with an initial condition at $\tau = t + 0$. From this latter point of view G represents the diffusion of a "heat pulse" released at $\mathbf{x} = \mathbf{y}$ at time $\tau = t$.

Green's function	"Heat pulse"
$-\dfrac{\partial G}{\partial t} - \nabla_x^2 G = \delta(\mathbf{x} - \mathbf{y})\delta(t - \tau)$	$\dfrac{\partial G}{\partial \tau} - \nabla_y^2 G = \delta(\mathbf{x} - \mathbf{y})\delta(t - \tau)$
$G(\mathbf{x}, t = \tau + 0; \mathbf{y}, \tau) = 0$	$G(\mathbf{x}, t; \mathbf{y}, \tau = t + 0) = \delta(\mathbf{x} - \mathbf{y})$
$G(\mathbf{x}, t; \mathbf{y}, \tau) = 0$ for $t > \tau$	$G(\mathbf{x}, t; \mathbf{y}, \tau) = 0$ for $\tau < t$

$$G(\mathbf{x}, t; \mathbf{y}, \tau) = G(\mathbf{y}, -\tau; \mathbf{x}, -t)$$

An eigenvalue problem like (16.1.46) would also be encountered in the search for the normal modes of oscillation of a system capable of supporting linear undamped waves. If one looks for simple oscillatory solutions of the form $v(\mathbf{x}, t) = u_n(\mathbf{x})e^{-i\omega_n t}$ of the wave equation

$$\frac{1}{c^2}\frac{\partial^2 v}{\partial t^2} - \nabla^2 v = 0, \tag{16.1.53}$$

one is led to (16.1.46) with $\lambda_n = \omega_n^2/c^2$. The Hilbert–Schmidt theorem then guarantees the existence of an infinity of normal modes of progressively increasing frequency; see §3.8 p. 78 for an example.

16.2 The diffusion equation

By combining the methods of §15.5 and §16.1 we can develop the Green's function theory for the non-homogeneous diffusion equation

$$\frac{\partial u}{\partial t} - \nabla_x^2 u = f(\mathbf{x}, t), \qquad \mathbf{x} \in V. \tag{16.2.1}$$

In this case the domain Ω has a space-time nature including both a spatial domain V and a time domain $0 < t < T$, $\Omega = V \times [0, T]$.

Following the general procedure outlined in §15.6 and §15.5, we start by determining the adjoint operator and the concomitant by multiplying (16.2.1) by a function $G(\mathbf{x}, t; \mathbf{y}, \tau)$ and integrating over V and over t from 0 to $\tau + 0 \leq T$

$$\int_0^\tau dt \int_V G\left(\frac{\partial u}{\partial t} - \nabla^2 u\right) d^d x = \int_0^\tau dt \int_V G(\mathbf{x}, t; \mathbf{y}, \tau) f(\mathbf{x}, t) d^d x, \tag{16.2.2}$$

where $d^d x$ is the volume element in the d-dimensional spatial domain V. Integration to τ rather than T is justified by the same causality argument given earlier in connection with (15.5.3) p. 388 and, as in §15.5, the specification $\tau + 0$ is added so as to have the full contribution of the $\delta(t - \tau)$ in (16.2.4) (cf. footnote p. 388). Integration by parts over t and Green's identity (16.1.7) give

$$\int_0^{\tau+0} dt \int_V \left(-\frac{\partial G}{\partial t} - \nabla^2 G \right) u(\mathbf{x}, t) \, d^d x \ = \ \int_0^{\tau} dt \int_V G f(\mathbf{x}, t) \, d^d x$$

$$+ \int_V [G(\mathbf{x}, 0; \mathbf{y}, \tau) \, u(\mathbf{x}, 0) - G(\mathbf{x}, \tau + 0; \mathbf{y}, \tau) \, u(\mathbf{x}, \tau)] \, d^d x$$

$$+ \int_0^{\tau} dt \int_S \mathbf{n} \cdot (G \nabla u - u \nabla G) \, dS. \tag{16.2.3}$$

We thus see that the appropriate form of the equation (15.6.6) p. 393 for G is

$$-\frac{\partial G}{\partial t} - \nabla_x^2 G \ = \ \delta(\mathbf{x} - \mathbf{y}) \, \delta(t - \tau). \tag{16.2.4}$$

For an initial-value problem data are prescribed at $t = 0$, but not at the final instant $t = \tau$. Thus $u(\mathbf{x}, \tau)$ is unknown and, as before in (15.5.6), we need to require that

$$G(\mathbf{x}, t = \tau + 0; \mathbf{y}, \tau) \ = \ 0, \tag{16.2.5}$$

so that (16.2.4) is to be solved backward in time.[6] Equation (16.2.3) becomes

$$u(\mathbf{y}, \tau) \ = \ \int_0^{\tau} dt \int_V G(\mathbf{x}, t; \mathbf{y}, \tau) \, f(\mathbf{x}, t) \, d^d x_x + \int_V G(\mathbf{x}, 0; \mathbf{y}, \tau) \, u(\mathbf{x}, 0) \, d^d x$$

$$+ \int_0^{\tau} dt \int_S (G \nabla_x u - u \nabla_x G) \cdot \mathbf{n}_x \, dS_x. \tag{16.2.6}$$

This formula is valid in any number of space dimensions. In one dimension, if the interval of interest extends for $a < x < b$, the surface integral must be interpreted as

$$\int_0^{\tau} dt \int_S dS_x \mathbf{n}_x \cdot (G \nabla_x u - u \nabla_x G) = \int_0^{\tau} dt \left[G \frac{\partial u}{\partial x} - u \frac{\partial G}{\partial x} \right]_{x=a}^{x=b} \tag{16.2.7}$$

where the square brakcets indicate the difference between the values of the enclosed quantity at $x = b$ and $x = a$.

As a justification one may think of a region in the form of a slab of unit cross section with the outward unit normals at $x = b$ and $x = a$ directed in the positive and negative x-directions respectively.

In §15.5 of the previous chapter we found, similarly to (16.2.4), that the dependence of the Green's function on its first time variable required an integration backward in time, while its dependence on the second time variable involved an integration forward in time.

---

[6] For normal parabolic equations this causes stability problems, but the equal signs of the two terms in the left-hand side of the diffusion equation (16.2.4) indicate that the diffusivity parameter is negative in this case (equal to -1 with our normalization). This is equivalent to solving a diffusion equation with positive diffusivity forward in time and no difficulties arise.

The situation is quite analogous here. If we apply the operator $\partial/\partial\tau - \nabla_y^2$ to (16.2.6) we find that, for this expression to be a solution of (16.2.1), it is necessary that

$$\frac{\partial G}{\partial\tau} - \nabla_y^2 G = \delta(\mathbf{x}-\mathbf{y})\,\delta(\tau - t). \tag{16.2.8}$$

Integration of (16.2.8) from $\tau = t - 0$ to $\tau = t + 0$ shows that the jump in G equals $\delta(\mathbf{x}-\mathbf{y})$. But $G = 0$ for $\tau < t$ by causality so that

$$G(\mathbf{x}, t; \mathbf{y}, \tau = t + 0) = \delta(\mathbf{x}-\mathbf{y}). \tag{16.2.9}$$

This relation also implies that the initial condition $u = u(\mathbf{x}, 0)$ at $\tau = 0+$ is satisfied. As before, we see then that, in its dependence on the second time variable, $G(\mathbf{x}, t; \mathbf{y}, \tau)$ represents the response of the system to a unit "heat pulse" occurring at $\mathbf{x} = \mathbf{y}$ and $\tau = t$. In this sense, the solution in the form (16.2.6) may be interpreted as the effect at time τ of heat pulses distributed over the volume of the medium and its surface at earlier times. Similarly to (16.1.17), the last term $u\,\mathbf{n} \cdot \nabla_x G$ may be interpreted as a "heat dipole." This physical picture justifies the name *fundamental solution* often used for the Green's function.

Table 16.1 summarizes and compares the two constructions of the Green's function for the diffusion equation according to (16.2.4), (16.2.5) or (16.2.8), (16.2.9). The symmetry property in the last line is proven similarly to (15.6.17) p. 395 by taking in (16.2.3) $u = G(\mathbf{x}, -t; \mathbf{y}, -\hat{t})$, $v = G(\mathbf{x}, t; \mathbf{y}, \tau)$ and using the fact that G vanishes for $t > \tau$.

In order to complete the specification of G we need to prescribe spatial boundary conditions which, as usual, will be chosen so as to turn (16.2.3) into an explicit result. There are various possibilities here; we consider a few cases which illustrate the general procedure.

16.2.1 Dirichlet problem

Suppose that u vanishes at $t = 0$ and is assigned on the boundary:

$$u(\mathbf{x}, t) = U(\mathbf{x}, t), \quad \mathbf{x} \in S \quad \text{and} \quad u(\mathbf{x}, 0) = 0. \tag{16.2.10}$$

Choose then

$$G(\mathbf{x}, t; \mathbf{y}, \tau) = 0 \quad \text{for} \quad \mathbf{x} \in S. \tag{16.2.11}$$

The formula (16.2.6) reduces to

$$u(\mathbf{y}, \tau) = -\int_0^\tau dt \int_S U(\mathbf{x})\,\mathbf{n}_x \cdot \nabla_x G\,dS_x + \int_0^\tau dt \int_V G(\mathbf{x}, t; \mathbf{y}, \tau)\,f(\mathbf{x}, t)\,d^d x. \tag{16.2.12}$$

Example 16.2.1 Consider the one-dimensional diffusion problem

$$\frac{\partial u}{\partial t} = \frac{\partial^2 u}{\partial x^2} \tag{16.2.13}$$

in $0 < x < \pi$, subject to the conditions

$$u(0, t) = U(t), \quad u(\pi, t) = 0, \quad u(x, 0) = 0. \tag{16.2.14}$$

The boundary condition (16.2.11) suggests to solve for G by using a Fourier sine series:

$$G(x, t; y, \tau) = \sum_{n=1}^{\infty} g_n(t; y, \tau) \sin nx. \tag{16.2.15}$$

Upon proceeding in the usual way (i.e., substituting into the one-dimensional form of (16.2.8) and taking scalar products) we are led to

$$\frac{\partial g_n}{\partial t} - n^2 g_n = -\frac{2}{\pi} \sin ny \, \delta(t - \tau), \tag{16.2.16}$$

which may be rewritten as

$$\frac{\partial}{\partial t} \left[\exp(-n^2 t) g_n \right] = -\frac{2}{\pi} \sin nx \, \exp(-n^2 \tau) \delta(t - \tau), \tag{16.2.17}$$

or, upon integration from t to $\tau + 0$,

$$\exp(-n^2 \tau) g_n(\tau + 0; x, t) - \exp(-n^2 t) g_n(t; x, \tau)$$
$$= -\frac{2}{\pi} \sin nx \, \exp(-n^2 t) \int_t^{\tau+0} \delta(\sigma - \tau) d\sigma. \tag{16.2.18}$$

The first term vanishes due to (16.2.5) and the integral of the δ equals $H(\tau - t)$ as the result would be zero unless $\tau + 0 > t$. Hence we find

$$g_n(t; x, \tau) = \frac{2}{\pi} \exp[-n^2(\tau - t)] \, H(\tau - t) \sin nx, \tag{16.2.19}$$

so that

$$G = \frac{2}{\pi} H(\tau - t) \sum_{n=1}^{\infty} \sin nx \, \sin ny \, e^{-n^2(\tau - t)}. \tag{16.2.20}$$

We can verify here explicitly the symmetry property in the last line of Table 16.1. When this expression is substituted into (16.2.12) and use is made of (16.2.7), the result is

$$u(y, \tau) = \int_0^{\tau} dt \, U(t) \left[\frac{2}{\pi} \sum_{n=1}^{\infty} n \, \sin nx \, e^{-n^2(\tau - t)} \right], \tag{16.2.21}$$

where the H function can be dropped as the integral is from 0 to τ. This is the same result as we found by using the Fourier series to solve (16.2.13) directly in §3.2.3 p. 62. □

If the boundary data are homogeneous, but $u(\mathbf{x}, 0) = u_0(\mathbf{x}) \neq 0$, we still take $G(\mathbf{x}, t; \mathbf{y}, \tau) = 0$ for $\mathbf{x} \in S$ and we find

$$u(\mathbf{y}, \tau) = \int_V G(\mathbf{x}, 0; \mathbf{y}, \tau) u_0(\mathbf{x}) d^d x + \int_0^{\tau} dt \int_V G(\mathbf{x}, t; \mathbf{y}, \tau) f(\mathbf{x}, t) d^d x. \tag{16.2.22}$$

Example 16.2.2 Consider the one-dimensional problem (16.2.13) on the infinite line $-\infty < x < \infty$, with conditions $u \to 0$ at infinity and $u(x, 0) = u_0(x)$. The solution of

(16.2.4) decaying at infinity is found by means of the Fourier transform in §4.4.1 and is given in (4.4.19) on p. 97:

$$G(x, t; y, \tau) = \frac{H(\tau - t)}{2\sqrt{\pi(\tau - t)}} \exp\left[-\frac{(x - y)^2}{4(\tau - t)}\right], \qquad (16.2.23)$$

which satisfies also the symmetry property in the last line of Table 16.1. The solution of the problem (16.2.31) takes then the well-known form already given in (4.4.13) p. 96:

$$u(y, \tau) = \frac{H(\tau)}{2\sqrt{\pi\tau}} \int_{-\infty}^{\infty} u_0(x) \exp\left[-\frac{(x - y)^2}{4\tau}\right] dx. \qquad (16.2.24)$$

□

As is evident from the expression (16.2.6), the solution of the combined initial-boundary value problem

$$u(\mathbf{x}, t) = U(\mathbf{x}, t), \qquad \mathbf{x} \in S \qquad \text{and} \qquad u(\mathbf{x}, 0) = u_0(\mathbf{x}) \qquad (16.2.25)$$

is given by the superposition of the two results (16.2.12) and (16.2.22):

$$\begin{aligned} u(\mathbf{y}, \tau) &= \int_V G(\mathbf{x}, 0; \mathbf{y}, \tau) u_0(\mathbf{x}) d^d x - \int_0^\tau dt \int_S U(\mathbf{x}) \mathbf{n}_x \cdot \nabla_x G dS_x \\ &+ \int_0^\tau dt \int_V G(\mathbf{x}, t; \mathbf{y}, \tau) f(\mathbf{x}, t) d^d x. \end{aligned} \qquad (16.2.26)$$

This is another example of the decomposition strategy mentioned on p. 65.

16.2.2 Neumann problem

If now

$$\mathbf{n} \cdot \nabla u(\mathbf{x}, t) = W(\mathbf{x}, t), \qquad \mathbf{x} \in S, \qquad u(\mathbf{y}, 0) = 0, \qquad (16.2.27)$$

we choose

$$\mathbf{n}_x \cdot \nabla_x G(\mathbf{x}, t; \mathbf{y}, \tau) = 0, \qquad \mathbf{x} \in S. \qquad (16.2.28)$$

This does not give rise to the difficulty encountered with the Poisson equation as integration of (16.2.4) for G over V gives

$$-\frac{\partial}{\partial t} \int_V G(\mathbf{x}, t; \mathbf{y}, \tau) d^d x = \delta(\tau - t), \qquad (16.2.29)$$

which poses no conflicting requirements. The formula (16.2.6) gives then

$$u(\mathbf{y}, \tau) = \int_0^\tau dt \int_S G(\mathbf{x}, t; \mathbf{y}, \tau) W(\mathbf{x}, t) dS_x + \int_0^\tau dt \int_V G(\mathbf{x}, t; \mathbf{y}, \tau) f(\mathbf{x}, t) d^d x. \qquad (16.2.30)$$

If the initial data are non-homogeneous

$$\mathbf{n} \cdot \nabla_x u(\mathbf{x}, t) = 0 \quad \text{for} \quad \mathbf{x} \in S, \qquad u(\mathbf{x}, 0) = v_0(\mathbf{x}), \qquad (16.2.31)$$

we retain the condition $\mathbf{n}_x \cdot \nabla_x G(\mathbf{x}, t; \mathbf{y}, \tau) = 0$ for $\mathbf{x} \in S$ and find from (16.2.6)

$$u(\mathbf{y}, \tau) = \int_V G(\mathbf{x}, 0; \mathbf{y}, \tau) v_0(\mathbf{x}) \, d^d x + \int_0^\tau dt \int_V G(\mathbf{x}, t; \mathbf{y}, \tau) f(\mathbf{x}, t) \, d^d x. \quad (16.2.32)$$

The solution corresponding to non-homogeneous boundary and initial data is found by superposition similarly to (16.2.26).

16.3 Wave equation

A Green's function theory for the wave equation

$$\frac{\partial^2 u}{\partial t^2} - \nabla_x^2 u = f(\mathbf{x}, t) \quad (16.3.1)$$

can be readily developed following the lines described for the diffusion equation. The procedure now gives

$$\int_0^\tau dt \int_V d^d x \, G \left[\frac{\partial^2 u}{\partial t^2} - \nabla^2 u \right] = \int_0^{\tau+0} dt \int_V d^d x \left[\frac{\partial^2 G}{\partial t^2} - \nabla_x^2 G \right] u$$
$$+ \int_V \left[\frac{\partial G}{\partial t} u - G \frac{\partial u}{\partial t} \right]_{t=0}^{t=\tau} d^d x + \int_0^\tau dt \oint_S (G \nabla_x u - u \nabla_x Gu) \cdot \mathbf{n} \, dS, (16.3.2)$$

which shows that the differential operator is formally self-adjoint. The Green's function satisfies therefore

$$\frac{\partial^2 G}{\partial t^2} - \nabla_x^2 G = \delta(\mathbf{x} - \mathbf{y}) \, \delta(t - \tau), \quad (16.3.3)$$

subject to the "final" conditions

$$G(\mathbf{x}, t = \tau + 0; \mathbf{y}, \tau) = 0, \qquad \left. \frac{\partial G}{\partial t} \right|_{t=\tau+0} = 0. \quad (16.3.4)$$

Upon integrating (16.3.3) from $t = \tau - 0$ to $t = \tau + 0$ we thus see that

$$G(\mathbf{x}, t = \tau - 0; \mathbf{y}, \tau) = 0, \qquad - \left. \frac{\partial G}{\partial t} \right|_{t=\tau-0} = \delta(\mathbf{x} - \mathbf{y}). \quad (16.3.5)$$

The relations (16.3.4) imply that $G = 0$ for $t > \tau$ as is necessary for causality. The arguments leading to these relations are quite similar to those given on p. 389 in §15.5 in connection with (15.5.6) and (15.5.8).

The explicit expression for u is then

$$u(\mathbf{y}, \tau) = \int_0^\tau dt \int_V G(\mathbf{x}, t; \mathbf{y}, \tau) f(\mathbf{x}, t) \, d^d x + \int_V \left[G \frac{\partial u}{\partial t} - u \frac{\partial G}{\partial t} \right]_{t=0} d^d x$$
$$+ \int_0^\tau dt \oint_S (G \nabla_x u - u \nabla_x G) \cdot \mathbf{n}_x \, dS_x. \quad (16.3.6)$$

In one space dimension the surface integral is to be interpreted similarly to (16.2.7), namely

$$\int_0^\tau dt \int_S dS_x \, (G\nabla_x u - u\nabla_x G) \cdot \mathbf{n}_x = \int_0^\tau dt \left[G\frac{\partial u}{\partial x} - u\frac{\partial G}{\partial x} \right]_{x=a}^{x=b}. \qquad (16.3.7)$$

Since, as described by the wave equation, there is a finite speed of propagation of disturbances, the actual spatial domain of integration in (16.3.6) and (16.3.7) may be smaller than indicated. This limitation will be an automatic consequence of the vanishing of the Green's function outside the domain of dependence of the solution.

We can subject (16.3.6) to an analysis similar to the one carried out for the initial-value problem in §15.5. In this way we find that

$$\frac{\partial^2 G}{\partial \tau^2} - \nabla_y^2 G = \delta(\mathbf{y} - \mathbf{x})\, \delta(\tau - t), \qquad (16.3.8)$$

and that

$$G(\mathbf{x}, t; \mathbf{y}, \tau = t + 0) = 0, \qquad \frac{\partial G}{\partial \tau}\bigg|_{\tau=t+0} = \delta(\mathbf{y} - \mathbf{x}). \qquad (16.3.9)$$

In its dependence on τ the Green's function therefore represents the effect at point $\mathbf{y}$ and time τ of a pulse emitted at the earlier time t from the point $\mathbf{x}$. For this reason, it is often referred to as *causal fundamental solution*. In the context of general linear hyperbolic equations the Green's function is often referred to as *Riemann function* (see e.g. Courant & Hilbert 1953, vol. 2, chapter 6; Garabedian 1964, chapter 5).

In this case, the symmetry relation satisfied by the Green's function is

$$G(\mathbf{x}, t; \mathbf{y}, \tau) = G(\mathbf{y}, -\tau; \mathbf{x}, -t), \qquad (16.3.10)$$

which is proven starting from (16.3.2) in the same way indicated on p. 412. Physically, this relation states that the effect at $(\mathbf{y}, \tau)$ of a pulse emitted at $\mathbf{x}$ at the earlier time t is the same as the effect at $(\mathbf{x}, -t)$ of a pulse emitted at $\mathbf{y}$ at time $-\tau$.

Example 16.3.1 As a simple example let us consider the one-dimensional case over the real line, assuming that $f \to 0$ at infinity sufficiently fast and that $u(x, 0) = u_0(x)$, $(\partial u/\partial t)(x, 0) = v_0(x)$. After Fourier-transforming over x, the equation (16.3.3) for G becomes

$$\frac{\partial^2 \tilde{G}}{\partial t^2} + k^2 \tilde{G} = \frac{1}{\sqrt{2\pi}} \delta(\tau - t) e^{iky}, \qquad (16.3.11)$$

subject to

$$\tilde{G}(k, t = \tau - 0; y, \tau) = 0, \qquad -\frac{\partial \tilde{G}}{\partial t}\bigg|_{t=\tau-0} = \frac{e^{iky}}{\sqrt{2\pi}}. \qquad (16.3.12)$$

The solution is

$$\tilde{G} = -\frac{H(\tau - t)}{2} \frac{1}{ik\sqrt{2\pi}} \left[e^{ik[y-(\tau-t)]} - e^{ik[y+(\tau-t)]} \right]. \qquad (16.3.13)$$

From Table 4.1 p. 85 $\mathscr{F}^{-1}\{\exp(ibk)f\}(k) = f(x-b)$ and, from Table 4.2 p. 86, $\mathscr{F}^{-1}\left\{-(ik\sqrt{2\pi})^{-1}\right\} = H(x) - 1/2$. Thus

$$G(x,t;y,\tau) = \frac{H(\tau-t)}{2}\left[H(x-(y-\tau+t)) - H(x-(y+\tau-t))\right], \quad (16.3.14)$$

which satisfies the symmetry relation (16.3.10) as is easily seen by using the property $H(-\xi) = 1 - H(\xi)$ of the step function. Substitution into (16.3.6) gives

$$
\begin{aligned}
u(y,\tau) &= \frac{1}{2}\int_0^\tau dt \int_{y-(\tau-t)}^{y+\tau-t} f(x,t)\,dx + \frac{1}{2}\left[u_0(y+\tau) + u_0(y-\tau)\right] \\
&+ \frac{1}{2}\int_{y-\tau}^{y+\tau} v_0(x)\,dx,
\end{aligned}
\quad (16.3.15)
$$

the last two terms of which embody d'Alembert's solution (4.5.9) of the homogeneous equation already derived on p. 101. Simple considerations show that the expression (16.3.14) can be written more compactly as

$$G(x,t;y,\tau) = \frac{1}{2}H(\tau - t - |x-y|). \quad (16.3.16)$$

□

17

Analytic Functions

It is very frequently the case that the true nature of a real function or a real differential equation can only be understood in full by extending it to the complex domain. Such extensions have many other applications, such as the exact calculation of integrals arising, e.g., in the use of transform techniques, or their approximate evaluation by means e.g. of the saddle-point method (§10.7 p. 276). Functions of a complex variable are also directly useful for the solution of a large number of two-dimensional field problems in many disciplines. These are just a few of the very many reasons which render the theory of analytic functions a powerful tool in applied mathematics.

17.1 Complex algebra

A complex number z with real part a and imaginary part b is written as

$$z = a + ib \quad \text{or} \quad z = r\,e^{i\theta}. \tag{17.1.1}$$

The second expression, in which r and θ are the *modulus* and *argument* of the complex number, is the *trigonometric* or *polar form*. The connection between the two forms is provided by *Euler's formula*

$$e^{i\theta} = \cos\theta + i\sin\theta, \tag{17.1.2}$$

so that $a = r\cos\theta$, $b = r\sin\theta$. The argument θ is only defined up to an arbitrary multiple of 2π, which is at the root of numerous subtle aspects of the theory of complex functions. On occasion it is convenient to limit this arbitrariness by restricting consideration to the *principal value* of the argument, written as Arg z; we take

$$-\pi < \text{Arg } z \leq \pi, \tag{17.1.3}$$

although the stipulation $0 \leq \text{Arg } z < 2\pi$ is also encountered in the literature. The *complex conjugate* $\bar{z}$ of z is defined by $\bar{z} = a - ib = re^{-i\theta}$.

The rules governing the algebra of complex numbers, summarized in Table 17.1, are a consequence of the fundamental relation $i^2 = -1$ and of the definition $z^{-1}z = 1$.

The indeterminacy of the argument forces us to write e.g.

$$\text{arg } z_1 z_2 = \text{arg } z_1 + \text{arg } z_2 \quad \text{mod } 2\pi \tag{17.1.4}$$

Table 17.1   Basic rules and properties of complex algebra.

$z = a + ib$ $\quad$ $i^2 = -1$		$\overline{z} = a - ib, \overline{(\overline{z})} = z$
$z_1 + z_2 = a_1 + a_2 + i(b_1 + b_2)$		$z_k = a_k + ib_k, k = 1, 2$
$z_1 z_2 = a_1 a_2 - b_1 b_2 + i(a_1 b_2 + a_2 b_1)$		$z\overline{z} \equiv |z|^2 = a^2 + b^2$ $\quad$ $|z|$ modulus of z $\quad$ $z^2 \equiv zz = a^2 - b^2 + 2iab$ $\quad$ $|z_1 z_2| = |z_1||z_2|$
$z^{-1} = \overline{z}/|z|^2 = (a - ib)/(a^2 + b^2)$		$|z^{-1}| = |z|^{-1}$
$\dfrac{z_1}{z_2} = \dfrac{z_1 \overline{z_2}}{|z_2|^2} = \dfrac{a_1 a_2 + b_1 b_2}{a_2^2 + b_2^2} + i\dfrac{a_2 b_1 - a_1 b_2}{a_2^2 + b_2^2}$		$\left|\dfrac{z_1}{z_2}\right| = \dfrac{|z_1|}{|z_2|}$ $\quad$ if $\quad z_2 \neq 0$
$\bigl||z_1| - |z_2|\bigr| \leq |z_1 + z_2| \leq |z_1| + |z_2|$		triangle inequality
$\sqrt{z} = \pm\left[\sqrt{\dfrac{1}{2}\left(a + \sqrt{a^2 + b^2}\right)}\right.$ $\left. + i\,(\mathrm{sgn}\,b)\sqrt{\dfrac{1}{2}\left(-a + \sqrt{a^2 + b^2}\right)}\right]$		If $b = 0$ $\quad$ $\sqrt{z} = \pm\sqrt{a}$ for $a > 0$ $\quad$ $\sqrt{z} = \pm i\sqrt{-a}$ for $a < 0$
$z = r e^{i\theta}, \overline{z} = r e^{-i\theta}$		$r = \sqrt{a^2 + b^2} \equiv |z|$ $\quad$ $\cos\theta = a/\sqrt{a^2 + b^2}$ $\quad$ $\sin\theta = b/\sqrt{a^2 + b^2}$
$z^n = r^n(\cos n\theta + i\sin n\theta)$		$n = 0, \pm 1, \ldots,$ De Moivre's formula
$\sqrt{z} = \sqrt{r}\,e^{i\theta/2}$ or $\sqrt{r}\,e^{i(\theta/2 + \pi)}$		
$\arg z_1 z_2 = \arg z_1 + \arg z_2$		$\mathrm{mod}\,2\pi$
$\arg z_1/z_2 = \arg z_1 - \arg z_2$		$\mathrm{mod}\,2\pi$
$-\pi < \mathrm{Arg}\,z \leq \pi$		principal value of the argument

signifying that the difference between the two sides of the equality is an arbitrary multiple of 2π. Alternatively, we may interpret the equality as relations between classes of numbers, in the sense that for each one of the infinite numbers ($\arg z_1 z_2$) on the left (all different from one another by an integer multiple of 2π), there is an equal number in the class of numbers ($\arg z_1 + \arg z_2$) on the right.

The visualization provided by the *stereographic projection* of the *Riemann sphere* is often a useful aid to intuition. The projection, illustrated in Figure 17.1, is a one-to-one correspondence between the points on the surface of a sphere of unit diameter and the points of the *compactified* or *extended complex plane*, i.e., the complex plane including the point at infinity. The "south pole" of the sphere coincides with the origin and each point

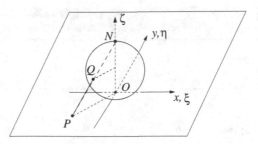

Stereographic projection of the Riemann sphere onto the compactified, or extended, complex plane, i.e., the complex plane argumented by the point at infinity.

Q of the sphere is projected onto a corresponding point P on the plane by prolongation of the straight line joining Q to the "north pole" N of the sphere. The north pole N itself is mapped onto the point at infinity of the complex plane. Note that there is only one such point, independent of the "direction of approach" to infinity.[1] Clearly, the extended complex plane has no boundary point while $z = \infty$ is the only boundary point of the (non-extended) complex plane. Consideration of the similar triangles in Figure 17.1 readily permits one to derive the following relations between the Cartesian coordinates $z = (x, y)$ of a point P on the plane and the Cartesian coordinates (ξ, η, ζ) of the corresponding point P on the unit sphere:

$$(\xi, \eta, \zeta) = \frac{(x, y, x^2 + y^2)}{1 + x^2 + y^2}, \qquad (x, y, x^2 + y^2) = \frac{(\xi, \eta, \zeta)}{1 - \zeta}. \qquad (17.1.5)$$

These relations permit an easy proof of the fact that circles and lines in the (x, y) plane are transformed into circles on the Riemann sphere and vice versa (see e.g. Hille 1959, p. 40).

There is a very large number of books on analytic function theory (for a recent compendium see Krantz 1999), and much of the material that follows can be found in most of them. Accordingly, we will make a sparing use of specific references.

---

[1] Thus, the point at infinity of the extended complex plane is different from the circle at infinity of the projective plane, which can be represented on the lower hemisphere of the Riemann sphere by projecting from the center of the sphere rather than from the north pole. There are significant differences between the complex plane and the extended complex plane. For example, the complex plane "punctured" by the deletion of a finite-size region or even a single point is not simply connected (definition §A.5), while the extended complex plane remains simply connected as the image of the punctured plane on the sphere is simply connected. Accordingly, the theory is developed differently in the complex plane and the extended complex plane; for our purposes the latter is more convenient (see e.g. Burckel 1979 p. 39 and *passim*).

17.2 Analytic functions

Given a pair of continuous functions of two variables $u(x, y)$ and $v(x, y)$, we can define the complex-valued function $w = u + iv$. With the definitions

$$z = x + iy, \qquad \overline{z} = x - iy, \tag{17.2.1}$$

from which

$$x = \frac{1}{2}(z + \overline{z}), \qquad y = \frac{1}{2i}(z - \overline{z}), \tag{17.2.2}$$

we can regard $w = w(z, \overline{z})$ as a function of z and $\overline{z}$ and in particular, if the indicated partial derivatives exist, the chain rule gives[2]

$$\frac{\partial w}{\partial z} = \frac{1}{2}\frac{\partial w}{\partial x} + \frac{1}{2i}\frac{\partial w}{\partial y}, \qquad \frac{\partial w}{\partial \overline{z}} = \frac{1}{2}\frac{\partial w}{\partial x} - \frac{1}{2i}\frac{\partial w}{\partial y}. \tag{17.2.3}$$

We say that a function w is *analytic* in a connected open region, or *domain*, of the complex plane[3] if

$$\frac{\partial w}{\partial \overline{z}} = 0 \tag{17.2.4}$$

which, from (17.2.3), requires that

$$\frac{\partial w}{\partial x} = -i\frac{\partial w}{\partial y} \qquad \Longleftrightarrow \qquad \frac{\partial u}{\partial x} = \frac{\partial v}{\partial y}, \quad \frac{\partial u}{\partial y} = -\frac{\partial v}{\partial x}. \tag{17.2.5}$$

These are the *Cauchy–Riemann relations*. When they are satisfied, w depends on the single variable z so that we may write $w'(z)$ in place of $\partial w/\partial z$.[4]

Definition 17.2.1 A function $w(z)$ defined for z in an open domain D is analytic in D if it is single-valued and continuous with a continuous derivative in D.

Continuity of the derivative actually follows from its existence, (Goursat's theorem, see e.g. Narasimhan & Nievergelt 2001, pp. 15 and 19) and, according to the monodromy theorem (see e.g. Ablowitz & Fokas 1997, p. 155; Henrici 1974, p. 167), single-valuedness is a consequence of the other conditions included in this definition provided the domain is simply connected. We have explicitly included these attributes in the definition for convenience.[5]

---

[2] To be sure, this is only a formal procedure in that it ignores the fact that the indicated partials have no clear definition as limits. It does however have the merit of stressing the difference between a general pair $[u(x, y), v(x, y)]$ of real functions of two variables, and an analytic function. A better definition would be built along the line mentioned as item 6 on p. 423.

[3] A region D of the complex plane is *open* if, for any point $z_1 \in D$, one can find an $\varepsilon > 0$ such that all the points in the circle $|z - z_1| < \varepsilon$ belong to D; see §A.1 in the Appendix for further details. A region is connected if any two of its points can be joined by a line entirely belonging to the region.

[4] It follows from these equalities that the derivative of a *real* function of a complex variable is either zero or does not exist.

[5] A definition with weaker requirements is the following: $w = u + iv$ is analytic in a domain D if $|w|$ is locally Lebesgue-integrable and the first partial derivatives of u and v exist, are finite, and satisfy the Cauchy–Riemann relations almost everywhere in D (see e.g. Boas 1987, p. 21).

The function $w(z)$ is *analytic at a point* if it is analytic in a neighborhood of that point. For the point at infinity we set $\zeta = 1/z$ and consider the resulting function in the neighborhood of $\zeta = 0$.

By use of (17.2.5) we have several equivalent expressions for the derivative; two compact ones are

$$w'(z) = \frac{\partial w}{\partial x} = -i\frac{\partial w}{\partial y}. \tag{17.2.6}$$

In the trigonometric representation of $z = re^{i\theta}$ the Cauchy–Riemann relations take the form

$$\frac{\partial u}{\partial r} = \frac{1}{r}\frac{\partial v}{\partial \theta}, \qquad \frac{1}{r}\frac{\partial u}{\partial \theta} = -\frac{\partial v}{\partial r}, \tag{17.2.7}$$

while, in the trigonometric representation of $w = R(x, y)\exp[i\Theta(x, y)]$, they are

$$\frac{\partial R}{\partial x} = R\frac{\partial \Theta}{\partial y}, \qquad \frac{\partial R}{\partial y} = -R\frac{\partial \Theta}{\partial x}. \tag{17.2.8}$$

In trigonometric form the derivative may be expressed e.g. as

$$w'(z) = e^{-i\theta}\left(\frac{\partial u}{\partial r} + i\frac{\partial v}{\partial r}\right) = \frac{e^{-i\theta}}{r}\left(\frac{\partial v}{\partial \theta} - i\frac{\partial u}{\partial \theta}\right). \tag{17.2.9}$$

We now list without proof some important properties:

1. An analytic function of an analytic function is analytic. In particular, $w_1 + w_2$, $w_1 w_2$ and w_1/w_2 are analytic, the last one at all points where $w_2 \neq 0$.
2. If u, v and their first partial derivatives are continuous at all points of D and the Cauchy–Riemann relations are satisfied, then $w = u + iv$ is analytic in D.
3. If $w = u + iv$ is analytic, the partials of any order of u and v exist and are continuous (see Theorem 17.3.4 p. 430). In particular, it then follows from the Cauchy–Riemann conditions that u and v are *harmonic*, i.e.

$$\frac{\partial^2 u}{\partial x^2} + \frac{\partial^2 u}{\partial y^2} = 0, \qquad \frac{\partial^2 v}{\partial x^2} + \frac{\partial^2 v}{\partial y^2} = 0. \tag{17.2.10}$$

Conversely, only if u and v are harmonic can $w = u + iv$ be analytic.
4. If u is harmonic, with the change of variables (17.2.1) we have

$$\frac{\partial^2 u}{\partial x^2} + \frac{\partial^2 u}{\partial y^2} = 4\frac{\partial^2 u}{\partial z \partial \bar{z}} = 0 \tag{17.2.11}$$

the most general solution of which is $u = f(z) + g(\bar{z})$. Since u is real by hypothesis, it must be that $g(\bar{z}) = \overline{f(z)}$, which requires that $\overline{f(z)} = f(\bar{z})$ (in this connection see (17.8.8) and Theorem 17.8.5 p. 458) and $u = f(z) + \overline{f(z)}$.
5. For a given u it follows from the Cauchy–Riemann relations that all the possible v's such that $u + iv$ is an analytic function are given by

$$v(x, y) = -\int_a^x \frac{\partial u}{\partial y}(\xi, b)\, d\xi + \int_b^y \frac{\partial u}{\partial x}(x, \eta)\, d\eta + v(a, b), \tag{17.2.12}$$

where $v(a, b)$ is the arbitrary value of v at the starting point of the integration. For example, if $u = xy$ and $a = b = 0$, we find $v = \frac{1}{2}(y^2 - x^2) + v(0, 0)$ so that $f = -\frac{1}{2}iz^2 + f(0)$. The imaginary part of an analytic function is the *harmonic conjugate* of the real part.

6. When the Cauchy–Riemann relations hold, the limit for $z \to z_0$ of $[w(z) - w(z_0)]/(z - z_0)$ is independent of the direction from which z approaches z_0. Conversely, by requiring that the limit be independent of direction, one finds the Cauchy–Riemann relations. To emphasize this property the word *monogenic* is also sometimes used synonymously with analytic. Because of this property the derivative of an analytic function can be calculated as the limit of the ratio $[w(z) - w(z_0)]/(z - z_0)$ from any direction and, as a consequence, the familiar differentiation rules of real functions apply.

7. If $w(z)$ is analytic at *some* point in every neighborhood of a point z_0, but not at z_0, z_0 is a *singular point* or singularity of w. The singular point is *isolated* if a region can be found, of which z_0 is an interior point, containing no other singularities; singular points are treated in §17.4.1.

8. It is often useful to consider the relation $w = w(z)$ as a mapping from the z-plane to the w-plane. As described in greater detail in §17.10, this mapping is *conformal* in the sense that the angle between two lines in the z-plane is preserved upon the transformation. The *open mapping theorem* states that the image of an open set under this mapping is open (see e.g. Narasimhan & Nievergelt 2001, p. 26).

9. By the Cauchy–Riemann relations, if $w = u + iv$, the families of level curves $u = $ const. and $v = $ const. are mutually orthogonal.

In conclusion, some terminology:

1. An analytic function is often also called *holomorphic*, *regular* or *monogenic* in the literature.

2. A function analytic at every point of the finite complex plane is *entire*.

3. A function whose only singularities in the finite plane are poles (see definition p. 435) is termed *meromorphic*.

4. A function is *univalent*, or *single-valued*, in a region if it assumes each one of its values just once in that region.

17.3 Integral of an analytic function

Integration of analytic functions is performed along paths in the complex plane. Precise definitions are given in §A.5 of the Appendix. Here we will refer to arcs and closed curves appealing to the intuitive meaning of the terms.

1. A *smooth curve* or *arc* has a continuously turning tangent.

2. *Contour* or *simple closed curve*: A continuous chain of a finite number of smooth arcs; the two end points of the chain coincide and there are no other intersections.

3. *Interior and exterior domains*: An essential property of a simple closed curve C is that it divides the plane in two regions, a bounded one, the *interior* of C, and an unbounded one, the *exterior* of C.

4. *Positive direction*: If a domain D is bounded by a contour C, the positive direction of traversal of C is defined as the direction which leaves D on the left. Thus, if the domain of interest is the interior of C, the positive direction is counter clockwise while it is clockwise if D is the exterior domain. On occasion it will be useful to consider the normal to the complex plane. We choose it to be oriented as the thumb of the right hand when the fingers are oriented in the positive sense of the contour.

5. *Simply connected domain*: Whenever the domain contains a closed curve, it also contains all the points in its interior. A domain containing a "hole" – even consisting of a single point – is not simply connected. The domain extending outside a hole all the way to infinity is not simply connected in the usual complex plane, but becomes simply connected in the extended complex plane as its image on the Riemann sphere is simply connected.

The *integral* of an analytic function w along a smooth arc $\mathscr{L}$ in the complex plane is defined analogously to the Riemann integral of a real function (see §A.4.1 p. 687), namely as the limit of the "discretized sum" $\sum w(\zeta_i)(z_{i+1} - z_i)$, with ζ_i a point of the arc of $\mathscr{L}$ joining z_i and z_{i+1} (see e.g. Conway 1978, p. 58). This integral is well defined if the arc has a finite length and the function w is continuous. Integrals along infinite paths, or of singular functions, are defined in the same way as for improper Riemann integrals (see p. 668).

Since, according to the definition of the integral, $w\,dz$ is legitimately interpreted as $u + iv$ times $dx + i\,dy$, the product rule gives

$$\int_{\mathscr{L}} w(z)\,dz = \int_{\mathscr{L}} (u\,dx - v\,dy) + i\int_{\mathscr{L}} (u\,dy + v\,dx). \qquad (17.3.1)$$

This relation reduces the complex integral to four real integrals.[6] In the actual evaluation it is often convenient to use a parameterization of the arc, e.g. $x = \phi(t)$, $y = \psi(t)$.

Example 17.3.1 Consider $\oint_C dz/z$ where the contour C is a circle of radius R centered at the origin. The quickest way to evaluate the integral is to use the trigonometric form $z = re^{i\theta}$, which effectively uses $\theta = \arg z$ to parameterize the contour. Since, along this contour, $dz = Rie^{i\theta}\,d\theta$, we have

$$\oint_C \frac{dz}{z} = \int_{-\pi}^{\pi} \frac{Rie^{i\theta}}{Re^{i\theta}}\,d\theta = 2\pi i. \qquad (17.3.2)$$

Alternatively we may set $x = R\cos t$, $y = R\sin t$, so that $dx = -R\sin t\,dt$, $dy = R\cos t\,dt$ and

$$\frac{1}{z} = \frac{x - iy}{x^2 + y^2} = \frac{1}{R}(\cos t - i\sin t). \qquad (17.3.3)$$

---

[6] It is perhaps worth noting explicitly that the real part of the integral is not the integral of the real part of the integrand.

Then, according to (17.3.1),

$$
\oint_{\mathscr{L}} \frac{dz}{z} = \int_{-\pi}^{\pi} \left[\frac{\cos t}{R} (-R \sin t \, dt) - \frac{(-\sin t)}{R} R \cos t \, dt \right]
$$
$$
+i \int_{-\pi}^{\pi} \left[\frac{\cos t}{R} (R \cos t \, dt) + \frac{(-\sin t)}{R} (-R \sin t \, dt) \right] = 2\pi i. \quad (17.3.4)
$$

$\square$

17.3.1 Cauchy's theorem

Given its similarity with the definition of the real Riemann integral, the integral of a complex function enjoys many of the same properties such as linearity (i.e., $\int (aw_1 + bw_2) \, dz = a \int w_1 \, dz + b \int w_2 \, dz$ for any numbers a and b), reversal of the sign upon interchange of the starting and end points of the arc $\mathscr{L}$, and concatenation (i.e., the sum of the integrals along two consecutive arcs equals the integral along the combined arc). Another property reminiscent of a similar one in the real case, which now follows from the triangle inequality (Table 17.1), is that, if $|w| < M$ on $\mathscr{L}$,

$$
\left| \int_{\mathscr{L}} w \, dz \right| \leq \int_{\mathscr{L}} |w| |dz| \leq ML, \quad (17.3.5)
$$

in which L is the length of $\mathscr{L}$. The most important property of the integral of an analytic function is expressed by (see e.g. Whittaker & Watson 1927, p. 85; Boas 1987, p. 45; Ablowitz & Fokas 1997, p. 81)

Theorem 17.3.1 [Cauchy] *If a function w is analytic at all points interior to and on a piecewise differentiable closed contour C,[7] then*

$$
\oint_C w \, dz = 0. \quad (17.3.6)
$$

A simple proof[8] can be based on Stokes's Theorem 16.1.2 p. 401 which, for the present purposes, can be stated as

$$
\oint_C \mathbf{V} \cdot \mathbf{t} \, d\ell = \int_S (\nabla \times \mathbf{V}) \cdot \mathbf{k} \, dS, \quad (17.3.7)
$$

in which $\mathbf{V}$ is a continuous vector field, $\mathbf{t}$ is the unit tangent to C and $\mathbf{k}$ the unit normal to the (x, y) plane, both in the positive direction; $d\ell$ is the arc length along C and S is the interior of C. To calculate the first integral in the right-hand side of (17.3.1) we take $\mathbf{V} = (u, -v, 0)$ to find

$$
\oint [u \, dx + (-v) \, dy] = \int_S \left(\frac{\partial v}{\partial x} - \frac{\partial u}{\partial y} \right) dS, \quad (17.3.8)
$$

---

[7] It is actually enough to require that w be continuous on and inside C.
[8] More general proofs are available, see e.g. Burckel 1979, p. 120.

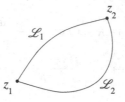
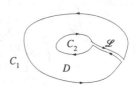

Fig. 17.2 Deformation of the contour of integration

which vanishes by the first of the Cauchy–Riemann relations (17.2.5); by taking $\mathbf{V} = (u, v, 0)$ we similarly prove that the imaginary part of the integral vanishes. This property actually characterizes analytic functions in view of:

Theorem 17.3.2 [Morera] *Let $u(x, y)$, $v(x, y)$ be continuous in a domain D. If $\oint (u + iv)\,dz = 0$ for every simple closed contour C whose interior lies in D, then $w = u + iv$ is analytic in D.*

A proof is given on p. 430.

A very important consequence of Cauchy's theorem is the possibility to *deform the path of integration* without changing the value of the integral; this is a major resource in many applications of analytic functions. Consider two arcs $\mathscr{L}_1$ and $\mathscr{L}_2$ having the same starting and end points z_1 and z_2 (Figure 17.2, left). If w is analytic on the arcs and inside the region they enclose, Cauchy's theorem says that

$$\oint_{\mathscr{L}_1 + \mathscr{L}_2} w\,dz = \int_{(1)\,z_1}^{z_2} w\,dz + \int_{(2)\,z_2}^{z_1} w\,dz = 0 \qquad (17.3.9)$$

where the first integral is along $\mathscr{L}_1$ and the second one along $\mathscr{L}_2$. As a consequence

$$\int_{(1)\,z_1}^{z_2} w\,dz = \int_{(2)\,z_1}^{z_2} w\,dz \qquad (17.3.10)$$

and the value of the integral is independent of the particular path used to evaluate it, provided the integrand is analytic in the region bounded by the original and the deformed path.

Example 17.3.2 Consider, for a and b real, the integral

$$I = \int_{-\infty}^{\infty} \exp(-a^2 x^2 + 2ibx)\,dx = e^{-b^2} \int_{-\infty}^{\infty} \exp[-(ax - ib)^2]dx. \qquad (17.3.11)$$

With the substitution $\zeta = ax - ib$ this becomes

$$I = \frac{e^{-b^2}}{a} \int_{-\infty - ib}^{\infty - ib} e^{-\zeta^2} d\zeta. \qquad (17.3.12)$$

On vertical lines $\zeta = \pm L + iy$, with $-b < y < 0$, $\exp(-\zeta^2) \to 0$ for $L \to \infty$ and the integrand is analytic inside the strip bound by the two horizontal lines Im $\zeta = 0$ and Im $\zeta = -b$. Thus the integral of $\exp(-\zeta^2)$ around the rectangle consisting of the portion of the strip bounded by the two vertical lines vanishes, with the vertical lines themselves giving no contribution in the limit $L \to \infty$. This argument shows that the integration path in (17.3.12) can be replaced by the real axis with the conclusion that

$$I = \frac{e^{-b^2}}{a} \int_{-\infty}^{\infty} e^{-\zeta^2} \, d\zeta = \sqrt{\pi} \, \frac{e^{-b^2}}{a}, \tag{17.3.13}$$

in agreement with one of the entries in the Fourier transform Table 4.2 p. 86. Since the imaginary part of the integral, being odd, vanishes, we also deduce

$$\int_{-\infty}^{\infty} e^{-a^2 x^2} \cos 2bx \, dx = \sqrt{\pi} \, \frac{e^{-b^2}}{a} \tag{17.3.14}$$

□

Cauchy's theorem enables us to introduce the notion of *indefinite integral* of a function analytic inside a simply connected domain D:

$$W(z) = \int_{z_0}^{z} w(z) \, dz. \tag{17.3.15}$$

This integral is an analytic function of its upper limit provided the path of integration is confined to the domain D and, furthermore, $W'(z) = w(z)$. All indefinite integrals differ by constants so that $\int_{z_1}^{z_2} w(z) \, dz = W(z_2) - W(z_1)$. A word of caution is in order here: if the domain is not simply connected (for instance, it is similar to a ring, or is "punctured" at a single point), the indefinite integral may become multi-valued (see e.g. Example 17.6.2 p. 449).

A similar property of path independence holds for the integrals along two closed curves C_1 and C_2 lying one inside the other if w is analytic in the annulus D bounded by them *but not necessarily inside the inner curve C_2*. Join the two curves arbitrarily by a line $\mathscr{L}$ entirely contained in D (Figure 17.2, right) and consider the closed contour that is described by going around C_1, turning into $\mathscr{L}$, going around C_2, and back to C_1 along $\mathscr{L}$ again, keeping the domain D on the left all the way. This closed contour encloses D, where w is analytic by hypothesis, and therefore the integral of w along the total path vanishes. Since the line $\mathscr{L}$ is traversed twice in opposite directions, its contribution cancels and one is left with

$$\oint_{C_1} w \, dz + \oint_{-C_2} w \, dz = 0 \tag{17.3.16}$$

where we write $-C_2$ to signify that the contour C_2 is traversed in the negative direction *with respect to the region that it encloses*, as is necessary if D is to be kept on the left. As a consequence, we find

$$\oint_{C_1} w \, dz = \oint_{C_2} w \, dz \tag{17.3.17}$$

where now both contours are described in the positive sense with respect to the regions that they enclose. The argument can be generalized to several contours $C_1, C_2, \ldots, C_n$

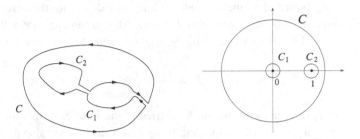

Fig. 17.3 Left: Illustration of the relation (17.3.18). Right: Evaluation of the integral (17.3.20)

entirely contained inside a contour C, not intersecting one other, and none of which lies inside any other one (see e.g. Figure 17.3, left):

$$\oint_C w\,\mathrm{d}z = \oint_{C_1} w\,\mathrm{d}z + \cdots + \oint_{C_n} w\,\mathrm{d}z. \tag{17.3.18}$$

This property affords a considerable simplification of the calculation of many integrals.

Example 17.3.3 Consider the following integral along the contour on the right in Figure 17.3:

$$\oint_C \frac{2z-1}{z^2-z}\,\mathrm{d}z = \oint_C \left(\frac{1}{z} + \frac{1}{z-1}\right)\,\mathrm{d}z. \tag{17.3.19}$$

The only singularities are at 0 and 1, and the integral can therefore be replaced by the sum of two integrals around arbitrarily small circles surrounding these singular points:

$$\oint_{C_1}\left(\frac{1}{z}+\frac{1}{z-1}\right)\mathrm{d}z + \oint_{C_2}\left(\frac{1}{z}+\frac{1}{z-1}\right)\mathrm{d}z = \oint_{C_1}\frac{\mathrm{d}z}{z} + \oint_{C_2}\frac{\mathrm{d}z}{z-1} = 4\pi i. \tag{17.3.20}$$

The first step is justified by the fact that $1/(z-1)$ does not contribute to the integral around C_1 as it is analytic inside C_1, and similarly for $1/z$ inside C_2. This example is a special case of the residue theorem given below in §17.9.   □

17.3.2 Cauchy's integral formula

Another key property of analytic functions extensively used in applications is (see e.g. Paliouras & Meadows 1990, p. 219; Conway 1978, p. 83; Ahlfors 1979, p. 114; Lang 1985)

Theorem 17.3.3 [Cauchy's integral formula] *Let $w(z)$ be analytic in an open domain D. Then the formula*

$$w(z) = \frac{1}{2\pi i} \oint_C \frac{w(\zeta)}{\zeta - z}\,\mathrm{d}\zeta \tag{17.3.21}$$

is valid for every simple positively oriented contour C and every point z in its interior, provided that C and its interior belong entirely to D.

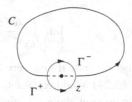

Fig. 17.4 Contours used in the calculation of the Cauchy principal value of an integral according to (17.3.23).

The remarkable fact is that once $w(z)$ is given on C, it is automatically determined inside C. This shows how "special" analytic functions are: every alteration of the value of w *inside* C must be accompanied by a change of its value *on* C if the function is to remain analytic. The proof is an easy consequence of (17.3.17) which enables us to shrink the contour of integration from C to an arbitrarily small circle C_ε centered at z.[9] We add and subtract $w(z)$ in the numerator to find

$$\frac{1}{2\pi i}\oint_C \frac{w(\zeta)}{\zeta - z}\mathrm{d}\zeta = \frac{w(z)}{2\pi i}\oint_{C_\varepsilon}\frac{\mathrm{d}\zeta}{\zeta - z} + \frac{1}{2\pi i}\oint_{C_\varepsilon}\frac{w(\zeta) - w(z)}{\zeta - z}\mathrm{d}\zeta. \quad (17.3.22)$$

The first integral has the value 1 as shown in Example 17.3.1. Thanks to the continuity of w, the second integral vanishes in the limit $\varepsilon \to 0$.

If z is outside C, the integrand in (17.3.21) is analytic on and inside C and the integral vanishes by Cauchy's theorem 17.3.1. It remains to consider z on C, in which case the integral (17.3.21) does not exist in the ordinary sense. It proves useful to define it as the *Cauchy principal value*:

$$\mathrm{PV}\oint \frac{w(\zeta)}{\zeta - z}\mathrm{d}\zeta = \lim_{\delta \to 0}\int_{C_\delta}\frac{w(\zeta)}{\zeta - z}\mathrm{d}\zeta, \quad (17.3.23)$$

where C_δ is obtained from C by removing all points ζ such that $|\zeta - z| < \delta$. This limit exists if w is continuous and reasonably smooth.[10] To calculate it we close the arc C_δ by a small semi-circle Γ^- of radius δ centered at z (see Figure 17.4), so that

$$\int_{C_\delta}\frac{w(\zeta)}{\zeta - z}\mathrm{d}\zeta + \int_{\Gamma^-}\frac{w(\zeta)}{\zeta - z}\mathrm{d}\zeta = 0, \quad (17.3.24)$$

since z is outside the contour $C_\delta + \Gamma^-$. Proceeding as in (17.3.22) we have

$$\int_{\Gamma^-}\frac{w(\zeta)\mathrm{d}\zeta}{\zeta - z} = w(z)\int_{\Gamma^-}\frac{\mathrm{d}\zeta}{\zeta - z} + \int_{\Gamma^-}\frac{w(\zeta) - w(z)}{\zeta - z}\mathrm{d}\zeta$$

$$= w(z)\int_\pi^0 \frac{i\delta\exp(i\theta)}{\delta\exp(i\theta)}\mathrm{d}\theta + O(\delta) \to -i\pi w(z), \quad (17.3.25)$$

---

[9] Considerably more general proofs are available for this theorem as well, see e.g. Burckel 1979, pp. 229, 264, 341; Narasimhan & Nievergelt 2001, p. 62.

[10] More precisely, the limit exists provided w satisfies a Lipschitz (or Hölder) condition (Definition A.3.5 p. 684) of order less ≤ 1.

so that

$$\lim_{\delta \to 0} \int_{C_\delta} \frac{w(\zeta)}{\zeta - z} d\zeta = \frac{1}{2} w(z). \tag{17.3.26}$$

The result would be the same if one were to close C_δ by the semicircle Γ^+ in Figure 17.4 as, in this case, (17.3.24) would equal $w(z)$ as the singularity of the integral is now inside the contour, but the integral along Γ^+ would equal $i\pi w(z)$ as the integration would be for $-\pi < \theta < 0$ rather than $\pi > \theta > 0$.[11] In conclusion we find therefore

$$\frac{1}{2\pi i} \oint \frac{w(\zeta)}{\zeta - z} d\zeta = \begin{cases} w(z) & z \text{ internal to } C \\ \frac{1}{2} w(z) & z \text{ on } C \\ 0 & z \text{ external to } C \end{cases} \tag{17.3.27}$$

which has an obvious relation with the Sokhotski–Plemelj formulae of Example 20.3.5 p. 574 and Example 20.8.1 p. 591.

A direct calculation of $\lim_{z \to z_0}[w(z) - w(z_0)]/(z - z_0)$ based on the representation (17.3.21) of w, shows that

$$w'(z) = \frac{1}{2\pi i} \oint_C \frac{w(\zeta)}{(\zeta - z)^2} d\zeta \tag{17.3.28}$$

and justifies therefore differentiation under the integral sign, provided z is interior to the contour C, which is taken to be simple and positively oriented. By induction, we then conclude that:

Theorem 17.3.4 *Let w be analytic in an open domain D. Then it has continuous derivatives of any order in D given by*

$$w^{(n)}(z) = \frac{n!}{2\pi i} \oint_C \frac{w(\zeta)}{(\zeta - z)^{n+1}} d\zeta \tag{17.3.29}$$

provided C and its interior belong entirely to D and z is interior to C.

Here is another striking difference with real functions for which the existence of a derivative does not imply anything about its nature or that of higher derivatives. Upon integration by parts, (17.3.29) gives

$$w^{(n)}(z) = \frac{1}{2\pi i} \oint \frac{w^{(n)}(\zeta)}{\zeta - z} d\zeta, \tag{17.3.30}$$

as one would expect from Cauchy's integral theorem applied to $w^{(n)}$.

We are now in a position to give a simple proof of Morera's theorem 17.3.2 p. 426. If w is continuous in a simply connected domain D and $\oint w(\zeta) d\zeta = 0$ along all closed contours lying wholly in D, we can define an indefinite integral $W(z)$ according to (17.3.15), which is analytic and for which $W' = w$. But if W is analytic, it also has a continuous second

---

[11] If at z the boundary has a corner with interior angle $\alpha\pi$ (i.e., tangents exist on the two sides of z, and they enclose an angle $\alpha\pi$), the same derivation shows that the integral along Γ^- would have the value $[-\alpha/2]w(z)$ and that along Γ^+ $[1 - \alpha/2]w(z)$ so that the 1/2 in the right-hand side would be replaced by $\alpha/2$.

derivative $W'' = w'$. Hence, being continuous by hypothesis and having a continuous first derivative, w is analytic by the definition 17.2.1 p. 421.

17.3.3 Further consequences

While reasonably well behaved, one cannot ask too much of an analytic function:

Theorem 17.3.5 [Liouville] *If w is entire (i.e., analytic at every finite point of the complex plane) and $|w|$ is bounded for all z, then w is a constant.*

Indeed, if M is the upper bound of $|w|$, the expression (17.3.28) for w' shows that, if the integration contour is chosen as a circle of radius R centered at z, then $|w'| \leq M/R$. which can be made as small as desired by taking R large enough. But this is only possible if $|w'| = 0$ and, since the argument can be repeated for any z, we find that $|w'| = 0$ everywhere, so that also $w' = 0$ and w is a constant.

Thus, a non-constant entire function must be singular at infinity. Furthermore, if w is entire $\exp(\pm w)$ also is by property 1 on p. 422. Since $|\exp(\pm w)| = \exp(\pm \mathrm{Re}\, w)$, it follows that there must be paths going to infinity in the complex plane along which $\mathrm{Re}\, w$ and $-\mathrm{Re}\, w$ tend to infinity.

This theorem enables us to give a very nice proof of

Theorem 17.3.6 [Fundamental theorem of algebra] *The equation*

$$P_m(z) \equiv a_0 + a_1 z + \cdots + a_m z^m = 0 \tag{17.3.31}$$

where P_m is a polynomial of degree m, has m roots (possibly not all distinct).

Indeed, suppose that $P_m(z) \neq 0$ for all z. Then $w = 1/P_m$ is analytic everywhere, and $|w| \to 0$ as $z \to \infty$, so that $|w|$ is bounded for all z. As a consequence, w is a constant, which is a contradiction. But if P_m has one root z_1, then we can form $P_m(z)/(z - z_1) = P_{m-1}$, find z_2, and so on, until we are left with a constant. Hence, any polynomial of degree m has m roots in the complex plane.

Another consequence is the *mean value theorem* for harmonic functions, already mentioned on p. 75. In Cauchy's integral formula (17.3.21) take the integration path to be a circle of arbitrary radius r centered at z. On this circle $\zeta = z + re^{i\theta}$ and the integral becomes

$$
\begin{aligned}
w(z) &= \frac{1}{2\pi i} \int_0^{2\pi} \frac{w(z + r \exp(i\theta))}{r \exp(i\theta)} ir \exp(i\theta) \, d\theta \\
&= \frac{1}{2\pi} \int_0^{2\pi} w(z + r \exp(i\theta)) \, d\theta
\end{aligned} \tag{17.3.32}
$$

or, equivalently,

$$w(z) = \frac{1}{2\pi r} \oint w(\zeta) \, d\ell, \tag{17.3.33}$$

where $d\ell$ is the arc length along the circle. Hence the value of an analytic function w at the center of a circle equals its average over the circle. Note that this result holds separately for u and v, i.e. any harmonic function in the plane. As mentioned on p. 75, this is a general property of harmonic functions in d-dimensional space: the value of these functions at a point equals their average over a $(d-1)$-dimensional sphere centered at that point. If $d = 1$, harmonic functions are straight lines and the result is obvious. If $d = 3$, the average of a harmonic function over the surface of the sphere equals the value at the center.

From (17.3.32) we have

$$|w(z)| \leq \int_0^{2\pi} |w(z + r \exp(i\theta))| \, d\theta . \tag{17.3.34}$$

Let $M(r) = \max |w(z + r \exp(i\theta))|$. If $|w| \neq$ const., there will be parts $\Delta\theta$ of the integration range which contribute less than $M \Delta\theta$ to the integral and, therefore, $|w(z)|$ will be *strictly less* than M. Thus, equality in (17.3.34) can only hold if $|w| =$ const. which, from the Cauchy–Riemann relations, implies also that $w(z) =$ const. By applying the same argument to a sequence of overlapping circles filling up any finite domain of analyticity D we conclude that (see e.g. Markusevich 1965, vol. 1 p. 297; Conway 1978, p. 79)

Theorem 17.3.7 [Maximum modulus principle] *If $w(z)$ is analytic in a bounded region D and $|w(z)|$ is continuous in $\overline{D}$, then $|w(z)|$ can only attain its maximum on the boundary of the region, unless w is a constant.*

A related result valid for harmonic functions in any number of space dimensions is the *maximum principle*, which asserts that a non-constant function harmonic inside a bounded connected closed domain cannot attain its maximum or minimum at an interior point of the domain, but only on its boundary. This results was supported by a physical argument on p. 11 and verified in a special case on p. 76; for a complete proof see e.g. Stakgold (1997), p. 51, Ablowitz & Fokas 1997, p. 97.

17.4 The Taylor and Laurent series

Some general results on power series have been presented in §8.4, where the notions of convergence and limit were defined and several convergence tests described. It was shown that a power series of the form

$$w(z) = \sum_{n=0}^{\infty} c_n (z - z_0)^n \tag{17.4.1}$$

is uniformly convergent in its *circle of convergence* of radius R centered at z_0, *inside* which it can be differentiated or integrated term-by-term any number of times. Repeated differentiation shows that

$$c_n = \frac{1}{n!} w^{(n)}(z_0). \tag{17.4.2}$$

Conversely, it is easy to prove from Cauchy's integral formula 17.3.3 p. 428 that

Theorem 17.4.1 [Taylor] *A function $w(z)$ analytic inside a circle $|z - z_0| < R$ may be represented inside the circle as a convergent power series of the form (17.4.1), the series being uniquely defined.*

Proof The proof is immediate: Take a circle C_ρ of radius $0 < \rho < R$ centered at z_0 and containing z (which is possible because z is assumed to be inside the circle of radius R, not on the boundary). Since w is analytic inside C_ρ, it can be expressed by Cauchy's integral formula as

$$w(z) = \frac{1}{2\pi i} \int_{C_\rho} \frac{w(\zeta)}{\zeta - z} \, d\zeta \, . \tag{17.4.3}$$

But, since $|(z - z_0)/(\zeta - z_0)| = |z - z_0|/\rho < 1$, we have as in Example 8.3.1 p. 225

$$\frac{1}{\zeta - z} = \frac{1}{\zeta - z_0 - (z - z_0)} = \frac{1}{\zeta - z_0} \sum_0^\infty \left(\frac{z - z_0}{\zeta - z_0} \right)^m . \tag{17.4.4}$$

Substitution into (17.4.3) and interchange of summation and integration (legitimate in view of the uniform convergence of the power series, see p. 229) gives

$$w(z) = \sum_0^\infty \left[\frac{1}{2\pi i} \int_{C_\rho} \frac{w(\zeta)}{(\zeta - z_0)^{n+1}} \, d\zeta \right] (z - z_0)^n \tag{17.4.5}$$

and, from (17.3.29), we recognize that the term in brackets is the same as (17.4.2). From Cauchy's theorem 17.3.1, the circle C_ρ may be replaced by any closed contour lying in the domain $|z - z_0| < R$ and containing z_0 as an interior point.   □

Uniqueness of the expansion is a consequence of

Theorem 17.4.2 *If a convergent power series vanishes for all z with $|z| < R$, and $R > 0$, then all the coefficients of the power series vanish.*[12]

If therefore $w(z) = \sum c_m (z - z_0)^m = \sum d_m (z - z_0)^m$, then $\sum (c_m - d_m)(z - z_0)^m \equiv 0$ from which $c_m = d_m$ follows.

Taylor's theorem establishes a one-to-one correspondence between a function analytic in the neighborhood of some point z_0 and a power series centered at z_0. Hence the concept of an analytic function as an infinitely differentiable function is equivalent to a function which can be represented as a power series. For this reason it is possible to base the entire theory of analytic functions on power series.

If the radius of the circle of convergence is not infinite it is natural to ask what prevents convergence of the series when $|z - z_0|$ equals or exceeds R. One might guess that singular points play a role and, indeed, one has the following

---

[12] The result remains true if the series vanishes for real x, or for an infinite sequence of points z_n converging to a limit internal to the circle of convergence.

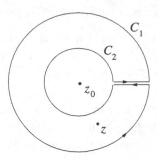

Fig. 17.5 For the proof of the Laurent series expansion.

Theorem 17.4.3 *On the boundary of the circle of convergence of a power series defining* $w(z)$ *there is at least one singular point of* $w(z)$.

From this theorem we deduce that the radius of convergence of a power series centered at z_0 is the distance from z_0 to the closest singular point of w; Example 17.5.4 p. 439 illustrates this aspect in a specific case.

17.4.1 The Laurent series and singular points

The Taylor series provides a representation of an analytic function w around a regular point z_0. A more general series representation exists in the neighborhood of a singular point z_0 of w.

Suppose that $w(z)$ is analytic and single-valued in an annulus $0 < R_2 < |z - z_0| < R_1$. Then, similarly to the procedure used to derive (17.3.10), we may write (see Figure 17.5)

$$
\begin{aligned}
w(z) &= \frac{1}{2\pi i} \oint_{C_1+C_2} \frac{w(\zeta)}{\zeta - z} \, d\zeta \\
&= \frac{1}{2\pi i} \oint_{C_1} \frac{w(\zeta)}{\zeta - z} \, d\zeta - \frac{1}{2\pi i} \oint_{C_2} \frac{w(\zeta)}{\zeta - z} \, d\zeta ,
\end{aligned} \tag{17.4.6}
$$

where, in the second line, both circles are described counterclockwise. We can now use in both integrals a series expansion as in (17.4.4) keeping in mind that, on C_1, $|\zeta - z_0| > |z - z_0|$ while, on C_2, $|\zeta - z_0| < |z - z_0|$. The result is:

$$
\begin{aligned}
w(z) &= \sum_{n=0}^{\infty} \left[\frac{1}{2\pi i} \oint_{C_1} \frac{w(\zeta)}{(\zeta - z_0)^{n+1}} \, d\zeta \right] (z - z_0)^n \\
&+ \sum_{k=0}^{\infty} \left[\frac{1}{2\pi i} \oint_{C_1} w(\zeta)(\zeta - z_0)^k \, d\zeta \right] \frac{1}{(z - z_0)^{k+1}} \\
&= \sum_{-\infty}^{\infty} c_n (z - z_0)^n ,
\end{aligned} \tag{17.4.7}
$$

in which, for both positive and negative n,

$$c_n = \frac{1}{2\pi i} \oint_C \frac{w(\zeta)}{(\zeta - z_0)^{n+1}} \, d\zeta \tag{17.4.8}$$

and C is any contour inside the annulus. The series (17.4.7) is the *Laurent series* expansion of w in the neighborhood of the singular point z_0 and is unique. The coefficient of the first negative power

$$c_{-1} = \frac{1}{2\pi i} \oint_C w(\zeta) \, d\zeta = \text{Res}\,[w, z_0] \tag{17.4.9}$$

is called the *residue* of w at z_0 and has a great importance as will be seen in §17.9. If z_0 is a regular point or a removable singularity, $c_{-1} = 0$ from Cauchy's theorem 17.3.1.

The Laurent series leads to the following characterization of the isolated singular points of a function $w(z)$:

1. The series does not contain negative powers of $(z - z_0)$. In this case the singularity is *removable*, since it can be seen that the series has a well-defined limit for $z \to z_0$ and it is therefore sufficient to redefine $w(z_0)$ as this limit. An example is $w(z) = (z^2 - z)/(1 - z)$, which is not defined at $z = 1$. However

$$\frac{z - z^2}{1 - z} = z \sum_0^\infty z^n - z^2 \sum_0^\infty z^n = \sum_0^\infty z^{m+1} - \sum_0^\infty z^{m+2} = z \to 1. \tag{17.4.10}$$

Generally, a singularity at z_0 is removable if and only if the function is bounded in the neighborhood of z_0.

2. The series has the form

$$w(z) = \sum_{-m}^\infty c_n(z - z_0)^n = \frac{c_{-m}}{(z - z_0)^m} + \cdots + \frac{c_{-1}}{z - z_0} + c_0 + \cdots \tag{17.4.11}$$

with a finite most negative power $-m$. In this case z_0 is a *pole of order* m. It can be shown that, if z_0 is a pole, $|w|$ increases without bound as $z \to z_0$ from any direction and vice versa, if $|w|$ increases without bound as $z \to z_0$ from any direction, and w is analytic in a neighborhood of z_0, but not at z_0 itself, then z_0 is a pole. A pole of order m for w is a zero of order m for $1/w$, and vice versa.

3. The Laurent series contains an infinity of negative powers of $(z - z_0)$. In this case the point z_0 is called an *essential singularity*. A striking difference with poles is that in the neighborhood of an essential singularity w takes all possible complex values except at most one (Picard's theorem, see e.g. Narasimhan & Nievergelt 2001, Chapter 4). A typical example is $z = 0$ for the function $\exp(1/z) = \sum_0^\infty 1/(n! z^n)$. It is easy to see that, for any a and b, one can find a $z = re^{i\theta}$ such that $\exp(1/z) = \exp(a + ib)$. Again unlike poles, if z_0 is an essential singularity for w, it is also an essential singularity for $1/w$. Essential singularities are by no means "rare," as the only analytic functions having only poles (including possibly at infinity) are rational functions.

In considering singularities one should not forget the point at infinity, which can be investigated by setting $\zeta = 1/z$ and studying the Laurent series expansion of $w(\zeta)$ in the neighborhood of $\zeta = 0$. Upon doing this and reverting to the variable z we find

$$w(z) = \cdots \frac{a_{-m}}{z^m} + \cdots + \frac{a_{-1}}{z} + a_0 + a_1 z + \cdots + a_n z^n + \cdots \qquad (17.4.12)$$

If the point at infinity is a regular point, there are no positive powers; if it is a pole of order k, only powers up to z^k appear while, if it is an essential singularity, there is an infinity of positive powers.

In general, near non-isolated singularities there are no expansions analogous to the Laurent series.

17.5 Analytic continuation I

It is very frequently the case that a function is defined by a specific representation with a certain domain of validity and one would like to extend this definition. This happens, for example, when we consider real functions such as e^x, $\sin x$ etc. which we would like to extend off the real axis into the complex domain. Or, a function may be defined as a power series with a certain radius of convergence, and we enquire whether there is a meaningful way to extend this definition beyond the radius of convergence of the given series. For a function defined by a real differential equation we might wonder whether the properties that follow from this definition hold when the variable is allowed to be complex.

This broad class of problems falls under the heading of *analytic continuation*. In this connection, a useful result is the following (see e.g. Henrici 1974, p. 150; Paliouras & Meadows 1990, p. 315):

Theorem 17.5.1 [Uniqueness theorem] *Let the functions $w_1(z)$ and $w_2(z)$ both be analytic in an open domain D and suppose that $w_1(z_m) = w_2(z_m)$ at all points of an infinite sequence $\{z_m\}$ converging to a point $z_0 \in D$. Then $w_1(z) = w_2(z)$ throughout D.*

Thus, if the two functions coincide in a neighborhood, however small, of a point z_0, or along a segment, however short, terminating at z_0, then $w_1 = w_2$ throughout their common domain of analiticity D. We illustrate the implications of this result by several examples.

Example 17.5.1 [Continuation of real functions] Let two complex functions $w_1(z)$ and $w_2(z)$ be analytic in two domains, not necessarily coincident, but both containing the same finite interval of the real axis, however short. Suppose further that, on this interval, they have the same value as a real function $f(x)$. The theorem then assures us that the two functions coincide in their common domain of analyticity. This uniqueness property makes it possible to use any convenient means to extend real functions to the complex domain as the extension obtained by any other means cannot be different. Thus we can use e.g. a power series (all that is needed being the replacement of x by z), or integration from a suitable real point to a point in the complex plane; the latter procedure is used for the logarithm in Example 17.6.2 p. 449.

Because of the strong constraint imposed by analyticity, the class of real functions which can be extended into the complex plane is severely limited. For example, even though they are infinitely differentiable, the test functions of the class $\mathcal{D}$ of the theory of distributions (§20.2) cannot be represented by a Taylor series and therefore no analytic continuation is possible. Nevertheless a bounded continuous function defined on the real axis and decaying, for $|x| \to \infty$, faster than $|x|^{-\alpha}$ with $\alpha > 0$, can be represented for all x in terms of a function $\hat{f}(z)$, analytic over the entire real plane except the real axis, as (see e.g. Bremermann 1965, p. 47; see also Titchmarsh 1948, p. 131)

$$\lim_{\varepsilon \to 0+} [\hat{f}(x + i\varepsilon) - \hat{f}(x - i\varepsilon)] = f(x). \tag{17.5.1}$$

The function $\hat{f}$ is given by

$$\hat{f}(z) = \frac{1}{2\pi i} \int_{-\infty}^{\infty} \frac{f(\xi)}{\xi - z} \, d\xi \tag{17.5.2}$$

and

$$\hat{f}(x + iy) - \hat{f}(x - iy) = \frac{1}{\pi} \int_{-\infty}^{\infty} \frac{y}{(\xi - x)^2 + y^2} f(\xi) \, d\xi. \tag{17.5.3}$$

We recognize here the solution of the Dirichlet problem given in §4.6 and the representation (20.6.8) p. 585 of the δ distribution. As shown in §20.8, a similar representation exists for distributions.   □

Example 17.5.2 [The function e^z] As a concrete illustration, consider the power series $\sum_{k=0}^{\infty} z^k / k!$ which, on the real axis, reduces to $\sum_{k=0}^{\infty} x^k / k! = e^x$. By virtue of the theorem, therefore,

$$e^z = \sum_{k=0}^{\infty} \frac{z^k}{k!} \tag{17.5.4}$$

uniquely defines the analytic continuation of the real exponential into the complex plane. The radius of convergence of the series is readily seen to be infinite by the ratio test (8.1.8) p. 218 or by the root test (8.1.7) p. 218 applied with the aid of Stirling's formula (Table 17.2 p. 444) for the factorial. The real exponential can be continued by other means (e.g. as that solution of the functional equation $f(z_1 + z_2) = f(z_1)f(z_2)$ which reduces to e^x for z real), but this would not lead to different functions in the common domain of analyticity if the values of the different continuations coincide for real $z = x$.

The known real expansions of $\sin y$ and $\cos y$ combined as $\cos y + i \sin y$ give a proof of Euler's formula (17.1.2). Manipulation of the power series also shows that (see e.g. Whittaker & Watson 1927 p. 581)

$$e^{z_1} e^{z_2} = e^{z_1 + z_2} \tag{17.5.5}$$

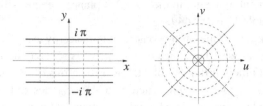

Fig. 17.6 The mapping of the fundamental domain $-\pi < \text{Im} z \leq \pi$ induced by the exponential function $u + iv = e^{x+iy}$.

from which, in particular,

$$e^z = e^{x+iy} = e^x (\cos y + i \sin y) \qquad (17.5.6)$$

so that

$$|e^z| = e^x, \qquad \arg e^z = y. \qquad (17.5.7)$$

By the definition of the argument of a complex number we could write $\arg e^z = y + 2k\pi$ for an arbitrary integer k, but then $\arg e^z$ would not tend to zero for $\text{Im} z \to 0$ and e^z would not be a continuation of the real exponential. Thus, $\arg e^z = y$ may be regarded as a definition.[13] More generally, if $u = u(x, y)$, $v = v(x, y)$,

$$|e^{u+iv}| = e^u, \qquad \arg e^{u+iv} = v. \qquad (17.5.8)$$

The relation (17.5.6) shows that the mapping $w = e^z$ is not one-to-one, as any two points separated by a multiple of 2π along a line $x = $ const. are mapped onto the same point in the w-plane. This feature is at the root of the difficulties associated with the complex logarithm (Example 17.6.2 p. 449). In order for the mapping $w = e^z$ to be one-to-one, we need to restrict the range of the variable z to an arbitrary horizontal strip of vertical extent 2π, i.e. $-\infty < \text{Re} z < \infty$, $y_0 < \text{Im} z \leq y_0 + 2\pi$, with y_0 arbitrary. A region of the complex plane which gives rise to all the possible values of $w(z)$ preserving the one-to-one quality of the mapping is called a *fundamental domain*. It is seen from (17.5.6) that, under the mapping $w = u + iv = e^{x+iy}$, lines $x = $ const. are mapped onto circles centered at the origin (i.e., at $w = 0$), while lines $y = $ const. are mapped onto rays emanating from the origin (Figure 17.6); as for the origin itself, there is no point z which is mapped onto it.

In spite of the familiar appearance, some properties of e^z differ in a striking way from those of e^x. For example, $(e^z)^\alpha \neq e^{\alpha z}$ in general. If true, this relation would permit one to "prove" that $e^{i\omega t} = 1$ through the following chain of relations: $e^{i\omega t} = e^{2\pi i f t} = \left(e^{2\pi i}\right)^{ft} = 1^{ft} = 1!$ (The result is of course correct if ft is an integer.) We resolve this paradox later in Example 17.8.2 p. 456. Some properties of the functions e^z obtained by continuation of the real exponential are summarized in Table 17.3 p. 449. $\qquad\square$

---

[13] As a consequence, the principal value of $\arg e^z$ equals y only if $|y| \leq \pi$, otherwise we must set $\text{Arg } e^z = y + 2N\pi$ with N chosen so that $|y + 2N\pi| \leq \pi$.

Example 17.5.3 [Trigonometric and hyperbolic functions] Again by the uniqueness theorem, the trigonometric and hyperbolic functions can be defined starting either from their own real power series or from the previous definition of e^z. Using the latter we have

$$\sin z = \frac{1}{2i} \left(e^{iz} - e^{-iz} \right), \qquad \cos z = \frac{1}{2} \left(e^{iz} + e^{-iz} \right), \qquad (17.5.9)$$

$$\sinh z = \frac{1}{2} \left(e^z - e^{-z} \right), \qquad \cosh z = \frac{1}{2} \left(e^z + e^{-z} \right). \qquad (17.5.10)$$

The related functions $\tan z = \sin z / \cos z$ etc. follow from these definitions and the uniqueness theorem. From the fundamental property (17.5.5) of the exponential, we can also prove the validity of the usual relations

$$\sin^2 z + \cos^2 z = 1, \qquad \sin(z_1 + z_2) = \sin z_1 \cos z_2 + \cos z_1 \sin z_2, \qquad (17.5.11)$$

and so forth. A more general way to prove these and similar relations is given later in Theorem 17.8.2 p. 458. Given the close relation between complex trigonometric and hyperbolic functions apparent from (17.5.9) and (17.5.10), it is clear that the limitations $|\sin x| \leq 1$, $|\cos x| \leq 1$ are not valid in the complex domain; as a matter of fact, both $\sin z$ and $\cos z$ take all complex values.   □

While we have succeeded in extending real functions to the complex domain, a question that we have not addressed is whether familiar real relations such as $x^\alpha x^\beta = x^{\alpha+\beta}$ etc. also hold for the complex continuations. We will see in the next two sections, where we show other examples of the continuation of real functions, that surprises are in store here when the analytic function of interest is multi-valued.

Example 17.5.4 [Continuation by power series] Suppose that, within a finite circle of convergence $|z - z_0| < R_0$, an analytic function w is represented by the power series

$$w(z) = \sum_0^\infty c_n (z - z_0)^n. \qquad (17.5.12)$$

Take a point z_1 inside the circle of convergence and write

$$\sum_0^\infty c_n (z - z_1 + z_1 - z_0)^n = \sum_0^\infty d_k (z - z_1)^k, \qquad (17.5.13)$$

where

$$d_k = \sum_{n=k}^\infty c_n \binom{n}{k} (z_1 - z_0)^{n-k}. \qquad (17.5.14)$$

This representation can be regarded both as a rearrangement of the terms of the original series, which leaves the sum of the series unchanged because of its absolute convergence (Theorem 8.1.6 p. 218), or as a Taylor series expansion of w around the regular point z_1 with a radius of convergence R_1. Since the two series have the same sum in a neighborhood of z_1, by Theorem 17.5.1 they must represent the same function in the common part of their

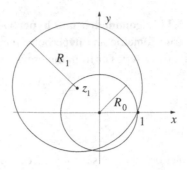

Fig. 17.7 Analytic continuation by power series.

circles of convergence. If the radius of convergence R_1 of the second series extends outside the circle of convergence of the original series, the second series provides a definition of the function beyond R_0, at least in a certain direction (Figure 17.7). It may be possible to repeat the process starting from a point z_2 inside the circle of radius R_1 and so on, thus progressively extending the definition of the function over a chain of domains.

A function known in closed form, so that no series expansion is really necessary, is a trivial example but serves to illustrate what happens without too much calculation. Consider

$$\frac{1}{1-z} = \sum_0^\infty z^n = \frac{1}{1-z_1+(z_1-z)} = \frac{1}{1-z_1}\frac{1}{1-\frac{z-z_1}{1-z_1}} = \sum_0^\infty \frac{(z-z_1)^n}{(1-z_1)^{n+1}}.$$

$$(17.5.15)$$

From Cauchy's root test (p. 218) the radius of convergence of the last series is $|1-z_1|^{-1}$ and can be greater than 1, the radius of the original circle of convergence. Whatever the point z_1, the boundary of the circle of convergence of the new series goes through the singular point $z = 1$ in agreement with Theorem 17.4.3 p. 434. Many special functions are generated in the process of solving certain differential equations by power series (see e.g. §2.5, §12.2, and §13.2). When these series have a finite radius of convergence, the above procedure is an effective way to extend their definition.

As shown in §8.5, the radius of convergence of power series can be extended by having recourse to generalized definitions of the sum of a series. Analytic continuation can also be effected by such methods.   □

Example 17.5.5 [Direct analytic continuation] Let $D = D_1 \cap D_2$ be the common part of two domains D_1 and D_2, and let $w_1(z)$ be analytic in D_1 and w_2 in D_2, with $w_1(z) = w_2(z)$ for $z \in D$. Then we set

$$w(z) = \begin{cases} w_1(z) & z \in D_1 \\ w_2(z) & z \in D_2 \end{cases}.$$

$$(17.5.16)$$

This function is analytic in the extended domain $D_1 + D_2$. We say that w_1 is the *direct analytic continuation* of w_2 into D_1 and similarly for w_1 into D_2. It is important to stress

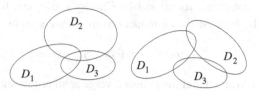

Fig. 17.8 Let w_3 defined in D_3 be obtained by direct analytic continuation of w_2 defined in D_2, itself a direct analytic continuation of w_1 defined in D_1. In the case on the left, it is guaranteed that $w_1 = w_3$ in $D_1 \cap D_3$, but not so in the case on the right.

that the theorem requires analyticity in the whole overlap region; in Example 17.6.2 p. 449 we illustrate the importance of this point. Generally, if the function can be extended over a sequence of domains such that the first and the last one overlap, there is no guarantee that the original function and the last continuation coincide in this overlap region unless all the domains overlap in a region common to all (Figure 17.8, left). In a situation such as that shown on the right of Figure 17.8, in which w_2 is a continuation of w_1 into the domain D_2, and w_3 is a continuation of w_2 into the domain D_3, arg z may be different between D_1 and D_3 and therefore w_1 may be different from w_3 in their overlap domain (see e.g. González 1991, p. 3). The situation is clarified by a consideration of the Riemann surface introduced in §17.7: two points z_1 and z_2 may be "on top of each other" but, if they belong to different Riemann sheets, they must be considered as distinct for the purposes of analytic continuation.   □

Example 17.5.6 [The equation $w(z) = a$] Let w be analytic in D. Then, unless w is a constant, the equation $w(z) = a$ has at most a finite number of roots in every bounded closed subregion D' of D as, otherwise, $w(z)$ would be equal to the constant function a at an infinite sequence of points in a closed domain, which would be convergent by the Bolzano–Weierstrass theorem (p. 216). Theorem 17.5.1 would then imply that w is a constant. Thus, if there is a limit point of an infinity of solutions of $w(z) = a$, this cannot be a point of regularity, but is necessarily a singular point for a non-constant w. For example, the equation $\sin(1/z) = 1/\sqrt{2}$ has the infinity of solutions $z_n = (\frac{1}{4}\pi + 2n\pi)^{-1}$ in any neighborhood of the origin, however small, and the sequence $\{z_n\}$ tends to 0, which is an essential singularity.   □

Example 17.5.7 [Lacunary functions] It is not always possible to carry out the process of analytic continuation. Consider for example

$$w(z) = \sum_0^\infty z^{2^n} = z + z^2 + z^4 + z^8 + z^{16} + \cdots . \qquad (17.5.17)$$

which converges for $|z| < 1$ by Cauchy's root test. It is clear that the series diverges for $z \to 1-$ because then it reduces to an infinite sum of 1's. It is readily established that

$$w(z) = z + w(z^2), \qquad w(z) = z + z^2 + w(z^4), \qquad (17.5.18)$$

and so on and therefore w must diverge at all points such that $z^2 = 1$, $z^4 = 1$, etc. But these points are dense on the circle $|z| = 1$ (i.e., any interval of non-zero length contains an infinity of them), which therefore implies that no convergent power series can be constructed by the method exemplified in Figure 17.7. In this case the circle of convergence is called the *limiting circle* or *natural boundary* for the *lacunary function* $w(z)$. Another example is $\sum_0^\infty z^{n!}$, which can be shown to diverge at any point $\exp(2\pi i p/q)$ with p, q integers.

Lacunary functions are by no means rare: one can prove that, given any connected open domain, there is an analytic function for which the boundary of the domain is a natural boundary (see e.g. Narasimhan & Nievergelt 2001, p. 124). $\qquad\qquad\qquad\qquad\square$

The result of performing all possible analytic continuations of a function is a *complete analytic function*. Each partial representation of the complete analytic function valid in a certain region is a *function element*.

17.5.1 The Gamma function

Another method of analytic continuation can be demonstrated in connection with Euler's Gamma function $\Gamma(z)$ which is defined by the integral

$$\Gamma(z) = \int_0^\infty t^{z-1} e^{-t} \, dt \qquad (17.5.19)$$

which only converges at the lower limit if $\mathrm{Re}\, z > 0$. Upon integration by parts we prove the essential relation

$$\Gamma(1+z) = \int_0^\infty t^z e^{-t} \, dt = \int_0^\infty z\, t^{z-1} e^{-t} \, dt, \qquad (17.5.20)$$

i.e.,

$$\Gamma(1+z) = z\,\Gamma(z). \qquad (17.5.21)$$

This functional relation can be used to continue $\Gamma(z)$ in the half-plane $\mathrm{Re}\, z < 0$. For example, if $-1 < \mathrm{Re}\, z < 0$, then $\mathrm{Re}\,(z+1) > 0$ and we can set

$$\Gamma(z) = \frac{\Gamma(z+1)}{z}. \qquad (17.5.22)$$

If $-2 < \mathrm{Re}\, z < -1$, we similarly have

$$\Gamma(z) = \frac{\Gamma(z+2)}{z(z+1)}, \qquad (17.5.23)$$

and so on. It is evident that this procedure breaks down if $z = 0$ or a negative integer, all of which are poles of $\Gamma(z)$.[14]

Another way to extend the definition (17.5.19), equivalent to the previous one by virtue of Theorem 17.5.1, is to rewrite (17.5.19) as

$$\Gamma(z) = \int_0^1 t^{z-1} e^{-t} \, dt + \int_1^\infty t^{z-1} e^{-t} \, dt. \qquad (17.5.24)$$

The second integral is evidently convergent for any finite z and it defines an entire function. Existence of the first integral has the same limitation as that of the original one (17.5.19) but we note that

$$\int_0^1 t^{z-1} \exp(-t) \, dt = \int_0^1 t^{z-1} \sum_0^\infty \frac{(-t)^k}{k!} \, dt = \sum_0^\infty \frac{(-1)^k}{k!(z+k)}. \qquad (17.5.25)$$

The series is absolutely and uniformly convergent for all z not equal to 0 or a negative integer, and we find in this way another equivalent analytic continuation of $\Gamma(z)$. This last representation also shows that the residue (§17.9) at the pole $z = -n$ equals $(-1)^n/n!$.

Analytic continuation is also useful to derive another important property of $\Gamma(z)$. Assume for the time being that $0 < \operatorname{Re} z < 1$; then we have from the defining integral (17.5.19)

$$\Gamma(z)\,\Gamma(1-z) = \int_0^\infty dt \int_0^\infty ds \; e^{-(s+t)} s^{-z} t^{z-1}. \qquad (17.5.26)$$

The change of variables $\sigma = s + t$, $\tau = t/s$ gives[15]

$$\Gamma(z)\,\Gamma(1-z) = \int_0^\infty d\tau \int_0^\infty d\sigma \, e^{-\sigma} \frac{\tau^z}{1+\tau} \frac{1}{} = \int_0^\infty \frac{\tau^{z-1}}{1+\tau} \, d\tau = \frac{\pi}{\sin \pi z}. \qquad (17.5.27)$$

By analytic continuation, this relation remains valid everywhere in the complex plane except at the poles.[16] By a similar procedure (see e.g. Lebedev 1965, p. 4) we establish Legendre's *duplication formula*

$$\Gamma(2z) = \frac{2^{2z-1}}{\sqrt{\pi}} \Gamma(z) \Gamma\left(z + \frac{1}{2}\right). \qquad (17.5.28)$$

The relation (17.5.27) provides us with a third way to continue $\Gamma(z)$ which rests on calculating the integral

$$\frac{1}{2\pi i} \int_C e^t t^{-z} \, dt \qquad (17.5.29)$$

around the path shown in Figure 17.9. On the lower and upper edges of the path $t^{-z} = \exp(-z \log|t| \pm i\pi z)$, respectively. If $\operatorname{Re} z < 1$, the small circle around the origin gives a

---

[14] The same analytic continuation would be found if t^{z-1} in the defining integral (17.5.19) is interpreted, in the sense of distributions, as $t^{z-1} H(t)$ for any z different from 0 or a negative integer, see (20.5.15) p. 582.

[15] The integral in the last step is the Mellin transform of $(1 + \tau)^{-1}$ and is calculated by contour integration as shown in the second line of Table 17.8 p. 470.

[16] A better justification of the argument may rely on an extension of Theorem 17.8.1 p. 456.

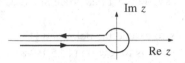

Fig. 17.9   Integration path for (17.5.29) and (17.5.31).

Table 17.2   Some properties of the function $\Gamma(z)$.

$$\Gamma(z) = \int_0^\infty \exp(-t)\, t^{z-1}\, dt \qquad \mathrm{Re}\, z > 0$$

$\Gamma(1+z) = z\,\Gamma(z)$	$\Gamma(n) = (n-1)!$
$\Gamma(2z) = 2^{2z-1}\,\Gamma(z)\,\Gamma\left(z+\tfrac{1}{2}\right)/\sqrt{\pi}$	$\Gamma\left(n+\tfrac{1}{2}\right) = \sqrt{\pi}\, 2^{-n}\,(2n-1)!!$
$\Gamma(z)\,\Gamma(1-z) = \pi/\sin \pi z$	$\Gamma\left(\tfrac{1}{2}\right) = \sqrt{\pi}$
$\Gamma(\tfrac{1}{2}+z)\,\Gamma(\tfrac{1}{2}-z) = \pi/\cos \pi z$	$\Gamma\left(\tfrac{1}{2}-n\right) = (-1)^n 2^n \sqrt{\pi}/(2n-1)!!$
$\Gamma(z) \simeq \sqrt{2\pi}\, e^{-z} z^{z-1/2}$ for $|z| \to \infty$	$n! \simeq \sqrt{2\pi n}\, e^{-n} n^n$ for $n \to \infty$
Poles at $z = -n$, $n = 0, 1, 2, \ldots$	Analytic continuation for $\mathrm{Re}\, z < 0$
with residue $(-1)^n/n!$	see (17.5.24) and (17.5.25)

vanishing contribution and we find

$$\frac{1}{2\pi i}\int_C e^t t^{-z}\, dt = \frac{1}{2\pi i}\int_{-\infty}^0 \left[e^{-z\,\log|t|+i\pi z} - e^{-z\,\log|t|-i\pi z}\right] dt$$

$$= \frac{e^{i\pi z} - e^{-i\pi z}}{2\pi i}\int_0^\infty e^{-t} t^{(1-z)-1}\, dt = \frac{\sin \pi z}{\pi}\Gamma(1-z). \qquad (17.5.30)$$

Comparing with (17.5.27) we deduce

$$\frac{1}{\Gamma(z)} = \frac{1}{2\pi i}\int_C e^t t^{-z}\, dt, \qquad (17.5.31)$$

which is the *Hankel integral representation* of $\Gamma(z)$. Although this result has been derived under the assumption $\mathrm{Re}\, z < 1$, both sides of the equality are entire functions and therefore the formula holds for all z.

The defining integral (17.5.19) is elementary for $z = 1$, with $\Gamma(1) = 1$, and for $z = 1/2$, with $\Gamma(1/2) = \sqrt{\pi}$. From these results and the fundamental relation (17.5.21) we readily see that

$$\Gamma(n+1) = n!, \qquad \Gamma\left(n+\frac{1}{2}\right) = \frac{\sqrt{\pi}}{2^n}\,(2n-1)!!, \qquad (17.5.32)$$

where $(2n-1)!! = (2n-1)(2n-3)\cdots 1$ is the semi-factorial of $2n-1$.

Table 17.2 summarizes some of the main properties of $\Gamma(z)$; derivations can be found in a large number of references (e.g. Whittaker & Watson 1927, chapter 12; Copson 1955, chapter 9; Lebedev 1965, chapter 1; Jeffreys & Jeffreys 1999; chapter 15). A particularly important relation is Stirling's formula, which gives the asymptotic behavior for $|z| \to \infty$. This result can be derived by means of the asymptotic method for Laplace-type integrals explained on p. 279. We write

$$\Gamma(x + 1) = \int_0^\infty \exp\left[-x\left(\frac{t}{x} - \log t\right)\right] dt. \tag{17.5.33}$$

The function $\phi = t/x - \log t$ has a minimum at $t = x$, where $\phi(x) = 1 - \log x$ and $\phi''(x) = 1/x^2$. Application of (10.7.17) p. 279 gives then $\Gamma(x + 1) \simeq \sqrt{2\pi x} \, (x/e)^x$. The same method can be used for complex argument and can be extended to give additional terms of the asymptotic expansion; the result is (see also Bender & Orszag 2005, p. 218)

$$\Gamma(z) \simeq \sqrt{2\pi} \, e^{-z} z^{z-1/2} \left[1 + \frac{1}{12z} + \frac{1}{288z^2} + O\left(\frac{1}{z^3}\right)\right], \qquad |\arg z| \le \pi - \delta, \tag{17.5.34}$$

with δ an arbitrarily small positive number.

A function intimately related to $\Gamma(z)$ is the derivative of its logarithm[17]

$$\psi(z) = \frac{d}{dz} \log \Gamma(z) = \frac{\Gamma'(z)}{\Gamma(z)}. \tag{17.5.35}$$

Among other properties one can show that

$$\psi(z + 1) = \psi(z) + \frac{1}{z}, \qquad \psi(z) = -\gamma + \sum_{n=0}^{\infty} \left(\frac{1}{n+1} - \frac{1}{n+z}\right) \tag{17.5.36}$$

so that $\psi(1) = -\gamma$ where $\gamma \simeq 0.5772156649\ldots$ is the Euler (or Euler–Mascheroni) constant.

Another function related to the Γ function is Euler's beta function defined, for $\mathrm{Re}\, z > 0$ and $\mathrm{Re}\, \zeta > 0$, by[18]

$$B(z, \zeta) = B(\zeta, z) = \int_0^1 t^{z-1}(1 - t)^{\zeta-1} dt. \tag{17.5.37}$$

The function is defined by analytic continuation for other values of z and ζ. For this purpose one may rely e.g. on its relation with the Γ function:[19]

$$B(z, \zeta) = \frac{\Gamma(z)\Gamma(\zeta)}{\Gamma(z + \zeta)}. \tag{17.5.38}$$

One defines *incomplete Gamma functions* $\gamma(z, a)$ and $\Gamma(z, a)$ by extending the integral in (17.5.19) between 0 and a or between a and ∞ respectively, so that $\gamma(z, a) + \Gamma(z, a) =$

---

[17] $\psi(z + 1)$ is sometimes called the digamma function and its successive derivatives polygamma functions.

[18] Here $t^{z-1} = e^{(z-1)\log t}$ and $(1 - t)^{\zeta-1} = e^{(\zeta-1)\log(1-t)}$, with the logarithms understood to have their principal value (p. 450).

[19] This relation is proven by noting that the product of the integrals defining $\Gamma(z)$ and $\Gamma(\zeta)$, $\left(\int_0^\infty t^{z-1} e^{-t} dt\right)\left(\int_0^\infty s^{\zeta-1} e^{-s} ds\right)$, can be interpreted as an area integral extended to the first quadrant of the (t, s) plane. The change of variables $s + t = \xi, s = \xi\eta$ in this area integral transforms it into the product of the two integrals defining $B(z, \zeta)$ and $\Gamma(z + \zeta)$; see e.g. Copson 1955, p. 212.

$\Gamma(z)$. The incomplete beta function $B_a(z, \zeta)$ is defined similarly by extending the integral in (17.5.37) from 0 to $a < 1$.

17.6 Multi-valued functions

The two possible values of a square root are a familiar example of multi-valuedness for real functions. This phenomenon acquires a much greater dimension and has more far-reaching implications with complex-valued functions due to the 2π-indeterminacy of the argument.

A simple example is $w = \sqrt{z}$, which is defined by $(\sqrt{z})^2 = z$. Thus, if $z = re^{i\theta}$ and $\sqrt{z} = se^{i\eta}$, we have $s^2 e^{2i\eta} = re^{i\theta}$, i.e. $s = \sqrt{r}$ and $2\eta = \theta + 2k\pi$, from which

$$\sqrt{z} = \sqrt{|z|} \exp i \frac{\arg z + 2k\pi}{2}. \tag{17.6.1}$$

As in the real case, there are two distinct values, or *branches*, or *determinations*, of $\sqrt{z}$ obtained by taking k even or odd, for example $k = 0$ and $k = 1$. However, unlike the real case, these two branches are connected to each other as, if we set $z = re^{i\theta}$ so that z goes around the origin once by varying θ, arg z increases by 2π so that $\arg(z/2)$ becomes $\arg(z/2) + \pi$ while $\arg(z/2) + \pi$ becomes $\arg(z/2) + 2\pi$ which gives for $\sqrt{z}$ the same value as $\arg(z/2)$. For this reason, the origin is called a *branch point* for the function $\sqrt{z}$. It is easily shown by setting $z = a(1 + \varepsilon e^{i\theta})$ with $|\varepsilon| < 1$ and a an arbitrary complex constant, that the function $\sqrt{z}$ has no other finite branch points. A similar study can be carried out for the function $\sqrt{z - z_0}$, concluding that its two branches are interchanged by executing a path which encircles the point z_0. These examples clarify the content of the following general definition:

Definition 17.6.1 A point z_0 is a *branch point* for a function $w(z)$ if there exists a neighborhood of z_0 such that one complete circuit around an arbitrary closed smooth curve containing z_0 in its interior and lying in the neighborhood carries every branch of $w(z)$ into another branch of $w(z)$.

For the n-th root $z^{1/n}$, defined so that $(z^{1/n})^n = z$, we have

$$z^{1/n} = |z|^{1/n} \exp i \frac{\arg z + 2k\pi}{n}. \tag{17.6.2}$$

There are now n branches of $z^{1/n}$ obtained by taking e.g. $k = 0, 1, \ldots, n - 1$ and the origin is a branch point of finite order $n - 1$. If z executes one loop around the origin the k-th branch becomes the $(k + 1)$-th, a second loop increase k by another unit and so on until, after n loops, each one of the n branches recovers its initial value. In a similar way

$$z^{m/n} = |z|^{m/n} \exp i \frac{m(\arg z + 2k\pi)}{n} \tag{17.6.3}$$

also has n distinct branches if m and n have no common factors.

For general real α we define z^α as suggested by (17.6.2) and (17.6.3):[20]

$$z^\alpha = |z|^\alpha \exp\left[i\alpha(\arg z + 2k\pi)\right], \qquad (17.6.4)$$

and this function has an infinity of distinct branches when α is irrational as $k\alpha$ cannot then be an integer for any k. In this case the origin is a branch point of infinite order, or logarithmic branch point; we return to this function in Example 17.6.3 later in this section and in Example 17.8.2 p. 456; some of the properties of the principal branch (i.e., $k = 0$, $\arg z$ limited to a 2π range containing 0) are summarized in Table 17.6 p. 452. Another important function with an infinity of distinct branches is the logarithm, which we consider in Example 17.6.2.

In studying the branch points of a function one should consider also the point at infinity, which can be examined by setting $\zeta = 1/z$ for large z. A neighborhood of the point at infinity is defined by considering, on the plane, the image of any portion of the Riemann sphere which contains the "north pole." Thus, for example, from $\sqrt{\zeta} = \sqrt{1/z}$, we see that the point at infinity is a branch point of order 1 for $\sqrt{z}$ while, since $\sqrt{z^2 - 1} = \sqrt{\zeta^{-2} - 1} \simeq 1/\zeta$, infinity is not a branch point for $\sqrt{z^2 - 1}$.

17.6.1 Branch cuts

A way to prevent the two branches of $w = \sqrt{z}$ from transforming into each other, thus retaining the same separation that exists between $+\sqrt{x}$ and $-\sqrt{x}$ in the real case, is to enforce a rule which prevents z from going around its two branch points $z = 0$ and ∞. It is convenient to visualize such a rule as a *branch cut* in the complex plane joining these points. For example, a cut extending between the origin and infinity along the negative real axis would enforce the limit $|\arg z| \leq \pi$, and neither of the two branches could be transformed into the other one. A similar procedure is essential in many applications, and notably in the evaluation of integrals by contour integration described in §17.9. Because of the jump in the argument, however, the function becomes discontinuous across a cut and, therefore, it ceases to be analytic in any region of the complex plane containing (part of) the cut in its interior. For example, with the cut described before, $\sqrt{z}$ is defined for $-\pi \leq \arg z \leq \pi$ but is only analytic for $|\arg z| < \pi$.

With such a cut, as z ranges over the complex plane, the branch of $w = \sqrt{z} = \sqrt{|z|}\exp\left[i(\arg z + 2k\pi)/2\right]$ with k even would range over the half-plane $-\frac{1}{2}\pi \leq \arg w \leq \frac{1}{2}\pi$ of the w-plane, while the branch with k odd would range over the half-plane $\frac{1}{2}\pi \leq \arg w \leq \frac{3}{2}\pi$ (Figure 17.10, left). Each one of these half planes of the w-plane is a *domain of univalence* of the multi-valued function $\sqrt{z}$, in the sense that there is a unique value of z which corresponds to each w. The same objective would be achieved by cutting the complex plane e.g. from 0 to infinity along the positive imaginary axis. In this

---

[20] This definition can be justified by taking the limit of a sequence z^{α_n}, where $\{\alpha_n\}$ is a rational sequence tending to α.

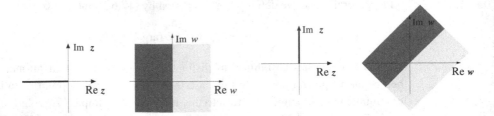

Fig. 17.10 Domains of univalence of the function $\sqrt{z}$ obtained by cutting the complex plane along the negative real axis (left figure) and along the positive imaginary axis.

case the two domains of univalence would be the half planes $-\frac{3}{4}\pi \le \arg w \le \frac{1}{4}\pi$ and $\frac{1}{4}\pi \le \arg w \le \frac{5}{4}\pi$ (Figure 17.10, right).[21]

Example 17.6.1 [Square roots of algebraic functions] For a multi-valued function such as $w = \sqrt{z^2 - 1} = \sqrt{(z-1)(z+1)}$ a "small" loop around $z = \pm 1$, obtained by setting $z = \pm 1 + \varepsilon e^{i\theta}$ and varying θ, increases $\arg(z \mp 1)$ by 2π and therefore $\arg \sqrt{z^2 - 1}$ by π. But a "large" loop which goes around *both* $z = 1$ and $z = -1$ increases $\arg \sqrt{z^2 - 1}$ by $\pi + \pi = 2\pi$ and leaves the value of the function unchanged. Thus, in this case, the proper cut needs to prevent encircling only one of $z = \pm 1$, and can therefore be taken along the real axis between -1 and $+1$. Another choice which achieves the same objective is cutting the real axis from $-\infty$ to -1 and from $+1$ to ∞, which can be viewed as a cut connecting -1 to $+1$ via the point at infinity (Figure 17.11). An analogous procedure is applicable to the square root of a quadratic polynomial $\sqrt{az^2 + bz + c}$, for which the cut needs to connect the two (distinct) roots.[22] A cubic polynomial $P_3(z)$ has three roots, each one of which is a branch point for $\sqrt{P_3}$; infinity is also a branch point as is easily seen. In this case

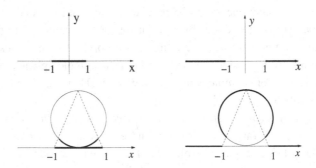

Fig. 17.11 Two possible cuts in the complex plane for the definition of $\sqrt{z^2 - 1}$ (upper row) and their images onto the Riemann sphere.

---

[21] Since $w = \sqrt{z}$ is by definition the inverse of the function $z = w^2$, the domain of univalence in the (direct) image plane $w = \sqrt{z}$ coincides with the fundamental domains (p. 438) in the image plane under the inverse mapping $z = w^2$.

[22] If the two roots coincide the function is evidently single-valued.

Table 17.3	Some properties of e^z, the complex continuation of the real exponential e^x.		
		restrictions	equation no.
$e^z = e^{x+iy} = e^x(\cos y + i \sin y)$		none	(17.5.6)
$\left\| e^{u(x,y)+iv(x,y)} \right\| = e^u$		none	(17.5.8)
$\arg e^{u(x,y)+iv(x,y)} = v$		none	(17.5.8)
$e^{z_1} e^{z_2} = e^{z_1+z_2}$		none	(17.5.5)
$\log e^z = z$		none	(17.6.8)
$e^{\log z} = z$		none	(17.6.8)
$(e^z)^n = e^{nz}$		n integer	
$(e^z)^\alpha = e^{\alpha z}$		$\operatorname{Im} z \in 2\pi\text{-range}$ including $\operatorname{Im} z = 0$	(17.8.1)

one needs two cuts connecting two pairs of branch points. In general, infinity is a branch point for the square root of a polynomial of odd degree, but not for one of even degree. Hence, polynomials always have an even number of branch points which can be connected pairwise by cuts.   □

Example 17.6.2 [The logarithm] A possible definition of the complex logarithm by continuation from the real axis is via the integral

$$\log z = \int_1^z \frac{d\zeta}{\zeta}.$$ (17.6.5)

As shown in Example 17.3.1 p. 424, however, this integral equals $2\pi i$ if the integration encircles the origin, and $2\pi N i$ if the origin is encircled N times. Thus, the origin is a branch point of infinite order, and so is the point at infinity as is found immediately by considering $\log 1/z$. Thus, (17.6.5) can only be used to extend $\log x$ to the complex domain by introducing a cut joining the two branch points 0 and ∞ in such a way that the value of the integral coincides with $\log x$ when $z = x$ is real; as usual, we take the cut along the negative real axis so that $\arg z = \operatorname{Arg} z$ with $|\operatorname{Arg} z| \le \pi$. By evaluating the integral along the real axis from 1 to $|z|$ and then along an arc of the circle of radius $|z|$ centered at the origin we find

$$\log z = \int_1^{|z|} \frac{dx}{x} + \int_0^{\operatorname{Arg} z} \frac{d\left(|z|e^{i\theta}\right)}{|z|e^{i\theta}} = \log|z| + i\operatorname{Arg} z.$$ (17.6.6)

In this cut plane $\log z$ is analytic except at 0 and at infinity for $|\arg z| < \pi$, but not for $\arg z = \pm\pi$ since it is discontinuous across the cut. An alternative definition of the complex logarithm can be based on the series

$$\log z = \sum_{n=1}^{\infty} \frac{(-1)^{n-1}}{n}(z-1)^n,$$ (17.6.7)

which however converges only for $|z| < 1$. The integral (17.6.5) in the cut plane can then be regarded as the analytic continuation of the function defined by (17.6.7), and the equivalence between the two is readily established by setting $z^{-1} = [1 + (z - 1)]^{-1}$, expanding, and integrating term by term. By applying (17.5.8) to (17.6.6) we conclude that

$$e^{\log z} = z, \qquad \log e^z = z, \qquad (17.6.8)$$

so that the function so defined is the inverse of the exponential as in the real case.

The need for a cut may impose restrictions on the validity of familiar relations such as

$$\log \frac{z_1}{z_2} = \log z_1 - \log z_2, \quad \log (z_1 z_2) = \log z_1 + \log z_2, \quad \log z^n = n \log z.$$
$$(17.6.9)$$

If $\arg z$ is limited to a 2π-range including 0 (i.e., the real axis), the first relation is always valid, the second one is valid up to $2N\pi i$, with $N = 0$ or ± 1 depending on the arguments of z_1 and z_2, and the last one is valid only if the argument of z^n is in the chosen range. Failure to respect these limitations may lead to paradoxical conclusions such as the following "proof", due to Jean Bernoulli, that $\log z = \log (-z)$:

$$\log (-z)^2 = \log z^2 \qquad \Longrightarrow \qquad 2 \log (-z) = 2 \log z \quad ! \qquad (17.6.10)$$

To better understand the root of the problem, consider a specific example, e.g. $z = e^{i\pi/4}$ so that $-z = e^{-3i\pi/4}$. Then $z^2 = e^{i\pi/2}$, but we must take $(-z)^2 = e^{i\pi/4}$ rather than $(-z)^2 = e^{3i\pi/2}$ to satisfy the argument limitation. With this the first equality is correct, but the second one does not follow from it because $\log(-z)^2 \neq 2 \log(-z)$.

With a different branch cut, e.g. along the positive imaginary semi-axis, we would have a different continuation of $\log x$, say $\log^* z$, defined in the domain $-\frac{3}{2}\pi < \arg z \leq \frac{1}{2}\pi$, $z \neq 0$. In order to apply the conclusion of Example 17.5.5 p. 440 and ascertain the uniqueness of analytic continuation, the two functions must be analytic in the overlap domain which therefore, in particular, cannot contain any branch cut. Thus, the overlap domain to which the result can be applied is only $-\pi < \arg z < \frac{1}{2}\pi$. In this domain the two functions are indeed equal. However, e.g. for $z = (-1/\sqrt{2}, 1/\sqrt{2})$, we have $\log z = \frac{3}{4}i\pi$, but $\log^* z = -\frac{5}{4}i\pi$.

We bring together some properties of the principal branch of the function $\log z$, namely the branch obtained by continuation of the real logarithm, in Table 17.4.

Another possible definition of the logarithm is as the inverse of the exponential. If we set $z = e^w$, and $z = |z|e^{i\theta}$, $w = u + iv$ we find

$$u = \log |z|, \qquad v = \theta + 2k\pi, \qquad (17.6.11)$$

so that

$$w = \log z = \log |z| + i(\arg z + 2k\pi), \qquad (17.6.12)$$

with $k = 0, \pm 1, \pm 2, \ldots$ The origin is therefore a branch point of infinite order for this function, which has an infinity of domains of univalence, each one being a horizontal strip of width 2π. Of all the infinite branches, the only one which is a legitimate continuation of $\log x$ is the one that is real when $z = x$ is real. In spite of this multi-valuedness, with the aid of (17.2.6) or (17.2.9), we readily establish that $d(\log z)/dz = 1/z$.

		restrictions	equation no.
	Some properties of $\log z$, the complex continuation of the real logarithm $\log x$. The principal value $\mathrm{Arg}\, z$, satisfying $-\pi < \mathrm{Arg}\, z \leq \pi$, is assumed throughout.		

Table 17.4 — Some properties of $\log z$, the complex continuation of the real logarithm $\log x$. The principal value $\mathrm{Arg}\, z$, satisfying $-\pi < \mathrm{Arg}\, z \leq \pi$, is assumed throughout.

	restrictions	equation no.
$\log z = \log |z| + i\,\mathrm{Arg}\, z$	none	(17.6.6)
$e^{\log z} = z$	none	(17.6.8)
$\log e^z = z$	none	(17.6.8)
$\log z_1/z_2 = \log z_1 - \log z_2$	none	(17.6.9)
$\log (z_1 z_2) = \log z_1 + \log z_2 + 2N\pi i$	$N = 0, \pm 1$	(17.6.9)
$\log z^n = n \log z$	$|\mathrm{Arg}\, z^n| \leq \pi$	(17.6.9)

With this more general definition of the logarithm, relations such as (17.6.9) are always valid if considered as equalities between classes of numbers or, alternatively, as equalities mod $2\pi i$ in the sense that the difference between the left- and right-hand sides is a multiple of $2\pi i$ (see e.g. Hille 1959, vol. 1 p. 143; Markusevich 1965, vol. 1 p. 224; Henrici 1974, p. 112). In this setting, the fallacy of (17.6.10) derives from the fact that

$$\log z^2 = \log z + \log z \neq 2 \log z. \qquad (17.6.13)$$

Indeed

$$\log z^2 = \log |z|^2 + i(\arg z^2 + 2k\pi) = 2(\log |z| + i \arg z) + 2k\pi i \qquad (17.6.14)$$

$$\log z + \log z = 2(\log |z| + i \arg z) + 2(m+n)\pi i \qquad (17.6.15)$$

while

$$2 \log z = 2(\log |z| + i \arg z + 2m\pi i). \qquad (17.6.16)$$

This argument illustrates another possible point of view under which (17.6.9) may be regarded, namely as equalities in the conventional sense provided the branch of one of the logarithms in a relation is suitably chosen in dependence of the choice made for the other ones, which remain arbitrary. Similarly, while $e^{\log z} = z$ for any branch by definition, the converse relation

$$\log e^z = z \qquad (17.6.17)$$

is only valid among classes, or mod $2\pi i$, unless one adopts the definition (17.5.7) $\arg e^z = y$. The situation is summarized in Table 17.5.

$\square$

Example 17.6.3 [Arbitrary powers z^α, a^z] With the definition (17.6.12) of $\log z$ we can express all the branches of z^α defined in (17.6.4) as

$$z^\alpha = e^{\alpha \log z}, \qquad (17.6.18)$$

Table 17.5	Properties of the multi-valued complex logarithm defined as the inverse of the exponential, equation (17.6.12).

$e^{\log z} = z$	always valid
$\log e^z = z$	mod $2\pi i$, otherwise unrestricted
$\log (z_1/z_2) = \log z_1 - \log z_2$	mod $2\pi i$, otherwise unrestricted
$\log (z_1 z_2) = \log z_1 + \log z_2$	two branches arbitrary, the third is uniquely determined
$\log z^n = n \log z$	mod $2\pi i$, otherwise unrestricted

Table 17.6	Some properties of z^α (α real), the complex continuation of the real arbitrary power x^α. The principal value $\operatorname{Arg} z$, satisfying $-\pi < \operatorname{Arg} z \le \pi$, is assumed throughout.

	restrictions	equation/example no.
$z^N = zz \cdots z$	N factors	(17.6.4)
$z^\alpha = |z|^\alpha e^{i\alpha \operatorname{Arg} z}$	none	(17.6.4), (17.6.18)
$z^\alpha z^\beta = z^{\alpha+\beta}$	none	17.8.2
$(z^\alpha)^N = z^{N\alpha}$	N integer	
$(z^\alpha)^\beta = z^{\alpha\beta}$	$|\alpha \operatorname{Arg} z| \le \pi$	17.8.2

which has a well-defined meaning also for complex α.[23] Similarly, for any real or complex constant $a \neq 0$, $a^z = e^{z \log a}$. As an example consider

$$i^i = e^{i \log i} = e^{i[i(\pi/2 + 2k\pi)]} = e^{-(2k+1/2)\pi}. \qquad (17.6.19)$$

Once again, complex multi-valuedness has some non-trivial consequences for the function z^α to which we return in Example 17.8.2. Some properties of the branch of z^α which is real for z real are summarized in Table 17.6.   □

Example 17.6.4 [Inverse trigonometric and hyperbolic functions] By writing $\cos z = \frac{1}{2}(e^{iz} + e^{-iz})$ and solving for z we find[24]

$$\cos^{-1} z = -i \log \left(z + \sqrt{z^2 - 1} \right) \qquad (17.6.20)$$

---

[23] In principle we could apply this definition to e^z interpreted as the number e raised to the power z. Since, by (17.6.12), $\log e = 1 + 2k\pi i$, we would then have $e^z = e^{z \log e} = e^z e^{2k\pi i z}$ (see e.g. Markusevich 1965, vol. 1 p. 230; Boas 1987, p. 93). We mention this point mainly to draw attention to the intricacies that arise when one continues real functions into the complex domain.

[24] Note that $z \pm \sqrt{z^2 - 1}$ are reciprocals of each other, $(z + \sqrt{z^2 - 1})(z - \sqrt{z^2 - 1}) = 1$.

and, in a similar way,

$$\sin^{-1} z = -i \, \log i \left(z + \sqrt{z^2 - 1} \right), \qquad \cosh^{-1} z = \log \left(z + \sqrt{z^2 - 1} \right). \quad (17.6.21)$$

An infinity of possible values of the logarithm correspond to each one of the two branches of the square root, so that these functions have a double infinity of different branches with two algebraic branch points at $z = \pm 1$ and a logarithmic branch point at infinity. In order to deal with this multi-valuedness, we introduce branch cuts joining the three branch points, e.g. along the real axis along $(-\infty, -1)$ and $(1, \infty)$, or along $(-1, \infty)$. Similar considerations apply to

$$\sinh^{-1} z = \log \left(z + \sqrt{z^2 + 1} \right). \quad (17.6.22)$$

□

The role of cuts is only to prevent circuits around branch points and therefore, in principle, they do not have to be straight lines, although there are few cases (if any) where a "wiggly" cut would be useful. Normally the specific cut to use is chosen on the basis of physical considerations or for convenience. For some multi-valued functions in widespread use, these considerations have led to the adoption of standard cuts as seen e.g. in §13.5 for the Legendre functions and in §12.2 for the Bessel functions.

17.7 Riemann surfaces

The introduction of cuts in the complex plane is one way to deal with multi-valued functions, but it suffers from the disadvantage that the single-valued function so obtained ceases to be continuous, and therefore analytic, in any region containing part of the cut in its interior. For example, for $\sqrt{z}$ with $|\arg z| \leq \pi$, just above the cut we have $z = |z| e^{i(\pi+\varepsilon)}$ so that $\sqrt{z} \to \sqrt{|z|} \exp(i\pi/2)$ while, just below the cut, $z = |z| e^{-i(\pi-\varepsilon)}$ and $\sqrt{z} \to \sqrt{|z|} \exp(-i\pi/2)$. On the cut, the function does not possess an ordinary derivative, but only one-sided derivatives. Another difficulty is that the use of relations such as those shown in Table 17.5 is restricted.

An alternative way to deal with the problem of multi-valuedness is to replace the complex plane by a finite or infinite set of complex planes "glued together" in a special way. This more general manifold is called a *Riemann surface* and each constituent complex plane a *Riemann sheet*.

As an example, consider once again $w = \sqrt{z}$. The point of view adopted earlier amounts to saying that the entire w-plane can be covered by considering the two branches $\pm\sqrt{|z|} \exp(i \arg z/2)$ together, with $-\pi \leq \arg z \leq \pi$. The alternative we now describe is based on the consideration of a *single* function $\sqrt{|z|} \exp(i \arg z/2)$ defined on a *pair* of complex planes connected in such a way that, on the upper one, $-\pi \leq \arg z \leq \pi$ while,

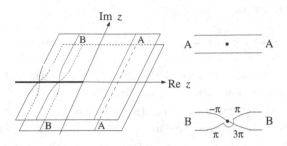

Fig. 17.12 Sketch of a Riemann surface for the function $\sqrt{z}$ with the two Riemann sheets connected along the negative semi-axis. The diagrams on the right indicate the trace that the Riemann surface would leave on planes cutting along AA and BB; the black dot is the Re z axis directed out of the page. The sketch in the lower right corner indicates that it is possible to go either along a "descending" or an "ascending" path from one sheet to the other, but there is no connection which would allow one to make a "U turn" $-\pi \rightarrow \pi$ or $\pi \rightarrow 3\pi$ from the upper to the lower sheet or vice versa; the real nature of this connection cannot be portrayed in ordinary space.

on the lower one, $\pi \leq \arg z \leq 3\pi$ (Figure 17.12).[25] These planes are joined in such a way that the line $\arg z = \pi$ belongs to both and the lines $\arg z = -\pi$ and $\arg z = 3\pi$ are on top of each other as indicated in the lower right corner of Figure 17.12. It must be imagined that these two lines have no common point, although this is impossible to represent in a drawing. With this construction, and in spite of its complicated global structure, near every point the local structure of a Riemann surface looks much like that of a portion of the ordinary complex plane. Corresponding arguments can be developed for similar manifolds obtained by connecting the two sheets along an arbitrary line joining 0 and ∞.

On these manifolds a closed path around the branch point $z = 0$ looks like the sketch in figure 17.13. Along this path we find

$$\oint \sqrt{z}\,dz = 0 \tag{17.7.1}$$

in agreement with Cauchy's theorem 17.3.1. Such relations are valid on a Riemann surface when the integration contour does not include a singularity as any function is single valued on its Riemann surface.

A possible Riemann surface for the function $\sqrt{z^2 - 1}$ (upper left corner of Figure 17.11 p. 448) is shown as another example in Figure 17.14. The Riemann surface for the function $z^{1/n}$ would consist of n sheets connected in such a way that points belonging to one of the top $n - 1$ sheets, after a 2π-circuit around the origin, are brought to the Riemann sheet immediately below, and a point initially located on the bottom-most n-th sheet is brought to the top sheet. For the logarithm, the Riemann surface has an infinity of sheets and, on this surface, there is no *closed* contour encircling the origin.

---

[25] This range has been chosen to establish a connection with the previous discussion on the cut plane, but on the Riemann surface any 4π-range for $\arg z$ would give equal results.

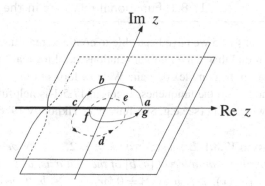

Fig. 17.13 A closed circuit around the branch point $z = 0$ on the Riemann surface of the previous figure for the function $\sqrt{z}$. The circuit starts at a, proceeds counter clockwise to b and then to c where it passes to the lower sheet, remaining there through d and e. The circuit returns to the upper sheet upon traversing f and joins its initial point at $g = a$.

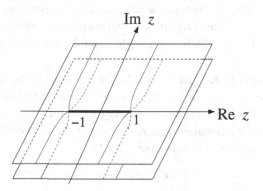

Fig. 17.14 The Riemann surface for the function $\sqrt{z^2 - 1}$ corresponding to the diagram in the upper left corner of Figure 17.11.

The idea of the Riemann surface, the embryo of which has been described here, has been developed in great depth in its topological and other aspects (see e.g. Weyl 1964; Farkas & Kra 1980; Jost 2006).

17.8 Analytic continuation II

In §17.5 we have seen several aspects and examples of analytic continuation. Here we consider several additional features of this topic, starting with conditions for the validity of real functional relations in the complex domain.

17.8.1 Functional relations in the complex domain

While in §17.5 we have been able to continue real functions into the complex domain, we have found that some familiar relations valid for real functions become either restricted or invalid in the complex domain. We now look at this issue in more general terms.

Once again the uniqueness theorem 17.5.1 is helpful as, on its basis, one can prove the following result (see e.g. Sveshnikov & Tikhonov 1978, p. 84):

Theorem 17.8.1 *Let $w_k(z)$ with $k = 1, 2, \ldots, N$ be analytic functions of z in a domain D containing an interval $[a, b]$ of the real axis, and let them satisfy a real relation of the form $F[w_1(x), \ldots, w_N(x)] = 0$ for $a \leq x \leq b$. It then follows that the analogous relation $F[w_1(z), \ldots, w_N(z)] = 0$ is valid for $z \in D$ provided the function F is analytic with respect to each one of its variables $w_1, \ldots, w_N$.*[26]

We readily recognize that a relation such as the first one of (17.5.11) is an immediate consequence of this theorem. If $w_1 = \log z$, $w_2 = z$ and $F = e^{w_1} - w_2$, with D_1 the complex plane cut along the negative real axis, this theorem gives another proof of the relation (17.6.8) $e^{\log z} = z$. Some properties of this continuation of the real exponential are summarized in Table 17.3 p. 449. In less straightforward cases some care is advisable.

Example 17.8.1 Let us consider the extension of the real relation $(e^x)^\alpha = e^{\alpha x}$ with α real. We set $w_1 = e^z$, $w_2 = e^{\alpha z}$, $F = w_1^\alpha - w_2$. The requirement that F be analytic and real for real w_1 and w_2 requires that $\arg w_1 = \operatorname{Im} z = y$ (see (17.5.7)) be suitably restricted to a 2π-range containing the real axis. For such z the relation $(e^z)^\alpha = e^{\alpha z}$ holds. In general however, from (17.6.18) and (17.6.17),

$$(e^z)^\alpha = e^{\alpha \log e^z} = e^{\alpha(z + 2k\pi i)} = e^{\alpha z} e^{2\alpha k\pi i}. \qquad (17.8.1)$$

The fallacy of the argument of p. 438 leading to the paradoxical result $e^{2\pi i f t} = 1$ is now clear as the relation $(e^{2\pi i/\lambda})^{\lambda\mu} = e^{2\pi i\mu}$ with λ real is valid only if $2\pi/\lambda$ satisfies the restriction imposed on the argument; in particular, it will never be satisfied if $\lambda = 1$. For example, with the complex plane cut along the negative real axis, we need $|\lambda| > 2$. □

Example 17.8.2 Let us consider now the validity of the real relation $(x^\alpha)^\beta = x^{\alpha\beta}$ with α and β both real. In order to apply Theorem 17.8.1 we take $w_1 = z^\alpha$ and $w_2 = z^{\alpha\beta}$. The real powers are only defined for $x > 0$, and therefore we must define $\arg z$ so that it has the value 0 for real and positive $z = x > 0$, e.g. by taking the principal value of $\arg z$. Failing this, the definition (17.6.4) of z^α would not give x^α for $z = x$ real. Thus we cut the complex plane so that, e.g., $|\arg z| \leq \pi$ (or, equivalently, set $\arg z = \operatorname{Arg} z$). With this

---

[26] A function of several complex variables $F(w_1, \ldots, w_N)$ defined for $w_k \in D_k$ is analytic with respect to each variable if, for any j, the corresponding function $\Phi_j(w_j) = F(w_1^0, \ldots, w_j, \ldots, w_N^0)$ of the single complex variable w_j obtained for arbitrary fixed values of the other variables is an analytic function of the single variable w_j.

prescription, $F = w_1^\beta - w_2 = 0$ for $z = x$ real. If β is not an integer, the requirement that F be analytic in w_1 requires that $\arg w_1 = \alpha \mathrm{Arg}\, z$ be suitably limited to a 2π-range and, with this provision, we conclude that $(z^\alpha)^\beta = z^{\alpha\beta}$ as is also apparent by comparing (17.8.2) and (17.8.3) below. If the argument is not restricted, according to the definition (17.6.4) of z raised to a real power, we have

$$
\begin{aligned}
\left(z^\alpha\right)^\beta &= |z|^{\alpha\beta} \exp i\beta \left(\arg z^\alpha + 2l\pi\right) = |z|^{\alpha\beta} \exp i\beta \left[\alpha(\arg z + 2k\pi) + 2l\pi\right] \\
&= |z|^{\alpha\beta} \exp i\left[\alpha\beta \arg z + 2\pi\beta(k\alpha + l)\right],
\end{aligned}
\tag{17.8.2}
$$

while

$$
z^{\alpha\beta} = |z|^{\alpha\beta} \exp i\left[\alpha\beta \arg z + 2\pi\alpha\beta n\right].
\tag{17.8.3}
$$

Thus, every value of $z^{\alpha\beta}$ is a value of $(z^\alpha)^\beta$ but not vice versa. Comparison of (17.8.2) and (17.8.3) shows that $z^{mn} = (z^m)^n$ with no restrictions if both exponents are integers.

We can study in a similar way the validity of the real relation $x^{\alpha+\beta} = x^\alpha x^\beta$. We apply Theorem 17.8.1 with $w_1(z) = z^\alpha$, $w_2(z) = z^\beta$, $w_3(z) = z^{\alpha+\beta}$ in the cut plane, and $F = w_3 - w_1 w_2$. The theorem then shows that $z^{\alpha+\beta} = z^\alpha z^\beta$. Without the limitation on $\arg z$ we have

$$
\begin{aligned}
z^\alpha z^\beta &= |z|^{\alpha+\beta} \exp i\left[(\alpha + \beta)\arg z + 2\pi(\alpha m + \beta n)\right], \tag{17.8.4} \\
z^{\alpha+\beta} &= |z|^{\alpha+\beta} \exp i\left[(\alpha + \beta)\arg z + 2\pi(\alpha + \beta)k\right]. \tag{17.8.5}
\end{aligned}
$$

Every value of $z^{\alpha+\beta}$ is also a value of $z^\alpha z^\beta$, but not the converse unless both exponents are integers.[27] More generally, we see that the relation holds if the same branch is taken for all powers so that $m = n = k$. $\square$

An easy corollary of Theorem 17.8.1 is that, if $w(x)$ satisfies the differential equation $F(w, w', w'', \ldots, x) = 0$, its analytic continuation $w(z)$ satisfies the same equation when the conditions of the theorem are fulfilled (see e.g. Sveshnikov & Tikhonov 1978, p. 86; González 1991, p. 28). In this case we can study differential equations in the complex plane, where the structure of the solutions is more apparent.

Example 17.8.3 The analyticity conditions of the theorem restrict the class of real differential equations which can be extended to the complex domain. For example, the solution of $w'(x) = 2|x|$ is $w(x) = x|x| + c$, but this equation is meaningless in the complex domain as $|z|$ is nowhere analytic and therefore cannot equal the derivative of an analytic function w. By a similar argument, the coefficients of the differential equation must be analytic functions of the independent variable for the differential equation to be extended into the complex plane (see e.g. Bender & Orszag 2005, p. 29). $\square$

An analog of Theorem 17.8.1 concerns multi-variable relations (see e.g. Sveshnikov & Tikhonov 1978, p. 86):

---

[27] It might be argued that a similar situation is encountered also in the real case since, for example, $3^{1/2+1/2} = 3$, while $3^{1/2}3^{1/2} = \pm 3$.

Theorem 17.8.2 *Let* $w_1(z_1), \ldots, w_N(z_N)$ *be analytic functions of the complex variables* z_k *in the respective domains* D_k, *each containing intervals* $[a_k, b_k]$ *of the real axis. Let the function* $F[w_1, \ldots, w_N]$ *be analytic with respect to each one of its variables in their range. Then, if the relation* $F[w_1(x_1), \ldots, w_N(x_N)] = 0$ *holds for* $x_k \in [a_k, b_k]$, *it follows that* $F[w_1(z_1), \ldots, w_N(z_N)] = 0$ *for* $z_k \in D_k$.

This theorem gives an alternative and more general proof of relations such as (17.5.5), the second one of (17.5.11) and others.

17.8.2 Continuation across a boundary

The process of direct analytic continuation described in Example 17.5.5 p. 440 can be applied also when the overlap of the two domains reduces to a portion of their boundaries; the following theorem is due to Painlevé.

Theorem 17.8.3 *Let the domains of analyticity* D_1 *and* D_2 *of* w_1 *and* w_2 *be joined along a piecewise smooth line* Γ, *and let* w_i *be continuous on* $D_i + \Gamma$. *Then the function*

$$w(z) = \begin{cases} w_1(z) & z \in D_1 + \Gamma \\ w_2(z) & z \in D_2 + \Gamma \\ w_1(z) = w_2(z) & z \in \Gamma \end{cases} \qquad (17.8.6)$$

is analytic in the domain $D_1 + D_2 + \Gamma$.

In the case of a multivalued function, it may be necessary for this continuation process to be carried out on a Riemann surface rather than on the ordinary complex plane.

While this theorem asserts that the combination of the two given functions gives rise to a single analytic function, in other cases we may know one of the two functions, and the task is then to determine the other one. In this connection the following result is useful (see e.g. Boas 1987, p. 140; Ablowitz & Fokas 1997, p. 346):

Theorem 17.8.4 [Reflection principle] *Let* $w(z)$ *be analytic in a domain* D *that lies on the upper half of the complex plane. Let* $\tilde{D}$ *be the image of* D *in the real axis, so that it lies in the lower half of the complex plane and consists of all the points* $\bar{z}$ *with* $z \in D$. *Then the function* $\overline{w(\bar{z})}$ *is analytic in* $\tilde{D}$.

The truth of this proposition is easily verified directly from the Cauchy–Riemann relations. This is actually a special case of a more general result, known as the *Schwarz reflection principle*, which enables one to construct an analytic function by reflection of a given function in a generic straight line or circular arc (see e.g. Paliouras & Meadows 1990, p. 550; Ablowitz & Fokas 1997, p. 379).[28] If the domain D of the theorem contains (a part of)

---

[28] That such generalizations should be possible is suggested by the fact that any straight line can be transformed into the real axis by a mapping of the form $\zeta = az + b$, and that mappings of the form $\zeta = (az + b)/(cz + d)$ transform circles into circles, see Example 17.10.4 below.

the real axis, the relation between D and $\tilde{D}$ becomes that of Theorem 17.8.3. In this case we can deduce that

Theorem 17.8.5 *If the analytic function $w(z)$ is real for z real, then*

$$\overline{w(\overline{z})} = w(z), \tag{17.8.7}$$

and conversely, if this relation holds, $w(z)$ is real for z real.

The first part is a consequence of the previous theorem; for the second one it is sufficient to note that, if $u(x, 0) - iv(x, 0) = u(x, 0) + iv(x, 0)$ then $v(x, 0) = 0$.

A useful consequence of this theorem relates to functions obtained from continuation off the real axis as, by taking the complex conjugate of (17.8.7), we deduce that

$$w(\overline{z}) = \overline{w(z)}. \tag{17.8.8}$$

Thus, for example, $\sin \overline{z} = \overline{\sin z}$, etc.

17.9 Residues and applications

We have already defined in (17.4.9) Res $[w, z_0]$, the residue of a function $w(z)$ at a point z_0. Residues are extremely useful in the calculation of integrals by virtue of the

Theorem 17.9.1 [Residue theorem] *Let w be analytic in a closed domain $\overline{D}$ except at a finite number of isolated singularities z_k, $(k = 1, \ldots N)$ lying inside D. Then*

$$\oint_C w(\zeta) \, d\zeta = 2\pi i \sum_{k=1}^{N} \text{Res}\,[w(z), z_k] \tag{17.9.1}$$

where C is the complete boundary of D traversed in the positive direction.

Proof In connection with Figure 17.3 and (17.3.18) we have already seen that, if the domain D is punctured by holes each one of which contains only one singularity, we have the equality

$$\oint_C w(\zeta) \, d\zeta = \sum_{k=1}^{N} \oint_{C_k} w \, d\zeta, \tag{17.9.2}$$

in which the C_k are non-intersecting contours surrounding each hole. In the neighborhood of each singularity w can be expanded in the Laurent series (17.4.7) which, being uniformly convergent, can be integrated term by term. A simple direct calculation then shows that only the term proportional to $(z - z_k)^{-1}$ gives a non-zero contribution to the integral along C_k (cf. Example 17.3.3 p. 428), which equals $2\pi i c_{-1}$ as shown before in Example 17.3.1; this proves the theorem. $\qquad\square$

	Residue of $w(z) = \phi(z)/\psi(z)$ at z_0
$m = 1$	$\dfrac{\phi(z_0)}{\psi'(z_0)}$
$m = 2$	$2\dfrac{\phi'(z_0)}{\psi''(z_0)} - \dfrac{2}{3}\dfrac{\psi'''(z_0)\phi(z_0)}{[\psi''(z_0)]^2}$

Table 17.7 Residue of $w = \phi/\psi$ at a zero of order $m = 1$ and 2 of $\psi(z)$; it is assumed that $\phi(z_0) \neq 0$.

It may be noted that the theorem is valid also for an infinite number of singular points provided the two sides of (17.9.1) converge as the contour C is enlarged to contain more and more singularities.

In order to define the residue of the point at infinity we use as a guide the requirement that the relation

$$\oint w \, dz = 2\pi i \operatorname{Res}[w, z = \infty] \qquad (17.9.3)$$

hold for a contour surrounding the point at infinity but no other singularity of w. Such a contour may be thought of as the image in the complex plane of a sufficiently small closed contour surrounding the "north pole" of the Riemann sphere (cf. p. 447). In order to traverse this contour keeping the neighborhood of ∞ on the left we must traverse it the clockwise direction, which would be the negative direction with respect to its interior. Furthermore, for no other singularity to be "enclosed," the contour must be so large as to encircle all the finite singularities. Choosing a contour around the point at infinity in this way, we find from the Laurent expansion (17.4.12)

$$\operatorname{Res}[w, z = \infty] = -a_{-1}. \qquad (17.9.4)$$

Note that $\operatorname{Res}[w, z = \infty]$ may be non-zero even though ∞ is a regular point. For example, $w = 1/z$ has a first-order zero at ∞ but, according to the definition, $\operatorname{Res}[1/z, \infty] = -1$. Similarly, the residue could be 0 even if the function has a pole at infinity, e.g., $w(z) = z$. A general rule is the following (Boas 1987, p. 76):

$$\operatorname{Res}[w(z), z = \infty] = \operatorname{Res}[-\zeta^2 w(\zeta), \zeta = 0], \qquad \zeta = \frac{1}{z}. \qquad (17.9.5)$$

Theorem 17.9.2 *If $w(z)$ is analytic except at isolated singular points, then the sum of all its residues, including the point at infinity, equals 0.*

The theorem is evident as the sum of the residues at all finite points equals the contour integral defining the residue of the point at infinity traversed in the opposite direction.

As will be seen presently, when coupled with Jordan's lemma, the residue theorem provides a powerful way to evaluate many integrals. It is therefore useful to be able to calculate residues easily without having to carry out the Laurent series expansion which defines them. A simple case is that of a pole of order m at z_0, so that the Laurent expansion takes the form (17.4.11) p. 435. If we multiply this expansion by $(z - z_0)^m$, take the derivative of order $(m - 1)$ and the limit $z \to z_0$, we find

$$c_{-1} = \frac{1}{(m-1)!} \lim_{z \to z_0} \frac{d^{m-1}}{dz^{m-1}}[(z - z_0)^m w(z)]. \tag{17.9.6}$$

Poles frequently arise when $w(z) = \phi(z)/\psi(z)$ and ψ has a zero of order m at z_0, while ϕ is analytic and non-zero at z_0; explicit results for this case with $m = 1$ and 2 are given in Table 17.7.

Example 17.9.1 Consider the integral of a rational function R of $\cos \theta$ and $\sin \theta$ over a 2π-range, which can be reduced to $[-\pi, \pi]$ by a simple translation. With the substitution $z = e^{i\theta}$ the integral becomes

$$\int_{-\pi}^{\pi} R(\cos \theta, \sin \theta) \, d\theta = i^{-1} \oint_{|z|=1} R\left(\frac{z + z^{-1}}{2}, \frac{z - z^{-1}}{2i}\right) \frac{dz}{z}, \tag{17.9.7}$$

to which the residue theorem is directly applicable. For example

$$I = \int_0^{2\pi} \frac{d\theta}{1 + a \cos \theta} = 2i^{-1} \oint_{|z|=1} \frac{dz}{az^2 + 2z + a}. \tag{17.9.8}$$

The only pole inside the unit circle is at $z = (-1 + \sqrt{1 - a^2})/a$ where the residue equals $2\sqrt{1 - a^2}$; thus, $I = 2\pi/\sqrt{1 - a^2}$ (cf. (13.3.13) p. 321). ☐

17.9.1 Jordan's lemma. Fourier and Laplace integrals

Many of the integrals that arise in practice are not contour integrals but integrals along a line, e.g. from $-\infty$ to ∞. The great usefulness of Jordan's lemma is that in many cases it permits us to close the path of integration by adding an arc at infinity which contributes nothing to the integral. In this case, the line integral can effectively be considered as a contour integral and evaluated by residues.

Let w be analytic in $\text{Im } z > 0$, save for a finite number of isolated singularities, and let it tend uniformly to zero as $|z| \to \infty$ for $0 \le \arg z \le \pi$. In other words, if $z = R \exp i\theta$, there is an $M(R)$ such that $|w(z)| < M(R)$ for $0 \le \theta \le \pi$, and $M(R) \to 0$ as $R \to \infty$. Jordan's lemma states that, for $k > 0$,[29]

$$\lim_{R \to \infty} \int_{C_R} e^{ikz} w(z) \, dz = 0, \tag{17.9.9}$$

where C_R is a semi-circle of radius R in the upper half-plane centered at the origin (Figure 17.15, left). As long as θ is not too close to 0 or π, the exponential has a negative real

[29] Not surprisingly, the result remains true if R tends to infinity through a discrete set of values rather than continuously.

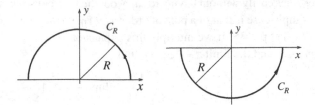

The semi-circles used in Jordan's lemma (17.9.9) for $k > 0$ (left) and $k < 0$ (right).

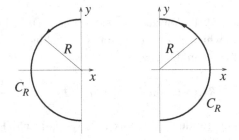

The semi-circles used in the form (17.9.10) of Jordan's lemma with $\kappa > 0$ (left) and $\kappa < 0$ (right).

part $\exp(-kR\sin\theta)$ which tends to zero for $R \to \infty$. The less than obvious part of the lemma is to show that the contributions to the integral from neighborhoods of the x axis (i.e., θ close to 0 or π) also vanish in the limit (see e.g. Henrici 1974, p. 255; Sveshnikov & Tikhonov 1978, p. 135; Ablowitz & Fokas 1997, p. 222).

This theorem has several very useful corollaries:

1. The same result holds for semi-circles in the lower half-plane if $k < 0$ and w satisfies corresponding hypotheses (Figure 17.15, right).

2. If $k = -i\kappa$, with $\kappa > 0$ or $\kappa < 0$, provided w satisfies analogous hypotheses, a similar result holds for integrals over semi-circles in the left or right half-planes respectively (Figure 17.16)

$$\lim_{R\to\infty} \int_{C_R} e^{\kappa\zeta}\, w(\zeta)\, d\zeta = 0. \qquad (17.9.10)$$

This form of Jordan's lemma is of particular relevance in connection with the inversion of the Laplace transform (§11.1).

3. By setting $\zeta = R\exp(i\theta) + iy_0$ in (17.9.9) the lemma can be proven also for the half-plane $\operatorname{Im} z \geq y_0$ with $y_0 > 0$ and also $y_0 < 0$ (figure 17.17). Similar generalizations are available for the lemma in the form (17.9.10).

The hypotheses under which Jordan's lemma holds can be weakened further (see e.g. Sveshnikov & Tikhonov 1978, p. 136).

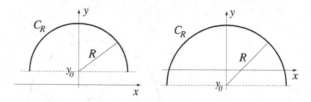

Examples of other situations in which Jordan's lemma holds.

In the first form stated the lemma is ideal for the evaluation of integrals that arise in the application of Fourier transforms (Chapters 4 and 10):

$$I = \int_{-\infty}^{\infty} \exp(iax)\, f(x)\, dx. \qquad (17.9.11)$$

Suppose $a > 0$, let the function $f(x)$ in the integrand have no singularities on the real axis and let it be possible to continue it into the upper half-plane $\text{Im } z \geq 0$. Let also the $f(z)$ thus obtained satisfy the conditions demanded by Jordan's lemma. Then one can close the path of integration along the real axis by a large semi-circle in the upper half-plane, which contributes nothing to the integral by Jordan's lemma, and apply the residue theorem to the path closed in this way. If $a < 0$ the same procedure can be followed by closing the path in the lower half-plane. In this case, however, the integration is carried out clockwise and an extra minus sign is needed. In conclusion

$$\int_{-\infty}^{\infty} \exp(iax)\, f(x)\, dx = 2\pi i\, (\text{sgn}\, a) \sum_{k} \text{Res}\,[\exp(iaz)\, f(z), z_k]. \qquad (17.9.12)$$

Example 17.9.2 To apply the method to

$$I = \int_{0}^{\infty} \frac{\cos ax}{x^2 + \beta^2}\, dx, \qquad a > 0,\ \beta > 0 \qquad (17.9.13)$$

we write

$$I = \frac{1}{2}\,\text{Re} \int_{-\infty}^{\infty} \frac{\exp(iax)}{x^2 + \beta^2}\, dx. \qquad (17.9.14)$$

The function $(z^2 + \beta^2)^{-1}$ tends to zero at infinity. Its only singularity in the upper half-plane is a pole at $i\beta$ and, from the first line of Table 17.7,

$$\text{Res}\left[\frac{\exp(iaz)}{z^2 + \beta^2}, ia\right] = \frac{\exp(-\beta a)}{2ia}. \qquad (17.9.15)$$

Hence

$$I = \frac{1}{2}\,\text{Re}\left[2\pi i\,\frac{\exp(-\beta a)}{2ia}\right] = \frac{\pi}{2a}\exp(-\beta a). \qquad (17.9.16)$$

□

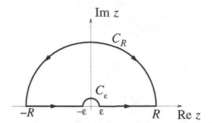

Fig. 17.18   Handling singularities on the real axis, Example 17.9.3

Example 17.9.3 The restriction on the absence of singularities of f on the real axis can be lifted in some cases. An example was discussed in §4.2 p. 90. As another example, consider

$$I = \int_0^\infty \frac{\sin ax}{x} \, dx = \frac{1}{2} \, \mathrm{Im} \int_{-\infty}^\infty \frac{\exp(iax)}{x} \, dx . \tag{17.9.17}$$

While the first integral is well defined, the second one exists only as a Cauchy principal value in the sense defined in (17.3.23):

$$I = \frac{1}{2} \, \mathrm{Im} \int_{-\infty}^\infty \frac{\exp(iax)}{x} \, dx = \lim_{\varepsilon \to 0, R \to \infty} \left(\int_{-R}^{-\varepsilon} + \int_\varepsilon^R \right) \frac{\exp(iax)}{2x} \, dx . \tag{17.9.18}$$

Take the integration contour shown in Figure 17.18 indented at $x = 0$. Then no poles are enclosed and

$$\left(\int_{-R}^{-\varepsilon} + \int_{C_\varepsilon} + \int_\varepsilon^R + \int_{C_R} \right) \frac{\exp(iaz)}{z} \, dz = 0. \tag{17.9.19}$$

The last term tends to zero by Jordan's lemma and

$$\int_{C_\varepsilon} \frac{\exp(iaz)}{z} \, dz = \int_\pi^0 \exp\left(ia\varepsilon \, e^{i\theta}\right) d\theta \to -i\pi. \tag{17.9.20}$$

In conclusion

$$\int_0^\infty \frac{\sin ax}{x} \, dx = \frac{\pi}{2} \qquad a > 0. \tag{17.9.21}$$

This calculation shows that, in effect, the pole can be considered as contributing only half of its residue. The calculation can also be carried out by using for C_ε a small half-circle in the lower half-plane. Now the pole is enclosed and the right-hand side of (17.9.19) has the value $2\pi i$. However the integral (17.9.20) is now from $-\pi$ to 0 and is therefore $i\pi$ giving the same result for the integral. The situation is quite similar to the one discussed in connection with Figure 17.4. If $a < 0$ Jordan's lemma would apply to a large half-circle in the lower half-plane. We then get the same result but with a minus sign as now the contour is traversed in the negative direction (i.e., leaving the domain on the right). We have thus found the following integral representation of the sign function:

$$\mathrm{sgn} \, a = \frac{2}{\pi} \int_0^\infty \frac{\sin ax}{x} \, dx . \tag{17.9.22}$$

$\square$

From this last example we can see the origin of the following rule for the calculation of the principal-value integral with $a > 0$

$$\text{PV} \int_{-\infty}^{\infty} f(x) e^{iax} \, dx = 2\pi i \sum_{\text{Im } z_k > 0} \text{Res}\,[f(z) e^{iaz}, z_k] + \pi i \sum_{\text{Im } z_j = 0} \text{Res}\,[f(x), x_j],$$

(17.9.23)

where the z_k are the singularities of f in the upper half-plane and the x_j its poles on the real axis, assumed simple. If $a < 0$, the same result holds with the opposite sign.

Some examples of applications to the Fourier transform are given in §4.2 and §4.8 and to the Laplace transform in §5.5 and §11.3.

17.9.2 Integrals of the form $\int_{-\infty}^{\infty} f(x) \, dx$

Similarly to (17.9.23), for such integrals the residue theorem gives

$$\text{PV} \int_{-\infty}^{\infty} f(x) \, dx = 2\pi i \sum_{\text{Im } z_k > 0} \text{Res}\,[f(z), z_k]$$

$$+ \pi i \sum_{\text{Im } z_j = 0} \text{Res}\,[f(x), x_j] - \pi i [\lim_{z \to \infty} z f(z)], \quad (17.9.24)$$

provided f can be continued into the upper half-plane and the limit exists. A similar relation with the last term preceded by the opposite sign holds if f can be continued into the lower half-plane. For example, in the upper half-plane, $(1 + z^4)^{-1}$ has poles at $e^{i\pi/4}$ and $e^{i3\pi/4}$ with residues, from Table 17.7, $(4z^3)^{-1} = \frac{1}{4} e^{-3i\pi/4}$ and $(4z^3)^{-1} = \frac{1}{4} e^{-9i\pi/4}$ respectively; thus

$$\int_0^{\infty} \frac{dx}{1 + x^4} = \frac{1}{2} \int_{-\infty}^{\infty} \frac{dx}{1 + x^4} = \frac{1}{2} 2\pi i \left(4 e^{i3\pi/4} + 4 e^{i9\pi/4} \right) = \frac{\pi}{2\sqrt{2}}. \quad (17.9.25)$$

17.9.3 Multivalued functions

Table 17.8 summarizes the application of contour integration to some typical cases involving multivalued functions. As an example of the procedure, let us consider the first case, namely

$$I = \int_0^{\infty} f(x) \log x \, dx \quad \text{with} \quad f(-x) = f(x). \quad (17.9.26)$$

Suppose that f has no singularities on the real axis and consider the integral $\oint f(z) \log z \, dz$ along the path shown in Figure 17.18. We have

$$\oint f(z) \log z \, dz = \left(\int_{-R}^{-\varepsilon} + \int_{C_\varepsilon} + \int_{\varepsilon}^{R} + \int_{C_R} \right) f(z) \log z \, dz$$

$$= 2\pi i \sum_k \text{Res}\,[f(z) \log z, z_k], \quad (17.9.27)$$

Analytic Functions

Table 17.8

Evaluation of the indicated integrals by contour integration. The diagrams show the path used to obtain the result and the last column lists sufficient conditions for its validity. It is assumed that the analytic continuation of f into the complex plane is a single-valued analytic function with a finite number of isolated singularities. The residues are evaluated at the poles of the indicated function.

Integral	Contour	Sufficient conditions
$\int_0^\infty f(x) \log x \, dx =$ $\pi i \sum_k \operatorname{Res}\left[f(z)\left(\log z - \frac{\pi}{2}i\right), z_k\right]$ $f(x) = f(-x)$		$\lim_{z\to 0} f(z) = 0$ $\|f\| < M/\|z\|^{1+\delta}, \delta > 0$ for $z \to \infty$
$\int_0^\infty f(x) x^{\alpha-1} \, dx =$ $\dfrac{2\pi i}{1 - e^{2\pi i \alpha}} \sum_k \operatorname{Res}\left[z^{\alpha-1} f(z), z_k\right]$ $0 < \alpha < 1$ (Mellin transform)		$f(x)$ regular for $0 \le x < \infty$ $\lim_{\varepsilon\to 0} \varepsilon^\alpha f(\varepsilon e^{i\theta}) = 0$ $\lim_{R\to\infty} R^\alpha f(R^\alpha e^{i\theta}) = 0$ both uniformly in θ
$\int_0^\infty f(x) \, dx =$ $-\sum_k \operatorname{Res}\left[f(z) \log(-z), z_k\right]$		$f(x)$ regular for $0 \le x < \infty$ $\lim z^{1+\delta} f(z) = 0, \delta > 0$ for $z \to 0$ and $z \to \infty$
$\int_0^1 x^{\alpha-1}(1-x)^{-\alpha} f(x) \, dx$ $= \dfrac{\pi [\lim_{z\to\infty} f(z)]}{\sin \pi \alpha} +$ $\dfrac{2\pi i}{1 - e^{2\pi \alpha i}} \sum_k \operatorname{Res}\left[z^{\alpha-1}(1-z)^{-\alpha} f(z), z_k\right]$		$\|f\| < M/\|z\|^{1+\delta}, \delta > 0$ for $z \to \infty$
$\int_0^\infty R(x^n) \, dx =$ $\dfrac{2\pi}{1 - e^{2\pi i/n}} \sum_k \operatorname{Res}\left[R(z^n), z_k\right]$		$R(x^n)$ rational function of x^n, n integer > 1

where the limits $R \to \infty$ and $\varepsilon \to$ will eventually be taken. For the integral around the small circle we have

$$\left| \int_{C_\varepsilon} f(z) \log z \, dz \right| \le \int_{C_\varepsilon} |f(z)| \, | \log \varepsilon + i\theta | \, \varepsilon \, d\theta \to 0, \qquad (17.9.28)$$

provided $f \to 0$ sufficiently fast and, similarly,

$$\left| \int_{C_R} f(z) \log z \, dz \right| \le R \int_0^\pi |f(z)| \, | \log R + i\theta | \, d\theta \qquad (17.9.29)$$

will also vanish e.g. if $|f| \le M/R^{1+\delta}$ with $\delta > 0$. On $-R < x < -\varepsilon$, $\log x = \log |x| e^{i\pi}$ and, therefore,

$$\int_{-R}^{-\varepsilon} f(x) \log x \, dx = \int_\varepsilon^R f(-x) \, [\log x + i\pi] \, dx$$

$$= \int_\varepsilon^R f(x) \log x \, dx + \pi i \sum_k \mathrm{Res}\,[f(z), z_k], \qquad (17.9.30)$$

where the last step is from (17.9.24) using the fact that f is even, has no singularities on the real axis and $z f(z) \to 0$ at infinity. Combining the various contributions, we find the result in the first column of the table. The identity

$$\int_0^\infty x^\alpha \log x \, f(x) \, dx = \frac{\partial}{\partial \alpha} \int_0^\infty x^\alpha f(x) \, dx \qquad (17.9.31)$$

provides an alternative way to calculate this type of integral.

With the substitution $y = e^{bx}$ we find

$$\int_{-\infty}^\infty e^{ax} f(e^{bx}) \, dx = \frac{1}{b} \int_0^\infty y^{a/b-1} f(y) \, dy \qquad (17.9.32)$$

which is of the type shown in the second entry of the table.

Many other methods to evaluate integrals by contour integration have been devised, some of which are of general applicability while others work only in special cases. An extensive coverage of the former can be found e.g. in Mitrinović & Kečkić 1984.

17.9.4 Summation of series

Many of the previous formulae still hold if the function f has an infinite number of poles and the corresponding residue series converges. This remark permits us to evaluate the sum of some series in closed form provided that the series can be interpreted as the result of a contour integral which can be evaluated by other means.

This idea can be implemented in several ways (see e.g. Henrici 1974, p. 265; Sveshnikov & Tikhonov 1978, p. 320; Mitrinović & Kečkić 1984, chapter 6). In applications to diffraction problems (see e.g. Sommerfeld 1950, p. 279; Nussenzweig 1972, chapter 7), it is sometimes referred to as the *Sommerfeld–Watson transformation* (Mathews & Walker 1971, p. 73). As an example, we describe its application to a series of the form $\sum_{-\infty}^\infty (-1)^n f(n)$.

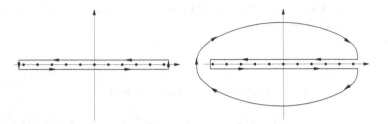

Fig. 17.19 Integration contours for the Sommerfeld–Watson transformation.

Let $f(z)$ be analytic except for isolated poles in the complex plane and let it have no singularities on the real axis. Then, the integral around a contour C surrounding the x-axis (Figure 17.19, left) gives

$$\oint \frac{f(z)}{\sin \pi z}\, dz \;=\; 2i \sum_{-\infty}^{\infty}(-1)^n f(n) \;=\; 2i\left(f(0) + \sum_{n=1}^{\infty}(-1)^n [f(n)+f(-n)]\right)$$

(17.9.33)

since $(\sin \pi z)^{-1}$ has poles at all the integers with residue $(-1)^n/\pi$. If $f(z)/\sin \pi z$ decays to 0 at infinity faster than $1/|z|$, the same integral can be calculated using the contour shown on the right in Figure 17.19. Only the singularities z_k of f contribute to this latter integral and, therefore, we find

$$\sum_{-\infty}^{\infty}(-1)^n f(n) \;=\; -\pi \sum_{k} \mathrm{Res}\left[\frac{f(z)}{\sin \pi z}, z_k\right],$$

(17.9.34)

where the minus sign is due to the traversal of the contour in the negative direction.

As an example, consider

$$S = \sum_{n=1}^{\infty}(-1)^{n+1}\frac{n \sin nx}{a^2 + n^2} = \frac{1}{2}\sum_{-\infty}^{\infty}(-1)^{n+1}\frac{n \sin nx}{a^2 + n^2}.$$

(17.9.35)

In the complex plane the function $z \sin xz/(a^2 + z^2)$ has poles at $z = \pm ia$ and the requirement of fast decay at infinity of $[z/(a^2 + z^2)](\sin xz/\sin \pi z)$ is met provided $|x| < \pi$. Upon evaluating the residues we then find

$$S = \frac{\pi}{2}\frac{\sinh ax}{\sinh a\pi}, \quad |x| < \pi.$$

(17.9.36)

The approach can also be applied to finite sums. For example, it leads to the following closed-form expression of the Gaussian sum (see e.g. Narasimhan & Nievergelt 2001, p. 82):

$$\sum_{k=0}^{n-1} e^{2\pi i k^2/n} = \frac{i + i^{1-n}}{1 + i}\sqrt{n}.$$

(17.9.37)

17.9.5 The logarithmic residue

If $f(z)$ has a pole of order n at z_0, near z_0 it has the form $f(z) \simeq (z - z_0)^{-n} g(z)$ with $g(z_0)$ regular. In this case

$$\frac{f'(z)}{f(z)} \simeq -\frac{n}{z - z_0} + \frac{g'(z)}{g(z)}, \tag{17.9.38}$$

with g'/g analytic in the neighborhood of z_0. Similarly, if f has a zero of order m at z_1, $f(z) \simeq (z - z_1)^m h(z)$ with $h(z_1)] \neq 0$ and

$$\frac{f'(z)}{f(z)} \simeq \frac{m}{z - z_1} + \frac{h'(z)}{h(z)}. \tag{17.9.39}$$

It is therefore evident that, if the only singularities of a function f inside a closed contour C are poles, and if f is regular and non-zero on C, the integral

$$\frac{1}{2\pi i} \oint_C \frac{f'(z)}{f(z)} \, dz = Z_f - P_f \tag{17.9.40}$$

equals the difference between the number of zeros Z_f and the number of poles P_f contained inside the contour C, each counted according to its order. This expression can be restated as follows by observing that the integrand is just $(d/dz) \log f(z)$:

$$\frac{1}{2\pi} \Delta_C \arg f(z) = Z_f - P_f \tag{17.9.41}$$

where $\Delta_C \arg f(z)$ is the change in the argument of f as z traverses the closed contour C so that $[\Delta_C \arg f(z)]/(2\pi)$ is the number of turns that the point $f(z)$ executes around the origin in the f-plane. This result is known as the *principle of the argument* and is a powerful tool to locate the zeros and poles of a function in the complex plane.

For example, a typical concern in stability theory is the establishment of conditions such that the characteristic equation $f(z) = 0$ does not have roots with a positive real part. Supposing that f has no poles, one may consider a contour in the z-plane consisting of the imaginary axis closed by a large semi-circle in the right half-plane (as in the right diagram in Figure 17.16), and look for conditions on the problem parameters such that the point $f(z)$ does not execute a path encircling the origin in the f-plane. In so doing, one would set $z = iy$, with $-R < y < R$ on the imaginary axis, $z = Re^{i\theta}$ with $\pi/2 > \theta > -\pi/2$ and study the image of this path in the f-plane as $R \to \infty$.

Example 17.9.4 As an example, let us investigate the conditions for which both roots of the equation

$$f(z) \equiv z^2 + az + b + ic = 0, \tag{17.9.42}$$

with a, b and c real, have a negative real part. We have, for large R

$$f\left(Re^{i\theta}\right) \simeq R^2 e^{2i\theta}. \tag{17.9.43}$$

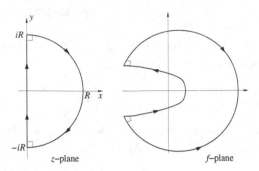

Fig. 17.20 Left: Contour used to determine the number of zeros with positive real part. Right: Image in the f-plane of the contour on the left for f given in (17.9.42) with $a > 0$ and $b - c^2/a^2 > 0$.

As θ goes from $\pi/2$ to $-\pi/2$, arg f goes from π to $-\pi$, i.e., f executes one loop around the origin. To avoid encircling the origin, the path of

$$f(iy) = b - y^2 + i(ay + c) \qquad (17.9.44)$$

must cancel this loop, which requires a situation like the one sketched in figure 17.20. Thus we find that

$$b - \frac{c^2}{a^2} > 0, \qquad a > 0. \qquad (17.9.45)$$

It is easy to convince oneself that any other possibility would give at least one full circle around the origin in the f-plane.[30]   □

In some cases, the application of the principle of the argument is facilitated by (see e.g. Hille 1959, vol. 1 p. 254)

Theorem 17.9.3 [Rouché] *Suppose that $f(z)$ in (17.9.41) can be written as $f(z) = g(z) + h(z)$, with $g(z) \neq 0$ and regular on C. Further, suppose that $|g(z)| > |h(z)|$ on C. Then*

$$\frac{1}{2\pi} \Delta_C \arg f(z) = \frac{1}{2\pi} \Delta_C \arg g(z) = Z_g - P_g = Z_f - P_f. \qquad (17.9.46)$$

An immediate application is to a polynomial of degree k, $f(z) = P_k(z) = z^k + P_{k-1}(z)$. Over a large circle $|z|^k > |P_{k-1}|$ and $\Delta_C z^k/(2\pi) = k$, which is the number of zeros of a polynomial of degree k.

---

[30] Generally speaking, if sketching the path it should be kept in mind that, just as one executes a $\pi/2$ turn to the right at $y = iR$ in going from the Im z axis onto the semi-circle (i.e., arg z start decreasing from $\pi/2$), one should also execute a $\pi/2$ turn to the right in the f-plane, so that arg f starts decreasing from π. This is a consequence of the conformal nature of the mapping from the z-plane onto the f-plane, see § 17.10.

An easy generalization of (17.9.41) is (see e.g. Hille 1959, vol. 1 p. 256)

$$\frac{1}{2\pi i} \oint_C g(z) \frac{f'(z)}{f(z)} \, dz = \sum_{j=1}^{n} g(a_j) - \sum_{k=1}^{m} g(b_k), \qquad (17.9.47)$$

where $g(z)$ is analytic inside C and each one of the zeros a_j and poles b_k of f (none of which is on C) is counted according to its multiplicity. By taking $g = \log z$ and carrying out the integral around the contour showed in the second entry of Table 17.8 one can prove *Jensen's formula*, which applies to a function f analytic in a disk $|z| < R$ and such that $f(z) \neq 0$ on $|z| = R$ (see e.g. Hille 1959, vol. 1 p. 256; Conway 1978, p. 280)

$$\frac{1}{2\pi} \int_0^{2\pi} \log \left| f(Re^{i\theta}) \right| \, d\theta = \log |f(0)| + \sum_{j=1}^{n} \log \frac{R}{|a_j|} - \sum_{k=1}^{m} \log \frac{R}{|b_k|}. \qquad (17.9.48)$$

For simplicity, in writing this result we have assumed $f(0) \neq 0$. If f has no poles or zeros, $\log |f|$ is harmonic and this relation reduces to the mean value theorem.

17.10 Conformal mapping

We have already pointed out that the relation $w = w(z)$ can be regarded as a mapping of an open domain D in the z-plane to an open domain $\mathfrak{D}$ in the w-plane. This mapping enjoys some special properties which make it useful for the solution of many two-dimensional problems involving harmonic functions.

From the expression of the derivative as the limit of the ratio $[w(z) - w(z_0)]/(z - z_0)$ (p. 423) we have

$$w(z) - w(z_0) \simeq w'(z_0)(z - z_0) = |w'(z_0)| \, |z - z_0| \, e^{i[\arg(z - z_0) + \arg w'(z_0)]}, \qquad (17.10.1)$$

which shows that the length $|z - z_0|$ of all (short) segments $z - z_0$ gets dilated by the same amount $|w'(z_0)|$ (provided $w'(z_0) \neq 0$), and that each one of them gets rotated by the same angle $\arg w'(z_0)$. As a consequence, the angle between the images of two segments $z_a - z_0$ and $z_b - z_0$ is the same as that between the original segments, a property known as *isogonality*. When the mapping is one-to-one, these properties characterize it as *conformal*.[31] If the image of a point z_0 is the point at infinity in the w-plane, the mapping is conformal if a neighborhood of z_0 is mapped conformally onto a neighborhood of $W = 1/w = 0$, and vice versa. Note also from (17.10.1) that $\arg w'(z_0) = \arg(w - w(z_0)) - \arg(z - z_0)$. Thus, $\arg w'(z_0)$ has the geometric meaning of the difference between the slopes of the complex vectors $w - w_0$ and $z - z_0$.

A conformal mapping maps an infinitesimal triangle in the z-plane onto a similar triangle in the w-plane and the image of a system of lines orthogonal to each other is a family of

---

[31] Generally speaking, if the direction of the rotation which brings $z_a - z_0$ onto that $z_b - z_0$ is the same as that of their images the mapping is *conformal of the first kind*; if it is opposite it is *conformal of the second kind*. If $w(z)$ effects a mapping of the first kind, $w_1(z) = \overline{w}(z)$ effects a mapping of the second kind.

lines with the same property. We show in §17.10.4 how this feature permits the generation of boundary fitted coordinates. Equation (17.10.1) also shows that, as z describes an open neighborhood of z_0, the image point $w(z)$ describes an open neighborhood of $w(z_0)$, a property known as *preservation of neighborhoods*.

The analytic nature of the function that effects the mapping is essential:

Theorem 17.10.1 *The conformal mapping of an open domain D in the z-plane onto an open domain $\mathfrak{D}$ in the w-plane is effected only by* single-valued *analytic functions $w(z)$ with $w'(z) \neq 0$ in D.*

Conformality of the mapping is violated at points such that $w' = 0$ or ∞; these are the *critical points* of the mapping. For example, two rays meeting at the origin $z = 0$ at an angle α are mapped by the function $w = z^n$ into rays meeting at the origin $w = 0$ at a different angle $n\alpha$ and, indeed, $w'(z = 0) = 0$.

In view of the central importance of boundary conditions in many problems, it is a quite useful feature of conformal mappings that (reasonably smooth) boundaries are mapped onto boundaries (see e.g. Henrici 1974, p. 383 and Chapter 14; Burckel 1979, p. 309):[32]

Theorem 17.10.2 [Osgood–Carathéodory] *Let D and $\mathfrak{D}$ be two regions bounded by piecewise simple closed contours C and $\mathfrak{C}$. Any analytic function $w(z)$ mapping D conformally and one-to-one onto $\mathfrak{D}$ can be extended to a one-to-one continuous mapping of $\overline{D} = D \cup C$ onto $\overline{\mathfrak{D}} = \mathfrak{D} \cup \mathfrak{C}$.*

The principle of the argument (17.9.41) gives an easy proof of the converse statement (*Darboux's Theorem*, see e.g. Burckel 1979, p. 310; Boas 1987, p. 181). Suppose that $w(z)$ is analytic inside C and maps C one-to-one onto $\mathfrak{C}$ preserving the direction of traversal. Then, since $w(z)$ has no poles inside D, the variation of $\arg(w(z) - w_1)$ as z traverses C is 2π if the point $w_1 \in \mathfrak{D}$ and 0 if $w_1 \notin \overline{\mathfrak{D}}$, which shows that D is mapped one-to-one onto $\mathfrak{D}$.[33]

Example 17.10.1 [Linear mapping] The mapping $w = z + b$ is a simple translation having ∞ as the only fixed point. The mapping $w = e^{i\alpha}z$, with α real, is a rotation by the angle α around the origin; the only fixed points are $z = 0$ and $z = \infty$. The mapping $w = az = |a|e^{i\alpha}z$ adds to the rotation a uniform dilation by $|a|$. The combination of these elementary components is the general linear mapping

$$w = az + b = |a|e^{i\alpha}z + b \tag{17.10.2}$$

---

[32] The open mapping theorem (p. 423) states that an open region is mapped onto an open region. Thus, the Riemann mapping theorem (Theorem 17.10.3 below) says nothing about whether the mapping extends to the boundaries. As a matter of fact, this is generally false without suitable restrictions on the nature of the boundary, e.g. that it be a Jordan curve as defined on p. 696 of the Appendix (see e.g. Gonzalez 1991, p. 99).

[33] That this result is also valid for unbounded domains is easily shown by applying it to the mapping of the corresponding bounded domain induced by the transformation $\zeta = 1/(z - z_0)$, with $z_0 \notin D$.

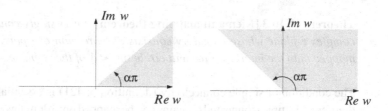

Fig. 17.21 Images of the upper half of the z-plane under the mapping $w = z^\alpha$ with $0 < \alpha < \frac{1}{2}$ (left) and $\frac{1}{2} < \alpha < 1$ (right).

which has as fixed points $z_0 = b/(1 - a)$ and ∞, the former one only if $a \neq 1$. The mapping may be considered as a rotation by α about z_0 followed by a dilation by $|a|$. $\square$

Example 17.10.2 [Inversion] The mapping

$$w = \frac{1}{z} = \frac{1}{|z|} e^{-i \arg z} \tag{17.10.3}$$

defines the inversion in the unit circle. It consists of a reflection in the unit circle, which changes $|z|$ into $1/|z|$, and a reflection in the real axis, which changes $\arg z$ to $- \arg z$. The interior of the unit circle in the z-plane is mapped onto the exterior of the unit circle in the w-plane and vice versa. In particular, $z = 0$ is mapped onto $w = \infty$ and $z = \infty$ onto $w = 0$. The only invariant points are $z = \pm 1$. $\square$

Example 17.10.3 [Powers] The function $w = z^2$ maps the first quadrant of the z plane $0 \leq \arg z \leq \frac{1}{2}\pi$ onto the upper half of the w-plane. The upper half of the z-plane is mapped onto the whole w plane, and the z-plane is mapped onto the two-sheeted Riemann surface of the inverse function $\sqrt{w}$.

To study the mapping of the upper half-plane induced by the real power

$$w = z^\alpha = |z|^\alpha e^{i\alpha \arg z} \qquad 0 < \alpha < 2, \qquad |\arg z| \leq \pi \tag{17.10.4}$$

we note that the real positive z-semi-axis is mapped onto the real positive w-semi-axis. On the real negative z-semi-axis, however, $\arg z = \pi$ and, therefore, $\arg w = \alpha\pi$, which corresponds to a ray issuing from $w = 0$ and enclosing an angle $\alpha\pi$ with the positive real w-semi-axis. Thus, under this mapping, the upper half of the z-plane is mapped onto a sector of angular aperture $\alpha\pi$ in the w plane. If $0 < \alpha < \frac{1}{2}$ the angle is acute, if $\frac{1}{2} < \alpha < 1$ it is obtuse and, as α grows from 1 to 2 it gradually encompasses more and more of the w-plane (Figure 17.21). $\square$

The results described apply to the mapping $w(z)$ generated by a given analytic function, but say nothing about whether such a function exists for an arbitrary domain. This matter is addressed by the fundamental

Theorem 17.10.3 [Riemann mapping theorem] *Every singly-connected domain D of the complex z-plane whose boundary consists of more than one point[34] can be conformally mapped onto the interior of the unit circle $|w| < 1$ of the w-plane.*

The condition of simple connectivity (definition p. 421) is essential for the validity of the theorem. Multiply-connected domains can be mapped onto domains with equal connectivity, but the generality of Riemann's theorem is lost; we briefly return on this topic in §17.10.2.

Since an analytic function of an analytic function is analytic (p. 422) and the mappings are one-to-one, it is evident that any two domains to which the theorem applies can be mapped into each other by combining the mapping of one onto the unit circle with the inverse mapping carrying the unit circle into the other one.

The fact that the unit circle can be mapped onto itself by a simple rotation shows that the mapping of Riemann's theorem cannot be unique without further conditions; some of the most useful ones are as follows:

Theorem 17.10.4 *The mapping mentioned in Riemann's theorem is unique if it is required that:*

(i) *$w(z_0) = 0$ and $\arg w'(z_0) = \alpha_0$, with $z_0 \in D$ and α_0 an arbitrary real number; or*
(ii) *$w(z_0) = 0$ with $z_0 \in D$ and $w(z_1) = w_1$ with $z_1 \in C$ and $w_1 \in \mathfrak{C}$; or*
(iii) *three boundary points $z_k \in C$, $k = 1, 2, 3$, are assigned to three points $w_k = w(z_k)$ on $\mathfrak{C}$, also numbered in the positive sense.*

It will be noted that, in each case, uniqueness is provided by the specification of three real numbers: in case (i) $\operatorname{Re} w(z_0)$, $\operatorname{Im} w(z_0)$, and $\arg w'(z_0)$; although the correspondence of two points (and, therefore, four real numbers) is prescribed in case (ii), there actually are only three degrees of freedom since $\operatorname{Re} w(z_1)$ and $\operatorname{Im} w(z_1)$ are not independent because the point is supposed to be on the boundary; and the same is true for case (iii).

Riemann's mapping theorem states the existence of the mapping, but does not provide a constructive method to find it. No such procedure exists in general and, in solving specific problems, one mostly has to have resort to intuition aided by the knowledge of a variety of existing mappings; an extended catalog is provided e.g. by Kober 1957 and Ivanov & Trubetskov 1995; Schinzinger & Laura 1991 also give many examples and applications, while Kythe 1998 focuses on computational aspects.

Example 17.10.4 [Möbius or fractional linear map] An important example of a conformal mapping is the Möbius, or fractional linear map:

$$w = \frac{az + b}{cz + d} = \frac{a}{c} + \frac{1}{c}\frac{bc - ad}{cz + d}. \tag{17.10.5}$$

---

[34] The only domain with no boundary points is the extended complex plane and, by Liouville's theorem 17.3.5 p. 431, the only analytic functions on such a domain are constants. A domain with one boundary point is the extended complex plane with one point deleted.

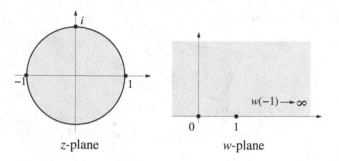

Fig. 17.22 The fractional linear mapping $w = i(1 - z)/(1 + z)$ transforms the interior of the unit circle in the z-plane onto the upper half of the w-plane with the points $z_1 = 1$, $z_2 = i$ and $z_3 = -1$ mapped onto $w_1 = 0$, $w_2 = 1$ and $w_3 = \infty$. The exterior of the unit circle is transformed into the lower half of the w-plane.

The cases $c = 0$ and $a = 0$ have already been essentially covered in Examples 17.10.1 and 17.10.2, respectively, and here we have assumed $c \neq 0$. The last equality in (17.10.5) shows that the mapping is the result of a translation, dilation and inversion (the latter two only provided $ad - bc \neq 0$). The inverse map is

$$z = \frac{b - dw}{cw - a},$$
(17.10.6)

and is therefore itself a Möbius transformation.[35] This is the only conformal map of the extended complex plane onto itself, with the point $z = -d/c$ mapped to the point $w = \infty$ while the point $w = a/c$ is the image of the point $z = \infty$. An important property of the Möbius map is that it transforms circles into circles (including straight lines, as circles with an infinite radius) and that points symmetric with respect to any circle are transformed into points symmetric with respect to the image of that circle.[36] These properties make the Möbius map very convenient to solve problems involving circles and half-planes.

Since there are essentially three independent constants in the definition (17.10.5) of the map, prescribing the position of the images w_k, $k = 1, 2, 3$, of three points z_k uniquely determines the map; a simple direct calculation shows that

$$\frac{w - w_1}{w - w_2} \frac{w_3 - w_2}{w_3 - w_1} = \frac{z - z_1}{z - z_2} \frac{z_3 - z_2}{z_3 - z_1}.$$
(17.10.7)

The three selected points uniquely identify a circle which, in conformity with point (iii) of the uniqueness theorem 17.10.4, is the boundary of the region in the z-plane mapped onto a corresponding region (circle) in the w-plane.

---

[35] If w is given by (17.10.5) and one sets $W = (Aw + B)/(Cw + D)$, the relation $W = W(z)$ retains the Möbius form. Thus, the Möbius transformations constitute a group in which (17.10.6) is the inverse transformation (see e.g. Hille 1959, p. 48).

[36] The symmetric point of a point P located at a distance r from the center C of a circle of radius R is the point on the line joining C and P lying at a distance R^2/r from the circle center.

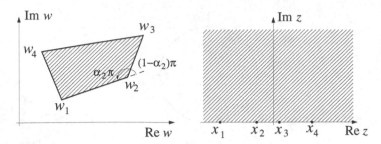

Fig. 17.23 The Schwarz–Christoffel transformation.

As an example, if the points $z_1 = 1, z_2 = i$ and $z_3 = -1$ are required to map onto $w_1 = 0$, $w_2 = 1$ and $w_3 = \infty$, then (17.10.7) readily gives $w = i(1 - z)/(1 + z)$ (Figure 17.22). Here the interior and exterior of the unit circle are mapped onto the upper and lower half-planes respectively. If we consider two points symmetric with respect to the line (i.e., circle with infinite radius) $\text{Re } w = 0$, for example $w_\pm = \pm i A$, from (17.10.6) we readily verify that the corresponding points satisfy $z_+ z_- = 1$, as expected. In general, if the correspondence between the points z_k and w_k is set so as to preserve the sense of traversal, then the interior of a circle is mapped into the interior of the corresponding circle while the opposite happens if the sense of traversal is inverted in the correspondence.

Another special case illustrates part (i) of the uniqueness theorem 17.10.4. Let it be required to map the unit circle into itself with the point z_2 mapped onto the center $w_2 = 0$. By the mentioned symmetry-preserving property, the point $w_3 = \infty$ will be the image of the symmetric point of z_2 in the circle, namely $1/\bar{z}_2$. From (17.10.7) we then find

$$w = \lambda \frac{z - z_2}{\bar{z}_2 z - 1} \qquad \text{with} \qquad \lambda = \frac{1 - \bar{z}_2 z_1}{z_2 - z_1} w_1, \qquad (17.10.8)$$

where z_1 and its image w_1 are still free. We should now enforce that the image circle have radius 1 according to the Riemann mapping theorem, i.e., that the image of all points $z = e^{i\alpha}$ satisfy $|w| = 1$. Enforcing this condition in (17.10.8) leads to $\lambda = e^{i\alpha_0}$ with α_0 undetermined. Prescribing a value for this parameter, then, uniquely specifies the mapping.[37] □

17.10.1 The Schwarz–Christoffel transformation

The Schwarz–Christoffel transformation maps polygons in the w-plane onto the upper half of the z-plane. As the examples to be given later show, the "polygons" to which the

---

[37] All the fractional linear mappings which transform the unit circle into itself have the form $w = (az + b)/(\bar{b}z + \bar{a})$ with $|a|^2 - |b|^2 > 0$.

transformation applies can be domains very different from what a narrow interpretation of the term would imply, and this feature makes this transformation particularly useful.

Consider an N-sided, non-self-intersecting polygon with interior angles $\alpha_1\pi$, $\alpha_2\pi$, ..., $\alpha_N\pi$ (Figure 17.23). For the time being we assume that $\alpha_j < 2$, so that the w_j is not the tip of a slit, and also that $\alpha_j > 0$. By an elementary property of polygons, we know that $\sum_{k=1}^{N} \alpha_k = N - 2$. In view of Darboux's Theorem, the vertices w_1, w_2, ..., w_N of the polygon, ordered in the positive sense, will be images of points $x_1 < x_2 < \cdots < x_N$ of the real axis of the z-plane; as we will see, some of the w_k and of the x_k may be infinite.

As the contour of the polygon is described in the positive direction, arg w remains constant along each side and jumps by an amount $(1 - \alpha_k)\pi$ as w traverses the k-th vertex (Figure 17.23, left). A function which mimics this behavior is the principal branch of $(z - x_k)^{\alpha_k-1}$ because, for $z = x > x_k$, its argument vanishes while, for $z = x < x_k$, it equals $(\alpha_k - 1)\pi$.[38] Along the two sides having the k-th vertex in common, therefore, the derivative of the function which effects the transformation can be taken to have the form

$$w' = (z - x_k)^{\alpha_k-1} f_k(z) \tag{17.10.9}$$

with arg f_k = const. Indeed, for z real to the right of x_k, arg $dw_+ = 0 + $ arg f_k while, to the left of x_k, arg $dw_- = \pi(\alpha_k - 1) + $ arg f_k so that the jump in the argument on traversing x_k is (arg $dw_+ - $ arg $dw_-) = \pi(1 - \alpha_k)$. The point x_k itself is a critical point of the mapping as $w' = 0$ or ∞ there.[39] Since the argument of a product is the sum of the arguments of the individual factors, it is clear that the function

$$w' = C(z - x_1)^{\alpha_1-1}(z - x_2)^{\alpha_2-1} \cdots (z - x_N)^{\alpha_N-1} \tag{17.10.10}$$

for any non-zero constant C fulfills the condition (17.10.9) at all the vertices of the polygon. Indeed, for a given $z = x$, all factors with $x_j < x$ have constant argument 0 while all factors with $x_k > x$ have constant argument $(\alpha_k - 1)\pi$. As the point x traverses the point x_k, arg w' therefore changes by the amount $(1 - \alpha_k)\pi$. Starting the traversal of the polygon at the vertex w_1 along a side with a certain inclination, when one returns to w_1 at the end of the traversal the argument will have rotated by an amount $\sum_{k=1}^{N}(1 - \alpha_k)\pi = \pi[N - (N - 2)] = 2\pi$ and will therefore have returned to the same slope.

Thus, if D designates an integration constant, the required transformation must have the form

$$w = C \int^z (\zeta - x_1)^{\alpha_1-1}(\zeta - x_2)^{\alpha_2-1} \cdots (\zeta - x_N)^{\alpha_N-1} \, d\zeta + D \tag{17.10.11}$$

which is the *Schwarz–Christoffel transformation*; we do not indicate the lower limt of integration which can be adjusted by changing D. On the basis of Example 17.10.1, it is clear that the constants C and D correspond to a dilation, rotation and translation of the mapping defined by the integral which, by itself, defines a mapping onto a polygon similar to the given one (see footnote 40 p. 480).

---

[38] We choose the principal branch for definiteness; any complex extension of $(x - x_k)^{\alpha_k-1}$ into the upper half of the complex plane defined so as to be real when $z - x$ is real and positive can be used.

[39] In order to remain on the chosen branch of $(z - x_k)^{\alpha_k-1}$ it is necessary to imagine going around the points x_k along an infinitesimal semi-circle in the upper half-plane, which corresponds to the interior of the polygon.

If, as we have assumed so far, $\alpha_j > 0$, the integral is finite for all z, even for $z = \infty$ because, since $\sum_{k=1}^{N}(\alpha_k - 1)\pi = -2$,

$$(\zeta - x_1)^{\alpha_1-1} \cdots (\zeta - x_N)^{\alpha_N-1} = \zeta^{-2}\left(1 - \frac{x_1}{\zeta}\right)^{\alpha_1-1} \cdots \left(1 - \frac{x_N}{\zeta}\right)^{\alpha_N-1}, \quad (17.10.12)$$

which converges at infinity. If we put $z = Re^{i\theta}$ in this relation and take R very large, we have, approximately,

$$w' = \frac{dw}{i Re^{i\theta} d\theta} \simeq \frac{C}{R^2}e^{-2i\theta}, \quad (17.10.13)$$

which can be integrated to find $w = w_\infty - (C/R)e^{-i\theta}$, with w_∞ the image of the point $z = \infty$. Thus, w describes a small semi-circle around the point w_∞. Before giving some examples, we make a few additional remarks.

1. Point $x_k \to \infty$. We have hitherto assumed that all the points x_k are finite. Suppose for example that it be required that the N-th vertex be the image of $z = \infty$. We can study this situation by introducing an intermediate mapping from a plane σ to the w-plane in which the N-th vertex corresponds to a finite point s_N, and then map this plane onto the upper z-plane by a fractional linear mapping governed by (17.10.7). This latter relation simplifies if we impose that two points s_a and s_b, with $s_a + s_b = s_N$ and $s_a s_b = -1$, retain their position. With this requirement, it is readily seen that (17.10.7) gives $\sigma = s_N + z^{-1}$. The mapping to the σ-plane has the standard form (17.10.11):

$$w = C\int^{\sigma}(s - s_1)^{\alpha_1-1}(s - s_2)^{\alpha_2-1} \cdots (s - s_N)^{\alpha_N-1} ds + D. \quad (17.10.14)$$

If we make the change of variables $\sigma = s_N + z^{-1}$ and recall that $\sum_{k=1}^{N}(\alpha_k - 1) = -2$, we find

$$w = C_1\int^{z}(\zeta - x_1)^{\alpha_1-1}(\zeta - x_2)^{\alpha_2-1} \cdots (\zeta - x_{N-1})^{\alpha_N-1-1} d\zeta + D, \quad (17.10.15)$$

where the x_j and C_1 depend on the s_k; the precise form of this dependence is of no particular interest in view of the purely auxiliary nature of the intermediate mapping. The point to note is that the final transformation (17.10.15) has precisely the same form as the general Schwarz–Christoffel one, except that the factor corresponding to the vertex at infinity is dropped and the corresponding angle does not appear. A similar result holds for any other point $x_k \to \infty$.

2. Point $w_N \to \infty$. The Schwarz–Christoffel formula can also be used to map an "open polygon," such as the one shown in Figure 17.24 onto the upper half-plane. For this purpose it is necessary to set $\alpha_4 = -\alpha_*$ as defined in Figure 17.24 (see e.g. Henrici 1974, p. 402). If the two sides "meeting" at ∞ are parallel, we would take the "enclosed angle" equal to 0 or π. It can be verified in the figure that, with this prescription, the relation $\sum_{k=1}^{N}\alpha_k = N - 2$ is verified. Since in a case such as this one $\alpha_4 < 0$, the integral diverges at $z = x_4$ so that this point is mapped to $w = \infty$.

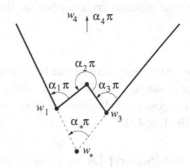

Fig. 17.24 Schwarz–Christoffel transformation of an open region.

Fig. 17.25 An interior angle equal to 2π (left) and to 0 (right); the interior of the "polygon" is shaded. The two sides are shown as separate for clarity, but in reality they are to be thought of as coinciding.

3. Slits. The Schwarz–Christoffel mapping works also if two sides of the region to be mapped are on top of each other as in the case of the slit shown on the left in Figure 17.25, where the interior of the region of interest is shaded. In this case we would take $\alpha = 2$. For example, applied to the mapping of the entire w-plane cut along the positive real axis, with $w = 0$ and $w = \infty$ going to $z = 0$ and $z = \infty$, (17.10.11) would give

$$w = C \int^{z} \zeta \, d\zeta = \frac{1}{2}Cz^{2}. \tag{17.10.16}$$

We can determine C by imposing the correspondence of a third point. In the opposite case of a line coming out of the boundary (Figure 17.25, right), we would take $\alpha = 0$.

4. Exterior domains. The Schwarz–Christoffel transformation can be applied also to the mapping onto the upper half-plane of the domain exterior to the polygon. In this case, the positive direction of traversal of the polygon is clockwise and the sign of the angles must be adjusted accordingly.

5. The parameter problem. According to the Riemann mapping theorem, the position of only three points along the boundary can be specified, which implies that, if $N > 3$, the position of $N - 3$ points in the z-plane is unknown and must be determined. This is the *Schwarz–Christoffel parameter problem*, which can be shown to admit a unique solution;

robust numerical methods are available for this purpose (see e.g. Driscoll & Trefethen 2002).[40]

6. Extensions. The same approach can be extended in various ways, e.g. to map a polygon onto different standard domains such as disks or others (see e.g. Driscoll & Trefethen 2002). It is also possible to "round" the corners of the polygon into smoother curves (see e.g. Henrici 1974, p. 422).

Example 17.10.5 [Sectors] Let us consider the mapping of the upper z-plane onto a two-vertex "polygon" consisting of an infinite sector centered at the point w_1 and extending over the angular range $\beta \leq \arg w \leq \beta + \alpha\pi$ with $0 < \alpha < 2$. Let us establish the correspondence $w(z_1 = 0) = w_1$, $w(z = \infty) = \infty$. The formula (17.10.15) gives[41]

$$ w = C \int^z \zeta^{\alpha-1}\, dz + D = \frac{C}{\alpha} z^\alpha + D. \tag{17.10.17} $$

By requiring that $w(0) = w_1$ we find $D = w_1$, and the condition $\arg w' = \beta$ for $z = x > 0$ requires $C = |C|e^{i\beta}$, so that $w = |C|e^{i\beta} z^\alpha/\alpha + w_1$. For a sector extending to the whole plane $\alpha = 2$ and we recover the earlier formula (17.10.16). As in that case, and for the same reason, there is a residual degree of freedom which corresponds to the scale factor $|C|$.

Let us take $\beta = 0$, $w_1 = -C/\alpha$ and consider the limit $\alpha \to 0$; in this case, the vertex moves to infinity and the sector degenerates to a horizontal strip. We find

$$ w = \frac{C}{\alpha}\left[e^{\alpha \log z} - 1\right] \to C \log z. \tag{17.10.18} $$

On the basis of our study of the complex logarithm (Example 17.6.2 p. 449), we see that this function maps the upper z-plane onto the infinite strip $0 \leq \operatorname{Im} w \leq \pi C$. ☐

Example 17.10.6 [Triangles] Since the factor corresponding to $x_N \to \infty$ drops out from the Schwarz–Christoffel integral (see comment 1 above), it is convenient to map one vertex, w_3, say, to ∞. We start by considering some degenerate "triangles" for which $w_3 = \infty$.

For the half-strip $0 \leq \operatorname{Re} w < \infty$, $0 \leq \operatorname{Im} w < b$ (Figure 17.26, left), the boundary is traversed in the positive direction if we take $w_1 = ib$, $w_2 = 0$ and $w_3 = \infty$, with $\alpha_1 = 1/2$, $\alpha_2 = 1/2$ (and $\alpha_3 = 0$, cf. comment 2 above). Reasons of symmetry suggest taking $w_1 = w(z = -1)$, $w_2 = w(z = 1)$, with which we have

$$ w = C \int^z \left(\zeta^2 - 1\right)^{-1/2} d\zeta + D = C \cosh^{-1} z + D. \tag{17.10.19} $$

---

[40] The constants C and D, which determine the scale and the orientation of the polygon, are also unknown, but it is clear that prescribing two vertices fixes the scale, and prescribing a third one gives the orientation of the polygon.

[41] According to comment 2 above, in this case the angle corresponding to the vertex at infinity would be taken as $-\alpha\pi$ and the sum of the α's would be 0 in keeping with the formula $\sum_{k=1}^N \alpha_k = N - 2$.

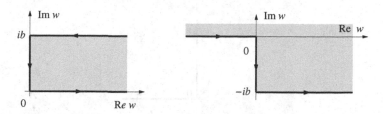

Fig. 17.26 Two degenerate triangles: a semi-infinite strip (left) and the domain above a step; the arrows show the positive sense of traversal of the boundaries of the shaded domains.

Imposing $w_2 = w(z = 1) = 0$ gives $D = 0$. Recall from (17.6.21) that $\cosh^{-1} z = \log\left(z + \sqrt{z^2 - 1}\right)$. For definiteness, we take the principal branches of both the square root and the logarithm. Since then $\log(-1) = i\pi$, requiring that $z = -1$ map to $w = ib$ gives $ib = i\pi C$ or $C = b/\pi$. Thus $w = (b/\pi)\cosh^{-1} z$ or $z = \cosh(\pi w/b)$.

If we wish to map the domain *exterior* to the semi-infinite strip, the positive direction of traversal is opposite and we now take $w_1 = 0 = w(-1)$ and $w_2 = ib = w(1)$. The positive direction of the angles is also inverted and, therefore, $\alpha_1 = \alpha_2 = 3/2$ so that the mapping is

$$w = C \int^z \left(\zeta^2 - 1\right)^{1/2} d\zeta + D = \frac{1}{2} C \left[z\sqrt{z^2 - 1} - \cosh^{-1} z\right] + D. \quad (17.10.20)$$

With the same choice of branches as before, $w(-1) = 0$ provided $D = \frac{1}{2}\pi C$ and $w(1) = ib$ provided $C = 2b/\pi$ so that

$$w = \frac{b}{\pi}\left[z\sqrt{z^2 - 1} - \cosh^{-1} z + i\pi\right]. \quad (17.10.21)$$

As another example, consider the domain above a step from $\operatorname{Im} w = 0$, $-\infty < \operatorname{Re} w < 0$, down to $\operatorname{Im} w = -b$, $0 < \operatorname{Re} w < \infty$ (Figure 17.26, right). Take as before $w_1 = w(z = -1) = 0$, $w_2 = w(z = 1) = -ib$ to find, since $\alpha_1 = 3/2$, $\alpha_2 = 1/2$,[42]

$$w = C \int^z \left(\frac{\zeta + 1}{\zeta - 1}\right)^{-1/2} d\zeta + D = C \left[(z^2 - 1)^{1/2} + \cosh^{-1} z\right] + D, \quad (17.10.22)$$

with C and D determined as before.

As a final example, let us consider the geometry of Figure 17.27. We map the point at "right infinity" to $z_1 = 0$ (with $\alpha_1 = 0$), the origin $w = 0$ to $z_2 = 1$ (with $\alpha_2 = 3/2$), and the point at "left infinity" to $z_3 = \infty$ (with $\alpha_3 = -1/2$). We now have

$$w = C \int^z \frac{\sqrt{\zeta - 1}}{\zeta} d\zeta + D = 2C \left[\sqrt{z - 1} - \frac{1}{2i}\log\frac{1 + i\sqrt{z - 1}}{1 - i\sqrt{z - 1}}\right] + D. \quad (17.10.23)$$

[42] According to the rule shown in Figure 17.24, the "angle" $\alpha_* \pi$ between the two horizontal lines should be taken as π, so that $\alpha_3 = -\alpha_* = -1$ and $\alpha_1 = \pi - \alpha_2 - \alpha_3 = 3/2$.

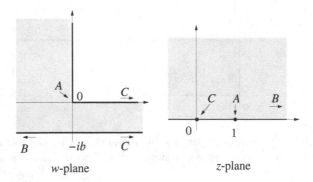

w-plane   z-plane

Fig. 17.27 The shading shows the region mapped in the final case of Example 17.10.6.

The condition that $z = 1$ correspond to $w = 0$ requires $D = 0$. Consider z real, large and negative. Then $\sqrt{z-1} \simeq i\sqrt{-x}$ and $w \simeq 2iC\left(\sqrt{-x} + \pi/2\right)$. This part of the domain should map on the horizontal line $\operatorname{Im} w = -ib$, which requires that $C = i\pi/b$. We then find

$$w = \frac{\pi}{b}\left[2i\sqrt{z-1} - \log\frac{1+i\sqrt{z-1}}{1-i\sqrt{z-1}}\right]. \tag{17.10.24}$$

If z is real, large and positive $w \simeq (2i\pi/b)\sqrt{x}$ so that the line $0 < x < \infty$ is mapped onto $\operatorname{Re} w = 0$, $0 < \operatorname{Im} w < \infty$. The segment $0 < \operatorname{Re} z < 1$ has been mapped on the line $0 < \operatorname{Re} w < \infty$, $\operatorname{Im} w = 0$. Thus, the part of the domain boundary having a corner has been mapped onto $\operatorname{Re} z > 0$ and the plane wall onto $\operatorname{Re} z < 0$, in both cases with $\operatorname{Im} z = 0$.

For a general triangle we would still map w_3 to $z = \infty$. If we map the other two vertices onto $z = 0$ and $z = 1$, (17.10.15) gives

$$w = C_1 \int^z \zeta^{\alpha_1 - 1}(1 - \zeta)^{\alpha_2 - 1}\, dz + D, \tag{17.10.25}$$

which is reducible to an incomplete beta function (p. 445).   □

Many other examples of applications of the Schwartz–Christoffel transformation can be found e.g. in Schinzinger & Laura 1991, section 3.3 and in Driscoll & Trefethen 2002.

17.10.2 Multiply connected domains

By Riemann's mapping theorem we can map a singly connected domain onto a circle. Imagine removing from the initial domain some internal areas, so that it becomes multiply connected. Because neighborhoods are preserved by the mapping (p. 472), each one of the removed areas will have an image internal to the circle, the removal of which endows the circle with the same connectivity as the original domain. This consideration suggests that it should be possible to map a multiply connected domain conformally onto a domain with the same connectivity. This is indeed true, although a result with the same generality as the Riemann mapping theorem does not exist.

In particular, every doubly connected domain having no isolated boundary points can be mapped conformally onto a circular annulus bounded by two concentric circles with radii R_1 and R_2 (see e.g. Bieberbach 1953, p. 209). A point to note, however, is that the ratio $\mu = R_2/R_1 > 1$ of the radii cannot be prescribed arbitrarily, but is an intrinsic attribute of the original doubly connected region called the *modulus* of the region (sometimes the modulus is defined as log R_2/R_1). This quantity is a conformal invariant in the sense that no conformal mapping will change it. Thus, all doubly connected domains domains fall into classes each member of which has the same value of μ. Domains belonging to the same class can be conformally mapped onto one another, but this is impossible for two domains having different moduli.

A similar result is valid for domains with a higher degree of connectivity: every finitely connected domain having no isolated boundary points can be mapped conformally onto a domain with the same connectivity bounded by a number of complete circles.

17.10.3 The Laplace equation

One of the major applications of analytic function theory is to the solution of the two-dimensional Laplace equation because the equation remains invariant in form under a conformal mapping, as we now show, while a suitable mapping can considerably simplify the shape of the domain (see Example 17.10.9) and help in other ways as well (see Example 17.10.10).

Consider a one-to-one coordinate transformation

$$\xi = \xi(x, y), \qquad \eta = \eta(x, y) \qquad (17.10.26)$$

as a consequence of which a function $U(x, y)$ becomes a function of the new variables ξ, η. We express the Laplacian in the new coordinates and ask under what conditions it is proportional to $\nabla^2_{\xi,\eta} U = \partial^2 U/\partial\xi^2 + \partial^2 U/\partial\eta^2$ so that the Laplace equation remains invariant in form. A patient application of the chain rule gives

$$\frac{\partial^2 U}{\partial x^2} + \frac{\partial^2 U}{\partial y^2} = (\xi_x^2 + \xi_y^2)\frac{\partial^2 U}{\partial\xi^2} + (\eta_x^2 + \eta_y^2)\frac{\partial^2 U}{\partial\eta^2} + 2(\xi_x\eta_x + \xi_y\eta_y)\frac{\partial^2 U}{\partial\xi\eta}$$

$$+ (\xi_{xx} + \xi_{yy})\frac{\partial U}{\partial\xi} + (\eta_{xx} + \eta_{yy})\frac{\partial U}{\partial\eta}. \qquad (17.10.27)$$

To meet our requirement it is therefore necessary that

$$\xi_x\eta_x + \xi_y\eta_y = 0, \qquad \xi_{xx} + \xi_{yy} = \eta_{xx} + \eta_{yy} = 0, \qquad (17.10.28)$$

while

$$\xi_x^2 + \xi_y^2 = \eta_x^2 + \eta_y^2 \neq 0. \qquad (17.10.29)$$

Using here the first relation of (17.10.28) we may rewrite this last condition as $(\xi_x/\eta_y)^2(\eta_x^2 + \eta_y^2) = \eta_x^2 + \eta_y^2$, from which $\xi_x = \pm\eta_y$. If we take the upper sign, the first relation of

(17.10.28) shows that $\xi_y = -\eta_x$ so that ξ and η satisfy the Cauchy–Riemann relations. This requirement implies that they are harmonic so that the last two relations of (17.10.28) are also satisfied.[43]

In conclusion we have found that, if the original (x, y)-plane is turned into the complex plane $z = x + iy$, the conformal transformation $\zeta(z) = \xi(x, y) + i\eta(x, y)$ leads to

$$\frac{\partial^2 U}{\partial x^2} + \frac{\partial^2 U}{\partial y^2} = |\zeta'(z)|^2 \left(\frac{\partial^2 U}{\partial \xi^2} + \frac{\partial^2 U}{\partial \eta^2} \right) = \frac{1}{|\psi'(\xi + i\eta)|^2} \left(\frac{\partial^2 U}{\partial \xi^2} + \frac{\partial^2 U}{\partial \eta^2} \right),$$
(17.10.30)

where ψ is the inverse of ζ and we have used the relation $\xi_x^2 + \xi_y^2 = \eta_x^2 + \eta_y^2 = |\zeta'(z)|^2$ which is an easy consequence of the Cauchy–Riemann relations (17.2.5). The Jacobian of the transformation (17.10.26) is readily found to be

$$J(\xi, \eta) = \frac{\partial(\xi, \eta)}{\partial(x, y)} = |\zeta'(z)|^2,$$
(17.10.31)

so that the mapping is conformal except at the critical points where $\zeta'(z) = 0$.

Example 17.10.7 [Laplace equation in a disk] Consider the problem of finding a function U harmonic inside the disk $x^2 + y^2 = 1$ reducing to a given $f(\theta)$ on its boundary. We have solved this problem by means of the Fourier series in §3.6. The same solution can be elegantly found as follows. The given disk in the (x, y)-plane is the the unit disk in the complex z-plane. Consider the linear fractional mapping given by (17.10.8) of Example 17.10.4, in which we write $\zeta(z)$ in place of $w(z)$ to conform to the present notation. This mapping transforms the unit z-disk onto the unit ζ-disk putting an arbitrary point $z_2 = r_2 e^{i\theta_2}$ at the center $\zeta = 0$. After this mapping has been determined, we can use the mean value theorem (17.3.32) to express the value of $U(\zeta)$ at $\zeta = 0$:

$$U(\zeta = 0) = \frac{1}{2\pi} \int_0^{2\pi} F(\phi) \, d\phi,$$
(17.10.32)

where $F(\phi)$ is the image of $f(\theta)$ under the mapping. The required mapping is given by (17.10.8), from which we have

$$d\phi = \frac{1 - r_2^2}{1 + r_2^2 - 2r_2 \cos(\theta - \theta_2)} \, d\theta,$$
(17.10.33)

so that

$$U(r_2, \theta_2) = \frac{1}{2\pi} \int_0^{2\pi} \frac{(1 - r_2^2) f(\theta)}{1 + r_2^2 - 2r_2 \cos(\theta - \theta_2)} \, d\theta$$
(17.10.34)

as found earlier.   □

---

[43] Taking $\xi_x = -\eta_y$, and consequently $\xi_y = \eta_x$, gives a conformal mapping of the second kind, cf. footnote on p. 471.

Example 17.10.8 [Green's function] It is briefly shown in §2.4 and, in greater detail, in Chapter 16, that the solution of the Dirichlet problem in a domain Ω can be written down explicitly once the Green's function G of the problem in known. In two dimensions, this function satisfies

$$\nabla^2_{x,y} G(x, y; x_0, y_0) = -2\pi \delta(x - x_0)\delta(y - y_0) \tag{17.10.35}$$

and vanishes on the boundary of Ω. We now show that the determination of G can be reduced to that of a suitable conformal mapping.

We know from (16.1.14) p. 402 that

$$\nabla^2_{\xi,\eta} \log \sqrt{(\xi - \xi_0)^2 + (\eta - \eta_0)^2} = 2\pi \delta(\xi - \xi_0)\,\delta(\eta - \eta_0). \tag{17.10.36}$$

Furthermore, $\log \sqrt{(\xi - \xi_0)^2 + (\eta - \eta_0)^2} = 0$ on the unit circle centered at (ξ_0, η_0). In terms of the complex variable z, the domain Ω of the (x, y)-plane becomes a domain of the complex z-plane. Let the function $\zeta(z) = \xi + i\eta$ map this domain onto the unit circle $|\zeta - \zeta_0| = 1$ where ζ_0 is the image of $z_0 = x_0 + iy_0$. By (17.10.30) we then have

$$\nabla^2_{x,y} G = |\zeta'(z)|^2 \nabla^2_{\xi,\eta} \log |\zeta - \zeta_0| = |\zeta'(z)|^2 \delta(\xi - \xi_0)\,\delta(\eta - \eta_0). \tag{17.10.37}$$

It is shown in §20.7 p. 588 that $\delta(x - x_0)\,\delta(y - y_0) = J\delta(\xi - \xi_0)\,\delta(\eta - \eta_0)$, where J is the Jacobian of the transformation equal here to $|\zeta'(z)|^2$ by (17.10.31). In conclusion we find therefore that

$$G(x, y; x_0, y_0) = \log \frac{1}{|\zeta - \zeta_0|}. \tag{17.10.38}$$

As a specific application, let us find the Green's function for the infinite strip $-\infty < x < \infty$, $0 < y < b$. This is also a good example to illustrate the flexibility achieved by combining a sequence of conformal mappings. To map the strip onto the unit disk we first map it onto the upper half-plane of an intermediate variable σ using the inverse of the mapping (17.10.18), namely $\sigma = e^{\pi z/b}$; by writing the argument as $\pi z/b$, we have $0 \le \arg \sigma \le \pi$. The point $z_0 = (x_0, y_0)$ is mapped to $\sigma_0 = e^{\pi z_0/b}$. Now we can use a fractional linear map to transform the upper σ-plane into the unit circle in the ζ-plane. It is convenient to prescribe that the image of σ_0 correspond to $\zeta = 0$, so that $\zeta_0 = 0$, which puts the general expression (17.10.5) in the form

$$\zeta = a \frac{\sigma - \sigma_0}{\sigma + d}, \tag{17.10.39}$$

where we have taken $c = 1$ without loss of generality. We want that the real axis of the w-plane be mapped onto $|\zeta| = 1$, which requires that

$$|a|^2 \left(\sigma^2 + |\sigma_0|^2 - 2\sigma \operatorname{Re}\sigma_0\right) = \sigma^2 + |d|^2 + 2\sigma \operatorname{Re}d. \tag{17.10.40}$$

To satisfy this equation identically for all real σ it is necessary that $|a| = 1$, $|\sigma_0| = |d|$ and $\operatorname{Re}\sigma_0 = -\operatorname{Re}d$. The last two conditions imply that $d = -\overline{\sigma}_0$ and we also may take $a = 1$

for simplicity.[44] Thus, finally,

$$\zeta = \frac{e^{\pi z/b} - e^{\pi z_0/b}}{e^{\pi z/b} - e^{\pi \bar{z}_0/b}} \qquad (17.10.41)$$

from which

$$G(x, y; x_0, y_0) = -\frac{1}{2} \log \frac{\cosh\left[\pi(x - x_0)/b\right] - \cos\left[\pi(y - y_0)/b\right]}{\cosh\left[\pi(x - x_0)/b\right] - \cos\left[\pi(y + y_0)/b\right]}. \qquad (17.10.42)$$

Upon inserting this formula in the general expression (16.1.19) p. 403, we recover the solution (4.6.7) p. 107 of the infinite-strip problem obtained by transform methods in §4.6. □

Example 17.10.9 Consider the geometry of Figure 17.27 and suppose that we are interested in the solution of the Laplace equation for a function U in the shaded domain, satisfying the boundary conditions $U = 0$ on the wall with a corner and $U = U_0$ on the plane wall. Let us rewrite the mapping (17.10.24) changing the symbol w to z and z to ζ, which are the original and the transformed coordinates in this case:

$$z = \frac{\pi}{b}\left[2i\sqrt{\zeta - 1} - \log \frac{1 + i\sqrt{\zeta - 1}}{1 - i\sqrt{\zeta - 1}}\right]. \qquad (17.10.43)$$

This mapping transforms the domain of interest in the z-plane into the upper ζ half-plane $\eta \geq 0$ (left and right parts of Figure 17.27, respectively) leaving the equation invariant. Thus, in the transformed coordinates, our problem amounts to finding the solution of the Laplace equation in a half plane, satisfying $U(\xi, 0) = 0$ for $\xi < 0$ and $U(\xi, 0) = U_0$ for $\xi > 0$. We have obtained a general solution of this type of problem by transform methods in §4.6. Adapting that solution, given in (4.6.9) p. 107, to the present situation we find

$$U(\xi, \eta) = \frac{U_0}{\pi} \int_0^\infty \frac{\eta}{(X - \xi)^2 + \eta^2} \, dX = \frac{U_0}{\pi}\left[\frac{\pi}{2} \pm \tan^{-1}\frac{\xi}{\eta}\right]. \qquad (17.10.44)$$

To resolve the sign ambiguity we recall that, when η is small, $U \simeq U_0$ for $\xi > 0$ and $U \simeq 0$ for $\xi < 0$. On this basis, we find $U(\xi, \eta) = (U_0/\pi)\,\theta$, where θ is the polar angle in the (ξ, η)-plane. An explicit form in the original coordinates (x, y) cannot be written down as it is not possible to invert the transformation (17.10.43) analytically, but the result obtained permits one to readily plot the solution and to investigate its nature by asymptotic methods (see e.g. Morse & Feshbach 1953, p. 450). □

Example 17.10.10 The previous problem can also be solved by noting that, in the new complex plane $W = U + iV$, the solution corresponds to the strip $0 \leq U \leq U_0$ parallel to the imaginary axis. Hence a mapping of the original domain (left in Figure 17.27) onto this

---

[44] The other possibility $d = -\sigma_0$ must be discarded because it would simply give $\zeta = a$. Note also that, by the uniqueness theorem of the mapping 17.10.4, we are at liberty to impose the value of arg $\zeta'(\sigma_0) = \arg[a/(\sigma_0 + d)]$; the choices made amount to imposing arg $\zeta'(\sigma_0) = 0$.

strip such that $z = z(W)$ implicitly provides the required solution. Such a mapping can be obtained by a combination of the mappings (17.10.43) and (17.10.41), the latter rotated by $\pi/2$. This method of solution is particularly effective when the given Dirichlet data are piece-wise constant on the boundaries of the physical domain, as e.g. with conductors held at prescribed potentials in electrostatics or free-streamline problems in inviscid irrotational flow.   □

17.10.4 Orthogonal coordinates

The analytic and numerical solution of many boundary value problems is greatly facilitated by the use of an orthogonal coordinate system having coordinate surfaces which conform with the shape of the boundaries. Due to their isogonality property, conformal mappings are useful for this purpose.

Let the problem be posed in the (x, y)-plane in which we introduce the complex variable $z = x + iy$. Our objective is to find a conformal mapping $z(\zeta)$ which transforms orthogonal coordinates in the ζ-plane into boundary-fitting orthogonal coordinates in the z-plane. If $\zeta = \xi + i\eta$, the mapping is usually such that the desired coordinate lines in the z-plane correspond to $\xi = $ const. and $\eta = $ const. The two-dimensional coordinate system thus generated can be transformed into a three-dimensional one either by translating the plane normal to itself, or by rotating it around the x- or y-axes.

By expressing the length of an infinitesimal segment ds referred to the coordinates (x, y) we find the expression of the *metric coefficients h_ξ and h_η*:

$$\mathrm{d}s^2 = \mathrm{d}x^2 + \mathrm{d}y^2 = \left(h_\xi \mathrm{d}\xi\right)^2 + \left(h_\eta \mathrm{d}\eta\right)^2, \qquad (17.10.45)$$

where

$$h_\xi = \sqrt{\left(\frac{\partial x}{\partial \xi}\right)^2 + \left(\frac{\partial y}{\partial \xi}\right)^2}, \qquad h_\eta = \sqrt{\left(\frac{\partial x}{\partial \eta}\right)^2 + \left(\frac{\partial y}{\partial \eta}\right)^2}. \qquad (17.10.46)$$

Clearly, $h_\xi \mathrm{d}\xi$ measures the displacement of a point in the direction $\eta = $ const. and similarly for $h_\eta \mathrm{d}\eta$. By using the Cauchy–Riemann relations (17.2.5) satisfied by $x(\xi, \eta)$ and $y(\xi, \eta)$:

$$\frac{\partial x}{\partial \xi} = \frac{\partial y}{\partial \eta}, \qquad \frac{\partial x}{\partial \eta} = -\frac{\partial y}{\partial \xi}, \qquad (17.10.47)$$

we readily show that $h_\xi = h_\eta$ as expected on the basis of the results of §17.10.3.

Example 17.10.11 [Plane polar coordinates] A very simple example is obtained from the mapping $z = e^\zeta$, illustrated in Figure 17.6 p. 438, which gives

$$x = e^\xi \cos \eta, \qquad y = e^\xi \sin \eta, \qquad (17.10.48)$$

with $h_\xi = h_\eta = e^\xi$. By using $r = e^\xi$ in place of ξ, we have the usual polar coordinates in the plane with

$$h_r = \sqrt{\left(\frac{\partial x}{\partial r}\right)^2 + \left(\frac{\partial y}{\partial r}\right)^2} = 1, \qquad h_\eta = r. \qquad (17.10.49)$$

□

Example 17.10.12 [Plane bi-polar coordinates] Consider the conformal mapping of the second kind (see footnote on p. 471) akin to the mapping (17.10.41) of Example 17.10.8:

$$\bar{z} = a\frac{e^{\zeta} + 1}{e^{\zeta} - 1} = a \coth \frac{\zeta}{2}. \qquad (17.10.50)$$

If a is real, this is

$$x = a\frac{\sinh \xi}{\cosh \xi - \cos \eta}, \qquad y = a\frac{\sin \eta}{\cosh \xi - \cos \eta}, \qquad (17.10.51)$$

which define bi-polar coordinates in the plane with

$$h_{\xi} = h_{\eta} = \frac{a}{\cosh \xi - \cos \eta}. \qquad (17.10.52)$$

The inverse relations are

$$(x - a \coth \xi)^2 + y^2 = \frac{a^2}{\sinh^2 \xi}, \qquad (y - a \cot \eta)^2 + x^2 = \frac{a^2}{\sin^2 \eta}. \qquad (17.10.53)$$

Thus, the lines of constant ξ are a family of circles centered on the x-axis at $a \coth \xi$ with radius $a/\sinh \xi$ and the lines of constant η are circles centered on the-y axis at $a \cot \eta$. It is seen from the second relation that the point $(x = a, \ y = 0)$ belongs to all the circles of the second family so that the mapping ceases to be one-to-one. This is indeed a critical point of the transformation corresponding to $\zeta \to \infty$. □

The transition to a three-dimensional system introduces a third coordinate and the corresponding metric coefficient. If the third coordinate arises by displacing the plane normally to itself by the amount Z, then h_{ξ} and h_{η} do not change and $h_Z = 1$. In addition to the planes of constant Z, the coordinate surfaces are infinite cylinders with the axis parallel to the axis of translation intersecting the $(x, \ y)$-plane along lines of constant ξ and η. In this way, for example, the plane polar coordinates of Example 17.10.11 generate the usual cylindrical coordinates in space (§6.1) and the bipolar coordinates of Example 17.10.12 generate bi-cylinder coordinates.

The third coordinate may also arise from a rotation by an angle ϕ, with $0 \le \phi < 2\pi$, of the $(x, \ y)$-plane around an axis. In this case one family of coordinate surfaces are the half-planes of constant ϕ. all of which go through the axis of rotation, or polar axis. Let us denote by Z the coordinate along this axis and by X and Y the coordinates in the plane $Z = 0$. If the rotation is around the y-axis we have (Figure 17.28)

$$X = x(\xi, \eta) \cos \phi, \qquad Y = x(\xi, \eta) \sin \phi, \qquad Z = y(\xi, \eta), \qquad (17.10.54)$$

and

$$h_X = h_Y = \sqrt{\left(\frac{\partial x}{\partial \xi}\right)^2 + \left(\frac{\partial y}{\partial \xi}\right)^2} = \sqrt{\left(\frac{\partial x}{\partial \eta}\right)^2 + \left(\frac{\partial y}{\partial \eta}\right)^2}, \qquad h_Z = x(\xi, \eta). \qquad (17.10.55)$$

If the polar axis is the x-axis, we similarly find

$$X = y(\xi, \eta) \cos \phi, \qquad Y = y(\xi, \eta) \sin \phi, \qquad Z = x(\xi, \eta), \qquad (17.10.56)$$

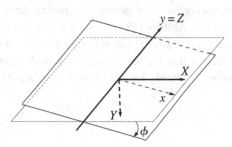

Fig. 17.28 Three-dimensional coordinate system obtain by rotating about the y-axis.

with

$$h_X = h_Y = \sqrt{\left(\frac{\partial x}{\partial \xi}\right)^2 + \left(\frac{\partial y}{\partial \xi}\right)^2} = \sqrt{\left(\frac{\partial x}{\partial \eta}\right)^2 + \left(\frac{\partial y}{\partial \eta}\right)^2}, \qquad h_Z = y(\xi, \eta).$$
(17.10.57)

Example 17.10.13 [Polar coordinates in space] The standard polar coordinates in space (§7.1) are obtained by rotating the plane polar coordinates around the x-axis. From (17.10.48) and (17.10.56) we have

$$X = e^\xi \sin \eta \cos \phi, \qquad Y = e^\xi \sin \eta \sin \phi, \qquad Z = e^\xi \cos \eta, \qquad (17.10.58)$$

and from (17.10.57) $h_X = h_Y = e^\xi, h_Z = e^\xi \cos \eta$. The substitution $r = e^\xi$ reduces these expressions to the usual form.

Rotation of the plane bi-polar coordinate system around the x-axis generates bi-spherical coordinates in which the coordinate surfaces are spheres, spindles and half-planes. Rotation around the y-axis generates toroidal coordinates in which the coordinate surfaces are tori and intersecting spheres.   □

This topic is treated in much greater detail e.g. in Morse & Feshbach 1953 section 5.1; Moon & Spencer 1961 and Lebedev 1965 where many more examples will be found.

A more recent development is the use of *quasi-conformal mappings* for grid generation. A quasi-conformal mapping differs from a conformal one in that the Cauchy–Riemann relations are generalized to the form (see e.g. Ahlfors 1987)

$$f\frac{\partial x}{\partial \xi} = \frac{\partial y}{\partial \eta}, \qquad \frac{\partial x}{\partial \eta} = -f\frac{\partial y}{\partial \xi}, \qquad (17.10.59)$$

with $f = f(\xi, \eta)$. Among other properties, quasi-conformal mappings are more flexible than conformal ones and can be better conditioned numerically. Furthermore, they can be used to map curvilinear or degenerate quadrilaterals onto other quadrilaterals of arbitrary *module* (see e.g. Ahlfors 1987).[45]

The recent interest in numerical grid generation has moved this field well beyond the classical results that we have briefly exposed; some recent references are the books by Thompson *et al.* (1999), Liseikin (1999, 2007) and others.

---

[45] The module of a rectangle is defined as the ratio of the lengths of the two sides. Conformal mappings can only map rectangles onto the class of quadrilaterals having the same module (see e.g. Henrici 1986), while quasi-conformal mappings are not subject to this limitation.

18

Matrices and Finite-Dimensional Linear Spaces

Matrix theory is a large subject which, among others, plays a central role in scientific computing. Here we will only deal with a few selected topics of direct relevance to classical field theory. In particular, we will focus on some aspects which are useful to illustrate the general ideas underlying the theory of linear operators, a topic which is taken up in Part III of this book.

It is next to impossible to discuss matrix theory without reference to the theory of linear vector spaces, which provides a deeper and ultimately much more illuminating view of the entire subject. We assume that the reader is familiar with the elementary aspects of this subject (finite-dimensional vectors, linear dependence and independence, dimension of a finite-dimensional space, scalar product and a few others). To avoid duplication, these concepts are only touched upon in §18.3. A more detailed description is provided in the more general context of infinite-dimensional spaces in Chapter 19 much of which – provided only that references to infinite dimensionality be deleted – could be placed at this point with but minor changes.

18.1 Basic definitions and properties

We assume that the reader is familiar with the basic notions of matrix theory; this section is meant mainly to establish the notation and to summarize some definitions and facts; Table 18.1 gives a synthetic presentation of the main definitions.

An $M \times N$ matrix A is represented as an array of numbers a_{ij}, complex in general, arranged in M rows and N columns:

$$
A = |a_{ij}| = \begin{vmatrix} a_{11} & a_{12} & a_{13} & \cdots & a_{1N} \\ a_{21} & a_{22} & a_{23} & \cdots & a_{2N} \\ \vdots & \vdots & \vdots & \vdots & \vdots \\ a_{M1} & a_{M2} & a_{M3} & \cdots & a_{MN} \end{vmatrix} .
\tag{18.1.1}
$$

When $N = 1$ there is only one column and we omit the second index; we refer to these special $M \times 1$ matrices as *vectors*. It is convenient at times to write the matrix in terms of column vectors

$$
A = |\mathbf{a}_1\ \mathbf{a}_2\ \mathbf{a}_3\ \cdots\ \mathbf{a}_N| \quad \text{where} \quad \mathbf{a}_k^T = |a_{1k}\ \cdots\ a_{Mk}|^T
\tag{18.1.2}
$$

Table 18.1	Some basic definitions of matrix theory; the symbol $\forall$ is read "for all" and $\emptyset$ denotes the null vector	

Operation	Definition	Notes
Addition	$A + B = \lvert a_{ij} + b_{ij} \rvert$	$A = \lvert a_{ij} \rvert,\ B = \lvert b_{ij} \rvert$
Adjoint	$A^* = \lvert \bar{a}_{ji} \rvert$	$(AB)^* = B^* A^*$
Diagonal	$D = \lvert d_i \delta_{ij} \rvert$	$D = \mathrm{diag}\,[d_1, \ldots, d_N]$
Exponential of a matrix	$\exp A = \sum_{n=0}^{\infty} (n!)^{-1} A^n$	if series converges ($\S 18.11$)
Inverse	$A^{-1} A = A A^{-1} = I$	$(AB)^{-1} = B^{-1} A^{-1}$
Hermitian	$A^* = A$	$a_{ij} = \bar{a}_{ji}$
Idempotent	$A^2 \equiv A A = A$	
Multiplication by number	$\alpha A = \lvert \alpha\, a_{ij} \rvert$	
Matrix multiplication	$AB = \lvert \sum_{k=1}^{N} a_{ik} b_{kj} \rvert$	no. of columns of A = no. of rows of B
Nilpotent with index m	$A^m = 0,\ A^{m-1} \neq 0$	m integer
Non-negative definite	$\mathbf{u}^* A \mathbf{u}$ real $\geq 0\ \forall\, \mathbf{u} \neq 0$	$(\mathbf{u}, A\mathbf{u}) \geq 0\ \forall\, \mathbf{u} \neq 0$
Normal	$A A^* = A^* A$	
Orthogonal	$O^T O = O O^T = I$	$O^T = O^{-1}$
Positive definite	$\mathbf{u}^* A \mathbf{u}$ real $> 0\ \forall\, \mathbf{u} \neq \emptyset$	$(\mathbf{u}, A\mathbf{u}) > 0\ \forall\, \mathbf{u} \neq \emptyset$
Similar	$B = T^{-1} A T$	T non-singular
Orthogonally similar	$B = O^T A O$	$O^T = O^{-1}$
Unitarily similar	$B = U^* A U$	$U^* = U^{-1}$
Singular	$\det A = 0$	A square
Skew-Hermitian	$A^* = -A$	$\bar{a}_{ij} = -a_{ji}$ a_{kk} pure imaginary
Skew-symmetric	$A^T = -A$	$a_{ij} = -a_{ji},\ a_{kk} = 0$
Symmetric	$A^T = A$	$a_{ij} = a_{ji}$
Trace	$\mathrm{Tr}\,A = \sum_{k=1}^{N} a_{kk}$	A square
Transpose	$A^T = \lvert a_{ji} \rvert$	$(AB)^T = B^T A^T$
Unitary	$U^* U = U U^* = I$	$U^* = U^{-1}$

in which the superscript T denotes the transpose. The *transpose* of an $M \times N$ matrix A, denoted by A^T, is the $N \times M$ matrix obtained by interchanging the rows and columns of A:

$$
A^T = \lvert a_{ji} \rvert =
\begin{vmatrix}
a_{11} & a_{21} & \cdots & a_{M1} \\
a_{12} & a_{22} & \cdots & a_{M2} \\
a_{13} & a_{23} & \cdots & a_{M3} \\
\vdots & \vdots & \vdots & \vdots \\
a_{1N} & a_{2N} & \cdots & a_{MN}
\end{vmatrix}
\tag{18.1.3}
$$

and, therefore, row and column vectors are the transpose of each other.

The matrix obtained by taking the complex conjugate of all the entries of the transpose is the *adjoint* of A, denoted by A^*:[1]

$$A^* = |\bar{a}_{ji}|. \tag{18.1.4}$$

If $A^T = A$ so that $a_{ji} = a_{ij}$ the matrix is *symmetric* while if $A^* = A$, so that $\bar{a}_{ij} = a_{ji}$, it is *Hermitian*. Given a matrix A, the definitions

$$H_1 = \frac{1}{2}(A + A^*), \qquad H_2 = \frac{1}{2i}(A - A^*) \tag{18.1.5}$$

permit one to write

$$A = H_1 + iH_2 \tag{18.1.6}$$

which is similar to the decomposition of a complex number into real and imaginary parts. It is evident that both H_1 and H_2 are Hermitian matrices.

The product of a matrix A by a number α and the sum of two $M \times N$ matrices A and B are defined, respectively, by

$$\alpha A = |\alpha\, a_{ij}|, \qquad A + B = |a_{ij} + b_{ij}|. \tag{18.1.7}$$

The product $C = AB$ between two matrices is defined only if the two matrices are *conformable*, i.e. the number of columns of the first factor A equals the number of rows of the second one B. The result of the multiplication of an $M \times L$ matrix A by an $L \times N$ matrix B is an $M \times N$ matrix C irrespective of the value of L; the entries c_{ik} of C are given by the familiar "row-by-column" rule

$$c_{ik} = \sum_{j=1}^{L} a_{ij}b_{jk}. \tag{18.1.8}$$

With this definition, the product Au of an $M \times N$ matrix A by an N-vector $u^T = |u_1\ u_2\ \ldots\ u_N|$ is an M-vector consisting of a linear combination of the columns of A:

$$Au = u_1 a_1 + u_2 a_2 + \cdots + u_N a_N. \tag{18.1.9}$$

Thus, the k-th column of $C = AB$ is just the matrix A applied to (i.e., multiplied by) the k-th column of B:

$$c_k = A\, b_k. \tag{18.1.10}$$

The multiplication rule also holds if a_{ij}, b_{jk} are formally replaced by conformable submatrices A_{ij} and B_{jk} obtained by partitioning the matrices A and B; in this case the matrix C consists of the submatrices $C_{ik} = \sum_j A_{ij}B_{jk}$. Even when AB and BA are both defined, they are not necessarily equal so that the matrix product is non-commutative in general. If $AB = BA$ the two matrices *commute*.

It is readily verified that, when the products are defined,

$$(AB)^T = B^T A^T, \qquad (AB)^* = B^* A^*. \tag{18.1.11}$$

---

[1] Also referred to as *conjugate transpose, Hermitian conjugate* or *Hermitian adjoint*. Sometimes the transpose matrix of cofactors in (18.2.16) is also referred to as the adjoint, but we prefer to use the term *adjugate* for this matrix to avoid confusion.

A matrix which commutes with its adjoint

$$AA^* = A^*A \tag{18.1.12}$$

is said to be *normal*. A matrix decomposed as in (18.1.6) is normal if and only if H_1 and H_2 commute.

The set of all vectors which can be written in the form Au defines the *range* $\mathscr{R}(A)$, or *column space*, of the matrix A. When the vector multiplies from the left, $u^T A$, the result is a linear combination of the rows of A, the *row space* of A.

When $M = N$ the matrix is *square of order* N. In this case we can define powers of A in an obvious way, e.g. $A^2 = AA$ and so on. A square matrix such that $A^2 = A$ is said to be *idempotent*. A square matrix A is *positive definite*, or *non-negative definite*, when, respectively,

$$u^*Au > 0 \quad \text{or} \quad u^*Au \geq 0 \quad \text{for all} \quad u \neq \emptyset. \tag{18.1.13}$$

in which $\emptyset$ denotes the null vector.

A matrix all the entries of which below or above the main diagonal vanish is called *upper* or *lower triangular*, respectively. A matrix $D = |d_i \delta_{ij}|$ with non-zero entries only on the main diagonal is called a *diagonal matrix* and is written as $D = \text{diag}[d_1, d_2, \ldots, d_N]$. The diagonal matrix $I = \text{diag}[1, 1, \ldots, 1]$ is the $N \times N$ *identity matrix*. This is the unique matrix such that $AI = IA = A$ for all $N \times N$ matrices A.

Matrices B and C such that

$$BA = I, \quad AC = I \tag{18.1.14}$$

(with the two identity matrices having different dimensionality in general) are the left- and right-inverses of A. A non-square matrix may have a left or right inverse, but not both.[2] For a square matrix A, however, a simultaneous left and right inverse square matrix A^{-1} may exist such that

$$A^{-1}A = I, \quad AA^{-1} = I; \tag{18.1.15}$$

A^{-1} is called simply the *inverse* of A and is unique.[3] When such an inverse exists, A is said to be non-singular, and *singular* otherwise. Equation (18.2.16) below gives an explicit expression for A^{-1}. It can be verified that

$$(AB)^{-1} = B^{-1}A^{-1}. \tag{18.1.16}$$

A square matrix O is *orthogonal* if its transpose equals its inverse:

$$O^T O = I = OO^T \quad \text{i.e.} \quad \sum_{k=1}^{N} o_{ki}o_{kj} = \delta_{ij}. \tag{18.1.17}$$

---

[2] The notion of inverse can be generalized so that any matrix possesses a generalized inverse, see e.g. Golub & van Loan 1989 or Piziak & Odell 2007.

[3] The explicit expression (18.2.16) shows that, if $A^{-1}A = I$, then also $AA^{-1} = I$ and vice versa so that, when it exists, A^{-1} is both a left and a right inverse. The same conclusion can also be reached directly by using (18.1.14): $B = BI = B(AC) = (BA)C = C$. A similar argument proves uniqueness. See §21.5 p. 647 for additional properties of the inverse of a linear operator.

When a similar relation holds for the adjoint

$$U^*U = I = UU^* \qquad \text{i.e.} \qquad \sum_{k=1}^{N} \bar{u}_{ki} u_{kj} = \delta_{ij} \tag{18.1.18}$$

the matrix is *unitary*. Either side of these relations follows from the other one by using the identity of the right and left inverses or the explicit representation of the inverse given in (18.2.16) below. It is evident that both orthogonal and unitary matrices are normal.

Two matrices A and B are *similar* if there exists a non-singular matrix T such that

$$B = T^{-1}AT \tag{18.1.19}$$

from which it also follows that $A = S^{-1}BS$ with $S = T^{-1}$. It is easily established that the similarity property is transitive; furthermore

$$(T^{-1}AT)^m = (T^{-1}AT) \cdots (T^{-1}AT) = T^{-1}A^m T. \tag{18.1.20}$$

When the matrix T intervening in the similarity relation (18.1.19) is orthogonal, or unitary, the two matrices are *orthogonally*, or *unitarily*, similar (or equivalent) respectively.

The *outer* or *tensor product* of two matrices A and B is the matrix obtained by replacing each entry a_{ij} of A by the block sub-matrix $a_{ij}B$. Clearly, this product is defined with no restriction on the number of rows and columns. In particular, the outer product of two vectors **a** and $\mathbf{b}^T$ of dimension M and N respectively, is the $M \times N$ matrix

$$\mathbf{ab}^T = \begin{vmatrix} a_1 b_1 & a_1 b_2 & \dots & a_1 b_N \\ a_2 b_1 & a_2 b_2 & \dots & a_2 b_N \\ \vdots & \vdots & \vdots & \vdots \\ a_M b_1 & a_M b_2 & \dots & a_M b_N \end{vmatrix}. \tag{18.1.21}$$

The *trace* of a matrix A is the sum of the diagonal terms:

$$\text{Tr}\,A = \sum_{k=1}^{N} a_{kk}. \tag{18.1.22}$$

If the entries of $A = |a_{ij}(t)|$ are differentiable functions of a parameter t, the derivative of A with respect to t is defined as the matrix having as entries the derivatives of the entries of A:

$$\frac{d}{dt}A = \left| \frac{da_{ij}}{dt} \right|. \tag{18.1.23}$$

Similarly, the integral of a matrix is defined as the matrix each entry of which is the integral of the corresponding entry of the integrand matrix:

$$\int A(t)\,dt = \left| \int a_{ij}(t)\,dt \right|. \tag{18.1.24}$$

Table 18.2	Some properties of determinants.
$\det A = (-1)^{k+1} a_{k1} M_{k1} + \cdots + (-1)^{k+N} a_{kN} M_{kN}$	expansion by row; M_{ij} minor
$\det A = (-1)^{j+1} a_{1j} M_{1j} + \cdots + (-1)^{j+N} a_{Nk} M_{Nj}$	expansion by column
$\det\lvert a_1 + b_1\, a_2\, \ldots\, a_N \rvert = \det\lvert a_1\, a_2\, \cdots\, a_N \rvert$	
$\qquad\qquad\qquad\qquad + \det\lvert b_1\, a_2\, \cdots\, a_N \rvert$	
$\det\lvert \alpha a_1\, \ldots\, a_N \rvert = \alpha\, \det\lvert a_1\, \cdots\, a_N \rvert$	
$\det\lvert \alpha A \rvert = \alpha^N \det A$	
$\det\lvert a_1\, \cdots\, b\, \cdots\, b\, \ldots\, a_N \rvert = 0$	also if two rows are equal
$\det A^T = \det A$	
$\det(AB) = (\det A)\,(\det B)$	
$\det(A^{-1}) = (\det A)^{-1}$	
$\det B = \det\left(T^{-1} A T\right) = \det A$	$B = T^{-1} A T$
$\det\lvert \mathrm{diag}[A_1, \ldots, A_k] \rvert = (\det A_1) \cdots (\det A_k)$	A block-diagonal

18.2 Determinants

Given a *square* $N \times N$ matrix A, consider all possible products of its entries of the form $a_{1j_1} a_{2j_2} \cdots a_{Nj_N}$ where $j_1, j_2, \ldots, j_N$ is a permutation p of $(1, 2, \ldots, N)$. Thus each such product contains one and only one entry from each row and one and only one entry from each column of A. The *determinant* of A is defined as

$$\det A = \sum_p (-1)^p a_{1j_1} a_{2j_2} \cdots a_{Nj_N}, \tag{18.2.1}$$

where the summation is extended to all the $N!$ permutations of 1, 2, ..., N and $(-1)^p$ is to be taken as $+1$ for an even permutation and -1 for an odd permutation.[4] Since every term of the summation in (18.2.1) contains an entry of the k-th column, for any k we can group the terms as

$$\det A = a_{1k} A_{1k} + a_{2k} A_{2k} + \cdots + a_{Nk} A_{Nk}. \tag{18.2.2}$$

The quantity A_{jk} appearing in this relation is the *cofactor* of a_{jk} and this formula gives the *expansion by column* of det A. A similar formula can be written down for the *expansion by row* of det A.

The *minor* M_{kl} of A is the determinant of the $(N-1) \times (N-1)$ matrix obtained by striking out the k-th row and l-th column of A, and it can be shown that

$$A_{kl} = (-1)^{k+l} M_{kl}, \tag{18.2.3}$$

---

[4] The determinant is a special case of a multilinear alternating form on the N-dimensional vector space; "multilinear" means that the form is linear in each one of its arguments, and "alternating" means that it changes sign upon an odd permutation of its arguments. The determinant is an N-form; one can similarly define k-forms with $k < N$ (see e.g. Halmos 1958 p. 52).

valid for both expansions by row and by column.

Some very well known properties of determinants (summarized in Table 18.2) are the following:

1. det A is an additive and homogeneous function of each row and each column, i.e.

$$\det|\mathbf{a}_1 \cdots \mathbf{a}_{k-1} (\alpha\mathbf{a}_k + \beta\mathbf{b}_k) \mathbf{a}_{k+1} \cdots \mathbf{a}_N| = \alpha \det|\mathbf{a}_1 \cdots \mathbf{a}_{k-1} \mathbf{a}_k \mathbf{a}_{k+1} \cdots \mathbf{a}_N|$$
$$+ \beta \det|\mathbf{a}_1 \cdots \mathbf{a}_{k-1} \mathbf{b}_k \mathbf{a}_{k+1} \cdots \mathbf{a}_N|. \tag{18.2.4}$$

2. The determinant of the transpose equals the determinant of the original matrix

$$\det A^T = \det A \tag{18.2.5}$$

so that the determinant of the adjoint is the complex conjugate of the original determinant.

3. The determinant changes sign upon interchanging two rows or columns. As a consequence, the determinant vanishes if two rows or two columns are equal and, by (18.2.4), also if they are proportional to each other. By the same argument, the determinant of a matrix the rows (or columns) of which are arbitrary linear combinations of those of another matrix equals that of the original matrix.

4. If $i \neq j$,

$$a_{j1}A_{i1} + a_{j2}A_{i2} + \cdots + a_{jN}A_{iN} = 0 \tag{18.2.6}$$
$$a_{1j}A_{1i} + a_{2j}A_{2i} + \cdots + a_{Nj}A_{Ni} = 0. \tag{18.2.7}$$

The first result is evident once it is observed that, by (18.2.2), it equals the determinant of a matrix whose i-th row has been replaced by the j-th row: with two equal rows, the determinant vanishes. The same argument applied to columns gives the second result.

5. The determinant of the product of two matrices equals the product of the determinants:[5]

$$\det(AB) = (\det A)(\det B). \tag{18.2.8}$$

As a consequence, since the determinant of the identity matrix is evidently equal to 1, we find that

$$\det(A^{-1}) = (\det A)^{-1}. \tag{18.2.9}$$

From these relations it also immediately follows that the determinants of two similar matrices are equal.

$$\det B = \det\left(T^{-1}AT\right) = \det A. \tag{18.2.10}$$

6. If the entries of a square matrix $A = |\mathbf{a}_1(t) \mathbf{a}_2(t) \cdots \mathbf{a}_N(t)|$ are differentiable functions of a parameter t, then

$$\frac{d}{dt} \det A = \det\left|\frac{d\mathbf{a}_1}{dt} \mathbf{a}_2 \cdots \mathbf{a}_N\right| + \cdots + \det\left|\mathbf{a}_1 \mathbf{a}_2 \cdots \frac{d\mathbf{a}_N}{dt}\right| \tag{18.2.11}$$

with a corresponding formula for differentiation by row.

---

[5] This is a special case of the Binet–Cauchy formula which relates the determinant of AB to products of all possible minors of order m of A with the corresponding ones of B.

7. *Schur's formulae.* Let a matrix of order $(2N) \times (2N)$ be partitioned into four $N \times N$ matrices:

$$M = \begin{vmatrix} A & B \\ C & D \end{vmatrix}. \qquad (18.2.12)$$

Then

$$\det M = \det \left| A(D - CA^{-1}B) \right| = \det \left| (A - BD^{-1}C)D \right| \qquad (18.2.13)$$

provided A or D be non-singular; $D - CA^{-1}B$ and $A - BD^{-1}C$ are, respectively, the *Schur complements* of A and D in M. Similar relations obtained by simultaneously replacing A and D by B and C also hold (see e.g. Gantmacher 1959 p. 46; Piziak & Odell 2007 p. 25).

8. If either B or C are zero in the previous formulae, $\det M = \det(AD) = (\det A)(\det D)$. Repeated application of this result shows that the determinant of a block-upper-triangular matrix is the product of the determinants of the individual diagonal blocks:

$$\det \begin{vmatrix} A_{11} & A_{12} & A_{13} & \cdots & A_{1N} \\ 0 & A_{22} & A_{23} & \cdots & A_{2N} \\ & & \vdots & & \\ 0 & 0 & 0 & 0 & A_{NN} \end{vmatrix} = (\det A_{11})(\det A_{22}) \cdots (\det A_{NN}), \quad (18.2.14)$$

and similarly for a block-lower-triangular matrix. If all the off-diagonal blocks are zero, the matrix is *block-diagonal*.

If we pre-multiply A by the transpose of the matrix of cofactors of its entries

$$\begin{vmatrix} A_{11} & A_{21} & \cdots & A_{N1} \\ A_{12} & A_{22} & \cdots & A_{N2} \\ \vdots & \vdots & \vdots & \vdots \\ A_{1N} & A_{2N} & \cdots & A_{NN} \end{vmatrix} \begin{vmatrix} a_{11} & a_{12} & \cdots & a_{1N} \\ a_{21} & a_{22} & \cdots & a_{2N} \\ \vdots & \vdots & \vdots & \vdots \\ a_{N1} & a_{N2} & \cdots & a_{NN} \end{vmatrix} \qquad (18.2.15)$$

because of the expansion rule (18.2.2) and the property (18.2.6) we obtain the identity matrix multiplied by det A. The same result is found by inverting the order of the multiplication. Thus the inverse of A is given by

$$A^{-1} = \frac{1}{\det A} \begin{vmatrix} A_{11} & A_{21} & \cdots & A_{N1} \\ A_{12} & A_{22} & \cdots & A_{N2} \\ \cdots & \cdots & \cdots & \cdots \\ A_{1N} & A_{2N} & \cdots & A_{NN} \end{vmatrix} \qquad (18.2.16)$$

provided $\det A \neq 0$. For example, the inverse of a 2×2 matrix

$$A = \begin{vmatrix} a & b \\ c & d \end{vmatrix} \quad \text{is} \quad A^{-1} = \frac{1}{ad - bc} \begin{vmatrix} d & -b \\ -c & a \end{vmatrix}. \qquad (18.2.17)$$

When $\det A = 0$, the inverse does not exist and the matrix is *singular*. From the explicit representation of the inverse one can verify that the transpose of the inverse equals the inverse of the transpose. The transpose of the matrix of co-factors in (18.2.16) is the *adjugate* of A (cf. footnote p. 493).

If we multiply from the left the linear algebraic system $A\mathbf{u} = \mathbf{b}$ by the adjugate of A, adj(A), we find

$$\mathbf{u} = \frac{1}{\det A}\,\mathrm{adj}(A)\,\mathbf{b}. \tag{18.2.18}$$

It can be shown that the j-th entry of the vector adj(A) $\mathbf{b}$ in the right-hand side equals the determinant of A_j, the matrix constructed by replacing the j-th column of A by $\mathbf{b}$. Thus we have *Cramer's rule* for the solution of a linear algebraic system:

$$u_j = \frac{\det A_j}{\det A}. \tag{18.2.19}$$

If $\mathbf{b} = 0$ and $\det A \neq 0$, the system $A\mathbf{u} = \emptyset$ evidently only has the $\emptyset$ solution, and conversely. Thus the homogeneous system $A\mathbf{u} = \emptyset$ can have a non-zero solution if and only if $\det A = 0$.

We have defined earlier what are often just called minors without further qualification. Actually, the definition given is that of the minors of order $N - 1$. More generally, a minor of order m is the determinant of the matrix obtained by striking out all but m rows and m columns of a matrix. Of particular importance are the *principal minors*, which are the determinants of the $m \times m$ square submatrices along the principal diagonal of A. Thus, the principal minors of order 1 are simply the entries along the diagonal while, for an $N \times N$ matrix, the minor of order N is the determinant $\det A$. Clearly, principal minors also exist for rectangular matrices.

The *determinant rank* of a matrix is the order of the largest non-vanishing principal minor. One also defines the *column rank* and the *row rank* as the maximum number of columns or rows that are linearly independent in the sense specified in connection with (18.3.2) below. The column, row and determinant ranks are all equal to each other (see e.g. Ortega 1987, p. 27; Piziak & Odell 2007 p. 100). According to these definitions, the rank of a non-singular matrix is N. The rank of the product of two matrices does not exceed that of either factor (see e.g. Gantmacher 1959, p 12; Ortega 1987, p. 64).

18.3 Matrices and linear operators

For the purposes of this book, a primary reason for our interest in matrices is that they are relatively simple examples of linear operators, their simplicity deriving from the fact that they operate on finite-dimensional spaces. A general theory of linear vector spaces is presented in §19.1 and a very abbreviated treatment is sufficient here.

An abstract linear vector space $\mathbb{S}$ over the field of real or complex numbers $\mathbb{R}$ or $\mathbb{C}$ is a set of entities $\mathbf{u}$, $\mathbf{v}$, ..., called vectors, or elements of the space, which can be combined in such a way that

$$\mathbf{w} = a\mathbf{u} + b\mathbf{v} \in \mathbb{S} \quad \text{for all} \quad \mathbf{u},\,\mathbf{v} \in \mathbb{S} \quad \text{and for all} \quad a,\,b \in \mathbb{C}. \tag{18.3.1}$$

In general we refer to the real or complex numbers a and b appearing here as *scalars*. The operations of sum of two vectors and multiplication by a scalar appearing in this definition enjoy the usual properties (see Tables 19.1 and 19.2 on p. 538). A simple example of an N-dimensional linear vector space is the set of all polynomials $p_N(x)$ of degree N, with the operations (18.3.1) defined in the usual way. The set of all $M \times N$ matrices also constitutes a linear vector space with the previous definitions (18.1.7) of addition and multiplication by scalars, see §19.1.

Vectors $\mathbf{u}_1$, $\mathbf{u}_2$, ..., $\mathbf{u}_K$ are *linearly independent* when the relation

$$\sum_{j=1}^{K} a_j \mathbf{u}_j = \emptyset \tag{18.3.2}$$

(in which the right-hand side is the zero vector) can only be satisfied if all the a_j are zero. The space is *N-dimensional* when no set of linearly independent vectors contains more than N elements.

Let $\mathbf{x}_1$, $\mathbf{x}_2$, ..., $\mathbf{x}_N$ be a set of N linearly independent vectors in an N-dimensional space. The addition of any other vector $\mathbf{u}$ will make the set linearly dependent so that a relation

$$\mathbf{u} = \sum_{i=1}^{N} u_i \mathbf{x}_i \tag{18.3.3}$$

must exist, with N (in general complex) numbers u_i, not all of which can vanish. This representation is unique due to the assumed linear independence of the $\mathbf{x}_i$. Any set $\mathbf{x}_1$, $\mathbf{x}_2$, ..., $\mathbf{x}_N$ which permits one to represent an arbitrary element of the space in this way is a *basis*, or base, for the N-dimensional space $\mathbb{S}^N$. In an N-dimensional space every basis contains precisely N linearly independent vectors and any set of linearly independent vectors $\mathbf{x}_1$, ..., $\mathbf{x}_k$, with $k < N$, can be augmented by vectors $\mathbf{x}_{k+1}$, ..., $\mathbf{x}_N$ so that the combination of the two sets forms a basis.[6]

The u_i in (18.3.3) are the *coordinates* of $\mathbf{u}$ in the basis $\mathbf{x}_1$, $\mathbf{x}_2$, ..., $\mathbf{x}_N$ and the representation (18.3.3) establishes a one-to-one correspondence (i.e., an isomorphism) between any N-dimensional linear vector space $\mathbb{S}^N$ and the space $\mathbb{C}^N$ (or $\mathbb{R}^N$) of N-dimensional sets of complex (or real) numbers. Thus, for example, each polynomial of degree $\leq N$ is uniquely specified by its coefficients. In this sense, all N-dimensional vector spaces are essentially equal and can be represented by the space of $N \times 1$ matrices (i.e., vectors) consisting of the coordinates of an element of the vector space with respect to a specific basis. However, committing oneself to a specific basis may obscure basis-independent properties that hold abstractly and which are often of interest in themselves.

The set of all possible linear combinations of K given vectors $\{\mathbf{u}_1, \mathbf{u}_2, ..., \mathbf{u}_K\} \in \mathbb{S}^N$, with $K \leq N$, itself a linear vector space, is a *subspace* of $\mathbb{S}^N$ *spanned*, or generated, by the K vectors. According to this definition, $\mathbb{S}^N$ and the $\emptyset$ element are also trivial subspaces of dimensions N and 0 respectively; any other subspace is called *proper*. It is evident from

---

[6] In general, the vectors of a basis need not be orthogonal or of unit length, concepts which, as a matter of fact, have not yet been introduced.

(18.3.1) that two subspaces must have at least the $\emptyset$ vector in common. When two subspaces $\mathbb{S}_1$ and $\mathbb{S}_2$ have *only* $\emptyset$ in common and, furthermore, every $\mathbf{u} \in \mathbb{S}$ can be represented as[7]

$$\mathbf{u} = \mathbf{u}_1 + \mathbf{u}_2, \qquad \mathbf{u}_j \in \mathbb{S}_j, \tag{18.3.4}$$

we say that $\mathbb{S}$ is *decomposed* into the sum of the two subspaces $\mathbb{S}_1$ and $\mathbb{S}_2$, or is the *direct sum* of $\mathbb{S}_1$ and $\mathbb{S}_2$, and we write[8]

$$\mathbb{S} = \mathbb{S}_1 \oplus \mathbb{S}_2. \tag{18.3.5}$$

Each one of $\mathbb{S}_1$ and $\mathbb{S}_2$ is the *algebraic complement* of the other. Any subspace has an infinity of algebraic complements. For example, in the plane, $\mathbb{S}_1$ could be the x-axis and $\mathbb{S}_2$ any other straight line intersecting it at the origin. All the algebraic complements of $\mathbb{S}_1$ have the same dimension, which is called the *co-dimension* of $\mathbb{S}_1$. The notion of direct sum is generalized to more than two subspaces in a straightforward way.

Let $\{u_j\}$ represent the vector $\mathbf{u}$ on a basis $\{\mathbf{x}_j\}$ of $\mathbb{S}^N$ as in (18.3.3), and let $\{\mathbf{y}_i\}$ be a basis in another space $\mathbb{S}^M$. An $M \times N$ matrix $\mathsf{A} = |a_{ij}|$ establishes a linear correspondence between the vector $\mathbf{u} \in \mathbb{S}^N$ and a vector $\mathbf{v} \in \mathbb{S}^M$ having coordinates $\{v_i\}$ in the basis $\{\mathbf{y}_i\}$ according to the rule

$$v_i = \sum_{j=1}^{N} a_{ij} u_j. \tag{18.3.6}$$

The correspondence between vectors of $\mathbb{S}^N$ and vectors of the image space $\mathbb{S}^M$ established by (18.3.6) defines a linear operator A the action of which may be indicated as

$$\mathbf{v} = \mathsf{A}\mathbf{u}. \tag{18.3.7}$$

We use the same symbol for the operator and the matrix that represents it in the given bases although these are of course different mathematical entities. In particular, (18.3.6) presupposes the choice of specific bases, while (18.3.7) denotes the correspondence in an abstract, basis-independent sense. The rules of matrix addition and multiplication induce corresponding rules for the addition and multiplication of linear operators.[9]

The set of vectors $\mathsf{A}\mathbf{u} \in \mathbb{S}^M$ obtained as the vector $\mathbf{u} \in \mathbb{S}^N$ ranges over the entire space $\mathbb{S}^N$ is the *range* $\mathscr{R}(\mathsf{A})$ of the operator A and is evidently the subspace spanned by the vectors $\mathsf{A}\mathbf{x}_1, \mathsf{A}\mathbf{x}_2, \ldots, \mathsf{A}\mathbf{x}_N$ (not all necessarily linearly independent) where $\{\mathbf{x}_j\}$ is any basis of the space $\mathbb{S}^N$. This definition clearly coincides with that given on p. 494 in terms of the columns of A. The set of vectors $\mathbf{u}$ such that $\mathsf{A}\mathbf{u} = \emptyset$ is the *null space* or *kernel* of A, denoted by $\mathscr{N}(\mathsf{A})$. The maximum number of linearly independent vectors in $\mathscr{N}(\mathsf{A})$ is the *nullity* of A. The range and the null space of an operator are independent of the particular basis chosen to represent it as will be clear from the following developments.

---

[7] Uniqueness of this representation follows from the assumption that the two subspaces have only $\emptyset$ in common.

[8] For a more detailed explanation of this concept see p. 541.

[9] The converse statements are also true: any linear operator over a finite-dimensional space can be represented by a matrix and the rules governing the addition and multiplication by scalars for operators correspond to those governing the same operations for matrices. In other words, there is a one-to-one correspondence between linear operators and matrices.

Let $\{\mathbf{x}_1, \ldots, \mathbf{x}_N\}$ and $\{\mathbf{y}_1, \ldots, \mathbf{y}_M\}$ be the chosen bases in $\mathbb{S}^N$ and $\mathbb{S}^M$ respectively. Then we can rewrite (18.3.7) as

$$\mathbf{v} = \sum_{i=1}^{M} v_i \mathbf{y}_i = \mathsf{A}\left(\sum_{j=1}^{N} u_j \mathbf{x}_j\right) = \sum_{j=1}^{N} u_j (\mathsf{A}\mathbf{x}_j) \qquad (18.3.8)$$

but also, using (18.3.6),

$$\mathbf{v} = \sum_{i=1}^{M}\left(\sum_{j=1}^{N} a_{ij} u_j\right) \mathbf{y}_i = \sum_{j=1}^{N} u_j \sum_{i=1}^{M} \mathbf{y}_i a_{ij}. \qquad (18.3.9)$$

From these two results, due to the uniqueness of the u_j and the arbitrariness of $\mathbf{u}$, we conclude that

$$\mathsf{A}\mathbf{x}_j = \sum_{i=1}^{M} \mathbf{y}_i a_{ij}. \qquad (18.3.10)$$

It should be noted that the summation involves the *first* index of the matrix entries. Upon comparing with (18.3.3) we see that a_{ij} is the i-th coordinate of the transformed basis vector $\mathsf{A}\mathbf{x}_j$ with respect to the basis in $\mathbb{S}^M$. In other words, the j-th column of the matrix A contains the coordinates of the transformed basis vector $\mathsf{A}\mathbf{x}_j$. It thus follows that the dimension of the range of an operator is equal to the column rank r (p. 499) of the matrix that represents it. For vectors $\mathbf{u}$ in the null space of A, (18.3.8) says that $\sum_{j=1}^{N} u_j (\mathsf{A}\mathbf{x}_j) = \emptyset$ and, since $\dim \mathscr{R}(\mathsf{A}) = r$, $N - r$ of the parameters $\{u_j\}$ are free or, in other words, $\dim \mathscr{N}(\mathsf{A}) = N - r$; thus we have the *rank plus nullity theorem*:

$$\dim \mathscr{R}(\mathsf{A}) + \dim \mathscr{N}(\mathsf{A}) = N. \qquad (18.3.11)$$

Note that this is a relation between the dimension of a set in the image space $\mathbb{S}^M$ and that of a set in the original space $\mathbb{S}^N$.

It may be useful to write in an expanded form (18.3.6) and (18.3.10); we have, respectively,

$$v_1 = a_{11}u_1 + a_{12}u_2 + \cdots + a_{1N}u_N, \quad \ldots, \quad v_M = a_{M1}u_1 + a_{M2}u_2 + \cdots + a_{MN}u_N \qquad (18.3.12)$$

and

$$\mathsf{A}\mathbf{x}_1 = a_{11}\mathbf{y}_1 + a_{21}\mathbf{y}_2 + \cdots + a_{M1}\mathbf{y}_M, \quad \ldots, \quad \mathsf{A}\mathbf{x}_N = a_{1N}\mathbf{y}_1 + a_{2N}\mathbf{y}_2 + \cdots + a_{MN}\mathbf{y}_M. \qquad (18.3.13)$$

Observe that the matrix that gives the image of the original basis vectors after the transformation is the *transpose* of that which transforms $\mathbf{u}$ into $\mathbf{v}$. This is a necessary consequence of requiring that the result BA of two successive transformations A and B be represented by the matrix formed by taking the row-by-column product BA of the two matrices B and A in the proper order.

Because of the linearity of the operator, the range and the null space are both subspaces, the former in the image space $\mathbb{S}^M$, the latter in the original space $\mathbb{S}^N$. When $N = M$, $\mathbb{S}^M$ and $\mathbb{S}^N$ may be taken to coincide, but $\mathscr{R}(\mathsf{A})$ and $\mathscr{N}(\mathsf{A})$ are not necessarily disjoint sets.

For example, a non-zero vector $\mathbf{u}$ may exist for which $A\mathbf{u} \neq \emptyset$, while $A^2\mathbf{u} = \emptyset$ so that $A\mathbf{u}$ belongs both to the range and to the null space of A. An example is a 2×2 matrix with a single non-zero entry a_{12}: the range and the null space both consist of vectors proportional to $|1\ 0|^T$ and, therefore, coincide. The relations among these subspaces are discussed further in connection with Theorem 18.5.2 below.

By definition, the operator A is invertible if, for any given $\mathbf{v} \in \mathbb{S}^M$, there is only one $\mathbf{u}$ such that $A\mathbf{u} = \mathbf{v}$. In general, from $A\mathbf{u}_1 = \mathbf{v}$ and $A\mathbf{u}_2 = \mathbf{v}$, we can deduce that $A(\mathbf{u}_1 - \mathbf{u}_2) = \emptyset$. For uniqueness it is necessary that $\mathbf{u}_1 = \mathbf{u}_2$, which is a legitimate conclusion only if the null space reduces to the zero vector.

Since there cannot be more than N linearly independent vectors in a space of dimension N, for any operator A mapping $\mathbb{S}^N$ into itself and any non-zero vector $\mathbf{u}$, the vectors $A^j\mathbf{u}$ with $j \geq N$ must be combinations linear combinations of $\mathbf{u}$, $A\mathbf{u}$, $A^2\mathbf{u}$, ..., $A^{N-1}\mathbf{u}$. The trivial example of the identity matrix shows that $A^j\mathbf{u}$ can be a linear combination of the preceding vectors also for j less than N. When, for some vector $\mathbf{u}$ and some integer $l < N - 1$, $A^{l+1}\mathbf{u}$ can be written as a linear combination of $\mathbf{u}$, $A\mathbf{u}$, ..., $A^l\mathbf{u}$, the subspace spanned by these vectors is left unchanged by the action of A and is called a *cyclic subspace*. This is an example of an *invariant subspace*, i.e. a subspace such that A applied to any of its elements produces a vector in the same subspace. The null space of A is another example of an invariant subspace under A. We can define the *restriction* of A to an invariant subspace as the operator acting only on vectors of that subspace.

18.4 Change of basis

Any set of N linearly independent vectors can be taken as a basis but, in general, the coordinates of the same vector $\mathbf{u}$ in the two different bases $\{\mathbf{x}_j\}$ and $\{\mathbf{x}'_j\}$ will be different. We wish to establish the relation between the old and new coordinates and the matrix representations of a linear operator A.

As suggested by (18.3.10) let us write

$$\mathbf{x}'_j = \sum_{i=1}^{N} \mathbf{x}_i t_{ij}. \tag{18.4.1}$$

As before, this relation shows that the j-th column of the matrix T contains the coordinates of the transformed basis vector $\mathbf{x}'_j$ with respect to the original basis. Thus, for the transformation to be invertible, T must be non-singular. If we construct two matrices the columns of which are the old and new basis vectors

$$X = |\mathbf{x}_1 \ \cdots \ \mathbf{x}_N|, \qquad X' = |\mathbf{x}'_1 \ \cdots \ \mathbf{x}'_N|, \tag{18.4.2}$$

(18.4.1) may be written

$$X' = XT \qquad \text{or} \qquad X = X'T^{-1} \tag{18.4.3}$$

from which

$$\mathbf{x}_j = \sum_{i=1}^{N} \mathbf{x}'_i (t^{-1})_{ij}. \tag{18.4.4}$$

With (18.4.1) we may write

$$\mathbf{u} = \sum_{j=1}^{N} u'_j \mathbf{x}'_j = \sum_{j=1}^{N} u'_j \sum_{i=1}^{N} \mathbf{x}_i t_{ij} = \sum_{i=1}^{N} \left(\sum_{j=1}^{N} t_{ij} u'_j \right) \mathbf{x}_i = \sum_{i=1}^{N} u_i \mathbf{x}_i \tag{18.4.5}$$

from which

$$u_i = \sum_{j=1}^{N} t_{ij} u'_j. \tag{18.4.6}$$

Conversely, the new coordinates of $\mathbf{u}$ are obtained by multiplying the old ones by the matrix T^{-1}.

We may also enquire what, in the new basis, is the matrix representation $\mathsf{A}' = |a'_{ij}|$ of the operator A which effects the transformation $\mathbf{v} = \mathsf{A}\mathbf{u}$ of the space $\mathbb{S}^N$ in itself. Noting that $\mathbf{v} = \sum_{i=1}^{N} v_i \mathbf{x}_i = \sum_{i=1}^{N} v'_i \mathbf{x}'_i$ and using the transformation relations (18.4.1) and (18.4.6) together with (18.3.6) we have

$$
\begin{aligned}
\sum_{i=1}^{N} v'_i \mathbf{x}'_i &= \sum_{i=1}^{N} \left(\sum_{j=1}^{N} a'_{ij} u'_j \right) \sum_{k=1}^{N} \mathbf{x}_k t_{ki} = \sum_{k=1}^{N} \mathbf{x}_k \left(\sum_{i,j=1}^{N} t_{ki} a'_{ij} u'_j \right) \\
&= \sum_{i,k=1}^{N} \mathbf{x}_k a_{ki} u_i = \sum_{k=1}^{N} \mathbf{x}_k \left(\sum_{i,j=1}^{N} a_{ki} t_{ij} u'_j \right),
\end{aligned} \tag{18.4.7}
$$

so that, by the arbitrariness of the u'_j, $\mathsf{T}\mathsf{A}' = \mathsf{A}\mathsf{T}$ or[10]

$$\mathsf{A}' = \mathsf{T}^{-1}\mathsf{A}\mathsf{T} \quad \text{and} \quad \mathsf{A} = \mathsf{T}\mathsf{A}'\mathsf{T}^{-1}. \tag{18.4.8}$$

Thus, A and A' are connected by a similarity transformation. The existence of a relation of this type supports the notion of an abstract operator which takes on different explicit matrix forms, related by similarity transformations, in dependence of the choice of specific bases. For this reason, no great harm is done referring to matrices and operators somewhat interchangeably.

---

[10] The same procedure shows that, if the operator transforms $\mathbf{u} \in \mathbb{S}^N$ into vectors of a different space $\mathbb{S}^M$, one finds $\mathsf{A}' = \mathsf{S}^{-1}\mathsf{A}\mathsf{T}$ where the matrix S plays in $\mathbb{S}^M$ the same role as T in $\mathbb{S}^N$. In this case S is $M \times M$ and T is $N \times N$.

18.5 Scalar product

By associating to each pair of vectors $\mathbf{u}$ and $\mathbf{v}$ a complex number, the *scalar product* $(\mathbf{v}, \mathbf{u})$ of $\mathbf{u}$ and $\mathbf{v}$, we turn the linear space into a scalar product space. The rules governing this mapping between vector pairs and complex numbers are discussed in some detail in §19.3 and summarized in Table 19.3 on p. 549. The scalar product must be defined in such a way as to satisfy the fundamental properties[11]

$$(\mathbf{u}, \mathbf{v}) = \overline{(\mathbf{v}, \mathbf{u})}, \qquad (\mathbf{v}, a\mathbf{u}_1 + b\mathbf{u}_2) = a(\mathbf{v}, \mathbf{u}_1) + b(\mathbf{v}, \mathbf{u}_2). \qquad (18.5.1)$$

As a consequence, $(a\mathbf{v}_1 + b\mathbf{v}_2, \mathbf{u}) = \bar{a}(\mathbf{v}_1, \mathbf{u}) + \bar{b}(\mathbf{v}_2, \mathbf{u})$. Two vectors the scalar product of which vanishes are said to be *orthogonal*. The only vector orthogonal to all the vectors of the space is the $\emptyset$ vector.

As a consequence of the first one of (18.5.1), $(\mathbf{u},\mathbf{u})$ is a real number which is required to be strictly positive except when $\mathbf{u} = \emptyset$, in which case it vanishes. The (Euclidean) *norm* of $\mathbf{u}$ induced by the scalar product (sometimes called the *natural norm* of the scalar-product space) is defined by[12]

$$\|\mathbf{u}\| = \sqrt{(\mathbf{u}, \mathbf{u})}. \qquad (18.5.2)$$

This is a non-negative quantity analogous to the length of an ordinary vector.

A set of vectors each one of which is orthogonal to all the others is an *orthogonal set*. If, in addition, each vector of the set has norm 1, the set is termed *orthonormal*. Thus, the vectors $\{\mathbf{e}_1, \mathbf{e}_2, \ldots, \mathbf{e}_N\}$ of an *orthonormal basis* satisfy the relation

$$(\mathbf{e}_j, \mathbf{e}_k) = \delta_{jk}. \qquad (18.5.3)$$

Orthonormal bases are particularly convenient as, by taking the scalar product of the relation

$$\mathbf{u} = \sum_{i=1}^{N} u_i \mathbf{e}_i \qquad (18.5.4)$$

by the generic basis vector $\mathbf{e}_k$ we have

$$(\mathbf{e}_k, \mathbf{u}) = u_k. \qquad (18.5.5)$$

The Gram–Schmidt orthogonalization procedure explained in §19.4.1 permits the systematic construction of an orthonormal set from an arbitrary set of linearly independent vectors. Thus, in particular, every N-dimensional scalar product space possesses an orthonormal basis and the same is true for its subspaces. For example, an orthonormal basis in an invariant subspace of A of dimension l can be generated by applying the Gram–Schmidt procedure to the l linearly independent vectors $\mathbf{v}_j = \mathsf{A}^j \mathbf{u}$ with $0 \le j \le l - 1$. By adding $N - l$ arbitrary linearly independent other vectors and again applying the procedure, one generates a

---

[11] Some authors use a different convention for the second rule according to which $(a\mathbf{u}_1 + b\mathbf{u}_2, \mathbf{v}) = a(\mathbf{u}_1, \mathbf{v}) + b(\mathbf{u}_2, \mathbf{v})$.

[12] See §19.2 for a general definition of norm and its properties.

new orthonormal basis e'_j in the range of A. This new basis is connected to the original one by a unitary transformation.

An explicit representation for the scalar product can be given in terms of the coordinates of the two vectors with respect to the same orthonormal basis. If $u = \sum_{i=1}^{N} u_i e_i$ and $v = \sum_{j=1}^{N} v_j e_j$, the fundamental property (18.5.3) of an orthonormal basis, together with the properties (18.5.1) of the scalar product, show that

$$(v, u) = \sum_{j=1}^{N} \sum_{k=1}^{N} \overline{v}_j u_k (e_k, e_j) = \sum_{j=1}^{N} \overline{v}_j u_j = v^* u, \qquad (18.5.6)$$

where in the last step we have used the notation of matrix product, v^* being the adjoint matrix of v, i.e., the row vector made up of the complex conjugates of the entries of v. As a consequence

$$\|u\| = \left[\sum_{j=1}^{N} |u_j|^2 \right]^{1/2} = \left[\operatorname{Tr} u u^* \right]^{1/2}. \qquad (18.5.7)$$

The last equality expresses $\|u\|$ as the trace of the matrix formed as the outer product (18.1.21) of u by u^*. Furthermore, if the vectors u and $v = Au$ are referred to the same orthonormal basis, one readily finds (see 18.3.6 and 18.3.7)

$$a_{ij} = (e_i, Ae_j). \qquad (18.5.8)$$

The same proof given on p. 549 can be used in the present case to show that

$$|(u, v)| \leq \|u\| \|v\| \qquad (18.5.9)$$

or, in component form,

$$\left| \sum_{j=1}^{N} \overline{u}_j v_j \right|^2 \leq \left(\sum_{j=1}^{N} |u_j|^2 \right) \left(\sum_{j=1}^{N} |v_j|^2 \right) \qquad (18.5.10)$$

a special case of an important relation known as the *Cauchy–Schwarz* (or Schwarz, or Cauchy–Schwarz–Buniakowsky) *inequality*.[13]

As a consequence of (18.3.6) we have the chain of relations

$$(w, Au) = \sum_{i=1}^{N} \overline{w}_i \sum_{j=1}^{N} a_{ij} u_j = \sum_{i,j=1}^{N} \overline{a_{ij} w_i}\, u_j = \sum_{j=1}^{N} \left(\sum_{i=1}^{N} \overline{a_{ji}^* w_i} \right) u_j = (A^* w, u) \qquad (18.5.11)$$

which defines the adjoint A^* of the operator A in terms of the adjoint of the matrix that represents it on given bases.

---

[13] The author referred to here is Hermann Amandus Schwarz (1843–1921) who proved the analogue of this inequality for integrals independently of Bunyakovsky who had found it earlier. The form for sums shown here was discovered by Cauchy. H. A. Schwarz should not be confused with Laurent Schwartz (1915–2002), the originator of distribution theory of Chapter 20. The inequality can be seen as a generalization of the fact that, for a and b vectors in ordinary two- or three-dimensional space, $a \cdot b = ab \cos\theta$, so that $|\cos\theta| = |a \cdot b|/ab \leq 1$.

For certain purposes it is convenient to make use of a generalized scalar product defined with the aid of a positive-definite matrix $B = |b_{ij}|$ by

$$(\mathbf{v}, \mathbf{u})_B \equiv \mathbf{v}^* B \mathbf{u} = \sum_{i,j=1}^{N} \bar{v}_i b_{ij} u_j. \tag{18.5.12}$$

We have introduced on p. 501 the notion of the complement of a subspace $\mathbb{S}_1$ of $\mathbb{S}$ and noted that many complements exist. The *orthogonal complement* $\mathbb{S}_1^{\perp}$ of $\mathbb{S}_1$, on the other hand, is unique. It consists of all the vectors of $\mathbb{S}$ which are orthogonal to all the vectors of $\mathbb{S}_1$; we write $\mathbb{S} = \mathbb{S}_1 \oplus \mathbb{S}_1^{\perp}$ to indicate the direct sum of $\mathbb{S}_1$ and its orthogonal complement. In particular, for a linear operator A, the image space $\mathbb{S}^M$ can be decomposed into the range of A and its orthogonal complement, $\mathbb{S}^M = \mathscr{R}(A) \oplus \mathscr{R}^{\perp}(A)$. It is a general property of linear operators that

Theorem 18.5.1 *If a subspace* $\mathbb{S}_1$ *is invariant under the action of* A, *its orthogonal complement* $\mathbb{S}_1^{\perp}$ *is invariant under* A^*.

Indeed, if $\mathbf{u}$ belongs to an invariant subspace $\mathbb{S}_1$, so does $A\mathbf{u}$ by definition of invariant subspace. Then, for any $\mathbf{v}$ in the orthogonal complement $\mathbb{S}_1^{\perp}$, we have $0 = (\mathbf{v}, A\mathbf{u}) = (A^*\mathbf{v}, \mathbf{u})$ so that $A^*\mathbf{v} \in \mathbb{S}_1^{\perp}$ as well. An important consequence of this simple fact is that

Theorem 18.5.2 *The orthogonal complement of the range of* A *is the null space of* A^*, $\mathscr{R}^{\perp}(A) = \mathscr{N}(A^*)$.

Proof Since the range of A is invariant under A, we conclude that its orthogonal complement $\mathscr{R}^{\perp}(A)$ is invariant under the action of A^*. Now, for any vector $\mathbf{w}$ in the null space of A^*, $(\mathbf{w}, A\mathbf{u}) = (A^*\mathbf{w}, \mathbf{u}) = 0$, so that $\mathbf{w} \in \mathscr{R}^{\perp}(A)$. Conversely, if $\mathbf{w} \in \mathscr{R}^{\perp}(A)$, $(A^*\mathbf{w}, \mathbf{u}) = (\mathbf{w}, A\mathbf{u}) = 0$ for all $\mathbf{u}$, because $A\mathbf{u} \in \mathscr{R}(A)$. As noted at the beginning of this section, the only vector orthogonal to all vectors of a space is the zero vector, and therefore $\mathbf{w} \in \mathscr{N}(A^*)$. We have thus shown that, if a vector belongs to $\mathscr{R}^{\perp}(A)$, it also belongs to $\mathscr{N}(A^*)$ and vice versa, which proves the theorem.   □

From this theorem we have the important relations

$$\mathscr{R}^{\perp}(A^*) = \mathscr{N}(A), \qquad \mathscr{R}(A^*) = \mathscr{N}^{\perp}(A), \qquad \mathscr{R}(A) = \mathscr{N}^{\perp}(A^*), \tag{18.5.13}$$

which derive from the decomposition $\mathbb{S}^M = \mathscr{R}(A) \oplus \mathscr{R}^{\perp}(A)$ and the relations $(A^*)^* = A$, $\mathscr{R}^{\perp\perp}(A) = \mathscr{R}(A)$, valid in a finite-dimensional space. As a consequence, if the dimension of $\mathscr{R}(A)$ is r, then $\dim \mathscr{R}^{\perp}(A) = M - r = \dim \mathscr{N}(A^*)$. It will be recalled from (18.3.11) that $\dim \mathscr{N}(A) = N - r$ which, then, is also the dimension of $\mathscr{R}^{\perp}(A^*)$. Summarizing,

$$\dim \mathscr{R}(A^*) = \dim \mathscr{R}(A) = r, \quad \dim \mathscr{N}(A) = N - r, \quad \dim \mathscr{N}(A^*) = M - r. \tag{18.5.14}$$

A very important consequence of the last one of (18.5.13) is:

Theorem 18.5.3 *The linear algebraic system* $A\mathbf{u} = \mathbf{b}$ *has solutions if and only if* $\mathbf{b}$ *is orthogonal to all solutions of* $A^*\mathbf{v} = \emptyset$.

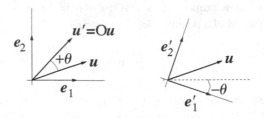

Fig. 18.1   The rotation of a vector by θ is equivalent to a rotation of the reference system by $-\theta$.

Indeed, if the system is solvable, then **b** must belong to the range of A and it must therefore be orthogonal to the null space of A^*. These ideas are also addressed in §15.7 and §21.7 in connection with the Fredholm Alternative Theorem.

A transformation $\mathbf{u}' = U\mathbf{u}$ of the space in itself will preserve scalar products when U is unitary (or orthogonal in the real case) since, then,

$$(\mathbf{v}', \mathbf{u}') = (U\mathbf{v}, U\mathbf{u}) = (U^*U\mathbf{v}, \mathbf{u}) = (\mathbf{v}, \mathbf{u}) \qquad (18.5.15)$$

for all pairs $\mathbf{u}$, $\mathbf{v}$. A typical simple example is a counterclockwise rotation by the angle θ in the plane, which is readily seen to be accomplished by the orthogonal matrix

$$O = \begin{vmatrix} \cos\theta & -\sin\theta \\ \sin\theta & \cos\theta \end{vmatrix}. \qquad (18.5.16)$$

Clearly a unitary (or orthogonal) transformation carries one orthonormal basis into another one. Conversely, two different orthonormal bases must be connected by a unitary (or, in the real case, orthogonal) transformation. Let us consider then an orthogonal transformation $O = |o_{ij}|$ from a basis $\{\mathbf{e}_j\}$ to a new basis $\{\mathbf{e}'_j\}$. According to (18.4.1), $\mathbf{e}'_j = \sum_{i=1}^{N} \mathbf{e}_i o_{ij}$. It was noted after (18.4.1) that in this case the j-th column of the matrix O is just the set of the old coordinates of the new basis vector $\mathbf{e}'_j$. Thus, upon using (18.5.5), we may write

$$O = |o_{ij}| = \left| (\mathbf{e}_i, \mathbf{e}'_j) \right|. \qquad (18.5.17)$$

For example, when a new basis in the plane is obtained by a counterclockwise rotation by θ of the old one, the orthogonal matrix connecting the new to the old basis vectors is

$$\begin{vmatrix} \cos\theta & \sin\theta \\ -\sin\theta & \cos\theta \end{vmatrix} \qquad (18.5.18)$$

which is the transpose (i.e., the inverse in this case) of that defined in (18.5.16). It is indeed obvious that, in a new basis rotated by $-\theta$, a vector will appear as the same vector rotated by $+\theta$ in the original basis (Figure 18.1).

Upon a change of basis, the matrices representing operators transform according to (18.4.8) which takes the form

$$A' = O^T A O, \qquad A = O A' O^T. \qquad (18.5.19)$$

18.6 Eigenvalues and eigenvectors

Any non-zero vector satisfying a relation of the form

$$A\mathbf{x} = \lambda\mathbf{x} \tag{18.6.1}$$

where A is a square matrix of order N, is called a (right) *eigenvector* of the matrix A and the number λ, in general complex and possibly zero, the *eigenvalue* of A corresponding to the eigenvector $\mathbf{x}$. Other common denominations are characteristic vector and characteristic value of A. If $\mathbf{x}$ is an eigenvector, so is any non-zero multiple of it.[14]

Regarded as a linear system, (18.6.1) is homogeneous and it will have non-zero solutions if and only if (p. 499)

$$C(\lambda) \equiv \det |\lambda\mathbf{I} - \mathbf{A}| = 0; \tag{18.6.2}$$

$C(\lambda)$ is an Nth order polynomial in λ which has the form

$$C(\lambda) = \lambda^N - (\mathrm{Tr}\,\mathbf{A})\,\lambda^{N-1} + \cdots + (-1)^N \det\mathbf{A}. \tag{18.6.3}$$

We call $C(\lambda)$ the *characteristic polynomial* of A, and (18.6.2) is the *characteristic* or *eigenvalue equation* of A. The coefficient of λ^k is the sum of all the principal minors (p. 499) of order $N - k$ multiplied by $(-1)^{N-k}$. Thus, in particular, the coefficient of λ^{N-1} is just the trace of A, while the last term is the determinant of A. If $\lambda = 0$ is an eigenvalue, $C(0) = (-1)^N \det A = 0$ and the matrix is singular, and vice versa. If A^{-1} exists, we have from (18.6.1)

$$A^{-1}\mathbf{x} = \lambda^{-1}\mathbf{x} \tag{18.6.4}$$

which shows that λ^{-1} is an eigenvalue of A^{-1} with the same eigenvector.

It was shown in §18.4 that the matrices representing an operator in two different bases are connected by a similarity transformation. Since, according to (18.2.10), the determinants of two similar matrices are equal, so are their eigenvalues.[15] Thus, all the matrices representing the same operator have the same eigenvalues, and we can therefore consider eigenvalues as attributes of the operator itself, independent of its matrix representation. The fact that the characteristic polynomial is invariant under a similarity transformation indicates that the sum of all the principal minors of any order of a matrix – and in particular its trace – is also invariant under such a transformation. Eigenvectors do depend on the matrix representation, however. A simple calculation shows that, if $\mathbf{x}$ is an eigenvector of A and $A' = T^{-1}AT$, then $T^{-1}\mathbf{x}$ is an eigenvector of A' corresponding to the same eigenvalue.

From the fundamental theorem of algebra, a polynomial with complex coefficients has exactly as many complex roots as its degree, if repeated roots are counted with their

---

[14] In a scalar product space, to deal with this lack of uniqueness it is sometimes useful to normalize the eigenvectors in such a way that their norm equal 1; this condition establishes uniqueness only up to a phase, i.e., a number of modulus 1.

[15] The converse, however, is not true: matrices having the same eigenvalues are not necessarily similar.

multiplicity. We therefore deduce that $C(\lambda)$ can be decomposed into factors as

$$C(\lambda) = (\lambda - \lambda_1)^{s_1}(\lambda - \lambda_2)^{s_2} \cdots (\lambda - \lambda_r)^{s_r}, \qquad s_1 + s_2 + \cdots + s_k = N. \quad (18.6.5)$$

The exponent s_j of the factor $\lambda - \lambda_j$ is the *algebraic multiplicity* of λ_j. If $s_j = 1$, the eigenvalue is *simple* while, if $s_j > 1$, it is *degenerate*. Upon comparing (18.6.5) with (18.6.3) we immediately deduce that

$$\operatorname{Tr} A = s_1\lambda_1 + \cdots + s_r\lambda_r, \qquad \det A = \lambda_1^{s_1} \cdots \lambda_r^{s_r}, \quad (18.6.6)$$

which complements the statement made after (18.6.3). In the case of a simple matrix, these formulae also follow directly from (18.7.8) below. We will see in §18.10 that, while there always are N eigenvalues, linearly independent eigenvectors in the sense of (18.6.1) may be fewer. However

Theorem 18.6.1 *Every square matrix has at least one eigenvector.*

Indeed, if λ is an eigenvalue of A, then the determinant of the matrix $A - \lambda I$ vanishes and the matrix is singular, which implies that the algebraic equation $(A - \lambda I)\mathbf{x} = \emptyset$ has non-zero solutions. By the same argument, there is at least one eigenvector associated to each distinct eigenvalue of A.

Since any multiple of $\mathbf{x}$ satisfies (18.6.1) if $\mathbf{x}$ does, the set of eigenvectors corresponding to an eigenvalue, augmented by the null vector which is not in eigenvector by definition, is a linear vector space with dimension at least 1. In the case of a degenerate eigenvalue λ_j, there may be $m_j \geq 1$ linearly independent vectors satisfying the eigenvalue equation (18.6.1). Since, evidently, any linear combination of these vectors also satisfies (18.6.1), in this case there is an m-dimensional *eigenspace* corresponding to λ_j and we call m_j the *geometric multiplicity* of λ_j. The algebraic multiplicity cannot be less than the geometric multiplicity, i.e, $s_j \geq m_j$ (see footnote p. 525), because, otherwise, the characteristic polynomial would have degree less than N. Matrices for which $s_j = m_j$ are called *simple* and will be considered in some detail in §18.7. Matrices for which $s_j > m_j$ are *defective* and will be considered in §18.10.

Eigenspaces are invariant subspaces for A and, in fact, A is a multiple of the identity when restricted to each one of its eigenspaces. By definition, the j-th eigenspace is the null space of $A - \lambda_j I$ and therefore the set of eigenvalues of A, the so-called *spectrum* of A, can be characterized as the set of numbers for which $A - \lambda I$ is not invertible.

Similar definitions hold for left eigenvectors and eigenvalues:

$$\mathbf{y}^*A = \mu\mathbf{y}^*. \quad (18.6.7)$$

Due to the second one of the properties (18.1.11), a left eigenvector of A is a right eigenvector of A^* corresponding to the complex conjugate eigenvalue. Therefore, in particular, the left and right eigenvectors of a Hermitian matrix are the same and its eigenvalues are real.

Theorem 18.6.2 *The set constituted by any number of eigenvectors, each one corresponding to a different eigenvalue, is linearly independent.*

For example, if $\mathbf{x}_1$ and $\mathbf{x}_2$ correspond to different eigenvalues λ_1 and λ_2, and the relation $\alpha\mathbf{x}_1 + \beta\mathbf{x}_2 = \emptyset$ were true, we could multiply it e.g. by $\lambda_1 I - A$ to find $\beta(\lambda_1 - \lambda_2)\mathbf{x}_2 = \emptyset$, which implies $\beta = 0$. The argument can be generalized to any number of right eigenvectors, none corresponding to equal eigenvalues, and to left eigenvectors as well.

Let B commute with A, and $\mathbf{x}$ be an eigenvector of A. Suppose that $\mathbf{x}$ and $B\mathbf{x}$ be linearly independent, while $B^2\mathbf{x}$ is a linear combination of them. Thus, the subspace spanned by $\mathbf{x}$ and $B\mathbf{x}$ is an invariant subspace for B. But a subspace invariant under the action of an operator must contain an eigenvector of the operator.[16] Thus, the subspace will contain an eigenvector of B, which is also an eigenvector of A since A and B commute. The argument is readily generalized to conclude that

Theorem 18.6.3 *Two commuting operators have at least one common eigenvector.*

A simple consequence of this theorem concerns the eigenvalues of a normal operator, for which $AA^* = A^*A$. Let $\mathbf{x}$ be a common eigenvector corresponding to the eigenvalue λ for A and μ for A^*, and consider $(\mathbf{x}, A\mathbf{x})$. This quantity equals $\lambda\|\mathbf{x}\|^2$, but it also equals $(A^*\mathbf{x}, \mathbf{x}) = \overline{\mu}\|\mathbf{x}\|^2$. Thus, for a normal operator, the eigenvalues of A^* are the complex conjugates of those of A (see also p. 667). This result holds true also for non-normal operators if A and A^* have a common eigenvector.

The eigenvalue problem (18.6.1) can be generalized to the form

$$A\mathbf{x} = \lambda B\mathbf{x} \qquad (18.6.8)$$

with suitable conditions on B. If B is non-singular, the problem may be phrased as a standard eigenvalue problem for the matrix $B^{-1}A$, but this approach suffers from the shortcoming that the latter matrix may lack properties, such as symmetry, which simplify the analysis of the original problem (18.6.8). This may be one instance in which the generalized scalar product (18.5.12) proves useful. The reader is referred e.g. to Golub & van Loan (1989, section 7.7) for details. Further properties of eigenvalues are discussed in §18.8.

18.7 Simple, normal and Hermitian matrices

The largest possible number of linearly independent eigenvectors of an operator in an N-dimensional space cannot exceed N. An operator (or the matrix that represents it) is *simple* if the number of its linearly independent eigenvectors is precisely equal to this maximum number; non-simple operators (or the matrices which represent them) are called *defective* and are considered in §18.10. Being simple does not imply that all eigenvalues are distinct, as shown by the trivial example of the identity matrix. However, if an eigenvalue of a simple matrix is repeated k times, we can always find k linearly independent vectors

---

[16] A heuristic justification is that one may consider the subspace and the restriction of the operator to it without reference to the original operator and bigger space. Theorem 18.6.1 then applies and guarantees the existence of an eigenvector in the subspace.

in its eigenspace. A matrix with N distinct eigenvalues is simple because, according to Theorem 18.6.2, its eigenvectors are linearly independent.

One of the main reasons of our interest in simple matrices is that their right eigenvectors $\mathbf{x}_1, \mathbf{x}_2, \ldots, \mathbf{x}_N$ are a maximal linearly independent set, and therefore can be used as a basis to express a generic vector $\mathbf{u}$ in the N-dimensional space as in (18.3.3). This choice of a basis simplifies, in a sense, the action of the operator A on the vectors of the space as, evidently,

$$\mathsf{A}\mathbf{u} = \sum_{i=1}^{N} \lambda_i u_i \mathbf{x}_i. \tag{18.7.1}$$

Thus, in principle, if we are interested in solving the equation

$$(\mathsf{A} - \lambda \mathsf{I})\mathbf{u} = \mathbf{f} \tag{18.7.2}$$

with λ a real or complex number, we can expand $\mathbf{u}$ and $\mathbf{f}$ on the basis $\{\mathbf{x}_i\}$:

$$\mathbf{u} = \sum_{i=1}^{N} u_i \mathbf{x}_i, \qquad \mathbf{f} = \sum_{i=1}^{N} f_i \mathbf{x}_i, \tag{18.7.3}$$

and substitute into (18.7.2) to find

$$\sum_{i=1}^{N} [(\lambda_i - \lambda)u_i - f_i]\,\mathbf{x}_i = 0. \tag{18.7.4}$$

From the linear independence of the basis vectors, if λ is not equal to any of the eigenvalues, we then have $u_i = f_i/(\lambda_i - \lambda)$ and

$$\mathbf{u} = \sum_{i=1}^{N} \frac{f_i}{\lambda_i - \lambda}\,\mathbf{x}_i. \tag{18.7.5}$$

Even when the basis vectors are known, this method is seldom useful as expanding $\mathbf{f}$ requires solving a linear system which may be just as expensive as solving (18.7.2) directly. As will be seen shortly, however, the method becomes quite practical if A is normal.

In the present special case of simple matrices, (18.7.1) permits us to develop a straightforward proof of an important general result:

Theorem 18.7.1 [Cayley–Hamilton] *Every square matrix satisfies its own characteristic equation, $C(\mathsf{A}) = 0$, the zero matrix.*

We will not prove the theorem for the general case. For a simple matrix, however, it is obviously true as, for any vector $\mathbf{u}$, we have $C(\mathsf{A})\mathbf{u} = \sum_{j=1}^{N} a_j C(\mathsf{A})\mathbf{x}_j = \sum_{j=1}^{N} a_j C(\lambda_j)\mathbf{x}_j = \emptyset$ so that $C(\mathsf{A})$ is the zero matrix. An immediate consequence which directly follows upon multiplying $C(\mathsf{A}) = \emptyset$ by A^{-1} when the inverse exists is the following representation of A^{-1} (cf. (18.6.3)):

$$\mathsf{A}^{-1} = \frac{(-1)^{N+1}}{\det \mathsf{A}}\left[\mathsf{A}^{N-1} - (\mathrm{Tr}\,\mathsf{A})\mathsf{A}^{N-2} + \cdots\right]. \tag{18.7.6}$$

There may be other polynomials $p(x)$ of degree smaller than N such that $p(A) = \emptyset$. For example, the first-degree polynomial $p(x) = x - 1$ evidently annihilates the identity matrix I. In this case, the matrix is labelled *derogatory*; more on these matrices will be said at the end of §18.10.

Arrange the eigenvalues of a simple matrix A along the main diagonal of a diagonal matrix Λ and construct a matrix X having as columns the corresponding eigenvectors:[17]

$$\Lambda = \text{diag}[\lambda_1, \lambda_2, \ldots, \lambda_N], \qquad X = |\mathbf{x}_1 \ \mathbf{x}_2 \ \cdots \ \mathbf{x}_N|. \qquad (18.7.7)$$

Then a simple calculation shows that $A X = X \Lambda$[18] and therefore

$$A = X \Lambda X^{-1}, \qquad \Lambda = X^{-1} A X, \qquad (18.7.8)$$

since the columns of X are linearly independent so that X is non-singular. We thus conclude that a simple matrix is similar to the diagonal matrix of its eigenvectors. Actually, this property characterizes simple matrices:

Theorem 18.7.2 *A matrix is simple if and only if it is similar to a diagonal matrix.*

The set of simple matrices is dense in the space of matrices, i.e., any matrix can be approximated with arbitrary accuracy by a simple matrix. On the basis of this property it is sometimes possible to prove a result for simple matrices and extend it by a continuity argument to non-simple matrices. We can also show that

Theorem 18.7.3 *If* A *is simple, so is* A* *and complete sets of eigenvectors* $\mathbf{x}_1, \ldots, \mathbf{x}_N$ *of* A *and* $\mathbf{y}_1, \ldots, \mathbf{y}_N$ *of* A* *can be chosen such that* $(\mathbf{y}_j, \mathbf{x}_k) = \delta_{jk}$.

The idea of the proof is simply illustrated by considering $N = 2$. Let $\mathbf{x}_j$, $j = 1, 2$ be linearly independent eigenvectors of A and $\mathbb{S}_j^\perp$ the orthogonal complements of the respective eigenspaces. Then, since each one of these eigenspaces is invariant under A* (Theorem 18.5.1 p. 507), it contains an eigenvector $\mathbf{y}_j$ of A* (footnote on p. 511), which proves that A* is simple as well. Now, the eigenvector $\mathbf{y}_1$ cannot be orthogonal to $\mathbf{x}_2$ without being the null vector because it is already orthogonal to $\mathbf{x}_1$, and similarly for $\mathbf{y}_2$ and $\mathbf{x}_1$. By adjusting the arbitrary multiplicative constants at our disposal and renumbering the eigenvectors, we can therefore make these eigenvector sets orthonormal.

This result can be strengthened if A is *normal* so that it commutes with A* as, in this case, all the eigenvectors of A are also eigenvectors of A* (Theorem 18.6.3). Therefore, after an appropriate normalization, $(\mathbf{e}_j, \mathbf{e}_k) = \delta_{jk}$, a property which characterizes normal matrices:

---

[17] A matrix so constructed is called a *modal matrix*. Since eigenvectors are not unique, modal matrices are not unique either.

[18] Note that $X\Lambda \neq \Lambda X$ in general: a diagonal matrix commutes with a symmetric matrix, but not a general unsymmetric one as is readily verified by a direct calculation.

Theorem 18.7.4 *A matrix is normal if and only if it has a complete orthonormal system of eigenvectors.*

In this case, as in (18.5.5) p. 505, the coefficients of the expansion of a vector $\mathbf{f}$ on the basis $\{\mathbf{e}_k\}$ of normalized eigenvectors of A are simply calculated as the scalar product $(\mathbf{e}_k, \mathbf{f})$. With this remark, the solution (18.7.5) of (18.7.2) becomes

$$\mathbf{u} = \sum_{k=1}^{N} \frac{(\mathbf{e}_k, \mathbf{f})}{\lambda_k - \lambda} \mathbf{e}_k. \tag{18.7.9}$$

Thanks to the previous theorem, the matrix X reducing a normal matrix to diagonal form can be taken as unitary by taking its columns equal to the normalized eigenvectors $\mathbf{e}_k$. Indeed, with this step X^*X is just the matrix having for entries $(\mathbf{e}_j, \mathbf{e}_k)$ and is therefore the identity matrix. This property permits another characterization of normal matrices:

Theorem 18.7.5 *A matrix is normal if and only if it can be reduced to diagonal form by a unitary similarity transformation.*

It should be noted that, while both normal and simple non-normal matrices can be diagonalized, the transforming matrix can be chosen unitary in the first case but not necessarily in the second.

Example 18.7.1   Consider, for real a and $b > 0$,

$$A = \begin{vmatrix} a & 1 \\ b & a \end{vmatrix} \qquad \text{for which} \qquad AA^* - A^*A = \begin{vmatrix} 1 - b^2 & 0 \\ 0 & b^2 - 1 \end{vmatrix} \tag{18.7.10}$$

so that A is not normal unless $b = 1$. The eigenvalues are $\lambda_\pm = a \pm \sqrt{b}$ with corresponding normalized eigenvectors

$$\mathbf{x}_\pm = (1 + b)^{-1/2} \begin{vmatrix} 1 \\ \pm\sqrt{b} \end{vmatrix} \qquad \text{with} \qquad (\mathbf{x}_-, \mathbf{x}_+) = \frac{1 - b}{1 + b}. \tag{18.7.11}$$

Thus, the matrix is simple with non-orthogonal eigenvectors if $b \neq 1$. In this case

$$X = \frac{1}{\sqrt{1 + b}} \begin{vmatrix} 1 & 1 \\ \sqrt{b} & -\sqrt{b} \end{vmatrix}, \qquad X^{-1} = \frac{1}{2}\sqrt{\frac{1 + b}{b}} \begin{vmatrix} \sqrt{b} & 1 \\ \sqrt{b} & -1 \end{vmatrix}, \tag{18.7.12}$$

from which

$$X^{-1}AX = \begin{vmatrix} a + \sqrt{b} & 0 \\ 0 & a - \sqrt{b} \end{vmatrix}, \tag{18.7.13}$$

as expected from Theorem 18.7.2. If $b = 1$, A is normal (in fact, symmetric), $X^{-1} = X^*$ and the eigenvectors are orthogonal (Theorem 18.7.4). The adjoint A^* has the same eigenvalues and, if its eigenvectors are normalized to

$$\mathbf{y}_\pm = \frac{1}{2}\sqrt{\frac{1 + b}{b}} \begin{vmatrix} \sqrt{b} \\ \pm 1 \end{vmatrix}, \tag{18.7.14}$$

we have $(\mathbf{y}_j, \mathbf{x}_i) = \delta_{ij}$ (Theorem 18.7.3). Note, however, that $\|\mathbf{y}_\pm\| \neq 1$. Furthermore – and not coincidentally – the eigenvectors of A are not parallel to those of A^* if $b \neq 1$.   □

The previous results become particularly powerful in the case of Hermitian matrices, which are obviously normal as $\overline{a}_{ji} = a_{ij}$ so that $A^* = A$. On the basis of the previous theorem, Hermitian matrices are therefore also simple. It was shown on p. 511 that, if A and A^* have a common eigenvector, the corresponding eigenvalues are complex conjugates. For Hermitian operators, this result implies that the eigenvalues are real, a fact that can also be proven in precisely the same way as done in §21.2.2 for Theorem 21.2.8 p. 630 in the more general case of Hermitian operators. Summarizing:

Theorem 18.7.6 *A Hermitian matrix has real eigenvalues, a complete set of orthonormal eigenvectors, and can be reduced to diagonal form by a unitary transformation.*

These properties characterize Hermitian matrices in the same sense as Theorem 18.7.4 characterizes normal ones. The importance of Hermitian matrices resides in their frequent appearance in applications which often reflects, in its turn, important symmetry properties of the underlying physics. Furthermore, Hermitian matrices are the simplest example of Hermitian operators, which will be studied in §21.2.2 and §21.6.1, and which lie at the basis of various eigenfunction expansions.

In the real case, all of the above applies equally well to real symmetric matrices which are a special case of Hermitian matrices.

18.8 Spectrum and singular values

The *spectrum* of an operator is the set of real or complex numbers λ for which the inverse of $A - \lambda I$ does not exist. This notion is independent of the particular basis chosen to represent the operator A.

The *spectral radius* $r_\sigma(A)$ of an operator A is the largest of the moduli of the eigenvalues:

$$r_\sigma(A) = \max_j |\lambda_j|. \tag{18.8.1}$$

This may be considered as a measure of the "magnitude" of A. An alternative measure can be based on the operator norm. Several norms are in use and a general theory is given in §19.2. For the present purposes we will mainly consider the norm *induced* by the vector norm (18.5.2), which is given by[19]

$$\|A\|_2 = \sup \frac{\|A\mathbf{u}\|}{\|\mathbf{u}\|} = \sup_{\|\mathbf{v}\|=1} \|A\mathbf{v}\|, \tag{18.8.2}$$

where *sup* denotes the supremum, or least upper bound. The norm $\|A\|_2$ is thus the maximum length of a unit vector after it is transformed by the action of A. The vectors $\mathbf{v}$ with unit

---

[19] The use of the subscript 2 is motivated in §19.3.

norm constitute the *unit sphere*, or *unit ball*, of the space. It is evident from the definition that, for any **u**,

$$\|A\mathbf{u}\| \leq \|A\|_2 \|\mathbf{u}\|. \tag{18.8.3}$$

Another useful norm is the *Frobenius norm*

$$\|A\|_F = \left(\sum_{i,j=1}^{N} |a_{ij}|^2 \right)^{1/2} = \left(\mathrm{Tr}\, A^*A \right)^{1/2} \tag{18.8.4}$$

also known as the *Euclidean* or *spectral* norm of A. These two norms are equivalent (in the sense made precise on p. 544) because (see e.g. Golub & van Loan 1989, p. 57)[20]

$$\|A\|_2 \leq \|A\|_F \leq \sqrt{N} \|A\|_2 \tag{18.8.5}$$

which implies, in particular, that (18.8.3) also holds if the Frobenius norm is used for A.

An interesting application of the notion of "magnitude" of a matrix is the *condition number*. Consider a slightly perturbed linear system $A(\mathbf{u} + \delta\mathbf{u}) = \mathbf{b} + \delta\mathbf{b}$, in which $\delta\mathbf{u} = A^{-1}\delta\mathbf{b}$ is the perturbation to the solution $\mathbf{u} = A^{-1}\mathbf{b}$ induced by a perturbation $\delta\mathbf{b}$ of the data. Then, by (18.8.3), we have, with either norm,

$$\frac{\|\delta\mathbf{u}\|}{\|\mathbf{u}\|} \leq \|A^{-1}\| \frac{\|\delta\mathbf{b}\|}{\|\mathbf{u}\|} \leq \left(\|A\| \|A^{-1}\| \right) \frac{\|\delta\mathbf{b}\|}{\|\mathbf{b}\|}. \tag{18.8.6}$$

This relation shows that the quantity $\kappa_A = \|A\| \|A^{-1}\|$, called the *condition number* of A, measures the "amplification of the error" that is incurred in dealing with imperfectly known data. This bound cannot be improved as it is always possible to find **b** and $\delta\mathbf{b}$ such that it is attained (see e.g. Ciarlet 1989, chapter 2). A similar argument shows that κ_A bounds also the change in the solution when the matrix A itself is perturbed. For a singular matrix A^{-1} does not exist and the condition number is considered infinite.

To address the relation between norm, eigenvalues and spectral radius let us begin by considering a Hermitian operator A and the *quadratic form* $(\mathbf{u}, A\mathbf{u})$ associated with it. Since $(\mathbf{u}, A\mathbf{u}) = (A^*\mathbf{u}, \mathbf{u}) = (A\mathbf{u}, \mathbf{u}) = \overline{(\mathbf{u}, A\mathbf{u})}$, this quantity is real and, by representing **u** on an orthonormal basis of eigenvectors of A, we have

$$(\mathbf{u}, A\mathbf{u}) = \sum_{j=1}^{N} \lambda_j |u_j|^2. \tag{18.8.7}$$

If λ_M and λ_m are the maximum and minimum eigenvalues of A, then evidently

$$\lambda_m \sum_{j=1}^{N} |u_j|^2 \leq (\mathbf{u}, A\mathbf{u}) \leq \lambda_M \sum_{j=1}^{N} |u_j|^2 \tag{18.8.8}$$

so that

$$\lambda_M = \max \frac{(\mathbf{u}, A\mathbf{u})}{(\mathbf{u}, \mathbf{u})}, \qquad \lambda_m = \min \frac{(\mathbf{u}, A\mathbf{u})}{(\mathbf{u}, \mathbf{u})}. \tag{18.8.9}$$

---

[20] Actually, all norms in a finite-dimensional space are equivalent, see e.g. Kress 1989, p.4.

The quantity $(\mathbf{u}, A\mathbf{u})/(\mathbf{u}, \mathbf{u})$ is known as the *Rayleigh quotient* and its values as $\mathbf{u}$ varies over the space are the *field of values* or *numerical range* of A. The maximum and the minimum are attained when $\mathbf{u}$ equals the eigenvectors of λ_M and λ_m, respectively; more generally, if $\mathbf{x}_j$ is the normalized eigenvector corresponding to λ_j, $\lambda_j = (\mathbf{x}_j, A\mathbf{x}_j)$. A direct calculation shows that the Rayleigh quotient is stationary in correspondence of each eigenvalue, in the sense that

$$\frac{\partial}{\partial u_k} \frac{(\mathbf{u}, A\mathbf{u})}{(\mathbf{u}, \mathbf{u})}\bigg|_{\mathbf{u}=\mathbf{x}_j} = 0 \qquad \text{for} \qquad k = 1, 2, \ldots, N. \tag{18.8.10}$$

Moreover, the following result, which generalizes (18.8.9), holds (see e.g. Lancaster & Tismenetsky 1985, Wilkinson 1988):

Theorem 18.8.1 [Courant–Fischer min-max principle] *If the eigenvalues of a Hermitian operator are arranged in ascending order $\lambda_1 \leq \lambda_2 \leq \ldots \leq \lambda_N$, then*

$$\lambda_k = \min_{\mathbb{S}^k} \max_{\mathbf{u} \in \mathbb{S}^k} \frac{(\mathbf{u}, A\mathbf{u})}{(\mathbf{u}, \mathbf{u})}, \qquad k = 1, 2, \ldots, N, \tag{18.8.11}$$

where the minimum is taken over all k-dimensional subspaces of $\mathbb{C}^N$.[21]

An equivalent way to state the theorem would be to consider only the values of $(\mathbf{v}, A\mathbf{v})$ as $\mathbf{v}$ ranges over the unit sphere of each subspace. For the largest eigenvalue λ_N there is only one subspace, namely $\mathbb{C}^N$ itself, there is no minimization to perform and this result is just the first one of (18.8.9). Conversely, for the smallest eigenvalue (assumed simple), only one-dimensional subspaces are allowed, the unit sphere contains essentially only one vector and there is no maximization to perform. To aid intuition, one may think of an ellipsoid in three-dimensional space. The traces of this ellipsoid on planes through the origin are ellipses, and it is evident that if one considers the major semi-axis of each one of these ellipses (the max part of the theorem), and takes the minimum value of all these major semi-axes, one is left with the intermediate semi-axis of the ellipsoid, i.e., the second eigenvalue.

An immediate consequence of (18.8.11) is that the eigenvalues of a positive-definite (non-negative definite) Hermitian matrix are positive (non-negative). In particular positive-definite or negative-definite matrices cannot be singular.

Again representing $\mathbf{u}$ on an orthonormal basis of eigenvectors of A, we have $\|A\mathbf{u}\|^2 = \sum_{j=1}^{N} |\lambda_j|^2 |u_j|^2$ from which we conclude that $\|A\|_2 = \max(|\lambda_M|, |\lambda_m|) = r_\sigma(A)$ if the norm is defined as in (18.8.2). Since, by (18.6.4), the eigenvalues of A^{-1} are the reciprocals of those of A, the condition number defined with the norm (18.8.2) equals the larger one of the ratios $|\lambda_M/\lambda_m|$ and $|\lambda_m/\lambda_M|$.

The stationarity property (18.8.10) also holds for simple but non-Hermitian matrices for which $\lambda_j = (\mathbf{y}_j, A\mathbf{x}_j)/(\mathbf{y}_j, \mathbf{x}_j)$, with $\mathbf{x}_j$ the eigenvector of A corresponding to λ_j and $\mathbf{y}_j$ the corresponding orthogonal eigenvector of A^* (cf. Theorem 18.7.3). In this more general case, however, there are no results analogous to (18.8.9) as there is no guarantee that the

---

[21] If the eigenvalues are numbered in the reverse order, the theorem should be phrased as the maximum over k-dimensional subspaces of the minimum of the Rayleigh quotient.

quadratic form $(\mathbf{u}, A\mathbf{u})$ is real. For operators which are neither Hermitian nor simple, these results need to be weakened further. In this more general case one can prove

Theorem 18.8.2 [Singular value decomposition] *Any $M \times N$ matrix* A *can be decomposed in the form*

$$A = X\Sigma Y^* \tag{18.8.12}$$

in which X *and* Y *are unitary (or real orthogonal if* A *is real)* $M \times M$ *and* $N \times N$ *matrices, respectively, and* Σ *is an* $M \times N$ *diagonal matrix with non-negative entries* σ_j *on the main diagonal.*

The non-negative numbers σ_j are the *singular values* of A and are equal to the singular values of A^*. For simplicity, we sketch the proof for a square matrix; only minor modifications are needed for the general case (see e.g. Golub & van Loan 1989, chapter 2; Ciarlet 1989, chapter 1; Horn & Johnson 1994, chapter 3).

Proof The operator A^*A is evidently Hermitian and non-negative definite as, for any vector $\mathbf{u}$, $(\mathbf{u}, A^*A\mathbf{u}) = (A\mathbf{u}, A\mathbf{u}) = \|A\mathbf{u}\|^2 \geq 0$. According to Theorem 18.7.5, there is therefore a unitary matrix Y such that

$$Y^*A^*AY = (AY)^*AY = \text{diag}[\sigma_1^2, \sigma_2^2, \ldots, \sigma_r^2, 0, \ldots, 0], \tag{18.8.13}$$

where $r \leq N$ is the rank of A, which can be proven to be equal to that of A^*A. If $\mathbf{y}_j$ is the j-th column vector of AY, $\mathbf{y}_i^*$ is the i-th row of $(Ay)^*$ so that, by the rule of matrix multiplication, (18.8.13) implies that $\mathbf{y}_i^*\mathbf{y}_j = \sigma_i^2\delta_{ij}$ for $i \leq r$, while $\mathbf{y}_k^*\mathbf{y}_k = \|\mathbf{y}_k\|^2 = 0$ for $r+1 \leq k \leq N$, so that $\mathbf{y}_k = 0$ for $k = r+1, \ldots, N$. Set $\mathbf{x}_j = \sigma_j^{-1}\mathbf{y}_j$ for $1 \leq j \leq r$. Then $\mathbf{x}_i^*\mathbf{x}_j = \delta_{ij}$ and these vectors form an orthonormal basis in an $r \times r$-dimensional subspace. This basis can be complemented by another arbitrary orthonormal set to form an orthonormal basis in the whole space (p. 500) and the matrix X having these vectors as columns will then be unitary. Furthermore,

$$(X^*AY)_{ij} = \mathbf{x}_i^*\mathbf{y}_j = \begin{cases} \sigma_i^{-1}\mathbf{y}_i^*\mathbf{y}_j = \sigma_i\delta_{ij} & 1 \leq i \leq r \\ 0 & r+1 \leq i \leq N \end{cases} \tag{18.8.14}$$

which is (18.8.12). In the course of the proof we have also shown that

$$A\mathbf{y}_j = \sigma_j\mathbf{x}_j, \qquad A^*\mathbf{x}_j = \sigma_j\mathbf{y}_j, \tag{18.8.15}$$

from which one readily deduces that the $\mathbf{x}_j$ and $\mathbf{y}_j$ are the eigenvectors of AA^* and A^*A, respectively, corresponding to the eigenvalues σ_j^2 or 0 (which are therefore the same for AA^* and A^*A). The $\mathbf{x}_j$ and $\mathbf{y}_j$ are, respectively, the left and right *singular vectors* of A. $\square$

For normal matrices the singular value decomposition is another way to state the factorization (18.7.8). The only difference is that Σ contains the moduli of the eigenvalues rather than the eigenvalues themselves, which is obtained by absorbing the sign (or the complex phase) of the eigenvalues in the columns of X (recall the phase indeterminacy of vectors of unit length, footnote on p. 509). Thus, for a normal matrix, $\sigma_i = |\lambda_i|$. For non-normal matrices the relation between eigenvalues and singular values is more complicated. In general, the following *Weyl inequalities* hold (see e.g. Horn & Johnson 1994 p. 171; Godunov 1998, chapter 6)

$$|\lambda_1| \leq \sigma_1, \quad |\lambda_1\lambda_2| \leq \sigma_1\sigma_2, \quad \ldots, \quad |\lambda_1\lambda_2\cdots\lambda_{N-1}| \leq \sigma_1\sigma_2\ldots\sigma_{N-1},$$
$$|\lambda_1\lambda_2\cdots\lambda_N| = \sigma_1\sigma_2\ldots\sigma_N, \tag{18.8.16}$$

with the last equality a consequence of the expression (18.6.6) of determinants in terms of eigenvalues. Here the singular values have been arranged in non-increasing order, $\sigma_1 \geq \sigma_2 \geq \cdots$, and the eigenvalues in non-increasing order of their modulus, $|\lambda_1| \geq |\lambda_2| \geq \cdots$.

Another consequence of the the singular value decomposition is a generalization of (18.7.1) expressing the action of A on a vector $\mathbf{u}$. Indeed, adopting the scalar product notation to express $\mathbf{y}_i^*\mathbf{u} = (\mathbf{y}_i, \mathbf{u})$, we have from (18.8.12)

$$\left(\mathsf{X}^*\mathsf{A}\mathbf{u}\right)_i = \sigma_i \delta_{ij}(\mathbf{y}_j, \mathbf{u}) = \sigma_i(\mathbf{y}_i, \mathbf{u}) \tag{18.8.17}$$

which implies that

$$\mathsf{A}\mathbf{u} = \sum_{j=1}^{r} \sigma_j(\mathbf{y}_j, \mathbf{u})\mathbf{x}_j. \tag{18.8.18}$$

From (18.8.4) and (18.8.2) we also have

$$\|\mathsf{A}\|_F^2 = \sum_{j=1}^{r} \sigma_j^2, \quad \|\mathsf{A}\|_2 = \max_j \sigma_j = \sqrt{r_\sigma(\mathsf{A}^*\mathsf{A})}. \tag{18.8.19}$$

The relation (18.6.6) between the trace and the eigenvalues of a matrix thus shows that, for a normal matrix,

$$\Delta^2(\mathsf{A}) = \|\mathsf{A}\|_F^2 - \sum_{j=1}^{N} |\lambda_j|^2 \tag{18.8.20}$$

vanishes. This quantity is a convenient measure of the *departure from normality* of the matrix. For the matrix in the example (18.7.10) we have $\Delta^2(\mathsf{A}) = (\sqrt{b} - 1)^2$ so that the matrix is normal if $b = 1$ as noted before.

18.9 Projections

According to the definition of p. 501, if the space $\mathbb{S}$ is the direct sum of subspaces $\mathbb{S}_1$ and $\mathbb{S}_2$, any vector $\mathbf{u} \in \mathbb{S}$ can be uniquely decomposed as $\mathbf{u} = \mathbf{u}_1 + \mathbf{u}_2$ with $\mathbf{u}_j \in \mathbb{S}_j$. On this basis we can define the *projection operator* (or projector), e.g. on $\mathbb{S}_1$, by

$$\mathsf{P}_1\mathbf{u} = \mathbf{u}_1. \tag{18.9.1}$$

It is evident from the definition that this operator behaves like the identity when acting on any vector of $\mathbb{S}_1$: $\mathsf{P}_1^2\mathbf{u} = \mathsf{P}_1\mathbf{u}_1 = \mathbf{u}_1$. Thus, the projection operator is idempotent, i.e. $\mathsf{P}_1^2 = \mathsf{P}_1$. This remark can be used to define a projection operator:

Definition 18.9.1 A linear operator P transforming a linear space into itself is said to be a *projection* if $P^2 = P$.

Furthermore, $P_2 = I - P_1$, so that the decomposition of $\mathbb{S}$ into the direct sum of $\mathbb{S}_1$ and $\mathbb{S}_2$ is mirrored by the decomposition of the identity $I = P_1 + P_2$. It also follows that $P_1 P_2 = P_2 P_1 = 0$. An alternative way to express these notions is to say that the range of P_1, for example, is $\mathbb{S}_1$ and its null space $\mathbb{S}_2$, so that $\mathscr{R}(P_1)$ and $\mathscr{N}(P_1)$ are algebraic complements of each other and $\dim \mathscr{R}(P_1) + \dim \mathscr{N}(P_1) = N$ in agreement with (18.5.14).

Since, evidently, any non-zero vector of $\mathscr{R}(P_1)$ or $\mathscr{N}(P_1)$ is an eigenvector of P_1, by combining the vectors of any orthonormal basis in each one of these subspaces we have a basis for the whole space. We thus conclude that a projection operator is simple. As the equation $P\mathbf{u} = \emptyset$ admits non-zero solutions, a projection operator is singular.

These definitions are extended in an obvious way to the decomposition of the space $\mathbb{S}$ into the direct sum of more than two subspaces. Thus, if $\mathbb{S} = \mathbb{S}_1 \oplus \mathbb{S}_2 \oplus \cdots \oplus \mathbb{S}_r$, we have the decomposition of the identity

$$I = P_1 + P_2 + \cdots + P_r. \tag{18.9.2}$$

We are particularly concerned with *orthogonal projection* operators, for which $\mathbb{S}_2 = \mathbb{S}_1^{\perp}$ and $\mathbb{S}_1 = \mathbb{S}_2^{\perp}$. In this case we have the analog of Pythagoras's theorem:

$$\|\mathbf{u}\|^2 = \|\mathbf{u} - P_j\mathbf{u}\|^2 + \|P_j\mathbf{u}\|^2, \qquad j = 1, 2. \tag{18.9.3}$$

If two vectors are orthogonally decomposed as $\mathbf{u} = \mathbf{u}_1 + \mathbf{u}_2$, $\mathbf{v} = \mathbf{v}_1 + \mathbf{v}_2$ we have $(\mathbf{v}, P_1\mathbf{u}) = (\mathbf{v}, \mathbf{u}_1) = (\mathbf{v}_1, \mathbf{u}_1) = (P_1\mathbf{v}, \mathbf{u}_1) = (P_1\mathbf{v}, \mathbf{u})$ which shows that orthogonal projectors are Hermitian. It is also clear that $(\mathbf{u}, P_1\mathbf{u}) = \|\mathbf{u}_1\|^2$ so that an orthogonal projector is non-negative definite. If $\mathbb{S}_1$ and $\mathbb{S}_2$ can be decomposed further we can iterate (18.9.3) to find

$$\|\mathbf{u}\|^2 = \sum_{k=1}^{r} \|P_k\mathbf{u}\|^2. \tag{18.9.4}$$

If $\mathbf{e}_1, \ldots, \mathbf{e}_k$ is an orthonormal basis in $\mathbb{S}_1$, the orthogonal projection of $\mathbf{u}$ on $\mathbb{S}_1$ is realized by

$$P_1\mathbf{u} = \sum_{j=1}^{k} (\mathbf{e}_j, \mathbf{u})\, \mathbf{e}_j. \tag{18.9.5}$$

An equivalent representation of this projector can be given in terms of the (rectangular) matrix $M = [\mathbf{e}_1\ \mathbf{e}_2\ \ldots\ \mathbf{e}_k]$:

$$P_1 = MM^* = \sum_{j=1}^{k} \mathbf{e}_j\, \mathbf{e}_j^*, \tag{18.9.6}$$

where the outer product defined in (18.1.21) is used in the last expression.

We have seen that eigenspaces are invariant subspaces, and that eigenvectors belonging to different eigenvalues are orthogonal if the operator is normal (Theorem 18.7.4). Thus, one can define an orthogonal projector P_j for each one of these subspaces. The set of these projectors constitutes a decomposition of the identity as in (18.9.2) and, therefore,

$$A = AI = A \sum_{j=1}^{N} P_j = \sum_{j=1}^{N} \lambda_j P_j \qquad (18.9.7)$$

which is called the *spectral decomposition* of the operator. For example, the matrix

$$A = \frac{1}{3} \begin{vmatrix} 4 & \sqrt{2} \\ \sqrt{2} & 5 \end{vmatrix} \qquad (18.9.8)$$

has eigenvalues $\lambda_1 = 1$ and $\lambda_2 = 2$ with corresponding normalized eigenvectors

$$\mathbf{u}_1 = \frac{1}{\sqrt{3}} \begin{vmatrix} \sqrt{2} \\ -1 \end{vmatrix}, \qquad \mathbf{u}_2 = \frac{1}{\sqrt{3}} \begin{vmatrix} 1 \\ \sqrt{2} \end{vmatrix}, \qquad (18.9.9)$$

so that, from (18.9.6),

$$P_1 = \frac{1}{3} \begin{vmatrix} 2 & -\sqrt{2} \\ -\sqrt{2} & 1 \end{vmatrix}, \qquad P_2 = \frac{1}{3} \begin{vmatrix} 1 & \sqrt{2} \\ \sqrt{2} & 2 \end{vmatrix}, \qquad (18.9.10)$$

with which both (18.9.2) and (18.9.7) are readily seen to be verified.

Since the eigenspaces are orthogonal to each other, we have $P_j P_k = \delta_{jk} P_j$ so that we find from (18.9.7), for example, $A^2 = \sum_{j=1}^{N} \lambda_j^2 P_j$ and more generally, if $p(x)$ is a polynomial,

$$p(A) = \sum_{j=1}^{N} p(\lambda_j) P_j. \qquad (18.9.11)$$

This remark immediately shows that, if A is normal with simple eigenvalues, the projector P_j can be represented by the polynomial[22]

$$P_j = \frac{\prod_{k=1, k \neq j}^{N} (A - \lambda_k I)}{\prod_{k=1, k \neq j}^{N} (\lambda_j - \lambda_k)}. \qquad (18.9.12)$$

The explicit analog of this relation for derogatory matrices is somewhat more complicated (see e.g. Horn & Johnson 1991 p. 401). However, we have the expression of general validity

$$P_j = \frac{1}{2\pi i} \oint_{C_j} (\lambda I - A)^{-1} \, d\lambda, \qquad (18.9.13)$$

where the contour C_j is a closed simple rectifiable curve (p. 697) containing only the eigenvalue λ_j in its interior.[23] The inverse of $\lambda I - A$ can be expressed as a polynomial in $\lambda I - A$ by use e.g. of (18.7.6). In particular, with this step and (18.9.11), one can establish the

---

[22] (18.9.11) with P_j given by (18.9.12) is sometimes referred to as *Sylvester's formula*.
[23] Recall that the integral of a matrix is simply the matrix whose entries are the integrals of the entries of the integrand matrix.

equivalence of this representation with (18.9.12) for normal matrices and simple eigen-values. The operator $R_\lambda = (A - \lambda I)^{-1}$ is called the *resolvent* of A (see §21.9) and is an example of an operator-valued analytic function of the complex variable λ (see e.g. Henrici 1974, vol. 1 p. 122; Taylor & Lay 1980, chapter 5); a little more on such mathematical entities will be said in §21.2.4 and §21.10. By using two nested contours and the resolvent equation (21.10.1) p. 669 it is not difficult to show that the representation (18.9.13) satisfies the fundamental relation $P_j^2 = P_j$.

If the integration contour in (18.9.13) encloses all the eigenvalues, the integral equals the identity operator in agreement with the decomposition (18.9.2):

$$ I = \frac{1}{2\pi i} \oint_C (\lambda I - A)^{-1} \, d\lambda \,. \tag{18.9.14} $$

By combining (18.9.13) with (18.9.7) and recalling the use of residues in complex contour integration (§17.9) we also have

$$ A = \frac{1}{2\pi i} \oint_C \lambda (\lambda I - A)^{-1} \, d\lambda \,, \tag{18.9.15} $$

where, again, the contour encloses all the eigenvalues. These relations are special cases of the general ones given on p. 670.

The relation (18.9.13) gives the correct projector also in the case of non-normal matrices. For example, in the case of the matrix (18.7.10) on p. 514, we have

$$ (\lambda I - A)^{-1} = \frac{1}{(\lambda - a)^2 - b} \begin{vmatrix} \lambda - a & 1 \\ b & \lambda - a \end{vmatrix} \tag{18.9.16} $$

and

$$ P_\pm = \frac{1}{2\pi i} \oint_{C_\pm} (\lambda I - A)^{-1} \, d\lambda = \frac{1}{2} \begin{vmatrix} 1 & \pm b^{-1/2} \\ \pm b^{1/2} & 1 \end{vmatrix} \,, \tag{18.9.17} $$

where $C_\pm$ are contours around $\lambda_\pm = a \pm \sqrt{b}$ respectively. It is readily verified that for a vector $\mathbf{u}$, $P_\pm \mathbf{u} = (\mathbf{y}_\pm, \mathbf{u})\mathbf{x}_\pm$, with $\mathbf{x}_\pm$ and $\mathbf{y}_\pm$ as given in (18.7.11) and (18.7.14). It is also evident that $P_+ + P_- = I$. An example of the application of (18.9.13) to a defective eigenvalue is given on p. 526.

18.10 Defective matrices

By definition, defective matrices do not have a full set of eigenvectors and several of the properties described in §18.7 for simple matrices fail. A theorem of general validity is the following:

Theorem 18.10.1 [Schur] *Any square matrix is unitarily similar to an upper triangular matrix.*

In other words, for any matrix A one can find a unitary matrix U such that U*AU is a matrix having non-zero entries only on the main diagonal and, possibly, the positions above it. This result can be considerably strengthened to put the upper triangular matrix in a particularly convenient form, namely the Jordan standard form. Rather than giving a full proof, we describe the main building blocks.

Theorem 18.10.2 [Fitting's lemma] *For any square matrix of order N, the space $\mathbb{S}$ can be decomposed as $\mathbb{S} = \mathbb{M} + \mathbb{N}$ so that A is non-singular when restricted to the subspace $\mathbb{M}$ and nilpotent when restricted to the subspace $\mathbb{N}$.*

Since the space is finite-dimensional, there must be a smallest finite k, called the *index* of A, such that $\mathscr{R}(A^{k+1}) = \mathscr{R}(A^k)$ (cf. e.g. the Cayley–Hamilton theorem). Then $\mathbb{M} = \mathscr{R}(A^k)$, the range of A^k, and $\mathbb{N} = \mathscr{N}(A^k)$, its null space; it is shown that the two subspaces are complements of each other in $\mathbb{S}$. Because of this circumstance, A is non-singular when restricted to $\mathbb{M}$ and nilpotent of index k when restricted to $\mathbb{N}$. If A is non-singular, then $k = 1$, $\mathbb{M} = \mathbb{S}$ and $\mathbb{N} = \{\emptyset\}$. For a complete proof the reader may consult e.g. Halmos 1958, p. 113; Cullen 1972, p. 195; Piziak & Odell 2007, p. 129.

If an operator N is nilpotent of index p, so that $N^p = 0$ while $N^{p-1} \neq 0$, then one can find a vector $\mathbf{u}$ such that $\mathbf{u}, N\mathbf{u}, \ldots, N^{p-1}\mathbf{u}$ are linearly independent.[24] The subspace spanned by these vectors, which we denote by $\mathbb{S}_p$, is invariant with respect to N, and so is its orthogonal complement. Take these vectors as the first p vectors of a basis for the N-dimensional space $\mathbb{S}$ and add $N - p$ linearly independent vectors from the orthogonal complement to form a basis for the whole space (this is always possible, cf. remark after Equation (18.3.3)). With respect to this basis the matrix N takes the form

$$N = \begin{vmatrix} N_1 & 0 \\ 0 & N_2 \end{vmatrix}. \tag{18.10.1}$$

A generic vector of the subspace $\mathbb{S}_p$ may be written in the form $\sum_{j=1}^{p} c_j N^{p-j}\mathbf{u}$ and, recalling that $N^p\mathbf{u} = \emptyset$, $N \sum_{j=1}^{p} c_j N^{p-j}\mathbf{u} = \sum_{j=1}^{p-1} c_{j+1} N^{p-j}\mathbf{u}$. Thus the submatrix N_1 must be $p \times p$ and have the form

$$N_1 = \begin{vmatrix} 0 & 1 & 0 & \ldots & 0 \\ 0 & 0 & 1 & \ldots & 0 \\ \vdots & \vdots & \vdots & \vdots & \vdots \\ 0 & 0 & 0 & \ldots & 1 \\ 0 & 0 & 0 & \ldots & 0 \end{vmatrix}. \tag{18.10.2}$$

---

[24] Take $p = 2$ as an example; we want to show that $a\mathbf{u} + b N\mathbf{u}$ cannot vanish for all $\mathbf{u}$ without a and b both vanishing. If $a = 0$ for all $\mathbf{u}$ but $b \neq 0$, $N\mathbf{u}$ would have to vanish for all $\mathbf{u}$ and then N would be 0 and the index of nilpotency would be 1, contrary to the hypothesis. Thus, there must be at least one vector for which $a \neq 0$, which also implies that $b N\mathbf{u} \neq \emptyset$, so that neither b nor $N\mathbf{u}$ can vanish. Apply N to find $a N\mathbf{u} + b N^2\mathbf{u} = \emptyset = a N\mathbf{u}$ from which $a = 0$ since $N\mathbf{u} \neq \emptyset$.

Indeed, a simple direct calculation shows that the non-zero elements of a matrix of this form move up to the next diagonal every time N_1 is multiplied by itself, until N_1^p is the 0 matrix. Clearly the same property remains true if some of the 1s are replaced by 0s.

Repeated application of Theorem 18.10.2 leads to the following conclusion:

Theorem 18.10.3 *Let* A *be a square* $N \times N$ *matrix with the characteristic polynomial* $C(\lambda) = \Pi_{k=1}^{r}(\lambda - \lambda_k)^{s_k}$, *where* $\sum_{k=1}^{r} s_k = N$. *Then there are* r *invariant subspaces* $\mathbb{S}_k$ *of* $\mathbb{S}$, *each one an eigenspace of* λ_k *of dimension* s_k, *such that* $\mathbb{S} = \mathbb{S}_1 + \cdots + \mathbb{S}_r$. *Restricted to each one of these subspaces* A *equals* $\lambda_k I + N_k$ *where* N_k *is nilpotent.*

Building on these results, one can prove the following theorem:

Theorem 18.10.4 [Jordan canonical form] *If the characteristic polynomial of a square matrix* A *of order* N *is written in the form* $C(\lambda) = \Pi_{k=1}^{r}(\lambda - \lambda_k)^{s_k}$, *where* $\sum_{k=1}^{r} s_k = N$, *a non-singular matrix* T *can be found such that*

$$T^{-1}AT = J = \begin{vmatrix} J_1 & 0 & 0 & \cdots & 0 \\ 0 & J_2 & 0 & \cdots & 0 \\ \vdots & \vdots & \vdots & \vdots & \vdots \\ 0 & 0 & \cdots & 0 & J_r \end{vmatrix}, \qquad (18.10.3)$$

where each $s_k \times s_k$ *block* J_k *has the form*

$$J_k = \lambda_k I + N_k = \begin{vmatrix} \lambda_k & * & 0 & \cdots & 0 \\ 0 & \lambda_k & * & \cdots & 0 \\ \vdots & \vdots & \vdots & \vdots & \vdots \\ 0 & 0 & 0 & \cdots & * \\ 0 & 0 & 0 & \cdots & \lambda_k \end{vmatrix} \qquad (18.10.4)$$

in which the asterisks are either 1 *or* 0.

A matrix reduced to the form (18.10.3) is said to be in the *Jordan canonical form*. Blocks for which some of the asterisks in (18.10.4) are 0 can be further subdivided into elementary sub-blocks. For example, the three blocks

$$\begin{vmatrix} \mu & 0 & 0 \\ 0 & \mu & 0 \\ 0 & 0 & \mu \end{vmatrix}, \qquad \begin{vmatrix} \mu & 0 & 0 \\ 0 & \mu & 1 \\ 0 & 0 & \mu \end{vmatrix}, \qquad \begin{vmatrix} \mu & 1 & 0 \\ 0 & \mu & 1 \\ 0 & 0 & \mu \end{vmatrix} \qquad (18.10.5)$$

all have the single eigenvalue μ. The first one can be subdivided into three 1×1 elementary blocks, the second one into one 1×1 and one 2×2 elementary blocks, while the third one cannot be subdivided further. For the first one of (18.10.5) one can find three linearly independent eigenvectors, for the second one two, and for the third one only one. This is a general property: the number of distinct elementary blocks containing the same eigenvalue equals the number of linearly independent eigenvectors corresponding to that eigenvalue.

Thus, the elementary Jordan blocks of simple matrices are all 1×1 and the Jordan normal form is purely diagonal in agreement with the results of §18.7.

When, as in (18.10.5), the same eigenvalue occurs in more than one elementary Jordan block, the matrix is termed *derogatory*. As noted on p. 512, the annihilating polynomial for a derogatory matrix of order N has a degree smaller than N (see below after Example 18.10.1). A derogatory matrix is only defective when the number of linearly independent eigenvectors is not equal to its order. Thus, in the examples (18.10.5), while the matrices are all derogatory, only the second and the third one are defective.

To develop this analysis further, let us consider for simplicity the case of a single eigenvalue λ of algebraic multiplicity N and geometric multiplicity 1 in an N-dimensional space.

In these hypotheses the matrix A can be reduced to the single Jordan block (18.10.4) in which all the asterisks are 1. In the basis in which the matrix takes this form, the vector $\mathbf{x}_1^T = |1\ 0\ \ldots\ 0|^T$ is clearly an ordinary eigenvector of λ since $(J - \lambda I)\mathbf{x}_1 = 0$. The vector $\mathbf{x}_2^T = |0\ 1\ \ldots\ 0|^T$ satisfies $(J - \lambda I)\mathbf{x}_2 = \mathbf{x}_1$ and, therefore, $(J - \lambda I)^2 \mathbf{x}_2 = 0$. Continuing in this way we construct a *Jordan chain* :

$$(J - \lambda I)\mathbf{x}_{k+1} = \mathbf{x}_k \quad \text{or} \quad (J - \lambda I)^{k+1}\mathbf{x}_{k+1} = 0. \tag{18.10.6}$$

A direct calculation shows that $(J - \lambda I)$ raised to the dimension of the block is identically zero, so that the Jordan chain stops when the number of its vectors equals the dimension of the block. Vectors satisfying (18.10.6) are *generalized eigenvectors*, or *principal vectors*, of the operator J. While these vectors are linearly independent, it is clear from the second of (18.10.6) that each $\mathbf{x}_k$ is determined only up to an arbitrary linear combination of $\mathbf{x}_1, \ldots, \mathbf{x}_{k-1}$.

If we replace J by $T^{-1}AT$ we find $T^{-1}(A - \lambda I)T\mathbf{x}_{k+1} = \mathbf{x}_k$ which shows that the generalized eigenvectors of A are $\mathbf{u}_k = T\mathbf{x}_k$ and

$$(A - \lambda I)\mathbf{u}_{k+1} = \mathbf{u}_k. \tag{18.10.7}$$

Since by construction the components of each $\mathbf{x}_k$ are all zero except the one in the k-position, $T\mathbf{x}_k$ is just the k-th column of T. Thus, we conclude that the columns of the matrix effecting the reduction to the Jordan form are the generalized eigenvectors of A. The subspace spanned by the ordinary plus generalized eigenvectors is a *generalized eigenspace* of A and, just like an ordinary eigenspace, it is invariant under A.[25]

Some insight into the origin of generalized eigenvectors can be gained from the following heuristic argument. As noted on p. 513, any $N \times N$ matrix can be approximated arbitrarily closely by a diagonalizable matrix. Let then $A + \varepsilon B$ be a diagonalizable approximation to a defective matrix A and let λ and $\lambda + \varepsilon\mu$ be two of its eigenvalues so that

$$(A + \varepsilon B)\mathbf{u} = (\lambda + \varepsilon\mu)\mathbf{u}, \qquad (A + \varepsilon B)\mathbf{w} = \lambda\mathbf{w}. \tag{18.10.8}$$

---

[25] Since $(J - \lambda I)$ raised to the dimension of the block is zero, the same is true for $(A - \lambda I)$, which makes this matrix nilpotent of the same order on its generalized eigenspace, in agreement with Theorem 18.10.3. The fact that the geometric multiplicity cannot exceed the algebraic multiplicity is a consequence of the fairly obvious chain of inclusions $\mathscr{N}(A - \lambda_j I) \subseteq \mathscr{N}\big((A - \lambda_j I)^2\big) \subseteq \ldots$ (see e.g. Dym 2006, pp. 67–68).

We can expect relations of this form to hold as the eigenvalues are continuous functions of the elements of a matrix, given that the roots of a polynomial depend continuously on the coefficients. Since the difference between $\mathbf{u}$ and $\mathbf{w}$ is of order ε, the limit $(\mathbf{u} - \mathbf{w})/(\varepsilon\mu)$ is finite for $\varepsilon \to 0$. If we let this limit be $\mathbf{v}$, upon subtracting the two equations, dividing by $\varepsilon\mu$ and taking the limit we find, similar to (18.10.7),

$$A\mathbf{v} = \lambda\mathbf{v} + \mathbf{u}. \tag{18.10.9}$$

Example 18.10.1 To illustrate the previous analysis, consider the matrix

$$A = \begin{vmatrix} 5 & 4 & 3 \\ -1 & 0 & -3 \\ 1 & -2 & 1 \end{vmatrix}. \tag{18.10.10}$$

The characteristic polynomial is $C(\lambda) = (\lambda + 2)(\lambda - 4)^2$. The eigenvalue -2 has an ordinary eigenvector which is an arbitrary multiple of $\mathbf{u}_1^T = |1 \ -1 \ -1|^T$. The eigenvalue 4 has only one ordinary eigenvector, an arbitrary multiple of $\mathbf{u}_2^T = |1 \ -1 \ 1|^T$. While $\mathbf{u}_1$ and $\mathbf{u}_2$ are linearly independent, they are not orthogonal as the matrix is not normal. To find the generalized eigenvector we solve

$$(A - 4I)\mathbf{u}_3 = c\mathbf{u}_2 = c|1 \ -1 \ 1|^T \tag{18.10.11}$$

where c is an arbitrary constant. The solution is

$$\mathbf{u}_3^T = c|0 \ 1 \ -1|^T + a\mathbf{u}_2^T, \tag{18.10.12}$$

where a is another arbitrary constant reflecting the arbitrariness in the determination of the generalized eigenvector. The transforming matrix and its inverse are therefore, with b a third arbitrary constant multiplying the ordinary eigenvector $\mathbf{u}_1$,

$$T = \begin{vmatrix} b & c & a \\ -b & -c & c-a \\ -b & c & a-c \end{vmatrix}, \quad T^{-1} = \frac{1}{2bc^2} \begin{vmatrix} 0 & -c^2 & -c^2 \\ 2b(c-a) & b(c-2a) & bc \\ 2bc & 2bc & 0 \end{vmatrix} \tag{18.10.13}$$

and one finds

$$T^{-1}AT = \begin{vmatrix} -2 & 0 & 0 \\ 0 & 4 & 1 \\ 0 & 0 & 4 \end{vmatrix} \tag{18.10.14}$$

irrespective of the values of a, b, c; a can be taken as 0, but this choice of course is not available for b or c.

It is also interesting to see what the formula (18.9.13) gives for the projector onto the eigenspace corresponding to the degenerate eigenvalue $\lambda = 4$. The result is more transparent (and quicker to calculate) if we start from

$$T^{-1}PT = \frac{1}{2\pi i} \oint_C (\lambda I - T^{-1}AT)^{-1} \, d\lambda, \tag{18.10.15}$$

where the integration contour surrounds $\lambda = 4$ but not $\lambda = -2$. We find

$$\frac{1}{2\pi i} \oint_C (\lambda I - T^{-1}AT)^{-1} \, d\lambda = \frac{1}{2\pi i} \oint_C d\lambda \begin{vmatrix} (\lambda + 2)^{-1} & 0 & 0 \\ 0 & (\lambda - 4)^{-1} & -(\lambda - 4)^{-2} \\ 0 & 0 & (\lambda - 4)^{-1} \end{vmatrix}$$

$$= \begin{vmatrix} 0 & 0 & 0 \\ 0 & 1 & 0 \\ 0 & 0 & 1 \end{vmatrix}, \tag{18.10.16}$$

from which P is readily found by pre-multiplying by T and post-multiplying by T^{-1}. $\square$

It was mentioned earlier in connection with the Cayley–Hamilton Theorem 18.7.1 that, while it is always true that an $N \times N$ matrix satisfies its own characteristic equation, there may be polynomials of degree smaller than N such that $p(A) = 0$. The *minimum polynomial*, $m_A(x)$, of A is the polynomial of the least degree such that $m_A(A) = 0$. This polynomial is unique if it is *monic*, i.e., if the coefficient of the highest power is unity. The minimum polynomial is a factor of the characteristic polynomial $(\lambda - \lambda_1)^{s_1}(\lambda - \lambda_2)^{s_2} \ldots (\lambda - \lambda_r)^{s_r}$ (cf. Equation 18.6.5), but it is not necessarily just $(\lambda - \lambda_1)(\lambda - \lambda_2) \ldots (\lambda - \lambda_r)$. Rather, it consists of the product of factors $(\lambda - \lambda_k)$, each one raised to the dimension of the largest elementary Jordan block associated with λ_k. Thus, for example, the three matrices (18.10.5) all have the same characteristic polynomial $(\lambda - \mu)^3$ but their minimum polynomials are, respectively, $A - \mu I$, $(A - \mu I)^2$ and $(A - \mu I)^3$. An extensive treatment of the minimum polynomial can be found in Gantmacher 1959, chapter 4; see also Lancaster & Tismenetsky 1985, Cullen 1972, Piziak & Odell 2007 and Ortega 1987.

The spectral decomposition of a defective matrix is somewhat more complicated than the form (18.9.7) valid for simple matrices. By using $A = TJT^{-1}$ it is evident from (18.10.3) that the decomposition will have the form

$$A = \sum_{j=1}^{N} (\lambda_j P_j + Q_j). \tag{18.10.17}$$

This relation can also be derived from (18.9.15) as

$$A = \frac{1}{2\pi i} \oint_C \lambda (\lambda I - A)^{-1} \, d\lambda = \frac{1}{2\pi i} \sum_j \oint_{C_j} \lambda (\lambda I - A)^{-1} \, d\lambda$$

$$= \frac{1}{2\pi i} \sum_j \left[\lambda_j \oint_{C_j} (\lambda I - A)^{-1} \, d\lambda + \oint_{C_j} (\lambda - \lambda_j)(\lambda I - A)^{-1} \, d\lambda \right]$$

$$= \sum_j \left[\lambda_j P_j + \frac{1}{2\pi i} \oint_{C_j} (\lambda - \lambda_j)(\lambda I - A)^{-1} \, d\lambda \right]. \tag{18.10.18}$$

The operator P_j is the projector in the direction of the ordinary eigenvector; it is idempotent and satisfies the decomposition of the identity (18.9.2). If the eigenvalue is simple, $(\lambda - \lambda_j)(\lambda I - A)^{-1}$ is analytic inside the contour of integration and $Q_j = 0$. By using nested contours and the resolvent equation (21.10.1) p. 669 it can be shown that

$$Q_j^n = \frac{1}{2\pi i} \oint_{C_j} (\lambda - \lambda_j)^n (\lambda I - A)^{-1} \, d\lambda \qquad (18.10.19)$$

and therefore, since $(\lambda I - A)^{-1}$ has only poles, Q_j must be nilpotent of an order equal to the order of the j-th pole.[26] By similar methods (or otherwise, see e.g. Lancaster & Tismenetsky 1985, chapter 5) one can also show that $P_j Q_k = Q_k P_j = \delta_{jk} Q_j$, $A P_k = P_k A = \lambda_k P_k + Q_k$ and $Q_j Q_k = 0$ if $j \neq k$.

18.11 Functions of matrices

Since the powers of a square matrix A are well defined, it is evident how one can give a meaning to the function $p(A)$ when $p(x)$ is a polynomial. Provided one can introduce the notion of convergence for a series of matrices, this remark opens the way to the definition of the function $f(A)$ for any function $f(x)$ which can be expanded in a power series, $f(x) = \sum_{k=0}^{\infty} c_k x^k$. The standard procedure is to consider the convergence of the sequence of partial sums $B^{(n)} = \sum_{k=0}^{n} c_k A^k$. This notion will be discussed for more general linear operators in §21.2. Let us just say here that the sequence converges to a matrix $B \equiv f(A)$ as $n \to \infty$ if and only if the (i, j) entry of $B^{(n)}$ converges to the corresponding entry of B in the sense of numerical series (see §8.1). This notion is equivalent to convergence in the sense of any of the operator norms such as those introduced in §18.8.

Let us start by considering simple matrices which, according to Theorem 18.7.2, can be diagonalized by a similarity transformation, $A = T \Lambda T^{-1}$, so that, as in (18.1.20), $A^k = T \Lambda^k T^{-1}$. In this case, for the partial sums $B^{(n)}$ we have

$$B^{(n)} = T \, \text{diag} \left[\sum_{k=0}^{n} c_k \lambda_1^k, \ \ldots, \ \sum_{k=0}^{n} c_k \lambda_N^k \right] T^{-1} \qquad (18.11.1)$$

and therefore $f(A)$ will exist provided all its eigenvalues are inside the circle of convergence of the power series which defines the function $f(x)$ (see §8.4 p. 228 for a definition of these terms). As an example one may consider the matrix exponential, which is of particular interest in applications (see §18.12). The power series for e^x has an infinite radius of convergence and, therefore,

$$e^A = T \begin{vmatrix} e^{\lambda_1} & 0 & \ldots & 0 \\ 0 & e^{\lambda_2} & \ldots & 0 \\ 0 & 0 & \ldots & 0 \\ 0 & 0 & \ldots & e^{\lambda_N} \end{vmatrix} T^{-1}. \qquad (18.11.2)$$

---

[26] Indeed, if n_j is the order of the j-th pole, $(\lambda - \lambda_j)^n (\lambda I - A)^{-1}$ is analytic for any $n \geq n_j$ so that its integral around C vanishes.

A point to note is that the matrix exponential so defined lacks some of the properties of the usual exponential function, such as the equality of $\exp(A + B)$ and $(\exp A)(\exp B)$, which however holds if A and B commute; in particular $\exp A \exp(-A) = I$.[27]

Relations such as (18.11.1) and (18.11.2) suggest a more direct way to define $f(A)$ for a simple matrix, which is applicable also when f cannot be defined in terms of a power series, namely

$$f(A) = T\, f(\Lambda)\, T^{-1} \quad \text{where} \quad f(\Lambda) = \text{diag}\,[f(\lambda_1), \ldots, f(\lambda_N)] \qquad (18.11.3)$$

provided the function f is defined on the spectrum of A. Recalling that the eigenvectors of A are the columns of the matrix T and the spectral decomposition (18.9.7), we have the equivalent expression

$$f(A) = \sum_{j=1}^{N} f(\lambda_j) P_j, \qquad (18.11.4)$$

in which P_j is the orthogonal projector in the direction of the j-th eigenvector. In this way we can define, e.g., the square root $A^{1/2}$ of a non-negative definite matrix. For the matrix (18.9.8), for example, this procedure gives

$$A^{1/2} = P_1 + \sqrt{2}P_2 = \frac{1}{3} \begin{vmatrix} 2 + \sqrt{2} & 2 - \sqrt{2} \\ 2 - \sqrt{2} & 2\sqrt{2} + 1 \end{vmatrix}. \qquad (18.11.5)$$

One can verify directly that $A^{1/2}A^{1/2}$ reproduces the original matrix.[28] As another example

$$(\lambda I - A)^{-1} = \sum_{j=1}^{N} \frac{1}{\lambda - \lambda_j} P_j. \qquad (18.11.6)$$

This definition fails when λ is an eigenvalue as then the operator $\lambda I - A$ does not have an inverse (§18.8).

Defective matrices cannot be diagonalized but can be brought into the Jordan normal form. To see how (18.11.3) extends to this case we start by noting that, if J is an elementary Jordan block of the form (18.10.4) with all the asterisks replaced by 1, then J^n can be written as[29]

$$J^n = \begin{vmatrix} \lambda^n & \frac{1}{1!}(\lambda^n)' & \frac{1}{2!}(\lambda^n)'' & \cdots & \frac{1}{(n-1)!}(\lambda^n)^{(n-1)} \\ 0 & \lambda^n & \frac{1}{1!}(\lambda^n)' & \cdots & \frac{1}{(n-2)!}(\lambda^n)^{(n-2)} \\ 0 & 0 & \cdots & \cdots & \cdots \\ 0 & 0 & \cdots & \lambda^n & \frac{1}{1!}(\lambda^n)' \\ 0 & 0 & \cdots & 0 & \lambda^n \end{vmatrix}, \qquad (18.11.7)$$

[27] If all the entries of A and B are algebraic numbers (i.e., roots of polynomial equations with rational coefficients), then $(\exp A)(\exp B) = \exp(A + B)$ if and only if A and B commute; if some of the entries are non-algebraic numbers (e.g., π), commutativity is sufficient but not necessary, see Horn & Johnson 1991 p. 437.

[28] The square of e.g. $P_1 - \sqrt{2}P_2$ would also reproduce the original matrix; the number of distinct branches of $f(A)$ is not necessarily equal to number of distinct branches of $f(x)$.

[29] This result can be proven by induction writing $J_{ij} = \lambda\delta_{ij} + \delta_{i+1,j}$.

where the superscripts $(\ldots)^{(k)}$ indicate differentiation. For a function f we therefore define

$$f(A) = T \operatorname{diag}(f(J_1) \, f(J_2) \ldots f(J_m)) \, T^{-1}, \qquad (18.11.8)$$

in which the J_k are the elementary Jordan blocks of A, each with dimension $n_k \times n_k$, and

$$f(J_k) = \begin{vmatrix} f(\lambda_k) & \frac{1}{1!}f'(\lambda_k) & \frac{1}{2!}f''(\lambda_k) & \cdots & \frac{1}{(n_k-1)!}f^{(n_k-1)}(\lambda_k) \\ 0 & f(\lambda_k) & \frac{1}{1!}f'(\lambda_k) & \cdots & \frac{1}{(n_k-2)!}f^{(n_k-2)}(\lambda_k) \\ 0 & 0 & \cdots & \cdots & \cdots \\ 0 & 0 & \cdots & f(\lambda_k) & \frac{1}{1!}f'(\lambda_k) \\ 0 & 0 & \cdots & 0 & f(\lambda_k) \end{vmatrix} \qquad (18.11.9)$$

provided the indicated derivatives of f exist on the spectrum of A. As an example, if A is non-singular, its inverse is given by (18.11.8) in which, for each elementary Jordan block,

$$J_k^{-1} = \begin{vmatrix} \lambda_k^{-1} & -\lambda_k^{-2} & \lambda_k^{-3} & \cdots \\ 0 & \lambda_k^{-1} & -\lambda_k^{-2} & \cdots \\ & & \cdots & \\ 0 & \cdots & 0 & \lambda_k^{-1} \end{vmatrix}. \qquad (18.11.10)$$

A third way of defining $f(A)$, applicable when $f(\lambda)$ is analytic inside a closed contour C which includes the spectrum of A, is by a Cauchy integral:

$$f(A) = \frac{1}{2\pi i} \oint_C f(\lambda) \, (\lambda I - A)^{-1} \, d\lambda. \qquad (18.11.11)$$

When A is diagonalizable, the truth of this statement follows from Cauchy's integral theorem (p. 428) applied to the integrals of the diagonal elements $1/(\lambda - \lambda_j)$. For the general case, one can have recourse to the continuity argument mentioned after Theorem 18.7.2 p. 513. Relations such as (18.9.14) and (18.9.15) clearly are special cases of (18.11.11).

The above definitions imply that two different functions $f(x)$ and $g(x)$ having, with the necessary derivatives, the same values on the spectrum of A, define the *same* function of A. In particular, therefore, any function of $f(A)$ can be represented by a polynomial interpolating $f(x)$ (and the requisite derivatives) on the spectrum of A. For example, use of the representation (18.9.12) for the projectors P_1 and P_2 in (18.11.5) expresses $A^{1/2}$ as a polynomial in A.

Much remains to be said on matrix functions; the reader is referred e.g. to Horn & Johnson, 1991 chapter 6 for an exhaustive treatment.

18.12 Systems of ordinary differential equations

A system of ordinary differential equations of any order can be reduced to a system of first-order equations simply by considering all the derivatives of the unknown function(s) except the highest one as auxiliary unknowns. As an example, consider the n-th order equation

$$\frac{d^n x}{dt^n} + c_1 \frac{d^{n-1}x}{dt^{n-1}} + \cdots + c_n x(t) = b(t) \qquad (18.12.1)$$

and, for $k = 0, 1, 2, \ldots, n-1$, let $x_k(t) = \mathrm{d}^k x/\mathrm{d}t^k$. Then the equation can be written as the first-order system

$$
\frac{\mathrm{d}}{\mathrm{d}t}
\begin{vmatrix} x_0 \\ x_1 \\ \vdots \\ x_{n-1} \end{vmatrix}
=
\begin{vmatrix}
0 & 1 & 0 & \cdots & 0 \\
0 & 0 & 1 & \cdots & 0 \\
& & \vdots & & \\
-c_n & -c_{n-1} & \cdots & -c_2 & -c_1
\end{vmatrix}
\begin{vmatrix} x_0 \\ x_1 \\ \vdots \\ x_{n-1} \end{vmatrix}
+
\begin{vmatrix} 0 \\ 0 \\ \vdots \\ b(t) \end{vmatrix}. \qquad (18.12.2)
$$

The matrix appearing here is the transpose of the so-called *companion matrix* of the polynomial $c(\lambda) = \lambda^n + c_1 \lambda^{n-1} + \cdots + c_n$.[30] The same procedure is readily extended to systems with more than one unknown function.

Thus, considering the linear homogeneous case, there is no loss of generality in restricting ourselves to first-order systems:

$$
\frac{\mathrm{d}x_i}{\mathrm{d}t} = \sum_{j=1}^{N} a_{ij}(t)\, x_j(t), \qquad i = 1, 2, \ldots, N. \qquad (18.12.3)
$$

In terms of the vector $\mathbf{x}^T(t) = |x_1(t) \ldots x_N(t)|$ and of the matrix $A(t) = |a_{ij}(t)|$ the system can be written compactly as

$$
\frac{d\mathbf{x}}{dt} = A(t)\,\mathbf{x} \qquad (18.12.4)
$$

or, arranging N arbitrary linearly independent solutions $\mathbf{x}_1(t), \ldots, \mathbf{x}_N(t)$ as the columns of a matrix $X(t) = |\mathbf{x}_1 \ldots \mathbf{x}_N|$, as

$$
\frac{dX}{dt} = A(t)\,X. \qquad (18.12.5)
$$

Any matrix X satisfying this relation is a *fundamental matrix* for the linear homogeneous system. Different initial conditions $X_0 = X(0)$ give rise to different fundamental matrices.

It is easy to check that, if X_0 is non-singular, so is $X(t)$ for all t. Indeed, by applying the analog of (18.2.11) to differentiation by row and using (18.12.3) we find

$$
\frac{\mathrm{d}}{\mathrm{d}t} \det X = \left[\sum_{k=1}^{N} a_{kk}(t) \right] \det X \qquad (18.12.6)
$$

from which we have *Jacobi's identity*

$$
\det X(t) = \exp\left[\int_0^t \mathrm{Tr}\, A(t')\, \mathrm{d}t' \right] \det X_0. \qquad (18.12.7)
$$

Any two non-singular matrix solutions $X(t)$ and $Y(t)$ of (18.12.5) satisfy a relation of the form $Y(t) = X(t)C$ where C is a constant matrix since, obviously, $X(t)C$ is a solution of the equation if X is, and choosing $C = X_0^{-1} Y(0)$ satisfies the initial conditions on $Y(t)$. This remark indicates the particular importance of the matrix solution satisfying the initial condition $X(0) = I$, which is known as the *fundamental solution*.

---

[30] So called because the characteristic equation of the companion matrix is exactly the polynomial.

Let us start from the simplest case by considering a non-homogeneous system

$$\frac{d\mathbf{x}}{dt} = A\mathbf{x} + \mathbf{b}(t) \qquad \mathbf{x}(0) = \mathbf{x}_0 \tag{18.12.8}$$

with A constant and simple and $\mathbf{b}(t) = |b_j(t)|$ a prescribed vector. A simple matrix can be reduced to diagonal form by a similarity transformation, so that $A = T\Lambda T^{-1}$. Let us then multiply the equation by T letting $\boldsymbol{\xi} = T^{-1}\mathbf{x}$ to find

$$\frac{d\boldsymbol{\xi}}{dt} = \Lambda\boldsymbol{\xi} + \boldsymbol{\beta} \tag{18.12.9}$$

in which $\boldsymbol{\beta}(t) = T^{-1}\mathbf{b}(t)$. The different components ξ_j of $\boldsymbol{\xi}$ are no longer coupled and each scalar component of this equation is just

$$\frac{d\xi_j}{dt} = \lambda_j\xi_j + \beta_j, \tag{18.12.10}$$

so that

$$\xi_j(t) = e^{\lambda_j t}\xi_j(0) + \int_0^t e^{\lambda_j(t-t')}\beta_j(t')\,dt'. \tag{18.12.11}$$

Returning to the original variable $\mathbf{x}$ one finds

$$\mathbf{x}(t) = e^{At}\mathbf{x}_0 + \int_0^t e^{A(t-t')}\mathbf{b}(t')\,dt'. \tag{18.12.12}$$

Since

$$\frac{d}{dt}e^{At} = T\,\mathrm{diag}\left[\frac{de^{\lambda_k t}}{dt}\right]T^{-1} = T\,\mathrm{diag}\left[\lambda_k e^{\lambda_k t}\right]T^{-1} = A\,e^{At} \tag{18.12.13}$$

and $e^{A0} = I$, we recognize that e^{At} is the fundamental solution of (18.12.5).

In some cases the system may be given in the form

$$B\frac{d\mathbf{x}}{dt} = A\mathbf{x} + \mathbf{b}(t). \tag{18.12.14}$$

If B is non-singular, the form (18.12.8) is recovered, at least in principle, by multiplication by B^{-1}. When $\det B = 0$, the system is *anomalous*. The vanishing of $\det B$ means that there is at least one left eigenvector $\boldsymbol{\mu}$ of B corresponding to the 0 eigenvalue, $\boldsymbol{\mu}^T B = \emptyset$. If we multiply the original system by $\boldsymbol{\mu}^T$ we therefore find

$$\boldsymbol{\mu}^T A\mathbf{x} + \boldsymbol{\mu}^T\mathbf{b}(t) = \emptyset \tag{18.12.15}$$

which is a linear relationship between the unknowns and the forcing $\mathbf{b}$. One possibility of dealing with this situation is to use these relations to eliminate some unknowns from the original system until a normal system remains. If the Laplace transform is used to solve the problem, the use of generalized functions may be convenient (see pp. 127 and 132). Evaluated at $t = 0$, (18.12.15) imposes a constraint on the initial data or on the values of $\mathbf{b}(0)$.

When A is constant, but not simple, it can be reduced to the Jordan form by a similarity transformation. From (18.11.9) we now have

$$\frac{d}{dt}\exp(Jt) = \frac{d}{dt}\begin{vmatrix} e^{\lambda t} & \lambda e^{\lambda t} & \frac{1}{2}\lambda^2 e^{\lambda t} & \cdots \\ 0 & e^{\lambda t} & \lambda e^{\lambda t} & \cdots \\ 0 & 0 & \cdots & \cdots \end{vmatrix}$$

$$= \begin{vmatrix} \lambda e^{\lambda t} & \lambda^2 e^{\lambda t} & \frac{1}{2}\lambda^3 e^{\lambda t} & \cdots \\ 0 & \lambda e^{\lambda t} & \lambda^2 e^{\lambda t} & \cdots \\ 0 & 0 & \cdots & \cdots \end{vmatrix} = J\exp(Jt) \qquad (18.12.16)$$

from which we deduce that $\exp(At) = T\,\mathrm{diag}[\exp(J_1 t), \ldots, \exp(J_m t)]T^{-1}$ is again a fundamental solution of (18.12.5). The solution of the non-homogeneous problem (18.12.8) is therefore expressed in the same form as (18.12.12).

The case in which the matrix A depends on the variable t is more complex as now the matrices T depend on t as well so that diagonalization and differentiation cannot be interchanged as was done in (18.12.13). To gain some insight into the situation we integrate both sides of (18.12.8) between 0 and t to find

$$\mathbf{x}(t) = \mathbf{x}(0) + \int_0^t \left[A(t')\mathbf{x}(t') + \mathbf{b}(t') \right] dt'. \qquad (18.12.17)$$

Iterating once, we have

$$\mathbf{x}(t) = \mathbf{x}(0) + \int_0^t \left\{ A(t') \left[\mathbf{x}(0) + \int_0^{t'} \left[A(t'')\mathbf{x}(t'') + \mathbf{b}(t'') \right] dt'' \right] + \mathbf{b}(t') \right\} dt'$$
$$(18.12.18)$$

and, if the iteration is continued,

$$\mathbf{x}(t) = \left[I + \int_0^t A(t')\,dt' + \int_0^t dt'A(t') \int_0^{t'} A(t'')\,dt'' + \cdots \right] \mathbf{x}(0)$$

$$+ \int_0^t \mathbf{b}(t')\,dt' + \int_0^t dt'A(t') \int_0^{t'} \mathbf{b}(t'')\,dt'' + \cdots \qquad (18.12.19)$$

Upon defining the *matrizant* (or matricant or, in the more recent literature, *state transition matrix*) M of A by

$$M(t,0) = I + \int_0^t A(t')\,dt' + \int_0^t dt'A(t') \int_0^{t'} A(t'')\,dt'' + \cdots \qquad (18.12.20)$$

the previous result may be written

$$\mathbf{x}(t) = M(t,0)\mathbf{x}(0) + M(t,0) \int_0^t M^{-1}(t',0)\mathbf{b}(t')\,dt'. \qquad (18.12.21)$$

When A is constant the iteration (18.12.19) just produces the power series expansion of $\exp At$, which is therefore seen to be the matrizant for constant matrices. It is evident that $M(t,t) = I$. From the definition one can prove that (see e.g. Gantmacher 1959, vol. 2 pp. 125–141; Pease 1965 chapter 7)

$$M(s,t')\,M(t',t) = M(s,t), \qquad M^{-1}(t,s) = M(s,t) \qquad (18.12.22)$$

and also that

$$\frac{\partial M(t, s)}{\partial t} = A(t)\, M(t, s), \qquad \frac{\partial M(t, s)}{\partial s} = M(t, s)\, A(s). \qquad (18.12.23)$$

The matrizant is not easy to use in general (see e.g. Moler & van Loan 1978). The expansion (18.12.20), though convergent under mild assumptions, usually converges only slowly. Approximate methods can, however, be based on this concept. For example, if $A(t)$ can be split into two matrices, $A(t) = A_0(t) + A_1(t)$, it can be shown that the matrizant may be written as $M = M_0 N$, where M_0 is the matrizant of A_0 and $N(t, s)$ satisfies

$$\frac{\partial N}{\partial t} = (M_0^{-1} A_1 M_0)\, N. \qquad (18.12.24)$$

If M_0 is known (e.g., when A_0 is constant) and A_1 is small in a suitable sense, one may approximate N by a few terms of the expansion (18.12.20); see Pease 1965 for an example.

PART III

Some Advanced Tools

19

Infinite-Dimensional Spaces

The similarity that the method of eigenfunction expansion bears to operations familiar from ordinary finite-dimensional spaces has been repeatedly emphasized in Part I and elsewhere in this book. In this chapter we begin to put this analogy on a firm mathematical basis by describing several aspects of the abstract theory of infinite-dimensional spaces. We introduce a notion of distance – the norm – which permits us to define convergence in a general way, we add the geometric notion of scalar product, which leads to the key property of orthogonality, and we describe linear functionals, the simplest example of linear operators and also a necessary premise for the definition of generalized functions in the next chapter.

19.1 Linear vector spaces

A linear vector space $\mathbb{S}$ is built from a non-empty set $u, v, w, \ldots$ of arbitrary entities, to which we refer as *elements*, *vectors* or *points*, and scalars (numbers) $a, b, c, \ldots$ from a field $\mathbb{F}$ (which we will take as the set $\mathbb{R}$ of real numbers or the set $\mathbb{C}$ of complex numbers), by defining two operations. The first one, called *addition* and denoted as $u + v$, is a rule which associates to any pair of elements of $\mathbb{S}$ an element of $\mathbb{S}$ (in other words, a mapping $\mathbb{S} \times \mathbb{S} \to \mathbb{S}$). The second operation, called *multiplication* (or multiplication by scalars) and denoted as $a \cdot u$ or, more simply, $a\,u$, is a rule which associates an element of $\mathbb{S}$ to the multiplication of any element of $\mathbb{S}$ by any number of $\mathbb{F}$ (i.e., a mapping $\mathbb{F} \times \mathbb{S} \to \mathbb{S}$). When $\mathbb{F} = \mathbb{R}$, we have a *real vector space* and a *complex vector space* if $\mathbb{F} = \mathbb{C}$.

In brief, the key requirement that these operations must satisfy for the collection of elements to constitute a linear vector space $\mathbb{S}$ over the given field is that, for any pair of numbers a, b of the field (including 0), and any pair of elements u, v of the set,

$$a \cdot u + b \cdot v \in \mathbb{S} \qquad \text{for all } u, v \in \mathbb{S}, \qquad \text{for all } a, b \in \mathbb{F}. \tag{19.1.1}$$

In other words, the set $\mathbb{S}$ must be *closed* under the operation of addition and multiplication. A detailed list of the properties that should be satisfied for $\mathbb{S}$ to qualify as a linear vector space is given in Tables 19.1 and 19.2. This is not the most concise set of axioms, in the sense that some of the properties indicated are consequence of others although we omit the proofs. These properties are of course all very familiar from the theory of ordinary vectors, which is one of the reasons why adopting the point of view of abstract linear vector spaces and operators is such a powerful aid to intuition.

	Properties of the addition operation (the symbol ∀ is read "for all"; the symbol ∃ is read "there exists").	
Table 19.1		

	Property	Definition/Consequences
(i)	Closure	$\forall\, u,\, v \in \mathbb{S},\, u + v \in \mathbb{S}$
(ii)	Commutativity	$\forall\, u,\, v \in \mathbb{S},\, u + v = v + u$
(iii)	Associativity	$\forall\, u,\, v,\, w \in \mathbb{S},\, (u + v) + w = u + (v + w)$
(iv)	Existence of neutral element	$\exists\, \emptyset \in \mathbb{S}$ such that, $\forall\, u \in \mathbb{S},\, u + \emptyset = u$
(v)	Uniqueness of neutral element	If, $\forall\, u \in \mathbb{S},\, u + \emptyset_1 = u,\, u + \emptyset_2 = u$ then $\emptyset_1 = \emptyset_2$
(vi)	Existence of the opposite	$\forall\, u \in \mathbb{S} \,\exists\, (-u) \in \mathbb{S}$ such that $u + (-u) = \emptyset$

	Properties of the operation of multiplication by scalars. For simplicity, the same symbol "+" is used to denote addition of scalars and vectors although these are, in fact, different notions.	
Table 19.2		

	Property	Definition/Consequences
(a)	Closure	$\forall\, u \in \mathbb{S}$ and $\forall\, a \in \mathbb{F},\, a \cdot u \in \mathbb{S}$
(b)	Distributive property I	$\forall\, a,\, b \in \mathbb{F}$ and $\forall\, u \in \mathbb{S},\, (a + b) \cdot u = a \cdot u + b \cdot u$
(c)	Distributive property II	$\forall\, a \in \mathbb{F}$ and $\forall\, u,\, v \in \mathbb{S},\, a \cdot (u + v) = a \cdot u + a \cdot v$
(d)	Associativity	$\forall\, a,\, b \in \mathbb{F}$ and $\forall\, u \in \mathbb{S},\, (ab) \cdot u = a \cdot (b \cdot u)$
(e)	Neutral element	For $1 \in \mathbb{F},\, \forall\, u \in \mathbb{S},\, 1 \cdot u = u$
(f)	Multiplication by 0	$\forall\, u \in \mathbb{S},\, 0 \in \mathbb{F},\, 0 \cdot u = \emptyset$
(g)	Denote $(-1) \cdot u$ by $-u$	$u - u = 1 \cdot u + (-1) \cdot u = (1 - 1) \cdot u = 0 \cdot u = \emptyset$
(h)	Equation $v + u = w$	uniquely solved by $v = w + (-1) \cdot u = w - u$
(i)	$a \cdot u = \emptyset,\, a \neq 0$	$\Longrightarrow u = \emptyset$
(j)	$a \cdot u = \emptyset,\, u \neq \emptyset$	$\Longrightarrow a = 0$

The following are simple examples of linear vector spaces, in which the operations of addition and multiplication by scalars are defined in the usual way:

(i) The sets $\mathbb{R}^N$ or $\mathbb{C}^N$ of all ordered N-tuples of real or complex numbers for fixed N; these are the usual vector spaces of elementary linear algebra

(ii) The set of all $M \times N$ matrices for fixed M and N

(iii) The set of polynomials of degree not greater than N

(iv) The set of functions of the form $a_1 g_1(x) + a_2 g_2(x) + \cdots + a_N g_N(x)$ where $g_1,\, g_2,\, \ldots,\, g_N$ are fixed given functions

(v) The set of all solutions of a given *linear* differential equation

(vi) The set of polynomials of arbitrary degree

(vii) The set whose elements are infinite sequences $(u_1, u_2, u_3, \ldots)$ of real or complex numbers

(viii) The set $C[a, b]$ of continuous functions $f(x)$ defined for $a \leq x \leq b$[1]

(ix) The set $C^1[a, b]$ of continuous functions $f(x)$ with a continuous first derivative defined for $a \leq x \leq b$

For some of the sets in these examples one can change the definitions of the operations of addition and multiplication in such a way that the basic axioms are not satisfied. Rather than presenting examples of this type, which are often somewhat artificial, we show three sets which *do not* constitute linear vector spaces with the usual definitions of addition and multiplication:

(i) The set of all polynomials taking a fixed value $c \neq 0$ at $x = x_0$. Indeed, if $p_1(x_0) = p_2(x_0) = c$, $(p_1 + p_2)(x_0) = 2c \neq c$ unless $c = 0$ and, furthermore, $ap(x_0) = ac \neq 0$ if $c \neq 0$.

(ii) For a similar reason, the set of all solutions of a non-homogeneous linear differential equation.

(iii) The set of all solutions of a non-linear differential equation as, in general, the sum of two solutions, or the product of a solution by a constant, are not solutions of the equation.

In all these cases the closure property (19.1.1) under addition and multiplication fails. Let us now consider in greater detail two more examples of linear vector spaces.

Example 19.1.1 Consider the set l^p of all sequences $u = \{u_n\} \equiv (u_1, u_2, \ldots)$ such that $\sum_{n=1}^{\infty} |u_n|^p < \infty$, with $p \geq 1$. This set is evidently closed under multiplication by scalars if the usual definition $a \cdot u = \{au_n\}$ is used. Closure under addition, defined as $u + v = \{u_n + v_n\}$, is a consequence of *Minkowski's inequality*

$$\left(\sum_{n=1}^{\infty} |u_n + v_n|^p \right)^{1/p} \leq \left(\sum_{n=1}^{\infty} |u_n|^p \right)^{1/p} + \left(\sum_{n=1}^{\infty} |v_n|^p \right)^{1/p}. \tag{19.1.2}$$

For $p = 1$, this is obvious. For general p, Minkowski's inequality follows from the *Hölder inequality*[2]

$$\sum_{n=1}^{\infty} |u_n v_n| \leq \left(\sum_{n=1}^{\infty} |u_n|^p \right)^{1/p} \left(\sum_{n=1}^{\infty} |v_n|^q \right)^{1/q} \tag{19.1.3}$$

in which $p^{-1} + q^{-1} = 1$. By noting that $|u_n + v_n|^p = |u_n + v_n| |u_n + v_n|^{p-1} \leq |u_n| |u_n + v_n|^{p-1} + |v_n| |u_n + v_n|^{p-1}$ and applying (19.1.3) with $q(p-1) = p$, one finds (19.1.2). All the axioms are then verified and the set l^p is a linear vector space. □

---

[1] Here and elsewhere in the book the symbol $C^k[\Omega]$ denotes the set of functions continuous together with their first k derivatives in and on the boundary of the domain Ω. Occasionally we write $C[\Omega]$ in place of $C^0[\Omega]$ to indicate continuous functions; $C^\infty[\Omega]$ denotes the set of infinitely differentiable functions.

[2] A proof of these standard inequalities can be found in a large number of books; see e.g. Debnath & Mikusiński 1990, p. 7; Boccara 1990, p. 71; Naylor & Sell 2000, Appendix A; Robinson 2001, p. 24.

Example 19.1.2 The set of all functions $u(x)$ defined on the (finite or infinite) interval (a, b) such that the Lebesgue integral is finite,

$$\int_a^b |u(x)|^p \, dx \ < \ \infty, \tag{19.1.4}$$

is also a linear space with the usual definitions of addition and multiplication. Closure under addition is a consequence of the continuous analog of Minkowski's inequality, namely

$$\left(\int_a^b |u(x) + v(x)|^p \, dx \right)^{1/p} \leq \left(\int_a^b |u(x)|^p \, dx \right)^{1/p} + \left(\int_a^b |v(x)|^p \, dx \right)^{1/p} \tag{19.1.5}$$

which, in turn, rests on the continuous analog of the Hölder inequality. We denote this linear vector space by L^p or $L^p(a, b)$. □

We conclude by defining a few concepts which will be useful in the following.

Linear dependence and independence. The set $u_1, u_2, \ldots, u_n$ of elements of $\mathbb{S}$ is *linearly dependent* if there are numbers $a_1, a_2, \ldots, a_n$, not all zero, such that

$$a_1 \cdot u_1 + a_2 \cdot u_2 + \cdots + a_n \cdot u_n = \emptyset. \tag{19.1.6}$$

In the opposite case the elements of the set are *linearly independent*. According to this definition, any set that contains the ∅ element (including the set consisting of the ∅ element only) is linearly dependent. A set consisting of an infinite number of elements is said to be linearly independent if every finite subset of them is linearly independent.

Dimension. A linear vector space is N-dimensional if it contains a set of N linearly independent vectors while any set of $N + 1$ vectors is linearly dependent. If, whatever N, an additional linearly independent vector can be found, the space is infinite-dimensional. The spaces of examples (i) to (v) on p. 538 are finite-dimensional, while those of examples (vi) to (ix), as well as those of Examples 19.1.1 and 19.1.2, have infinite dimension.

Linear Manifold. Consider a finite or infinite subset $u_1, u_2, \ldots$ of elements of $\mathbb{S}$. The set of all *finite* linear combinations of these elements, i.e. the set all elements of $\mathbb{S}$ of the form $a_1 \cdot u_1 + a_2 \cdot u_2 + \cdots + a_n \cdot u_n$ constitutes the *linear manifold M* spanned by the elements of the set and we write

$$M = \text{span} \ (u_1, u_2, \ldots, u_n, \ldots); \tag{19.1.7}$$

M is the *algebraic span*, or simply *span*, or *linear hull* of the u_k.[3] It is evident that the *same* linear manifold can be generated, or spanned, by many different subsets of elements

---

[3] A spanning set is not necessarily a basis even in a finite-dimensional space (see definition on p. 547) as the spanning vectors need not be linearly independent. The restriction to finite linear combinations is necessary as at present we have no concept which would permit us to give a meaning to an infinite sum.

of $\mathbb{S}$. A linear manifold must contain the neutral element $\emptyset$ because, by definition, it contains e.g. $u_1 - u_1$. The validity of the other properties of Tables 19.1 and 19.2 under the same rules inherited from $\mathbb{S}$ is readily ascertained so that a linear manifold is itself a linear vector space over the same field of scalars. For this reason, linear manifolds are often also referred to as subspaces; we prefer to reserve the latter denomination for complete and closed linear manifolds of normed spaces (p. 548). Familiar examples of linear manifolds (in fact, subspaces) are lines and planes through the origin in three-dimensional space.

Direct sum. If $\mathbb{S}_1$ and $\mathbb{S}_2$ are two linear vector spaces, we define their *product space* (or Cartesian product space) $\mathbb{S}$ as the set of all ordered pairs $\langle u_1, u_2 \rangle$ with $u_1 \in \mathbb{S}_1$ and $u_2 \in \mathbb{S}_2$. An elementary instance of this notion is a two-dimensional Cartesian plane constructed from two lines, not necessarily orthogonal. Another example is the tensor product of two matrices defined on p. 495. This product space acquires the structure of a linear vector space with the rule

$$a \cdot \langle u_1, u_2 \rangle + b \cdot \langle v_1, v_2 \rangle = \langle a \cdot u_1 + b \cdot v_1, a \cdot u_2 + b \cdot v_2 \rangle. \tag{19.1.8}$$

With this definition the spaces $\mathbb{S}_1$ and $\mathbb{S}_2$ can be identified with the linear manifolds of $\mathbb{S}$ consisting of elements of the form $\langle u_1, \emptyset \rangle$ and $\langle \emptyset, u_2 \rangle$ respectively, and the space $\mathbb{S}$ thus constructed becomes the direct sum of $\mathbb{S}_1$ and $\mathbb{S}_2$; we write[4]

$$\mathbb{S} = \mathbb{S}_1 \oplus \mathbb{S}_2. \tag{19.1.9}$$

19.2 Normed spaces

A crucial procedure in applied mathematics is the construction of the solution of a problem by the systematic refinement of an approximation. This is what it means, for example, that an infinite series or an iterative process converge to the exact solution of a problem. In order to judge how good the approximation is, and to define convergence, we need to introduce a notion that can be interpreted as a *distance* between the true solution and the approximation. This leads us to the concept of normed spaces.[5] We have already briefly considered norms in finite-dimensional vector spaces in §18.5 and §18.8.

---

[4] This construction synthesizes in a convenient (if not logically precise) way two different ideas, the *internal* and *external* direct sums (see e.g. Naylor & Sell 2000, pp. 196–200 and 340–344). Let $\mathbb{S}_1$ and $\mathbb{S}_2$ be two linear manifolds of the same space $\mathbb{S}$ with no common element other than $\emptyset$. The set consisting of elements of the form $u_1 + u_2$, with $u_j \in \mathbb{S}_j$, is the *internal direct sum* of $\mathbb{S}_1$ and $\mathbb{S}_2$ and is the smallest linear manifold of $\mathbb{S}$ containing both $\mathbb{S}_1$ and $\mathbb{S}_2$; we denote this set by $\mathbb{S}_1 \oplus \mathbb{S}_2$. When $\mathbb{S} = \mathbb{S}_1 \oplus \mathbb{S}_2$, this is the same concept introduced on p. 501 for finite-dimensional spaces. If $\mathbb{S}_1$ and $\mathbb{S}_2$, instead, are two different linear vector spaces, we define their product space $\mathbb{S}$ as the set of all ordered pairs $\langle u_1, u_2 \rangle$ with $u_j \in \mathbb{S}_j$. This product space acquires the structure of a linear vector space with the rule (19.1.8) and it is then called the *external direct sum* of the spaces $\mathbb{S}_1$ and $\mathbb{S}_2$. If $\mathbb{S}_1$ is identified with the linear manifold of $\mathbb{S}$ consisting of elements $\langle u_1, \emptyset \rangle$ and $\mathbb{S}_2$ with the linear manifold of $\mathbb{S}$ consisting of elements $\langle \emptyset, u_2 \rangle$ the distinction between external and internal direct sums becomes unnecessary and we can simply refer to the direct sum of the two spaces or linear manifolds.

[5] A set of elements for which a distance function is defined constitutes a *metric space*, which does not presuppose the algebraic structure (i.e., the notions of addition and scalar multiplication) of a linear vector space (see e.g.

A *normed space* is a linear real vector space $\mathbb{S}$ over a field $\mathbb{F}$ equipped with a rule which associates to each element $u \in \mathbb{S}$ a non-negative number, denoted by $\|u\|$ and called the *norm* of u, in such a way that the following axioms are satisfied:

(i) For all $u \in \mathbb{S}$, $\|u\| \geq 0$
(ii) $\|u\| = 0$ if and only if $u = \emptyset$
(iii) For all $a \in \mathbb{F}$ and all $u \in \mathbb{S}$, $\|a \cdot u\| = |a| \, \|u\|$
(iv) *Triangle inequality*: For all $u, v \in \mathbb{S}$, $\|u + v\| \leq \|u\| + \|v\|$.

By writing $u = (u - v) + v$ and $v = (v - u) + u$ and using the triangle inequality it is easy to establish the inequalities

$$\big| \, \|u\| - \|v\| \, \big| \leq \|u \pm v\| \leq \|u\| + \|v\|. \tag{19.2.1}$$

Example 19.2.1 Consider the spaces $\mathbb{R}^N$ or $\mathbb{C}^N$ of all ordered N-tuples of real or complex numbers for a certain fixed N with the usual definitions of addition and scalar multiplication. On these spaces we can define norms by

$$\|u\|_p = \left(\sum_{n=1}^{N} |u_n|^p \right)^{1/p} \qquad \text{or} \qquad \|u\|_\infty = \max_{1 \leq n \leq N} |u_n|. \tag{19.2.2}$$

Axioms (i) to (iii) are evidently satisfied. For the first norm, axiom (iv) is satisfied in view of Minkowski's inequality (19.1.2); for the second one it is obviously verified. The label ∞ of the second norm is justified by the fact that it coincides with the first one in the limit $p \to \infty$ (see e.g. Wouk 1979, p. 69; Ablowitz & Fokas 1997, p. 434). The case $p = 2$ is called the *Euclidean norm* and is of special importance, as will be seen later. $\square$

Example 19.2.2 Consider the set of all real polynomials $p(x) = a_0 + a_1 x + \cdots + a_n x^n$ of arbitrary order n with the usual definitions of addition and multiplication. The reader can readily verify that this set becomes a normed space with the norm

$$\|p\| = |a_0| + |a_1| + \cdots + |a_n|. \tag{19.2.3}$$

$\square$

Example 19.2.3 Let us return to the space l^p of Example 19.1.1 p. 539 of all sequences u such that $\sum_{n=1}^{\infty} |u_n|^p < \infty$, with $p \geq 1$. Generalizing (19.2.2) to an infinity of terms, we define the norm

$$\|u\|_p = \left(\sum_{n=1}^{\infty} |u_n|^p \right)^{1/p}. \tag{19.2.4}$$

All the norm axioms are satisfied as in the first example. $\square$

Naylor & Sell 2000, chapter 3; Kubrusly 2001, chapter 3). However, since we are really interested in a metric that recognizes the operations inherent in the notion of linear vector space, we skip the specific treatment of metric spaces and proceed directly to normed spaces which fulfill both of our requirements.

Example 19.2.4 In the set of all bounded infinite sequences $u \equiv \{u_n\} = (u_1, u_2, \ldots)$ we can define a norm similar to the second one of (19.2.2):

$$\|u\|_\infty = \sup_{1 \leq n < \infty} |u_n| \tag{19.2.5}$$

which trivially satisfies all the axioms.[6] This new normed space is denoted by l^∞. Clearly the elements of the space l^p are all elements of this new space, and in fact they constitute a proper subspace of it (definition on p. 548).   □

Example 19.2.5 In the space $C[a, b]$ of continuous functions $u(x)$ defined on the finite interval $[a, b]$ we define a norm, known as the L^p-norm, by

$$\|u\|_p = \left(\int_a^b |u(x)|^p \, dx \right)^{1/p} . \tag{19.2.6}$$

The validity of the norm axioms is readily verified, with (19.1.5) proving the triangle inequality. The case $p = 2$ gives the L^2-norm

$$\|u\|_2 = \left(\int_a^b |u(x)|^2 \, dx \right)^{1/2} \tag{19.2.7}$$

which is of particular importance, as will become clear in the following section. The same space can be equipped with different norms; an important one is

$$\|u\|_\infty = \max_{a \leq x \leq b} |u(x)|. \tag{19.2.8}$$

We refer to this as the C^∞-norm or maximum norm.   □

Example 19.2.6 If, in the space $C^1[a, b]$ of continuous functions with continuous first derivative, we attempted to define a norm by $\|u\| = \max_{a \leq x \leq b} |u'(x)|$, we would violate axiom (ii) as this norm would vanish for constants. This would be an acceptable norm, however, in the subspace $C_0^1[a, b]$ of functions which vanish at a point (cf. § 19.5).   □

Other examples are the Sobolev spaces of §19.5 and the space of linear functionals of §19.6.

We have seen that the same linear space S can be equipped with different norms, some of which may induce a very different "landscape" in the space, in the sense that the same elements of S can be close in one norm and far apart in another one. For example, in the space $C[0, 1]$, a function having only a very narrow and tall peak will have a large norm with the definition (19.2.8), while it might have a small norm according to the definition (19.2.6). To distinguish norms that do not "change the landscape" too much, we introduce the notion of *equivalent norms*:

---

[6] The supremum, *sup*, is the least upper bound; we use this notion in place of *max* as, for an infinite sequence, there might be an upper bound which is never quite attained, see §A.6.

Definition 19.2.1 Two norms $\|\cdot\|_1$ and $\|\cdot\|_2$ are equivalent if there are two positive constants α and β such that, for all $u \in \mathbb{S}$,

$$\alpha \|u\|_1 \leq \|u\|_2 \leq \beta \|u\|_1. \tag{19.2.9}$$

For example, let $s(x)$ be a strictly positive continuous function over $a \leq x \leq b$. Then the expression

$$\|u\|_s = \left(\int_a^b s(x) |u(x)|^2 \, dx \right)^{1/2} \tag{19.2.10}$$

defines a norm equivalent to the L^2-norm defined in (19.2.7) as, evidently,[7]

$$\left(\min_{a \leq x \leq b} s(x) \right)^{1/2} \|u\|_2 \leq \|u\|_s \leq \left(\max_{a \leq x \leq b} s(x) \right)^{1/2} \|u\|_2. \tag{19.2.11}$$

In a finite-dimensional space, all norms are equivalent (see e.g. Kress 1989, p. 3), but not so if the number of dimensions is infinite.

19.2.1 Convergence in a normed space

We are now in a position to define convergence in a normed space.

Definition 19.2.2 The sequence $\{u_n\}$ of elements of a normed space $\mathbb{S}$ is *strongly convergent* to a limit $u \in \mathbb{S}$ when

$$\lim_{n \to \infty} \|u - u_n\| = 0. \tag{19.2.12}$$

The notions of convergence in the mean and L^1-convergence already encountered in §8.2.3 coincide with this definition when use is made of the L^2 and L^1 norms, respectively.

The triangle inequality shows that the limit is unique when it exists as, if $\|U_1 - u_n\| \to 0$ and $\|U_2 - u_n\| \to 0$, then $\|U_2 - U_1\| = \|(U_2 - u_n) - (U_1 - u_n)\| \leq \|U_2 - u_n\| + \|U_1 - u_n\| \to 0$. Furthermore, (19.2.1) implies that, if $\|u - u_n\| \to 0$, then $\|u_n\| \to \|u\|$. The triangle inequality can also be used to prove continuity of the norm as, if $\|u - u_n\| \to 0$, then $\|v - u_n\| \leq \|v - u\| + \|u - u_n\|$ so that $\|v - u_n\| \to \|v - u\|$. In other words, the limit of the sequence of the norms equals the norm of the limit of the sequence.

In the case of numerical sequences we saw that the Cauchy property was an important test to decide whether the sequence was convergent or not (cf. §8.1). In the present context the definition of a Cauchy sequence is the following:

---

[7] $s(x)$ cannot be negative in order not to violate the positivity of the norm. It could however vanish, or be singular, at isolated points without changing sign. These special situations need to be considered case by case (cf. the variety of behaviors encountered in connection with singular Sturm–Liouville problems in §15.4 p. 375). In such cases one or both inequalities might not be satisfied and the norm would therefore not be necessarily equivalent to the L^2-norm.

Definition 19.2.3 [Cauchy sequence] In a normed space the sequence $\{u_n\}$ is said to be a *Cauchy sequence* if, for any $\varepsilon > 0$, it is possible to find an $N = N(\varepsilon)$ such that

$$\|u_n - u_m\| \leq \varepsilon \qquad \text{for all} \qquad n, m > N(\varepsilon). \tag{19.2.13}$$

A simple application of the triangle inequality shows that, if a sequence is strongly convergent, then it is also a Cauchy sequence since

$$\|u_n - u_m\| = \|u_n - u + u - u_m\| \leq \|u_n - u\| + \|u_m - u\| \leq 2\varepsilon. \tag{19.2.14}$$

We know that, e.g. for general sequences of rational numbers, the converse is not true: one can write down many rational Cauchy sequences which approximate π, without π being a rational number. This is, in fact, a powerful motivation for the introduction of irrational numbers. Real numbers – i.e. the combination of rational and irrational numbers – enjoy the property of *completeness*, in the sense that any Cauchy sequence of real numbers is convergent to a real number. This completeness property of real numbers renders the finite-dimensional spaces of Example 19.2.1 complete. The infinite-dimensional spaces of Examples 19.2.3 and 19.2.4 are also complete. In general, however, completeness does not necessarily hold for an arbitrary normed space $\mathbb{S}$. In other words, it is not necessarily true that all Cauchy sequences formed with elements of $\mathbb{S}$ possess a limit belonging to $\mathbb{S}$. This point is immediately evident for Example 19.2.2 p. 542, as increasing indefinitely the order of polynomials leads to infinite series, which are not polynomials. A more interesting illustration can be given with the space of Example 19.2.5 equipped with the L^2 norm.

Example 19.2.7 In the space $C[0, 1]$ of continuous functions defined for $0 \leq x \leq 1$ consider the sequence

$$u_n = \begin{cases} n^{1/3} & 0 \leq x \leq 1/n \\ x^{-1/3} & 1/n \leq 1 \end{cases}. \tag{19.2.15}$$

With the L^2 norm (19.2.7) we have, for $m < n$,

$$\int_0^1 |u_n - u_m|^2 \, dx = m^{-1/3} \left[1 - \left(\frac{m}{n}\right)^{1/3} \right] \left[1 - \left(\frac{m}{n}\right)^{1/3} + \left(\frac{m}{n}\right)^{2/3} \right] \tag{19.2.16}$$

which tends to 0 as $m \to \infty$ if $m/n \leq 1$ as we have assumed. Thus, this is a Cauchy sequence with this norm. It is also evident that

$$\int_0^1 \left| x^{-1/3} - u_n \right|^2 dx = n^{-1/3} \to 0 \tag{19.2.17}$$

so that $u_n \to x^{-1/3}$ in the L^2 sense. However, $x^{-1/3} \notin C[0, 1]$, which is therefore seen not to be complete if equipped with the L^2 norm. The problem does not arise with the maximum norm (19.2.8) with which (19.2.15) is not a Cauchy sequence as is readily shown.   □

This lack of completeness of the space $C[0, 1]$ equipped with the L^2 norm may be remedied by enlarging it with the addition of functions such as $x^{-1/3}$ which, though singular, are integrable. Other examples can be constructed where sequences of continuous functions converge, with the L^2 norm, to a discontinuous function, so that such functions also should be added, and so forth. It is found that this process terminates once all the functions such that $|u|^2$ is integrable in the sense of Lebesgue have been included.[8] This enlarged space of functions, to which we refer as the L^2 space, is then complete. We similarly define the space L^p, with $p \geq 1$, as the completion of the set of functions such that $|u|^p$ is Lebesgue-integrable.

The process of enlargement described in the context of functions can always be carried out in a normed space. Just as in the case of real numbers, it can be effected in such a way that the new members of the class can be approximated arbitrarily well (in the sense of the norm) by the original members:

Theorem 19.2.1 *It is always possible to enlarge a normed linear space $\mathbb{S}$ to a new complete space $\mathbb{S}' \supseteq \mathbb{S}$ in such a way that*

- *If an element u belongs to both $\mathbb{S}$ and $\mathbb{S}'$, then $\|u\|$ is the same in the two spaces;*
- *$\mathbb{S}$ is dense in $\mathbb{S}'$ i.e., for all $u' \in \mathbb{S}'$ and all $\varepsilon > 0$, it is possible to find a u in the original space $\mathbb{S}$ such that $\|u' - u\| < \varepsilon$.[9]*

Every element of $\mathbb{S}'$ can be represented as the limit of a Cauchy sequence $\{u_n\}$ of elements of $\mathbb{S}$,[10] and $u + v = \lim (u_n + v_n)$, $au = \lim au_n$, $\|u\| = \lim \|u_n\|$. With reference to the previous example, we would have $\mathbb{S} = C[0, 1]$ and $\mathbb{S}' = L^2(0, 1)$. As another example we may cite the space of polynomials, which is not complete with the maximum norm. According to a result of Weierstrass, its completion in this norm gives the space of continuous functions.

In view of this theorem, we can always consider a normed space as complete; a normed linear space complete in its norm is called a *Banach space*. Normed spaces "small enough" to contain a *countable* dense set are called *separable*. Among the previous examples, the norm (19.2.5), together with its continuous analog (19.2.8), lead to non-separable spaces (see e.g. Wouk 1979, p. 77).

It should be noted that, while in enlarging $C[0, 1]$ to $L^2(0, 1)$ with the L^2 norm we have succeeded in making every Cauchy sequence a convergent sequence, we have lost something in the sense that we have been forced to add many functions which are not equal to zero in a point-wise sense, although $\int_0^1 |u|^2 \, dx = 0$. An example would be functions that vanish everywhere except at a countable set of points. In keeping with the first axiom of the norm, we must regard functions such as these, which vanish *almost everywhere* (i.e., everywhere except for a set of measure zero), as equal and as the neutral element $\emptyset$ of the space. Similarly, any two functions u, v such that $\int_0^1 |u - v|^2 \, dx = 0$ must be considered

---

[8] See §A.4.3 in the Appendix for a very brief description of the Lebesgue integral.

[9] Generally speaking, a set $\mathscr{E} \subseteq \mathbb{S}$ of a normed space $\mathbb{S}$ is said to be *dense* in $\mathbb{S}$ if, for every $u \in \mathbb{S}$ and every $\varepsilon > 0$, there is an element $v \in \mathscr{E}$ such that $\|u - v\| < \varepsilon$. Further information on this notion may be found in §A.1 of the Appendix.

[10] An element u belonging to $\mathbb{S}'$ because it already belongs to $\mathbb{S}$ is the limit of the trivial Cauchy sequence $u, u, \ldots, u, \ldots$.

as the same function irrespective of their point-wise behavior.[11] We already encountered the problem in Chapter 9 when we declared the Fourier series of a discontinuous function convergent, but at the price of losing point-wise convergence. Thus, strong convergence does not imply pointwise convergence, but point-wise convergence does not imply strong convergence either. For example, a sequence of functions $\{u_n(x)\}$ each one of which is non-zero only in a neighborhood of $x = n$, converges to 0 in the point-wise sense, but not necessarily strongly.

We have addressed the convergence of sequences. That of series can be discussed along similar lines by considering the convergence of the sequence of the partial sums. For series, however, one may consider another mode of convergence:

Definition 19.2.4 [Absolute convergence] A series $\sum_{n=1}^{\infty} u_n$ is said to converge absolutely when $\sum_{n=1}^{\infty} \|u_n\|$ converges.

The obvious question then arises of whether there is an element v which is the strong limit of the series, $v = \sum_{n=1}^{\infty} u_n$. In a Banach space the answer is in the affirmative as, evidently, if $M < N$, $\|\sum_{j=1}^{N} u_j - \sum_{j=1}^{M} u_k\| \le \sum_{j=M+1}^{N} \|u_j\|$ and the latter sum can be made as small as desired given that the sum of the norms is a convergent numerical sequence. We conclude that $\sum_{n=1}^{\infty} u_n$ is a Cauchy sequence and therefore convergent. The converse is also true: A normed space is a Banach space if and only if every absolutely convergent series converges also in the sense of the norm (see e.g. Kubrusly 2001, p. 201).

The notion of convergence permits us to define a basis, or base, for the Banach space:

Definition 19.2.5 [Basis] A set of vectors $\{h_1, h_2, \ldots, h_m, \ldots\}$ is called a basis for a Banach space if and only if any u in the space can be uniquely represented as

$$u = \sum_{1}^{\infty} \alpha_k h_k. \tag{19.2.18}$$

If the space is finite-dimensional the sum is finite and, by the definition of dimension (p. 540), all bases have the same number of elements; if the space is infinite-dimensional and separable, all bases have a countable infinity of elements.

In general, even if the set $\{h_k\}$ is infinite, contains no linearly dependent elements and is dense in H (so that any element of H can be approximated arbitrarily closely by a finite linear combination of h_k's) it is still not necessarily a basis for the space. For example, the set of monomials $\{x^k\}$ is infinite, linearly independent and, by Weirstrass's theorem (p. 546), complete in the space $C[0, 1]$ equipped with the maximum norm. However, it is not a basis as only analytic functions possess a representation of the form $f(x) = \sum_{n=0}^{\infty} c_n x^n$ and there are plenty of continuous functions which are not analytic, e.g. continuous functions with a discontinuous first derivative. We return on this point on p. 560.

---

[11] Thus, rather than with individual functions, we are forced to deal with *equivalence classes* of functions, each one constituted by all the functions that differ from each other over a set of zero measure. If we insisted in considering two such functions as distinct, we would have a *semi-norm* (also called *pseudonorm*) instead of a norm, which satisfies all the axioms of a norm except for the second one.

We have already encountered the notion of subspace on p. 500 in connection with finite-dimensional spaces. A more general definition in the present context can be given starting with linear manifolds (p. 540). Consider the linear manifold generated by the elements of a subset of a linear vector space $\mathbb{S}$. This manifold, complemented with all the limit points of all the strongly convergent sequences that can be constructed with elements of the manifold, is a *linear subspace*, or simply *subspace*, of the linear vector space $\mathbb{S}$. Since by definition a subspace contains the limit points of all its convergent sequences, it is closed,[12] and it is also complete. A subspace of a Banach space is evidently itself a Banach space with respect to the two operations of addition and multiplication by scalars inherited from $\mathbb{S}$.

According to the definition, the element $\emptyset$ of $\mathbb{S}$ and $\mathbb{S}$ itself constitute (trivial) subspaces of $\mathbb{S}$; any other subspace is a *proper subspace*. For example, the space l^p of Example 19.1.1 is a proper subspace of the space of all infinite sequences.

The intersection $\mathbb{S}_1 \cap \mathbb{S}_2$ of $\mathbb{S}_1$ and $\mathbb{S}_2$ (i.e., the set of elements that belong to both $\mathbb{S}_1$ and $\mathbb{S}_2$) is also a subspace (possibly containing only the element $\emptyset$). If $\mathbb{S}_1$ and $\mathbb{S}_2$ are subspaces of $\mathbb{S}$, so is their union (or internal direct sum, p. 541) $\mathbb{S}_1 \cup \mathbb{S}_2$, i.e. all the elements of $\mathbb{S}$ of the form $u_1 + u_2$ with $u_1 \in \mathbb{S}_1$ and $u_2 \in \mathbb{S}_2$.

19.3 Hilbert spaces

We are now in a position to introduce the crucial notions of scalar product and orthogonality in a general infinite-dimensional space.

Starting from a real or complex linear vector space $\mathbb{S}$, we introduce a rule that associates to each ordered pair u, v of elements u and v of $\mathbb{S}$ a real or complex number which we denote by (u, v). If the rule satisfies the properties listed in Table 19.3, we call (u, v) the *scalar product* of u and v and the space so equipped a *scalar product space*. For fixed u, (u, v) is a functional of v (definition on p. 563) and axiom (ii) states that this functional is linear so that $(u, a \cdot v) = a(u, v)$. Upon combining with axiom (i), it then follows that the scalar product is *antilinear* in its first element, $(a \cdot u, v) = \bar{a}(u, v)$.[13] The two properties are both implied by the statement that the scalar product is a *sesquilinear form*; when the scalars are real numbers $(a \cdot u, v) = a(u, v)$ and the scalar product is then a *bilinear form*.

In view of axioms (ii), (iv) and (v), it is obvious that properties (i), (ii) and (iii) of the norm (p. 542) are satisfied by the definition

$$\|u\| = \sqrt{(u, u)}. \tag{19.3.1}$$

For this relation to define a proper norm, we must make sure that the triangle inequality (property (iv) p. 542) is also satisfied, namely that

$$\sqrt{(u + v, u + v)} \leq \sqrt{(u, u)} + \sqrt{(v, v)}, \tag{19.3.2}$$

---

[12] For a summary of the notions and properties of closed and open sets see §A.1 of the Appendix.

[13] In the mathematics community many authors use the opposite convention so that $(u, a \cdot v) = \bar{a}(u, v)$; our convention is the one prevailing in physics.

| | Table 19.3 | The axioms governing the (in general, complex) scalar product and some of their consequences; overline denotes the complex conjugate. |

	Axiom	Consequences
(i)	$\forall\, u,\, v \in \mathbb{S},\, (u, v) = \overline{(v, u)}$	(u, u) real
(ii)	$\forall\, a \in \mathbb{C},\, (u, a \cdot v) = a\,(u, v)$	$(a \cdot u, v) = \bar{a}\,(u, v)$
		$(a \cdot u, a \cdot u) = |a|^2 (u, u)$
		$(\emptyset, u) = (u, \emptyset) = 0$
(iii)	$(u, v + w) = (u, v) + (u, w)$	$(u + v, w) = (u, w) + (v, w)$
		$(u + v, u + v) = (u, u) + 2\mathrm{Re}(u, v) + (v, v)$
(iv)	$(u, u) \geq 0$	
(v)	$(u, u) = 0$ if and only if $u = \emptyset$	

or

$$(u + v, u + v) \leq (u, u) + (v, v) + 2\sqrt{(u, u)(v, v)}. \tag{19.3.3}$$

Proof A direct calculation using the properties of Table 19.3 shows that

$$(u + v, u + v) = (u, u) + (v, v) + (u, v) + (v, u) = (u, u) + (v, v) + 2\mathrm{Re}\,(u, v) \tag{19.3.4}$$

so that (19.3.3) is satisfied if

$$\mathrm{Re}\,(u, v) \leq \sqrt{(u, u)(v, v)}. \tag{19.3.5}$$

This inequality follows immediately from the *Cauchy–Schwarz inequality* (see footnote p. 506)

$$|(u, v)|^2 \leq (u, u)\,(v, v), \tag{19.3.6}$$

with equality holding only when $u = a \cdot v$, with a a real or complex number. Note first that this relation is trivially satisfied if $v = \emptyset$. Assume therefore that $v \neq \emptyset$ and note the relation $(u - w, u - w) \geq 0$, which is a consequence of axiom (iv). Calculating the scalar product by using (i) and (iii) we find

$$\begin{aligned} 0 &\leq (u - w, u - w) = (u, u) - (u, w) - (w, u) + (w, w) \\ &= (u, u) - 2\,\mathrm{Re}\,(u, w) + (w, w) \end{aligned} \tag{19.3.7}$$

from which

$$(u, u) \geq 2\,\mathrm{Re}\,(u, w) - (w, w), \tag{19.3.8}$$

with equality holding only if $u = w$ by axiom (v). Take now $w = [(v, u)/(v, v)] \, v$ to find

$$
\begin{aligned}
(u, u) \;\geq\; & 2\,\mathrm{Re}\left(u, \frac{(v, u)}{(v, v)} v\right) - \frac{|(u, v)|^2}{(v, v)^2}(v, v) \\
\geq\; & 2\,\mathrm{Re}\left[\frac{(v, u)}{(v, v)}(u, v)\right] - \frac{|(u, v)|^2}{(v, v)} = \frac{|(u, v)|^2}{(v, v)},
\end{aligned}
\qquad (19.3.9)
$$

from which (19.3.6) follows. The fact that the equal sign only holds for $u = w$ implies that it only holds for $u = [(v, u)/(v, v)] \cdot v$ so that u and v are linearly dependent. The Cauchy–Schwarz inequality (19.3.6) is satisfied and therefore so is the triangle inequality. □

We thus conclude that (19.3.1) defines a proper norm which is the *natural norm* of the scalar product space. With this definition, a scalar product space becomes a normed space, which can be rendered complete by Theorem 19.2.1. A real or complex linear scalar product space complete in its natural norm is called a *Hilbert space*.[14] As before in the case of Banach spaces, linear subspaces of a Hilbert space will be assumed to be closed and complete.

In the space l^2 of square-summable sequences, the norm (19.2.4) with $p = 2$ is the natural norm induced by the scalar product

$$
(u, v) = \sum_{n=1}^{\infty} \bar{u}_n v_n. \qquad (19.3.10)
$$

Similarly, in the space $L^2(a, b)$ of square integrable functions over a finite or infinite interval (a, b) the norm (19.2.7) is the natural norm corresponding to the scalar product

$$
(u, v) = \int_a^b \bar{u}(x)\, v(x)\, dx. \qquad (19.3.11)
$$

However, no scalar product can be defined in the L^p spaces for $p \neq 2$. In other words, while it is always possible to turn a scalar-product space into a normed space, the converse is not necessarily true. As a matter of fact (see e.g. Kubrusly 2001, p. 316):

Theorem 19.3.1 [Jordan–von Neumann] *A normed space can be turned into a scalar product space if and only if the* parallelogram law *holds:*

$$
\|u + v\|^2 + \|u - v\|^2 = 2\|u\|^2 + 2\|v\|^2. \qquad (19.3.12)
$$

The validity of this relation in a scalar product space is readily established by a direct calculation. The real content of the theorem is the converse statement. When the parallelogram law is satisfied, the scalar product can be defined in terms of the norm by the so-called

---

[14] Some authors add the further stipulation that the space be infinite-dimensional.

polarization procedure:

$$(u, v) = \frac{1}{4}\left(\|u+v\|^2 - \|u-v\|^2\right) + \frac{i}{4}\left(\|u+iv\|^2 - \|u-iv\|^2\right)$$

$$= \frac{1}{4}\sum_{n=1}^{4} i^n \|u + i^n v\|. \tag{19.3.13}$$

It can be verified directly that the scalar product axioms of Table 19.3 are satisfied by this relation. When applied to the L^2-norm, for example, we recover the definition (19.3.11) of scalar product.

There are many Banach spaces for which the parallelogram law does not hold. An example is $L^1(a, b)$ with

$$\|f\|_1 = \int_a^b |f| \, dx. \tag{19.3.14}$$

If two functions f and g are such that they are simultaneously non-zero at no point, then

$$\|f \pm g\|_1 = \|f\|_1 + \|g\|_1, \tag{19.3.15}$$

the parallelogram law is not satisfied and no scalar product can be defined.

It is in some sense unfortunate that $C[a, b]$ is "too small" to be turned into a Hilbert space: If we equip it with the $\| \ \|_2$ norm, we have an incomplete space (Example 19.2.7) while, if we equip it with the $\| \ \|_\infty$ norm, we get a Banach space whose norm is not induced by a scalar product.

The same latitude encountered in the definition of the norm exists for the definition of the scalar product. For example, in dealing with the Sturm–Liouville expansions in §15.3, we have used a scalar product defined by (15.3.2) p. 366, namely

$$(u, v)_s = \int_a^b \overline{u}(x)\, v(x)\, s(x)\, dx, \tag{19.3.16}$$

provided the weighting function $s(x)$ is positive over the finite interval $a \le x \le b$ (see footnote p. 544); (19.2.10) is the natural norm induced by this definition. Another example is the definition (18.5.12) given on p. 507 for the case of finite-dimensional vector spaces.

Theorem 19.3.2 *The scalar product depends continuously on its arguments i.e., if* $\|u_m - u\| \to 0$, *then*

$$\lim_{m \to \infty} (v, u_m) = (v, \lim u_m) = (v, u), \tag{19.3.17}$$

and similarly if $\|v_n - v\| \to 0$.

This result follows from the Schwarz inequality as, e.g.,

$$|(v, u_m) - (v, u_n)| = |(v, u_m - u_n)| \le \|v\| \ \|u_m - u_n\| \tag{19.3.18}$$

and the right-hand side converges to zero by hypothesis.[15]

---

[15] If $u_m \to u$ and $v_n \to v$ simultaneously we have $(u_m, v_n) - (u, v) = (u_m - u, v_n) + (u, v_n - v)$, so that the result also holds in this case.

19.3.1 Weak convergence

Since Hilbert spaces are inherently Banach spaces, strong and absolute convergence can be defined as above on pp. 544 and 547. The continuity of the scalar product however enables us to introduce yet another type of convergence:

Definition 19.3.1 A sequence $\{u_n\}$ of elements of a Hilbert space H is said to be *weakly convergent* to a limit $u \in H$ if and only if, for all $v \in H$,[16]

$$\lim_{n\to\infty} (v, u_n) = (v, u). \tag{19.3.19}$$

It is evident from the relation (19.3.18) that a strongly convergent sequence is also weakly convergent. The converse is also true in a finite-dimensional space,[17] but not in an infinite-dimensional one. For example, the Fourier coefficients $(v, \sin nx)$ tend to zero according to the Riemann–Lebesgue lemma (p. 249), without the sequence $\{\sin nx\}$ having any limit, strong or point-wise. As a matter of fact, $\| \sin nx - \sin mx \|_2^2 = 2\pi$ if $m \neq n$ so that the sequence $\{\sin nx\}$ is not even a Cauchy sequence in $L^2(-\pi, \pi)$. It is however true in general that *weakly convergent sequences are bounded* so that, if $\|u_n\| \leq C$, then also $\|u\| < C$ (see e.g. Riesz & Sz.-Nagy 1955, p 61). The converse is also true: every bounded sequence in L^2 has a weak limit (see e.g. Lieb & Loss 1997, p. 62).[18]

With the additional hypothesis that $\|u_n\| \to \|u\|$, (19.3.19) implies that $u_n \to u$ strongly as well, as readily follows passing to the limit in the relation $\|u - u_n\|^2 = \|u\|^2 + \|u_n\|^2 - (u_n, u) - (u, u_n)$ in view of the continuity of the scalar product.

Relations such as (19.3.19), in which an equality does not hold between the primary quantities, but between the scalar products of these primary quantities with a sufficiently large subset of elements of the space, are qualified by the adjective "weak." Thus, for example, the *weak derivative* of a function $f(x)$ in an interval (a, b) is a function $g(x)$ such that

$$(v, g) = -(v', f) \tag{19.3.20}$$

for all the differentiable functions v which vanish at $x = a$ and $x = b$. Evidently, this relation does not require f to be differentiable in the ordinary sense, but only integrable.

Example 19.3.1 In $L^2(0, 2)$ consider the continuous function

$$f(x) = \begin{cases} x & 0 \leq x \leq 1 \\ 1 & 1 \leq x \leq 2 \end{cases}. \tag{19.3.21}$$

---

[16] Weak convergence in a normed space is defined by demanding that $\ell(u_n) \to \ell(u)$ for all continuous linear functionals ℓ (definition in § 19.6).

[17] The easiest proof requires the use of an orthogonal basis (§19.4.3) and of Parseval's equality (19.4.23). Indeed $\|u - u_n\|^2 = \sum_{k=1}^{N} |(u - u_n, e_k)|^2$ which can be made as small as desired as $(u - u_n, e_k) \to 0$ for all k because of the hypothesized weak convergence.

[18] Thus, one may say that, in a Hilbert space, a set is weakly pre-compact if and only if it is bounded. For the definition of pre-compactness see Definition 21.3.1 p. 636.

Then, upon integration by parts,

$$-(v', f) = -\int_0^1 \overline{v}'x\,dx - \int_1^2 \overline{v}'dx = \int_0^1 \overline{v}dx \qquad (19.3.22)$$

since $v(2) = 0$. Thus, we conclude that the weak derivative g is given by

$$g(x) = \begin{cases} 1 & 0 \leq x \leq 1 \\ 0 & 1 \leq x \leq 2 \end{cases}. \qquad (19.3.23)$$

However, if, we re-define the function f to equal 2 for $1 < x \leq 2$, a similar calculation gives

$$-(v'f) = \int_0^1 \overline{v}dx + \overline{v}(1) \qquad (19.3.24)$$

and there is no function g capable of reproducing this result. Thus, even though with this latter definition f is differentiable almost everywhere, it does not admit a weak derivative. In spite of its similarity with the definition of the derivative of a distribution (p. 575), the weak derivative is still an ordinary function and, therefore it is different from the distributional derivative. As another example, the step function $H(x)$ admits the distributional derivative $\delta(x)$ which is not a function and, therefore, not a weak derivative. $\qquad\square$

The definition is extended to higher-order derivatives as $\left(v, f^{(k)}\right) = (-1)^k \left(v^{(k)}, f\right)$.[19]

In a domain Ω of a space of dimension N we would define the weak first partial derivatives of a function f by requiring that

$$(v, \partial_i f) = -(\partial_i v, f) = -\int_\Omega \frac{\partial v}{\partial x_i} f(\mathbf{x})\,d\Omega \qquad (19.3.25)$$

be satisfied for all differentiable $v \in H_0(\Omega)$, i.e., all $v \in H$ vanishing on the boundary of Ω. For example, if $N \geq 3$, the function defined by $f(\mathbf{x}) = r^{-(N-2)}$ for $\mathbf{x} \neq 0$, with $r = (\mathbf{x} \cdot \mathbf{x})^{1/2}$, does not possess a derivative at $r = 0$ no matter how it is defined there. However it does have the weak partial derivatives $\partial_i f = (2 - N)x_i/r^N$ since the volume element $d\Omega$ in the integral brings in a factor r^{N-1} and x_i/r is well defined.

The notion is extended to the concept of *weak solution* of a differential equation. For example, a weak solution of the Poisson equation $\nabla^2 u = -f$ in a domain Ω is a function w such that

$$(\nabla v, \nabla w) = (v, f) \qquad (19.3.26)$$

for all differentiable $v \in H_0(\Omega)$. With this notion the regularity requirements on w are milder than those that would apply to the *classical solution* u.

We return to these ideas and amplify them in the context of the theory of distributions in Chapter 20.

---

[19] In all cases, when it exists, the connection between the ordinary notion and the weak notion is provided by an integration by parts with the integrated terms vanishing due to the vanishing of the auxiliary functions v on the boundary (cf. p. 686). Thus, whenever derivatives exist in an ordinary sense, the ordinary and weak definitions coincide except possibly on a set of measure zero. More precisely, if a continuous function admits continuous weak derivatives up to and including order k, then it is k times continuously differentiable in the ordinary sense (see e.g. Edmunds & Evans 1987, p. 219).

19.4 Orthogonality

Two elements u and v of a real or complex Hilbert space are said to be *orthogonal* when

$$(u, v) = 0. \tag{19.4.1}$$

A direct calculation shows then that

$$\|u + v\|^2 = \|u\|^2 + \|v\|^2, \tag{19.4.2}$$

a generalization of Pythagoras's theorem. A vector is orthogonal to a subset of H when it is orthogonal to all the elements of that subset. The $\emptyset$ vector is orthogonal to the whole space and, conversely, the only vector orthogonal to all the vectors of the space – or just to all the vectors of a set dense in the space – is the $\emptyset$ vector. Two subsets of H are orthogonal when each vector of one is orthogonal to all the vectors of the other. A set of mutually orthogonal vectors is an *orthogonal set* and, if all the vectors have unit norms, an *orthonormal set*.

Much as in ordinary three-dimensional space, the notion of orthogonality is at the basis of the geometric properties of Hilbert spaces.

Definition 19.4.1 [Orthogonal manifold] Let $M \subset H$ be a linear manifold of a Hilbert space H. The set $M^\perp$ of elements of H all of which are orthogonal to all the elements of M is the *orthogonal manifold* of M.

The continuity of the scalar product implies that $M^\perp$ is a closed set as the limit of any convergent sequence of elements of $M^\perp$ will also be orthogonal to the elements of M. Thus, $M^\perp$ is always closed whether M is closed or not and is, therefore, a subspace of H.

The following theorem is of fundamental importance in finding approximate solutions to many problems:

Theorem 19.4.1 [Projection theorem] *Let $E \subset H$ be a closed linear subspace of a Hilbert space H. Each $u \in H$ can be written in one and only one way as*

$$u = v + w \tag{19.4.3}$$

where $v \in E$ and w is orthogonal to all the elements of E; v is called the orthogonal projection *of u onto E, and is the element of E closest (in the sense of the norm) to u.*

Since w is orthogonal to all the elements of E, it is orthogonal to v and (19.4.2) holds. The association of v to u defines a mapping called *orthogonal projection*, realized by a *projection operator* P; we write $v = \mathsf{P}u$.

Proof Consider the set of all "distances" $\|u - v\|$ of u from all elements of E, let d be their greatest lower bound[20] and consider a sequence $\{v_n\}$ such that $\|u - v_n\| \to d$. From

---

[20] Such a lower bound exists and is finite as $\|u - v\| \geq 0$ and, since $\emptyset \in E$, $\|u - \emptyset\| = \|u\| < \infty$ so that $d \leq \|u\|$.

the parallelogram law (19.3.12) applied to $u - v_n$ and $u - v_m$ we have

$$\|v_n - v_m\|^2 \leq 2\|u - v_n\|^2 + 2\|u - v_m\|^2 - \|2u - (v_n + v_m)\|^2. \tag{19.4.4}$$

But $v_n + v_m$ is also in E and, therefore, the right-hand side tends to 0 as $n, m \to \infty$, which shows that the sequence $\{v_n\}$ is a Cauchy sequence and, therefore, convergent since subspaces are complete (p. 548). Let v be the limit of this sequence; we can then show that $w = u - v$ is orthogonal to all elements of E by noting that, for any $v' \in E$, $\|w\|^2 \leq \|w - \lambda v'\|^2$ since $\|w\|$ is the minimum distance. Upon taking $\lambda = (v', w)/\|v'\|^2$ and calculating $\|w - \lambda v'\|^2$ explicitly we are led to

$$0 \leq -\frac{(v', w)(w, v')}{\|v'\|^2} \tag{19.4.5}$$

which can only be satisfied if $(w, v') = 0$. The closure of E plays an essential role here in that it ensures the existence of a limit $v \in E$.   □

By virtue of the projection theorem, any Hilbert space can be decomposed into the direct sum (p. 541) of two mutually orthogonal subspaces, $H = E \oplus E^\perp$. For this reason, $E^\perp$ is called the *orthogonal complement* of E. In general, $(E^\perp)^\perp$ equals the closure of the set E (p. 681) but, since E is a subspace, it is already closed and therefore $(E^\perp)^\perp = E$ so that E is the orthogonal complement of $E^\perp$.[21] Since E and $E^\perp$ are subspaces, they may be regarded as spaces in their own right and further decomposed (perhaps repeatedly) on the basis of the theorem.

Conversely, given two Hilbert spaces H_1 and H_2, one can construct a larger Hilbert space by forming their direct sum $H = H_1 \oplus H_2$ (p. 541) and endowing it with the scalar product

$$(u, v)_H = (u_1, v_1)_{H_1} + (u_2, v_2)_{H_2}. \tag{19.4.6}$$

It is evident that, with this definition, H_1 and H_2 are each other's orthogonal complement in H and that the norm induced by (19.4.6) is

$$\|u\|_H^2 = \|u_1\|_{H_1}^2 + \|u\|_{H_2}^2. \tag{19.4.7}$$

A similar construction holds for any finite or countably infinite set of "component" Hilbert spaces.[22]

While we have proven that the orthogonal decomposition (19.4.3) exists, we have not given a procedure to actually calculate it. We will do this in §19.4.2, where we need the Gram–Schmidt orthogonalization procedure.

19.4.1 Gram–Schmidt orthogonalization procedure

The Gram–Schmidt procedure is a systematic method to construct an orthonormal set of vectors (i.e., a set of pairwise orthogonal vectors with unit norm) from a finite or infinite

---

[21] Orthogonality plays an important role as, in general, the direct sum of two non-orthogonal subspaces is not closed and, therefore, not a subspace (see e.g. Kubrusly 2001, pp. 327 and 414).

[22] In the latter case, the elements $u = \langle u_1, u_2, \ldots \rangle$ of the resultant space are formed by those elements of the component spaces such that $\sum \|u_k\|_{H_k}^2$ is finite.

set of vectors u_0, u_1, u_2, ..., u_n, ... in a Hilbert space. Starting from any one element, e.g. u_0, we construct a vector with unit norm

$$e_0 = \frac{u_0}{\|u_0\|}. \tag{19.4.8}$$

We then select another vector, for example u_1, and subtract from it the component parallel to e_0 by defining

$$v_1 = u_1 - (e_0, u_1)e_0. \tag{19.4.9}$$

If u_1 is parallel to u_0, $v_1 = \emptyset$ and we move on to another vector. Otherwise, we turn v_1 into a unit vector by defining

$$e_1 = \frac{v_1}{\|v_1\|}. \tag{19.4.10}$$

Evidently, $(e_0, e_1) = 0$ by construction. The procedure continues in this way by constructing

$$v_k = u_k - (e_0, u_k)e_0 - (e_1, u_k)e_1 - \cdots - (e_{k-1}, u_k)e_{k-1}, \quad e_k = \frac{v_k}{\|v_k\|}. \tag{19.4.11}$$

It is readily verified that the vectors e_k constitute an orthonormal set.

Example 19.4.1 As an example, consider $u_k = x^k$ over the interval $-1 \leq x \leq 1$, with the scalar product (19.3.11). A simple calculation shows that

$$e_0 = \frac{1}{\sqrt{2}}, \quad e_1 = \sqrt{\frac{3}{2}} x, \quad e_2 = \sqrt{\frac{5}{2}} \frac{1}{2} (3x^2 - 1) \tag{19.4.12}$$

and so forth. Up to the multiplicative constant $\sqrt{(2\ell + 1)/2}$, these will be recognized as the Legendre polynomials of §13.3. In the same way, by considering other intervals $a < x < b$ and weighting functions $w(x)$, one can construct families of orthogonal polynomials; a brief description of some these families and of the various relations they satisfy is given in §13.9 p. 334. $\qquad\square$

19.4.2 Best Approximation

We can now address the problem of constructing the orthogonal projection of a vector on a subspace. Let us begin by considering a finite-dimensional subspace spanned by vectors v_1, v_2, ... v_N. By the Gram–Schmidt procedure we construct from the $\{v_j\}$ an orthonormal set $\{e_j\}$ and we want then to determine N complex numbers $\{\beta_j\}$ such that

$$w_N = \sum_{j=1}^{N} \beta_j e_j \tag{19.4.13}$$

is closest to u in the sense of the norm. Since $(e_j, e_k) = \delta_{jk}$, the norm of the difference $u - w_N$ is

$$\|u - \sum_{j=1}^{N} \beta_j e_j\|^2 = (u, u) - \sum_{j=1}^{N} \overline{\beta_j} (e_j, u) - \sum_{j=1}^{N} \beta_j (u, e_j) + \sum_{j=1}^{N} |\beta_j|^2 \tag{19.4.14}$$

or, upon adding and subtracting $\sum_{j=1}^{N} |(e_j, u)|^2$,

$$\|u - \sum_{j=1}^{N} \beta_j \, e_j\|^2 = (u, u) + \sum_{j=1}^{N} |\beta_j - (e_j, u)|^2 - \sum_{j=1}^{N} |(e_j, u)|^2. \qquad (19.4.15)$$

The minimum will evidently be obtained by choosing $\beta_j = (e_j, u)$; hence we find that

$$w_N = \sum_{j=1}^{N} (e_j, u) \, e_j, \qquad (19.4.16)$$

is the best approximation to u and (19.4.15) gives the square of the "distance" between u and the span of the v_j as

$$\|u - \sum_{j=1}^{N} (e_j, u) e_j\|^2 = \|u\|^2 - \sum_{j=1}^{N} |(e_j, u)|^2. \qquad (19.4.17)$$

This relation implies that

$$\sum_{j=1}^{N} |(e_j, u)|^2 \le \|u\|^2. \qquad (19.4.18)$$

If the space has dimension N, every u can be written as a linear combination of the $\{e_j\}$ and, therefore, this relation holds with the equal sign for any u; in this case, the same coefficients may also be found by writing

$$u = \sum_{j=1}^{N} \alpha_j e_j \qquad (19.4.19)$$

and taking scalar products with $e_1, e_2, \ldots, e_N$. A more interesting situation is encountered when the space is infinite dimensional. We start with a simple but important result.

Theorem 19.4.2 [Riesz–Fischer] *Let $\{e_k\}$ be an orthonormal set in an infinite-dimensional Hilbert space, and $\{\alpha_k\}$ a sequence of complex numbers. Then*

$$\sum_{1}^{\infty} \alpha_k \, e_k \quad and \quad \sum_{1}^{\infty} |\alpha_k|^2 \qquad (19.4.20)$$

converge or diverge together.

Proof Let us consider two elements of the sequence $u_k = \sum_{1}^{k} \alpha_j e_j$ corresponding to m and n, with $n > m$. Then

$$\|u_n - u_m\|^2 = \|\sum_{m+1}^{n} \alpha_j \, e_j\|^2 = \sum_{m+1}^{n} |\alpha_j|^2 = \sum_{1}^{n} |\alpha_j|^2 - \sum_{1}^{m} |\alpha_j|^2. \qquad (19.4.21)$$

This relation shows that, if $\{u_k\}$ is a Cauchy sequence, then also the numerical sequence $\sum |\alpha_j|^2$ is a Cauchy sequence and therefore converges. Conversely, if $\sum |\alpha_j|^2$ converges, and is therefore a Cauchy sequence, so is $\{u_k\}$ and therefore $\{u_k\}$ converges given the completeness of the space. □

The argument leading to the best approximation result shows that adding to $\sum_{k=1}^{N}(e_k, u)e_k$ a new term $(e_{N+1}, u)e_{N+1}$ (where e_{N+1} has unit norm and is orthogonal to all the other e_k) cannot make the approximation worse.[23] In other words, $\sum_{k=1}^{N}|(e_k, u)|^2$ is a non-decreasing sequence always smaller than, or at most equal to, $\|u\|^2$. If we now consider an infinite set of orthonormal vectors, according to Theorem 8.1.3 p. 216, we conclude that the infinite series is convergent and we have the *Bessel inequality*

$$\sum_{k=1}^{\infty}|(e_k, u)|^2 \leq \|u\|^2. \tag{19.4.22}$$

By the Riesz–Fischer theorem, the first series in (19.4.20) is then convergent.

Even with an infinite number of terms, the inequality cannot necessarily be replaced by an equality. For example, the sine functions $\{\pi^{-1/2}\sin kx\}$ over the interval $(-\pi, \pi)$ are an infinite orthonormal set for which there will be a strict inequality for any continuous function $u(x)$ such that $u(-x) \neq -u(x)$. On the other hand, if the functions $(2\pi)^{-1/2}$ and $\{\pi^{-1/2}\cos kx\}$ are added to this set, we do have an equality as we know from the theory of the Fourier series in Chapter 9.

19.4.3 Orthonormal basis

When (19.4.22) holds with an equal sign,

$$\sum_{1}^{\infty}|(e_j, u)|^2 = \|u\|^2 \tag{19.4.23}$$

we have the *Parseval equality*, which generalizes the relation by the same name of §9.8 p. 257. This case is of particular interest as then the Riesz–Fischer theorem shows that, when it is satisfied, there is an element v of the Hilbert space such that $v = \sum_{j=1}^{\infty}(e_j, u)e_j$ in the sense of the norm. However, since strong convergence implies weak convergence (p. 552),

$$(e_k, v) = \lim_{n\to\infty}\sum_{j=1}^{n}(e_j, u)(e_k, e_j) = (e_k, u) \tag{19.4.24}$$

and we conclude that $(e_k, v - u) = 0$ for all k. Suppose that the set $\{e_k\}$ is dense in the space, i.e., that every element of the space can be approximated arbitrarily closely by finite linear combinations of elements of the set. Then, from the fact that $(e_k, v - u) = 0$ for

---

[23] We see here the advantage of executing the Gram–Schmidt orthogonalization prior to calculating the minimum distance. Had we not done so, we would have found
$$\|u - \sum_{j=1}^{N}\gamma_j v_j\|^2 = (u, u) - \sum_{j=1}^{N}\overline{\gamma}_j(v_j, u) - \sum_{j=1}^{N}\gamma_j(u, v_j) + \sum_{j=1}^{N}\sum_{k=1}^{N}\overline{\gamma}_j\gamma_k(v_j, v_k)$$
the minimization of which leads to $\sum_{k=1}^{N}\gamma_k(v_\ell, v_k) = (v_\ell, u)$, $\ell = 1, 2, \ldots$ This is a linear system that must be solved for the coefficients γ_k. If we were to add another vector, the work already done to solve this system would be useless and the new system would need to be solved *ex novo*. On the other hand, after orthogonalization, the coefficients (e_j, u) already determined do not change.

all e_k, we deduce that $v - u$ is orthogonal to all the elements of the Hilbert space and, in particular, to itself, so that $\|u - v\|^2 = 0$. Axiom (v) of the scalar product (Table 19.3 p. 549) then implies that

$$u = \sum_{j=1}^{\infty} (e_j, u) \, e_j. \tag{19.4.25}$$

We say that the $\{e_k\}$ constitute an *orthonormal basis* or a *complete orthonormal set* in the Hilbert space. The same argument also proves that, if $(e_k, f) = (e_k, g)$ for all e_k, then $f = g$.

Since, by definition, every separable Hilbert space contains a countable dense set (p. 546), which can be turned into an orthonormal set by the Gram–Schmidt procedure, we deduce that[24]

Theorem 19.4.3 *Every separable Hilbert space possesses an orthonormal basis.*

By writing (19.4.23) for $u \pm v$, $u \pm iv$ and applying the polarization relation (19.3.13), the Parseval equality can be generalized to

$$\sum_{j=1}^{\infty} (v, e_j) \, (e_j, u) = (v, u), \tag{19.4.26}$$

with the series converging absolutely since $|(v, e_j) \, (e_j, u)| \leq \frac{1}{2}[|(v, e_j)|^2 + |(e_j, u)|^2]$ and the series $\sum |(v, e_j)|^2$ and $\sum |(e_j, u)|^2$ converge. This relation extends to infinite-dimensional spaces the familiar rule for calculating the scalar product of two vectors by avoiding the products of their components on the same base.

The characterization of the sets $\{e_k\}$ which constitute a basis is evidently a matter of great importance. We give some criteria without proof:

Theorem 19.4.4 *The orthonormal set $\{e_k\}$ constitutes a basis for the Hilbert space H so that, for any $u \in H$,*

$$u = \sum_{k=1}^{\infty} (e_k, u) e_k : \tag{19.4.27}$$

1. *if it is dense in H; or*
2. *the set is maximal, in the sense that it is impossible to find another non-zero vector orthogonal to all the e_k; or*
3. *the only element of H orthogonal to all the e_k is the $\emptyset$ vector; or*
4. *for any $u \in H$ Parseval's equality (19.4.23) is satisfied.*

It should be stressed that the equality in (19.4.27) should be interpreted in the sense of the norm. In particular, there is no implication e.g. of point convergence when u and the e_k's are ordinary functions.

---

[24] There are non-separable Hilbert spaces with uncountably infinite orthonormal bases, see Kubrusly 2001, p. 364.

As in the case of Banach spaces, any two different bases have the same number of elements or are both countably infinite according as the space has a finite or infinite dimensionality. The relation (19.4.27) is a generalization of the Fourier series and, indeed, it is frequently referred to by this name in the literature.

In connection with the general definition of a basis given on p. 547 we have pointed out that an *infinite, linearly independent dense set* is not necessarily a basis for a Banach space. Theorem 19.4.4 states, however, that such a set is a basis provided it is orthonormal (or, more generally, orthogonal).

To gain some insight into this somewhat puzzling situation let us consider again the monomials x^k, $k = 0, 1, 2, \ldots$ in the interval $-1 \leq x \leq 1$. The Legendre polynomials are an orthogonal basis over this interval, and each one of them is a finite linear combination of monomials. Thus, an expansion of the type $\sum_0^\infty c_n P_n(x)$ is a rearrangement of the terms of a power series $\sum_0^\infty a_k x^k$ and we know from §8.1.2 (p. 217) that such a rearrangement may well change the sum of the series. Theorem 8.1.6 states that this does not happen if the series converges absolutely, which is the case for power series inside their circle of convergence (Theorem 8.3.6 p. 227). In this case, $\sum_0^\infty c_n P_n(x)$ and $\sum_0^\infty a_k x^k$ have the same sum, but this is not true in general.

Once a basis has been chosen, by (19.4.27), the knowledge of the coefficients $\alpha_k = (e_k, u)$ is equivalent to a knowledge of u and conversely, given u, the coefficients are uniquely determined. As in the finite-dimensional case, there is therefore an isomorphism, namely, a one-to-one correspondence, between separable Hilbert spaces and the space l^2 of sequences $(\alpha_1, \alpha_2, \ldots, \alpha_k, \ldots)$, all square-summable by Parseval's equality. Thus, all separable Hilbert spaces are isomorphic to each other and, by (19.4.26), the correspondence preserves scalar products. In other words, there exists essentially only one infinite-dimensional separable Hilbert space and l^2 serves as a "universal model" for all of its realizations. Since, in an intuitive sense, l^2 is a fairly straightforward extension of the usual finite-dimensional Euclidean space, we have another indication of why the concept of Hilbert space proves a powerful aid to intuition.

19.5 Sobolev spaces

For boundary value problems one is interested in finding the solution to an equation which satisfies certain specific boundary conditions. This solution must therefore be sought in a function space where the notion of boundary value is well defined. We touched on this problem in §15.4 in connection with singular Sturm–Liouville problems. The issue may exhibit unexpected subtleties. For example, a function belonging to $L^2(-\infty, \infty)$ does not necessarily tend to 0 at infinity as shown e.g. by $x^2 \exp(-x^8 \sin^2 x)$, which is square integrable without being even bounded (see e.g. Richtmyer 1978, p. 85; Boccara 1990, p. 280). Similarly, over a finite interval, a function $f \in L^p(a, b)$ may not be defined at the end points so that it would be impossible to impose boundary conditions there. Spaces L^p are therefore too large for boundary-value problems and one needs to consider particular subspaces of them. In more than one dimension, these difficulties are compounded as then the solution of the problem must become a prescribed function on the boundary, which may not be smooth. Sobolev spaces are useful in this respect as, on the boundary of the domain of interest, functions that belong to them reduce to well-defined functions (or possibly distributions).

Let us consider a function $u(x_1, x_2, \ldots, x_N)$ defined in a region $\Omega \in \mathbb{R}^N$. The Sobolev space $W^{1,2}$ consists of those functions u such that u itself and all its distributional first-order partial derivatives $\partial u / \partial x_k$ belong to L^2:

$$W^{1,2} = \{ u \in L^2(\Omega), \partial u / \partial x_k \in L^2(\Omega), k = 1, 2, \ldots, N \}. \tag{19.5.1}$$

This space is not a closed linear manifold with the L^2 norm, but it becomes a complete separable Hilbert space with respect to the inner product

$$(v, u)_1 = (v, u) + (\nabla v, \nabla u) = \int_\Omega \left(\overline{v} u + \sum_{k=1}^N \frac{\partial \overline{v}}{\partial x_k} \frac{\partial u}{\partial x_k} \right) d\Omega \tag{19.5.2}$$

and the norm induced by it:

$$\| u \|_{W^{1,2}}^2 = \| u \|_2^2 + \| \nabla u \|_2^2. \tag{19.5.3}$$

Evidently $\| u \|_2 \leq \| u \|_{W^{1,2}}$ and, therefore, a sequence converging in $W^{1,2}$ will also converge in L^2. Often the norm (19.5.3) has the physical meaning of an energy, with the first term related to the kinetic energy and the second one to the potential energy of the system.

The idea can be generalized by using one of the p-norms with $p \geq 1$ (see e.g. Renardy & Rogers 1993, p. 205; Edmunds & Evans 1987, p. 320; Robinson 2001, chapter 5; Adams & Fournier 2003):

$$\| u \|_{W^{1,p}} = \left(\| u \|_p^p + \| \nabla u \|_p^p \right)^{1/p} \tag{19.5.4}$$

in which $\| \cdot \|_p$ is the norm defined in (19.2.6). These spaces are complete separable Banach spaces in these norms. A further extension is obtained by including distributional derivatives of order up to k, giving rise to spaces $W^{k,p}$ which are also complete normed separable spaces. The norm is then defined by summing over all derivatives of order $\leq k$. It is evident that, with increasing k, the elements of these spaces become smoother and smoother.

When $p = 2$ the space $W^{k,2}$ becomes a Hilbert space, denoted by H^k, with an inner product defined by the obvious extension of (19.5.2) obtained by including in the right-hand sides the L^2 inner products of all the derivatives up to order k. Another way to describe these spaces is as the completion, under the norm induced by this scalar product, of the set $C^k(\Omega)$ of functions continuous with their derivatives up to order k. Hence, any function in H^k can be approximated by functions whose derivatives exist in the classical sense. This statement cannot be extended to functions continuous on the closed domain $\overline{\Omega}$ (i.e. consisting of Ω and all its boundary points) without assuming some smoothness properties of the boundary $\partial \Omega$ of Ω (see e.g. Renardy & Rogers 1993, p. 208). We will not pursue this aspect here, but we will simply state that, if $\partial \Omega$ is sufficiently smooth, the space $C^\infty(\overline{\Omega})$ of infinitely differentiable functions defined on the closed domain $\overline{\Omega}$ is also dense in $H^k(\Omega)$.

We have already made use of functions which vanish on the boundary of the domain of interest in connection with the notion of weak derivative (p. 552). The subset of $C^k(\Omega)$ consisting of such functions is denoted by $C_0^k(\Omega)$ and the completion of this set with the above norm generates the Hilbert space H_0^k, in general different from H^k.[25] When the domain is bounded, one can prove the *Poincaré–Friedrichs inequality*

$$\|u\|_{L^2} \leq c(\Omega)\|\nabla u\|_{L^2} \qquad \text{for all} \qquad u \in H_0^1(\Omega), \tag{19.5.5}$$

where $c(\Omega)$ is a constant solely dependent on the domain. As a consequence, in H_0^1 we can introduce the scalar product

$$(v, u)_{H_0^1} = \int \nabla \bar{v} \cdot \nabla u \, d\Omega \tag{19.5.6}$$

and the norm that it induces is equivalent to the norm (19.5.3) in the sense of (19.2.9).

In a bounded domain Ω, the space $H_0^1(\Omega)$ has a striking connection with $L^2(\Omega)$ (see e.g. Ladyzhenskaya 1985, p. 25; Edmunds & Evans 1987, p. 230; Robinson 2001, p. 143):

Theorem 19.5.1 [Rellich–Kondrachov] *A bounded set* $S \in H_0^1(\Omega)$ *is pre-compact in* $L^2(\Omega)$, *i.e., every infinite sequence of elements of S contains a subsequence converging strongly (i.e., in the sense of the L^2-norm) to an element of* $L^2(\Omega)$.

Since a bounded sequence is (or contains) a weakly convergent (sub-) sequence (p. 552), this theorem implies that a weakly convergent sequence in H_0^1 (definition on p. 552) is (or contains) a strongly convergent sequence in L^2.[26]

Example 19.5.1 The set $\{\sqrt{2/\pi} \sin nx\}$ is an orthonormal basis in the space $L^2(0, \pi) = H_0(0, \pi)$. The corresponding orthonormal basis in $H_0^k(0, \pi)$ is

$$e_n^k(x) = \sqrt{\frac{2}{\pi}} \frac{\sin nx}{\sqrt{1 + n^2\pi^2 + \ldots + (n\pi)^{2k}}}. \tag{19.5.7}$$

By the Riemann–Lebesgue lemma, the sequence $\{e_n^1\}$ converges weakly in the H_0^1 norm, but converges strongly in the L^2 norm. $\square$

The simplest connection between functions belonging to a Sobolev space and continuous functions arises in one space dimension. It can be shown that, if $u \in H^1(a, b)$, then u is continuous on the closed (finite) interval $[a, b]$.[27] More generally,

Theorem 19.5.2 [Sobolev embedding theorem] *Let* $s > N/2$. *Then every element* $u \in H^s(\mathbb{R}^N)$ *is a bounded continuous function on* $\mathbb{R}^N$ *and there is a constant c, independent of u, such that*

$$\|u\|_\infty \equiv \max |u(x_1, \ldots, x_N)| \leq c\|u\|_{H^s}. \tag{19.5.8}$$

[25] The set $C_0^k(\Omega)$ contains the set $C_c^\infty(\Omega)$ of infinitely differentiable functions which vanish outside a compact (i.e., closed and bounded) domain. This latter set is dense in $H_0^k(\Omega)$, even if $\Omega = \mathbb{R}^N$.

[26] In view of the importance of compact operators (§21.3), it may be worth while recasting the previous statement as saying that the operator which embeds H^1 in L^2 is compact.

[27] More precisely, if $u \in H^1(a, b)$, there is a continuous function defined on $a \leq x \leq b$ which is equal to u except possibly on a set of measure zero.

The analogous statement for bounded domains requires suitable smoothness properties of the boundary of the domain.

For $N = 1$ this theorem implies that a function belonging to $L^2(-\infty, \infty)$ with its derivative is continuous and tends to zero at infinity.[28] For $N = 2$ and 3, we need to have $s \geq 2$ to draw the same conclusion.

The embedding theorem mentioned before is one of several embedding theorems concerning Sobolev spaces. For example, one can show that

$$\cdots \subset H_0^{n+1} \subset H_0^n \subset \cdots \subset H_0^0 \subset H^0 = L^2. \tag{19.5.9}$$

An important point is that, in removing elements from L^p to form a Sobolev space, one is left with a complete separable space, rather than just a subset of the original space. In the case of the spaces H^k this new smaller space has a full Hilbert space structure permitting the application of Hilbert space results which are not applicable to L^2. For example, differential operators which are not defined for all elements of L^2 may happen to be defined and bounded when operating on all elements of H^k. This circumstance permits the application of the Riesz representation theorem 19.6.3 introduced in the next section; an example of the use of this feature is provided on p. 660.

19.6 Linear functionals

Definition 19.6.1 Let M be a linear manifold in a Banach space. A *linear functional* defined over M is a relation which associates to each element $v \in M$ a number $\ell(v)$ (in general complex) in such a way that

$$\ell(a \cdot v + b \cdot w) = a\ell(v) + b\ell(w) \tag{19.6.1}$$

for any $v, w \in M$ and any real or complex numbers a and b; M is the domain of definition of the linear functional.

Since $\ell(v - v) = \ell(\emptyset) = \ell(v) - \ell(v) = 0$, for all linear functionals $\ell(\emptyset) = 0$.

A typical and important example is the scalar product with a fixed element of the space, $\ell_f(v) = (f, v)$. For example, on the space $L^2(a, b)$,

$$\ell_f(v) = \int_a^b \overline{f}(x)\, v(x)\, dx, \tag{19.6.2}$$

where f is any fixed element of $L^2(a, b)$. But note that (v, f) and $\|v\|$ are not linear functionals according to the definition (19.6.1) as $(a \cdot v, f) = \overline{a}(v, f)$ and $\|a \cdot v\| = |a|\,\|v\|$. Another important example of a linear functional defined on the space of functions continuous at 0 is

$$\ell_\delta(v) = v(0). \tag{19.6.3}$$

---

[28] For an elementary proof see e.g. Boccara 1990, p. 280 or Richtmyer 1978, p. 86.

The set of all linear functionals defined on a linear vector space B is itself a linear space if the functional $\ell = \ell_1 + \ell_2$ obtained by the addition of two functionals ℓ_1 and ℓ_2 is defined as $\ell(u) = \ell_1(u) + \ell_2(u)$ for all u, and multiplication by a number as $(a \cdot \ell)(u) = a\ell(u)$. This is the *algebraic dual space* or *algebraic conjugate space* of the original space and is usually denoted by B^f.

A linear functional is *continuous at* v if, for any sequence $\{v_n\}$ such that $\|v - v_n\| \to 0$, $\ell(v) - \ell(v_n) \to 0$.[29]

Theorem 19.6.1 *A linear functional continuous at any point is continuous everywhere in its domain of definition.*

This is a simple consequence of linearity as, if ℓ is continuous at v and $\{w_n\} \to w \neq v$, $\ell(w) - \ell(w_n) = \ell(w - w_n + v) - \ell(v)$. Since $w - w_n \to 0$, by continuity at v, $\ell(w - w_n + v) \to \ell(v)$ and thus we conclude that $\ell(w) - \ell(w_n) \to 0$ so that ℓ is also continuous at any w. Thus, it is sufficient to check continuity at a point.

A linear functional is *bounded* in M if and only if, for some fixed non-negative number c,

$$|\ell(v)| \leq c\|v\| \qquad \text{for all} \qquad v \in M. \tag{19.6.4}$$

There is a strong connection between boundedness and continuity:

Theorem 19.6.2 *For a linear functional boundedness and continuity imply each other.*

Continuity readily follows from boundedness since, if $\|v_n - v\| \to 0$, then $|\ell(v_n) - \ell(v)| \leq c\|v_n - v\| \to 0$ as well. If ℓ were continuous but not bounded one could construct a sequence $\{v_n\}$ such that $\ell(v_n) \geq n\|v_n\|$, i.e. $\ell(v_n/n\|v_n\|) \geq 1$. But this construction produces a sequence $v_n/n\|v_n\| \to 0$ without $\ell(v_n/n\|v_n\|)$ tending to 0, which violates continuity.

The set of all continuous linear functionals defined on the Banach space B is a subset of the algebraic dual space B^f; we refer to it simply as the *dual space* B' of the original Banach space B without further qualifiers. In the literature, it is also referred to as the *continuous dual space*.

We can re-write the definition (19.6.4) of bounded linear functional as $|\ell(v/\|v\|)| \leq c$. It is evident then that the smallest possible value of c coincides with the largest value of $|\ell(v/\|v\|)|$. We define this smallest c as the *norm* of the functional:

$$\|\ell\| = \sup_{\|e\|=1} |\ell(e)| \tag{19.6.5}$$

where the sup is taken over all vectors of norm 1, the so-called *unit ball* in the Banach space. It can be shown that this is indeed a norm in the dual space in the sense that it satisfies the norm axioms on p. 542. The norm in the dual space evidently depends on that in the original

---

[29] Alternatively, for any $\varepsilon > 0$ there is a $\delta > 0$ such that $|\ell(v) - \ell(w)| < \varepsilon$ whenever $\|v - w\| < \delta$.

space but, with this norm, the dual space is complete even when the original space is not.[30] Let us consider some examples.

Example 19.6.1 By adopting the L^2 norm, the Banach space becomes a Hilbert space and we can apply the Cauchy–Schwarz inequality to the functional (19.6.2). We then deduce that $\|\ell_f\|_2 \leq \|f\|_2$. Upon taking $v = f$, we have $|\ell_f(f)| = \|f\|_2^2$ and, therefore, $\|\ell_f\|_2 = \|f\|_2$. In the special case $f = 1$, $\|\ell_1\|_2 = \sqrt{b-a}$. Continuity of the functional follows from that of the scalar product. $\qquad\square$

Example 19.6.2 For the linear functional $\ell_I = \int_a^b u(x)\,dx$, with u continuous on the finite interval $a \leq x \leq b$, we have

$$|\ell_I(u)| \leq (b-a) \max_{a \leq x \leq b} |u(x)| \qquad (19.6.6)$$

and, therefore, $\|\ell_I\|_\infty \leq b - a$. Taking $u = 1$ proves that the norm equals $b - a$. Comparison with the previous example illustrates the different norms induced in the dual space by different norms in the original space. The functional is evidently continuous. $\qquad\square$

Example 19.6.3 In the space l^p of Example 19.2.3 (p. 542) consider the linear functional

$$\ell_\alpha(u) = \sum_1^\infty \overline{\alpha}_i u_i \qquad (19.6.7)$$

with the l^p norm. It is found that

$$\|\ell_\alpha\| = \left(\sum_{j=1}^\infty |\alpha_i|^q \right)^{1/q} \qquad \text{with} \quad q^{-1} + p^{-1} = 1. \qquad (19.6.8)$$

More generally, the dual space of l^p is isomorphic to l^q and all the bounded linear functionals on l^p can be put in the form (19.6.7). $\qquad\square$

Example 19.6.4 It is a remarkable fact, which will prove important later in Chapter 20, that the functional ℓ_δ defined in (19.6.3) is not continuous with the L^2 norm. For the proof consider any continuous function f positive for $-1 < x < 1$, vanishing for $1 \leq |x| < \infty$, and define the sequence $w_n = f(nx)$; because of the choice of f, $w_n = 0$ for $1/n < |x|$ and, as is easily shown, $\|w_n\|_2 = (1/\sqrt{n})\|f\|_2 \to 0$, while $\ell_\delta(w_n) = f(0) \neq 0$. As expected from Theorem 19.6.2, this functional is not bounded either. To prove this directly, define $v_n = \sqrt{n}\,w_n$. Then $\|v_n\|_2 = \|f\|_2$, while $\ell_\delta(v_n) = \sqrt{n}\,f(0) \to \infty$. $\qquad\square$

---

[30] The norm of the linear functional can be given a geometrical interpretation as the inverse of the distance from the origin to the hyperplane such that $\ell(u) = 1$ (see e.g. Kolmogorov & Fomin 1957, chapter 4).

Example 19.6.5 Every element of a finite-dimensional Banach space can be expressed in terms of a basis, $v = \sum_{k=1}^{N} \alpha_k h_k$, and therefore, by the linearity property (19.6.1),

$$\ell(v) = \ell\left(\sum_{k=1}^{N} \alpha_k h_k\right) = \sum_{k=1}^{N} \ell_k \alpha_k \qquad \text{where} \quad \ell_k = \ell(h_k). \tag{19.6.9}$$

The action of ℓ on all the vectors of the basis then completely determines the functional via the N numbers ℓ_k. Conversely, N arbitrary numbers $\ell_1, \dots, \ell_N$ specify a linear functional. It is evident therefore that all linear functionals on a finite-dimensional space are bounded so that the algebraic dual space coincides with the dual space, $B^f = B'$. If the basis is orthogonal, the linear functional can be expressed as a scalar product $\ell(v) = (\lambda, v)$ with the vector $\lambda = \sum_{k=1}^{N} \overline{\ell(e_k)}\, e_k$. Thus, in a finite-dimensional Hilbert space, any linear functional can be represented as a scalar product with some fixed vector so that the dual space coincides with the original space.[31]   □

In an infinite-dimensional space the second step in (19.6.9) requires continuity of the functional but, even with this property, the last summation in (19.6.9) might be divergent. It is therefore remarkable that in a Hilbert space, even infinite-dimensional, a linear functional can always be represented as a scalar product

Theorem 19.6.3 [Riesz representation theorem] *To each continuous (and therefore bounded) linear functional ℓ defined over the entire Hilbert space H corresponds a uniquely defined element $\lambda \in H$ such that $\ell(v) = (\lambda, v)$ for all $v \in H$; in other words, $H' = H$.*

The specification that ℓ should be defined over the whole space is actually redundant as, by the Hahn–Banach theorem 21.2.3, which is proven in the more general setting of linear operators on p. 626, the domain of a bounded linear functional over a normed space can be extended to the entire space without increasing its norm. A proof of Theorem 19.6.3 can be found in most of the books cited in this chapter. An implication worthy of explicit mention is that, in the case of functions, knowledge of $\ell(v)$ for all v's is equivalent to the knowledge of $\lambda(x)$ in a point-wise sense except for a set of zero measure.

As an application of this theorem we introduce the notion of *dual* or *reciprocal basis* in a Hilbert space. Consider a basis $\{h_n\}$ for a Hilbert space H consisting of vectors of norm 1 but not necessarily orthogonal to each other. How are the coefficients of the expansion (19.2.18) of a general vector u on this basis to be calculated? The Riesz representation theorem ensures that one can find a set of unit vectors $\{h_m^*\}$, the dual basis, such that $(h_m^*, h_n) = \delta_{mn}$, after which the components α_n readily follow as $\alpha_n = (h_n^*, u)$. Indeed, for each m, define a linear functional ℓ_m such that $\ell_m(h_n) = \delta_{mn}$ as n ranges over all the vectors of the basis. This defines ℓ_m for all the elements of the basis and, therefore, for the entire space. By the Riesz representation theorem, we conclude that there is a vector h_m^* which represents the linear functional.

---

[31] More precisely, the dual space consists of row vectors $|\ell_1, \ell_2, \dots, \ell_N|^T$ each one of which can be made to correspond to a column vector in the original space. This one-to-one correspondence establishes an isomorphism between the original space and its dual so that the two can be identified with each other.

Since the dual space B' is itself a normed linear space, one can consider its own dual space B'' – the *second dual*, or *bi-dual*, of the original space B. Hilbert spaces are examples of *reflexive spaces* for which $H'' = H$. The Banach space l^p is also reflexive for any $p > 1$. For a general normed space B one finds that there is an isometric isomorphism (i.e., a one-to-one correspondence which preserves the norm) between B and a linear manifold of B'', but not necessarily with the entire B'' (see e.g. Kubrusly 2001, p. 266). This isomorphism defines the *natural embedding* of B into B''.

Theory of Distributions

20.1 Introduction

The immense usefulness of the "δ-function", amply documented in this book, joint to its "non-existence" in the conventional sense in which a function is understood, would be motivation enough to develop a framework generalizing in a mathematically satisfactory way the concept of function. But a long list of additional justifications for such an extension is readily developed:

1. A naive limiting process suggests that the d'Alembert general solution of the wave equation in the form $f(x - ct) + g(x + ct)$ (p. 8) should also make sense when f and g are not twice differentiable functions. It should be possible to decide whether a proposed non-differentiable function is indeed a solution of the equation (for example, whether it propagates with the right speed), without the cumbersome apparatus required by standard analysis.

2. The problem is even more acute for non-linear waves, which frequently "break," i.e., develop discontinuities, in a finite time. Again, it makes sense that information on the speed of propagation of the discontinuous solution be contained in the equation, but this information is not easily extracted within the standard framework.

3. In the application of the Fourier and Laplace transforms often one can apply formally standard rules, e.g. for differentiation, convolution, etc., finding correct results even though it would be difficult to justify some of the intermediate steps within the bounds of the standard theory. For example, by a formal use of the convolution theorem one might write

$$f(x) = \frac{1}{\sqrt{2\pi}} \int_{-\infty}^{\infty} e^{-ikx}\, \tilde{f}(k)\, \mathrm{d}k \; = \; \frac{1}{2\pi} \int_{-\infty}^{\infty} \mathscr{F}^{-1}\{1\}(\xi)\, f(x - \xi)\, \mathrm{d}\xi \qquad (20.1.1)$$

even though the inverse Fourier transform of 1 does not even exist.

This list could easily be made much longer, but suffice it to say that the theory of distributions, or generalized functions, outlined in this chapter provides a very effective tool to deal with these and other issues as well as with a variety of applications which would otherwise require a great deal of conventional analysis or, sometimes, lie beyond its reach.

The theory of distributions is a vast topic on which many good references, at a variety of different levels, exist (see e.g. Lighthill 1958, Gelfand & Shilov 1964, Schwartz 1966a, 1966b, Jones 1966, Hörmander 1983, Richards & Youn 1990, Strichartz 1994, Saichev &

Woyczyński 1997, Kanwal 2004, Vladimirov 2002). Most books devoted to partial differential equations also have chapters on this topic (see e.g. Richtmyer 1978, Zauderer 1989, Griffel 2002, Renardy & Rogers 1993, Stakgold 1997). Here we provide a synthetic treatment focusing mostly on the δ and related distributions which are the distributions that appear in the vast majority of practical problems.

20.2 Test functions and distributions

While one ordinarily thinks of a function as a set of values $\{f(x)\}$, each one associated to a particular point x, Riesz's representation theorem 19.6.3 p. 566 affords the possibility of conceiving a function f in a more "global" sense as the set of values that the linear functional $\ell_f(u) \equiv (f, u)$ takes in correspondence with each element $u \in L^2$. This is the key observation on which a generalization of the notion of function is built.

While Riesz's theorem provides a new way to look at ordinary functions, to go beyond this concept we evidently need to weaken the hypotheses on which the theorem rests. Those hypotheses are: (i) that the functional ℓ_f is defined *for all* $u \in L^2$, and (ii) that it is continuous in the sense of the L^2-norm (which is equivalent to stating that it is bounded, see Theorem 19.6.2 p. 564).

Concerning the first requirement, it is evident that, if we were to restrict the class of functions to which the functional can be applied, we might end up with a richer set of functionals not all of which, presumably, would be ordinary functions. It is equally evident that the broader the allowed class of functions, the narrower the set of resulting functionals, and conversely. One must strike a delicate balance here as, for example, if the class of allowed functions $\{u\}$ were too limited, it might happen that $(f, u) = (g, u)$ even though f and g were to differ on a set of non-zero measure. The second requirement, L^2-continuity, must be weakened as well, but not so much as to destroy the basis on which limit considerations, which play such a vital role in judging convergence of series and others, rest.

The French mathematician Laurent Schwartz showed how to achieve both objectives by introducing a class of *test functions* ϕ, defined on the real line $\mathbb{R}$, which

1. are infinitely differentiable, and
2. have compact support, i.e., vanish outside a closed bounded interval (p. 683).

This class of functions is usually denoted by $\mathscr{D}(\mathbb{R})$ or by $C_0^\infty(\mathbb{R})$, the superscript referring to the first property and the subscript to the second one.[1] A standard textbook example is

$$\phi_1(x) = \begin{cases} \exp\left[-1/(1-x^2)\right] & \text{for } |x| < 1 \\ 0 & \text{for } 1 \le |x| \end{cases} \qquad (20.2.1)$$

---

[1] These functions do have unusual properties. For example, since they vanish with all their derivatives at the boundary points of their support, they cannot be represented as Taylor series. Indeed, the only analytic functions with compact support is identically 0.

and its translates/scaled versions $\phi_a(x; x_0) = \phi_1((x - x_0)/a)$, but an infinite number of other ones exist. As a matter of fact, a crucial property of this class of functions is that it is dense in $L^2(-\infty, \infty)$ in the L^2 norm so that, for any $u \in L^2$ and any positive ε, one can find a ϕ such that $\|u - \phi\|_{L^2} < \varepsilon$. Furthermore, any absolutely integrable *continuous* function f can be *uniformly* approximated within an arbitrary positive ε by a test function $\phi \in \mathcal{D}$.[2] In other words, the space $\mathcal{D}$ is dense in the space of absolutely integrable continuous functions with the maximum norm. An important aspect of these properties lies in the fact that, by continuity, they permit one to extend many operations carried out on the test functions to other functions.

By allowing linear functionals to operate only on test functions we have weakened the first hypothesis of Riesz's theorem. In order to affect the second one – continuity – we restrict the class of sequences used to test this property as follows:[3] We stipulate that a sequence $\{\phi_n\}$ converges to ϕ *in the sense of* $\mathcal{D}$ if:

1. The sequence $\{\phi_n\}$ converges to ϕ uniformly.
2. The sequence of derivatives of any order $\{\phi_n^{(k)}\}$ of all the $\{\phi_n\}$ converges to $\phi^{(k)}$ uniformly for all k (but not necessarily uniformly in k).[4]
3. There is a finite interval which contains all the supports of all the functions of each sequence.

If, to test continuity, we are only allowed to use sequences of functions of $\mathcal{D}$ satisfying the previous criteria, one may expect that there will be linear functionals continuous in the $\mathcal{D}$ sense, but *not* in the L^2 sense. These functionals therefore, cannot be represented by ordinary L^2-functions.

At this point we are ready for the definition:

Definition 20.2.1 A *distribution* or *generalized function* is a continuous linear functional $d(\phi)$ defined on the space of test functions $\mathcal{D}$.[5] Continuous means that $d(\phi_n) \to d(\phi)$ if $\phi_n \to \phi$ *in the sense of* $\mathcal{D}$. The set of distributions (which is the dual space of $\mathcal{D}$ as defined on p. 564) is denoted by $\mathcal{D}'$.

To avoid non-essential complications, for the time being we restrict our considerations to the real case.

---

[2] Specifically, let A be such that $|f(x)| < \varepsilon$ for $A < |x|$ and, for an arbitrary $a > 0$, consider $\phi_a(x) = \phi_1(x/a)/C_a$ where $C_a = \int_{-\infty}^{\infty} \phi_a(x)\,dx$. Then it can be shown that the functions $\psi_a(x) = \int_{-A}^{A} \phi_a(x - \xi) f(\xi)\,d\xi$ are test functions and that, for any $\varepsilon > 0$, one can choose an a_0 such that $|f(x) - \psi_{a_0}| < \varepsilon$.

[3] On p. 564 continuity of a linear functional $\ell(\cdot)$ at a point v was defined as $\ell(v_n) \to \ell(v)$ for *any* sequence $\{v_n\} \to v$. The essential ingredient of this definition is that convergence is understood in the L^2 sense. In other words, to test continuity, we must establish that the equality holds *for any* sequence $\{v_n\}$ tending to v in the L^2 norm. In order to kill the link between continuity and boundedness, we must restrict the class of sequences used to test continuity.

[4] In other words, the value of N such that $|\phi_n^{(k)} - \phi^{(k)}| < \varepsilon$ for $n \geq N$ may be different for different k.

[5] By use of the axiom of choice, it is possible to prove the existence of discontinuous linear functionals over the space $\mathcal{D}$. It is not easy to give an explicit example, and it appears doubtful that such functionals would arise in practice.

It should be explicitly noted that a distribution is fully known, or fully specified, when its action on *all* the test functions of $\mathcal{D}$ – i.e., the value of $d(\phi)$ for all ϕ – is known.

The usefulness of the theory of distributions would be very limited if we were restricted only to the class $\mathcal{D}$, as very few functions that arise in practice belong to this class. The power of this new concept lies in the fact that the action of many functionals may be extended beyond this class by using continuity and the approximation results mentioned before. Thus, for example, since a function u continuous at 0 can be approximated as closely as desired by test functions with support near 0, we can define $\ell_\delta(u) = u(0)$ even though, in principle, only $\ell_\delta(\phi)$ is specified. For such a u, however, the linear functional corresponding to the derivative of l_δ (see below) may fail to be reasonably defined. One might say that the importance of the space $\mathcal{D}$ lies in the fact that it is the space over which *all* the continuous functionals can be defined or, alternatively, that it is the intersection of all the domains of definition of all possible linear functionals continuous in the sense given above.

An important – if expected – fact is that, if f is an ordinary L^2 function, the scalar product (f, ϕ) is a continuous linear functional in the sense of $\mathcal{D}$ so that f generates – and therefore can be identified with – a distribution. Distributions generated in such a way are called *regular distributions*. Distributions that are not regular are called *singular*.

A consequence of Riesz's theorem is that knowledge of the functional $\ell_f(u) = (f, u)$ for all $u \in L^2$ fully determines f in the point-wise sense up to a set of zero measure. The same property holds also if ℓ_f is known only for all $\phi \in \mathcal{D}$. Furthermore, if $(f, \phi) = (g, \phi)$ for all test functions ϕ, one can show that $f = g$ except for a set of zero measure and, conversely, that if f and g are two continuous functions which differ on a set of non-zero measure, there is a test function ϕ such that $(f, \phi) \neq (g, \phi)$. These are further illustration of the fact that the ϕ's are "numerous enough" to preserve important features of analysis. We also note the following

Definition 20.2.2 A point $\overline{x}$ is a singular point for a distribution $d(x)$ if $d(x)$ equals an infinitely differentiable function in (finite or infinite) intervals $x < \overline{x}$ and $\overline{x} < x$ on either side of $\overline{x}$.

The close relation between scalar products and regular distributions suggests to express the duality between distributions and test functions by a notation reminiscent of the scalar product. Thus one writes

$$d(\phi) \equiv \langle d(x), \phi(x) \rangle \tag{20.2.2}$$

or even

$$d(\phi) \equiv \int_{-\infty}^{\infty} d(x)\,\phi(x)\,\mathrm{d}x \tag{20.2.3}$$

to denote the action of the distribution d on a test function ϕ even though the symbol $d(x)$ appearing in these expressions, in general, is not a function in any sense and is only defined by its action on the test function. A powerful reason to adopt this suggestive notation is that, as will be seen later, most of the manipulations that can be carried out on integrals in the case of ordinary functions are also legitimate with distributions.

The space $\mathscr{D}(\mathbb{R})$ is not the only test-function space on which continuous linear functionals can be usefully defined. By *enlarging* the class of test functions, we obtain a more *restricted* set of generalized functions. Some examples, which will prove important later, are:

1. The space $C_0^\infty[a, b]$ of infinitely differentiable functions whose support is contained in $[a, b]$; this test function space is used to define distributions over a finite interval (a, b).

2. The space $\mathscr{S}(\mathbb{R})$ of C^∞ functions which, together with all their derivatives, vanish faster than any power of $1/x$ for $|x| \to \infty$. This is the space of test functions on which the definition of the Fourier transform of distributions is based (§20.10); $\mathscr{D}(\mathbb{R})$ is a proper subset of $\mathscr{S}(\mathbb{R})$ and is dense in it. The set $\mathscr{S}'$ of continuous linear functionals on $\mathscr{S}$ are the so-called *tempered distributions*. Continuity is with respect to the topology of the space $\mathscr{S}$: a sequence $\{\phi_k\} \in S$ converges to $\phi \in S$ if all ϕ_k as well as all their derivatives converge uniformly to ϕ and the corresponding derivatives.

3. The space $\mathscr{E}(\mathbb{R})$ of C^∞ functions with arbitrary support; a sequence $\{\phi_k\} \in \mathscr{E}$ is said to converge to $\phi \in \mathscr{E}$ if all ϕ_k as well as all their derivatives converge uniformly to ϕ and the corresponding derivatives on every compact subset. The set of continuous linear functionals defined on these test functions is denoted by $\mathscr{E}'$ (§20.8). The space $\mathscr{D}$ is dense in $\mathscr{E}$ and, therefore, the distributions in $\mathscr{E}'$ are a subset of the distributions defined on $\mathscr{D}$. They also all have compact support as defined on p. 575.

4. Test functions defined in the complex plane $\mathbb{C}$ can be used in place of test functions defined on $\mathbb{R}$ (§20.8).

5. The requirement of infinite differentiability can be replaced by differentiability up to and including a finite order m, with the usual definition of convergence except that only derivatives up to order m are included. The smallest m necessary to make a linear functional continuous is the *order* of the distribution defined by that functional (see also p. 583).

All of these spaces are complete with the topology specified for each one of them, and yet many others can be introduced (see e.g. Pathak 1997, chapter 1).

In each case we can define complex-valued distributions by allowing the test functions to be complex, and N-dimensional distributions – further considered in §20.7 – by allowing the test functions to be defined on an N-dimensional space $\mathbb{R}^N$ or $\mathbb{C}^N$.

20.3 Examples

It is now time to give some important examples of distributions. In each case it must be assured that the action of the distribution on *all* test functions is unambiguously specified, and that this action is continuous in the sense of $\mathscr{D}$. We will usually omit the proof of the latter requirement.

Example 20.3.1 The "δ-function":

$$\langle \delta(x), \phi(x) \rangle = \int_{-\infty}^{\infty} \delta(x)\,\phi(x)\,\mathrm{d}x \equiv \phi(0). \qquad (20.3.1)$$

Continuity of this functional is immediate as, according to the definition of convergence in the $\mathscr{D}$ sense, $\phi_n \to \phi$ implies uniform convergence, i.e. $|\phi_n(x) - \phi(x)| < \varepsilon$ and, therefore, also $|\phi_n(0) - \phi(0)| < \varepsilon$. □

Example 20.3.2 Heaviside's step function $H(x)$ is defined, as a distribution, by

$$\langle H(x), \phi(x) \rangle = \int_{-\infty}^{\infty} H(x)\,\phi(x)\mathrm{d}x = \int_0^{\infty} \phi(x)\mathrm{d}x. \qquad (20.3.2)$$

The integral is certainly finite as ϕ is continuous and vanishes outside a finite interval. It is evident that

$$H(x) + H(-x) = 1, \qquad (20.3.3)$$

the unit distribution. A related distribution is the *sign* (or *signum*) function sgn x

$$\operatorname{sgn} x = H(x) - H(-x) = \begin{cases} 1 & x > 0 \\ -1 & x < 0 \end{cases}. \qquad (20.3.4)$$

Upon adding (20.3.3) and (20.3.4) we also find

$$H(x) = \frac{1}{2}(1 + \operatorname{sgn} x). \qquad (20.3.5)$$

□

Example 20.3.3 The distribution x_+^α is defined by

$$\langle x_+^\alpha, \phi \rangle = \int_0^\infty x^\alpha \phi(x)\,\mathrm{d}x \qquad (20.3.6)$$

and, similarly, x_-^α is defined by

$$\langle x_-^\alpha, \phi \rangle = \int_{-\infty}^0 |x|^\alpha \phi(x)\,\mathrm{d}x. \qquad (20.3.7)$$

These integrals exist for all ϕ for $\alpha > -1$. We will see in §20.5 how the definition can be extended to other α. As will be seen later (p. 577) the definition of the product of two distributions is not a straightforward matter, but we can use $x_\pm^\alpha$ to define the products

$$x^\alpha H(x) = x_+^\alpha, \qquad |x|^\alpha H(-x) = x_-^\alpha, \qquad (20.3.8)$$

provided $\alpha > -1$. By using these relations we can also set

$$|x|^\alpha \operatorname{sgn} x = x_+^\alpha - x_-^\alpha = |x|^\alpha [H(x) - H(-x)], \qquad |x|^\alpha = x_+^\alpha + x_-^\alpha. \qquad (20.3.9)$$

□

Example 20.3.4 The integral $\int_{-\infty}^{\infty} [\phi(x)/x] \, dx$ has a meaning for all the test functions that vanish at $x = 0$. However, in order to consider $1/x$ as a distribution, it must be given a meaning also for those ϕ such that $\phi(0) \neq 0$. This is the problem of regularization of divergent integrals which has a rich literature (see e.g. Gelfand & Shilov 1964, vol. 1; Saichev & Woyczyński 1997; Kanwal 2004) which goes back to Hadamard (1923) and, before him, to Cauchy. One possibility is to define the integral as a Cauchy principal value:

$$\left\langle \frac{1}{x}, \phi \right\rangle = \lim_{\varepsilon \to 0} \left[\int_{-\infty}^{-\varepsilon} \frac{\phi(x)}{x} \, dx + \int_{\varepsilon}^{\infty} \frac{\phi(x)}{x} \, dx \right]$$

$$= \int_{0}^{\infty} \frac{\phi(x) - \phi(-x)}{x} \, dx . \tag{20.3.10}$$

It is readily seen that this integral exists if ϕ is a test function as[6]

$$\frac{\phi(x) - \phi(-x)}{x} = \int_{-1}^{1} \phi'(sx) \, ds \tag{20.3.11}$$

and, as a function of x, the integral has bounded support and is infinitely differentiable. [7] An infinity of possible regularizations exists differing from each other by a multiple of $\delta(x)$ (see p. 610). This particular one recommends itself on the basis of the fact that, as shown in §20.5, it is the distributional derivative of $\log |x|$. We refer to this particular regularization of $1/x$ as the *pseudo-function* $(1/x)$. We return on this concept in Example 20.5.4 p. 582.

□

Example 20.3.5 The *Sokhotski–Plemelj* formulae are given by

$$\lim_{\varepsilon \to 0} \frac{1}{x \pm i\varepsilon} = \frac{1}{x \pm i0} = \mp i\pi \delta(x) + \frac{1}{x}, \tag{20.3.12}$$

where $1/x$ is as defined in (20.3.10). It will be sufficient to establish one of these relations. Consider

$$\int_{-\infty}^{\infty} \frac{\phi(x)}{x - i\varepsilon} \, dx = \int_{0}^{\infty} \left[\frac{\phi(x)}{x - i\varepsilon} - \frac{\phi(-x)}{x + i\varepsilon} \right] dx$$

$$= \int_{0}^{\infty} \frac{x}{x^2 + \varepsilon^2} [\phi(x) - \phi(-x)] \, dx + i\varepsilon \int_{0}^{\infty} \frac{\phi(x) + \phi(-x)}{x^2 + \varepsilon^2} \, dx . \tag{20.3.13}$$

The first integral exists in the ordinary sense as $\varepsilon \to 0$ and is the same as (20.3.10) defining the principal value of $1/x$. As for the second integral, we write

$$\int_{0}^{\infty} \frac{\phi(x) + \phi(-x)}{x^2 + \varepsilon^2} \, dx = \int_{0}^{\infty} \frac{\phi(x) + \phi(-x) - 2\phi(0)}{x^2 + \varepsilon^2} \, dx + \int_{0}^{\infty} \frac{2\phi(0)}{x^2 + \varepsilon^2} \, dx . \tag{20.3.14}$$

---

[6] Here $\phi'(sx)$ should be interpreted as $\phi'(z)|_{z=sx}$, rather than $(d/ds)\phi(sx)$.

[7] In general, any function $f(x)$ can be decomposed into the sum of an even, $f_+(x)$, and an odd, $f_-(x)$, component, $f = f_+ + f_-$, where $f_\pm = \frac{1}{2}[f(x) \pm f(-x)]$. The Cauchy principal-value integral of f_+/x vanishes by symmetry. Thus, for the Cauchy principal-value integral of f/x to exist it is necessary that f_-/x be integrable.

As shown later in (20.5.10), the first term is finite as $\varepsilon \to 0$ and therefore, since it is multiplied by ε in (20.3.13), contributes nothing in the limit. As for the second term we have

$$\int_0^\infty \frac{\mathrm{d}x}{x^2 + \varepsilon^2} = \frac{\pi}{2\varepsilon} \tag{20.3.15}$$

and the desired result follows. The Sokhotski–Plemelj formulae are related to the Cauchy integral in the theory of analytic functions (§17.3.2 p. 430) and they establish a link between analytic functions and distributions (see §20.8 and, e.g., Bremermann 1965; Roos 1969; Carmichael & Mitrović 1989; Vladimirov 2002). An application to the one-dimensional Helmholtz equation was shown in §4.2. $\qquad\square$

20.4 Operations on distributions

Since the concept of value at a point has no meaning for distributions, to define the notion of equality we cannot use the same criterion we would use for ordinary functions.

Definition 20.4.1 Two distributions d_1 and d_2 are *equal* in an open set Ω if and only if $\langle d_1, \phi \rangle = \langle d_2, \phi \rangle$ for all test functions ϕ having support in Ω.

It can be shown that, if d_1 and d_2 are continuous functions, this definition also implies equality in the point-wise sense.

The *support* of an ordinary function is the closure of the set of points where the function is non-zero (p. 683). Again, we cannot use this definition for distributions, but the previous definition of equality offers an obvious alternative. We start by considering all the open sets in which $d = 0$ in the previous sense; the *support* of the distribution is then the (closed) complement of this set. As an example, the distribution $\delta(x)$ applied to any test function the support of which does not contain the origin has the same action as the 0 distribution. Thus $\delta(x)$ equals the 0 distribution – defined by $\langle 0, \phi \rangle = 0$ for all ϕ – in the open sets $-\infty < x < 0$ and $0 < x < \infty$ and its support reduces to the single point $x = 0$. In this sense one might say that "$\delta(x)$ vanishes everywhere except at the origin."

Since all L^2 functions are distributions, it should be possible to extend to distributions many of the familiar operations that are possible for functions. As a general guideline to this end, we define the result of these manipulations so that, for regular distributions (i.e., L^2 functions), they reduce to the usual forms. In each case, we must make sure that the definition gives unambiguously the action of the distribution on all the test functions and that this action is continuous. As before, the first property will be fairly self-evident, and we omit proof of the second one.

1. Addition. The sum of two distributions d_1 and d_2 is defined by the obvious relation

$$\langle d_1 + d_2, \phi \rangle = \langle d_1, \phi \rangle + \langle d_2, \phi \rangle. \tag{20.4.1}$$

The right-hand side is known for any ϕ because, by hypothesis, d_1 and d_2 are given, which implies that their action on any ϕ is known. This relation then specifies uniquely the action of the linear functional $d_1 + d_2$ on any test function, and this is what is required to define unambiguously a distribution.

2. Multiplication by a number. We define the product of a given distribution d by a constant c by simply setting

$$\langle c\, d, \phi \rangle = c \langle d, \phi \rangle. \tag{20.4.2}$$

The set of distributions equipped with the definition (20.4.1) of addition and (20.4.2) of multiplication by a number forms a linear vector space. With the definition of convergence given later in (20.6.1), this becomes a topological vector space denoted by D'.

3. Multiplication by a function. The situation complicates considerably when we come to the definition of the product of a function $\alpha(x)$ and a distribution d. The natural definition would be

$$\langle \alpha\, d, \phi \rangle = \langle d, \alpha\, \phi \rangle, \tag{20.4.3}$$

but we face the difficulty that, unless $\alpha\, \phi$ is a test function, $\langle d, \alpha\, \phi \rangle$ in general has no meaning as only the action of d on test functions is known. Since ϕ is a test function, $\alpha\phi$ will have a compact support, but for it to be infinitely differentiable, α must enjoy this property. If we restrict the admissible functions in this way, then (20.4.3) gives a well-defined rule associating a number to the expression $\langle \alpha\, d, \phi \rangle$ for any ϕ, and therefore it properly defines the distribution $\alpha\, d$.[8] As an example of the application of this and of the previous rule, together with the definition of equality, we can write

$$\alpha(x)\, \delta(x) = \alpha(0)\, \delta(x) \tag{20.4.4}$$

in the sense that the distributions in the two sides of this equation have the same action on all test functions. A particular case, to which we return on p. 609, is

$$x\, \delta(x) = 0. \tag{20.4.5}$$

By exploiting the fact that infinitely differentiable functions are dense in the space of continuous functions, one might take (20.4.4) to hold for any function α continuous at 0. The validity of relations derived from (20.4.4) by differentiation k times (§20.5), however, would require α to be k times continuously differentiable in the neighborhood of $x = 0$.

---

[8] The set of all functions $\alpha(x)$ such that $\alpha(x)\phi(x) \in \mathscr{D}$, equipped with a suitable norm, is a linear space called the *space of multipliers* (see e.g. Friedman 1963, p. 28; Bremermann 1965, pp. 14, 98, 164; Pathak 1997, p. 21).

4. Multiplication of two distributions The previous considerations show the difficulties one encounters in an attempt to define the product of two distributions. It is impossible to formulate a definition that always coincides with the usual product of two functions in the case of regular distributions and is continuous, commutative and associative. For example, it is easy to check that $(1/x)\,x = 1$. Then we see that $[(1/x)\,x]\,\delta(x) = \delta(x)$, while $(1/x)\,[x\delta(x)] = 0$. Again, this impossibility refers to the *general* case as, for example, $x\,d(x)$ is not only the product of the infinitely differentiable function x by the distribution d, but may also be regarded as the product of the two distributions x and d. This example shows that it is possible to define the product of two distributions d_1 and d_2 if the singular points (Definition 20.2.2 p. 571) of d_2 are at points in the neighborhood of which d_1 equals an infinitely differentiable function and vice versa (see e.g. Hörmander 1983, p. 55). In some cases it is possible to define the product of two distributions also when both are singular at the same point. Examples are the distributions $x^\alpha H(x)$, $|x|^\alpha H(-x)$ and $|x|^\alpha \operatorname{sgn} x$ defined in Example 20.3.3 p. 573 when $-1 < \alpha < 0$. Another instance is

$$
\langle \operatorname{sgn} x \, \log|x|, \phi \rangle = \int_{-\infty}^{0} -\log(-x)\,\phi(x)\,dx + \int_{0}^{\infty} \log x\,\phi(x)\,dx
$$

$$
= \int_{0}^{\infty} \log x\,[\phi(x) - \phi(-x)]\,dx .
\tag{20.4.6}
$$

A similar procedure would give good definitions of products such as $|x|^\alpha \log|x|$, $|x|^\alpha \log|x| \operatorname{sgn} x$ and $x^\alpha H(x) \log x$ which are considered later in Example 20.6.3 p. 585. No straightforward interpretation is possible, however, for an expression like $\delta^2(x)$.[9]

An important situation in which the issue of the product of distributions arises is in the evaluation of finite integrals such as

$$
I_d = \int_{a}^{b} d(x)\,\phi(x)\,dx .
\tag{20.4.7}
$$

If the support of the distribution d is entirely contained in the interval (a, b), the integral can be extended over the entire real line and no difficulties arise. More generally, one can rewrite (20.4.7) formally as

$$
I_d = \int_{-\infty}^{\infty} [H(x - a) - H(x - b)]\,d(x)\phi(x)\,dx
\tag{20.4.8}
$$

and the product of distributions occurring here may or may not have a meaning. For example, if $d(x) = \delta(x - c)$, with $a < c < b$, $H(x - a) - H(b - x)$ is regular at the singular point of $\delta(x - c)$ and we can safely deduce that

$$
\int_{a}^{b} \delta(x - c)\,\phi(x)\,dx = [H(c - a) - H(c - b)]\,\phi(c).
\tag{20.4.9}
$$

---

[9] One may attempt to define this product by writing one of the δ as the limit of a suitable sequence (see §20.6). The difficulty is that the result will depend on the particular sequence chosen or, one might say, on the assumed "inner structure" of the distribution. Additional information on the product of distributions can be found e.g. in Jones 1966, chapter 6.

However, ambiguities arise if $c = a$ or b wich may be interpreted e.g. as

$$\int_c^b \delta(x-c)\phi(x)\,dx = \phi(c) \qquad \text{or} \qquad \int_c^b \delta(x-c)\phi(x)\,dx = \frac{1}{2}\phi(c). \quad (20.4.10)$$

Some comments on this situation can be found on p. 38.

5. Translation. In order to define the action of $d(x-a)$ when that of $d(x)$ is known, we make the change of variable $y = x - a$ in the "integral" representing the action of d according to the notation introduced in (20.2.3):

$$\langle d(x-a), \phi(x)\rangle = \int_{-\infty}^{\infty} d(x-a)\phi(x)\,dx = \int_{-\infty}^{\infty} d(y)\phi(y+a)\,dy$$

$$= \langle d(x), \phi(x+a)\rangle. \qquad (20.4.11)$$

In particular

$$\langle \delta(x-a), \phi(x)\rangle = \langle \delta(x), \phi(x+a)\rangle = \phi(a). \qquad (20.4.12)$$

6. Change of scale. In order to define the action of $d(\lambda x)$ when $d(x)$ is known, we make the change of variable $y = \lambda x$ in the "integral" representing the action of d to find

$$\langle d(\lambda x), \phi(x)\rangle = \frac{1}{|\lambda|}\langle d(x), \phi(x/\lambda)\rangle. \qquad (20.4.13)$$

The modulus is due to the fact that, when $\lambda < 0$, the integration interval $(-\infty, \infty)$ is mapped into $(\infty, -\infty)$ which gives rise to an overall minus sign. A particular case is

$$\delta(\lambda x) = \frac{\delta(x)}{|\lambda|}. \qquad (20.4.14)$$

7. Change of variable. The most common case is that involving the δ, which we address by considering the example $\langle \delta(x^2 - a^2), \phi(x)\rangle$ (Figure 20.1). If instead of δ we had an ordinary function f, we would write

$$\left\langle f(x^2 - a^2), \phi(x)\right\rangle = \int_{-\infty}^0 f(x^2 - a^2)\phi(x)\,dx + \int_0^{\infty} f(x^2 - a^2)\phi(x)\,dx \quad (20.4.15)$$

splitting the integral into two parts in such a way that the change of variable $y = x^2 - a^2$ is one-to-onc in cach interval. With $y = x^2 - a^2$, in the first integral we have $x = -\sqrt{y + a^2}$ and therefore

$$\int_{-\infty}^0 f(x^2 - a^2)\phi(x)\,dx = \int_{-a^2}^{\infty} f(y)\phi(-\sqrt{y+a^2})\frac{dy}{2\sqrt{y+a^2}}, \qquad (20.4.16)$$

while in the second integral $x = \sqrt{y + a^2}$ and

$$\int_0^{\infty} f(x^2 - a^2)\phi(x)\,dx = \int_{-a^2}^{\infty} f(y)\phi(\sqrt{y+a^2})\frac{dy}{2\sqrt{y+a^2}}. \qquad (20.4.17)$$

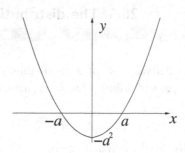

Fig. 20.1 Example for the change of variable calculation

Proceeding in the same way for the δ distribution we are led to[10]

$$
\langle \delta(x^2 - a^2), \phi(x) \rangle = \left\langle \delta(y), \frac{\phi(-\sqrt{y + a^2})}{2\sqrt{y + a^2}} \right\rangle + \left\langle \delta(y), \frac{\phi(\sqrt{y + a^2})}{2\sqrt{y + a^2}} \right\rangle
$$

$$
= \frac{\phi(-a)}{2|a|} + \frac{\phi(a)}{2|a|} \tag{20.4.18}
$$

and thus to the result

$$
\delta(x^2 - a^2) = \frac{\delta(x + a) + \delta(x - a)}{2|a|}. \tag{20.4.19}
$$

An application of this relation is encountered in some aspects of the theory of linear non-dispersive waves in the form of the relation:

$$
\delta \left(t^2 - \frac{|\mathbf{x}|^2}{c^2} \right) = \frac{c}{2|\mathbf{x}|} \left[\delta \left(t - \frac{|\mathbf{x}|}{c} \right) + \delta \left(t + \frac{|\mathbf{x}|}{c} \right) \right]. \tag{20.4.20}
$$

The rule is easily generalized to find:

$$
\delta \left(g(x) \right) = \sum_i \frac{\delta(x - x_i)}{|g'(x_i)|}, \tag{20.4.21}
$$

where the $x = x_i, i = 1, 2, \ldots$ are the zeros of the function $g(x)$. The g' in the denominator arises from the change of variable $y = g(x)$, the sum from the necessity to break up the integral in pieces in each one of which the mapping is one-to-one, and the modulus from the fact that the integration is in the negative direction in each piece in which $g' < 0$. It should be noted that $\delta(g(x))$ is undefined at zeros of g of order 2 and higher, as is evident, e.g., from (20.4.19) when $a = 0$.

---

[10] Strictly speaking, knowledge of the action of a distribution on a test function requires knowledge of the integral over the entire real axis; an integral extending from $-a^2$ to infinity is actually undefined. For the reason noted before in connection with (20.4.9), this is however a moot point in the present case.

20.5 The distributional derivative

We define the derivative d' of a distribution d by a rule justified by the procedure of integration by parts of ordinary functions, namely[11]

$$\langle d', \phi \rangle = -\langle d, \phi' \rangle. \tag{20.5.1}$$

More generally, for the k-th derivative,

$$\langle d^{(k)}, \phi \rangle = (-1)^k \langle d, \phi^{(k)} \rangle. \tag{20.5.2}$$

With this we see immediately that *any distribution is differentiable any number of times.* This is the primary reason to require that test functions be infinitely differentiable.

It is easy to verify the validity of the ordinary Leibniz rule for the derivative of a product and, in particular, that

$$(d_1 d_2)' = d_1' d_2 + d_1 d_2' \tag{20.5.3}$$

whenever the indicated products exist (see p. 577).[12]

Example 20.5.1 The application of the differentiation rule to $H(x)$ and $\delta(x)$ was shown earlier on p. 36. In a similar way we have

$$\frac{\mathrm{d}}{\mathrm{d}x} \operatorname{sgn} x = 2\delta(x). \tag{20.5.4}$$

By differentiating the relation $\alpha(x)\delta(x) = \alpha(0)\delta(x)$ using (20.5.3) we find

$$\alpha'(x)\delta(x) + \alpha(x)\delta'(x) = \alpha(0)\delta'(x) \tag{20.5.5}$$

and, therefore,

$$\alpha(x)\delta'(x) = \alpha(0)\delta'(x) - \alpha'(0)\delta(x) \tag{20.5.6}$$

and so on for the higher derivatives provided α is infinitely differentiable. In particular, one has

$$x^n \delta^{(m)}(x) = \begin{cases} (-1)^n \frac{m!}{(m-n)!} \delta^{(m-n)}(x) & m \geq n \\ 0 & m < n \end{cases}, \tag{20.5.7}$$

from which $x^n \delta^{(n)}(x) = (-1)^n n!\, \delta(x)$ and $x\delta'(x) = -\delta(x)$.   □

---

[11] Since $\phi = 0$ at $\pm\infty$, in the case of an ordinary function the integrated term vanishes.

[12] If the products $d_1 d_2$ and $d_1 d_2'$ exist, this relation can be used to *define* the product $d_1' d_2$ (see Jones 1966, p. 145).

Example 20.5.2 The distributional derivative of $\log |x|$ is calculated as

$$\left\langle \frac{d}{dx} \log |x|, \phi \right\rangle = -\langle \log |x|, \phi' \rangle$$

$$= -\int_{-\infty}^{0} \log(-x)\,\phi'(x)\,dx - \int_{0}^{\infty} \log x\,\phi'(x)\,dx$$

$$= -\int_{0}^{\infty} \log x\,[\phi'(x) + \phi'(-x)]\,dx. \qquad (20.5.8)$$

Since the integral exists in the ordinary sense, it can also be evaluated as the limit for $\varepsilon \to 0$ of the integral from ε to infinity. In this integral, we can then use integration by parts to find

$$\int_{0}^{\infty} \log x\,[\phi'(x) + \phi'(-x)]\,dx$$

$$= \lim_{\varepsilon \to 0} \left\{ [\phi(x) - \phi(-x)] \log x |_{\varepsilon}^{\infty} - \int_{\varepsilon}^{\infty} \frac{\phi(x) - \phi(-x)}{x}\,dx \right\}.$$

The first term vanishes in the limit and we thus conclude that, in a distributional sense,

$$\frac{d}{dx} \log |x| = \frac{1}{x}, \qquad (20.5.9)$$

in which $1/x$ is the pseudo-function defined as a Cauchy principal-value in (20.3.10). □

Example 20.5.3 The same procedure applied to the derivative of $1/x$ gives

$$-\left\langle \frac{d}{dx}\left(\frac{1}{x}\right), \phi \right\rangle = \left\langle \frac{1}{x}, \phi' \right\rangle = \lim_{\varepsilon \to 0} \int_{\varepsilon}^{\infty} \frac{\phi'(x) - \phi'(-x)}{x}\,dx$$

$$= \lim_{\varepsilon \to 0} \left\{ \left[\frac{\phi(x) + \phi(-x)}{x} \right]_{\varepsilon}^{\infty} + \int_{\varepsilon}^{\infty} \frac{\phi(x) + \phi(-x)}{x^2}\,dx \right\}$$

$$= \lim_{\varepsilon \to 0} \left\{ \left[\frac{2\phi(0)}{x} \right]_{\varepsilon}^{\infty} + \int_{\varepsilon}^{\infty} \frac{\phi(x) + \phi(-x)}{x^2}\,dx \right\}$$

$$= \int_{0}^{\infty} \frac{\phi(x) + \phi(-x) - 2\phi(0)}{x^2}\,dx = \langle \frac{1}{x^2}, \phi \rangle, \qquad (20.5.10)$$

which defines the pseudo-function $1/x^2$ and agrees with the finite part of the integral as defined by Hadamard. Continuing we find the pseudo-functions

$$\langle x^{-2m}, \phi \rangle = \int_{0}^{\infty} x^{-2m} \left[\phi(x) + \phi(-x) \right. \qquad (20.5.11)$$

$$\left. -2\left(\phi(0) + \frac{x^2}{2!}\phi''(0) + \cdots + \frac{x^{2m-2}}{(2m-2)!}\phi^{(2m-2)}(0) \right) \right]\,dx,$$

$$\langle x^{-2m-1}, \phi \rangle = \int_{0}^{\infty} x^{-2m-1} \left[\phi(x) - \phi(-x) \right. \qquad (20.5.12)$$

$$\left. -2x\left(\phi'(0) + \frac{x^2}{3!}\phi'''(0) + \cdots + \frac{x^{2m-2}}{(2m-1)!}\phi^{(2m-1)}(0) \right) \right]\,dx,$$

□

Example 20.5.4 Since ϕ/x^α is integrable for $\alpha > -1$, the distributions $|x|^\alpha$, $|x|^\alpha \mathrm{sgn}\, x$ and $x^\alpha H(x)$ are well defined for $\alpha > -1$. The following relations are then an immediate consequence of the definition (20.5.1):

$$\frac{\mathrm{d}|x|^\alpha}{\mathrm{d}x} = \alpha |x|^{\alpha-1} \mathrm{sgn}\, x, \qquad \frac{\mathrm{d}}{\mathrm{d}x}\left(|x|^\alpha \mathrm{sgn}\, x\right) = \alpha |x|^{\alpha-1}, \qquad (20.5.13)$$

$$\frac{\mathrm{d}}{\mathrm{d}x}\left[x^\alpha H(x)\right] = \alpha x^{\alpha-1} H(x), \qquad (20.5.14)$$

and they suggest a way to define the distributions $|x|^\alpha$, $|x|^\alpha \mathrm{sgn}\, x$ and $x^\alpha H(x)$ also for $\alpha < -1$ as

$$\left.\begin{array}{c} |x|^\alpha \\ |x|^\alpha \mathrm{sgn}\, x \\ x^\alpha H(x) \end{array}\right\} = \frac{1}{(\alpha+1)(\alpha+2)\dots(\alpha+n)} \frac{\mathrm{d}^n}{\mathrm{d}x^n} \left\{\begin{array}{c} |x|^{\alpha+n}(\mathrm{sgn}\, x)^n \\ |x|^{\alpha+n}(\mathrm{sgn}\, x)^{n+1} \\ x^{\alpha+n} H(x) \end{array}\right\}, \qquad (20.5.15)$$

where n is any integer such that $\alpha + n > -1$. These formulae fail for α a negative integer, which is a special case considered in Examples 20.5.5 and 20.6.4 p. 586. These relations lead to results that are identical to the conventional definition of the Hadamard finite part of the corresponding integrals. Entirely similar definitions can be applied to $|x-a|^\alpha$ etc., and to functions possessing more than one singularity of this type. The related distributions $|x|^\alpha \log|x|$, $|x|^\alpha \log|x| \,\mathrm{sgn}\, x$ and $x^\alpha H(x) \log x$ are addressed in Example 20.6.3 p. 585.

□

Example 20.5.5 The Cauchy principal value permits a good definition of $1/x$ but a similar approach is not useful for $1/|x|$. In order to define this distribution we note that $x/|x| = \mathrm{sgn}\, x$ and observe that

$$x \frac{\mathrm{d}}{\mathrm{d}x}(\mathrm{sgn}\, x \,\log|x|) = \mathrm{sgn}\, x, \qquad (20.5.16)$$

as is readily proven by applying the definition of derivative and integrating by parts.[13] Since the general solution of the equation $xd(x) = 0$ is $d(x) = C\delta(x)$ (see (20.4.5) and p. 609), (20.5.16) suggests the definition (see p. 611)

$$\frac{1}{|x|} = \frac{\mathrm{d}}{\mathrm{d}x}(\mathrm{sgn}\, x \,\log|x|) + c\delta(x), \qquad (20.5.17)$$

where the constant c, by the very nature of the approach, remains undetermined (see e.g. Lighthill 1958, p. 41; Jones 1966, p. 93).[14] Hadamard's definition of the pseudo-function $1/|x|$ corresponds to taking $c = 0$ (see e.g. Zemanian 1965, p. 18). The distribution

---

[13] The expression $\mathrm{sgn}\, x \,\log|x|$ is the product of two distributions for both of which $x = 0$ is a singular point and should, therefore, be properly interpreted; see comment after (20.4.6).

[14] Other choices for the definition of $1/|x|$ do remove the ambiguity of this definition; see e.g. Gelfand & Shilov 1964, vol. 1 section 3.5.

x^{-m} sgn x is defined by

$$x^{-m} \operatorname{sgn} x = \frac{(-1)^{m-1}}{(m-1)!} \frac{\mathrm{d}^{m-1}}{\mathrm{d}x^{m-1}} \frac{1}{|x|} \qquad (20.5.18)$$

with which, by means of (20.3.5), one can define $x^{-m} H(x)$. A similar ambiguity arises in the definition of $|x|^{-1} \log|x|$ which we define by $x|x|^{-1} \log|x| = \operatorname{sgn} x \, \log|x|$. In this case one finds (Jones 1966, p. 102)

$$|x|^{-1} \log|x| = \frac{\mathrm{d}}{\mathrm{d}x} \left[\frac{1}{2} (\log|x|)^2 \operatorname{sgn} x \right] + C\delta(x), \qquad (20.5.19)$$

from which we have

$$
\begin{aligned}
x^{-m} \log|x| \operatorname{sgn} x \;=\;& \frac{(-1)^{m-1}}{(m-1)!} \frac{\mathrm{d}^{m-1}}{\mathrm{d}x^{m-1}} \left[\left(|x|^{-1} \log|x| \right) \right] \\
& + \left(1 + \frac{1}{2} + \cdots + \frac{1}{m-1} \right) x^{-m} \operatorname{sgn} x.
\end{aligned}
\qquad (20.5.20)
$$

$\square$

The definition of order of a distribution given on p. 572 can equivalently be stated by saying that a distribution has order m if it can be obtained by differentiating in the sense of distributions a regular distribution (i.e., an ordinary function) m times, but cannot be expressed as a derivative of a regular distribution of lower order.

Theorem 20.5.1 *Every distribution has a finite order, i.e., it is either regular or can be obtained by differentiating a finite number of times a regular distribution.*[15]

Thus, just as rational numbers are as far as one can go enlarging the field of integers by "liberalizing" division, distributions are as far as one can go in the process of enlarging functions by liberalizing differentiation.

We conclude with a brief note on the *integral* of a distribution d, which is defined as a distribution D such that $D' = d$, i.e., such that $\langle D', \phi \rangle = -\langle D, \phi' \rangle$ for all ϕ. The problem with this seemingly straightforward definition is that, since $\int_{-\infty}^{\infty} \phi' \, \mathrm{d}x = 0$, the equality defines D only on the *subset* of test functions the integral of which vanishes. To circumvent this difficulty we introduce an arbitrary test function χ_0 such that $\int_{-\infty}^{\infty} \chi_0 \, \mathrm{d}x = 1$ and, for any test function ψ, we write $\psi = (\psi - \Psi \chi_0) + \Psi \chi_0$, in which $\Psi = \int_{-\infty}^{\infty} \psi \, \mathrm{d}x$. Then, evidently, $\int_{-\infty}^{\infty} (\psi - \Psi \chi_0) \, \mathrm{d}x = 0$ so that the action of D on this function is well defined. As for the other term, $\langle D, \Psi \chi_0 \rangle = \Psi c$, in which $c = \langle D, \chi_0 \rangle$ is an arbitrary constant. We will not have much use for this notion for which the reader is referred to most textbooks on the subject.

---

[15] For a general distribution the statement is valid only locally, while for distributions with compact support or tempered distributions it is valid globally, see e.g. Bremermann 1965, p. 30; Strichartz 1994.

20.6 Sequences of distributions

Not surprisingly, the sequence of distributions $\{d_n\}$ is said to converge to the distribution d if, for all $\phi \in \mathscr{D}$,

$$\lim_{n\to\infty} \langle d_n, \phi \rangle = \langle d, \phi \rangle. \tag{20.6.1}$$

This mode of convergence is similar to what was termed weak convergence in Chapter 19. An analogous definition applies to the convergence of distributions d_s depending on a continuous parameter s in place of the discrete one n. For example, the superficially trivial result

$$\lim_{s\to 0} |x|^s = 1 \tag{20.6.2}$$

is proven by showing that the limit of $\langle |x|^s - 1, \phi \rangle$ vanishes (Jones 1966, p. 68). Another interesting example is (see Lighthill 1958; Saichev & Woyczyński 1997)

$$\lim_{s\to 0+} \frac{1}{2} s |x|^{s-1} = \delta(x), \tag{20.6.3}$$

which is proven by differentiating the result $\lim_{s\to 0} |x|^s \operatorname{sgn} x = \operatorname{sgn} x$.

It is a remarkable fact that if, for every ϕ, $\langle d_n, \phi \rangle$ or $\langle d_s, \phi \rangle$ admit a finite limit, this limit defines a distribution. In other words, the set of distributions is *complete*: it is not possible to extend this set further by applying a limit process in the above sense.

In view of the definition of derivative, if $\langle d_n, \phi \rangle$ converges for any ϕ, so does $-\langle d_n, \phi' \rangle = \langle d_n', \phi \rangle$. Thus, the derivative of any order of a converging sequence of distributions converges to the corresponding derivative of the limit. It is also remarkable that

Theorem 20.6.1 *Every singular distribution is the limit of a sequence of regular distributions.*

A particularly important illustration of this property is offered by the δ.

Example 20.6.1 [δ-families] There are many families of functions which approximate any given distribution. In the case of the δ the most standard one is

$$\lim_{n\to\infty} \sqrt{\frac{n}{\pi}} e^{-n^2 x^2} = \delta(x). \tag{20.6.4}$$

More generally, given any non-negative locally integrable function $f(x)$ such that $\int_{-\infty}^{\infty} f(x)\,dx = 1$ it can be shown that

$$\lim_{n\to\infty} n f(nx) = \delta(x) \tag{20.6.5}$$

and, similarly, that

$$\lim_{s\to\infty} s\, f(sx) = \delta(x) \quad \text{and} \quad \lim_{\sigma\to 0} \frac{1}{\sigma} f(x/\sigma) = \delta(x). \tag{20.6.6}$$

Using this result, we readily establish the validity of relations such as

$$\lim_{t \to 0} \frac{\exp(-x^2/4t)}{\sqrt{4\pi t}} = \delta(x), \tag{20.6.7}$$

$$\lim_{y \to 0} \frac{y}{\pi(x^2 + y^2)} = \delta(x) \quad \text{and} \quad \lim_{n \to \infty} \frac{1}{\pi n(x^2/n^2 + 1)} = \delta(x), \tag{20.6.8}$$

$$\lim_{s \to \infty} \frac{1 - \cos sx}{\pi s x^2} = \lim_{s \to \infty} \frac{\sin^2 sx}{\pi s x^2} = \delta(x) \tag{20.6.9}$$

and similar ones. The first one, (20.6.7), shows, for example, that the solution (4.4.13) of the one-dimensional diffusion equation does indeed satisfy the initial condition as we have shown by a different argument on p. 96. The convergence of the family of functions in (20.6.8), sometimes called the Lorentz family, proves e.g. that the solution (4.6.9) of the Laplace equation in a strip satisfies the boundary condition at $y = 0$. The first form of the last relation arises, among others, in the Fejér theory of the Fourier series (p. 264).

The Fourier inversion formula implies that

$$\lim_{n \to \infty} \frac{\sin nx}{\pi x} = \lim_{n \to \infty} \int_{-n}^{n} \frac{e^{ikx}}{2\pi} \, dk = \delta(x), \tag{20.6.10}$$

from which

$$\lim_{n \to \infty} \int_{-n}^{n} e^{ikx} \, dk = 2\pi \, \delta(x). \tag{20.6.11}$$

We have thus obtained the important formulae

$$\int_{-\infty}^{\infty} e^{ikx} \, dk = 2\pi \, \delta(x) \quad \text{and} \quad \int_{0}^{\infty} \cos kx \, dk = \pi \, \delta(x). \tag{20.6.12}$$

A proof of these relations will be given in §20.10.

All the δ-families of these examples were constructed with functions which decay to 0 at infinity and extend on both sides of the origin, but neither of these conditions is necessary. The real part of $\sqrt{i/2\pi\varepsilon} \exp(-ix^2/2\varepsilon)$, for example, converges to $\delta(x)$ as $\varepsilon \to 0$, and a one-sided example is provided by taking the function f of (20.6.5) and (20.6.6) as $f(x) = H(x) x^{-2} \exp(-1/x)$ (see Saichev & Woyczyński 1997). $\square$

Example 20.6.2 It can be shown that any distribution can be written as the limit of distributions of the form

$$\sum_{k} \sum_{n} a_k \delta(x - b_n). \tag{20.6.13}$$

$\square$

Example 20.6.3 The analog of definition (20.6.1) for distributions dependent on a parameter permits us to define the derivative with respect to that parameter in an obvious way:

$$\left\langle \frac{\partial}{\partial s} d_s(x), \phi(x) \right\rangle = \lim_{h \to 0} \frac{1}{h} \left[\langle d_{s+h}(x), \phi(x) \rangle - \langle d_s(x), \phi(x) \rangle \right] \qquad (20.6.14)$$

provided the limit exists. For example

$$\langle \frac{\delta(x-s-h) - \delta(x-s)}{h}, \phi \rangle = \frac{\phi(s+h) - \phi(s)}{h} \to \phi'(s) \qquad (20.6.15)$$

which shows that

$$\frac{\partial}{\partial s} \delta(x-s) = -\delta'(x-s). \qquad (20.6.16)$$

Differentiation with respect to x and with respect to s commute when $\partial d_s / \partial s$ exists (see e.g. Jones 1966, p. 72). By using this definition and property one can define distributions such as

$$|x|^\alpha \log |x| = \frac{\partial}{\partial \alpha} |x|^\alpha, \qquad |x|^\alpha (\operatorname{sgn} x) \log |x| = \frac{\partial}{\partial \alpha} \left(|x|^\alpha \operatorname{sgn} x \right), \qquad (20.6.17)$$

$$x^\alpha H(x) \log x = \frac{\partial}{\partial \alpha} \left[x^\alpha H(x) \right], \qquad (20.6.18)$$

which exist in the ordinary sense of functions if $\alpha > -1$ or $\operatorname{Re} \alpha > -1$, and can be defined using differentiation for other values of α (see e.g. Lighthill 1958, p. 34; Jones 1966, p. 101). For example, the relation

$$\begin{aligned} \frac{\partial}{\partial x} \left(|x|^\alpha \log |x| \right) &= \frac{\partial}{\partial x} \frac{\partial |x|^\alpha}{\partial \alpha} = \frac{\partial}{\partial \alpha} \left(\alpha |x|^{\alpha-1} \operatorname{sgn} x \right) \\ &= |x|^{\alpha-1} (1 + \alpha \log |x|) \operatorname{sgn} x \end{aligned} \qquad (20.6.19)$$

with the second relation in (20.6.17), gives a definition of $|x|^{\alpha-1} \log |x| \operatorname{sgn} x$. □

Example 20.6.4 It follows from the definition (20.6.17) that

$$\lim_{\alpha \to 1} \langle |x|^{-\alpha} \operatorname{sgn} x, \phi(x) \rangle = \lim_{\alpha \to 1} \int_0^\infty \frac{\phi(x) - \phi(-x)}{x^\alpha} \, dx = \langle x^{-1}, \phi \rangle \qquad (20.6.20)$$

and, more generally, that (see e.g. Jones 1966, p. 97)

$$\lim_{\alpha \to 1} \frac{(\operatorname{sgn} x)^{m+1}}{|x|^{\alpha+m}} = \frac{1}{x^{m+1}}. \qquad (20.6.21)$$

On a similar basis we can define

$$x^{-m} \log |x| = \lim_{\alpha \to m} \frac{\log |x|}{|x|^\alpha} (\operatorname{sgn} x)^m. \qquad (20.6.22)$$

□

Example 20.6.5 We have seen that the Taylor formula is not necessarily valid for test functions of $\mathscr{D}$. However a family of test functions can be found for which the formula is valid and, using them, one can prove a "Taylor series expansion" for the δ (see e.g. Saichev & Woyczyński 1997, p. 34):[16]

$$\delta(x+a) = \lim_{N\to\infty} \sum_{k=0}^{N} \frac{a^k}{k!} \delta^{(k)}(x). \qquad (20.6.23)$$

$\square$

It is perhaps unexpected that point-wise convergence of regular distributions does not ensure distributional convergence. A simple example is given by

$$d_n'(x) = -\frac{2}{\pi} \frac{n^3 x}{(1+n^2x^2)^2} \qquad (20.6.24)$$

which, for $x = 0$, converges to 0 in a point-wise sense while, as is evident from the second of (20.6.8), it converges to $\delta'(x)$ in the sense of distributions. A uniformly convergent sequence of locally summable functions converging uniformly to a locally summable function $f(x)$ in every bounded region does, however, converge distributionally as well.[17]

As usual, the convergence of a series can be reduced to that of a sequence by considering the partial sums. We thus readily establish the validity of the remarkable property that any distributionally convergent series can be differentiated term by term with no restrictions:

$$\frac{\mathrm{d}}{\mathrm{d}x} \sum_{k=1}^{\infty} d_n(x) = \sum_{k=1}^{\infty} \frac{\mathrm{d}d_n}{\mathrm{d}x}. \qquad (20.6.25)$$

One must carefully note the real meaning of a relation of this type, namely

$$\left\langle \frac{\mathrm{d}}{\mathrm{d}x} \sum_{k=1}^{\infty} d_n(x), \phi \right\rangle = \sum_{k=1}^{\infty} \left\langle \frac{\mathrm{d}d_n}{\mathrm{d}x}, \phi \right\rangle = -\sum_{k=1}^{\infty} \langle d_n, \phi' \rangle. \qquad (20.6.26)$$

As usual in the theory of distributions, one can apply this relation to a broader class of functions when they can be approximated by test functions. This point is of great importance in connection with the convergence of the Fourier series and is addressed more fully in §20.11.

---

[16] This formula holds, in particular, for test functions belonging to the space Z introduced in the footnote on p. 595.

[17] Distributional convergence follows from point-wise convergence in other cases as well, for example if $f_n \to f$ almost everywhere, and $|f_n|$ is bounded by a fixed constant or a locally integrable function, or if $f_n \to f$ monotonically increasing or decreasing and f is locally summable.

20.7 Multi-dimensional distributions

In an N-dimensional space we introduce test functions $\phi(x_1, \ldots, x_N)$, infinitely differentiable and with compact support, and define multi-dimensional distributions as continuous linear functionals acting on these functions. The notion of support of a one-dimensional distribution given on p. 575 generalizes in a straightforward way to this multi-dimensional situation.

A simple special case is the product of one-dimensional distributions acting on different variables, for example $d_1(x)\, d_2(y)$. It can be proven that both $\psi(x) = \langle d_2(y), \phi(x, y)\rangle$ and $\chi(y) = \langle d_1(x), \phi(x, y)\rangle$ are one-dimensional test functions of x and y, respectively, so that $\langle d_1(x), \psi(x)\rangle$ and $\langle d_2(y), \chi(y)\rangle$ have their usual one-dimensional meaning; the fact they are equal can be used to define $\langle d_1(x)d_2(y), \phi(x, y)\rangle$. It is fairly obvious that the support of the distribution $d_1(x)d_2(y)$ is the direct product of the supports of d_1 and d_2, namely the set of points (x, y) such that x belongs to the support of $d_1(x)$ and y to that of $d_2(y)$.

The notions of derivative, convergence etc. are obtained by an obvious extension of the corresponding definitions in the one-dimensional case. For example

$$\int \frac{\partial^p}{\partial y_i \partial y_j \cdots} \delta(\mathbf{x} - \mathbf{y})\, \phi(\mathbf{y})\, \mathrm{d}y_1\, \mathrm{d}y_2 \cdots = (-1)^p \frac{\partial^p}{\partial x_i \partial x_j \cdots} \phi(\mathbf{x}). \qquad (20.7.1)$$

Since test functions are infinitely smooth, this definition shows that, in the distributional sense, the equality of mixed derivatives $\partial^2 d/\partial x_i \partial x_j = \partial^2 d/\partial x_j \partial x_i$ is *always* true.

One new aspect, which we encountered in a much simpler context in the one-dimensional case, arises upon a change of variables. Let us again confine ourselves to the δ distribution and, for simplicity, let us consider the two-dimensional case. In the original Cartesian coordinates we have

$$\int_{-\infty}^{\infty} \int_{-\infty}^{\infty} \delta(x - x_0)\, \delta(y - y_0)\, \phi(x, y)\, \mathrm{d}x\, \mathrm{d}y = \phi(x_0, y_0). \qquad (20.7.2)$$

If we effect a change of variables

$$\xi = \xi(x, y), \qquad \eta = \eta(x, y), \qquad (20.7.3)$$

with $\xi_0 = \xi(x_0, y_0)$, $\eta_0 = \eta(x_0, y_0)$, the surface element $\mathrm{d}x\, \mathrm{d}y$ becomes $|J(\xi, \eta)|\, \mathrm{d}\xi\, \mathrm{d}\eta$, where $J(\xi, \eta) = \frac{\partial(x, y)}{\partial(\xi\eta)}$ is the Jacobian of the transformation. On the other hand, the δ in the new variables must be defined so that the result of the integration would be $\phi(\xi_0, \eta_0)$. It is therefore evident that the proper rule is

$$\delta(x - x_0)\, \delta(y - y_0) = \frac{\delta(\xi - \xi_0)\, \delta(\eta - \eta_0)}{|J(\xi, \eta)|} = \frac{\delta(\xi - \xi_0)\, \delta(\eta - \eta_0)}{|J(\xi_0, \eta_0)|}. \qquad (20.7.4)$$

For example, for polar coordinates in the plane, $J = r$ so that

$$\delta(x - x_0)\, \delta(y - y_0) = \frac{1}{r} \delta(r - r_0)\, \delta(\theta - \theta_0) = \frac{1}{r_0} \delta(r - r_0)\, \delta(\theta - \theta_0). \qquad (20.7.5)$$

Table 20.1	Different expressions for $\delta(\mathbf{x} - \mathbf{x}_0)$ in plane polar coordinates; see Tables 6.3, p. 148, and 7.4, p. 173, for analogous expressions in cylindrical and spherical coordinates.	
Pole position		$\delta(\mathbf{x} - \mathbf{x}_0)$
r_0, θ_0		$\dfrac{1}{r_0} \delta(r - r_0) \delta(\theta - \theta_0)$
$r_0 = 0$		$\dfrac{\delta(r)}{2\pi r}$

One aspect worthy of special attention is that the coordinate transformation may have singular points, such as the origin in plane polar coordinates, where $J = 0$. At such points one or more of the coordinates become ignorable, in the sense that the test function cannot depend on them. For example, in plane polar coordinates, the value of a function at the origin cannot depend on the angular variable so that the integration over θ can be carried out directly. It is therefore evident that, in this case, the proper rule is

$$\delta(x)\,\delta(y) = \frac{\delta(r)}{2\pi r} \tag{20.7.6}$$

so that

$$\phi(x = 0, y = 0) = \int_{-\infty}^{\infty} \int_{-\infty}^{\infty} \delta(x)\delta(y)\phi(x, y)\,\mathrm{d}x\,\mathrm{d}y = \int_{0}^{\infty} \int_{0}^{2\pi} r\,\frac{\delta(r)}{2\pi r}\,\phi(r = 0)\,\mathrm{d}r\,\mathrm{d}\theta$$

$$= \int_{0}^{\infty} r\,\frac{\delta(r)}{2\pi r}\,\phi(r = 0)\,\mathrm{d}r \int_{0}^{2\pi} \mathrm{d}\theta = \phi(r = 0). \tag{20.7.7}$$

The general rule readily follows (see e.g. Stakgold 1997, p. 120; Kanwal 2004, section 3.1): at a singular point where e.g. the coordinate η is ignorable, the denominator $|J|$ in (20.7.4) should be replaced by

$$\left| \int_{D_\eta} J(\xi, \eta)\,\mathrm{d}\eta \right|, \tag{20.7.8}$$

where the integration is extended to the range D_η of the ignorable coordinate. These special situations are shown explicitly in Table 20.1 for the two-dimensional δ in plane polar coordinates; analogous expressions for cylindrical and spherical coordinates are given in Tables 6.3 p. 148 and 7.4 p. 173 respectively.

We have considered up to now the δ localized on a zero-dimensional manifold, i.e., a point. We can also consider the δ localized on a $(N - 1)$-dimensional manifold $S(\mathbf{x}) = 0$ in an N-dimensional space:

$$I(\phi) = \int \delta\left(S(\mathbf{x})\right) \phi(\mathbf{x})\,\mathrm{d}^N x. \tag{20.7.9}$$

For $N = 3$ we have the surface $S(x, y, z) = 0$ and for $N = 2$ the line $S(x, y) = 0$. We treat this situation by transforming to a system of coordinates consisting of the family of surfaces

$S = $ const. and the normal direction n. The result is[18]

$$I(\phi) = \int_S \frac{\phi(\mathbf{x}_S)}{|\nabla S(\mathbf{x}_S)|} \, dS, \tag{20.7.10}$$

where the integration is over the surface (or line) $S = 0$ and $\mathbf{x}_S$ is the generic point on this surface. For example, if $S = x^2 + y^2 - R^2$, $|\nabla S| = 2\sqrt{x^2 + y^2}$ which, evaluated on $S = 0$, is just $|\nabla S| = 2R$, and $dS = R \, d\theta$. Thus

$$\iint \delta\left(x^2 + y^2 - R^2\right) \phi(\mathbf{x}) \, dx \, dy = = \frac{1}{2} \int_0^{2\pi} \phi(R, \theta) \, d\theta \tag{20.7.11}$$

which can be confirmed by expressing the integral in polar coordinates. Equation (20.7.10) may be seen as a generalization of (20.4.21).

As a second application of (20.7.10), let us give a brief derivation of a generalized form of the divergence (or Gauss) theorem 16.1.1 p. 400 by considering

$$\int_V \frac{\partial \phi}{\partial x_i} \, dV = \int H(S) \frac{\partial \phi}{\partial x_i} \, dV, \tag{20.7.12}$$

where the boundary of the integration volume V is the surface $S = 0$, with $S > 0$ inside V and $S < 0$ outside; in the second integral integration is over all space. By using the definition of distributional derivative and (20.7.10) we find

$$\int_V \frac{\partial \phi}{\partial x_i} \, dV = \int H(S) \frac{\partial \phi}{\partial x_i} \, dV = -\int \phi \frac{\partial H(S)}{\partial x_i} \, dV = -\int \phi \delta(S) \frac{\partial S}{\partial x_i} \, dV$$

$$= -\int \phi \frac{\partial S}{\partial x_i} \frac{dS}{|\nabla S|} = \int \phi n_i \, dS, \tag{20.7.13}$$

where $n_i = -(\partial S/\partial x_i)/|\nabla S|$ is the outward normal to the integration volume V.[19] In the same way it is possible to prove a generalized form of Green's identity (16.1.7) p. 401.

The notion of δ-families of Example 20.6.1 p. 584 extends to an N-dimensional space. If we let

$$\rho_\varepsilon(\mathbf{x}) = \frac{1}{\varepsilon^N} f\left(\frac{\mathbf{x}}{\varepsilon}\right), \tag{20.7.14}$$

where the function f is spherically symmetric with support in the unit sphere and normalized so that $\int f(\mathbf{y}) \, d^N y = 1$, then the relation

$$d_\varepsilon(\mathbf{y}) = \langle d(\mathbf{x}), \rho_\varepsilon(\mathbf{y} - \mathbf{x}) \rangle = \int d(\mathbf{x}) \rho_\varepsilon(\mathbf{y} - \mathbf{x}) \, d^N x \tag{20.7.15}$$

defines a smoothing of the distribution d and the corresponding operator is called a *mollifier*. As $\varepsilon \to 0$, $d_\varepsilon \to d$ in the sense of distributions.[20] Furthermore, $\partial d_\varepsilon(\mathbf{y})/\partial y_j = \langle \partial d(\mathbf{x})/\partial x_j, \rho_\varepsilon(\mathbf{y} - \mathbf{x}) \rangle$ so that the derivatives of d_ε converge to those of d. Equation (20.7.15) is an instance of a convolution, which we consider in some detail in §20.9.

---

[18] Note the similarity with (20.4.21) p. 579 in the one-dimensional case.

[19] The minus sign arises from the fact that ∇S is directed in the direction in which the function S increases, which is inward in this case.

[20] Cf. footnote 2 on p. 570.

20.8 Distributions and analytic functions

The representation of a real continuous function as the jump across the real axis of an analytic function shown on p. 437 has an analog for distributions. In this case we consider the space $\mathcal{E}$ of infinitely differentiable complex test functions $\psi(x)$ with arbitrary (i.e., not necessarily compact) support (p. 572). Since this is a larger space than $\mathcal{D}$, the set $\mathcal{E}'$ of distributions defined on these test functions is smaller than the set $\mathcal{D}'$ considered until now. In particular, all these distributions have *compact support*; distributions of $\mathcal{D}'$ with compact support are also distributions of $\mathcal{E}'$.

Definition 20.8.1 [Cauchy representation] Let $e(x) \in \mathcal{E}'$ be a distribution over the test functions of $\mathcal{E}$. The function

$$\hat{e}(z) = \frac{1}{2\pi i} \left\langle e(x), \frac{1}{x - z} \right\rangle \tag{20.8.1}$$

is the Cauchy representation of the distribution e.

It can be shown that $\hat{e}(z)$ is an analytic function of z in the complex plane except for the support of $e(x)$ and that (see e.g. Bremermann 1965, p. 44)

$$\frac{\mathrm{d}^n \hat{e}}{\mathrm{d}z^n} = \frac{n!}{2\pi i} \left\langle e(x), \frac{1}{(x - z)^{n+1}} \right\rangle. \tag{20.8.2}$$

Furthermore $\hat{e}$ has a Laurent series expansion

$$\hat{e}(z) = e_0 + \frac{e_1}{z} + \frac{e_2}{z^2} + \cdots \tag{20.8.3}$$

from which one can deduce that, provided ψ is bounded with all its derivatives,[21]

$$\lim_{\varepsilon \to 0} \int_{-\infty}^{\infty} \left[\hat{e}(x + i\varepsilon) - \hat{e}(x - i\varepsilon) \right] \psi(x)\, \mathrm{d}x = \langle e, \psi \rangle. \tag{20.8.4}$$

This relation generalizes to distributions the representation of continuous functions mentioned earlier. For distributions in $\mathcal{D}$ a similar relation holds, although in general it cannot be written in terms of the Cauchy representation (20.8.1) (see e.g. Bremermann 1965, p. 48; Roos 1969, section 5.5; Hörmander 1983, vol. 1 p. 65). If $\hat{e}$ in (20.8.4) is replaced by $\hat{e} + g(z)$, with g an entire function, the relation still holds as $g(x + i\varepsilon) - g(x - i\varepsilon) \to 0$; thus, the representation (20.8.4) is not unique.

Example 20.8.1 The δ has compact support and we have

$$\hat{\delta}(z) = -\frac{1}{2\pi i z}, \qquad \hat{\delta}^{(n)}(z) = (-1)^{n+1} \frac{n!}{2\pi i z^{n+1}}. \tag{20.8.5}$$

---

[21] Thus, this relation does not hold for all the test functions of $\mathcal{E}$, but only for a subset.

In this case (20.8.4) is just (20.6.8) and it is valid for all functions of $\mathcal{D}$. By an argument similar to that used on p. 574 to establish the Sokhotski–Plemelj formulae one can show that also

$$\hat{\delta}_+(z) = \begin{cases} -\frac{1}{2\pi i z} & \text{Im } z > 0 \\ 0 & \text{Im } z < 0 \end{cases}, \qquad \hat{\delta}_-(z) = \begin{cases} 0 & \text{Im } z > 0 \\ -\frac{1}{2\pi i z} & \text{Im } z < 0 \end{cases} \qquad (20.8.6)$$

are analytic representations of distributions and, evidently,

$$\hat{\delta}(z) = \hat{\delta}_+(z) + \hat{\delta}_-(z). \qquad (20.8.7)$$

On the other hand, since $\hat{\delta}_\pm(x \mp i\varepsilon) = 0$ by definition,

$$i\pi \int_{-\infty}^{\infty} \left[(\hat{\delta}_- - \hat{\delta}_+)(x + i\varepsilon) - (\hat{\delta}_- - \hat{\delta}_+)(x - i\varepsilon) \right] \psi(x)\,dx$$

$$= i\pi \int_{-\infty}^{\infty} \left[-\hat{\delta}_+(x + i\varepsilon) - \hat{\delta}_-(x - i\varepsilon) \right] \psi(x)\,dx = \left\langle \frac{1}{x}, \psi \right\rangle, \qquad (20.8.8)$$

where the last expression is the Cauchy principal value defined in Example 20.3.4 p. 574. We thus find that

$$\widehat{\left(\frac{1}{x}\right)}(z) = i\pi \left[\hat{\delta}_-(z) - \hat{\delta}_+(z) \right]. \qquad (20.8.9)$$

By combining with (20.8.7) we have the following analytic representation of the Sokhotski–Plemelj formulae (20.3.12) p. 574:

$$\mp 2\pi i \hat{\delta}_\pm(z) = \widehat{\left(\frac{1}{x}\right)}(z) \mp \pi i \hat{\delta}(z) \qquad (20.8.10)$$

which can be differentiated to any order. $\square$

Example 20.8.2 Since the distribution $H(x)$ does not have a compact support, it does not admit a Cauchy representation. Nevertheless, a representation of the form (20.8.4) does exist in terms of the function

$$\hat{H}(z) = -\frac{1}{2\pi i} \log(-z) = -\frac{1}{2\pi i} \log z + \frac{1}{2}\text{sgn}(\text{Im } z) \qquad (20.8.11)$$

(see Bremermann 1965, p. 67) so that

$$-\frac{1}{2\pi i} \int_{-\infty}^{\infty} \left[\log(-x - i\varepsilon) - \log(-x + i\varepsilon) \right] \psi(x)\,dx \rightarrow \int_0^{\infty} \psi(x)\,dx. \qquad (20.8.12)$$

This relation may be compared with the entry shown in Table 17.8 p. 470 for the evaluation of the integral $\int_0^\infty \psi(x)\,dx$ by contour integration. $\square$

20.9 Convolution

We have already introduced in (10.3.6) p. 271 the convolution of two integrable functions $f(x)$ and $g(x)$, denoted by $f * g$:

$$h(x) = (f * g)(x) = \int_{-\infty}^{\infty} f(x - \xi)g(\xi)\, d\xi = \int_{-\infty}^{\infty} f(\xi)g(x - \xi)\, d\xi = (g * f)(x).$$
(20.9.1)

The scalar product of $h = f * g$ (assumed real) and a test function ϕ is

$$
\begin{aligned}
(h, \phi) &= \int_{-\infty}^{\infty} h(x)\phi(x)\, dx = \int_{-\infty}^{\infty} \left[\int_{-\infty}^{\infty} f(x - \xi)g(\xi)\, d\xi \right] \phi(x)\, dx \\
&= \int_{-\infty}^{\infty} d\xi \int_{-\infty}^{\infty} d\eta\, f(\xi)\, g(\eta)\, \phi(\xi + \eta),
\end{aligned}
$$
(20.9.2)

where Fubini's theorem A.4.3 p. 690 has been used to interchange the order of the integrations. This relation suggests a possible route to the definition of the convolution of two distributions d_1 and d_2, namely

$$\langle (d_1 * d_2)(x), \phi(x) \rangle = \langle d_1(x)\, d_2(y), \phi(x + y) \rangle.$$
(20.9.3)

The difficulty that arises is that $\phi(x + y)$ is not a test function because, while it is infinitely smooth, its support lies in an unbounded strip of the (x, y) plane parallel to the line $x + y = 0$. However, if the support of $d_1(x)d_2(y)$ intersects this strip in a bounded set σ, and if $\alpha(x, y)$ is infinitely differentiable and equal to 1 in a neighborhood of σ, then the functions $\alpha(x, y)\phi(x + y)$ are test functions which can be used for a good definition of the convolution (see e.g. Zemanian 1965, p. 122; Kanwal 2004, section 7.4). As noted in the previous section, the support of $d_1(x)d_2(y)$ is the direct product of the supports of the two distributions and, therefore, sufficient conditions for the existence of the convolution product are:

1. Either one (or both) of d_1 and d_2 has a bounded support; in this case, the support of $d_1(x)d_2(y)$ is a horizontal or vertical strip of finite width which will intersect the "diagonal" strip with the support of $\phi(x + y)$ in a bounded region.
2. The supports of both d_1 and d_2 are bounded from the same side; for example, if the support of $d_1(x)$ is contained in the half-plane $x > a$ and that of $d_2(y)$ in the half-plane $y > a$, the support of $d_1(x)d_2(y)$ is in a quarter-plane $a < x < \infty$, $a < y < \infty$ which will have a bounded intersection with the strip containing the support of $\phi(x + y)$.

An important instance of the first possibility is the δ:

$$\langle (\delta * d)(y), \phi(y) \rangle = \langle \delta(y)\, d(x), \phi(x + y) \rangle = \langle d(x), \phi(x) \rangle.$$
(20.9.4)

Thus, for any distribution d,

$$(\delta * d)(x) = d(x)$$
(20.9.5)

which generalizes to distributions the symbolic relation

$$\int_{-\infty}^{\infty} \delta(x - y)\, d(y)\, dy = d(x).$$
(20.9.6)

Table 20.2	Some properties of the convolution of distributions.

	Notes
$d_1 * d_2 = d_2 * d_1$	
$(\delta * d)(x) = d(x)$	
$\delta(x - a) * d(x) = d(x - a)$	
$\delta^{(n)}(x) * d(x) = d^{(n)}(x)$	
$(D\delta * d)(x) = Dd(x)$	D arbitrary differential operator
$D(d_1 * d_2) = (Dd_1) * d_2 = d_1 * (Dd_2)$	D arbitrary differential operator
$d_1(x + h) * d_2(x) = (d_1 * d_2)(x + h)$	
$d_1(-x) * d_2(-x) = (d_1 * d_2)(-x)$	
$d_1 * (d_2 * d_3) = (d_1 * d_2) * d_3$	Any two supports bounded on both sides, or all three bounded on the same side (conditions only sufficient)
$d_n * g \to d * g$	if $\{d_n\} \to d$ for $n \to \infty$ [a]
$d_t * g \to d * g$	if $\{d_t\} \to d$ for $t \to t_0$ [b]
$(\partial/\partial t)(d_t * d) = (\partial d_t/\partial t) * d$	see footnote b

[a] Provided the supports of all the d_n are contained in the same bounded set, or the support of g is bounded, or the supports of d_n and g are bounded on the same side independently of n.
[b] Provided the supports of all the d_t are contained in the same bounded set, or the support of g is bounded, or the supports of d_t and g are bounded on the same side independently of n.

With the aid of the relation $d_1' * d_2 = d_1 * d_2'$ (see Table 20.2), we find a counter example related to the failure of the hypothesis of case 2 as $H' * 1 = \delta * 1 = 1$, while $H * 1' = H * 0 = 0$. Thus $H' * 1 \neq H * 1'$ because the supports of H and 1 are not bounded from the same side. That the two conditions cited are only sufficient can be illustrated by noting that

$$\left\langle \frac{1}{x}, \phi(x + y) \right\rangle = \int_0^\infty \frac{\phi(y + x) - \phi(y - x)}{x} \, dx = \left(\int_y^\infty + \int_{-\infty}^y \right) \frac{\phi(\eta)}{\eta - y} \, d\eta,$$

(20.9.7)

the support of which evidently lies in the smallest interval $-a < y < a$ containing the support of ϕ. Thus, the convolution $(1/x) * d$ is well defined for any d.

It is interesting to note that

$$(d * \phi)(x) = \langle d(y), \phi(x - y) \rangle \in C^\infty(\mathbb{R}).$$

(20.9.8)

If the integral of ϕ is normalized to 1, this relation defines a *regularization* of d (cf. (20.8.15) p. 590). In particular, for a δ-family, in virtue of (20.9.5), the left-hand side tends to d, which shows that any distribution is a weak limit (definition on p. 552) of its own regularization. Remembering that the space $\mathscr{D}$ is dense in C^∞ (p. 570), we may also say that the space $\mathscr{D}$ is dense in the space of distributions $\mathscr{D}'$.[22] Furthermore

$$(d' * \phi) = \langle d'(y), \phi(x - y)\rangle = -\left\langle d(y), \frac{\partial}{\partial y}\phi(x - y)\right\rangle = \frac{\partial}{\partial x}\langle d(y), \phi(x - y)\rangle.$$

(20.9.9)

In Chapter 15 we have seen the role played by the operation of convolution in expressing a particular solution of an equation $\mathsf{L}u = f(x)$. We return on this topic in §20.13. An important application of the convolution of distributions is to the Hilbert transform described in §20.14.

20.10 The Fourier transform of distributions

In the previous sections we have seen many examples of a general procedure which reduces the definition of any new operation on a distribution to a known operation on test functions. The primary guide in this step is to consider properties and manipulations which are known to be valid in the case of regular functions. In applying this approach to the definition of the Fourier transform of a distribution, we recall the generalized Parseval equality (10.2.10) p. 269 of the standard Fourier transform theory:

$$\left(\tilde{f}, \tilde{g}\right) = \int_{-\infty}^{\infty} \overline{\tilde{f}(k)}\,\tilde{g}(k)\,\mathrm{d}k = \int_{-\infty}^{\infty} \overline{f(x)}\,\mathscr{F}^{-1}\{\tilde{g}\}(x)\,\mathrm{d}x = \left(f, \mathscr{F}^{-1}\{\tilde{g}\}\right).$$

(20.10.1)

If we were assured that both $\tilde{g}$ and $\mathscr{F}^{-1}\{\tilde{g}\}$ are test functions, we could use this relation to define the Fourier transform $\tilde{d}$ of the distribution d setting[23]

$$\langle \tilde{d}(k), \overline{\tilde{\phi}}(k)\rangle = \langle d(x), \overline{\mathscr{F}^{-1}\{\tilde{\phi}\}(x)}\rangle.$$

(20.10.2)

It is found, however that, if $\phi \in \mathscr{D}$, the same is not necessarily true for its Fourier transform.[24]

The way out of this dilemma is to adopt a new, more suitable set of test functions, referred to as functions of rapid decay, or "good functions." The functions in this class, denoted by

---

[22] In other words, any distribution is the distributional limit of a sequence of elements of $\mathscr{D}$. Conversely, since $\mathscr{D}$ is complete, every weak limit of locally summable functions is a distribution. For this reason it is possible to construct a theory of distributions based on weakly convergent sequences of locally summable ordinary functions. Conceptually there is an analogy here with the construction of the system of real numbers on the basis of sequences of rational numbers.

[23] Consistently with the convention employed hitherto, in using the notation $\langle \cdot, \cdot \rangle$ we indicate explicitly the complex conjugate.

[24] Test functions vanish outside a finite interval and, therefore, their Fourier transform exists for complex k and the defining integral can be differentiated any number of times. Thus $\tilde{\phi}(k)$ possesses all derivatives with respect to k and is therefore an analytic (in fact, entire) function of this variable. Since, according to Liouville's theorem (p. 431), the only analytic functions bounded for all values of the argument are constants, the Fourier transforms of $\mathscr{D}$ functions cannot belong to $\mathscr{D}$. Gelfand & Shilov (1964) base their theory on a different space of test functions, denoted by $\mathscr{Z}$, which consists of entire functions of the complex variable z which, with their derivatives, satisfy a certain bound as $|z| \to \infty$. The Fourier transforms of $\mathscr{D}$ functions belong to the space $\mathscr{Z}$.

$\mathscr{S}(\mathbb{R})$ (for Schwartz class), are infinitely differentiable functions which, for suitable finite constants C_{MN}, satisfy

$$\left| x^M \frac{\mathrm{d}^N \phi}{\mathrm{d}x^N} \right| < C_{MN} \tag{20.10.3}$$

for all M, $N \geq 0$. In other words, these functions, together with their derivatives, decay to 0 faster than any power of x as $|x| \to \infty$. A typical example would be $\exp(-x^2)$.

Due to the duality between x and $\mathrm{d}/\mathrm{d}x$ vis-a-vis $\mathrm{d}/\mathrm{d}k$ and k, it is evident that, if ϕ satisfies (20.10.3), its Fourier transform satisfies

$$\left| k^N \frac{\mathrm{d}^M \tilde{\phi}}{\mathrm{d}k^M} \right| < C_{MN} \tag{20.10.4}$$

and, therefore, belongs to the same class. Hence the Fourier transform maps the space $\mathscr{S}$ onto itself.

Definition 20.10.1 The continuous linear functionals on the space $\mathscr{S}$ of test functions are called *tempered distributions* or *distributions of slow growth*. The set of such functionals is denoted by $\mathscr{S}'$.

Continuity is to be judged with respect to the topology of the space $\mathscr{S}$: a sequence $\{\phi_k\} \in \mathscr{S}$ converges to $\phi \in \mathscr{S}$ if all the ϕ_k as well as all their derivatives converge uniformly to ϕ and the corresponding derivatives; furthermore, for each function ϕ_k, the inequality (20.10.3) is satisfied with constants C_{MN} independent of k. The space $\mathscr{S}$ is closed with respect to this definition of convergence. A characterization of the distributions in $\mathscr{S}'$ is contained in the following theorem (see e.g. Friedman 1963, p. 91; Beltrami & Wohlers 1966, p. 36):

Theorem 20.10.1 *Any distribution $d(x) \in \mathscr{S}'$ can be written as the M-th order distributional derivative of $(1 + x^2)^{N/2} f(x)$ for some integers M and N, with f a bounded continuous function. The same representation holds with f a continuous L^2 function, possibly with a different value of N.*

In dealing with the Fourier transform it is natural to allow for complex-valued test functions of the *real variable* x, which are just pairs of real functions of x, each one of which belongs to $\mathscr{S}$. Continuous linear functionals on these test functions are, in general, complex-value tempered distributions.[25] When we want to refer to the action of the complex-conjugate distribution $\bar{d}(x)$, rather than $d(x)$, we will write explicitly $\langle \bar{d}(x), \phi(x) \rangle$, which is defined by

$$\langle \bar{d}(x), \phi(x) \rangle = \overline{\langle d(x), \overline{\phi}(x) \rangle}. \tag{20.10.5}$$

Clearly all $\phi \in \mathscr{D}$ are also in $\mathscr{S}$ and are, in fact, dense in this space. Therefore, in going from $\mathscr{D}$ to $\mathscr{S}$, we have extended the set of available test functions, which results in a *restriction* of the admissible functionals, i.e., distributions. For example, e^{x^2} and $e^{\pm x}$ generate

---

[25] In the same way one can introduce complex-valued distributions using complex-valued functions in $\mathscr{D}$.

(regular) distributions in $\mathscr{D}$ but not in $\mathscr{S}$. Furthermore, convergence according to the $\mathscr{D}$ criterion of p. 570 implies convergence in $\mathscr{S}$, but not vice versa. Therefore, the latter is more stringent than the former: functionals which are continuous according to the $\mathscr{D}$ criterion may not be according to the criterion in $\mathscr{S}$.[26]

The definitions of multiplication by a constant, change of variable and differentiation given above in §20.4 and §20.5 are also valid for tempered distributions, and so is the notion of convergence of §20.6, all with the obvious modification that the test functions involved belong to $\mathscr{S}$ rather than to $\mathscr{D}$. On the other hand, not all multiplications by smooth functions allowable for distributions are legitimate for tempered distributions. For example, $\langle e^{x^2} d(x), e^{-x^2}\rangle = \langle d(x), 1\rangle$ may fail to exist, e.g. for $d = x$.[27]

Example 20.10.1 As a typical example of the calculation of the Fourier transform of tempered distributions, we consider $\delta(x - a)$. Upon applying (20.10.2) we have

$$
\begin{aligned}
\langle \tilde{\delta}, \overline{\tilde{\phi}}(k)\rangle &= \langle \delta(x - a), \overline{\mathscr{F}^{-1}\{\tilde{\phi}\}(x)}\rangle = \overline{\mathscr{F}^{-1}\{\tilde{\phi}\}(a)} \\
&= \frac{1}{\sqrt{2\pi}} \int_{-\infty}^{\infty} e^{ika} \overline{\tilde{\phi}}(k)\,\mathrm{d}k = \left\langle \frac{e^{ika}}{\sqrt{2\pi}}, \overline{\tilde{\phi}}(k)\right\rangle
\end{aligned}
\qquad (20.10.6)
$$

which shows that

$$
\mathscr{F}\{\delta(x - a)\} = \frac{e^{ika}}{\sqrt{2\pi}},
\qquad (20.10.7)
$$

i.e. the same result that one would find by evaluating formally the integral $(2\pi)^{-1/2} \int_{-\infty}^{\infty} \delta(x - a) e^{ikx}\,\mathrm{d}x$. This latter procedure may actually be justified by observing that, while e^{ikx} is not a test function, it can be approximated arbitrarily closely by a (complex) test function in the neighborhood of $x = a$. Another application of the definition (20.10.2) is given in Example 20.11.6 p. 604.   □

Reading the definition (20.10.2) from right to left we have the rule for the inversion of the Fourier transform:

$$
\langle \mathscr{F}^{-1}\{\tilde{d}\}(x), \overline{\tilde{\phi}}(x)\rangle = \langle \tilde{d}(k), \overline{\mathscr{F}\{\phi\}}(k)\rangle.
\qquad (20.10.8)
$$

For example

$$
\left\langle \frac{e^{ika}}{\sqrt{2\pi}}, \overline{\tilde{\phi}}\right\rangle = \int_{-\infty}^{\infty} \frac{e^{ika}}{\sqrt{2\pi}} \overline{\tilde{\phi}}(k)\,\mathrm{d}k = \overline{\phi}(x - a) = \langle \delta(x - a), \overline{\phi}(x)\rangle
\qquad (20.10.9)
$$

so that $\mathscr{F}^{-1}\{e^{ika}/\sqrt{2\pi}\} = \delta(x - a)$. Special cases of this result and of (20.10.7) are $\mathscr{F}^{-1}\{1/\sqrt{2\pi}\} = \delta(x)$ and $\mathscr{F}\{\delta(x)\} = 1/\sqrt{2\pi}$. Using the same convention used before

---

[26] Lighthill (1958) introduced the descriptive denomination "good functions" and based his entire exposition of distribution theory on functions in $\mathscr{S}(\mathbb{R})$, rather than $\mathscr{D}$. Thus, his theory includes only a subset of the set of distributions defined here which, however, is large enough to accommodate most practical needs.

[27] Multiplication by infinitely smooth functions of slow growth (also called "fairly good" functions in Lighthill's terminology) is however always allowed. These functions are infinitely differentiable any number of times and, with all their derivatives, are of order $|x|^N$ for some (positive, zero or negative) N as $|x| \to \infty$.

Table 20.3	Some operational rules for the exponential Fourier transform of tempered distributions.	
	x-domain	k-domain
1	$d^n d/dx^n$	$(-ik)^n \tilde{d}(k)$
2	$x^n d(x)$	$(-i d/dk)^n \tilde{d}(k)$
3	$e^{iax} d(x)$	$\tilde{d}(k+a)$
4	$d(ax-b)$	$\exp(ibk/a)\tilde{d}(k/a)/|a|$

of writing $\langle \cdot, \cdot \rangle$ in the form of an integral, we express them as

$$\int_{-\infty}^{\infty} e^{\pm ikx} \, \delta(x) \, dx = 1, \qquad \int_{-\infty}^{\infty} e^{\pm ikx} \, dk = 2\pi \, \delta(x). \qquad (20.10.10)$$

The latter result was already given before in (20.6.12).

As noted, the space of tempered distributions is indicated by $\mathscr{S}'$. Given the character of the space $\mathscr{S}$ of test functions, it is clear that the Fourier transform of a tempered distribution is a tempered distribution. The Fourier transform of a distribution with compact support is actually an entire analytic function of the conjugate variable k in the entire complex k-plane and is therefore a regular distribution.

The validity of the standard properties of the Fourier transform given in §10.3 can be proven for the Fourier transform of distributions as well. For example, for the differentiation rule,

$$\begin{aligned}
\langle \mathscr{F}\{d'\}, \bar{\tilde{\phi}} \rangle &= \langle d', \overline{\mathscr{F}^{-1}\{\tilde{\phi}\}} \rangle = -\langle d, (d/dx)\overline{\mathscr{F}^{-1}\{\tilde{\phi}\}} \rangle \\
&= -\langle d, ik\,\overline{\mathscr{F}^{-1}\{\tilde{\phi}\}} \rangle = \langle -ikd, \bar{\tilde{\phi}} \rangle
\end{aligned} \qquad (20.10.11)$$

so that $\mathscr{F}\{d'\} = -ik\mathscr{F}\{d\}$. Various rules are summarized in Table 20.3.

These rules permit us to derive several useful results. For example

$$\mathscr{F}\{x^n\}(k) = \mathscr{F}\{x^n \cdot 1\}(k) = \left(-i\frac{d}{dk}\right)^n \mathscr{F}\{1\} = (-i)^n \sqrt{2\pi}\,\delta^{(n)}(k). \qquad (20.10.12)$$

To calculate the Fourier transform of $H(x)$ we use the fact that $H'(x) = \delta(x)$, so that

$$\frac{1}{\sqrt{2\pi}} = \mathscr{F}\{\delta(x)\} = \mathscr{F}\{H'(x)\} = -ik\mathscr{F}\{H(x)\}. \qquad (20.10.13)$$

The solution of this equation is

$$\mathscr{F}\{H(x)\} = \frac{i}{k\sqrt{2\pi}} + c\delta(k). \qquad (20.10.14)$$

To determine c we use the transform of $H(x) + H(-x) = 1$ and entry 4 in the table to find

$$
\begin{aligned}
\mathscr{F}\{H(x)\} + \mathscr{F}\{H(-x)\} &= \frac{i}{\sqrt{2\pi}k} + c\delta(k) + \frac{i}{\sqrt{2\pi}(-k)} + c\delta(-k) \\
&= 2c\delta(k) = \sqrt{2\pi}\,\delta(k), \qquad\qquad (20.10.15)
\end{aligned}
$$

in which we have used the second one of (20.10.10) for the Fourier transform of 1. Thus we conclude that

$$
\mathscr{F}\{H(x)\} = \frac{1}{\sqrt{2\pi}}\left[\frac{i}{k} + \pi\delta(k)\right] = \frac{i}{\sqrt{2\pi}}\frac{1}{k+i0}, \qquad\qquad (20.10.16)
$$

where the last step follows from the Sokhotski–Plemelj formulae (20.3.12) p. 574. Several Fourier transforms of tempered distributions are summarized in Table 4.3 p. 87. As in the case of the ordinary Fourier transform, the tables can be readily extended by using the operational rules of Table 20.3 and the result

$$
\mathscr{F}^2\{d(x)\} = \mathscr{F}\{\tilde{d}(k)\} = d(-x) \qquad\qquad (20.10.17)
$$

valid also for the ordinary Fourier transform. For example, from (20.10.7),

$$
\mathscr{F}\left\{\frac{e^{iak}}{\sqrt{2\pi}}\right\} = \mathscr{F}^2\{\delta(x-a)\} = \delta(-x-a) = \delta(x+a). \qquad\qquad (20.10.18)
$$

Similarly to the case of functions, the Fourier transform of the convolution product of two distributions d_1 and d_2, when it is defined, is

$$
\mathscr{F}\{d_1 * d_2\} = \sqrt{2\pi}\,\tilde{d}_1\tilde{d}_2. \qquad\qquad (20.10.19)
$$

The product in the right-hand side defines a distribution because, as noted on p. 598, the Fourier transform of a distribution with compact support is actually in C^∞.

The continuity of tempered distributions has the consequence that the Fourier transform of a converging sequence converges to the Fourier transform of the limit, and that the Fourier transform of a series is the series of the Fourier transform of the individual terms.

The Fourier transform can be extended to more than one space dimension by generalizing the previous approach. The usual properties of the ordinary Fourier transform carry over. Table 4.4 p. 89 shows some Fourier transform pairs in N-dimensional space.

It is possible to define the Fourier transform also for non-tempered distributions with the aid of another class of test functions denoted by $\mathscr{Z}$ (see footnote 25 p. 595); the reader may consult several references for details (see e.g. Gelfand & Shilov 1964, vol. 1, chapter 2; Pandey 1996, p. 49; Pathak 1997, chapters 2 and 3; Zemanian 1965, p. 198). The set of continuous linear functionals on the space $\mathscr{Z}$ constitutes the space of *ultradistributions*.

20.11 The Fourier series of distributions

A distribution is said to be *periodic* with period L if

$$\langle d(x + L), \phi(x) \rangle = \langle d(x), \phi(x - L) \rangle = \langle d(x), \phi(x) \rangle \qquad (20.11.1)$$

for all $\phi \in \mathcal{D}(\mathbb{R})$. As we did in the case of the Fourier series in Chapter 9, we develop the theory for $L = 2\pi$. Rather than limiting ourselves to a fundamental interval, it is convenient to work over the entire real line using periodic extensions where appropriate.

Theorem 20.11.1 *Any trigonometric series*

$$d(x) = \sum_{-\infty}^{\infty} a_m \frac{e^{imx}}{\sqrt{2\pi}} \qquad (20.11.2)$$

the coefficients of which, as $|m| \to \infty$, do not grow faster than $|m|^N$ (with $N \geq 0$ a finite integer) converges distributionally to a periodic distribution $d(x)$.

The proof is immediate noting that, under the stated hypotheses, the series with coefficients $a_m/(im)^{N+2}$ converges uniformly, and hence distributionally (p. 587). The distributional convergence will not be destroyed by the repeated differentiation necessary to return to (20.11.2).

The set of coefficients $\{a_m/(im)^{N+2}\}$ is uniquely related to the function that they represent as a Fourier series. For this reason, the relation between a distribution and its Fourier coefficients is also unique.

Example 20.11.1 The Fourier series of the function

$$f(x) = \frac{1}{4}(|x| - \pi)^2, \qquad -\pi \leq x \leq \pi \qquad (20.11.3)$$

is readily calculated to be

$$f(x) = \frac{\pi^2}{12} + \sum_{k=1}^{\infty} \frac{\cos kx}{k^2} \qquad (20.11.4)$$

and, since $|\cos kx|/k^2 \leq 1/k^2$, it converges uniformly by Weierstrass's M-test (p. 227). Upon differentiating (20.11.3) and (20.11.4) twice in the sense of distributions, the latter term-by-term, noting that $|x|' = \operatorname{sgn} x$, $(\operatorname{sgn} x)' = 2\delta(x)$, and equating, we have

$$f''(x) = \frac{1}{2}[1 - 2\pi\delta(x)] = -\sum_{k=1}^{\infty} \cos kx, \qquad -\pi < x < \pi, \qquad (20.11.5)$$

a relation which is valid distributionally even though the indicated series does not converge classically. Upon rearranging we find

$$2\pi\delta(x) = 1 + 2\sum_{k=1}^{\infty} \cos kx = \sum_{m=-\infty}^{\infty} e^{imx}, \qquad -\pi < x < \pi, \qquad (20.11.6)$$

or also[28]

$$\delta(x - \xi) = \frac{1}{2\pi}\left[1 + 2\sum_{k=1}^{\infty}\cos k(x - \xi)\right], \qquad 0 < |x - \xi| < \pi. \tag{20.11.7}$$

By applying this relation to a test function with support in the interval $|x| \leq \pi$ we have

$$\phi(x) = \frac{1}{2\pi}\int_{-\pi}^{\pi}\phi(\xi)\,d\xi + \frac{1}{\pi}\sum_{k=1}^{\infty}\left[\cos kx\int_{-\pi}^{\pi}\phi(\xi)\cos k\xi\,d\xi\right.$$

$$\left. + \sum_{k=1}^{\infty}\sin kx\int_{-\pi}^{\pi}\phi(\xi)\sin k\xi\,d\xi\right] \tag{20.11.8}$$

which is the Fourier series expansion of ϕ. The distributional derivative of (20.11.6) is

$$2\pi\delta'(x) = -2\sum_{k=1}^{\infty}k\sin kx = \sum_{m=-\infty}^{\infty}ime^{imx}, \qquad -\pi < x < \pi, \tag{20.11.9}$$

and other relations of this type may be obtained by further differentiation.   □

Example 20.11.2 [Poisson summation formula] The right-hand side of (20.11.6) is already periodic; by extending periodically the left-hand side to the entire real axis we have the Fourier series for the so-called "Dirac comb":

$$2\pi\sum_{n=-\infty}^{\infty}\delta(x - 2n\pi) = 1 + 2\sum_{k=1}^{\infty}\cos kx = \sum_{m=-\infty}^{\infty}e^{imx}, \qquad -\infty < x < \infty \tag{20.11.10}$$

or, with the change of variable $y = x/2\pi$

$$\sum_{n=-\infty}^{\infty}\delta(y - n) = 1 + 2\sum_{k=1}^{\infty}\cos 2\pi ky, \qquad -\infty < y < \infty. \tag{20.11.11}$$

Applied to a test function ϕ, this relation gives

$$\sum_{n=-\infty}^{\infty}\phi(n) = \int_{-\infty}^{\infty}\phi(y)\,dy + \sum_{k=1}^{\infty}\int_{-\infty}^{\infty}\phi(y)\cos 2\pi ky\,dy. \tag{20.11.12}$$

More generally, upon setting $y = \beta x/(2\pi)$, we find

$$\beta\sum_{n=-\infty}^{\infty}\phi(n\beta) = \int_{-\infty}^{\infty}\phi(y)\,dy + 2\sum_{k=1}^{\infty}\int_{-\infty}^{\infty}\phi(y)\cos\frac{2\pi ky}{\beta}\,dy. \tag{20.11.13}$$

By the usual continuity arguments, both relations will hold not only for test functions but for any function F for which the individual terms in the series make sense and the

---

[28] Formally, this relation can be derived by writing $\delta(x - \xi) = \sum_{-\infty}^{\infty}(e_n, \delta)\,e_n(x)$, with e_n the Fourier basis functions. For $|x - \xi|$ outside the indicated range, the result of the sum is the periodic extension of the left-hand side as in (20.11.10).

series themselves converge. By setting $F(x) = \frac{1}{2}[\phi(x) + \phi(-x)]$, we recover the Poisson summation formulae (8.7.32) and (8.7.33) p. 238. For this reason the distributions in the left-hand sides of (20.11.10) and (20.11.11) are sometimes called Poisson distributions.

$\square$

The derivation of (20.11.6) shows one possible way to calculate the Fourier coefficients of a distribution. A useful remark is that, if $d(x)$ is the distribution corresponding to a normal periodic function (continuous or not), the standard rules of the Fourier series apply (see e.g. Jones 1966, p. 134).

Example 20.11.3 For ξ fixed, we can consider $H(x - \xi)$ as a normal function defined in $|x| \leq \pi$. As such, it can be expanded in a Fourier series in the standard way, $H(x - \xi) = \sum_{-\infty}^{\infty} (e_n, H) e_n(x)$ with $e_n(x)$ the Fourier basis functions. If $\xi < -\pi$, $H(x - \xi) = 1$ in the range of integration defining the scalar product while, if $\xi > \pi$, $H = 0$ in the range. The interesting case is $|\xi| < \pi$, to which we limit ourselves. Upon calculating the Fourier coefficients with the usual rule we find

$$H(x - \xi) = \frac{\pi - \xi}{2\pi} + \frac{1}{\pi} \sum_{k=1}^{\infty} \frac{\sin k(x - \xi) - (-1)^k \sin kx}{k}, \qquad -\pi < x, \xi < \pi$$

(20.11.14)

and in particular, for $\xi = 0$,

$$H(x) = \frac{1}{2} + \frac{2}{\pi} \sum_{\ell=1}^{\infty} \frac{\sin(2\ell + 1)x}{2\ell + 1}, \qquad -\pi < x < \pi.$$

(20.11.15)

With the identity $\sum_{k=1}^{\infty}(-1)^k (\sin kx)/n = -x/2$ valid for $|x| < \pi$ (Table 3.2 p. 56), the general result (20.11.14) may be written as

$$H(x - \xi) = \frac{1}{2} + \frac{x - \xi}{2\pi} + \frac{1}{\pi} \sum_{k=1}^{\infty} \frac{\sin k(x - \xi)}{k}, \qquad -\pi < x < \pi.$$

(20.11.16)

The same expression can be recovered by integrating the δ-expansion (20.11.7) provided $x \neq \xi$ (see p. 578):

$$\int_{-\pi}^{x} \delta(x' - \xi)\, dx' = H(x - \xi) = \frac{x + \pi}{2\pi} + \frac{1}{\pi} \sum_{k=1}^{\infty} \frac{\sin k(x - \xi) + \sin k(\xi + \pi)}{k}.$$

(20.11.17)

Since $\sin k(\xi + \pi) = (-1)^k \sin k\xi$, the second summation equals $-\xi/2$ and (20.11.16) follows. If $x = \xi$, (20.11.16) gives $H(0) = 1/2$, which is what one would expect for a normal function on the basis of the behavior of the Fourier series at a point of discontinuity. However, as remarked in connection with (20.4.10) p. 578, in the sense of distributions this result is ambiguous and therefore, while use of (20.11.16) as a distribution (i.e., applied to test functions) is safe, the correctness of a point-wise interpretation of it depends on the specific context.

$\square$

Example 20.11.4 The same procedure can also be followed in other cases; for example, $d(x) = [\sin x]^{-1}$, must be understood in the principal value sense and one has

$$a_m = \frac{1}{\sqrt{2\pi}} \lim_{\varepsilon \to 0} \left[\int_{-\pi+\varepsilon}^{-\varepsilon} + \int_{\varepsilon}^{\pi-\varepsilon} \right] \frac{e^{-imx}}{\sin x} \, dx = \begin{cases} -i\sqrt{2\pi} \operatorname{sgn} m & m = 2k+1 \\ 0 & m = 2k \end{cases}$$
(20.11.18)

so that, in the sense of distributions,[29]

$$\frac{1}{\sin x} = 2 \sum_{k=0}^{\infty} \sin(2k+1)x.$$
(20.11.19)

□

In attempting to calculate the coefficients a_m of a general distribution d according to the ordinary rule $a_m = (e_m, d)$, one encounters the difficulty that the integral of a distribution between finite limits is not always well defined as was noted before on p. 577. The reader is referred to several references for a discussion of this point (Lighthill 1958; Zemanian 1965; Jones 1966; Richards & Youn 1990; Vladimirov 2002). A useful and general rule is (e.g. Richards & Youn 1990, p. 84)

$$a_m = \lim_{b \to \infty} \left\langle \sqrt{2\pi} e^{-imx} d(x), b^{-1} \phi(x/b) \right\rangle \quad \text{with} \quad \int_{-\infty}^{\infty} \phi(x) \, dx = 1.$$
(20.11.20)

An intuitive justification of this relation is that, for very large b, $\phi(x/b)$ is nearly constant over every interval of length 2π so that (20.11.20) amounts to the weighted sum of integrals of $e^{-inx} d(x)$, each one of which is similar to the calculation of the Fourier coefficients of a normal function, and the sum of the weights equals 1 in view of the hypothesis on the integral of ϕ.

Example 20.11.5 According to (20.11.20) the Fourier coefficients of the translated Dirac comb are

$$\begin{aligned}
a_m &= \lim_{b \to \infty} \left\langle \sqrt{2\pi} e^{-imx} \sum_{n=-\infty}^{\infty} \delta(x - \xi - 2\pi n), b^{-1} \phi(x/b) \right\rangle \\
&= \sqrt{2\pi} \lim_{b \to \infty} \sum_{n=-\infty}^{\infty} \left\langle \delta(x - 2\pi n), e^{-im(x+\xi)} b^{-1} \phi((x+\xi)/b) \right\rangle \\
&= \sqrt{2\pi} e^{-im\xi} \lim_{b \to \infty} \left[b^{-1} \sum_{n=-\infty}^{\infty} \phi(2\pi n/b + \xi/b) \right] = \frac{e^{-im\xi}}{\sqrt{2\pi}}. \text{(20.11.21)}
\end{aligned}$$

The last step follows from (20.11.13) with $\beta = 2\pi/b$ recalling that ϕ integrates to 1 by hypothesis and appealing to the Riemann–Lebesgue lemma (p. 249) to discard the

[29] For a "verification" of this result, multiply both sides by $\sin x$ to find $\sum_{k=0}^{\infty} [\cos kx - \cos(k-1)x] = 1$. A better procedure would be based on the derivative of the Fourier expansion of $-\sum_{k=0}^{\infty} \cos[(2k+1)x]/(2k+1) = \frac{1}{2} \log|\tan(x/2)|$, similar to the derivation of (20.11.6).

summation in the right-hand side of this latter relation.[30] Thus we find

$$\sum_{n=-\infty}^{\infty} \delta(x - \xi - 2\pi n) = \frac{1}{2\pi} \sum_{m=-\infty}^{\infty} e^{im(x-\xi)}, \qquad -\infty < x < \infty \qquad (20.11.22)$$

which, restricted to the range $|x - \xi| \leq 2\pi$, is (20.11.7). $\square$

Yet another way to calculate the Fourier coefficients of a distribution is based on the observation that, because of the rapid decay of the $\mathscr{S}$-type test functions and the bound on the growth of the coefficients a_m, periodic distributions are tempered. By periodicity, the series (20.11.2) defines then the distribution $d(x)$ over the entire real line and, as mentioned on p. 599, the Fourier transform of a series can be calculated term by term. Thus, taking the Fourier transform of (20.11.2) and using (20.10.18) we have

$$\tilde{d} = \mathscr{F}\left\{\sum_{-\infty}^{\infty} a_m \frac{e^{imx}}{\sqrt{2\pi}}\right\} = \sum_{-\infty}^{\infty} a_m \mathscr{F}\{\frac{e^{imx}}{\sqrt{2\pi}}\} = \sum_{-\infty}^{\infty} a_m \delta(k + m). \qquad (20.11.23)$$

The coefficients a_m are then calculated by reducing $\tilde{d}$ to the form (20.11.23) (see e.g. Richards & Youn 1990, p. 80).

Example 20.11.6 Let us calculate by this method the Fourier expansion of $H_p(x)$, the periodic extension of $H(x)$, defined by

$$H_p(x) = \begin{cases} 1 & \text{for} & 2n\pi < x < (2n+1)\pi \\ 0 & \text{for} & (2n+1)\pi < x < 2(n+1)\pi \end{cases} \qquad (20.11.24)$$

with $-\infty < n < \infty$. By applying the definition (20.10.2) of the Fourier transform, we readily find

$$\begin{aligned}
\langle \tilde{H}_p(x), \overline{\tilde{\phi}(k)} \rangle &= \int_{-\infty}^{\infty} \left(\sum_{n=-\infty}^{\infty} \int_{2n\pi}^{(2n+1)\pi} \frac{e^{ikx}}{\sqrt{2\pi}} \, dx \right) \overline{\tilde{\phi}(k)} \, dk \\
&= \int_{-\infty}^{\infty} \sum_{n=-\infty}^{\infty} \frac{e^{(2n+1)\pi ik} - e^{2n\pi ik}}{ik\sqrt{2\pi}} \, \overline{\tilde{\phi}} \, dk , \qquad (20.11.25)
\end{aligned}$$

from which

$$\tilde{H}_p(x) = \sum_{n=-\infty}^{\infty} \frac{e^{(2n+1)\pi ik} - e^{2n\pi ik}}{ik\sqrt{2\pi}} = \frac{e^{ik\pi} - 1}{ik} \sum_{n=-\infty}^{\infty} \frac{e^{2\pi ikn}}{\sqrt{2\pi}}. \qquad (20.11.26)$$

This can be reduced to the form (20.11.23) by using the δ-expansion (20.11.11)

$$\sum_{m=-\infty}^{\infty} \delta(x + m) = \sum_{n=-\infty}^{\infty} e^{2\pi inx} \qquad (20.11.27)$$

---

[30] A better justification of this last step can be based on the asymptotic estimate of the Fourier cosine transform given in §10.8.

to find, upon using the property $f(k)\delta(k+n) = f(-n)$,

$$\tilde{H}_p(x) = \sqrt{\frac{\pi}{2}} + \sum_{n=-\infty,n\neq 0}^{\infty} \frac{e^{in\pi}-1}{in\sqrt{2\pi}}\delta(k+n), \qquad (20.11.28)$$

from which, upon inversion of the transform, it follows that

$$H_p(x) = \frac{1}{2} + \sum_{n=-\infty,n\neq 0}^{\infty} \frac{e^{in\pi}-1}{in\sqrt{2\pi}}\frac{e^{inx}}{\sqrt{2\pi}}. \qquad (20.11.29)$$

In the fundamental interval $-\pi < x < \pi$ this result can be verified by reducing the summation and using the Fourier expansion (9.5.8) p. 250 of $\operatorname{sgn} x$ to find

$$H(x) = \frac{1}{2} + \sum_{n=-\infty,n\neq 0}^{\infty} \frac{e^{in\pi}-1}{2\pi in}e^{inx} = \frac{1}{2} + \frac{2}{\pi}\sum_{\ell=0}^{\infty} \frac{\sin(2\ell+1)\pi x}{2\ell+1} = \frac{1}{2}\left(1+\operatorname{sgn} x\right),$$

$$(20.11.30)$$

which indeed equals $H(x)$ in $-\pi < x < \pi$ and agrees with (20.11.15) found earlier. □

20.12 Laplace transform of distributions

There are several ways to define the Laplace transform of distributions (see e.g. Zemanian 1965, p. 222; Kanwal 2004, p. 200; Pathak 1997, p. 121; Vladimirov 2002). Probably the most direct one is via the relation (11.1.2) p. 285 between Fourier and Laplace transforms already used to define the Laplace transform of ordinary functions; as in that case we denote the variable by t rather than x as done until now.

Let $d(t) \in \mathcal{D}'$ be a distribution with support contained in $[0, \infty)$ and suppose that $d(t)e^{-\sigma t} \in \mathcal{S}'$ for all real numbers $\sigma > \sigma_c$. Then we define the Laplace transform of d by

$$\hat{d} = \mathcal{L}\{d\} \equiv \sqrt{2\pi}\,\mathcal{F}\{d(t)\,e^{-\sigma t}\} \qquad (20.12.1)$$

or, from (20.10.2),

$$\langle \hat{d}, \bar{\tilde{\phi}} \rangle = \sqrt{2\pi}\,\langle d(t)e^{-\sigma t}, \overline{\mathcal{F}^{-1}\{\tilde{\phi}\}} \rangle = \sqrt{2\pi}\,\langle d(t), e^{-\sigma t}\overline{\mathcal{F}^{-1}\{\tilde{\phi}\}} \rangle. \qquad (20.12.2)$$

According to this definition, for example,

$$\langle \mathcal{L}\{\delta(t-a)\}, \bar{\tilde{\phi}} \rangle = \sqrt{2\pi}\,\langle \delta(t-a), e^{-\sigma t}\overline{\mathcal{F}^{-1}\{\tilde{\phi}\}} \rangle = \int_{-\infty}^{\infty} e^{-(\sigma+i\omega)a}\bar{\tilde{\phi}}(\omega)\,d\omega \qquad (20.12.3)$$

and, therefore,

$$\mathcal{L}\{\delta(t-a)\} = e^{-as}, \qquad s = \sigma + i\omega. \qquad (20.12.4)$$

A similar calculation shows that

$$\mathcal{L}\{\delta(t)\} = 1. \qquad (20.12.5)$$

Table 20.4	Some operational rules for the Laplace transform of distributions.

$$\mathscr{L}\{d^{(k)}(t)\} = s^k \hat{d}(s) \qquad\qquad \mathscr{L}\{t^k d(t)\} = (-1)^k \hat{d}^{(k)}(s)$$

$$\mathscr{L}\{d(t-a)\} = e^{-as}\hat{d}(s) \qquad\qquad \mathscr{L}\{e^{-bt}d(t)\} = \hat{d}(s+b)$$

$$\mathscr{L}\{d(at)\} = a^{-1}\hat{d}(s/a) \qquad\qquad \mathscr{L}\{\int_0^t d_1(t)\,d_2(t-\tau)\,d\tau\} = \hat{d}_1(s)\,\hat{d}_2(s)$$

Furthermore, from (20.10.16),

$$\mathscr{L}\{H(t)\} = \frac{1}{s} + \pi\delta(is) \tag{20.12.6}$$

so that

$$\mathscr{L}\{H(t)\} = \begin{cases} 1/s & \sigma \neq 0 \\ -i/\omega + \pi\delta(\omega) & \sigma = 0 \end{cases}. \tag{20.12.7}$$

Since

$$\sqrt{2\pi}\,e^{-\sigma t}\overline{\mathscr{F}^{-1}\{\tilde{\phi}\}} = \overline{\int_{-\infty}^{\infty} e^{-(\sigma+i\omega)t}\tilde{\phi}(\omega)\,d\omega} \tag{20.12.8}$$

and the application of $d(t)$ to $e^{-\sigma t}\overline{\mathscr{F}^{-1}\{\tilde{\phi}\}}$ removes the dependence on the variable t, we see that the proper variable for $\mathscr{L}\{d\}$ is $s = \sigma + i\omega$ as shown explicitly in the previous examples.

According to the general rule for the differentiation of distributions we have

$$\langle\mathscr{L}\{d'\},\overline{\tilde{\phi}}\rangle = -\sqrt{2\pi}\,\langle d(t), \frac{d}{dt}\left[e^{-\sigma t}\overline{\mathscr{F}^{-1}\{\tilde{\phi}\}}\right]\rangle = \langle d(t), se^{-\sigma t}\overline{\mathscr{F}^{-1}\{\tilde{\phi}\}}\rangle \tag{20.12.9}$$

and, therefore,

$$\mathscr{L}\{d'\} = s\,\hat{d}(s). \tag{20.12.10}$$

This rule is only superficially different from that applicable to ordinary absolutely continuous functions according to which $\mathscr{L}\{f'\} = s\,\hat{f}(s) - f(0)$ (p. 287). Indeed, if $f(0)$ exists and $f(t) = 0$ for $t < 0$, the distributional derivative of f is $H(t)f'(t) + f(0)\delta(t)$ the Laplace transform of which is $[s\hat{f}(s) - f(0)] + f(0) = s\hat{f}(s)$ in agreement with (20.12.10). This and other operational rules are summarized in Table 20.4, which may be compared with those valid for ordinary functions given in Table 5.3 p. 126; proofs can be found in the references cited at the beginning of this section. Worthy of notice in this table are the validity of the convolution theorem and the fact that $\tilde{d}$ admits any number of ordinary derivatives so that, in fact, it is an *analytic function* of s in its region of convergence.

Example 20.12.1 It is known that (see (11.4.7) p. 297 and (17.5.36) p. 445)

$$\mathscr{L}\{H(t)\log t\} = \int_0^{\infty} e^{-st}\log t\,dt = -\frac{\gamma + \log s}{s}, \tag{20.12.11}$$

in which γ is Euler's constant. By (20.3.5), (20.5.9) and (20.5.17) we have

$$\frac{d}{dt}[H(t)\log t] = \frac{1}{2}\frac{d}{dt}[(1+\operatorname{sgn} t)\log|t|] = \frac{1}{2}\left(\frac{1}{t}+\frac{1}{|t|}-C\delta(t)\right)$$

$$= \frac{H(t)}{t} - \frac{1}{2}C\delta(t) \qquad\qquad (20.12.12)$$

and therefore, upon using the differentiation formula, we find

$$\mathscr{L}\left\{\frac{H(t)}{t}\right\} = -\gamma - \log s + C \qquad\qquad (20.12.13)$$

upon renaming the arbitrary constant. A short list of the Laplace transform of distributions is given in Table 5.2 p. 125. $\qquad\square$

For the inversion of the Laplace transform we will rely on a theorem due to Paley, Wiener and Schwartz, the proof of which uses Theorem 20.10.1 p. 596 (see e.g. Zemanian 1965, p. 236; Beltrami & Wohlers 1966, p. 47; Friedlander & Joshi 1998, p. 132):

Theorem 20.12.1 *Let d be a distribution in $\mathscr{D}$ with support in $[0,\infty)$ and let $d(t)e^{-\sigma t} \in \mathscr{S}'$ for $\sigma > \sigma_c$. Then*

$$|\hat{d}(s)| \le |P(s)| \qquad \text{for} \qquad \operatorname{Re} s > \sigma_c \qquad\qquad (20.12.14)$$

where P is a polynomial. Conversely, if $\hat{d}(s)$ is analytic for $\operatorname{Re} s > \sigma_c$ and it satisfies (20.12.14), then it is the Laplace transform of a distribution in $\mathscr{D}$ with support in $[0,\infty)$ such that $d(t)e^{-\sigma t} \in \mathscr{S}'$.

In particular, therefore, every rational function of s is the Laplace transform of a distribution if σ_c is taken to the right of all the poles of the function.

This theorem assures us that, for any $\hat{d}(s)$, there is an integer N such that $\hat{d}(s)/s^N \to 0$ fast enough for it to be the Laplace transform of a classical L^1-function. To find $d(t)$, then, we can use the differentiation rule (20.12.10) N times:

$$d(t) = \mathscr{L}^{-1}\{\hat{d}\} = \frac{d^N}{dt^N}\left[\mathscr{L}^{-1}\left\{\frac{\hat{d}}{s^N}\right\}\right]. \qquad\qquad (20.12.15)$$

Before taking the derivatives, it is necessary to restore the factor $H(t)$ in the inverse Laplace transform of $\hat{d}(s)/s^N$, which is often understood with ordinary functions but which is essential in dealing with distributions as the comment following (20.12.10) implies.

Example 20.12.2 Consider the inversion

$$\mathscr{L}^{-1}\left\{\frac{s^3}{s^2+1}\right\} = \frac{d^3}{dt^3}\left[\mathscr{L}^{-1}\left\{\frac{1}{s^2+1}\right\}\right] \qquad\qquad (20.12.16)$$

and, since $\mathscr{L}^{-1}\{(s^2+1)^{-1}\} = H(t)\sin t$ (Table 5.1 p. 124),

$$\mathscr{L}^{-1}\left\{\frac{s^3}{s^2+1}\right\} = [\delta'(t) - H(t)]\cos t = \delta'(t) - H(t)\cos t, \qquad\qquad (20.12.17)$$

where we have used the relation $\delta(t) \sin t = 0$ and (20.5.6) in the last step. We may check the result by noting that

$$\mathscr{L}^{-1} \left\{ \frac{s^3}{s^2 + 1} \right\} = \mathscr{L}^{-1} \left\{ s - \frac{s}{s^2 + 1} \right\} = \delta'(t) - H(t) \cos t. \qquad (20.12.18)$$

$\square$

20.13 Miscellaneous applications

Chapters 15 and 16 show the role of the δ distribution in relation to the Green's function and the inverse of differential operators. Additional examples of this type of calculation are given in several of the early chapters of this book. We now show a few other typical and useful applications.

20.13.1 Approximate evaluation of integrals

A very powerful application of the Fourier transform of generalized functions is to the asymptotic evaluation of direct and inverse Fourier transforms (Lighthill 1958; Jones 1966, chapter 9). We begin with a definition:

Definition 20.13.1 A distribution $d(x)$ is said to be well-behaved at infinity if for some X the distribution $d(x) - F(x)$ is absolutely integrable in the intervals $-\infty < x < -X$, $X < x < \infty$, where $F(x)$ is a linear combination of functions of the type

$$e^{iax}|x|^\lambda, \quad e^{iax}|x|^\lambda \operatorname{sgn} x, \quad e^{iax}|x|^\lambda \log |x|, \quad e^{iax}|x|^\lambda \log |x| \operatorname{sgn} x \qquad (20.13.1)$$

for different values of a and $\lambda > -1$.

The application under consideration rests on the following result:

Theorem 20.13.1 *If the distribution $d(x)$ has a finite number of singular points $x_1, x_2, \ldots, x_J$ and if $d(x) - F_j(x)$ has absolutely integrable M-th derivative in an interval including x_j, $j = 1, 2, \ldots, J$, where F_j is a linear combination of functions of the type*

$$|x - x_j|^\lambda, \qquad |x - x_j|^\lambda \operatorname{sgn}(x - x_j), \qquad |x - x_j|^\lambda \log |x - x|,$$
$$|x - x_j|^\lambda \operatorname{sgn}(x - x_j) \log |x - x_j|, \qquad \delta^{(p)}(x - x_j) \qquad (20.13.2)$$

for different values of λ and p, and if $d^{(M)}$ is well-behaved at infinity, then

$$\tilde{d}(k) = \sum_{j=1}^{J} \tilde{F}_j(k) + O(|k|^{-M}) \qquad \text{for} \qquad |k| \to \infty. \qquad (20.13.3)$$

This theorem is a consequence of a generalization of the Riemann–Lebesgue lemma (p. 249) which states that the Fourier transform of a distribution which is well-behaved at infinity and absolutely integrable on any finite interval tends to zero as $|k| \to \infty$ (Lighthill 1958, p. 49; Jones 1966, p. 316). The Fourier transforms of the distributions in (20.13.2) can be found from the entries in Tables 4.3 p. 87 combined with the translation rule (last entry in Table 20.3).

Example 20.13.1 We apply this theorem to the asymptotic evaluation of the Fourier transform of $d(x) = \sqrt{|x + 1|/|x - 1|}$. Near $x = 1$ we isolate the most singular part by writing identically

$$d(x) = \sqrt{\frac{2}{|x - 1|}} + \left[\sqrt{\frac{|x + 1|}{|x - 1|}} - \sqrt{\frac{2}{|x - 1|}} \right] \tag{20.13.4}$$

and note that

$$\frac{d}{dx} \left[\sqrt{\frac{|x + 1|}{|x - 1|}} - \sqrt{\frac{2}{|x - 1|}} \right] = -\frac{\text{sgn}\,(x - 1)}{\sqrt{2}|x - 1|^{3/2}} \left[\sqrt{\frac{2}{|x + 1|}} - 1 \right]$$

$$\simeq \frac{\text{sgn}\,(x - 1)}{4\sqrt{2}|x - 1|^{1/2}} \left[1 + O(x - 1)^2 \right] \tag{20.13.5}$$

which is integrable. Thus, for the singularity at $x = 1$, we have $F_1(x) = \sqrt{2/|x - 1|}$; d itself is integrable near $x = -1$ so that $F_{-1} = 0$. Thus

$$\tilde{d}(k) = \sqrt{2} e^{ik} \mathscr{F} \left\{ |x|^{-1/2} \right\} + O(k^{-1}) = \sqrt{\frac{2}{|k|}} e^{ik} + O(k^{-1}). \tag{20.13.6}$$

This estimate can be improved by adding to $F_{\pm 1}$ more terms of the Taylor series expansion of d in the neighborhood of the singularities to render integrable its derivatives of order 2 etc. With one more subtraction one finds (Jones 1966, p. 318)

$$\tilde{d}(k) = \sqrt{\frac{2}{|k|}} e^{ik} + \frac{i e^{ik} \text{sgn}\,k - e^{-ik}}{2\sqrt{2}|k|^{3/2}} + O(k^{-2}), \tag{20.13.7}$$

which shows that the error estimate in (20.13.6) turns out to be conservative in this particular example.   □

20.13.2 Algebraic equations

We have already seen in (20.4.5) that $c\delta(x)$, with c an arbitrary constant, satisfies the distributional equation

$$x\, d(x) = 0. \tag{20.13.8}$$

It can actually be proven that all solutions of this equation have the form $c\delta(x)$.[31] Similarly, the general solution of e.g. $(x - a)(x - b) d(x) = 0$ is $d(x) = c_1\delta(x - a) + c_2\delta(x - b)$. The general solution of the equation

$$x^{n+1} d(x) = 0 \tag{20.13.9}$$

is

$$d(x) = c_0\delta(x) + c_1\delta'(x) + \cdots + c_n\delta^{(n)}(x), \tag{20.13.10}$$

as can be readily verified from (20.5.7). Equation (20.13.9) implies that the support of $d(x)$ must be at the origin. It can be shown that a distribution with support at a single point a must be a linear combination of $\delta(x - a)$ and its derivatives. Furthermore, a relation of the form

$$a_0\delta(x) + a_1\delta'(x) + \cdots + a_n\delta^{(n)}(x) = 0 \tag{20.13.11}$$

can only be satisfied if $a_0 = a_1 = \cdots = a_n = 0$.

The general solution of

$$x \, d(x) = f(x) \tag{20.13.12}$$

is a distribution which acts on test functions according to

$$\langle d(x), \phi \rangle = \int_{-\infty}^{\infty} f(x) \frac{\phi(x) - \phi(0)\eta(x)}{x} \, dx + C\phi(0), \tag{20.13.13}$$

where $\eta(x)$ is a function belonging to the space $\mathscr{D}$ or $\mathscr{S}$ which is arbitrary except for being equal to 1 in a neighborhood of the origin. The integral is a particular solution of the equation and the second term the general solution of the homogeneous equation. Evidently, all solutions differ from each other by a multiple of $\delta(x)$. For example, if $f(x) = \delta(x)$, since $[\phi(x) - \eta(x)\phi(0)]/x \to \phi'(0)$, we have

$$\langle d(x), \phi \rangle = \phi'(0) + C\phi(0) \tag{20.13.14}$$

so that

$$d(x) = -\delta'(x) + C\delta(x), \tag{20.13.15}$$

as can be verified by taking $\alpha(x) = x$ in (20.5.6), or by taking the Fourier transform of the equation $x \, d(x) = \delta(x)$.

---

[31] This may be a good example to demonstrate a typical procedure of distribution theory an instance of which was already sketched on p. 583 in the context of the indefinite integral of a distribution. Equation (20.13.8) implies $\langle x \, d, \phi \rangle = 0$ which defines the action of d only on test functions which can be written as the product of x and other test functions. Evidently there are many more test functions that cannot be expressed in this way and, for such a test functions χ, (20.13.8) does not assign a value to $\langle d, \chi \rangle$. The trick is to show that any test function χ can be written in the form $\chi(x) = \chi(0)\psi(x) + x\phi(x)$ where $x\phi(x) = [\chi(x) - \chi(0)]\psi(x)$ and ψ is a test function such that $\psi(0) = 1$. This function is arbitrary but, once it has been chosen, it is not to be changed as ϕ ranges over all the functions of $\mathscr{D}$. With this decomposition, we have $\langle x \, d(x), \chi \rangle = \chi(0)\langle x \, d(x), \psi \rangle + \langle x \, d(x), \phi \rangle$ $= \chi(0)\langle x \, d(x), \psi \rangle = c\chi(0)$ where $c = \langle x \, d(x), \psi \rangle$ is an arbitrary constant.

The function η needs to be included to render the subtracted term integrable. Take, for example, $f = 1$. Then, since the integral in (20.13.13) exists in an ordinary sense, we are at liberty to calculate it however we please; let us rewrite it then as

$$\int_{|x|>\varepsilon} \frac{\phi(x)}{x}\,dx - \phi(0)\int_{|x|>\varepsilon} \frac{\eta(x)}{x}\,dx + \int_{-\varepsilon}^{\varepsilon} \frac{\phi(x)-\phi(0)\eta(x)}{x}\,dx. \qquad (20.13.16)$$

If we take the limit $\varepsilon \to 0$ the first term reproduces the pseudo-function $1/x$ defined in (20.3.10) p. 574, while the remaining terms become

$$\left[-\int_{|x|>\varepsilon} \frac{\eta(x)}{x}\,dx + \int_{-\varepsilon}^{\varepsilon} \frac{1-\eta(x)}{x}\,dx \right]\phi(0) \qquad (20.13.17)$$

which would not correspond to the addition of a finite multiple of $\delta(x)$ to $d(x)$ if the factor η were replaced by 1.

As another example we consider the equation $|x|\,d(x) = 1$ or, equivalently,

$$x\,d(x) = \operatorname{sgn} x. \qquad (20.13.18)$$

By inserting $\operatorname{sgn} x$ for f into (20.13.13) and integrating by parts, we find

$$\langle d(x), \phi \rangle = -\int_{-\infty}^{\infty} (\operatorname{sgn} x)\,\log|x|\phi'(x)\,dx + \phi(0)\int_{-\infty}^{\infty} (\operatorname{sgn} x)\,\log|x|\,\eta'(x)\,dx \qquad (20.13.19)$$

so that, with C suitably defined,

$$d(x) = \frac{d}{dx}\left[(\operatorname{sgn} x)\,\log|x|\right] + C\delta(x), \qquad (20.13.20)$$

as noted earlier in Example 20.5.5 p. 582.

In applications it is important to keep in mind the possible existence of non-zero solutions with support on a set of measure zero. We have already seen an instance in (20.10.13) in connection with the calculation of the Fourier transform of $H(x)$. As another very simple example consider

$$\frac{d^2 u}{dx^2} = 1 \qquad (20.13.21)$$

the general solution of which is $u(x) = \frac{1}{2}x^2 + c_1 x + c_2$. If we were to use the Fourier transform to solve the equation we would be led to

$$(-ik)^2 \tilde{u} = \sqrt{2\pi}\,\delta(k). \qquad (20.13.22)$$

By an argument similar to the one given above we see that a particular solution is $-\sqrt{\pi/2}\,\delta''(k)$ so that $\tilde{u} = -\sqrt{\pi/2}\,\delta''(k) + c_1\delta'(k) + c_2\delta(x)$ which, transformed back, gives the correct solution. It is only by including the solution (20.13.10) of the homogeneous equation $(-ik)^2 \tilde{u} = 0$ that we can recover the correct general solution.

20.13.3 Generalized solutions of differential equations

In considering the ordinary differential equation

$$\mathsf{L}u \equiv \sum_{k=0}^{N} a_m(x) \frac{\mathrm{d}^m u}{\mathrm{d}x^m} = f(x), \qquad a < x < b, \tag{20.13.23}$$

where the $a_m(x)$ are infinitely differentiable, we are led to three possible notions of solution:

1. If f is a continuous function and u has continuous derivatives up to order N included, then u is the classical (or strong) solution. If $a_N(x)$ is always non-zero in the domain of interest, this is the only possible solution of the equation (see e.g. Stakgold 1997, p. 172; Kanwal 2004, p. 215).

2. If u is an ordinary function, but it does not possess all the necessary derivatives, then the equation may be satisfied in the sense that

$$(\mathsf{L}^* \chi, u) = (\chi, f) \tag{20.13.24}$$

for all functions $\chi \in C_0^N[a, b]$, i.e., all functions continuous with their first N derivatives vanishing at a and b. Here L^* is the formal adjoint of L (Chapter 15) obtained by multiplying (20.13.23) by χ and integrating by parts N times. In this case u is a *weak solution*, a notion which predates the introduction of generalized functions.

3. If the equation is satisfied in a distributional sense, so that $\langle \mathsf{L}u, \phi \rangle = \langle f, \phi \rangle$ for all test functions, and u is a singular distribution, we have a *distributional solution*.

The last two possibilities extend the classical notion of solution of a differential equation. A simple example is the equation $x^k u' = 0$, the only classical solution of which is $u = c_1 =$ const. For all k this equation has the weak solution $u = c_1 + c_2 H(x)$, which equals the distributional solution for $k = 1$. For $k = 2$ the weak solution is the same, but the distributional solution becomes $u = c_1 + c_2 H(x) + c_3 \delta(x)$, with additional arbitrary multiples of derivatives of the δ for $k > 2$.[32] It will be observed that the generalized solution of the equation may depend on more arbitrary constants than one might expect on the basis of the classical theory.

These notions extend readily to partial differential equations (see e.g. Renardy & Rogers 1993). To give just one example, we consider the non-linear hyperbolic equation

$$\frac{\partial u}{\partial t} + \frac{\partial}{\partial x} q(u) = 0 \tag{20.13.25}$$

the solutions of which, for smooth u, coincide with those of

$$\frac{\partial u}{\partial t} + q'(u) \frac{\partial u}{\partial x} = 0. \tag{20.13.26}$$

---

[32] In the theory of distributions the integrations constants arise by a mechanism similar to the one described in the footnote on p. 610.

It is however well known that this latter equation can develop discontinuities, or shocks, across which the solution undergoes a finite jump (see e.g. Whitham 1974, p. 26). If there is only one discontinuity at $x = s(t)$ we write

$$u(x, t) = u_+(x, t)H(x - s(t)) + u_-(x, t)H(s(t) - x), \qquad (20.13.27)$$

decompose $q(u(x, t))$ similarly and substitute into (20.13.25). The terms not involving singular distributions just show that $u_\pm$ satisfy (20.13.25), or (20.13.26), away from the discontinuity. The terms proportional to $\delta(x - s)$, which arise upon differentiating H, give

$$\frac{ds}{dt} = \frac{[q(u_+) - q(u_-)]_{x=s}}{[u_+ - u_-]_{x=s}} \qquad (20.13.28)$$

which is the correct speed of propagation of the discontinuity.

20.13.4 Convolution and fundamental solutions

In Chapter 15 we have seen that, over $-\infty < x < \infty$, a particular solution u_p of the ordinary differential equation with constant coefficients

$$\mathsf{L}u_p \equiv \sum_{m \leq N} c_m \frac{d^m u_p}{dx^m} = f(x) \qquad (20.13.29)$$

could be written in terms of a Green's function G as

$$u_p(x) = \int G(x - y)f(y)\,dy = G * f. \qquad (20.13.30)$$

In the language of the theory of distributions, since $\delta^{(n)}(x) * u(x) = u^{(n)}(x)$ (Table 20.2), we may write the equation as

$$d * u_p = f \qquad (20.13.31)$$

if we introduce the distribution

$$d(x) = \sum_{m \leq N} c_m \frac{d^m}{dx^m} \delta(x). \qquad (20.13.32)$$

We then recover (20.13.30) by solving

$$d * G = \delta(x) \qquad (20.13.33)$$

since[33]

$$d * (G * f) = (d * G) * f = \delta * f = f. \qquad (20.13.34)$$

The same developments apply to more general equations, e.g.

1. Partial differential equations with constant coefficients, with d given by an obvious generalization of (20.13.32).

---

[33] As noted in Table 20.2, the convolution of three distributions may not be defined, or may not satisfy the associative rule used here. These steps are however legitimate in the case of the δ.

2. Linear constant-coefficients difference equations, for which

$$d(x) = \sum_m b_m \delta(x - x_m) \tag{20.13.35}$$

since $\delta(x - x_m) * u = u(x - x_m)$.

3. Integral equations of the first or second kind of the form

$$\int d(x - y) u(y) \, dy = f(x) \qquad \text{or} \qquad u(x) + \int g(x - y) u(y) \, dy = f(x) \tag{20.13.36}$$

with d a locally integrable regular distribution (i.e., a function) or $d = \delta(x) + g(x)$, with g a locally integrable function.

In all these cases, when the *fundamental solution* G of (20.13.33) is known, we can find a particular solution of the non-homogeneous equation by means of the convolution $G * f$. We do not need to worry about the uniqueness of the fundamental solution if we are only interested in a particular solution.

Example 20.13.2 [Ordinary differential equations] In §15.5 p. 388 we have studied the Green's function for the initial-value problem of a general second-order ordinary differential equation. Let us consider here the initial-value problem for an equation of the form (20.13.23) of arbitrary order with constant coefficients. We find that $G = H(x)u_c(x)$ where u_c is the classical solution of the homogeneous equation subject to $u_c(0) = u_c'(0) = \ldots = u_c^{N-1}(0) = 0, u_c^{(N)}(0) = 1/a_N$ is the required fundamental solution. Indeed, $[H(x)u_c(x)]' = H(x)u_c'(x) + \delta(x)u_c(0) = H(x)u_c'(x)$ and so on for all the first $N - 1$ derivatives, while $[H(x)u_c(x)]^{(N)} = H(x)u_c^{(N)} + \delta(x)/a_N$. Thus

$$d * [H(x)u_c(x)] = H(x) d * u_c(x) + \delta(x) = \delta(x) \tag{20.13.37}$$

since u_c is a solution of the homogeneous equation. A particular solution of the non-homogeneous equation is then given by

$$u_p(x) = \int_0^\infty u_c(\xi) \, f(x - \xi) \, d\xi , \tag{20.13.38}$$

which coincides with the first term of (15.5.7) p. 389 since, with constant coefficients, $G(x, \xi) = G(x - \xi)$. It will also be observed that, thanks to this dependence, the initial conditions on u_c mentioned before are the same ones imposed on G in (15.5.11) p. 390.

$\square$

Example 20.13.3 [Diffusion equation] The solution of

$$\frac{\partial G}{\partial t} - D\nabla^2 G = \delta(\mathbf{x}) \delta(t) \tag{20.13.39}$$

in N-dimensional space can be found by taking a Fourier transform with respect to the spatial variables:

$$\frac{\partial \tilde{G}}{\partial t} + D\mathbf{k} \cdot \mathbf{k}\tilde{G} = (2\pi)^{-N/2} \delta(t). \tag{20.13.40}$$

A particular solution of this equation is

$$\tilde{G} = (2\pi)^{-N/2} H(t) \exp(-D\mathbf{k} \cdot \mathbf{k} t), \qquad (20.13.41)$$

which can be inverted with one of the entries of table 4.4 p. 89 to find

$$G = \frac{H(t)}{(4\pi Dt)^{N/2}} \exp\left(-\frac{\mathbf{x} \cdot \mathbf{x}}{4Dt}\right). \qquad (20.13.42)$$

The convolution

$$u_p(\mathbf{x}, t) = \int_0^t d\tau \int_{-\infty}^{\infty} d\xi_1 \cdots \int_{-\infty}^{\infty} d\xi_N G(\mathbf{x} - \boldsymbol{\xi}, t - \tau) f(\boldsymbol{\xi}, \tau) \qquad (20.13.43)$$

reproduces the solutions obtained by different means in §4.4 p. 95.   □

Example 20.13.4 [Poisson equation] The solution of

$$-\nabla^2 G = \delta(\mathbf{x}) \qquad (20.13.44)$$

can be found by using the Fourier transform:

$$\mathbf{k} \cdot \mathbf{k} \tilde{G} = (2\pi)^{-N/2}. \qquad (20.13.45)$$

For $N \geq 3$ we can use one of the entries in Table 4.4 p. 89 to find

$$G = \mathscr{F}^{-1}\left\{(2\pi)^{-N/2} k^{-2}\right\} - \frac{\Gamma(N/2 - 1)}{4\pi^{N/2}} r^{2-N}. \qquad (20.13.46)$$

For $N = 3$ this reduces to $G = (4\pi r)^{-1}$. For $N = 2$ the formula is not directly applicable. A "fast and dirty" method, the results of which can be checked a posteriori, is to let $N = 2(1 + \varepsilon)$ and to take the limit $\varepsilon \to 0$. Upon so doing, and using the relation $\Gamma(\varepsilon) = \Gamma(1 + \varepsilon)/\varepsilon \simeq \varepsilon^{-1}$ (p. 442), we find

$$G \simeq \frac{1}{2\pi} \left[\frac{1}{2\varepsilon} - \log r + O(\varepsilon)\right]. \qquad (20.13.47)$$

The first term diverges in the limit and neglecting it amounts to choosing a particular regularization of the divergent integral that arises in the attempt to invert the transform; this step may be regarded as neglecting a particular solution of the homogeneous equation. Thus, we recover the known result in two dimensions, $G = (2\pi)^{-1} \log(1/r)$ which can be proven directly by integrating $\nabla^2 G$ over a small disk centered at the origin and using the divergence theorem:

$$\frac{1}{2\pi} \int_{r \leq \varepsilon} \nabla^2 \log r \, dS_\varepsilon = \frac{1}{2\pi} \oint_{r=\varepsilon} \left(\frac{\partial}{\partial r} \log r\right) d\ell_\varepsilon = 1 \qquad (20.13.48)$$

which implies (20.13.44). It is also interesting to note that, for $N = 1$, (20.13.46) gives $-\frac{1}{2}r$, which is the correct result in one dimension if r is interpreted as $|x|$.

By integrating the diffusion equation (20.13.39) over $-\infty < t < \infty$ we recover the Poisson equation. It is easy to verify that, for $N \geq 3$ (which is necessary for convergence), integration of (20.13.42) gives (20.13.46).

For the biharmonic equation $\nabla^4 G = \delta(\mathbf{x})$ we can proceed in a similar way finding e.g., for $N = 3$, $G = r/(8\pi)$. $\square$

Example 20.13.5 [Schrödinger equation] We are now interested in solving

$$i\frac{\partial G}{\partial t} + \frac{1}{2m}\nabla^2 G = \delta(\mathbf{x})\,\delta(t) \tag{20.13.49}$$

which is formally the same as the diffusion equation (20.13.39) with $D = i/(2m)$ and the right-hand side multiplied by $-i$. We cannot use the formula (20.13.42) directly as the function to be inverted now oscillates at infinity. To improve the situation, we give a small positive imaginary part to m, so that $\mathrm{Re}\,D = \mathrm{Re}\,i/[2(m+i\varepsilon)] > 0$ and, using the continuity of the Fourier transform operator, we take $\varepsilon = 0$ after the inversion. In this way we find

$$
\begin{aligned}
G &= -\frac{iH(t)}{(4\pi Dt)^{N/2}}\exp\left(-\frac{\mathbf{x}\cdot\mathbf{x}}{4Dt}\right) \\
&= -iH(t)\left(-\frac{im}{2\pi t}\right)^{N/2}\exp\left(\frac{imr^2}{2t}\right) \\
&= -iH(t)\left(\frac{m}{2\pi t}\right)^{N/2}\exp\left(i\frac{mr^2}{2t} - i\frac{\pi N}{4}\right).
\end{aligned}
\tag{20.13.50}
$$

$\square$

20.14 The Hilbert transform

The Hilbert transform of a function $f(x)$ is defined by the convolution[34]

$$\check{f}(x) \equiv \mathscr{H}\{f\}(x) = -\frac{1}{\pi}(x^{-1}) * f(x) = \frac{1}{\pi}\mathrm{PV}\int_{-\infty}^{\infty}\frac{f(\xi)}{\xi-x}\,d\xi, \tag{20.14.1}$$

provided the principal value of the integral exists (see e.g. Titchmarsh 1948, chapter 5; Bracewell 1986, p. 269; Pandey 1996).[35] The function $\check{f}(x)$ is said to be the conjugate of f.

The same formula defines the Hilbert transform of a distribution d with compact support (see e.g. Gelfand & Shilov 1964, vol. 1 section 3.8; Bremermann 1965, p. 72; Pandey 1996):

$$\mathscr{H}\{d\}(x) = -\frac{1}{\pi}(x^{-1}) * d(x). \tag{20.14.2}$$

[34] In the definition used by some authors the denominator in the integral is written as $x - \xi$.
[35] Sufficient conditions are that f and its first derivative tend to 0 faster than $1/|x|$, see Titchmarsh 1948, chapter 5. A weaker condition is that $f \in L^p(-\infty, \infty)$ with $p > 1$ and satisfies a Lipschitz condition of order $\alpha < 1$ (Titchmarsh 1948, p. 145).

As shown earlier in (20.9.7) this convolution product exists. Using an entry from Table 4.3, we find

$$\mathscr{F}\{(x^{-1}) * d(x)\} = i\pi \, (\text{sgn}\,k) \, \tilde{d}(k) \tag{20.14.3}$$

and, therefore,

$$
\begin{aligned}
(x^{-1}) * \left[(x^{-1}) * d(x) \right] &= \sqrt{2\pi}\,\mathscr{F}^{-1} \left\{ \sqrt{\pi/2}\,i \, (\text{sgn}\,k) \, \mathscr{F}\{(x^{-1}) * d(x)\} \right\} \\
&= -\pi^2 \mathscr{F}^{-1}\{\tilde{d}\} = -\pi^2 d(x). \tag{20.14.4}
\end{aligned}
$$

We have thus proven the *inversion theorem* for the Hilbert transform:

$$\mathscr{H}^{-1}\{\check{d}\} = d(x) = -\frac{1}{\pi}(x^{-1}) * (\mathscr{H}\{d\}) = -\mathscr{H}\{\mathscr{H}\{d\}\} = -\mathscr{H}\{\check{d}\} \tag{20.14.5}$$

so that

$$\mathscr{H}^{-1} = -\mathscr{H} \tag{20.14.6}$$

and the relation between a function and its Hilbert transform is skew-reciprocal; for an ordinary function, we write

$$f(x) = -\frac{1}{\pi}\text{PV} \int_{-\infty}^{\infty} \frac{\check{f}(\xi)}{\xi - x} \, \mathrm{d}\xi \,. \tag{20.14.7}$$

If the transform (20.14.2) is applied to a function $g(x)$ which is the boundary value of a function analytic in the upper half-plane, by using the representation (20.8.9) of $1/x$ we see that

$$\mathscr{H}\{g\} = -\left(\frac{1}{\pi x}\right) * g = -i\,(\delta_- - \delta_+) * g = ig \tag{20.14.8}$$

which is an important property of the Hilbert transform. An alternative derivation of this result is given on p. 111, where a connection with the Kramers–Kronig relations is established.

For functions $f, g \in L^2(-\infty, \infty)$, when the order of integration can be interchanged,[36] we have

$$
\begin{aligned}
\int_{-\infty}^{\infty} f(x)\check{g}(x)\,\mathrm{d}x &= \frac{1}{\pi} \int_{-\infty}^{\infty} \mathrm{d}x \, f(x) \int_{-\infty}^{\infty} \frac{g(\xi)}{\xi - x}\,\mathrm{d}\xi \\
&= \frac{1}{\pi} \int_{-\infty}^{\infty} \mathrm{d}\xi \, g(\xi) \int_{-\infty}^{\infty} \frac{f(x)}{\xi - x}\,\mathrm{d}x = -\int_{-\infty}^{\infty} \check{f}(x)g(x)\,\mathrm{d}x, \tag{20.14.9}
\end{aligned}
$$

i.e.

$$(f, \mathscr{H}\{g\}) = -(\mathscr{H}\{f\}, g) \tag{20.14.10}$$

so that

$$(f, g) = -(f, \mathscr{H}\{\mathscr{H}\{g\}\}) = (\mathscr{H}\{f\}, \mathscr{H}\{g\}) = \left(\check{f}, \check{g}\right) \tag{20.14.11}$$

---

[36] See p. 690; a better proof is given in Titchmarsh 1948, p. 123.

| Table 20.5 | Some Hilbert transform pairs and properties. |

$f(x) = -(1/\pi)\int_{-\infty}^{\infty}[\breve{f}(\xi)/(\xi-x)]\,d\xi$	$\breve{f}(x) = (1/\pi)\int_{-\infty}^{\infty}[f(\xi)/(\xi-x)]\,d\xi$
const.	0
$x^{-1}H(x-a)\ (a>0)$	$[1/(\pi x)]\log|a/(a-x)|$
$x^{-2}H(x-a) \quad a>0$	$\dfrac{1}{\pi x^2}\log\left|\dfrac{a}{a-x}\right| - \dfrac{1}{\pi a x} \quad x\neq 0, a$
$\cos ax\ (a>0)$	$-\sin ax$
$\sin ax\ (a>0)$	$\cos ax$
$(\sin ax)/x\ (a>0)$	$(\cos ax - 1)/x$
$1/(a^2+x^2)\ (\mathrm{Re}\,a>0)$	$-(x/a)/(a^2+x^2)$
$H(x-a)\,H(b-x)$	$(1/\pi)\log|(b-x)/(a-x)|$
$\delta(x)$	$-1/(\pi x)$
$|x|^{\nu-1}\ (0<\mathrm{Re}\,\nu<1)$	$-\cot(\pi\nu/2)(\mathrm{sgn}\,x)|x|^{\nu-1}$
$(\mathrm{sgn}\,x)|x|^{\nu-1}\ (0<\mathrm{Re}\,\nu<1)$	$\tan(\pi\nu/2)|x|^{\nu-1}$
$(\mathrm{sgn}\,x)\sin(a|x|^{1/2})\ (a>0)$	$\cos(a|x|^{1/2})+\exp(-a|x|^{1/2})$
$f(ax)$	$(\mathrm{sgn}\,a)\breve{f}(ax)$
$f(x+b)$	$\breve{f}(x+b)$
$x\,f(x)$	$x\breve{f}(x)+(1/\pi)\int_{-\infty}^{\infty}f(\xi)\,d\xi$
$f'(x)$	$\breve{f}'(x)$
$f*g$	$\mathcal{H}\{f\}*g = f*\mathcal{H}\{g\}$
$\int_{-\infty}^{\infty}f(x)g(x)\,dx = \int_{-\infty}^{\infty}\breve{f}(x)\breve{g}(x)\,dx$	
$\int_{-\infty}^{\infty}f(x)\,g(y-x)\,dx = -\int_{-\infty}^{\infty}\breve{f}(x)\,\breve{g}(y-x)\,dx$	

which is Parseval's relation for the Hilbert transform. Several other properties are collected in Table 20.5; for proofs see e.g. Titchmarsh 1948, chapter 5; Bracewell 1986, p. 269; Pandey 1996. An extended table of Hilbert transforms is given in Erdélyi *et al.* 1954, vol. 2, chapter 15.

Example 20.14.1 [A singular Fredholm integral equation] Consider the singular integral equation of the Fredholm type

$$u(x) = f(x) + \frac{\lambda}{\pi} \text{PV} \int_{-\infty}^{\infty} \frac{u(\xi)}{\xi - x} \, d\xi = f(x) + \lambda \mathcal{H}\{u\}(x) \qquad (20.14.12)$$

where $f(x)$ is given and λ is a parameter. Apply the Hilbert transform to both sides to find

$$\lambda \mathcal{H}\{\mathcal{H}\{u\}\} + \mathcal{H}\{f\} = \mathcal{H}\{u\} \qquad (20.14.13)$$

or, using the inversion theorem (20.14.5) and the original equation,

$$-\lambda u(x) + \mathcal{H}\{f\} = \frac{1}{\lambda} [u(x) - f(x)] \qquad (20.14.14)$$

so that

$$u(x) = \frac{f(x) + \lambda \mathcal{H}\{f\}(x)}{1 + \lambda^2}. \qquad (20.14.15)$$

The solution fails when $\lambda = \pm i$, which are eigenvalues of the Hilbert transform operator; indeed it can be verified that

$$\mathcal{H}\left\{(x + a \pm ib)^{-1}\right\} = \pm i(x + a \pm ib)^{-1}, \qquad b > 0. \qquad (20.14.16)$$

$\square$

21 Linear Operators in Infinite-Dimensional Spaces

Many of the examples in Part I of this book are built around the idea of solving a linear equation by expanding the solution on a basis of eigenfunctions of a suitable operator. At a superficial glance, even integral transform techniques appear to be based on the same general idea although function such as e^{ikx}, for example, cannot be regarded as a true eigenfunction of a differential operator.

Anyone familiar with the analogous method for dealing with a linear algebraic finite-dimensional system, $Au = f$, cannot but be struck by the similarity of the approaches. There is an important difference, however, and that resides in the infinite dimensionality of the space in which the solution of most differential and integral equations is to be sought. Nevertheless, the finite-dimensional ideas do carry over to some extent to the infinite-dimensional case, thus providing a great transparency to the "machinery" behind the mathematical manipulations and a powerful aid to intuition. In this chapter we consider similarity and differences, and we explore to what extent ideas familiar from the finite-dimensional realm carry over to bigger spaces. Two notions arise repeatedly in this chapter:[1]

- A *closed set* is a set which contains all its limit points;
- A normed space, or subspace, is *complete* when every Cauchy sequence (definition p. 545) has a limit in that space.

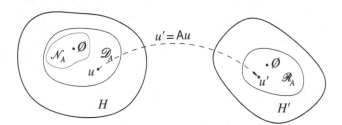

Fig. 21.1 An operator A maps each element u of a set $\mathscr{D}_A \subseteq H$, the domain of A, into an element $u' = Au$ of a set $\mathscr{R}_A \subseteq H'$, the range of A. All the elements of $\mathscr{D}_A$ mapped into the Ø element of H' form the null space $\mathscr{N}_A$ of A.

---

[1] See §A.1 for further information on these concepts.

21.1 Linear operators

An *operator* A from a linear vector space H to another, not necessarily distinct, linear vector space H' is a rule which associates to each element of a certain set $\mathcal{D}_A \subseteq H$ of elements of H, called the *domain* of A, a *unique* element u' of H'; we write $u' = Au$ and call u' the *image* of u under A (Figure 21.1). The rule that defines the action of the operator on the elements of its domain (e.g., differentiation) is referred to as the *formal operator*. The domain is considered an integral part of the definition of the operator: two operators are equal not only when the corresponding formal operators are equal, but also their domains. The operator is *linear* if H and H' are linear vector spaces over the same scalar field of real or complex numbers and the rule that defines the action of A is such that

$$A(a \cdot u + b \cdot v) = a \cdot Au + b \cdot Av \qquad (21.1.1)$$

for all elements $u, v \in \mathcal{D}_A$ and all scalars (i.e., numbers) a, b. In this chapter we deal exclusively with linear operators and the adjective "linear" will mostly be omitted.

In stating the results that follow we have in mind normed (Banach) spaces or scalar-product (Hilbert) spaces, even though in some cases extensions to more general spaces are possible. Since a Hilbert space equipped with its natural norm is also a Banach space, results stated for Banach spaces are also valid for Hilbert spaces. We generically use the symbol H to denote Banach or Hilbert spaces without further specification when the context is such that no ambiguity can arise. For brevity, we also often write "linear operator" in place of "linear operator on a Banach space" or "linear operator on a Hilbert space".

The set of all u' for which there is a $u \in \mathcal{D}_A$ such that $u' = Au$ is called the *range* of A and is denoted by $\mathcal{R}_A$. Just as in the case of linear functionals in §19.6, the linearity property implies that $\emptyset \in \mathcal{D}_A$ and that $A\emptyset = \emptyset$ so that also $\emptyset \in \mathcal{R}_A$. Thus, the definition (21.1.1) shows that $\mathcal{D}_A$ and $\mathcal{R}_A$ are linear manifolds (§19.1) of H and H' respectively. The dimension of $\mathcal{R}_A$ is defined as the *rank* of the operator A. When H' is a Hilbert space, we define the *deficiency* def A of A as the co-dimension of $\mathcal{R}_A$, namely the dimension of $\mathcal{R}_A^{\perp}$, the orthogonal complement of $\mathcal{R}_A$ (§19.4). Thus, def A characterizes the extent to which A "fails" to fill out H' with its action on the elements of H. We will see on pp. 630 and 655 that, for closed (and, in particular, bounded) operators $\mathcal{R}_A^{\perp}$ coincides with the null space of the adjoint operator. In these cases, therefore, def A may be thought of as the dimension of the null space of the adjoint.

Besides $\emptyset$, there may be other elements of $\mathcal{D}_A$ which are mapped into the $\emptyset$ element of H'. The set of all such elements is the *null space* of A and is denoted by $\mathcal{N}_A$; the dimension of $\mathcal{N}_A$ is the *nullity* nul A of A.[2]

The *index* of A is defined as

$$\text{ind A} = \text{def A} - \text{nul A}. \qquad (21.1.2)$$

The index of the operators with which we will be chiefly concerned is 0.

---

[2] According to the convention on p. 541 $\mathcal{N}_A$ is a true subspace, rather than simply a linear manifold, only when it is closed, which is the case e.g. for closed (and, in particular, bounded) operators.

The *graph* $\Gamma(A)$ of an operator A is constituted by all the ordered pairs $\langle u, u' \rangle$, with $u' = Au$, considered as elements of the direct sum space $H \oplus H'$ (p. 541). If the sum space is endowed with a Hilbert space structure, u is the orthogonal projection of $\langle u, u' \rangle$ on H and u' the orthogonal projection on H' (p. 555).

Example 21.1.1 [Zero operator] The zero operator is uniquely defined as $0u = \emptyset$ for all u. Evidently $\mathscr{D}_0 = \mathscr{N}_0 = H, \mathscr{R}_0 = \{\emptyset\}$. ☐

Example 21.1.2 [Identity operator] The identity operator I, uniquely defined by $Iu = u$, maps H onto itself. Obviously $\mathscr{D}_I = \mathscr{R}_I = H$ and $\mathscr{N}_I = \{\emptyset\}$. Since the range of I is H itself, the rank of the identity operator equals the dimension of H. ☐

Example 21.1.3 [Projection operator] Let H_1 be a (closed) subspace of H and let $v = Pu$ be the orthogonal projection of u on H_1 (§19.4 p. 554). In this case $\mathscr{D}_P = H, \mathscr{R}_P = H_1$, $\mathscr{N}_P = H_1^\perp$, where $H_1^\perp$ is the orthogonal complement of H_1, i.e., the set of all elements of H orthogonal to all elements of H_1. The rank of P is the dimension of H_1 and can therefore be finite or infinite. We note also that, as is intuitively clear, $P(Pu) = Pu$, which we can express by writing $P^2 = P$, an equality which defines *idempotent operators*. ☐

Example 21.1.4 [Linear functionals.] The linear functionals discussed in §19.6 are linear operators with range in the space of real or complex numbers. ☐

Example 21.1.5 [Integral operator.] Let $k(x, \xi)$ be square integrable in the rectangle $[a, b] \times [a, b]$. Then the integral operator

$$v(x) \equiv (Ku)(x) = \int_a^b k(x, \xi)\, u(\xi)\, d\xi \qquad (21.1.3)$$

maps functions $u \in L^2(a, b)$ into other functions $v \in L^2(a, b)$ (see Example 21.2.1 p. 625); k is called the *kernel* of the integral operator. Depending on k, an integral operator may have a finite or infinite rank. For example, if $k(x, \xi) = \sum_{j=1}^{N} h_j(x)k_j(\xi)$, the range is the N-dimensional subspace spanned by the N functions $h_j(x)$ if they are linearly independent and the rank is then also equal to N; dimension and rank will be smaller than N if the h_j are not linearly independent. ☐

Example 21.1.6 [Differentiation operator] The operator $D = d/dx$ can be considered as mapping elements of $u(x) \in \mathscr{D}_D = C^1[a, b]$ to elements $u'(x) = du/dx \in \mathscr{R}_D = C^0[a, b]$. The null space $\mathscr{N}_D$ is the set of constant functions. With the L^2 norm, both $\mathscr{D}_D$ and $\mathscr{R}_D$ can be considered as subspaces of $L^2(a, b)$, so that in this case $H =$

$H' = L^2(a, b)$. With the maximum norm, C^1 would be a subspace of C^0 so that again $H = H' = C^0[a, b]$, or C^1 and C^0 may be considered as distinct (normed) spaces. Since $C^0[a, b]$ is infinite-dimensional, the rank is infinite. ☐

The sum of two linear operators A and B and the product of a linear operator A by a number a are linear operators defined, respectively, as

$$(A + B)u = Au + Bu, \qquad (a\,A)u = a\,(Au). \qquad (21.1.4)$$

The domain of the sum is evidently the intersection $\mathscr{D}_A \cap \mathscr{D}_B$ of the two domains (p. 679), namely the set of elements belonging to both $\mathscr{D}_A$ and $\mathscr{D}_B$, while $\mathscr{D}_{aA} = \mathscr{D}_A$.

Definition 21.1.1 An operator A' is an *extension* of an operator A, which is expressed by writing $A \subset A'$, if $\mathscr{D}_A \subset \mathscr{D}_{A'}$, and $A'u = Au$ for all $u \in \mathscr{D}_A$.

Essentially, therefore, an extension of a given operator amounts to the definition of a new, broader domain for the same formal operator.

The *product* BA of the operators B and A is defined as $(BA)u = B(Au)$. When the domain and the range of both operators are in the same space, the domain of BA is generally smaller than $\mathscr{D}_A$ as it consists only of those $u \in \mathscr{D}_A$ such that $Au \in \mathscr{D}_B$. The product AB is different from BA unless the operators *commute*.[3] In a similar way one can define A^2 and higher powers.

Example 21.1.7 For the operator $D^2 = d^2/dx^2$, $\mathscr{D}_{D^2} = C^2[a, b] \subset C^1[a, b] = \mathscr{D}_D$, and so on as the order of differentiation increases. In order to avoid this progressively shrinking domain, it would be more convenient to define the differentiation operator on the space $C^\infty[a, b]$ of infinitely differentiable functions. The problem that one encounters is that infinitely differentiable functions are a relatively "small" set. Distributions (Chapter 20) offer the possibility of considering the differentiation operator as a continuous operator over a much larger space where it can be applied any number of times. ☐

21.2 Bounded operators

We have discussed bounded linear functionals in §19.6. Many of the notions covered there transfer directly to the case of linear operators.

---

[3] Operators cannot be considered separately from their domain. By saying that two operators commute to express the fact that AB = BA, we imply that AB and BA have a common domain on which they are equal. A weaker notion is that of *permuting operators*: B permutes with A (in this order) if and only if $BA \subset AB$, i.e., AB is an extension of BA. This terminology is not universally adhered to; some authors use "commute" in the sense of "permute".

Table 21.1	Some properties of bounded operators.	
Property		**Theorem/equation no./page**
norm $\|B\| = \sup_{\|e\|=1} \|Be\|$		Eq. (21.2.2)
B continuous at any u $\Rightarrow$ continuous everywhere		Th. 21.2.1
B bounded $\Leftrightarrow$ B continuous		Th. 19.6.2
$\mathscr{D}_B = H$		Th. 21.2.3
Null space $\mathscr{N}_B$ is closed		p. 626
$\|B_1 + B_2\| \le \|B_1\| + \|B_2\|$		
$\|aB\| = |a|\,\|B\|$		
$\|B_1 B_2\| \le \|B_1\|\,\|B_2\|$		Eq. (21.2.7)
B^{-1} bounded when it exists		
$B_n \to B$ uniformly $\Leftrightarrow \|B - B_n\| \to 0$		
$B_n \to B$ strongly $\Leftrightarrow \|(B - B_n)u\| \to 0$ for all $u \in H$		
uniform convergence $\Rightarrow$ strong convergence		
If $\sum_{n=1}^{\infty} |c_n|\,\|B\|^n < \infty$, then $S = \sum_{n=1}^{\infty} c_n B^n$ exists		
$(\mu I - B)^{-1} = (1/\mu)\sum_{j=0}^{\infty}(B/\mu)^j$ if $\|B\| < \mu$		
$(B_1 B_2)^\dagger = B_2^\dagger B_1^\dagger$		Th. 21.2.1

A linear operator B on a normed space H is bounded if, for all $u \in \mathscr{D}_B$, there is a constant c such that

$$\|Bu\| \le c\,\|u\|, \tag{21.2.1}$$

where the norm on the right is in the original space H and that on the left in the image space H'. All $M \times N$ matrices are bounded operators from the N-dimensional spaces $\mathbb{R}^N$ or $\mathbb{C}^N$ to the M-dimensional spaces $\mathbb{R}^M$ or $\mathbb{C}^M$. Much of the material covered in Chapter 18 therefore can be used to illustrate the more general ideas introduced here and will not be repeated.

The smallest c for which the inequality (21.2.1) is satisfied is designated as the *norm* of the operator. This is a true norm in the linear space consisting of all bounded operators as will be seen shortly. Similarly to the characterization (19.6.5) of the norm of linear functionals, the norm $\|B\|$ can equivalently be defined as

$$\|B\| = \sup_{\|e\|=1} \|Be\| \tag{21.2.2}$$

namely as the supremum over all the unit vectors e of the vector norm of Be. The dependence of the norm assigned to an operator upon the norms of the spaces in which both its range and its domain lie is evident from the definition; for this reason, this norm is sometimes called the *induced norm*. Just as in the case of a matrix, the norm of an operator may be thought of as a measure of the distortion, amplification, or compression that the elements of its domain undergo.

For the zero operator, $\|0\| = 0$ while, for the identity operator, $\|I\| = 1$. For a projection operator on the subspace H_1, evidently $\|Pu\| \le \|u\|$, with $Pu = u$ for all $u \in H_1$; thus, $\|P\| = 1$.

Example 21.2.1 By applying the Schwarz inequality (19.3.6) p. 549 to the integral operator (21.1.3) considered in $L^2(a, b)$ we have

$$\left| \int_a^b k(x, \xi)\, u(\xi)\, d\xi \right|^2 \le \|u\|^2 \int_a^b |k(x, \xi)|^2\, d\xi \tag{21.2.3}$$

and, therefore,

$$\|v\|^2 = \int_a^b |v|^2\, dx \le \|u\|^2 \int_a^b dx \int_a^b |k(x, \xi)|^2\, d\xi \tag{21.2.4}$$

which shows that

$$\|K\|^2 \le \int_a^b dx \int_a^b |k(x, \xi)|^2\, d\xi. \tag{21.2.5}$$

Integral operators such that $\|K\| < \infty$ are called *Hilbert–Schmidt* integral operators. Not all bounded integral operators have a square-integrable kernel; an example is the Fourier transform over the real line, which is a unitary operator considered in §21.4 □

A linear operator is strongly *continuous* at u if, for any sequence $\{w_n\}$ such that $\|u - w_n\| \to 0$, $\|Bu - Bw_n\| \to 0$. The following two theorems are analogous to the corresponding ones for linear functionals, Theorems 19.6.1 and 19.6.2 respectively (p. 564):

Theorem 21.2.1 *A linear operator strongly continuous at a point is continuous everywhere in its domain of definition.*

Theorem 21.2.2 *For a linear operator boundedness and strong continuity imply each other or, in other words, a linear operator is strongly continuous (respectively, bounded) if and only if it is bounded (respectively, strongly continuous).*

For this reason we use the denominations bounded operator and continuous operator somewhat interchangeably.

Example 21.2.2 Let us consider the differentiation operator D with domain $C^1[0, 1] \subset L^2(0, 1)$ and let us apply it to the family of functions $u_n(x) = \sqrt{2}\, \sin n\pi x$. All these functions have norm 1 but $\|Du_n\| = n\pi$ grows without bound. Thus, we conclude that the differentiation operator is *not bounded* with the L^2 norm and, therefore, it is not continuous.

If the same operator is considered in $C^1[0, 1]$ equipped with the norm

$$\|u\|_1 = \sup_{0 \le x \le 1} \left(|u| + \left| \frac{du}{dx} \right| \right) \tag{21.2.6}$$

while the range $C^0[0, 1]$ is equipped with the maximum norm, the operator becomes bounded as $\sup_x |du/dx| \leq \sup_x (|u| + |du/dx|)$ so that $\|D\| \leq 1$. In order to achieve boundedness for this operator in this way, we have given up the idea of scalar product and the Hilbert space structure of its domain.   □

It is a simple consequence of continuity that the null space of a bounded operator on H is a closed subspace of H.

Theorem 21.2.3 [Hahn–Banach] *The domain of a bounded linear operator* B *with range in a Banach space can be extended to the entire Banach space* H *without changing the norm of the operator.*

This is a deep theorem with many ramifications (see e.g. Taylor & Lay 1980, pp. 125, 134; Kubrusly 2001, p. 260, and many others).[4] For our limited purposes its importance lies in the fact that the domain of a bounded operator may be considered to be the entire Banach space.

The set of bounded linear operators on H all having their range in the same space H' can be turned into a linear space by defining the addition of two operators and the multiplication by a scalar as in (21.1.4) of the previous section. It is easy to show that (21.2.2) defines a proper norm in this space, i.e., that the axioms of p. 542 defining a norm are satisfied.

To show that this space is a Banach space we need to show that it is complete, i.e., that any Cauchy sequence of operators is convergent to an operator. Before addressing this aspect, it is necessary to explain various modes of convergence of operator sequences.

In general, in the case of sequences $\{B_n\}$ of bounded operators we distinguish three types of convergence:

1. *Uniform convergence*: $\lim_{n \to \infty} \|B - B_n\| \to 0$; this is also called "convergence in the norm". The denomination "uniform convergence" is justified by the fact that, in this case, $\|B_n u - Bu\|$ can be bounded by an arbitrarily small $\varepsilon > 0$ independent of u for all u in a bounded set.
2. *Strong convergence*: $\lim_{n \to \infty} \|Bu - B_n u\| \to 0$ for all $u \in H$; this is a direct generalization of the point-wise convergence of a sequence of functions as the sequence $B_n u$ converges to Bu at each "point" u.
3. *Weak convergence*: $\lim_{n \to \infty}[(v, B_n u) - (v, Bu)] \to 0$ for all $u, v \in H$.

By the Cauchy–Schwarz inequality we have $|(v, (B_n - B)u)| \leq \|v\|\|(B_n - B)u\|$ and, therefore, strong convergence implies weak convergence to the same limit, but not conversely.

Since $\|(B - B_n)u\| \leq \|B - B_n\| \|u\|$, a uniformly convergent sequence is also strongly convergent to the same limit. As for the convergent proposition, in general it may well happen that the sequence $\{B_n\}$ corresponding to a strongly convergent sequence $\{Bu_n\}$ is not even a Cauchy sequence in the operator norm, and therefore it cannot converge uniformly.

---

[4] In a sense, the theorem states that the action of a linear bounded operator over the entire space is essentially determined by its action over a "relatively small" set of elements of the space.

However, if $\{B_n\}$ *is* a Cauchy sequence in the operator norm, then strong convergence of $\{Bu_n\}$ does imply uniform convergence of $\{B_n\}$. Indeed, in this case $\|B_n - B_m\|$ can be made smaller than any ε by taking n, m sufficiently large. Thus, by continuity of the (vector) norm, from $\|(B_n - B_m)u\| \leq \varepsilon\|u\|$ we also have $\|(B_n - B)u\| \leq \varepsilon\|u\|$. Since this relation is valid for any u and therefore, in particular, for $\sup_{u \neq \emptyset} \|(B_n - B)(u/\|u\|)\|$, we conclude that $\|B_n - B\| \leq \varepsilon$, i.e., B is not only the strong, but also the uniform limit of $\{B_n\}$. Thus, we conclude that a sequence $\{B_n\}$ of bounded operators is uniformly convergent if and only if it is a strongly convergent Cauchy sequence.

We can now prove the completeness of the space of bounded linear operators on H with range in a complete space H'. If $\{B_n\}$ is a Cauchy sequence in the operator norm, for any $u \in H$, $\{B_n u\}$ is also a Cauchy sequence as $\|(B_n - B_m)u\| \leq \|B_n - B_m\|\|u\|$. Since H' is complete, we are assured that this sequence $\{B_n u\}$ will be convergent to some $u' \in H'$ so that, for each $u \in H$, there is a well-defined $u' \in H'$. This correspondence defines an operator B from H to H' which is the strong limit of the sequence B_n. It is not difficult to show that this limit operator is linear, bounded and unique. By the previous remark, we know that the strong convergence of $\{B_n u\}$ implies the uniform convergence of $\{B_n\}$. We thus conclude that the space of bounded linear operators defined in H with range in a complete Banach space H' is complete and, therefore, itself a Banach space.

We discuss more generally the inverse operator in §21.5, but we note here

Theorem 21.2.4 [Banach's theorem on the inverse] *When it exists, the inverse operator* B^{-1} *of a bounded operator B is bounded.*

We will see later (§21.7) the importance of the closure of the range of an operator. Thus we take note of the fact that

Theorem 21.2.5 *The range of an invertible bounded operator is closed.*

This result is true because a Cauchy sequence in $\mathscr{R}_B$ can be made to correspond to a Cauchy sequence in $\mathscr{D}_B$ through the inverse operator. But since this latter sequence converges, its image must also converge by continuity of the operator, which makes $\mathscr{R}_B$ closed.

Since the domain of bounded operators mapping a Banach space into itself is the entire Banach space, their product is always well defined. An easy deduction is that, if B_1 and B_2 are bounded, $\|(B_1 B_2)u\| \leq \|B_1\|\|B_2 u\| \leq \|B_1\|\|B_2\|\|u\|$ and, therefore,

$$\|B_1 B_2\| \leq \|B_1\|\|B_2\|. \tag{21.2.7}$$

In particular, $\|B^2\| \leq \|B\|^2$. This relation holds with the equal sign, among others, for projection operators for which, as noted before (Example 21.1.3), $P^2 = P$.

In addition to sequences of operators, we can also define *series* of operators by considering the sequence of their partial sums. In particular, for a bounded operator B from H into H, we can formally define a *power series*

$$S = \sum_{n=1}^{\infty} c_n B^n, \tag{21.2.8}$$

Table 21.2	Some properties of self-adjoint bounded operators.	
Property		Theorem/equation no.
adjoint operator $B^\dagger$: $(v, Bu) = (B^\dagger v, u)$ for all u, v		
$\|B\| = \|B^\dagger\|$		Th. 21.2.6
$\|B\| = \sup_{\|u\|=1} |(u, Bu)|$		Eq. (21.2.20)
if $Bu = \lambda u$, $\lambda = \bar{\lambda}$		Th. 21.2.8
$Bu = \lambda u$, $Bu' = \lambda' u'$, $\Rightarrow (u, u') = 0$ if $\lambda \neq \lambda'$		Th. 21.2.8
If $Bu = \lambda u$, $|\lambda| \leq \|B\|$		Th. 21.2.9

in which $\{c_n\}$ is a numerical sequence. A sufficient criterion for uniform convergence of the series is that the numerical series $\sum_{n=1}^{\infty} |c_n| \|B\|^n < \infty$ because, in this case,

$$\| \sum_{k=1}^{n} c_k B^k - \sum_{k=1}^{m} c_k B^k \| = \| \sum_{k=m+1}^{n} c_k B^k \| \leq \sum_{k=m+1}^{n} |c_k| \|B^k\| \leq \sum_{k=m+1}^{n} |c_k| \|B\|^k.$$

(21.2.9)

The last term can be made as small as desired by the assumed convergence of the numerical series which shows that the sequence of partial sums of the operator series is a Cauchy sequence.

A special class of bounded operators are *isometric operators* for which

$$(Uu, Uv) = (u, v) \tag{21.2.10}$$

for all u and v. From this relation $\|Uu\| = \|u\|$ for all u so that $\|U\| = 1$ and an isometric operator is bounded. More on this class of operators will be said in §21.5 and §21.9.

21.2.1 The adjoint of a bounded operator

Let now H and H' be Hilbert spaces and consider, for fixed $v \in H'$, the linear functional $\ell(u) = (v, Bu)$. If B is bounded, the functional is also bounded and, by the Riesz representation theorem 19.6.3 p. 566, there must be an element $w \in H$ such that $\ell(u) = (w, u) = (v, Bu)$. This argument shows that, for every $v \in H'$, one can identify a corresponding element $w \in H$. The linearity of this correspondence is obvious. Uniqueness is also easy to prove. Thus, we have a linear bounded operator, which we designate by B^*, with domain in H' and range in H, for which $w = B^* v$ i.e.

$$(v, Bu) = (B^* v, u). \tag{21.2.11}$$

This operator is the uniquely defined *adjoint* of B.[5] Note that the scalar product in the left-hand side is in the space H', while that on the right-hand side is in the space H.

Clearly, the notion of adjoint matrix and adjoint operator defined in §18.5 for finite-dimensional spaces precisely conform to the definition (21.2.11). In Chapter 15 we have referred to the adjoint of a differential operator, which is a closely related concept although different in detail as shown below in §21.6.1.

Example 21.2.3 For a Hilbert–Schmidt integral operator, defined in (21.1.3), we have

$$(v, \mathsf{K}u) = \int_a^b \mathrm{d}x \, \overline{v}(x) \int_a^b k(x, \xi) \, u(\xi) \, \mathrm{d}\xi. \qquad (21.2.12)$$

Since the double integral exists for a Hilbert–Schmidt kernel, we can interchange the order of integration (Fubini's theorem, p. 690). In so doing we also set $x' = \xi, \xi' = x$ to distinguish the variable used for the scalar product, x', from ξ' used for the action of the operator. Then we have

$$(v, \mathsf{K}u) = \int_a^b \mathrm{d}\xi' \overline{\left[\int_a^b \mathrm{d}x' \, \overline{k}(\xi', x') \, v(\xi') \right]} u(x) \, \mathrm{d}x = \left(\mathsf{K}^* v, u \right) \qquad (21.2.13)$$

which shows that

$$\left(\mathsf{K}^* v \right)(x) = \int_a^b \overline{k}(\xi, x) \, v(\xi') \, \mathrm{d}\xi. \qquad (21.2.14)$$

Similarly to the case of matrices (p. 493), the kernel of the adjoint operator is obtained by switching the variables and taking the complex conjugate.

As a specific example, consider $k(x, \xi) = H(x - \xi)$, with H the Heaviside step function, in $0 \le x, \xi \le 1$. Then K^* has kernel $H(\xi - x)$ and

$$\mathsf{K}u = \int_0^x u(\xi) \, \mathrm{d}\xi, \qquad \mathsf{K}^* v = \int_x^1 v(\xi) \, \mathrm{d}\xi. \qquad (21.2.15)$$

as is easily confirmed directly by interchanging the order of integration in

$$(v, \mathsf{K}u) = \int_0^1 \overline{v}(x) \int_0^x u(\xi) \, \mathrm{d}\xi. \qquad (21.2.16)$$

$\square$

One can readily prove that

Theorem 21.2.6 *A bounded operator* B *and its adjoint* B* *have the same norm,* $\|\mathsf{B}\| = \|\mathsf{B}^*\|$ *and, furthermore,* $\mathsf{B}^{**} \equiv (\mathsf{B}^*)^* = \mathsf{B}$.

---

[5] The adjoint can also be defined in a Banach space without a scalar product structure. Consider a linear functional ℓ_g in H'; this is an element of H'^*, the dual space of H'. As u ranges over H, $\ell_g(\mathsf{B}u)$ defines a linear functional $\ell_f(u)$ on H, which is therefore an element of H^*, the dual space of H. Thus we have a linear correspondence between the dual spaces H^* and H'^* which defines the Banach adjoint, also called the conjugate or dual operator.

As in the finite-dimensional case, the adjoint of a product is the product of the adjoints in reverse order, $(AB)^* = B^*A^*$.

The following theorem is a restatement of Theorem 18.5.1 p. 507 for operators in finite-dimensional spaces:

Theorem 21.2.7 *If an operator* B *maps all the elements of a subspace H_1 of H into elements of the same subspace (which we express by saying that H_1 is invariant with respect to* B*), the orthogonal complement $H_1^\perp$ of H_1 in H is invariant with respect to* B*.*

The analog of Theorem 18.5.2 p. 507 is also valid for bounded operators, so that

$$\mathscr{R}_A^\perp = \mathscr{N}_{A^*}, \qquad \mathscr{R}_{A^*}^\perp = \mathscr{N}_A. \tag{21.2.17}$$

The proof is the same as given later for closed operators on p. 655. Since the orthogonal complement of the orthogonal complement of a set is the closure of the set (p. 681), from these relations we have

$$\overline{\mathscr{R}}_A = \mathscr{N}_{A^*}^\perp, \qquad \overline{\mathscr{R}}_{A^*} = \mathscr{N}_A^\perp. \tag{21.2.18}$$

In dealing with the finite-dimensional case we have seen the importance of normal operators (§18.7). We define this concept in the same way for operators in an infinite-dimensional space: an operator N is *normal* when it commutes with its adjoint, $NN^* = N^*N$. More on these operators will be said in §21.6.2 and §21.9.

21.2.2 Self-adjoint bounded operators

A bounded operator such that $B = B^*$ is called *Hermitian* or *self-adjoint* and, in many respects, behaves like a Hermitian matrix. In fact, many of the results to be described here have their analogue in §18.7 and this similarity with the finite-dimensional case makes such operators particularly simple.

Theorem 21.2.8 *The eigenvalues of a self-adjoint operator are real and eigenvectors corresponding to two different eigenvalues are orthogonal.*

Proof Let u_1 and u_2 be the eigenvectors corresponding to the two eigenvalues λ_1 and λ_2 respectively. Take the scalar product from the left of the j-th eigenvalue equation by u_k ($j, k = 1, 2$), the scalar product of the k-th eigenvalue from the right by u_j, and subtract:

$$(u_k, Bu_j) - (Bu_k, u_j) = (u_k, Bu_j) - (u_k, Bu_j) = 0 = (\lambda_j - \overline{\lambda}_k)(u_k, u_j). \tag{21.2.19}$$

If $j = k$, this relation proves that the eigenvalues are real while, if $j \neq k$, it proves the eigenvectors to be orthogonal. $\qquad\square$

The set of eigenvectors corresponding to the same eigenvalue λ (plus the $\emptyset$ vector which, by definition, is not an eigenvector), constitutes a linear space, the *eigenspace* of the eigenvalue λ. The dimension of the eigenspace is the number of linearly independent eigenvectors and is the *geometric multiplicity* of the eigenvalue. If the multiplicity is 1, the eigenvalue is *simple* while, if it is greater than 1, the eigenvalue is *multiple* or *degenerate*. In this latter case, by using the Gram–Schmidt orthogonalization procedure, one can generate a set of orthogonal vectors in the subspace corresponding to each eigenvalue. Thus, without loss of generality, we can assume that all the eigenvectors of a Hermitian operator are orthogonal to each other.

As in the case of matrices (p. 525), for a general operator A (not necessarily bounded) it is possible that, while $(A - \lambda I)v \neq 0$, $(A - \lambda I)^k v = 0$ for some integer k. All such vectors are generalized eigenvectors corresponding to the eigenvalue λ and their set (over all k) constitutes the algebraic eigenspace corresponding to λ. The dimension of this eigenspace is the *algebraic multiplicity* of the eigenvalue and it is always greater than, or equal to, the geometric multiplicity.

For a self-adjoint operator $(u, Bu) = (Bu, u)$ and, therefore, (u, Bu) is a real number. Similarly to (18.8.9) for the finite-dimensional case, we can say that

Theorem 21.2.9 *The norm* $\|B\|$ *of a linear, bounded self-adjoint operator* B *satisfies the relation*

$$\|B\| = \sup_{\|e\|=1} |(e, Be)| \qquad (21.2.20)$$

and, as a consequence, the modulus of all the eigenvalues of a bounded operator B *is bounded above by the norm of* B, $|\lambda| \leq \|B\|$.

For the first part of the theorem see e.g. Griffel 2002 p. 224. The second part is straightforward as, if $\lambda u = Bu$, then $|\lambda| \leq \|Bu\|/\|u\| \leq \|B\|$. More specifically, all the eigenvalues of a self-adjoint bounded operator satisfy

$$\inf_{\|v\|=1} (v, Bv) \leq \lambda \leq \sup_{\|v\|=1} (v, Bv). \qquad (21.2.21)$$

A natural question is whether there are eigenvalues which actually equal the bounds or, in other words, whether the *sup*, which denotes the supremum, or least upper bound, can be changed to *max*, as in the corresponding theorem for the finite-dimensional case, and similarly for the *inf*, the greatest lower bound. In general the answer is no, although this property is guaranteed for compact operators as will be seen in the course of the proof of Theorem 21.3.4. Even if they are not necessarily eigenvalues, and therefore they do not belong to the point spectrum, the bounds in (21.2.21) do belong to other parts of the spectrum (see §21.9).

Since (u, Bu) is real for a self-adjoint operator, we can define a positive, or non-negative, operator as in the case of matrices as an operator having the property that, for all u, $(u, Bu) > 0$ or $(u, Bu) \geq 0$, respectively.[6] For example, B^*B and BB^* are non-negative. With this notion we can introduce a partial ordering in the space of self-adjoint operators by saying

---

[6] Sometimes operators for which $(u, Bu) \geq 0$ are just termed *positive*, dispensing with the notion of non-negative operators.

that $B_2 < B_1$ when $B_1 - B_2$ is positive. For non-negative operators the following analog of the Cauchy–Schwarz inequality holds[7]

$$\|(u, Av)\|^2 \leq (u, Au)(v, Av).$$ (21.2.22)

Example 21.2.4 [Projection operators] If P is the orthogonal projection on a subspace $H_1 \subset H$ and $Pu = u_1$, $Pv = v_1$, we have $(v, Pu) = (v, u_1) = (v_1, u_1) = (Pv, u_1) = (Pv, u)$ so that P is self-adjoint (note that orthogonality is essential in this argument). Furthermore, since P is idempotent so that $P^2 = P$, $(u, Pu) = (u, P^2u) = (Pu, Pu) = \|Pu\|^2$ and P is non-negative definite. ☐

21.2.3 Contraction mapping

Iteration is a standard procedure by which one may try to solve an equation of the form

$$Bu = u,$$ (21.2.23)

in which B is an operator mapping a Banach space H into itself. The idea is to start from a first guess u_0 and generate a sequence $u_{n+1} = Bu_n$ which will hopefully converge to the solution of (21.2.23). It is easy to construct examples where the method is successful and counter examples where it is not.

A bounded operator for which

$$\|B\| \leq 1$$ (21.2.24)

is called a *contraction operator*. If the relation holds with a strict inequality, B is a *strict contraction* operator. A bounded operator B defines a *contraction mapping* if its norm is strictly less than 1:

$$\|B(u - v)\| \leq \|B\| \|u - v\|, \qquad \|B\| < 1.$$ (21.2.25)

Theorem 21.2.10 *Every linear contraction mapping from a complete normed space H into itself admits one and only one fixed point.*

Proof Starting from an arbitrary u_0, define $u_1 = Bu_0$, $u_2 = Bu_1 = B^2u_0$ and so on. If $m < n$, evidently $u_n - u_m = B^m(u_{n-m} - u_0)$ so that $\|u_n - u_m\| \leq \|B\|^m \|(u_{n-m} - u_{n-m-1}) + (u_{n-m-1} - \cdots + (u_1 - u_0)\|$ from which $\|u_n - u_m\| \leq \|B\|^m (\|u_{n-m} - u_{n-m-1}\| + \|u_{n-m-1} - u_{n-m-2}\| + \cdots + \|u_1 - u_0\|)$. However $\|u_{k+1} - u_k\| \leq \|B\| \|u_k - u_{k-1}\|$ and, therefore, $\|u_n - u_m\| \leq \|B\|^m (\|B\|^{n-m} + \|B\|^{n-m-1} + \cdots + 1) \|u_1 - u_0\| \leq \|B\|^m \|u_1 - u_0\|/(1 - \|B\|)$ which tends to 0 as $m \to \infty$. Thus, the sequence $\{u_n\}$ is a Cauchy sequence which admits a limit point since the space is complete. By appealing once again to the boundedness of B it is easy to show that this limit point is unique. ☐

---

[7] This is most easily seen by noting that, for a non-negative operator, (u, Av) has all the properties of a scalar product except positivity, which is not necessary to prove the Cauchy–Schwarz inequality (p. 549).

21.2.4 The Neumann series

Let B be a continuous bounded linear operator from a Banach (or Hilbert) space H into itself. We are interested in solving the equation

$$(\mathsf{B} - \lambda \mathsf{I})\, u = f, \tag{21.2.26}$$

in which $f \in H$ is given and λ is a given complex number. To this end, let us consider the power series (assumed uniformly convergent)

$$\mathsf{R} = -\frac{1}{\lambda} \sum_{j=0}^{\infty} \frac{\mathsf{B}^j}{\lambda^j} \tag{21.2.27}$$

and the associated partial sums R_n. A simple calculation shows that

$$\mathsf{I} - (\mathsf{B} - \lambda \mathsf{I})\, \mathsf{R}_n = \frac{\mathsf{B}^{n+1}}{\lambda^{n+1}} \tag{21.2.28}$$

and similarly for $\mathsf{I} - \mathsf{R}_n\, (\mathsf{B} - \lambda \mathsf{I})$. We thus see that these differences tend to zero in the operator norm provided $\|\mathsf{B}\|/|\lambda| < 1$, which is also a sufficient condition for the uniform convergence of the series (21.2.27).[8] This argument shows that the *Neumann series* (21.2.27) is the inverse of the operator $\mathsf{B} - \lambda \mathsf{I}$; R is the *resolvent* of $\mathsf{B} - \lambda \mathsf{I}$ and, when the series converges, we say that $\lambda \in \rho(\mathsf{B})$, the *resolvent set* of the operator B. The solution of (21.2.26) is then given by

$$u = \mathsf{R}f = -\frac{1}{\lambda} \left(I + \frac{\mathsf{B}}{\lambda} + \frac{\mathsf{B}^2}{\lambda^2} + \cdots \right) f. \tag{21.2.29}$$

With $u_n = -\sum_{j=0}^{n} \mathsf{B}^j f/\lambda^{j+1}$, this expression may be written as the recurrence

$$u_{n+1} = \frac{1}{\lambda}\,(\mathsf{B}u_n - f), \qquad u_0 = -\frac{1}{\lambda} f. \tag{21.2.30}$$

The solution of the problem can then be seen as the (unique) fixed point of the mapping $u = (\mathsf{B}u - f)/\lambda$.

While carrying the recurrence to infinity will give the exact solution, stopping after a finite number of steps will produce an approximation the quality of which will depend on the magnitude of the ratio $\|\mathsf{B}\|/|\lambda|$. In particular, stopping after one step gives

$$u_1 = -\frac{1}{\lambda^2}\,(\mathsf{B}f + \lambda f) \tag{21.2.31}$$

which, e.g. in scattering theory, is known as the *Born approximation* (see e.g. Sakurai 1994, p. 386).

Example 21.2.5 [Integral equations] Consider an integral equation of the Fredholm type

$$u(x) = g(x) + \mu \int_a^b k(x, \xi)\, u(\xi)\, d\xi \tag{21.2.32}$$

---

[8] The series may also converge for some points of the circle $\|\mathsf{B}\|/|\lambda| = 1$, although it diverges if $\|\mathsf{B}\|/|\lambda| > 1$.

with, e.g., g and k both continuous. By setting $\lambda = 1/\mu$ and $f(x) = -\lambda g(x)$, the equation takes the form (21.2.26) and the recurrence relation (21.2.30) is

$$u^{n+1}(x) = g(x) + \mu \int_a^b k(x,\xi) u^n(\xi)\,d\xi. \tag{21.2.33}$$

We have

$$u^0(x) = g(x), \qquad u^1(x) = g(x) + \mu \int_a^b k(x,\xi)\,g(\xi)\,d\xi, \tag{21.2.34}$$

$$u^2(x) = g(x) + \mu \int_a^b k(x,\xi) u^1(\xi)\,d\xi$$
$$= g(x) + \mu \int_a^b k(x,\eta) \left[g(\eta) + \mu \int_a^b k(\eta,\xi)\,g(\xi)\,d\xi \right] d\eta \tag{21.2.35}$$

It is therefore seen that the final solution will have the form

$$u(x) = g(x) + \int_a^b K(x,\xi;\mu)\,g(\xi)\,d\xi \tag{21.2.36}$$

with

$$K(x,\xi;\mu) = \sum_{n=1}^{\infty} \mu^n k_n(x,\xi), \qquad k_n(x,\xi) = \int_a^b k(x,\eta)\,k_{n-1}(\eta,\xi)\,d\eta. \tag{21.2.37}$$

We can adapt this development to an equation of the Volterra type

$$u(x) = g(x) + \mu \int_a^x h(x,\xi)\,u(\xi)\,d\xi, \qquad a < x < b, \tag{21.2.38}$$

by setting $k(x,\xi) = h(x,\xi)\,H(x-\xi)$ with H the step function. For example, if the equation of interest is

$$u(x) = g(x) + \mu \int_a^x u(\xi)\,d\xi = g(x) + \mu \int_a^b H(x-\xi)u(\xi)\,d\xi, \tag{21.2.39}$$

we find

$$k_2(x,\xi) = \int_a^b H(x-\eta)\,H(\eta-\xi)\,d\eta = H(x-\xi)\int_\xi^x d\eta = (x-\xi)\,H(x-\xi). \tag{21.2.40}$$

Proceeding by induction one can easily prove that

$$k_n(x,\xi) = \frac{(x-\xi)^{n-1}}{(n-1)!} H(x-\xi), \tag{21.2.41}$$

so that

$$K(x,\xi;\mu) = \sum_{n=1}^{\infty} \mu^n \frac{(x-\xi)^{n-1}}{(n-1)!} H(x-\xi) = \exp[\mu(x-\xi)]\,H(x-\xi) \tag{21.2.42}$$

and

$$u(x) = f(x) + \int_a^x e^{\mu(x-\xi)} f(\xi)\,d\xi, \tag{21.2.43}$$

as can be found by the method of §5.6 p. 143 or, much more simply, by differentiating (21.2.38) and solving the resulting differential equation.   □

The series $\sum_{j=0}^{\infty} \|B\|^j / |\lambda|^{j+1}$ converges to $(|\lambda| - \|B\|)^{-1}$ and, since $\|B^j\| \leq \|B\|^j$, we have

$$\| (B - \lambda I)^{-1} \| \leq \frac{1}{|\lambda| - \|B\|} \tag{21.2.44}$$

in agreement with Theorem 21.2.4 p. 627 which states that the inverse of a bounded operator is bounded.

As we will see in §21.9, the spectrum $\sigma(A)$ of an operator A is the complement of the resolvent set $\rho(A)$, i.e. the set of complex numbers λ for which the operator $(A - \lambda I)^{-1}$ does not exist, is unbounded or does not have a dense domain. An immediate consequence of the previous development is therefore that the spectrum of a bounded operator must lie inside or on the boundary of the circle $|\lambda| = \|B\|$, in agreement with (21.2.21) for the self-adjoint case. The *spectral radius* $r_\sigma(A)$ of an operator is the number

$$r_\sigma(A) = \sup_{\lambda \in \sigma(A)} |\lambda|. \tag{21.2.45}$$

Since the spectrum is a closed set (§21.9) and the modulus a continuous function, for a bounded operator we can actually write

$$r_\sigma(B) = \max_{\lambda \in \sigma(B)} |\lambda| \leq \|B\|. \tag{21.2.46}$$

Equality holds when B is normal, i.e. $BB^* = B^*B$.

If we consider the series (21.2.27) as a power series in $1/\lambda$, application of Cauchy's root test for convergence (p. 218) gives the *Gelfand–Beurling formula*:[9]

$$r_\sigma(B) = \lim_{n \to \infty} \|B^n\|^{1/n}. \tag{21.2.47}$$

Since B and B^* have the same norm (Theorem 21.2.6 p. 629),

$$r_\sigma(B^*) = r_\sigma(B). \tag{21.2.48}$$

The Neumann series can be generalized to the operator $B - (\lambda + \mu)I$ observing that

$$B - (\lambda + \mu)I = -\mu (B - \lambda I) \left[(B - \lambda I)^{-1} - \mu^{-1}I \right] = \sum_{j=0}^{\infty} \mu^n (B - \lambda I)^{-(n+1)}, \tag{21.2.49}$$

provided that

$$\| (B - \lambda I)^{-1} \| < \frac{1}{|\mu|}. \tag{21.2.50}$$

This argument proves in particular that the resolvent set is open.

---

[9] A full justification requires use of the Spectral Mapping Theorem 21.9.1 plus a bound on $r_\sigma(B^n)$ (see e.g. Taylor & Lay 1980, p. 280).

The Neumann series (21.2.27) is reminiscent of the Taylor series expansion of the analytic function $(1 - z)^{-1}$ in the circle $|z| < 1$. This similarity is not superficial as it is possible to construct a theory of analytic operator-valued functions $f(\lambda)$ which is quite similar to the usual one (Chapter 17) except that moduli are replaced by norms (see e.g. Taylor & Lay 1980, p. 265; Locker 2000, p. 9). The contour-integral representations of some finite-dimensional operators given in §18.9 and §18.11 are other manifestations of this general theory. An interesting application is to show that the spectrum of a bounded operator cannot be empty. If this were not so, by the operator-function analog of Liouville's theorem, according to which only the constant function is analytic over the entire complex plane (p. 431), R would have to be a constant operator, and this constant should vanish as (21.2.27) shows that $R \to 0$ as $\lambda \to \infty$.

21.3 Compact operators

The simplest class of operators for which the technique of eigenfunction expansion can be justified is that of *compact operators*, a special class of bounded operators. We start with two definitions (see also p. 680 in the Appendix):

Definition 21.3.1 A subset S of a normed space is *compact* (or *sequentially compact*) if every infinite sequence of elements of S converges to an element of S. A subset S of a normed space B is *relatively compact* (or *pre-compact*) if every infinite sequence of elements of S converges to an element of B.[10]

Of course, here and in the following convergence is understood in the sense of the norm. The difference between a compact and a relatively compact set is that, in the latter, the limit point does not have to belong to S. An example might be the set of numbers 1, 1/2, 1/3, ..., which converges to 0 in the compact interval [0,1] but not in the relatively compact interval (0,1). Obviously a compact set is also relatively compact, and the closure of a relatively compact set is compact. Hence, all closed relatively compact sets are compact.

In a finite-dimensional space, all bounded and closed sets are compact: this is the *Bolzano–Weierstrass theorem* (p. 216), a cornerstone of analysis. In an infinite-dimensional space this is not true: even bounded and closed sets may be "too large" to be compact. For example, the set of all $u \in H$ with $\|u\| \leq 1$ is evidently bounded and also closed (definition 21.6.1, p. 650), as the limit of any converging sequence with norm less than or equal to 1 also belongs to the set. However, in this space there are sequences, such as the set $\{e_n\}$ of unit basis vectors, which are not convergent to an element of the set. By definition, this

---

[10] More properly, in both definitions the words "every infinite sequence of elements of S converges" should be replaced by "every infinite sequence contains a non-repeating subsequence converging." With this modification, the definition would apply also to (rather uninteresting) sequences in which, for example, the same element showed up every now and then an infinite number of times. The adjective "sequentially" is used to distinguish this notion of compactness from the corresponding topological one. For normed separable spaces the two notions coincide and therefore we will drop the specification "sequential."

phenomenon cannot occur in a compact set in an infinite-dimensional space. In this sense, such sets are the closest analog to closed and bounded sets in a finite-dimensional space. Since in the eigenfunction expansion approach we generate infinite series of functions, the reasons for our interest in properties and criteria which ensure convergence are obvious.

It is evident that a bounded operator transforms a bounded set into a bounded set as, for all the vectors $u \in H$ such that $\|u\| \leq C$ for some constant C, we have $\|Au\| \leq C\|A\|$. Compact (or completely continuous) linear operators do more than this:

Definition 21.3.2 An operator C is *compact* if and only if it transforms a bounded set into a relatively compact set.

It is intuitively clear that a compact operator must be bounded. Several important linear operators are compact. For example, every bounded linear operator of finite rank (i.e., with a finite-dimensional range) enjoys this property by the Bolzano–Weierstrass theorem. Integral operators of the Hilbert–Schmidt type (Example 21.2.3) are compact, as will be shown momentarily in Example 21.3.1. A simple example of a bounded non-compact operator is the identity in an infinite-dimensional space as, for example, no convergent subsequence can be extracted from the sequence $I e_n = e_n$ of the identity I applied to the unit vectors of a basis, as already noted.

Loosely speaking (in the sense of the footnote on p. 636), we may say that a compact operator C transforms a bounded sequence $\{u_n\}$ into a strongly convergent sequence $C u_n$, i.e., convergent in the sense of the norm. Since a weakly convergent sequence is bounded (p. 552), a compact operator C in a Hilbert space transforms a weakly convergent sequence into a strongly convergent one. In particular, if $\{e_n\}$ is an orthonormal basis, then $C e_n \to \emptyset$. Indeed, if $C e_n \to v \neq \emptyset$, then $(v, C e_n) = (C^* v, e_n)$ would not tend to zero, which is impossible as $\overline{(C^* v, e_n)}$ is a coefficient of the expansion of the vector $C^* v$ on a basis in the space. Here are three important properties of compact operators:

Theorem 21.3.1

  (i) *Every compact operator is bounded.*
 (ii) *The adjoint of a compact operator is compact.*
(iii) *If C is compact and B is bounded, then CB and BC are compact. As a consequence, the product of two compact operators is compact.*

An important deduction from part (iii) of this theorem is that the inverse C^{-1} of a compact operator in an infinite-dimensional space cannot be compact or even bounded because, if it were, $C^{-1}C = I$ would be compact which, as we have just noted, it is not. From some point of view this is a disadvantage of compact operators. On the other hand, in some cases we can get a handle on unbounded operators by studying their inverse when it is compact.

A theorem useful for proving the compactness of an operator is the following:

Theorem 21.3.2 *Let* C_1, C_2, ... *be a sequence of compact operators between two Banach spaces. If the sequence is uniformly convergent to an operator* C, *i.e.* $\lim \| C - C_n \| = 0$, *then* C *is compact.*

A partial converse valid for a Hilbert, but not a Banach space, is that every compact operator is the uniform limit of operators of finite rank as illustrated in the example that follows.

Example 21.3.1 An interesting application of this theorem shows that an integral operator K with a Hilbert–Schmidt kernel $k(x, \xi)$ (i.e., square-integrable in the Lebesgue sense in the rectangle $a \leq x, \xi \leq b$, p. 625) is compact with the L^2-norm. The proof relies on expanding k into a double series. The fact that the square integral of k is finite by hypothesis shows that the series converges in the L^2 sense (the Parseval equality). On the other hand, the integral operator in which k is replaced by the series truncated after a finite number of terms has finite rank, and is therefore compact. Thus, K can be approximated by a converging sequence of compact operators and is therefore itself compact. In particular, this theorem shows that the Green's function for regular Sturm–Liouville problems (Chapter 15) generates a compact integral operator.

The same conclusion holds with the *sup* norm if k is continuous in $a \leq x, \xi \leq b$ as a consequence of the Ascoli–Arzelá theorem (p. 693). The restrictions on k may also be weakened somewhat without compromising the compact nature of the operator (see e.g. Taylor & Lay 1980, p. 296; Akhiezer & Glazman 1961, vol. 1 p. 54 and 58).   □

With an eye toward eigenfunction expansions, we are interested in the eigenvalues of a compact operator.

Theorem 21.3.3 *The eigenspace of a non-zero eigenvalue of a compact operator* C *in a Banach space is finite-dimensional.*

The null space of $C - \lambda I$ is invariant under C and, in this null space, C reduces to λI. Thus, if the null space were infinite dimensional, it would be impossible to extract a convergent subsequence from the sequence Ce_n of C applied to each vector e_n of a basis in the subspace.

If there is an infinity of non-zero eigenvalues λ_n, each one with a corresponding eigenvector e_n with $\|e_n\| = 1$, form the sequence $\{e_n/\lambda_n\}$. Unless $\lambda_n \to 0$, this sequence is bounded, but then $\{C(e_n/\lambda_n)\} = \{e_n\}$ from which no converging subsequence can be extracted. In this case the operator would fail to transform a bounded set into a relatively compact one, and would not be compact, contrary to the hypothesis. This argument shows that the sequence must be unbounded, which requires $\lambda_n \to 0$. We thus conclude that only a finite number of eigenvalues can exceed in modulus any finite positive number so that it must be possible to arrange the eigenvalues of a compact operator in a non-increasing sequence $|\lambda_1| \geq |\lambda_2| \geq \cdots \to 0$. If the number of non-zero eigenvalues is finite, and $\lambda = 0$ is an eigenvalue, it must be infinitely degenerate (see e.g. Riesz & Sz.-Nagy 1955, p. 232; Kubrusly 2001, p. 479).

We now come to a central theorem which justifies for self-adjoint compact operators the same eigenvector expansion which is possible, for example, for Hermitian matrices (Theorem 18.7.6):

Theorem 21.3.4 [Hilbert–Schmidt] *For every compact self-adjoint operator* C *in a Hilbert space* H *there is an orthogonal set* $\{e_n\}$ *of eigenvectors corresponding to eigenvalues* λ_n *such that every element* $u \in H$ *can be uniquely written as*

$$u = \sum_k (e_k, u)\, e_k + u_0, \qquad \text{where} \quad C u_0 = 0. \tag{21.3.1}$$

Furthermore

$$C u = \sum_k \lambda_k (e_k, u)\, e_k \tag{21.3.2}$$

and, if the system $\{e_n\}$ *is infinite,*

$$\lim_{n \to \infty} \lambda_n = 0; \tag{21.3.3}$$

$\lambda = 0$ *belongs to the spectrum of* C, *but it may or may not be an eigenvalue.*

Given the centrality of this theorem to the eigenfunction expansion procedure, it is interesting to sketch a proof for the case in which there is an infinity of eigenvalues $\lambda_n \to 0$; more complete proofs are given in most of the books mentioned in this chapter.

Proof The first step is to show that, if C is compact, $\|C\| = \sup_{\|e\|=1} |(e, Ce)|$ is not just an upper limit but is actually attained for a vector with norm 1. Indeed, if the sequence $\{u_n\}$ with $\|u_n\| = 1$ is such that $|(u_n, Cu_n)| \to \|C\|$, due to the hypothesis of compactness, one can extract from it a converging subsequence the limit of which can be proven to have norm 1 and to satisfy $\|C\| = \sup_{\|e\|=1} |(e, Ce)|$. This limit point can then be taken as the first eigenvector e_1 as, if the equality $C e_1 = \lambda_1 e_1$ were not satisfied, $|(e_1, Ce_1)|$ could not be the maximum value of $|(e, Ce)|$. The subspace H_1 generated by the vector e_1 is invariant with respect to C in the sense of Theorem 21.2.7, and, by the same theorem, $H_1^\perp = H - H_1$, the orthogonal complement of H with respect to e_1, is invariant with respect to C^* or, since C is self-adjoint, with respect to C itself. The same argument can then be applied to $H_1^\perp$ to find a second eigenvector and so on. Since, by Theorem 21.3.3, every non-zero eigenvalue can only have a finite number of eigenvectors, either the process terminates after a finite number of steps, or the eigenvectors so constructed correspond to an infinite number of eigenvalues such that $\lambda_n \to 0$. In the first case, in an infinite-dimensional space, the orthogonal complement of the subspace consisting of the union of all the eigenspaces of the non-zero eigenvalues is infinite dimensional. In the second case, the process generates a sequence of orthonormal eigenvectors which, as any such sequence, converges weakly to $\emptyset$.[11] Since a compact operator transforms a weakly convergent sequence into a strongly convergent one, we deduce that the sequence $C e_n = \lambda_n e_n$ converges strongly to $\emptyset$, i.e., $|\lambda_n| = \|Ce_n\| \to 0$.

---

[11] For any vector u, the numbers (e_n, u) are the coefficients in the representation of u on a basis and therefore they must tend to 0.

We now show that the set of vectors e_n so constructed is a basis by proving that the only vector orthogonal to all of them is the vector $\emptyset$ (cf. Theorem 19.4.4 p. 559). For a vector w orthogonal to all the e_n, we would have $|(w, Cw)| \leq |\lambda_n| \|w\|^2$ for all n because, if this were not true, e_n would not maximize $|(e_n, Ce_n)|$ in the appropriate subspace. Since $\lambda_n \to 0$, the only possibility is then that $(w, Cw) = 0$. But, since we may take w to have norm 1, this implies that $Cw = \emptyset$, as C must have norm 0 when restricted to the subspace orthogonal to all the eigenspaces corresponding to the non-zero eigenvalues. For the last statement of the theorem see p. 665.

In the course of the proof we have constructed an orthonormal basis in the union of eigenspaces corresponding to the non-zero eigenvalues. If there is only a finite number of them, we can add an orthonormal basis in the remainder of the space and therefore, in either case, we have generated a basis in the Hilbert space which justifies the representation (21.3.1).

In order to prove (21.3.2), we note that the coefficients of the expansion of Cu on the basis $\{e_n\}$ are given by

$$(e_n, Cu) = (Ce_n, u) = \lambda_n(e_n, u) \tag{21.3.4}$$

where we have used the self-adjointness of C in the first step. The last statement of the theorem follows from the fact that, as noted before, any orthonormal basis $\{e_j\}$ converges weakly to $\emptyset$ so that $Cu_j \to 0$ strongly, which proves that 0 is a regular or generalized eigenvalue of C (see p. 665; Helmberg 1969, p. 195; Roman 1975, p. 617).

$\square$

Upon substituting the expansion (21.3.2) into the equation $(C - \lambda I)\, u = f$, we find

$$\sum_k \lambda_k(e_k, u)\, e_k - \lambda u = f \tag{21.3.5}$$

from which, upon taking the scalar product with each e_j in turn, by the continuity of the scalar product, we have

$$(\lambda_j - \lambda)(e_j, u) = (e_j, f). \tag{21.3.6}$$

If λ is not one of the eigenvalues this equation can be solved for (e_j, u) and the result substituted into $\lambda u = Cu - f$ to find (Taylor & Lay 1980, p. 356)

$$u = \frac{1}{\lambda} \sum_j \frac{\lambda_j}{\lambda_j - \lambda}(e_j, f)e_j - \frac{f}{\lambda} \tag{21.3.7}$$

from which, upon nothing that $u = (C - \lambda I)^{-1}$ and calculating the norm, it also follows that

$$\| (C - \lambda I)^{-1} \| \leq \frac{1}{|\lambda|} \left(1 + \sup_j \left| \frac{\lambda_j}{\lambda_j - \lambda} \right| \right). \tag{21.3.8}$$

If the set of eigenvectors forms a basis in the Hilbert space, we also have $f = \sum_j (e_j, f)e_j$ and the previous equation becomes

$$u = \sum_j \frac{(e_j, f)}{\lambda_j - \lambda}\, e_j, \tag{21.3.9}$$

a result similar to (18.7.9) p. 514 for the finite-dimensional case and identical in form to (15.3.12) p. 368 for Sturm–Liouville operators. If λ equals the eigenvalue λ_n, then the previous procedure shows that $(C - \lambda I)\, u = f$ has no solution unless f is orthogonal to the n-th eigenspace so that both sides of (21.3.6) vanish for $j = n$. If this condition is satisfied, the solution of the equation is given by

$$u = \frac{1}{\lambda} \sum_{j \neq n} \frac{\lambda_j}{\lambda_j - \lambda_n} (e_j, f) e_j - \frac{f}{\lambda} + u_n, \qquad (21.3.10)$$

where all the terms with $\lambda_j = \lambda_n$ are excluded from the summation and u_n is an arbitrary vector of the n-th eigenspace. In essence, this is Fredholm's alternative theorem 21.7.3 to which we return on p. 658.

We have seen in §18.9 p. 519 that the expansion of a finite-dimensional vector over an orthonormal set can be expressed in terms of a family of projection operators which constitute a resolution of the identity. For the present case of compact self-adjoint operators we can readily derive a similar decomposition. We define orthogonal projectors P_k onto the k-th eigenspace of dimension i_k by

$$P_k u = \sum_{i=1}^{i_k} (e_{k_i}, u)\, e_{k_i} \qquad (21.3.11)$$

and similarly define the projector P_0 on the orthogonal complement of all the subspaces corresponding to non-zero eigenvalues. Then we see that the expansions (21.3.1) and (21.3.2) can be recast as

$$I = P_0 + \sum_k P_k, \qquad C = \sum_k \lambda_k P_k, \qquad (21.3.12)$$

where the summations extend over all distinct eigenvalues and the corresponding eigenspaces. The first summation is a *resolution of the identity*. When the summations contain an infinity of terms, the second one is understood in the sense of the operator (uniform) norm (which implies strong convergence, p. 626), while the first one does not converge in the uniform sense and must be understood in the sense of strong convergence. Since each eigenspace is orthogonal to all the others, we have $P_k P_n = \delta_{kn} P_k$ as in the finite-dimensional case.

We can express (21.3.7) in terms of the same entities P_0 and P_k as

$$(C - \lambda I)^{-1} = \frac{1}{\lambda} \sum_k \frac{\lambda_k}{\lambda_k - \lambda} P_k - \frac{1}{\lambda} I \qquad (21.3.13)$$

provided $\lambda \neq 0$ and $\lambda \neq \lambda_k$. This relation shows that λ_k is a first-order pole of the operator-valued function $(C - \lambda I)^{-1}$. Similarly to (18.11.4) p. 529 for matrices, we can also rely on (21.3.12) to define a function $F(C)$ of the operator C starting from a general complex-valued function $F(z)$ defined on the spectrum of C (and, if needed, for $z = 0$):

$$F(C) = \sum_k F(\lambda_k) P_k + F(0) P_0. \qquad (21.3.14)$$

It can be shown that (see e.g. Conway 1985, p. 58)

$$\|F(C)\| = \sup_{\lambda \in \sigma(C)} |F(\lambda)|, \tag{21.3.15}$$

where $\sigma(C)$ is the spectrum of the operator C.

21.3.1 Self-adjoint operators with a compact resolvent

We have seen that compact self-adjoint operators share some significant similarities with self-adjoint operators in finite-dimensional spaces. The analogy is not as simple or as complete for more general operators as will be seen (in part) in §21.9, but it does extend to self-adjoint operators with a compact resolvent. This result is of particular interest here as, among others, it applies to Sturm–Liouville operators. The considerations that follow are amplified in Theorem 21.11.2 p. 672.

Let then C be compact and $C^{-1} = A$ exist, which in particular implies that 0 is not an eigenvalue of C (p. 648). We know from Theorem 21.3.1 that A is unbounded. From $Ce_n = \lambda_n e_n$ it follows that $C^{-1}e_n = \lambda_n^{-1}e_n$ so that A has the same eigenvectors as C with eigenvalues $\mu_k = 1/\lambda_k$; since $\lambda_k \to 0$, $|\mu_k| \to \infty$.

By definition, a vector v belongs to the range $\mathscr{R}_A$ if and only if there is a u such that $Au = v$, which implies that $u = Cv$ or, from (21.3.2),

$$u = \sum_{k=0}^{\infty} \lambda_k(e_k, v)\, e_k. \tag{21.3.16}$$

But since C is self-adjoint, its eigenvalues are real so that $\lambda_k(e_k, v) = (\lambda_k e_k, v) = (Ae_k, v) = (e_k, Av) = (e_k, u)$. Thus we have the representation

$$u = \sum_{k=0}^{\infty}(e_k, u)\, e_k \tag{21.3.17}$$

with a corresponding resolution of the identity of the form (21.3.12) without the term P_0. Furthermore

$$Au = \sum_{k=0}^{\infty} \lambda_k \mu_k(e_k, v)\, e_k = \sum_{k=0}^{\infty}(e_k, v)\, e_k = \sum_{k=0}^{\infty} \frac{(e_k, u)}{\lambda_k}\, e_k = \sum_{k=0}^{\infty} \mu_k(e_k, v)\, e_k, \tag{21.3.18}$$

which is similar to (21.3.2) and the second one of (21.3.12). We also find, using the expansion (21.3.17) and the fact that the e_k are eigenvectors of C,

$$(A - \mu I)^{-1} u = \frac{1}{\mu}\left(\frac{A}{\mu} - I\right)^{-1} AA^{-1}u = \frac{1}{\mu}\left(\frac{I}{\mu} - C\right)^{-1} Cu = \sum_{k=0}^{\infty} \frac{(e_k, u)\, e_k}{\mu_k - \mu} \tag{21.3.19}$$

provided $\mu \neq \mu_k$. The relations (21.3.17) and (21.3.18) are formally the same as (21.3.1) and (21.3.2), and (21.3.19) has the same form as the expansion (15.3.11) p. 368 of the Green's function theory for Sturm–Liouville operators. Since these relations have been derived for convergence according to the L^2 norm, they are insufficient to examine more

subtle convergence properties such uniformity, point-wise convergence, term-by-term differentiability and others. For this purpose, it is necessary to suitably refine this theorem.
Some of the ideas involved in this additional work have been described in §9.3 for the case
of the Fourier series.

21.3.2 Non-self-adjoint compact operators

The eigenfunction expansion (21.3.1) and the decompositions (21.3.12) are also valid for
compact non-self-adjoint operators which are normal, so that $CC^* = C^*C$ (see e.g. Naylor
& Sell 2000, p. 459). In this case, however, the eigenvalues λ_k are not real numbers. Again,
this is in perfect analogy with normal finite-dimensional operators described in §18.7 p. 511.

If the operator is compact, but not normal, we are in a situation similar to that encountered
in the case of non-normal simple matrices (§18.7). As in Theorem 18.7.3 p. 513, we can find
two orthonormal sets $\{e_k\}$ and $\{e'_j\}$ which permit one to express the action of the operator
in a simple form.

It is interesting to look at the proof of this decomposition, which is similar to the singular-
value decomposition of §18.8. We start by considering the operator C^*C which is compact,
self-adjoint and non-negative definite. Its eigenvalues μ_1^2, μ_2^2, ... are real and non-negative
and we may write, as in (21.3.2),

$$C^*Cu = \sum_k \mu_k^2 (e_k, u)e_k, \tag{21.3.20}$$

where the e_k are the eigenvectors belonging to the eigenvalue μ_k^2 and each eigenvalue is
repeated according to its multiplicity. Since C^*C is compact and self-adjoint, we know that
its eigenvectors are mutually orthogonal. Now, for $\mu_j \neq 0$, define e'_j by

$$Ce_j = \mu_j e'_j \tag{21.3.21}$$

and note that

$$(e'_j, e'_k) = \frac{1}{\mu_j \mu_k} (Ce_j, Ce_k) = \frac{1}{\mu_j \mu_k} (e_j, C^*Ce_k) = \frac{\mu_k}{\mu_j}(e_j, e_k) = \delta_{jk}, \tag{21.3.22}$$

so that the $\{e'_k\}$ form an orthonormal system. Even if there is only a finite number of non-zero
μ_k, each one of the two orthonormal systems $\{e_k\}$, $\{e'_j\}$ can be extended to an orthonormal
basis by adding an orthonormal basis in the remainder of the space. Thus we can write

$$u = \sum_{k=1}^{\infty} (e_k, u)e_k \tag{21.3.23}$$

and, by (21.3.21),

$$Cu = \sum_{k=1}^{\infty} (e_k, u)Ce_k = \sum_{k=1}^{\infty} \mu_k(e_k, u)e'_k, \tag{21.3.24}$$

in which the sum will terminate if there is only a finite number of non-zero eigenvalues.

21.4 Unitary operators

It is easily seen that an isometric operator U (defined on p. 628) is non-singular. If its domain and range are the entire spaces, than it is invertible and is called *unitary*. In this case, U^{-1} is also isometric and equals U^* so that[12]

$$UU^* = I, \qquad U^*U = I, \tag{21.4.1}$$

which shows that a unitary operator with domain and range in the same space is normal since it commutes with its adjoint. If λ is an eigenvalue of U with eigenvector u so that $Uu = \lambda u$, then

$$|\lambda|^2(u, u)\,(Uu, Uu) = (U^*Uu, u) = \left(U^{-1}Uu, u\right) = (u, u) \tag{21.4.2}$$

so that all the eigenvalues of a unitary operator lie on the unit circle. If, in particular, U is self-adjoint, then $\lambda = \pm 1$.

If A is self-adjoint, it can be shown that the operator

$$U(\xi) = \exp(i\xi A), \tag{21.4.3}$$

with ξ real and the exponential defined analogously to (21.3.14) (or, more precisely, as in (21.11.35)), is unitary. Furthermore, $U(\xi + \eta) = U(\xi)\,U(\eta)$ so that operators of the form (21.4.3) form a one-parameter *unitary group*. The converse result is contained in *Stone's theorem* which asserts that, with the added hypothesis of continuity, all unitary groups have the form (21.4.3) (see e.g. Conway 1985, p. 337). For example, the translation operator $f(x) \to f(x + \xi)$ defined on $f \in L^2(-\infty, \infty)$ is clearly unitary with an inverse defined by $f(x) \to f(x - \xi)$. The corresponding self-adjoint operator is $A = -i\,d/dx$.[13]

An important example of a unitary operator is the Fourier transform. The standard definition for an absolutely integrable function u given in Chapter 10 is

$$(Fu)(k) = \tilde{u}(k) = \frac{1}{\sqrt{2\pi}} \int_{-\infty}^{\infty} e^{ikx}u(x)\,dx, \tag{21.4.4}$$

but this definition is not necessarily applicable if $u \in L^2(-\infty, \infty)$ without being absolutely integrable. We now describe a procedure to extend the operator to L^2-functions. We present only the logic of the arguments omitting details (see e.g. Helmberg 1969, p. 108; Roman 1975, p. 556).

Let us start by considering the operator defined by (21.4.4) over a set $\mathscr{D}_F$ of absolutely integrable functions dense in L^2. We can for example take the functions $e^{-x^2/2}x^n$ linear combinations of which – namely Hermite polynomials multiplied by $e^{-x^2/2}$ – are proportional to an orthonormal set in the space (§13.9 p. 334) and therefore dense. A direct

[12] In the first relation I is the identity operator in the image space while, in the second one, it is the identity operator in the original space.
[13] Compare with the translation rule (10.3.5) p. 271 of the Fourier transform.

calculation shows that F applied to these functions is isometric and that $\tilde{u} \in \mathscr{D}_F$. On $\mathscr{D}_F$ F has an inverse given by the standard formula

$$(F^{-1}\tilde{u})(x) = (F^*\tilde{u})(x) = u(x) = \frac{1}{\sqrt{2\pi}} \int_{-\infty}^{\infty} e^{-ikx}\tilde{u}(k)\,dk. \tag{21.4.5}$$

Since F and F^{-1} are bounded and are defined over a dense set in L^2, we expect to be able to extend them by continuity to the entire space $L^2(-\infty, \infty)$. We also expect that these extensions will preserve the norm and define unitary operators.

In the first place, the extensions are isometric as, for example, $\|Fu\| = \|\lim Fu_n\| = \lim \|Fu_n\| = \lim \|u_n\| = \|u\|$ due to continuity of the norm and to the isometry of F on $\mathscr{D}_F$. Secondly, the domain of the extended F is the whole space as we can see by proving that FF^{-1} and $F^{-1}F$ are the identity operator for any $u \in L^2(-\infty, \infty)$. To show this we note that both these operators are bounded as they are the product of bounded operators, and they therefore are both continuous. Thus, for example, $F^{-1}Fu = \lim F^{-1}Fu_n = \lim u_n = u$.

While these arguments prove that the operator can be extended to the entire space, they do not give a concrete representation of this extension. Of course, if $u \in \mathscr{D}_F$, (21.4.4) and (21.4.5) apply, but what about more general L^2 functions? A possible procedure is to construct the sequence $\{u_n\}$ by truncating u so that

$$(Fu_n)(k) = \tilde{u}_n(k) = \frac{1}{\sqrt{2\pi}} \int_{-n}^{n} e^{ikx} u(x)\,dx \tag{21.4.6}$$

and similarly for the inverse. The limit is of course in the sense of the norm

$$\|F(u - u_n)\|^2 = \int_{-\infty}^{\infty} |\tilde{u} - \tilde{u}_n|^2\,dk \to 0, \tag{21.4.7}$$

so that Fu is the limit in the mean of Fu_n.[14] An alternative characterization of the extended Fourier operator is given by Plancherel's formulae, valid almost everywhere (see e.g. Titchmarsh 1948, p. 69; Riesz & Sz.-Nagy 1955, p. 294):

$$(Fu)(k) = \frac{1}{\sqrt{2\pi}} \frac{d}{dk} \int_{-\infty}^{\infty} \frac{e^{ikx} - 1}{ix} u(x)\,dx,$$

$$(F^{-1}\tilde{u})(x) = \frac{1}{\sqrt{2\pi}} \frac{d}{dx} \int_{-\infty}^{\infty} \frac{e^{-ikx} - 1}{-ik} \tilde{u}(k)\,dk. \tag{21.4.8}$$

For $u \in L^1(-\infty, \infty)$ one can differentiate under the integral sign (p. 691) and these relations reduce to (21.4.4) and (21.4.5). Equivalence with the earlier definition (10.2.3), p. 268 for $u \in L^2(-\infty, \infty)$ can also be proven (see Titchmarsh 1948, p. 74). One may note an analogy between Plancherel's formulae and the regularization of divergent integrals sketched on p. 232.

In the case of finite-dimensional spaces we have seen that unitary operators transform orthonormal bases into orthonormal bases, and that they establish equivalence

---

[14] Recall from §19.5 p. 560 that a function belonging to $L^2(-\infty, \infty)$ does not necessarily tend to 0 at infinity. As a consequence, (21.4.7) does not necessarily imply that $Fu = \lim Fu_n$ almost everywhere.

relations between operators, $B = U^*AU$ and $A = UBU^*$. Similar properties hold for infinite-dimensional spaces. An example is the operator of multiplication by $-ik$, which is unitarily equivalent to the operator d/dx via the Fourier transform:

$$(\tilde{v}, -ik\tilde{u}) = \left(\tilde{v}, F\frac{d}{dx}F^*\tilde{u}\right) = \left(F^*\tilde{v}, \frac{d}{dx}F^*\tilde{u}\right) = \left(v, \frac{du}{dx}\right). \tag{21.4.9}$$

This will be recognized as one of the operational rules of the Fourier transform, Equation (10.3.1) p. 270.

It is an immediate consequence of (21.4.4) and (21.4.5) that $F^2u(x) = u(-x)$ so that $F^4 = I$. The eigenvalues of F are therefore the fourth roots of 1, namely ± 1 and $\pm i$. The corresponding eigenvectors are proportional to $e^{-x^2/2}H_n(x)$, where the H_n are the Hermite polynomials (§13.9 p. 334).

21.4.1 Bochner's theorem

The Fourier–Plancherel transform is a particular example of a general result known as *Bochner's theorem* (see e.g. Riesz & Sz.-Nagy 1955 p. 291) which states the following:

Theorem 21.4.1 *To every unitary transformation $v = Uu$ of the space L^2 one can connect two functions $k(\xi, x)$ and $h(\xi, x)$, defined on a finite or infinite rectangle $a < \xi, x < b$, belonging to $L^2(a, b)$ for every fixed value of ξ, for which*

$$\int_0^\xi v(x)\,dx = \int_a^b \overline{k(\xi, x)}u(x)\,dx, \qquad \int_0^\xi u(x)\,dx = \int_a^b \overline{h(\xi, x)}v(x)\,dx. \tag{21.4.10}$$

These functions satisfy the relations

$$\int_a^b \overline{k(\xi, x)}\,k(\eta, x)\,dx = \int_a^b \overline{h(\xi, x)}\,h(\eta, x)\,dx = \begin{cases} \min(|\xi|, |\eta|) & \text{if } \xi\eta \geq 0 \\ 0 & \text{if } \xi\eta \leq 0 \end{cases} \tag{21.4.11}$$

and

$$\int_0^\eta k(\xi, x)\,dx = \int_0^\xi \overline{h(\eta, x)}\,dx. \tag{21.4.12}$$

Conversely, every pair of functions satisfying these conditions generates by (21.4.10) a unitary transformation U of the space L^2 with inverse $U^{-1} = U^*$.

In particular, the choice

$$k(\xi, x) = \frac{\overline{\chi(\xi x)}}{x}, \qquad h(\xi, x) = \frac{\chi(\xi x)}{x}, \tag{21.4.13}$$

with $\chi(x)/x \in L^2(a, b)$ and such that (21.4.11) holds, satisfies (21.4.12) automatically and we have the transform pair

$$v(k) = \frac{d}{dk}\int_a^b \frac{\chi(kx)}{x}u(x)\,dx, \qquad u(x) = \frac{d}{dx}\int_a^b \frac{\overline{\chi(kx)}}{x}v(k)\,dk. \tag{21.4.14}$$

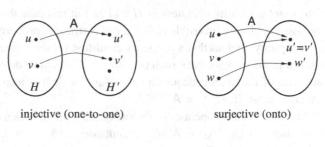

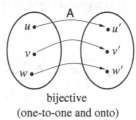

Fig. 21.2 For an injective operator A from H to H' there is one and only one element of the domain which is mapped onto each element of the range, which is not necessarily the entire space H'. If the operator is surjective, every element of H' is the image of some element of H, not necessarily unique.

The Fourier–Plancherel transform corresponds to $a \to -\infty$, $b \to \infty$, $\chi = (e^{ix} - 1)/(i\sqrt{2\pi})$.

21.5 The inverse of an operator

The solution of a problem $Au = f$ hinges on the existence of an inverse operator such that, symbolically, $u = A^{-1}f$. Here we collect some general information on the inverse of an operator, getting into further details for specific cases in later sections.

When, for each element $u' \in \mathcal{R}_A$, there is only one element $u \in \mathcal{D}_A$ such that $u' = Au$, the operator is said to be *injective*, or *one-to-one*, or *non-singular* (Figure 21.2). In this case, $\mathcal{N}_A = \{\emptyset\}$ i.e., the null space reduces to the $\emptyset$ element. One can introduce an operator A^{-1} with domain $\mathcal{D}_{A^{-1}} = \mathcal{R}_A$ and range $\mathcal{R}_{A^{-1}} = \mathcal{D}_A$ such that

$$A^{-1}u' = A^{-1}Au = u, \qquad Au = A(A^{-1}u') = u'. \qquad (21.5.1)$$

The first relation shows that $A^{-1}A$ is the identity operator in H *restricted* to $\mathcal{D}_A$ and the second one that AA^{-1} is the identity operator in H' *restricted* to $\mathcal{R}_A$. It is easy to show that A^{-1} is also linear, unique and that $(A^{-1})^{-1} = A$.

If $\mathcal{R}_A = H'$, i.e., the range of the operator is the entire space H', the operator is *surjective* or *onto*. An operator which is both surjective and injective is *bijective* and if, in addition, $\mathcal{D}_A = H$, (21.5.1) are valid for all $u \in H$ and $u' \in H'$ so that $A^{-1}A = I$ and $AA^{-1} = I'$, the *unrestricted* identity operators in H and H'; in this case the operator is termed *invertible* or *regular*. Being invertible is thus a much stronger property than being injective.[15]

It is pretty obvious that a necessary condition for the existence of A^{-1} is that $\mathcal{N}_A = \{\emptyset\}$ as, if this were not so, the u such that $Au = u'$ would be determined only up to the general solution of the homogeneous equation $Au_0 = \emptyset$. It is also easy to show that, when the inverses exist, $(BA)^{-1} = A^{-1}B^{-1}$.

Since the identity operator is obviously self-adjoint, when the inverse operators exist we have that $I^* = (A^{-1}A)^* = A^*(A^{-1})^*$, but also $A^*(A^*)^{-1} = I$, and similarly for the products in the reverse order. This argument proves the first statement of the following

Theorem 21.5.1

(i) *If* A *is a bounded operator and* A^{-1} *and* $(A^*)^{-1}$ *exist,*

$$(A^{-1})^* = (A^*)^{-1}. \tag{21.5.2}$$

(ii) A^{-1} *exists and is bounded if and only if* $(A^*)^{-1}$ *exists and is bounded.*
(iii) *If* A *and* A^{-1} *are closed and densely defined then* A^* *has an inverse and (21.5.2) holds.*

The notion of densely defined closed operator is introduced in §21.6.

Theorem 21.5.2 A *has a bounded inverse if and only if it is bounded away from* 0, *i.e., there is a positive constant* c *such that, for all* $u \in \mathcal{D}_A$, $u \neq \emptyset$,

$$\|Au\| \geq c\,\|u\|. \tag{21.5.3}$$

Proof Suppose A is bounded away from 0. Then $Au = \emptyset$ has only the zero solution and therefore A^{-1} exists. It is also bounded since, if $v = Au$, we have $\| v \| = \| Au \| \geq c \| u \|$, but since $u = A^{-1}v$, it follows that

$$\| A^{-1}v \| \leq c^{-1} \| v \| \qquad \Rightarrow \qquad \| A^{-1} \| \leq c^{-1} \tag{21.5.4}$$

which shows that the largest possible value of c is the reciprocal of the norm of A^{-1}. Suppose now that A has a bounded inverse. This means that A^{-1} exists and, for some C, $\| A^{-1}v \| \leq C \| v \|$. But $u = A^{-1}v$ and, therefore, $\| Au \| \geq C^{-1}\|u\|$ so that A is bounded away from 0. $\qquad\square$

We will see in §21.7 the crucial importance of a closed range for the solvability of a problem. If one can construct the operator A^{-1} it is therefore useful to know that

Theorem 21.5.3 *Let* A *be a closed operator, and let* A^{-1} *exist. Then* $\mathcal{R}_A$ *is closed if and only if* A^{-1} *is bounded.*

---

[15] Neither $A^{-1}A = I$ nor $AA^{-1} = I'$, by themselves, guarantee invertibility. For example, the differentiation of the integral of a function is the original function, but differentiation is not one-to-one, while integration is not onto and therefore neither is invertible.

The notion of closed operator is introduced in §21.6; in particular, bounded operators are closed.

Example 21.5.1 In Chapter 15 we have considered the Sturm–Liouville operator

$$\mathsf{L} = -\frac{d}{dx}\left[p(x)\frac{d}{dx}\right] + q(x) \tag{21.5.5}$$

with homogeneous boundary conditions. Since this operator involves derivatives, it is unbounded. By solving the equation $\mathsf{L}u = f$ we have obtained

$$u = \int_a^b G(x;\xi)\,f(\xi)\,d\xi \tag{21.5.6}$$

which evidently specifies the inverse operator L^{-1}. As we noted in Example 21.3.1 p. 638, the Green's function G for a regular Sturm–Liouville problem is square-integrable so that the integral operator L^{-1} is Hilbert–Schmidt, which makes the operator both bounded and compact (Example 21.3.1 p. 638). □

21.6 Closed operators and their adjoint

When, as in many examples in this book, we attempt to solve a problem $\mathsf{A}u = f$ by a series expansion, what we are really doing is generating a sequence of approximations $\{u_n\}$, each one the n-th partial sum of the series, hoping that, at the same time, $u_n \to u$ and $\mathsf{A}u_n \to f$ as $n \to \infty$. This may or may not happen. For example, the Fourier series of a continuous but non-smooth function may converge point-wise but, if the operator A contains derivatives of a sufficiently high order, $\mathsf{A}u_n$ may not converge (at least in the point-wise sense). We are thus faced with two questions:

(i) Given that all the elements $u_n \in \mathscr{D}_A$, does u also belong to $\mathscr{D}_A$, the domain of A? In other words, does it "make sense" to apply A to u?

(ii) Does the sequence $\mathsf{A}u_n$ converge, and if it does, does it converge to $\mathsf{A}u$?

The Hahn–Banach theorem, 21.2.3 p. 626, answers both these questions in the affirmative for bounded (and therefore continuous) operators. We now consider them more generally.

Question (i) can be rephrased in the following way: Consider all the possible convergent sequences $\{u_n\}$ made up of points of $\mathscr{D}_A$. Some of these sequences will converge to points that belong to the boundary of $\mathscr{D}_A$, without belonging to $\mathscr{D}_A$ itself. The question is whether these points can be added to $\mathscr{D}_A$ (in other words, whether it is possible to define the action of A on these points in a sensible way) so as to make the domain of the operator closed. In this way one would obtain an extension of the original operator (Definition 21.1.1 p. 623).

A difficulty we encounter in trying to extend the domain of an unbounded operator is that, if two sequences $\{u_n\}$ and $\{v_m\}$ both tend to the same element u, we cannot rely on an inequality of the type $\|\mathsf{A}(u_n - v_m)\| \leq \|\mathsf{A}\|\,\|u_n - v_m\|$ to guarantee that $\mathsf{A}u_n$ and $\mathsf{A}v_m$ tend to the same limit as the norm of the operator is not finite. Failing this property, different sequences all approaching the same u may well give rise to different and incompatible definitions of Au, which would make the operator ill-defined. As a matter of fact, the very unboundedness of the operator tells us that that there will be null sequences $\{w_m\}$ (i.e., sequences such that $\|w_m\| \to 0$) for which Aw_m does not tend to Ø, since otherwise A would be continuous at Ø and therefore continuous on all of $\mathcal{D}_A$ (Theorem 21.2.1 p. 625) and, hence, bounded (Theorem 21.2.2). For these operators an extension as complete as that achievable for bounded operators is not possible and we have to settle for less. For this purpose, let us introduce the notion of *closed operator*, which is in some sense a weaker version of a bounded operator. Here and in the rest of this section convergence is understood in the strong sense, i.e., in the sense of the vector norm.

Definition 21.6.1 An operator A is *closed* if, for any sequence $\{u_n\}$ such that $u_n \in \mathcal{D}_A$ for all n, $u_n \to u$ *and* A$u_n \to f$, it happens that $u \in \mathcal{D}_A$ and A$u = f$ or, in other words, that A$(\lim u_n) = \lim (Au_n)$.

The point to note is that one gives up proving that, from $u_n \to u$, it follows that A$(\lim u_n) = \lim(Au_n)$: this may or may not happen. The operators for which it *does happen* (in a sense to be made precise presently) are those that we call closed, a denomination justified by the fact that the graph of the operator (p. 621) is then closed as it contains all its limit points.

It is obvious that all bounded operators are closed. However, this definition includes a broader class of operators as, unlike the case of bounded operators, for operators that are only closed the existence of a limit for the sequence $\{u_n\}$ does not automatically imply the existence of a limit for the sequence Au_n. However, as we now show, it is still true that, if two different sequences $\{u_n\}$, $\{v_n\}$ tend to the same limit, and if both the sequences Au_n and Av_n converge, they converge to the same limit.

As noted before, for an unbounded operator we are guaranteed that there will be *at least some* null sequences $\{w_n\}$ such that $\lim$ Aw_n fails to tend to zero. This can happen in either one of two ways: (*i*) either $\lim$ Aw_n does not exist, or (*ii*) $\lim$ Aw_n exists but is not zero. Operators for which the first possibility applies to some null sequences and the second one to other null sequences do not exist.

As an example of an operator for which possibility (*i*) applies, let A $= \mathrm{d}/\mathrm{d}x$ and consider $w_n = \sin nx/n$, for example in $C^1[0, \pi] \subset L^2(0, \pi)$. Then clearly $\| w_n \| \to 0$ (i.e., $\{w_n\}$ is a null sequence), while A$w_n = \cos nx$ has no limit.[16] As an example of possibility (*ii*), consider

$$Au = \int_{-\infty}^{\infty} u(x)\, \mathrm{d}x, \qquad (21.6.1)$$

with domain in the space of $L^2(-\infty, \infty)$ functions with compact support. In this space consider the sequence

$$w_n = \begin{cases} 1/n & |x| \le n \\ 0 & |x| > n \end{cases}. \qquad (21.6.2)$$

---

[16] The sequence $\cos nx$ tends to zero in the weak and distributional sense, but here we are concerned with strong convergence in the L^2 norm. Of course, there are plenty of null sequences for which the limit of Aw_n is zero, for example $\sin nx/n^2$.

Then $\| w_n \|_{L^2} = \sqrt{2/n} \to 0$ while $Aw_n = 2$ for all n.

We are particularly interested in operators for which the first possibility (i) occurs:

Definition 21.6.2 An operator A is called *closable* or *pre-closed* if, for all null sequences $\{w_n\}$, with $w_n \in \mathscr{D}_A$, the fact that $Aw_n \to f$ implies $f = \emptyset$. Alternatively: for all null sequences, either $Aw_n \to \emptyset$ or it has no limit.

These are the unbounded operators that arise most frequently in the problems of classical field theory.

Example 21.6.1 Consider $D = d/dx$ with domain in $C^1[0, 1] \subset L^2(0, 1)$ as before (Example 21.1.6). We have just exhibited the example $\{\sin nx/n\}$ of a null sequence for which Dw_n has no limit. We now show that the operator is closable, i.e. that, *if* the null sequence $\{w_n\}$ is such that Dw_n has a limit f, then $f = 0$. Indeed, if $Dw_n \to f$, then $(v, Dw_n) \to (v, f)$ for any v by continuity of the scalar product (Theorem 19.3.2 p. 551). Take $v \in C^1[0, 1]$ with $v(0) = v(1) = 0$, which is a dense set in $L^2(0, 1)$. Then

$$(v, Dw_n) = \int_0^1 \overline{v} \frac{dw_n}{dx} \, dx = \overline{v} w_n \big|_0^1 - \int_0^1 \frac{d\overline{v}}{dx} w_n \, dx = -(v', w_n) \to 0 \qquad (21.6.3)$$

since $w_n \to 0$. Hence $(v, f) \to 0$, but since the v are dense in $L^2(0, 1)$, it follows that $f = 0$ (p. 554). □

The reason for calling such operators closable is that their domain can be extended in such a way that they become closed simply by defining $Au = f$ for any sequence $u_n \to u$ such that $\lim Au_n$ exists. No ambiguity in the definition of the action of A on u can arise because, if both $\{u_n\}$ and $\{v_n\}$ tend to u, $\{u_n - v_n\}$ is a null sequence. Then, either $A(u_n - v_n) \to \emptyset$, which means that the two limits are equal, or either Au_n or Av_n, or both, have no limit, since $A(u_n - v_n)$ has no limit. We may say that there are "good" and "bad" directions of approach to u, and that Au can be properly defined using only the good directions. By adding to $\mathscr{D}_A$ all points of $\overline{\mathscr{D}_A}$ for which a good direction of approach exists, we form an extension $\tilde{A}$ of A such that $\tilde{A} = A$ on $\mathscr{D}_A$.

Theorem 21.6.1 *The above extension uniquely defines a closed operator $\tilde{A}$ which is the minimal closed extension of A in the sense that, for any other closed A' such that $A = A'$ on $\mathscr{D}_A$, we would have $\mathscr{D}_{A'} \supset \mathscr{D}_{\tilde{A}}$.*

Not only cannot the domain of $\tilde{A}$ be extended to the entire space if the operator is unbounded, but it cannot even be a closed set as it can be shown that a closed linear operator on a closed domain in a Banach space is bounded (Banach's famous *closed graph theorem*).[17] This

---

[17] The closed graph theorem is equivalent to two other theorems (see e.g. Renardy & Rogers 1993, p. 240). The first one is the *Open Mapping Theorem*: If A is bounded and surjective (onto), the image of every open set is an open set. The second one is Banach's theorem 21.2.4 on the inverse operator.

implies that, for an unbounded operator, $\mathscr{D}_{\tilde{A}}$ cannot contain all the points of $\overline{\mathscr{D}}_A$ or, equivalently, that there are points of $\overline{\mathscr{D}}_A$ for which no good direction of approach exists. Therefore a closed, unbounded operator cannot be defined on the entire space, and, conversely, an operator whose domain cannot be extended beyond a proper dense subspace of H is necessarily unbounded. In particular, differential operators can only be defined on domains at most dense in the space. Of course, the norm of an operator depends on the norm adopted for the space to which its domain belongs and therefore, in particular, an operator may be bounded in one space and not in another one. We have seen two instances in §19.5 and in Example 21.2.2.

Example 21.6.2 We have seen in the previous example that the differentiation operator in $C^1[0, 1] \subset L^2(0, 1)$ is closable. The domain of its closure consists of all the *absolutely continuous* functions (definition p. 685) which are continuous and have the property that there is an L^2 function v such that

$$u(x) = \int_0^x v(\xi)\,\mathrm{d}\xi + u(0). \tag{21.6.4}$$

For such functions, $\tilde{D}u = v$. Clearly $\tilde{D}$ defined in this way coincides with the usual definition of derivative when v is continuous, but it is more general as (21.6.4) makes sense also when v is only integrable. For example, v may be undefined on a set of measure zero.　　□

More generally, all differential operators with smooth coefficients are closable (see e.g. Renardy & Rogers 1993, p. 238). This includes all the classical differential equations of mathematical physics and, therefore, we may safely assume to be dealing with closed operators. A general result of this nature is very useful because, usually, it is no easy matter to determine the actual closure of a closable operator, i.e. to fully characterize all – and only those – functions that belong to the domain of the closure.

We have seen that the null space of a bounded operator is closed (p. 626) and that the inverse of a bounded operator is bounded (p. 627); similarly:

Theorem 21.6.2 *The null space of a closed operator is closed. If* A *is closed and* A^{-1} *exists on* $\mathscr{R}_A$*, then* A^{-1} *is closed.*

21.6.1 Adjoint of a closed operator

In defining the adjoint of a bounded operator we could rely on the Riesz representation theorem as in that case the operator could be considered to be defined over the entire Hilbert space H. For unbounded operators this is not possible as their domain is at most dense in H. But we can still take an element $v \in H'$ and consider (v, Au) as u varies over $\mathscr{D}_A$. Again this is a linear functional, for which it may happen that some $g_v \in H$ exists such that $(g_v, u) = (v, Au)$ for all $u \in \mathscr{D}_A$. The difference with the bounded case is that the existence of g_v is not guaranteed for all v.

In order for the correspondence $v \leftrightarrow g_v$ to be well defined g_v should be unique. Suppose that there were two different g's, g_1 and g_2, for both of which $(g, u) = (v, Au)$. Then, subtracting, $(g_1 - g_2, u) = 0$ for all $u \in \mathscr{D}_A$. If we want to deduce from this relation that $g_1 = g_2$, we must make the assumption that $\mathscr{D}_A$ is at least dense in H, which is expressed by labelling the operator as *densely defined*. This makes the correspondence unique and we may introduce the adjoint operator by $g_v = A^* v$. Thus

$$(v, Au) = (A^* v, u) \qquad \text{for all} \qquad u \in \mathscr{D}_A, \quad v \in \mathscr{D}_{A^*}. \qquad (21.6.5)$$

The linearity of A^* is immediately established.[18] Given the continuity of the scalar product, it is not surprising that if A is closable, A and its closure have the same adjoint A^*.

When A^{-1} exists and is densely defined, one can define its adjoint $(A^{-1})^*$. As stated in Theorem 21.5.1, $(A^{-1})^* = (A^*)^{-1}$.

In general, existence of the adjoint of the adjoint $(A^*)^* = A^{**}$ is not guaranteed as $\mathscr{D}_{A^*}$ may not be dense in $\mathscr{H}$. However (see e.g. Riesz & Sz.-Nagy 1955, p. 305; Mlak 1991, p. 122; Locker 2000, pp. 6 and 7):

Theorem 21.6.3 *A densely defined operator* A *is closed if and only if* A^* *is densely defined. In this case,* A^{**} *exists and* $A^{**} = A$.

For the type of problems we are dealing with the requirement that $\mathscr{D}_A$ be dense is not particularly stringent as, for example, the set of infinitely differentiable functions with compact support in any number of variables is dense in the corresponding L^2 space.

We have illustrated in §18.7 p. 511 the importance of self-adjoint operators mapping a finite-dimensional spaces into itself. Infinite dimensions render necessary a finer distinction on the basis of the domain:

Definition 21.6.3

(i) An operator A is *symmetric* (or formally self-adjoint) if $\mathscr{D}_A \subseteq \mathscr{D}_{A^*}$ and $A^* = A$ for all the elements common to $\mathscr{D}_A$ and $\mathscr{D}_{A^*}$.
(ii) An operator A is *self-adjoint*, or *Hermitian*, if $\mathscr{D}_A = \mathscr{D}_{A^*}$ and $A = A^*$ on their common domain.

For a symmetric operator (u, Au) is real. Every self-adjoint operator is evidently symmetric.

If A is bounded, $\mathscr{D}_A = H$ and therefore there is no difference between self-adjointness and symmetry. However unbounded operators exist which are symmetric but not self-adjoint since $\mathscr{D}_{A^*}$ can be larger than $\mathscr{D}_A$. In fact, it is a general property that (see e.g. Roman 1975, p. 541)[19]

$$\mathscr{D}_A \subseteq \mathscr{D}_{A^{**}} \subseteq \mathscr{D}_{A^*}. \qquad (21.6.6)$$

---

[18] Actually, somewhat unexpectedly, this argument shows that A^* is a linear operator even when A is not. What we have defined here is the Hilbert adjoint, which is not the same as the adjoint of an operator acting between normed spaces (see e.g. Renardy & Rogers 1993, p. 252; Taylor & Lay 1980, p. 242).

[19] In general, if $\mathscr{D}_A$ is dense and $A \subset B$, so that B is an extension of A, then $B^* \subset A^*$ (see e.g. Riesz & Sz.-Nagy 1955, p. 301; Mlak 1991, p. 121).

It is a straightforward consequence of Theorem 21.6.3 that a self-adjoint operator is closed. A symmetric operator is closable and its closure is also symmetric.

Example 21.6.3 Consider A defined, over the subset of absolutely continuous functions (Example 21.6.2) belonging to $L^2(a, b)$, by

$$Au = -i \frac{du}{dx} \qquad (21.6.7)$$

with

$$\mathscr{D}_A = \{u \in L^2 : u' \in L^2, u(a) = u(b) = 0\}. \qquad (21.6.8)$$

The operator A is symmetric since, if v is also absolutely continuous so that integration by parts is legitimate,

$$(v, Au) = \int_a^b \overline{v}(-iu')\,dx = -ivu\big|_a^b + \int_a^b i\,\overline{v}'u\,dx = (-iv', u) = (Av, u). \qquad (21.6.9)$$

To find the adjoint, given a v, we look for a w such that $(v, Au) = (w, u)$ for all $u \in \mathscr{D}_A$ or, in other words, such that

$$\int_a^b \overline{v}\left(-iu'\right)\,dx = \int_a^b \overline{w}u\,dx. \qquad (21.6.10)$$

Upon integrating by parts we find

$$\int_a^b \left(\overline{w + iv'}\right) u\,dx = -i\overline{v}\,u\big|_a^b = 0, \qquad (21.6.11)$$

where the last step follows from the boundary conditions satisfied by u which are specified by the definition (21.6.8) of the domain of A. This equality can be satisfied for all absolutely continuous u only if $w = -iv'$ so that $A^* = -i\,d/dx$. As for the domain of A^*, we see that the equality (21.6.10) will be true irrespective of the values of $v(a)$ and $v(b)$ so that

$$\mathscr{D}_{A^*} = \{v \in L^2 : v' \in L^2\}, \qquad (21.6.12)$$

with no additional conditions at $x = a, b$.[20] Thus, the *formal operators* of A and A^* (i.e., the operations corresponding to A and A^*) are the same, but the operators – consisting of operations plus domain – are not equal and $\mathscr{D}_A \subset \mathscr{D}_{A^*}$ in conformity with (21.6.6). While A is symmetric, it is not self-adjoint and, in fact, A^* can be seen as an extension of A. A similar calculation shows that $A^{**} = A$ in conformity with Theorem 21.6.3, which implies that A^* is not self-adjoint either.

Just as in the finite-dimensional case, the operator A^*A is symmetric (actually self-adjoint) for any closed A even if unbounded. In the present example $A^*Au = -d^2u/dx^2$. As for the domain of this operator, we must require that $u \in \mathscr{D}_A$ (21.6.8), and also that $Au \in \mathscr{D}_{A^*}$ defined in (21.6.12). □

---

[20] Strictly speaking, this argument does not prove that all possible pairs v, w have been found. Nevertheless this is, indeed, true (see e.g. Stakgold 1997, p. 317.)

Example 21.6.4 Define the operation of A as in the previous example but change the definition of its domain to

$$\mathscr{D}_A = \{u \in L^2 : u' \in L^2, \ u(a) = 0\} \tag{21.6.13}$$

while $u(b)$ is arbitrary. An argument similar to the previous one gives then $A^* = -i \, d/dx$, with

$$\mathscr{D}_{A^*} = \{v \in L^2 : v' \in L^2, \ v(b) = 0\}. \tag{21.6.14}$$

In this case there is only partial overlap between the domains of A and A^* and neither operator is symmetric. □

Example 21.6.5 As before, but now with

$$\mathscr{D}_A = \{u \in L^2(a, b) : u' \in L^2, \ u(b) = \beta u(a)\} \tag{21.6.15}$$

with β a constant. The same integration-by-parts argument shows that

$$\mathscr{D}_{A^*} = \{v \in L^2(a, b) : v' \in L^2, \ v(a) = \overline{\beta} v(b)\}. \tag{21.6.16}$$

This is equal to $\mathscr{D}_A$ if and only if $\beta = 1/\overline{\beta}$ or $|\beta|^2 = 1$. Any β with unit modulus gives rise to a self-adjoint operator. In particular, with $\beta = 1$, we have the operator on the eigenfunctions of which the Fourier series is constructed. □

The relations (21.2.17), namely

$$\mathscr{R}_A^\perp = \mathscr{N}_{A^*}, \qquad \mathscr{R}_{A^*}^\perp = \mathscr{N}_A, \tag{21.6.17}$$

also hold for a densely defined closed operator. Indeed, if $v \in \mathscr{R}_A^\perp$, then $(v, Au) = 0 = (A^*v, u)$ which, by the density of the u, implies that $A^*v = \emptyset$ so that $\mathscr{R}_A^\perp \subset \mathscr{N}_{A^*}$. The converse relation $\mathscr{N}_{A^*} \subset \mathscr{R}_A^\perp$ is proven similarly starting with $v \in \mathscr{N}_{A^*}$. The second one of (21.6.17) follows by interchanging the roles of A and A^* (see e.g. Taylor & Lay 1980, p. 244; Renardy & Rogers 1993, p. 253). As before, by the closure of the orthogonal complement of the orthogonal complement of a set (p. 554), these relations imply

$$\overline{\mathscr{R}_A} = \mathscr{N}_{A^*}^\perp, \qquad \overline{\mathscr{R}_{A^*}} = \mathscr{N}_A^\perp. \tag{21.6.18}$$

Example 21.6.6 Let $H^k[a, b]$ denote the subspace of $L^2(a, b)$ consisting of all functions belonging to $C^{k-1}[a, b]$ with $(k-1)$-st derivative absolutely continuous on $[a, b]$ and k-th derivative in $L^2(a, b)$;[21] define the differential operator

$$A = \sum_{j=0}^{n} a_j(x) \frac{d^j}{dx^j} \tag{21.6.19}$$

---

[21] With the norm (19.5.4) p. 561, these functions constitute the Sobolev space $W^{k,2} = H^k$.

with the real coefficients $a_j(x) \in H^j[a, b]$ for $j = 0, 1, \ldots, n$, and $a_n(x) \neq 0$ on $[a, b]$. Define the domain of the operator as consisting of functions $u \in H^n[a, b]$ satisfying the ℓ linearly independent boundary conditions

$$B_k(u) = \sum_{j=0}^{n-1} \alpha_{kj} u^{(j)}(a) + \sum_{j=0}^{n-1} \beta_{kj} u^{(j)}(b) = 0, \qquad k = 1, 2, \ldots, \ell \le 2n, \quad (21.6.20)$$

in which the superscript (j) indicates the derivative of order j and α_{kj}, β_{kj} are numbers. When all these numbers vanish the functions in $\mathscr{D}_A$ are not required to satisfy any boundary conditions and we have the *maximal operator* A_M. At the opposite extreme, when the boundary conditions are $u^{j-1}(a) = u^{j-1}(b) = 0$, with $j = 1, 2, \ldots, n$, we have the *minimal operator* A_m.

By forming a scalar product analogous to (21.6.9) and integrating by parts, the adjoint operator can be reduced to the form

$$A^* = \sum_{j=0}^{n} b_j(x) \frac{d^j}{dx^j}, \qquad (21.6.21)$$

with the $b_j \in H^j[a, b]$ given by linear combinations of the a_j and their derivatives. The domain of A^* consists of functions $u \in H^n[a, b]$ satisfying the boundary conditions

$$B_k^*(u) = \sum_{j=0}^{n-1} \alpha_{kj}^* u^{(j)}(a) + \sum_{j=0}^{n-1} \beta_{kj}^* u^{(j)}(b) = 0, \qquad k = 1, 2, \ldots, 2n - \ell, \quad (21.6.22)$$

where the constants α_{kj}^* and β_{kj}^* are found by requiring that the boundary values arising from the repeated integration by parts vanish, after accounting for the boundary conditions (21.6.20). In particular, it is easily seen that $A_m^* = A_M$ and $A_M^* = A_m$; Example 21.6.3 illustrates a simple special case. By (21.6.17) the dimension of $\mathscr{R}_A^\perp$ equals that of $\mathscr{N}_{A^*}$. It can be shown that the index of the operator (definition on p. 621) is given by

$$\dim \mathscr{N}(A) - \dim \mathscr{N}(A^*) = n - \ell. \qquad (21.6.23)$$

When the number ℓ of independent boundary conditions equals the order n of the differential operator the index vanishes, and this is the only case in which a well-structured spectral theory for operators of this type can be developed (see e.g. Locker 2000, p. 58); it is also the case most frequently occurring in applications.

The operator A is (*i*) densely defined and closed in $L^2(a, b)$,[22] (*ii*) its range is a closed subspace of $L^2(a, b)$ and (*iii*) the null spaces $\mathscr{N}(A)$ and $\mathscr{N}(A^*)$ have finite dimension. Operators enjoying these properties are called *Fredholm operators*. $\qquad\qquad\qquad\Box$

Several other examples of the calculation of the adjoint operator can be found in Chapters 15 and 16.

---

[22] Note that these operators are closed in the subspaces H^k but not in the subspaces C^k.

21.6.2 Normal operators

We have defined a normal operator N as an operator which commutes with its adjoint, which implies that the range and the domain of a normal operator belong to the same space. When the domain of the operator is not the entire space, it is necessary to specify that the equality $NN^* = N^*N$ must be understood as also implying equality of the domains of NN^* and N^*N. By noting that

$$\|Nu\|^2 = (Nu, Nu) = (u, N^*Nu) = (u, NN^*u) = (N^*u, N^*u) = \|N^*u\|^2 \quad (21.6.24)$$

we see that, if an operator is normal, then $\|Nu\| = \|N^*u\|$ for all vectors $u \in \mathscr{D}_N, \mathscr{D}_{N^*}$. The converse of this proposition is also true if the domain is dense in the space as, from the previous relations, we then have

$$\big(u, (NN^* - N^*N)u\big) = 0 \quad (21.6.25)$$

for all u, which implies that $NN^* - N^*N$ is the zero operator. An important property of normal operators is:

Theorem 21.6.4 *If* N *is normal and densely defined, then* $\mathscr{N}_N$ *and* $\overline{\mathscr{R}}_N$ *are orthogonal complements,* $H = \mathscr{N}_N \oplus \overline{\mathscr{R}}_N$.

This result is an immediate consequence of (21.6.17) once it is observed that, by (21.6.24), $N^*u = \emptyset$ if $Nu = \emptyset$, i.e., $\mathscr{N}_N = \mathscr{N}_{N^*}$.

21.7 Solvability conditions and the Fredholm alternative

In the treatment of Green's functions in §15.7 p. 396 we have already considered the solvability of a problem of the form

$$Au = f, \quad (21.7.1)$$

with A a linear differential operator. We now return to the same question in a more general setting. The operator can be bounded or unbounded, but we require that it be closed and densely defined in a Hilbert space H so that it admits an adjoint A^*.

A first obvious remark is that, unless f is in the range of A, the equation does not have a solution. Consider the homogeneous adjoint equation

$$A^*v = \emptyset \quad (21.7.2)$$

and form the scalar product

$$(v, f) = (v, Au) = (A^*v, u) = 0. \quad (21.7.3)$$

This relation shows that the condition $(v, f) = 0$ is *necessary* for the solvability of the original equation (21.7.1). In other words, the right-hand side f must be orthogonal to all the solutions of the homogeneous adjoint equation (cf. (15.7.4) p. 396). Whenever the equation $A^* v = \emptyset$ admits non-zero solutions, (21.7.3) imposes non-trivial conditions on the solvability of (21.7.1).

While necessary, the relation (21.7.3) is not sufficient because, by the continuity of the scalar product and as (21.6.18) shows, it would be satisfied by an $f \notin \mathscr{R}_A$ while still $f \in \overline{\mathscr{R}_A}$. Thus, a proper statement of the solvability condition must include the hypothesis of closed range, so that $\mathscr{R}_A = \overline{\mathscr{R}_A}$:

Theorem 21.7.1 *If* A *is densely defined operation with a closed range,* $Au = f$ *has solutions if and only if* f *is orthogonal to all solutions of* $A^* v = \emptyset$.

By Theorem 21.5.3 p. 648, a closed operator with a bounded inverse has a closed range and a conclusion similar to that of Theorem 21.7.1 therefore applies. A related useful criterion is the following:

Theorem 21.7.2 *Let* A *be a closed linear operator, and let* A *be bounded away from 0, in the sense of (21.5.3), on* $\mathscr{N}_A^\perp \cap \mathscr{D}_A$. *Then* $\mathscr{R}_A$ *is closed.*

The conditions of the theorem ensure the existence of a bounded inverse on the range of the operator similarly to Theorem 21.5.2.

Another important class of operators with a closed range is that of compact operators (§21.3), a class which in particular includes operators on a finite-dimensional space (Theorem 18.5.3 p. 507) or with finite-dimensional range (p. 637), but also others, such as integral operators with a Hilbert–Schmidt kernel. A famous result for such operators is

Theorem 21.7.3 [Fredholm alternative theorem] *Let* C *be a compact operator on a Hilbert space H and consider the equation*

$$(C - \lambda I)u = f \tag{21.7.4}$$

and the associated homogeneous equation

$$(C - \lambda I)v = \emptyset \tag{21.7.5}$$

with λ a given real or complex number.

(a) *Either the homogeneous equation (21.7.5) has only the the solution $v = \emptyset$ (i.e., λ is not an eigenvalue of C), in which case (21.7.4) has a unique solution for every $f \in H$, or*

(b) *Equation (21.7.5) has non-zero solutions $v_1, v_2, \ldots, v_K$, in which case the adjoint homogeneous equation $(C^* - \bar{\lambda} I)w = \emptyset$ also has an equal number of non-zero solutions $w_1, w_2, \ldots, w_K$, and (21.7.4) is solvable if and only if $(w_j, f) = 0$ for $1 \leq j \leq K$. In this case the solution is given by*

$$u = u_p + \sum_{j=1}^{K} c_j v_j, \tag{21.7.6}$$

where u_p is any particular solution of (21.7.4) and $c_1, c_2, \ldots, c_K$ are arbitrary constants.

The proof of this theorem in part is a direct consequence of the closure of the range of a compact operator. For the other part of the theorem, namely that the dimension of $\mathcal{N}_{C-\lambda I}$ equals that of $\mathcal{N}_{C^*-\bar{\lambda}I}$ so that the index vanishes (see e.g. Helmberg 1969, p. 196; Kubrusly 2001, p. 479 and also §21.9, p. 668).

From this theorem we can derive a similar conclusion for operators which admit a compact inverse on their range. As in §21.3.1 p. 642, we let $C = A^{-1}$, $\mu = 1/\lambda$, $g = -A^{-1}f/\lambda$. The equation $(A - \lambda I)u = f$ becomes $(C - \mu I)u = g$ to which the theorem applies.

Example 21.7.1 In $0 < x < \ell$ consider $A = d^2/dx^2$ with domain consisting of $L^2(0, \ell)$-functions with absolutely continuous first derivative and such that $u(0) = 0$, $u(\ell) = 0$. To determine the inverse we solve the equation $Aw = g$ finding

$$w = \int_0^\ell G(x, \xi) g(\xi) \, d\xi \quad \text{with} \quad G(x, \xi) = \frac{x\xi}{\ell} - xH(\xi - x) - \xi H(x - \xi).$$

(21.7.7)

The integral operator is evidently compact and, since its kernel is symmetric in x and ξ, it is also self-adjoint (Example 21.2.3 p. 629). Let $\lambda = -k^2$. The solution of $(A - \lambda I)u = f$ vanishing at $x = 0$ is

$$u = \frac{1}{k} \left[B \sin kx + \int_0^x \sin k(x - \xi) f(\xi) \, d\xi \right].$$

(21.7.8)

If $\sin k\ell \neq 0$, this equation can be solved to give a value of B such that $u(\ell) = 0$. However, if $\sin k\ell = 0$, which occurs when the homogeneous equation $(A - \lambda I)v = \emptyset$, which, here, is the same as $(A^* - \bar{\lambda}I)v = \emptyset$, has the non-trivial solution $v \propto \sin kx$, the condition $u(\ell) = 0$ requires that

$$\int_0^\ell \sin k(\ell - \xi) f(\xi) \, d\xi = 0 \quad \text{or} \quad \int_0^\ell \sin k\xi \, f(\xi) \, d\xi = 0,$$

(21.7.9)

which is precisely the orthogonality condition of the theorem. This problem was considered in the context of modified Green's functions in Example 15.7.1 p. 398.

Often this orthogonality condition has a transparent physical meaning. Consider for example a string of length ℓ vibrating under the action of a distributed force density $F(x, t) = f(x) \cos \omega t$ so that (p. 18)

$$\frac{1}{c^2} \frac{\partial^2 \eta}{\partial t^2} - \frac{\partial^2 \eta}{\partial x^2} = f(x) \cos \omega t$$

(21.7.10)

where η is the displacement of the string. The power exerted by the force on the element dx of the string is proportional to the product of the force $F \, dx$ times the velocity of the string element $\partial \eta/\partial t$. The total instantaneous power is therefore

$$\mathscr{P} \propto \int_0^\ell F \frac{\partial \eta}{\partial t} \, dx.$$

(21.7.11)

If we look for a steady solution of the form $\eta(x, t) = u(x) \cos(\omega t)$, the equation for $u(x)$ is the one we just solved in the previous example. If the parameters are such that $\sin k\ell = 0$, so that a normal mode of the string is resonant with the forcing, it is immediately verified that the orthogonality condition is equivalent to requiring that no net work be performed on the resonant mode for the obvious reason that an undamped oscillator forced at resonance does not have a steady state. □

The power of Fredholm's alternative theorem can be substantially increased by recasting the problem in a suitable Hilbert space. The general idea can be described on an example (see e.g. Ladyzhenskaya 1985, chapter 2). Consider the homogeneous Dirichlet problem

$$-\nabla^2 u - \lambda u = f, \qquad u|_{\partial\Omega} = 0, \tag{21.7.12}$$

in a bounded smooth domain Ω; λ is a complex number. The Laplace operator is unbounded in L^2 but we have seen in §19.5 that, if u and its first-order derivatives are square integrable over Ω and $u = 0$ on the boundary, then u belongs to the Hilbert space $H_0^1(\Omega)$ in which (p. 562)

$$(v, u)_{H_0^1} = \int \nabla\bar{v} \cdot \nabla u \, d\Omega \tag{21.7.13}$$

defines a proper scalar product and induced norm. By multiplying (21.7.12) by a function $v \in H_0^1$ and integrating by parts the first term we have

$$(v, u)_{H_0^1} - \lambda(v, u)_{L^2} = (v, f)_{L^2}. \tag{21.7.14}$$

For a given $u \in H_0^1$, $(v, u)_{L^2}$ is a bounded linear functional defined for all $v \in H_0^1$ and therefore, by the Riesz representation theorem 19.6.3 (p. 566), there is an element $U \in H_0^1$ such that $(v, u)_{L^2} = (v, U)_{H_0^1}$. This correspondence defines an operator $U = Au$ with domain and range in H_0^1, which is clearly self-adjoint and positive definite, and can be shown to be compact (Theorem 19.5.1). Similarly, $(v, f)_{L^2}$ defines a vector $F \in H_0^1$ such that $(v, f)_{L^2} = (v, F)_{H_0^1}$. The problem is then reduced to

$$Au - \frac{1}{\lambda}u = -\frac{1}{\lambda}F \tag{21.7.15}$$

which is the form (21.7.4) to which the Fredholm alternative theorem applies.

21.8 Invariant subspaces and reduction

The consideration that follow refer to operators having domain and range in the same space.

A subspace $H_1 \subset H$ is invariant under the action of an operator A when $Au \in H_1$ whenever $u \in H_1 \cap \mathscr{D}_A$. The vector $\emptyset$ and the entire space H are trivial invariant subspaces; the null space and the closure of the range of a bounded operator are some non-trivial invariant subspaces.

An eigenspace of a closed linear operator is closed (Theorem 21.6.2) and is an invariant subspace on which the operator itself reduces to a simple multiplication. In Chapter 18 and elsewhere in this book we have seen the use of such eigenspaces as essential ingredients in the practical solution of problems. To invariant subspaces is also connected the most important, and still unsolved, theoretical open problem of operator theory: does every operator on an infinite-dimensional separable Hilbert space have a non-trivial invariant subspace?[23] We summarize here a few important facts related to invariant subspaces and the reduction of an operator.

The following theorem is similar to a statement made in the proof of Theorem 18.5.2 for operators in finite-dimensional spaces (see e.g. Kubrusly 2001, p. 392):

Theorem 21.8.1 *A subspace H_1 of a Hilbert space H is invariant with respect to a densely defined operator* A *if and only if the orthogonal complement $H_1^\perp$ of H_1 in H is invariant with respect to* A^*. *Thus* A *has a non-trivial invariant subspace if and only if* A^* *does.*

A non-trivial *reducing subspace* H_1 for an operator A on a Hilbert space is a subspace (other than $\emptyset$ and H itself) such that both H_1 and $H_1^\perp$ are invariant under A. An operator is *reducible* if it has a non-trivial reducing subspace. If A_1 and $A_1^\perp$ are the restrictions of A to H_1 and $H_1^\perp$, respectively, on $H = H_1 \oplus H_1^\perp$ we can represent the operator as

$$A = \begin{vmatrix} A_1 & 0 \\ 0 & A_1^\perp \end{vmatrix}, \tag{21.8.1}$$

in which it is understood that a vector u is represented as $u = |u_1 \;\; u_1^\perp|^T$, with $u_1 = P_1 u$, $u_1^\perp = (I - P_1)u$. Thus, the study of A can be reduced to that of A_1 and $A_1^\perp$, which explains the denomination. It is evident that $A_1 = P_1 A P_1$ and similarly for $A_1^\perp$.

The definition extends in an obvious way to more than two reducing subspaces when the restriction of A to a reducing subspace can be reduced further. If the reduction can be continued so that all the reducing subspaces have dimension 1, the analog of (21.8.1) becomes a complete diagonalization of the operator, as in the case of a simple matrix (p. 513). In an infinite-dimensional space, this reduction can be attained only for restricted classes of operators, e.g. self-adjoint and compact.

The domain, range, null space and inverse (if it exists) of A can be decomposed similarly as sums of the domains, ranges, null spaces and inverses of A_1 and $A_1^\perp$.

A consequence of Theorem 21.8.1 is that H_1 reduces A if and only if it reduces A^*. Thus, if the operator is symmetric, every invariant subspace is a reducing subspace.

21.9 Resolvent and spectrum

In §21.2.2 we have briefly considered the eigenvalues of a bounded self-adjoint operator and in §21.3 we have addressed in some detail the eigenvalues and eigenvectors of compact operators. It is now time to consider the spectral properties of more general operators.

We consider an arbitrary linear operator A with domain $\mathscr{D}_A$ and associate with it the family of operators

$$A_\lambda = A - \lambda I \tag{21.9.1}$$

---

[23] A famous theorem proven by Lomonosov in 1973 states that: A linear operator in a Banach space has a non-trivial closed invariant subspace if it commutes with an operator which commutes with a non-zero compact operator (none of the operators mentioned being zero or a multiple of the identity).

where λ is a real or complex number. When it exists, the inverse operator

$$R_\lambda = (A - \lambda I)^{-1} \tag{21.9.2}$$

is called the *resolvent*. Existence and properties of this operator are of course matters of primary interest when attempting to solve the equation

$$(A - \lambda I)\, u = f. \tag{21.9.3}$$

Definition 21.9.1 The *resolvent set* $\rho(A)$ of the linear operator A consists of all the real or complex numbers λ for which A_λ^{-1} exists, has a dense domain $\mathcal{D}_{A_\lambda^{-1}} = \mathcal{R}_{A_\lambda}$ and is bounded, and therefore continuous. Any $\lambda \in \rho(A)$ is a *regular value* of A.

The domain of a bounded operator can be extended by continuity to the entire space (Hahn–Banach theorem 21.2.3 p. 626) but, in the previous definition, we refer only to the natural domain of A_λ^{-1}, namely the range of $\mathcal{R}_{A_\lambda}$ of $A - \lambda I$.

Definition 21.9.2 The *spectral set* or *spectrum* $\sigma(A)$ of a linear operator A consists of all the real or complex numbers λ which are not in the resolvent set.

The set $\rho(A)$ is open; this was shown explicitly for the special case of bounded operators in §21.2.4, but it is a general property (p. 670). In other words, there are sequences $\{\lambda_k\} \in \rho(A)$ which converge to $\lambda \notin \rho(A)$. Being the complement of an open set, the spectrum is closed and it contains therefore all its limit points.

   Considering the qualifications which specify the resolvent set, we see that the spectrum of a closed operator consists of:

1. Those λ for which $(A - \lambda I)^{-1}$ does not exist because the equation $(A - \lambda I)u = \emptyset$ has non-trivial solutions. In other words, λ is an eigenvalue with one or more linearly independent eigenvectors u; this is the *point spectrum* $P\sigma(A)$.
2. Those λ for which $A - \lambda I$ has a dense range (and therefore A_λ^{-1} a dense domain) on which it is one-to-one, but the inverse defined on this range is not bounded; this is the *continuous spectrum* $C\sigma(A)$.
3. Those λ for which $A - \lambda I$ is one-to-one on its range, which however is not dense. In this case, the inverse (bounded or not) exists on the range of A_λ, but it is not densely defined; this is the *residual spectrum* $R\sigma(A)$.

According to these definitions, the same $\lambda \in \sigma(A)$ cannot belong simultaneously to more than one component of the spectrum.[24]

   The eigenvalues of linear operators in a finite-dimensional space are discrete points and these operators therefore only have a point spectrum (§18.6). In an infinite-dimensional space, however, the point spectrum does not necessarily consist of isolated points, as will

---

[24] The possibility that $(A - \lambda I)^{-1}$ exist, be defined on the entire space but be unbounded does not occur for closed operators.

be seen in some of the following examples. Thus, the denomination *discrete spectrum* sometimes used for the point spectrum does not necessarily reflect the true nature of this set.[25]

As a general rule, establishing that λ belongs to the point spectrum requires solving the corresponding eigenvalue problem, i.e., finding the associated eigenvector(s).

Example 21.9.1 Consider the bounded self-adjoint operator

$$Au = x\,u(x) + \int_0^1 u(x)\,dx \qquad (21.9.4)$$

with domain $\mathscr{D}_A = L^2(0, 1)$. The eigenvalue problem $Au = \lambda u$ is readily solved to find the single eigenvalue $\lambda_0 = e/(e-1)$ with the corresponding eigenvector $u_0 \propto 1/(\lambda_0 - 1)$. Since $\lambda_0 > 1$, $u_0 \in L^2(0, 1)$. By showing that $A - \lambda I$ is not bounded away from 0 when $0 \le \lambda \le 1$, we prove that each $\lambda \in [0, 1]$ belongs to the continuous spectrum (Theorem 21.5.2). To solve $A_\lambda u = f$ with $f \in L^2(0, 1)$, we introduce the unknown constant $U = \int_0^1 u(x)\,dx$ to find $u = (f - U)/(x - \lambda)$. Upon substituting into the definition of U, we determine this constant as

$$U = \frac{\int_0^1 [f(x)/(x - \lambda)]\,dx}{1 + \log|(\lambda - 1)/\lambda|}. \qquad (21.9.5)$$

The denominator vanishes if $\lambda = \lambda_0$ and the integral in the numerator does not exist if $0 \le \lambda \le 1$. Thus, this solution belongs to $L^2(0, 1)$ only if $\lambda \neq \lambda_0$ and $\lambda \notin [0, 1]$ so that all complex numbers except these belong to the resolvent set. $\qquad\qquad \square$

It is seen in (21.9.5) that, the closer λ is to the eigenvalue λ_0, the larger is U and, therefore, the norm of the solution. It is a general property of the resolvent of a normal operator (already suggested by the bound (21.2.50) for a bounded operator) that its norm is always bounded from below by the inverse of the distance $d(\lambda, \sigma(A))$ between λ and the closest point of the spectrum (see e.g. Richtmyer 1978, vol. 1 p. 148; Kubrusly 2001, p. 501)

$$\|(A - \lambda I)^{-1}\| \ge \frac{1}{d(\lambda, \sigma(A))}. \qquad (21.9.6)$$

In particular, the closer λ is to the spectrum of A, the greater the effect of errors in the data on the quality of the solution of $(A - \lambda I)u = f$ (cf. the discussion of the condition number of matrices in §18.8).

Example 21.9.2 As already mentioned, the point spectrum does not necessarily consist of isolated points. As an example, consider the operator $A = -d^2/dx^2$ acting on functions $C^1[-1, 1]$ with u' absolutely continuous and $u'' \in L^2(-1, 1)$.[26] We complete the specification of the domain of the operator by requiring that

$$u(-1) - u(1) = 0, \qquad u'(-1) + u'(1) = 0. \qquad (21.9.7)$$

---

[25] Some authors define the discrete spectrum as the subset of the point spectrum consisting of isolated eigenvalues with finite-dimensional eigenspaces.

[26] In the language of Example 21.6.6 p. 655 these functions belong to the subspace H^2 of L^2.

It is evident that the equation $Au - \lambda u = \emptyset$ has the non-trivial solution $u(x) \propto \cos(\sqrt{\lambda}\, x)$ whatever λ. Thus the resolvent set is empty and the entire complex plane belongs to the point spectrum with the null space $\mathcal{N}_{A-\lambda I}$ having dimension 1.

The adjoint operator is readily found to be $A^* = -d^2/dx^2$ acting on functions having the same features as those of A, but satisfying the boundary conditions

$$v(-1) + v(1) = 0, \qquad v'(-1) - v'(1) = 0, \tag{21.9.8}$$

so that $\mathcal{D}_A \neq \mathcal{D}_{A^*}$. The equation $A^*v - \lambda v = \emptyset$ has the non-trivial solution $v(x) \propto \sin(\sqrt{\lambda}\, x)$ whatever λ. Again, the resolvent set is empty and the entire complex plane belongs to the point spectrum with the null space $\mathcal{N}_{A^*-\lambda I}$ having dimension 1. $\qquad\qquad\square$

Example 21.9.3 A seemingly slight change in the previous example produces an operator with an empty spectrum. Consider again $A = -d^2/dx^2$ acting on C^1-functions with absolutely continuous u' and $u'' \in L^2$. If we now impose

$$u(0) = 0, \qquad u'(0) = 0, \tag{21.9.9}$$

we readily find that $Au - \lambda u = 0$ has only the 0 solution. This is of course in keeping with the existence and uniqueness theorem for ordinary differential equations. $\qquad\square$

As already noted, the *geometric multiplicity* of an eigenvalue λ is the dimension of the null space $\mathcal{N}_{A_\lambda}$, i.e. the number of linearly independent solutions of $(A - \lambda I)u = \emptyset$. The eigenvalue is simple if the eigenspace is one-dimensional and degenerate otherwise.

The eigenspace of the eigenvalue of a closed operator A is closed (Theorem 21.6.2) and is an invariant subspace of A. If the space is separable, the eigenspace will possess a finite or at most a countably infinite orthonormal basis. The same argument used in the finite-dimensional case (p. 510) shows that a finite or countably infinite set of eigenvectors, each one belonging to a different eigenvalue, is linearly independent.

In spite of the name, the continuous spectrum need not be uncountable and it may, in fact, consist of a single point as in the following example.

Example 21.9.4 Let $\{e_k\}$ be an orthonormal basis in a Hilbert space and consider a sequence of real numbers λ_k all smaller than 1 and such that $\lambda_k \to 1$ from below. For any vector $u = \sum_k u_k e_k$ define the operator A by $Au = \sum_k \lambda_k u_k e_k$. Since $(A - \lambda I)u = \sum_k (\lambda_k - \lambda)u_k e_k$, the only eigenvalues are the λ_k with corresponding eigenvectors e_k because, for any other λ, the equation $(A - \lambda I)u = \emptyset$ would imply that all $u_k = \emptyset$ so that $u = \emptyset$. Since all the e_k are in the domain $\mathcal{D}_{A_\lambda^{-1}} = \mathcal{R}_{A_\lambda}$, the domain of $(A - \lambda I)^{-1}$ is certainly dense and, if $v = \sum_k v_k e_k$, clearly,

$$(A - \lambda I)^{-1}v = \sum_k \frac{1}{\lambda_k - \lambda} v_k e_k. \tag{21.9.10}$$

This relation shows that, if $\lambda = 1$, the norm of $(A - \lambda I)^{-1}v$ can be made arbitrarily large by selecting v so that it has non-zero components v_k corresponding to λ_k close to 1. In this case the continuous spectrum of A consists of the single point $\lambda = 1$ and all other $\lambda \neq \lambda_k$ are in the resolvent set.   □

In addition to the classification on p. 662, there are several other ways to subdivide the spectrum of an operator. For example, the *approximate point spectrum* is defined as the set of λ's such that a sequence of unit-norm vectors $\{w_n\}$ exists with the property that

$$\lim_{n \to \infty} (A - \lambda I)w_n = \emptyset. \tag{21.9.11}$$

In this case, λ is called a *generalized, asymptotic* or *approximate eigenvalue*. A regular eigenvalue is also a generalized eigenvalue (just set w_n equal to an eigenvector for all n), but the approximate point spectrum is larger. Indeed, if (21.9.11) is satisfied, it is obvious that the operator $A - \lambda I$ is not bounded away from $\emptyset$ and, therefore, its inverse is unbounded. Thus, the approximate spectrum includes the point spectrum, the continuous spectrum and that part of the residual spectrum for which the inverse exists but is unbounded (see e.g. Mlak 1991, p. 110; Stakgold 1997, p. 325; Kubrusly 2001, p. 453).

It is an immediate consequence of this definition that $\lambda = 0$ must be an approximate eigenvalue for a compact operator C. Indeed, by considering the sequence of normalized eigenvectors, we see that $Ce_n - 0\,e_n = \lambda_n e_n$, and this sequence terminates if 0 is a true eigenvalue, or tends to zero if it is not according to the results of the Hilbert–Schmidt Theorem 21.3.4 p. 639. Another instance of this concept is offered by the previous example 21.9.4, for which we see that $\lambda = 1$ belongs to the approximate point spectrum by taking $w_n = e_n$. Setting $(A - \lambda I)e_n = \eta_n$, we have $e_n = R_\lambda \eta_n$, with $\|e_n\| = 1$ for all n, while $\|\eta_n\| \to 0$, which suggests that the resolvent is unbounded, as can be seen from (21.9.10) (see e.g. Mlak 1991, p. 110).

Example 21.9.5 The operator $Au = -i\,du/dx$, with both u and du/dx in $L^2(-\infty, \infty)$, is self-adjoint. The formal solution of the eigenvalue equation $Au = \lambda u$ is $u \propto e^{i\lambda x}$, which is not an eigenvector as it is not square integrable over the real line. However, all elements of the sequence $u_n = n^{-1/2}\pi^{-1/4}\exp[-x^2/(2n^2) + i\lambda x]$, with λ real, have norm 1 and $\|(A - \lambda I)u_n\| = (2n^2)^{-1} \to 0$. Thus, for this operator, all real λ are generalized eigenvalues. In this case, the general solution of (21.9.3) is

$$u(x) = ce^{i\lambda x} + \int_0^\infty e^{i\lambda(x-\xi)} f(\xi)\,d\xi. \tag{21.9.12}$$

If $\operatorname{Im}\lambda > 0$, u will vanish as $x \to -\infty$ provided

$$c = \int_{-\infty}^0 e^{-i\lambda\xi} f(\xi)\,d\xi \qquad \text{so that} \qquad u(x) = \int_{-\infty}^x e^{i\lambda(x-\xi)} f(\xi)\,d\xi. \tag{21.9.13}$$

If $\operatorname{Im}\lambda < 0$, a similar argument gives

$$u(x) = -\int_x^\infty e^{i\lambda(x-\xi)} f(\xi)\,d\xi. \tag{21.9.14}$$

Thus, all non-real λ belong to the resolvent set. If λ is real, $u(x) \in L^2(-\infty, \infty)$ if and only if

$$\int_{-\infty}^{\infty} e^{-i\lambda\xi} f(\xi) \, d\xi = 0. \tag{21.9.15}$$

The set of such functions is dense in L^2 (the condition just requires that the Fourier transform of f vanish at $-\lambda$), but evidently it does not exhaust L^2.

It is interesting to note that, if the same operator were considered on the finite interval $-a < x < a$ with boundary conditions that would make it self-adjoint (e.g., periodicity), one would find a point spectrum with eigenvalues $\lambda_k = \pi k/a$. These eigenvalues "coalesce" into the real line as $a \to \infty$ and the point spectrum turns into a continuous spectrum; we have seen several examples of this phenomenon in dealing with singular Sturm–Liouville problems in §15.4 p. 375 (see also e.g. Stakgold 1979, p. 430).   □

As noted in §21.4, two operators A_1, A_2 are said to be *unitarily equivalent* when they are connected by a similarity transformation, $A_2 = UA_1U^*$, with a unitary operator U. As in the finite-dimensional case (p. 509), unitarily equivalent operators have the same spectrum (see e.g. Mlak 1991, p. 110).

Example 21.9.6 Consider the self-adjoint multiplication operator $Au(x) = x\,u(x)$ defined for functions u such that $xu(x) \in L^2(-\infty, \infty)$. We know from the theory of the Fourier transform that $\mathscr{F}(xu) = -ik d\tilde{u}/dk$ (p. 270), and we have already remarked on p. 646 that $xu = \mathscr{F}^{-1}\{(-ik d/dk)\mathscr{F}(u)\}$. The operators x of this example and $-ik d/dk$ of the previous example are thus unitarily equivalent and indeed have the same spectrum, $C\sigma(A) = (-\infty, \infty)$, $P\sigma(A) = 0$, $R\sigma(A) = 0$.   □

We now summarize some basic features of the spectrum of several types of linear operators, in some cases repeating results already presented. For proofs see many of the books cited in this chapter, and in particular Roman 1975, Stakgold 1979, Taylor & Lay 1980, Mlak 1991 and Kubrusly 2001.

21.9.1 Bounded operators

We have dealt with several aspects of the theory of bounded operators in §21.2 and §21.3.

As noted on p. 636, the spectrum of a bounded operator B is not empty, although this feature does not imply that eigenvalues necessarily exist. By (ii) of Theorem 21.5.1 (p. 648), λ is in the resolvent set $\rho(B)$ if and only if $\bar{\lambda} \in \rho(B^*)$. Since $\sigma(B) + \rho(B) = \mathbb{C}$, we then conclude that the spectrum of B is the complex conjugate of the spectrum of B*. If $\lambda \in C\sigma(B)$, then $\bar{\lambda} \in C\sigma(B^*)$ and $\sigma(B^{-1}) = [\sigma(B)]^{-1}$.

As on p. 516, we define the *numerical range* $W(B)$ of an operator as

$$W(B) = \{\lambda \in \mathbb{C} : \lambda = (u, Bu) \text{ for some } \|u\| = 1\}. \tag{21.9.16}$$

Then we have that $\sigma(B) \subseteq \overline{W}$. Since for a bounded operator $\lambda \in \sigma(B)$ if and only if $|\lambda| \le \|B\|$, it follows that the spectral radius $r_\sigma(B) \le \|B\|$, with equality holding when B is normal, $BB^* = B^*B$. The spectral radius may be 0, for example in the case of a nilpotent operator.[27]

Theorem 21.9.1 [Spectral mapping theorem] *If $p(x)$ is any polynomial with real or complex coefficients and B a bounded operator, $\sigma(p(B)) = p(\sigma(B))$.*

Normal bounded operators. If N is normal, so is $N_\lambda = N - \lambda I$ and therefore, from (21.6.24), $\|(N - \lambda I)u\| = \|(N^* - \bar{\lambda}I)u\|$. Thus, an eigenvector u of N corresponding to λ, is also an eigenvector of N^* corresponding to $\bar{\lambda}$ so that $P\sigma(N^*) = \overline{P\sigma(N)}$ with the same eigenspaces; we have already seen this property in the case of finite-dimensional spaces (p. 511).

Eigenvectors u_1 and u_2 belonging to distinct eigenvalues λ_1 and λ_2 are orthogonal as

$$(u_2, Nu_1) = \lambda_1(u_2, u_1) = (N^*u_2, u_1) = \lambda_2(u_2, u_1) \tag{21.9.17}$$

and, if $\lambda_2 \neq \lambda_1$, this equality can only hold provided $(u_2, u_1) = 0$.

From Theorem 21.6.4 we know that $H = \mathcal{N}_{N-\lambda I} \oplus \overline{\mathcal{R}}_{N-\lambda I}$. If λ is not an eigenvalue, then $\mathcal{N}_{N_\lambda} = \{\emptyset\}$ and $H = \overline{\mathcal{R}}_{N_\lambda}$. Since the range of a bounded operator is closed, it follows that $\lambda \in \rho(A)$ so that $\sigma(A) = P\sigma(A)$.

Unitary operators. (§21.4). Unitary operators are bounded and normal and, therefore, their residual spectrum is empty. If $Uu = \lambda u$, then $(Uu, Uu) = |\lambda|^2(u, u)$, but also $(Uu, Uu) = (u, U^*Uu) = (u, u)$. Thus $|\lambda| = 1$ and all the eigenvalues lie on the unit circle; the same is true of the continuous spectrum. All other λ are in the resolvent set.

Bounded self-adjoint operators. For a bounded self-adjoint operator B (§21.2.2), consider the equation $(B - \lambda I)u = f$. Using the fact that (u, Bu) is real, upon taking the scalar product with u, we find

$$-(\text{Im}\,\lambda)\|u\|^2 = \text{Im}\,(u, f) \tag{21.9.18}$$

and, therefore, if $\text{Im}\,\lambda \neq 0$, f cannot vanish without u vanishing as well. This argument shows that all eigenvalues are real. By the Schwarz inequality, the previous relation gives $|\text{Im}\,\lambda|\|u\| \leq \|f\|$ from which, using $u = (B - \lambda I)^{-1}f$, we have

$$\|(B - \lambda I)^{-1}f\| \leq \frac{1}{|\text{Im}\,\lambda|}\|f\| \tag{21.9.19}$$

which is a weaker form of the previous relation (21.9.6). This relation also shows that $C\sigma$, if not empty, is confined to the real axis as well. Since B is bounded, so is $B_\lambda = B - \lambda I$ and therefore its range is closed. If λ is not real and therefore not an eigenvalue of B, it is not an eigenvalue of $B^* = B$ either, so that $\mathcal{N}_{B_\lambda^*} = \{\emptyset\}$. By (21.6.18), we therefore conclude

---

[27] An operator is nilpotent of order m if $A^m = 0$, while $A^{m-1} \neq 0$; cf. the definition for matrices in Table 18.1 p. 492.

that the range of $B - \lambda I$, which is the domain of $(B - \lambda I)^{-1}$, is the entire H, so that $R\sigma$ is empty and any non-real λ belongs to the resolvent set.

If λ is an eigenvalue, define

$$m(B) \equiv \inf_{\|u\|=1} (u, Bu) \le \lambda \le \sup_{\|u\|=1} (u, Bu) \equiv M(B). \tag{21.9.20}$$

Then $\sigma(B)$ lies in the interval $[m, M]$ and

$$\|B\| = \sup_{\|u\|=1} |(u, Bu)| = \max(|m|, |M|). \tag{21.9.21}$$

As a consequence of this relation, every $|\lambda| > \|B\|$ belongs to the resolvent set. When m and M are not eigenvalues, they belong to the continuous spectrum. Eigenspaces corresponding to distinct eigenvalues are orthogonal. If B is positive, or non-negative, the spectrum lies in $0 < \lambda < \infty$ or $0 \le \lambda < \infty$, respectively.

Compact operators. (§21.3). A compact (not necessarily self-adjoint) operator C has at most a countable set of eigenvalues $\lambda_1, \lambda_2, \ldots$, each one with a finite multiplicity. If there are infinitely many eigenvalues, then $\lim_{n \to \infty} \lambda_n = 0$ and there is no other accumulation point. As for $\lambda = 0$, it belongs either to the point, continuous or residual spectrum.[28] Since the continuous spectrum can at most only contain $\lambda = 0$, any non-zero λ for which $\mathcal{N}_{C-\lambda I} = \{\emptyset\}$ must belong to the resolvent. The point spectrum may be empty but, if the operator is also self-adjoint, there is at least one non-zero eigenvalue; the modulus of at least one eigenvalue equals $\|C\|$. Whether $\lambda \in \rho(C)$ or $\lambda \in \sigma(C)$, the dimensions of $\mathcal{M}(C - \lambda I)$ and $\mathcal{M}(C^* - \bar{\lambda} I)$ are equal so that the index of $C - \lambda I$ vanishes (see e.g Kubrusly 2001, p. 479).

Projection operators. We have seen that an orthogonal projection operator is idempotent (Example 21.1.3), self-adjoint and non-negative definite (Example 21.2.4). These operators only have a point spectrum with eigenvalues 0 and 1. In an infinite-dimensional space, at least one of them is infinitely degenerate. Every other λ is in the resolvent set.

21.9.2 Unbounded closed operators

Let A be a *closed* operator with a dense domain in a Hilbert space. As in the bounded case, if $\lambda \in C\sigma(A)$, then $\bar{\lambda} \in C\sigma(A^*)$; if $\lambda \in \rho(A)$, then $\bar{\lambda} \in \rho(A^*)$; and if $\lambda \in P\sigma(A)$, then either $\bar{\lambda} \in P\sigma(A^*)$ or $\bar{\lambda} \in R\sigma(A^*)$.

Normal unbounded operators. As in the bounded case, N and N^* have the same eigenspaces with complex conjugate eigenvalues. Furthermore, eigenspaces corresponding to distinct eigenvalues are orthogonal. If $\mathcal{R}_{N_\lambda} = H$, λ is in the resolvent set, while

---

[28] $\lambda = 0$ may belong to the resolvent set when $\mathcal{D}_A$ is finite dimensional.

$\lambda \in C\sigma(\mathsf{N})$ if $\mathscr{R}_{N_\lambda} \neq H$ but is dense. If A is normal, closed, densely defined and $\mathscr{D}_A = \mathscr{D}_{A^*}$, $R\sigma(\mathsf{A})$ is empty.

Symmetric unbounded operators. All the developments in the paragraph containing (21.9.19) except the last one apply also to a general unbounded symmetric operator A, and we therefore conclude that $P\sigma(\mathsf{A})$ and $C\sigma(\mathsf{A})$ are confined to the real axis. We cannot however deduce that $R\sigma(\mathsf{A})$ is empty because now $\mathsf{A} \neq \mathsf{A}^*$ (their domains may be different) and therefore $\mathscr{N}_{A_\lambda^*}$ does not have to be just the $\emptyset$ vector. As a consequence, we cannot use (21.6.18) to conclude that $\mathscr{R}_{A_\lambda} = H$, which means that the inverse operator $(\mathsf{A} - \lambda \mathsf{I})^{-1}$ is not necessarily defined over the entire space either. However, if $R\sigma(\mathsf{A})$ is not empty, it must be complex and, if it contains a complex point, then it contains also the entire upper or lower semi-plane to which that complex value belongs (it is also possible that it contains both semi-planes). Eigenspaces corresponding to distinct eigenvalues are orthogonal.

If the operator is unbounded and not only symmetric, but *self-adjoint*, the conclusion that the residual spectrum is empty however does apply and, actually, it can be shown that the operator is self-adjoint if and only if all non-real complex numbers are in the resolvent set, i.e., the entire spectrum is real. Due to the unboundedness of the operator, in this case, one or both of m and M defined in (21.9.20) will be infinite.

An operator is *essentially self-adjoint* when its closure is self-adjoint. For such operators the entire spectrum is real as well and the residual spectrum empty.

21.10 Analytic properties of the resolvent

As defined in (21.9.2), the resolvent R_λ associated to an operator A is $R_\lambda = (\mathsf{A} - \lambda \mathsf{I})^{-1}$, when the inverse exists. This operator satisfies the *resolvent equation*

$$\frac{R_\lambda - R_\mu}{\lambda - \mu} = R_\lambda R_\mu = R_\mu R_\lambda, \tag{21.10.1}$$

as is readily proven by multiplying both sides on the left by $(\mathsf{A} - \lambda \mathsf{I})$ or on the right by $(\mathsf{A} - \mu \mathsf{I})$; it is also clear that R_λ and R_μ commute. An explicit example of this general relation for the Sturm–Liouville problem was given in (15.3.47) p. 373.

For bounded operators, we mentioned in §21.2.4 that the resolvent can be considered an operator-valued function of the complex variable λ, and we have mentioned that several properties of ordinary analytic functions extend to such operator-valued functions. Related comments and results were given in §18.9 and §18.11 for the case of matrices.

For an unbounded operator A, the Neumann series (21.2.27)

$$R_\lambda = -\frac{1}{\lambda} \sum_{j=0}^{\infty} \frac{A^j}{\lambda^j} \tag{21.10.2}$$

may not converge uniformly, but it can still converge strongly and can be used to solve the equation $(\mathsf{A} - \lambda \mathsf{I})u = f$ when it does. A similar comment applies to the more general expansion (21.2.49) which may be rewritten as

$$R_\mu = \sum_{n=0}^{\infty} (\mu - \lambda)^n R_\lambda^{n+1}.$$

(21.10.3)

As mentioned earlier, this expansion proves that the resolvent set is open.

The connection between operator-valued analytic functions and ordinary analytic functions can be clarified by applying both sides of this relation to a vector u such that the series converges strongly, and taking the scalar product with another vector v to find

$$(v, R_\mu u) = \sum_{n=0}^{\infty} (\mu - \lambda)^n (v, R_\lambda^{n+1} u).$$

(21.10.4)

This relation shows that $(v, R_\mu u)$ is an analytic function of μ. Upon taking the derivative and evaluating at $\lambda = \mu$ we have

$$\frac{d}{d\mu}(v, R_\mu u)\bigg|_{\mu=\lambda} = (v, R_\lambda^2 u),$$

(21.10.5)

which may also be written as

$$\frac{d}{d\mu}R_\mu\bigg|_{\mu=\lambda} = R_\lambda^2$$

(21.10.6)

and is the same result as found from the resolvent equation (21.10.1) upon taking the limit $\mu \to \lambda$. Thus, the resolvent R_λ is an analytic operator-valued function of λ on the resolvent set. The eigenvalues of A give rise to poles of R_λ and the continuous spectrum generates branch cuts. Equation (21.10.2) shows that R_λ is analytic at infinity, where it has a zero of order 1.

Suppose that A has a pure point spectrum which is bounded. If we integrate (21.10.2) term by term around a contour which encloses the spectrum, we have from the analog of Cauchy's integral formula (Theorem 17.3.3 p. 428)

$$I = -\frac{1}{2\pi i}\oint_C R_\lambda \, d\lambda.$$

(21.10.7)

If the point spectrum can be separated into n disjoint sets, the contour C can be replaced by n non-intersecting contours C_i, each one encircling one set, and

$$I = \sum_{i=1}^{N} P_i \quad \text{where} \quad P_i = -\frac{1}{2\pi i}\oint_{C_i} R_\lambda \, d\lambda$$

(21.10.8)

is the projector on the eigenspace corresponding to the eigenvalues encircled by the i-th contour. Continuing this line of argument one can show that

$$A^n = -\frac{1}{2\pi i}\oint_C \lambda^n R_\lambda \, d\lambda$$

(21.10.9)

and, more generally

$$f(A) = -\frac{1}{2\pi i}\oint_C f(\lambda) R_\lambda \, d\lambda,$$

(21.10.10)

where the integration contour can be replaced by a set of smaller contours as before; we have already seen relations of this type in the case of matrices on p. 522. Some applications of these ideas can be found e.g. in Taylor & Lay 1980, section V.8. Explicit implementations of these general expressions are given in connection with Green's functions in §15.3.3 p. 373 and 15.4.3 p. 385.

21.11 Spectral theorems

We mentioned in §21.3.2 that, in the case of a compact normal operator with an infinite number of eigenvalues, there is a resolution of the identity

$$I = \sum_{k=1}^{\infty} P_k \tag{21.11.1}$$

with a spectral representation, or decomposition, of the operator

$$A = \sum_{k=1}^{\infty} \lambda_k P_k \quad \text{or} \quad Au = \sum_{k=1}^{\infty} \lambda_k P_k u, \tag{21.11.2}$$

where P_k is the orthogonal projector on the eigenspace corresponding to the eigenvalue λ_k, satisfying $P_k P_n = \delta_{kn} P_k$. At the end of the previous section we have seen that these relations remain valid when A has a bounded pure point spectrum. The question we turn to now is whether one can establish similar relations for other classes of operators. This issue requires considerable new developments to be addressed properly. We will limit ourselves to some very heuristic considerations.

The crucial property of compact normal operators that underlies (21.11.1) and (21.11.2) is that they have a non-empty point spectrum and sufficiently many eigenspaces to span H. More general operators may have an empty point spectrum, or an insufficient number of eigenspaces. Nevertheless, generalizations of (21.11.1) and (21.11.2) can be found for some classes of operators, at the price of replacing the summations by integrals of projectors over the continuous spectrum. The proper definition of these mathematical entities requires the use of measure theory and will not be pursued here.

Let us start with the following preliminary result (see Riesz & Sz.-Nagy 1955, p. 314):

Theorem 21.11.1 *Let* H_1, H_2, ... *be a finite or infinite set of mutually orthogonal subspaces which span the whole* H, *and let* A_1, A_2, ... *be a sequence of linear bounded self-adjoint operators each one transforming the subspace with the same index into itself. Then there is one and only one linear operator* A, *in general unbounded, which reduces to* A_k *on* H_k. *Its domain consists of all the* $u \in H$ *such that*

$$\sum_k \|A_k u_k\|^2 < \infty, \tag{21.11.3}$$

where $u_k = P_k u$ is the orthogonal projection of u onto H_k, and, for these u,

$$Au = \sum_k A_k u_k. \qquad (21.11.4)$$

The operator A will not be bounded unless the A_k have a common bound.

The similarity between (21.11.4) and (21.11.2) may be misleading, as the subspaces H_k of the theorem are not necessarily eigenspaces corresponding to discrete eigenvalues. The analogy is precise only if the entire spectrum of A consists of eigenvalues with the corresponding eigenspaces mutually orthogonal and spanning H, as in this case one may take $A_k = \lambda_k P_k$. A sufficient condition is contained in the following theorem, which extends the results of §21.3.1 p. 642 (see Naylor & Sell 1982, p. 486; Richtmyer 1978, p. 249):

Theorem 21.11.2 *Let A be a densely defined, linear operator on a Hilbert space. If there is a complex number λ_0 in the resolvent set of A such that $(A - \lambda_0 I)^{-1}$ is compact and normal and 0 is not an eigenvalue, then A can be expressed as a weighted sum of projections as in (21.3.18) and the corresponding projectors constitute a resolution of the identity.*

Proof Let μ_k be the eigenvalues of $(A - \lambda_0 I)^{-1}$ and P_k the projectors on the corresponding eigenspaces. Then, as in the second one of (21.3.12) p. 641, we have the expansion

$$(A - \lambda_0 I)^{-1} = \sum_{k=1}^{\infty} \mu_k P_k \qquad (21.11.5)$$

from which, since the eigenvalues of the inverse are the inverse of the eigenvalues,

$$A - \lambda_0 I = \sum_{k=1}^{\infty} \frac{1}{\mu_k} P_k. \qquad (21.11.6)$$

From the first one of (21.3.12) the P_k generate a resolution of the identity as in (21.11.1) and, therefore, this equation becomes

$$A = \sum_{k=1}^{\infty} \left(\lambda_0 + \frac{1}{\mu_k} \right) P_k \qquad (21.11.7)$$

which is (21.11.2) with $\lambda_k = \lambda_0 + 1/\mu_k$. If $\mu_k \to 0$, $\lambda_k \to \infty$. $\qquad \square$

This theorem only expresses a sufficient condition as an expansion of the form (21.3.18) exists also for other classes of operators.

It is also interesting to note that, in the hypotheses of the theorem, R_λ is compact for any $\lambda \in \rho(A)$ as from the resolvent equation (21.10.1) p. 669 we have

$$R_\lambda = R_{\lambda_0} [I + (\lambda - \lambda_0) R_\lambda]. \qquad (21.11.8)$$

The resolvent R_λ is bounded as can be seen by applying to $A - \lambda I$ the steps leading from (21.11.7) to (21.11.5) in the reverse order. Thus, the term in brackets is bounded and, from Theorem 21.3.1 p. 637 on the product of a compact by a bounded operator, we conclude that $R_\lambda = (A - \lambda I)^{-1}$ is compact for any λ in the resolvent set.

21.11.1 Continuous spectrum

We now provide some very heuristic considerations that indicate how the spectral decomposition is extended to the case of a continuous spectrum; the reader may turn to most references on functional analysis for a proper treatment (e.g., Riesz & Sz.-Nagy 1955, Akhiezer & Glazman 1961, Liusternik & Sobolev 1961, Taylor & Lay 1980, Berezansky *et al.* 1996 and many others).

Let us return to the family of projectors appearing in (21.11.1) for the self-adjoint compact case. Let us arrange for notational simplicity the terms of the sum in order of increasing λ_k, which can be done without altering the series due to its definition as the strong limit of the partial sums. We define the *spectral family* of projection operators

$$\mathsf{E}_\lambda = \sum_{\lambda_k < \lambda} \mathsf{P}_k. \qquad (21.11.9)$$

If m and M, defined as in (21.9.20), are the left and right boundaries of the segment of the real axis containing the spectrum, we evidently have $\mathsf{E}_\lambda = 0$ for $\lambda < m$ and $\mathsf{E}_\lambda = \mathsf{I}$ for $M \leq \lambda$. Furthermore $\mathsf{E}_\lambda \mathsf{E}_\mu = \mathsf{E}_\lambda$ for $\lambda \leq \mu$ and $\mathsf{E}_{\lambda+\varepsilon} \to \mathsf{E}_\lambda$ for $\varepsilon \to 0+$. If the interval between λ and $\lambda + \Delta\lambda$ contains only one eigenvalue λ_j, we also have

$$\Delta\mathsf{E}_\lambda \equiv \mathsf{E}_{\lambda+\Delta\lambda} - \mathsf{E}_\lambda = \mathsf{P}_j \qquad (21.11.10)$$

while, if it contains no eigenvalue, $\Delta\mathsf{E}_\lambda = 0$.

Given two vectors u, $v \in H$, one can consider the function $F(\lambda) = (v, \mathsf{E}_\lambda u)$. For general u and v, this function jumps every time λ increases past an eigenvalue, is continuous from the right and evidently, due to orthogonality,

$$F(\lambda) = \sum_{k=1}^{K} (v_k, u_k) \qquad \text{when} \qquad \lambda_K \leq \lambda < \lambda_{K+1} \qquad (21.11.11)$$

with $u_k = \mathsf{P}_k u$, $v_k = \mathsf{P}_k v$. This function is not quite a distribution because, by definition, it has a well-defined value at the points of discontinuity, but provided we keep this difference in mind, we can think of it as such for the present purposes. In terms of a "differential" $dF = F' \, d\lambda$, with F' the distributional derivative, we may re-express (21.11.2) as

$$(v, \mathsf{A}u) = \int_{-\infty}^{\infty} \lambda \, dF(\lambda) \qquad (21.11.12)$$

and, more generally,

$$(v, \Phi(\mathsf{A})u) = \int_{-\infty}^{\infty} \Phi(\lambda) \, dF(\lambda) \qquad (21.11.13)$$

for any function Φ such that $\Phi(\mathsf{A})$ is well defined. A better way to understand this integral is in terms of a Riemann–Stieltjes (or Lebesgue–Stieltjes) integral (see §A.4.2 p. 688 and, e.g., Kolmogorov & Fomin 1957, chapter 6; Berezansky *et al.* 1996, chapter 3). Just as the ordinary Riemann integral of Φ over an interval $a \leq x \leq b$ is defined as the limit of sums

$$\sum_i \Phi(\xi_i)(x_{i+1} - x_i), \qquad x_i \leq \xi_i \leq x_{i+1}, \qquad (21.11.14)$$

as the partition of the interval becomes infinitely fine, the Riemann–Stieltjes integral is defined as the limit of sums of the form

$$\sum_i \Phi(\xi_i)[F(x_{i+1}) - F(x_i)], \qquad x_i \le \xi_i \le x_{i+1}, \qquad (21.11.15)$$

where F is a monotonic non-decreasing, right-continuous function. If F is absolutely continuous, then the Stieltjes integral $\int \Phi(x)\, dF(x)$ becomes the ordinary Riemann integral $\int \Phi(x)\, F'(x)\, dx$. The major difference between the two notions – which is quite apparent from the previous considerations – is that, unlike the Riemann integral, the values of the Stieltjes integral over $[a, b)$, $(a, b]$ and $[a, b]$ are different in general.

We are now ready to extend the preceding ideas to unbounded self-adjoint operators. We define a spectral family as a family of orthogonal projectors depending on a real parameter λ, with $-\infty < \lambda < \infty$, having the following properties:

(a) $E_\lambda E_\mu = E_\lambda$ for $\lambda \le \mu$
(b) $E_{\lambda+\varepsilon} \to E_\lambda$ for $\varepsilon \to 0+$
(c) $E_\lambda \to 0$, the zero operator, for $\lambda \to -\infty$
(d) $E_\lambda \to I$, the identity operator, for $\lambda \to \infty$.

This definition makes sense whether E_λ is discrete, as in the previous case, continuous, or partly discrete and partly continuous, and whether the limits 0 and I are attained for finite or infinite values of λ. This spectral family gives rise to a resolution of the identity in the sense that, for any vector $u \in H$,

$$(u, u) = \int_{-\infty}^{\infty} d(u, E_\lambda u) = \int_{-\infty}^{\infty} d\|E_\lambda u\|^2, \qquad (21.11.16)$$

where the second step is justified by the fact that projectors are idempotent and self-adjoint so that $(u, E_\lambda u) = (u, E_\lambda^2 u) = (E_\lambda u, E_\lambda u)$. With these concepts, we can now state a generalization of the decompositions (21.11.2) (see e.g. Roman 1975, p. 636).

Theorem 21.11.3 *Let E_λ be a resolution of the identity in a Hilbert space H. The family E_λ defines a self-adjoint operator A in H with domain $\mathscr{D}_A$ consisting of all vectors u for which*

$$\int_{-\infty}^{\infty} \lambda^2 d\|E_\lambda u\|^2 < \infty. \qquad (21.11.17)$$

The action of the operator A is uniquely defined by[29]

$$(v, Au) = \int_{-\infty}^{\infty} \lambda\, d(v, E_\lambda u). \qquad (21.11.18)$$

---

[29] In general $(v, E_\lambda u)$ may be complex. This situation is reduced to the sum of four real integrals by expressing the scalar product in terms of the polarization identity (19.3.13), p. 551, namely as $(v, w) = \frac{1}{4}\left(\|w + v\|^2 - \|w - v\|^2\right) + \frac{i}{4}\left(\|w - iv\|^2 + \|w + iv\|^2\right)$, with $w = E_\lambda u$.

While the theorem is illuminating, its real power lies in the converse statement, namely that *every* self-adjoint operator can be represented in the form (21.11.18). With the interpretation suggested by this theorem in mind, we can re-express the content of (21.11.18) as

$$Au = \int_{-\infty}^{\infty} \lambda \, d(E_\lambda u), \qquad u \in \mathcal{D}_A, \tag{21.11.19}$$

or even

$$A = \int_{-\infty}^{\infty} \lambda \, dE_\lambda. \tag{21.11.20}$$

In the first integral the integrand is a Hilbert space vector and in the second one a linear operator; for a proper definition of such expressions see e.g. Liusternik & Sobolev 1961, p. 173 or Berezansky *et al.* 1996, chapter 13. In a similar vein we may write

$$u = \int_{-\infty}^{\infty} d(E_\lambda u), \qquad I = \int_{-\infty}^{\infty} dE_\lambda. \tag{21.11.21}$$

These equations and (21.11.20) represent the extension of (21.11.1) and (21.11.2) to the case of general self-adjoint operators.

Example 21.11.1 Let us consider the operator $Au = x\,u(x)$ of Example 21.9.6, which has a purely continuous spectrum consisting of the entire real line, and the scalar product

$$(v, x\,u) = \int_{-\infty}^{\infty} \overline{v}(\lambda)\,\lambda\,u(\lambda)\,d\lambda = \int_{-\infty}^{\infty} \lambda\,d\left(\int_{-\infty}^{\lambda} \overline{v}(x)\,u(x)\,dx\right). \tag{21.11.22}$$

Define the projectors of the spectral family as

$$(E_\lambda u)(x) = \begin{cases} u(x) & \text{if } x \le \lambda \\ 0 & \text{if } x > \lambda \end{cases}. \tag{21.11.23}$$

It is easily seen that, as a function of λ, $\|E_\lambda u\|^2$ is continuous and therefore, in particular, continuous from the right, and the other properties defining a spectral family are also easily checked. Then

$$\int_{-\infty}^{\lambda} \overline{v}(x)\,u(x)\,dx = \int_{-\infty}^{\infty} \overline{v}(x)\,E_\lambda u(x)\,dx = (v, E_\lambda u) \tag{21.11.24}$$

and (21.11.22) becomes

$$(v, x\,u) = \int_{-\infty}^{\infty} \lambda\,d(v, E_\lambda u). \tag{21.11.25}$$

It is suggestive to note that, in a heuristic sense, $d\,(E_\lambda u)(x) = \delta(\lambda - x)u(\lambda)\,d\lambda$ because, then, the resolution of the identity (21.11.21) becomes

$$u(x) = \int_{-\infty}^{\infty} \delta(\lambda - x)\,u(\lambda)\,d\lambda \tag{21.11.26}$$

which may be interpreted as the expansion of u on the "generalized eigenfunctions" $\delta(\lambda - x)$ of the operator A. $\qquad\square$

Example 21.11.2 Let us consider the operator $Au = -i\,du/dx$ of Example 21.9.5, which has a purely continuous spectrum consisting of the entire real line, and the scalar product

$$\left(v, -i\frac{du}{dx}\right) = \int_{-\infty}^{\infty} \overline{v}(\lambda)\,\lambda\left(-i\frac{du}{d\lambda}\right) d\lambda. \tag{21.11.27}$$

Define the projectors of the spectral family as

$$(\mathsf{E}_\lambda u)(x) = \frac{1}{2\pi}\int_{-\infty}^{\lambda} dv \int_{-\infty}^{\infty} d\xi\, e^{iv(x-\xi)}u(\xi). \tag{21.11.28}$$

Again this definition can be shown to satisfy the requisites of a spectral family. Now[30]

$$
\begin{aligned}
(v, \mathsf{E}_\lambda u) &= \frac{1}{2\pi}\int_{-\infty}^{\infty} dx\,\overline{v}(x)\int_{-\infty}^{\lambda} dv \int_{-\infty}^{\infty} d\xi\, e^{iv(x-\xi)}u(\xi) \\
&= \int_{-\infty}^{\lambda} \overline{\tilde{v}}(v)\,\tilde{u}(v)\,dv,
\end{aligned}
\tag{21.11.29}
$$

where the tilde denotes the Fourier transform. From this result $d(v, \mathsf{E}_\lambda u) = \overline{\tilde{v}}(\lambda)\,\tilde{u}(\lambda)\,d\lambda$ and (21.11.27) becomes

$$\left(v, -i\frac{du}{dx}\right) = \int_{-\infty}^{\infty} \lambda\,\overline{\tilde{v}}(\lambda)\,\tilde{u}(\lambda)\,d\lambda, \tag{21.11.30}$$

in agreement with the Plancherel–Parseval relation (10.2.10) p. 269 of the Fourier transform together with the operational rule (10.3.2) for the derivative. In this case the first one of (21.11.21) is simply

$$u(x) = \frac{1}{2\pi}\int_{-\infty}^{\infty} d\lambda \int_{-\infty}^{\infty} d\xi\, e^{i\lambda(x-\xi)}u(\xi), \tag{21.11.31}$$

which embodies the inversion formula of the Fourier transform and may suggestively be written as

$$u(x) = \int_{-\infty}^{\infty} \tilde{u}(\lambda)\frac{e^{i\lambda x}}{\sqrt{2\pi}}\,d\lambda, \tag{21.11.32}$$

in which $\tilde{u}(\lambda) = (e_\lambda, u)$ is the "coefficient" of the the expansion of u on the basis of the "generalized eigenfunctions" $e_\lambda(x) = e^{i\lambda x}/\sqrt{2\pi}$ of the operator $A = -i\,d/dx$ "normalized" according to

$$(e_\mu, e_\lambda) = \int_{-\infty}^{\infty} \frac{e^{-i\mu x}}{\sqrt{2\pi}}\frac{e^{i\lambda x}}{\sqrt{2\pi}}\,dx = \delta(\lambda - \mu). \tag{21.11.33}$$

$\square$

Accepting this heuristic use of "generalized eigenfunctions," (21.11.23) and (21.11.28) suggest that the spectral family can be constructed as

$$(\mathsf{E}_\lambda u)(x) = \int_{-\infty}^{\lambda} (e_\mu, u)\,e_\mu(x)\,d\mu. \tag{21.11.34}$$

---

[30] The inversion of the order of integration necessary to obtain this result is justified by Fubini's theorem, p. 690.

Proceeding with further suggestive generalizations, and again making no attempt at rigor, we define a function $\Phi(A)$ of the operator A by

$$\Phi(A) = \int_{-\infty}^{\infty} \Phi(\lambda) \, dE_\lambda, \tag{21.11.35}$$

of which (21.11.20) and (21.11.21) will be recognized as special cases. Similarly to (21.11.17), a condition for the existence of this function is that

$$\int_{-\infty}^{\infty} |\Phi(\lambda)|^2 d\|E_\lambda u\|^2 < \infty, \tag{21.11.36}$$

which defines the domain of $\Phi(A)$. Of particular interest is the resolvent $R_\mu = (A - \mu I)^{-1}$ which, according to (21.11.35), is given by

$$R_\mu = \int_{-\infty}^{\infty} \frac{1}{\lambda - \mu} \, dE_\lambda. \tag{21.11.37}$$

From this formula, it is possible to show (see e.g. Roman 1975, p. 651) that the resolvent set consists of non-real values of μ, and of real values of μ such that, in a closed interval of which μ is an interior point, $dE_\lambda = 0$ so that the spectral family has a point of constancy. If μ is real and is a point of discontinuity for E_λ, then μ is an eigenvalue. Finally, if $dE_\lambda \neq 0$ in an interval containing μ, then μ belongs to the continuous spectrum (recall that the residual spectrum of a self-adjoint operator is empty).

Appendix

We collect here several definitions and properties which have been referred to repeatedly in the previous pages. We try to avoid unnecessary generality focusing on the aspects of specific relevance to our subjects.

A.1 Sets

A set is any well-defined collection of objects which, in our cases, have a mathematical nature. A set is typically given by enumerating all its elements or by formulating a membership criterion. Here are some properties and definitions:

1. *Empty set*: This is the unique set containing no elements and is denoted by $\emptyset$.
2. *Countable set:* Contains either a finite or a countably infinite number of elements.
3. *Equal sets*: Both sets contain exactly the same elements.
4. *Disjoint sets*: The two sets have no elements in common.
5. *Subset*: A set A all the elements of which are also elements of another set B. If the set B is "larger" than A, in the sense that there are elements of B which are not in A, one writes $A \subset B$. If the possibility $A = B$ is not excluded, one writes $A \subseteq B$. For every non-empty set A one has $\emptyset \in A$. A subset $A \subset B$ is *proper* if it is not empty and does not contain all the elements of B.
6. *Power set*: The set $P(A)$ of all the subsets of A, including the empty set and A itself.
7. *Ordered pair*: A pair of objects x and y one of which is designated as first and the other one as second; written as $\langle x, y \rangle$. In a similar way one defines an ordered n-tuple.
8. *Cartesian product*: If A and B are two sets, the set made up of all the ordered pairs $\langle a, b \rangle$, with $a \in A$ and $b \in B$ is the Cartesian product $A \times B$ of A and B; note that $A \times B \neq B \times A$ unless $A = B$. This definition is extended in an obvious way to the Cartesian product of more than two sets.
9. *Difference of two sets*: The difference of two sets A and B, denoted by $A \backslash B$ or $A - B$, is the set of all elements x such that $x \in A$ while $x \notin B$.
10. *Complement of a set*: The complement of a set $A \subseteq S$ is the set $S \backslash A$.

Two elementary operations on sets are:

(i) *Union*: The union of two sets A and B, denoted by $A \cup B$ or $B \cup A$, is the set formed by all the elements that belong to either A or B or to both; the definition is extended in an obvious way to the union of any number of sets (finite or countably or uncountably infinite).

(ii) *Intersection*: The intersection of two sets A and B, denoted by $A \cap B$ or $B \cap A$, is the set formed by all the elements that belong to both A and B. Two sets are disjoint if $A \cap B = \emptyset$. The intersection of a collection of sets consists of the elements that belong to all sets of the collection.

Theorem A.1.1 [De Morgan's laws] *Let $A \subseteq S$ and $B \subseteq S$.*

(i) *The complement of the union equals the intersection of the complements:*

$$S \backslash (A \cup B) = (S \backslash A) \cap (S \backslash B).$$
(A.1.1)

(ii) *The complement of the intersection equals the union of the complements:*

$$S \backslash (A \cap B) = (S \backslash A) \cup (S \backslash B).$$
(A.1.2)

The properties hold also for more than two subsets.

This theorem is at the basis of the *duality principle*: from any equality related to subsets of S one can automatically obtain a dual equality replacing the sets by their complements, unions by intersections and intersections by unions.

Passage to the limit is one of the most important operations of analysis. This operation is based on the notion of distance which is introduced in the theory of metric spaces. Normed spaces are special cases of metric spaces, with the norm playing the role of distance; use of this particular metric is sufficient for our purposes. The sets to which we refer in the rest of this section are sets of elements of a normed space and the word "convergence" is understood in the sense of the norm (definition p. 544).

Definition A.1.1 A set A in a normed space is *bounded* if and only if there is a positive constant M such that $\|u - v\| < M$ for all pairs of elements $u, v \in A$.

Definition A.1.2 The *open ball* $B(u, r)$ of radius r centered at u is the set of all elements v of the space such that $\|v - u\| < r$. The set $B(u, r)$ is an *open neighborhood* of the point u.

Definition A.1.3 A set is *open* if it contains all the elements of a sufficiently small open ball centered at each one of its points. A set is *closed* if its complement (i.e., the set of all the elements of the space not belonging to the set in question) is open.

According to this definition, "closed" and "open" are dual of each other: A is open in S if and only if its complement is closed in S and B is closed if and only if its complement is open. Note however that these categories are neither exclusive (a set in a metric space may be both open and closed) nor exhaustive (a set in a metric space may be neither open nor closed). For example, the empty set $\emptyset$ and the whole space S are both open and closed.

Theorem A.1.2 *The union of an arbitrary collection of open sets and the intersection of a finite collection of open sets is open.*

By applying the duality principle we deduce that

Theorem A.1.3 *The intersection of any collection of closed sets and the union of a finite collection of closed sets is closed.*

Definition A.1.4 A point $u \in S$ is called a *point of accumulation* of a set $A \subseteq S$ if any neighborhood of u contains an infinity of points of A. A point $u \in A$ is called an *isolated point* of a set A if there is a neighborhood of u which contains no other point of A. A point $u \in A$ is called a *boundary point* of a set A if any non-zero neighborhood of u contains both points that belong to A and points that do not. The collection of all boundary points forms the *boundary* of A.

A point of accumulation of a set may or may not belong to the set.

Definition A.1.5 The *closure* $\overline{A}$ of a set A is the closed set formed by adding to the set all its points of accumulation.[1]

Thus, the closure of a set A is constituted by all the isolated points of A and all the points of accumulation of A. In particular, a closed set contains all the limit points of all converging sequences of elements of A. Equivalently: A set is *closed* when it contains all its boundary points.

In Chapters 19 and 21 we have repeatedly used the notion of denseness, which can be defined in several equivalent ways:[2]

Definition A.1.6 A set $A \subseteq S$ of a normed space S is said to be *dense* in S if

- every element $u \in S$ can be approximated arbitrarily closely by an element $a \in A$, i.e., for every $u \in S$ and every $\varepsilon > 0$ there is an element $a \in A$ such that $\|u - a\| < \varepsilon$, or
- for every $u \in S$ there is a sequence $\{v_n\}$ of elements of A such that $v_n \to u$ in the sense of the norm, $\|u - v_n\| \to 0$, or
- the closure of A coincides with S.

In §21.3 we have referred to compact and relatively compact sets (p. 636). In a finite-dimensional space all closed and bounded sets are also compact (Bolzano–Weierstrass theorem, p. 216), but not necessarily so in an infinite-dimensional space. For example, the set $C[0, 1]$ of continuous functions $f(x)$ with norm $\|f\|_\infty = \max |f(x)| \leq 1$ is bounded and closed since it contains the limit points of all sequences that converge in this norm (Weierstrass theorem, p. 546) but, as noted on p. 636, it is not compact. As another example, it is impossible to extract a converging subsequence from the sequence $u_n = \sin n\pi x$ although $\|u_n\|_\infty = 1$ for all n.

An additional property that arises in passing from normed to scalar-product spaces is that the orthogonal complement $A^\perp$ of a set A (definition on p. 555) is always closed, whether

---

[1] Alternatively, the closure of an open set is the intersection of all the closed sets containing the given set.
[2] With obvious changes, the definition can be extended to sets dense in other sets.

A itself is closed or not (p. 554). This property, which is a consequence of the continuity of the scalar product (Theorem 19.3.2 p. 551), implies that

$$(A^\perp)^\perp = \overline{A}. \tag{A.1.3}$$

Many of the books on advanced mathematics and functional analysis cited in the bibliography include some elements of set theory (see e.g. Kolmogorov & Fomin 1957, Haaser & Sullivan 1991, Roman 1975, Boccara 1990, Naylor & Sell 2000, Kubrusly 2001). In addition, there are many books specifically devoted to the subject such as Halmos 1960, Hrbacek & Jech 1999, Hajnal & Hamburger 1999 and many others.

A.2 Measure

There is a deep mathematical theory of measure, but here we are only interested in the simplest notion which is necessary to obtain a glimpse of the Lebesgue integral in §A.4.3; the reader is referred to a large number of available books giving short (e.g. Kolmogorov & Fomin 1957, Roman 1975, Boccara 1990, Naylor & Sell 2000) or detailed (e.g. Halmos 1974; Doob 1994, Rudin 1987, Berezansky *et al.* 1996, Rao 2004) treatments of the subject for further information.

We start with the measure of an open interval (a, b), which is simply defined as $b - a$. The measure of a finite collection of disjoint open intervals is the sum of the measures of each one of them. This definition is extended to an infinite collection of disjoint intervals: when the series so defined is convergent the measure is finite, when it is not, it is infinite. The result is independent of the order of summation since measures are inherently non-negative so that the series either converges absolutely or diverges (Theorem 8.1.6, p. 218).

For a set E of a more general nature we proceed as follows. We consider all the countable collections of disjoint intervals which contain ("cover") E. To each such collection we can apply the previous definition of measure. The greatest lower bound, or infimum (§A.6) of all these measures is the *outer measure* $m_o(E)$ of E. We then consider all the countable collections of disjoint intervals which are contained in E. The least upper bound (or supremum) of the measures of all these collections is the *inner measure* $m_i(E)$ of E. The set E is *measurable* when $m_o(E) = m_i(E)$ and its measure $m(E)$ is this common value, which may be infinite. Thus, essentially, a set is measurable if it can be "approximated to arbitrary accuracy," both from the inside and the outside, with a set of disjoint segments the total length of which is the same as the inside segments are made as long as possible and the outside segments as short as possible. In brief, this is the concept of the *Lebesgue measure*.

Theorem A.2.1 [Additivity of the Lebesgue measure] *If* $\{A_n\}$ *is a collection of pairwise disjoint measurable sets, then*

$$m\left(\cup_n A_n\right) = \sum_n m(A_n). \tag{A.2.1}$$

Any point can be covered by an interval as short as desired, so that $m_o = 0$. The point cannot contain an interval of positive measure, so that $m_i = 0$ as well. A single point is therefore a set of measure 0. The same conclusion holds for any finite set of points. An infinite denumerable set of points can also be covered by a collection of intervals the total length of which can be made arbitrarily small.[3] This shows in particular that the set of rational numbers in any finite interval, which is a countable set,[4] has measure 0. Since measures are additive, and since the set of irrational plus rational numbers in (0, 1) is the entire interval and has measure 1, we deduce that $m(\text{irrational}) = 1 - m(\text{rational}) = 1$.

Analogous constructions can be carried out in more than one space dimensions. For example, in the plane one can start with rectangles $(a_1, b_1) \times (a_2, b_2)$ the measure of which is defined as $(b_1 - a_1)(b_2 - a_2)$, and then proceed as before.

The words *almost everywhere* and *for almost all x* encountered many times in this book have the specific meaning of "except for a set of measure zero" in the above sense. The same concept can also be expressed by the word "essential" and related ones. For example, a function $f(x)$ defined almost everywhere on a finite or infinite interval is said to be *essentially bounded* if there is a number $A > 0$ such that $|f(x)| \le A$ for almost all x. The least possible choice for the constant A is then the *essential supremum* or *essential upper bound* of f.

A.3 Functions

One of the important realizations of mathematics in the late nineteenth and early twentieth century has been how relatively "few" continuous and differentiable functions really exist. The functions that most earlier mathematicians regarded as "normal" turned out to be relatively isolated specimens in a huge sea of more or less pathological functions (see e.g. Riesz & Sz.-Nagy 1955, p. 1; Medvedev 1991, p. 228; Kharazishvili 2006).[5] The vast majority of these functions cannot be written down in a simple way, but nevertheless they are much more "numerous" than the "nice" functions. Because of the bewildering variety of possible behaviors, it is often very difficult to characterize with precision ("if and only if") the specific features of functions for which any given mathematical statement is valid. For this reason several characterizations have been developed which permit weaker statements ("if"), but which are often sufficient and have appeared repeatedly in this book. Here we collect and define several such attributes; excellent treatments can be found e.g. in Riesz & Sz.-Nagy 1955, Kolmogorov & Fomin 1957, Rudin 1987 and many others. We start with a very basic one:

---

[3] For example, take an interval of length ε, divide it into two parts, divide one of the remaining parts into two parts and so on ad infinitum, and then place a little piece of the collection so obtained on top of each one of the points.

[4] Rational numbers can be arranged e.g. in order of increasing denominator: 0, 1, 1/2, 1/3, 2/3, 1/4, 3/4, 1/5, 2/5, 3/5, 4/5, 1/6,..., and therefore they are a countable set.

[5] Medvedev's book offers an exceedingly interesting historical survey of the developments of function theory.

Definition A.3.1 [Support] The support of a function is the closure of the set of points where the function is non-zero.

Since the support is a closed set by definition and, in a finite-dimensional space, compact sets are closed and bounded (Bolzano–Weierstrass Theorem, p. 636), "compact support" and "bounded support" can be used interchangeably.

Definition A.3.2 [Continuity] A function $f(x)$ defined at a point x_0 is continuous at that point if and only if, for any $\varepsilon > 0$, one can find a $\delta(\varepsilon) > 0$ such that

$$|f(x) - f(x_0)| < \varepsilon \qquad \text{for all} \qquad |x - x_0| < \delta, \tag{A.3.1}$$

which is also written as

$$\lim_{x \to x_0} f(x) = f(x_0) \qquad \text{or} \qquad \lim_{\eta \to 0} f(x_0 + \eta) = f(x_0). \tag{A.3.2}$$

A function which is defined at x_0 but has a jump there may be continuous from the right or from the left at x_0 when, respectively,

$$f(x_0 + 0) \equiv \lim_{x \to x_0+} f(x) = f(x_0) \qquad \text{or} \qquad f(x_0 - 0) \equiv \lim_{x \to x_0-} f(x) = f(x_0). \tag{A.3.3}$$

Here $x \to x_0\pm$ means that x approaches x_0 from the right or from the left. Of course, the value taken by f at x_0 may also be different from either limit and, in this case, the function is not continuous at x_0 from either side.

If a function is continuous in a closed interval $a \le x \le b$ (continuity being from the right at $x = a$ and from the left at $x = b$), then there are points $x \in [a, b]$ where the function attains its maximum and its minimum values (both of which are finite), and there are points where it attains any value in between. A stronger property is:

Definition A.3.3 [Uniform continuity] A function f defined on an open or closed, finite or infinite interval is uniformly continuous in that interval if and only if, for any $\varepsilon > 0$, there is a $\delta(\varepsilon) > 0$ such that $|f(y) - f(x)| < \varepsilon$ for any pair of points x and y belonging to the interval such that $|y - x| < \delta$.

Ordinary continuity at every point of a *closed interval* implies uniform continuity, but not for an *open interval*, as shown e.g. by the function $1/x$ on the interval $(0, 1)$. Thus, while uniformly continuous functions are continuous, there are continuous functions which are not uniformly continuous.

Definition A.3.4 [Differentiability] A function is differentiable at x_0 if and only if the following two limits exist and are equal:

$$\lim_{\delta \to 0+} \frac{f(x_0 + \delta) - f(x_0)}{\delta} = \lim_{\delta \to 0+} \frac{f(x_0) - f(x_0 - \delta)}{\delta} = f'(x_0). \tag{A.3.4}$$

If only the first limit exists, the function is differentiable from the right, and from the left if only the second limit exists. A *continuously differentiable* function is continuous and differentiable with a continuous derivative. A theorem due to Lebesgue asserts that every monotonic function possesses a finite derivative almost everywhere.

Definition A.3.5 [Lipschitz condition] A function is said to satisfy a Lipschitz condition (or to be Lipschitz-continuous) of order α at x_0 if there is a $\delta > 0$ such that

$$|f(x) - f(x_0)| < c|x - x_0|^\alpha \qquad \text{for all} \qquad |x - x_0| < \delta, \qquad (A.3.5)$$

where the constants $\alpha > 0$ and c are independent of x. If c and α are independent of x_0 in a certain interval, the function is said to satisfy a uniform Lipschitz condition in that interval.

A function satisfying a Lipschitz condition at x_0 is evidently continuous at x_0 but, even if $\alpha = 1$, the function need not be differentiable at x_0 (e.g. $|x|$ at 0) or have bounded variation (e.g., $f(x) = x \sin 1/x$ with $f(0) = 0$, see below). In practice, only the range $0 < \alpha \leq 1$ is of interest as, if $\alpha > 1$, the function is a constant.

The Lipschitz condition is intermediate between continuity and differentiability. It may be sufficient for the derivation of some results in cases where continuity is insufficient and differentiability is sufficient but not necessary. A condition formally similar to Lipschitz's, but concerning either multi-dimensional functions or vector-valued functions in abstract spaces, in which the absolute value is replaced by a suitable norm, is often referred to as the *Hölder condition*.[6]

Definition A.3.6 [Bounded variation] A function f defined at all points of a finite closed interval $[a, b]$ has bounded variation on $[a, b]$ if and only if a single number A exists such that

$$\sum_{n=1}^{N-1} |f(x_{n+1}) - f(x_n)| < A \qquad (A.3.6)$$

for every finite set of numbers $a \leq x_1 < x_2 < \ldots < x_N \leq b$.

If the relation (A.3.6) holds for *any* finite set of points with $x_1 < x_2 < \ldots < x_N$, the function is of bounded variation on $(-\infty, \infty)$.

As an example, in [0,1] consider the function $f(x) = x \sin^2(1/x)$, $f(0) = 0$, which is continuous, and take the points $x_n = 2/(n\pi)$. Then $f(x_n) = 0$ for $n = 2k$ even, while $f(x_n) = 2/[(2k + 1)\pi]$ for $n = 2k + 1$ odd and the sum (A.3.6) becomes

$$\sum_{n=1}^{N-1} |f(x_n) - f(x_{n-1})| = |f(x_1)| + 2|f(x_3)| + \cdots + 2|f(x_{2k+1})| + \cdots$$

$$= \frac{2}{\pi}\left[1 + \frac{2}{3} + \cdots + \frac{2}{2k+1} + \cdots\right] \qquad (A.3.7)$$

---

[6] Terminology is not uniquely established here. Sometimes the Lipschitz condition is referred to as the Hölder condition, with the denomination Lipschitz reserved for special cases (e.g. $\alpha = 1$ or $x_0 = 0$). Other authors (e.g. Schwartz 1970, p. 109) refer to the "Lipschitz or Hölder condition."

which is unbounded with increasing N. This example shows that continuity (even together with differentiability) is not sufficient for a function to have bounded variation. On the other hand, the function $f(x) = x^2 \sin^2(1/x)$, $f(0) = 0$ does have bounded variation. Continuity is not necessary either as, evidently, a step function has bounded variation. The quantity

$$V_a^b[f] = \sup \sum_{n=1}^{N-1} |f(x_{n+1}) - f(x_n)|, \tag{A.3.8}$$

where the supremum is taken over all the possible finite subdivisions of the interval $[a, b]$ is the *total variation* of f on $[a, b]$. If f is continuously differentiable on $[a, b]$, then

$$V_a^b[f] = \int_a^b |f'(x)| \, dx. \tag{A.3.9}$$

Some properties of a function f of bounded variation on an interval $[a, b]$ are (see e.g. Kolmogorov & Fomin 1957, chapter 6; Natanson 1961, vol. 1 chapter 8; Haaser & Sullivan 1991, p. 232):

1. f is continuous except at most on a countable set.
2. f has one-sided limits everywhere, from the left everywhere in $(a, b]$, from the right everywhere in $[a, b)$.
3. f can be written as the difference of two monotonic non-decreasing functions and therefore, by the Lebesgue theorem on differentiability of monotonic functions mentioned before (p. 684), it admits a finite derivative almost everywhere.
4. A function of bounded variation is not necessarily continuous. If f is continuous and differentiable in $[a, b]$ and its derivative is bounded for all $x \in [a, b]$, than f has bounded variation in $[a, b]$.
5. Every function f of bounded variation can be written as the sum of a step function, an absolutely continuous function and a singular function. Thus, the derivative of f equals the derivative of the absolutely continuous component almost everywhere. For this reason, the integral of f' can only reconstruct the absolutely continuous component, but not the entire function f.

Definition A.3.7 [Absolute continuity] A function for which the relation

$$F(x) = F(a) + \int_a^x f(\xi) \, d\xi \tag{A.3.10}$$

holds on $x \in [a, b]$ is absolutely continuous on $[a, b]$.[7]

---

[7] This is actually the most significant property (for our purposes) of absolutely continuous functions which are more properly defined as follows: A real-valued function F defined on a finite closed interval $[a, b]$ is absolutely continuous if, for any $\varepsilon > 0$, there exists a $\delta(\varepsilon) > 0$ such that $\sum_{i=1}^n |F(y_i) - F(x_i)| < \varepsilon$ for every finite pairwise disjoint collection $\{(x_i, y_i) : i = 1, \ldots, n\}$ of open intervals in $[a, b]$ with $\sum(y_i - x_i) < \delta$. In other words, the variation of the function becomes small on systems of intervals of sufficiently small total length.

The integral in this definition is to be understood in the Lebesgue sense as there are functions with bounded derivatives which are not Riemann-integrable. A function absolutely continuous on any finite interval is said to be absolutely continuous on $(-\infty, \infty)$.

Absolute continuity is stronger than mere continuity. For example, the function $f(x) = x \cos(\pi/2x)$, $f(0) = 0$ is continuous but not absolutely continuous (see e.g. Natanson 1961, vol. 1 p. 246).

The crucial property of absolutely continuous functions which follows from (A.3.10) is that an absolutely continuous function F is differentiable almost everywhere with $F' = f$ almost everywhere. Note that this property does not necessarily hold if F is simply continuous and differentiable. For example, there are continuous non-decreasing functions differentiable almost everywhere with zero derivative (see e.g. Riesz & Sz.-Nagy 1955, p. 48; Haaser & Sullivan 1991, p. 243). Some further properties of absolutely continuous functions are:

1. An absolutely continuous function is uniformly continuous, but not the converse (Kolmogorov & Fomin 1957, p. 337).
2. An absolutely continuous function has bounded variation (Kolmogorov & Fomin 1957, p. 338).
3. As a consequence, an absolutely continuous function can be written as the difference of two non-decreasing functions, which are also absolutely continuous (Haaser & Sullivan 1991, p. 239).
4. If f and g are absolutely continuous, integration by parts is valid, namely (see e.g. Haaser & Sullivan 1991, p. 244; Bary 1964, vol. 1 p. 78)

$$\int_a^b f(x)g'(x)\,\mathrm{d}x \;=\; f(b)g(b) - f(a)g(a) - \int_a^b f'(x)g(x)\,\mathrm{d}x. \qquad (A.3.11)$$

The fact that, of all the functions with bounded variation, only those belonging to the relatively restricted class of absolutely continuous functions can be reconstructed from their derivative gives another indication of the need to generalize the notion of function and derivative as done in the theory of distributions. Of all the functions with bounded variation, only for those which are absolutely continuous does the usual derivative coincide with the distributional derivative.

Definition A.3.8 [Equicontinuity] A family of functions $f_\alpha(x)$ is equicontinuous on an interval $[a, b]$ if and only if, for every $\varepsilon > 0$, there is a $\delta(\varepsilon) > 0$, independent of α, such that $|f_\alpha(x) - f_\alpha(y)| < \varepsilon$ whenever $|x - y| < \delta$ for all $x, y \in [a, b]$ and all functions of the family.

Every finite set of continuous functions on a closed interval is equicontinuous.

A.4 Integration

Integration plays a crucial role in the methods used in this book. The reader will be familiar with the Riemann integral which, however, proves too limiting for many developments and requires the extension provided by the Lebesgue integral. The point is not so much that one encounters in practice functions which are not Riemann integrable, but that the Lebesgue integral provides a necessary inner coherence to the analysis. For example, there are convergent sequences of Riemann-integrable functions whose point-wise limit is not Riemann integrable. In other cases, the limit is Riemann-integrable, but interchanging integral and limit may lead to a false result.

The situation is similar to that which requires the introduction of irrational numbers: most practical calculations are done with rational approximations to irrational numbers, but without irrational numbers the theory lacks the consistency provided by the completeness of the real number system.

A further extension of the notion of integral which was necessary to describe the spectral theory of unbounded operators in §21.11 is that of the Stieltjes integral, which is also briefly described.

A.4.1 The Riemann integral

The reader will be familiar with the *proper* Riemann integral between the finite limits a and b of a function $f(x)$, which may be visualized as the area under the graph of the function. This area is evaluated by dividing it into vertical strips, approximating the area of each strip by the product of its width and the value of the function at some arbitrary point within the strip:

$$\sum_k f(\xi_k)(x_k - x_{k-1}), \qquad x_{k-1} \le \xi_k \le x_k, \tag{A.4.1}$$

and taking the limit as the number of strips goes to infinity and the maximum width goes to zero.[8] For the limit to be well defined, it must be independent of the choice of the points ξ_k. For example, in the case of the famous *Dirichlet function* $d(x)$ defined in [0, 1] by

$$d(x) = \left\{ \begin{array}{ll} 0 & x \text{ irrational} \\ 1 & x \text{ rational} \end{array} \right\} = \lim_{m \to \infty} \lim_{n \to \infty} \cos^{2n}(m!\pi x) \tag{A.4.2}$$

the sum (A.4.1) would equal 1 if the ξ_k are taken rational and 0 if irrational no matter how fine the subdivision of the interval. Such a function, therefore, is not Riemann integrable. More generally one has the (see e.g. Apostol 1974; Champeney 1987, p. 9; Natanson 1961, vol. 1, p. 132):

Theorem A.4.1 *A necessary and sufficient condition for a function to have a proper Riemann integral over* [a, b] *is that it be bounded and continuous almost everywhere.*

---

[8] More precisely, the x-axis is divided into a finite number of intervals on each one of which one constructs the largest rectangle which "fits under" the function and the smallest rectangle that "contains" the function One then forms the sums of the areas of these rectangles and calculates the infimum (smallest lower bound) of the sums of the "outer" rectangles and the supremum (largest upper bound) of the sums of the "inner" rectangles over all possible partitions. If these two quantities are equal, their common value defines the Riemann integral of the function.

When the function is not bounded or the integration interval is infinite the Riemann integral is termed *improper* and is defined by an additional limit process (see e.g. Whittaker & Watson 1927, p. 69); for example:

$$\int_a^\infty f(x)\,dx = \lim_{b\to\infty} \int_a^b f(x)\,dx \tag{A.4.3}$$

and

$$\int_{-\infty}^\infty f(x)\,dx = \lim_{a\to\infty,\, b\to\infty} \int_{-a}^b f(x)\,dx, \tag{A.4.4}$$

with the two limits taken in any order and independently of each other.

Two fundamental properties of Riemann integration are (see e.g. Haaser & Sullivan 1991, p. 219):

Theorem A.4.2 *If $f(x)$ is continuous and $F(x) = \int_a^x f(\xi)\,d\xi$, then $F'(x) = f(x)$, and $\int_a^b f(x)\,dx = F(b) - F(a)$.*

A.4.2 The Riemann–Stieltjes integral

A generalization of the Riemann integral is the *Riemann–Stieltjes* integral: Let $\alpha(x)$ be of bounded variation on $a < x < b$ (Definition A.3.6). For all partitions of (a, b) and points x_k, consider

$$\sum_{k=1}^n f(\xi_k)[\alpha(x_k) - \alpha(x_{k-1})], \qquad x_{k-1} \le \xi_k \le x_k. \tag{A.4.5}$$

When it exists, the limit of this sum as the largest of the intervals $x_k - x_{k-1} \to 0$ is the Riemann–Stieltjes integral (see e.g. Wouk 1979, p. 35; Haaser & Sullivan 1991, chapter 10; Jeffreys & Jeffreys 1999, p. 26; Stroock 1999, p. 7); we say that f is integrable with respect to α and write

$$\int_a^b f(x)\,d\alpha(x). \tag{A.4.6}$$

It is evident that taking $\alpha = x$ gives the Riemann integral.

As an example, if $\alpha(x) = H(x)$ is the Heaviside step function, it is pretty evident that $\int_{-A}^A f(x)\,dH(x) = f(0)$ (see e.g. Kanwal 2004, section 3.6).

If f is Riemann-integrable on $[a, b]$ and α is differentiable (or only absolutely continuous) on $[a, b]$, then f is integrable with respect to α on $[a, b]$ and (Haaser & Sullivan 1991, p. 255)

$$\int_a^b f(x)\,d\alpha(x) = \int_a^b f(x)\,\alpha'(x)\,dx. \tag{A.4.7}$$

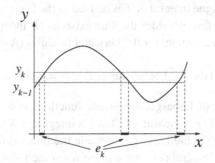

Fig. A.1   For the definition of the Lebesgue integral in (A.4.9).

If f is integrable with respect to g, then g is integrable with respect to f and we have the formula of integration by parts for the Riemann–Stieltjes integral:

$$\int_a^b f \, \mathrm{d}g = [fg]_a^b - \int_a^b g \, \mathrm{d}f. \tag{A.4.8}$$

A.4.3 The Lebesgue integral

Let us now give a rapid and far from rigorous description of one of several possible approaches to the Lebesgue integral. For a detailed treatment the reader may refer to a number of available references (see e.g. Riesz & Sz.-Nagy 1955, Kolmogorov & Fomin 1957, Haaser & Sullivan 1991, Roman 1975, Boccara 1990, Stroock 1999, Naylor & Sell 2000 and many others).

We say that a function f is *measurable* over a finite or infinite interval if the set of points where $|f(x)| < A$ is (Lebesgue-)measurable for all A in the sense of §A.2.[9] The sum, difference and product of two measurable functions is measurable, and so is their ratio provided the denominator does not vanish.

Lusin's theorem casts an interesting light on the notion of measurable functions over a finite interval: A function $f(x)$ defined over a segment $[a, b]$ is measurable there if and only if, for any $\varepsilon > 0$, there is a function $\phi(x)$ continuous over $[a, b]$ such that the measure of the set where $f(x) \neq \phi(x)$ is less than ε (see e.g. Kolmogorov & Fomin 1957, chapter 5; Haaser & Sullivan 1991, p. 151).

If a function is measurable, then the set of points e_k where $y_{k-1} \leq f(x) < y_k$ (Figure A.1) is also measurable.[10] To define the Lebesgue integral, divide the y axis by means of points $y_0 < y_1 < \ldots$ and form sums

$$y_1 \, m(e_1) + y_2 \, m(e_2) + \cdots y_n \, m(e_n). \tag{A.4.9}$$

---

[9] Recall that a measurable set does not necessarily have a finite measure. Note also that if the set of points where $|f(x)| < A$ is measurable, so are the sets where $|f(x)| \leq A$, $|f(x)| > A$ and $|f(x)| \geq A$.

[10] Conversely, a function f is measurable when the inverse image under f of every interval is measurable.

The Lebesgue integral of f is defined as the limit of these sums as the subdivision becomes arbitrarily fine, provided the limit exists and is independent of how the subdivision is carried out. As an example, for the Dirichlet function (A.4.2), we have

$$\int_0^1 d(x)\, dx \;=\; 1 \times m(\text{rational}) + 0 \times m(\text{irrational}) \;=\; 1 \times 0 + 0 \times 1 \;=\; 0. \quad \text{(A.4.10)}$$

The set of Lebesgue-integrable functions is a *subset* of all measurable functions. For example, $1/x$ is measurable, but not integrable. We express the fact that f is integrable (or summable) in the Lebesgue sense in the interval (a, b) by writing $f \in L(a, b)$. Since the Lebesgue integral does not change when the value of a function is altered on a set of zero measure, the integrals over (a, b) and over $[a, b]$ are the same.

Any function bounded on a finite interval is Lebesgue integrable, but a simple necessary and sufficient condition for Lebesgue integrability is not known.

The definition of the Lebesgue integral entails several properties most of which (e.g., $\int (cf)\, dx = c \int f\, dx$ for any constant c and integrable f, $\int (f + g)\, dx = \int f\, dx + \int g\, dx$ for integrable f and g, and others) are familiar from the Riemann integral. However it is important to mark explicitly the following ones:

1. Either *both* f and $|f|$ are integrable, or *neither one* is.
2. If $\int |f(x)|\, dx = 0$, then $f = 0$ almost everywhere.
3. If f is integrable in $[a, b]$, it is also integrable on any interval $[a, x] \subset [a, b]$. The indefinite integral of f is defined as $F(x) = \int_a^x f(\xi)\, d\xi$. From the fact that this function has bounded variation it follows that $F' = f$ almost everywhere in $[a, b]$ similarly to Theorem A.4.2 (see e.g. Haaser & Sullivan 1991, p. 234). For example, the indefinite integral of $d(x)$, where $d(x)$ is the Dirichlet function (A.4.2), vanishes and $F' = 0$ almost everywhere.

Definition A.4.1 A function f is *locally integrable*, $f \in L_{loc}$, when $f \in L(a, b)$ whenever a and b are finite.

For example, the function $1/x$ is measurable but not locally integrable, while $1/|x|^{1/2}$ is both. Boundedness without measurability does not ensure local integrability.

The definition of the Lebesgue integral can be extended to several dimensions. A question that arises in this case is the relation of the multidimensional integral to the iterated integrals, especially in connection with an interchange of the order of integration. We quote two results (see e.g. Whittaker & Watson 1927, p. 68; Natanson 1961, vol. 2 p. 86; Haaser & Sullivan 1991, p. 134; Stroock 1999, p. 69):

Theorem A.4.3 [Fubini] *If $f(x, y)$ is integrable over a finite or infinite region of the x, y plane, then the following integrals exist and are equal:*

$$\int f(x, y)\, dS \;=\; \int dx \int f(x, y)\, dy \;=\; \int dy \int f(x, y)\, dx. \quad \text{(A.4.11)}$$

Note that the statement of the theorem implies that the inner integrals exist almost everywhere.

Unfortunately the existence of one of the iterated integrals does not guarantee the existence of either one of the other integrals although, if $f \geq 0$ in S, the finiteness of one of the repeated integrals implies the summability of f in S and the equality (A.4.11) (see e.g. Natanson 1961, vol. 2 p. 92; Bary 1964, vol. 1 p. 31). For example, the function $xy/(x^2 + y^2)$ integrates to 0 between -1 and 1 with respect to both x and y (except for the sets of zero measure $y = 0$ and $x = 0$, respectively), but it is not integrable in the square $[-1, 1] \times [-1, 1]$ (see e.g. Kolmogorov & Fomin 1957, p. 313; Natanson 1961, vol. 2, p. 91). A sufficient condition is the following (see e.g. Champeney 1987, p. 19):

Theorem A.4.4 [Tonelli] *Let $f(x, y)$ be defined almost everywhere and measurable in a region of the x, y plane and suppose that at least one of the iterated integrals $\int dx \int |f| dy$ or $\int dy \int |f| dx$ exists. Then it follows that f is integrable in the region so that the conditions of the preceding theorem are satisfied and the equalities (A.4.11) valid.*

A.4.4 Comparison of the Riemann and Lebesgue integrals

As noted before, a function may fail to be Riemann-integrable because it oscillates so violently that its range of variation does not decrease as the strips are made thinner and thinner. This cannot happen for the Lebesgue integral, because the range of variation of f in any subdivision is controlled by the values of f itself, rather than by intervals of the x-axis, as we have seen in deriving (A.4.10).

Any function admitting a *proper* Riemann integral is also Lebesgue-integrable, but a difference may arise in the case of functions admitting an improper Riemann integral. In this case, if the function is not absolutely integrable in the Riemann sense, it will not be Lebesgue integrable as may be expected from the first property of the Lebesgue integral mentioned before. For example, the improper Riemann integral $\int_0^\infty (\sin x/x) dx$ exists according to the definition (A.4.3) as the positive and negative areas balance and give the finite result $\pi/2$ as the upper limit goes to infinity. In the Lebesgue integration theory, however, each side of the relation (A.4.3) has its own definition and the equality shown in the equation is not true as the positive and negative portions have "infinite areas" and the Lebesgue integral in the left-hand side does not exist. Similarly, the integral $\int_0^1 x^{-1} \sin x^{-1} dx$ exists as an improper Riemann integral, but not in the Lebesgue sense.

However, the conditions expressed in Theorem A.4.1 for the existence of the Riemann integral are sufficient (although not necessary) for the existence of the Lebesgue integral. In particular, therefore, when the improper Riemann integral exists, so will the corresponding limit of the Lebesgue integral. Thus, the integral $\int_0^b (\sin x/x) dx$ exists in the Lebesgue sense for any finite b and tends to $\pi/2$ in the limit as $b \to \infty$.

A.4.5 Differentiation under the integral sign

If $f = f(x, y)$, then in principle $\int_a^b f(x, y) dy = F(x)$ and one may inquire how the properties of F depend on those of f. If f is integrable over a finite or infinite region

$\alpha \le x \le \beta, a \le y \le b$, then $F(x)$ exists almost everywhere for $\alpha \le x \le \beta$ and, by Fubini's theorem, it is integrable there (see e.g. Whittaker & Watson 1927, p. 67; Natanson 1961, vol. 2 p. 86; Haaser & Sullivan 1991, p. 134).

If f is continuous in its two variables, $F(x)$ is also continuous if b is finite. If $b \to \infty$, continuity is guaranteed if the integral converges uniformly in the sense of Riemann: for any $\varepsilon > 0$, there is a $\Xi(\varepsilon)$ independent of x such that $| \int_\xi^\infty f(x, y) \, \mathrm{d}y | < \varepsilon$ for all $x \in [\alpha, \beta]$ and all $\xi > \Xi$. Just as in the case of series (p. 226), uniform convergence is also necessary when the integrand is positive (see e.g. Whittaker & Watson 1927, p. 67; de La Valleée Poussin 1949, vol. 2 p. 31).

Let f and $\partial f/\partial x$ be integrable over the same finite or infinite region and let $f(x, y)$ be absolutely continuous in x. Then, from Fubini's theorem, we may write

$$I(x) \equiv \int_\alpha^x \mathrm{d}x' \int_a^b \frac{\partial f}{\partial x'} \, \mathrm{d}y = \int_a^b \mathrm{d}y \int_\alpha^x \frac{\partial f}{\partial x'} \, \mathrm{d}x'$$

$$= \int_a^b [f(x, y) - f(\alpha, y)] \, \mathrm{d}y = F(x) - F(a), \qquad (A.4.12)$$

from which

$$\frac{\mathrm{d}F}{\mathrm{d}x} = \frac{\mathrm{d}}{\mathrm{d}x} \int_\alpha^x \mathrm{d}x' \int_a^b \frac{\partial f}{\partial x'} \, \mathrm{d}y = \int_a^b \frac{\partial f}{\partial x'}\bigg|_{x'=x} \mathrm{d}y = \int_a^b \frac{\partial f}{\partial x} \, \mathrm{d}y, \qquad (A.4.13)$$

for almost all $x \in [\alpha, \beta]$ (see e.g. Widder 1961, p. 358). If b is finite, equality holds at all x provided $\partial f/\partial x$ exists and is continuous. If $b \to \infty$, equality holds at each point x in the neighborhood of which the integral converges uniformly; (A.4.13) is (a special case of) the *Leibniz rule*.

For analytic functions we have the following results (see e.g. Burckel 1979, pp. 48, 228):

Theorem A.4.5 *Let C be a piecewise smooth curve (not necessarily closed) and $f(z)$ continuous on C. Then, provided $z \notin C$, the function*

$$F(z) = \int_C \frac{f(\zeta)}{\zeta - z} \, \mathrm{d}\zeta \qquad (A.4.14)$$

is infinitely differentiable and its derivatives can be calculated by differentiating under the integral sign.

Theorem A.4.6 *For each t in a finite interval $[a, b]$, let $f(z, t)$ be analytic for z in an open domain D of $\mathbb{C}$. Let f be bounded in every closed and bounded subset of D, and let it be Riemann-integrable in t for each $z \in D$. Then the function*

$$F(z) = \int_a^b f(z, t) \, \mathrm{d}t \qquad (A.4.15)$$

is analytic in D and $F'(z)$ can be calculated by differentiating under the integral sign.

A.4.6 Sequences

To the theorems on the convergence of sequences of §8.2 we add the following results (see e.g. Riesz & Sz.-Nagy 1957, Kolmogorov & Fomin 1957, Lang 1983, Rudin 1987, Haaser & Sullivan 1991):

Theorem A.4.7 *If $F_n(x) \to F(x)$ almost everywhere, and if all the F_n are measurable, then F is also measurable.*

Definition A.4.2 [Convergence in measure] A sequence of measurable functions $F_n(x)$ converges in measure to $F(x)$ if and only if the measure of the set where $|F_n(x) - F(x)| \geq \varepsilon$ tends to 0 as $n \to \infty$ for any $\varepsilon > 0$.

Theorem A.4.8 *Convergence almost everywhere implies convergence in measure.*

Theorem A.4.9 *Let $F_n(x) \to F(x)$ in measure. Then one can extract from $\{F_n\}$ a subsequence converging almost everywhere to $F(x)$.*

A famous result which, among others, provides a criterion for compactness in infinite-dimensional spaces, is (see e.g. Lang 1983, p. 55; Rudin 1987, p. 245; Haaser & Sullivan 1991, p. 87);

Theorem A.4.10 [Ascoli–Arzelà] *Let $\{F_n(x)\}$ be an equicontinuous sequence of functions defined on a closed and bounded set I such that $\max_{x \in I} |F_n(x)| < M$ for all n. Then it is possible to extract from $\{F_n\}$ a subsequence which converges uniformly to a function F continuous on I, and vice versa.*[11]

If the set I is not closed or not bounded, the result still holds except that convergence may not be uniform.

Example A.4.1 By appealing to this theorem we can give a proof of the compactness of a Hilbert-Schmidt integral operator K which does not rely on a series expansion as in the argument given in Example 21.3.1 p. 638. Let us suppose that the kernel $k(x, \xi)$ of K is continuous in $[a, b] \times [a, b]$ and consider a bounded sequence $\{u_n(x)\} \in C[a, b]$, with $[a, b]$ finite. Since k is continuous in a closed domain it is uniformly continuous (definition

[11] More concisely: A set of functions in $C[I]$ with the sup norm is relatively compact if and only if it is uniformly bounded and equicontinuous on I.

A.3.2 p. 675) and, therefore,

$$|v_n(x_2) - v_n(x_1)| \leq \int_a^b |k(x_2, \xi) - k(x_1, \xi)| d\xi \sup_{a \leq x \leq b} |u_n| \leq \varepsilon(b-a) \sup_n \|u_n\|$$

(A.4.16)

where the existence of an arbitrarily small ε follows from the uniform continuity of k and that of $\sup_n \|u_n\| = \sup_n(\sup_x |u_n|)$ is guaranteed by the assumed boundedness of the sequence. Since this bound on $|v_n(x_2) - v_n(x_1)|$ is independent of n, the family $\{v_n\}$ is equicontinuous. By the Ascoli–Arzelá theorem it is therefore possible to extract from it a convergent subsequence, which proves the compactness of the operator K.   □

The following theorems complement Lebesgue's dominated convergence theorem given on p. 222 presenting additional results on the interchange of integration and passage to the limit.

Theorem A.4.11 [B. Levi] *Let $\{F_n(x)\}$ be a set of functions integrable on a finite or infinite interval I, such that*

$$F_1(x) \leq F_2(x) \leq \ldots \leq F_n(x) \leq \cdots$$

(A.4.17)

and $\int_I F_n(x)\,dx \leq A$ for all n. Then the sequence $\{F_n(x)\}$ admits almost everywhere a finite limit $F(x)$ which is also integrable on I and

$$\lim_{n \to \infty} \int_I F_n(x)\,dx = \int_I F(x)\,dx .$$

(A.4.18)

Theorem A.4.12 [Fatou's lemma] *If a sequence of non-negative integrable functions $F_n(x)$ converges to $F(x)$ almost everywhere on a finite or infinite interval I and $\int_I F_n(x)\,dx \leq A$ for all n, then F is also integrable over I and $\int_I F(x)\,dx \leq A$.*[12]

A.4.7 Series

Some results on the integration and differentiation of series were presented in §8.3 p. 224 and §9.8 p. 255. We collect here some additional theorems.

The first one extends Theorem 8.3.3 p. 226 to an infinite interval (see e.g. Bromwich 1926, pp. 500, 502):

Theorem A.4.13 *If $\sum_k f_k(x)$ converges uniformly in any finite interval $a \leq x \leq b$, with b arbitrary, and if $g(x)$ is continuous for all finite values of x, then*

$$\sum_{k=1}^n \int_a^\infty f_k(x)\,g(x)\,dx = \int_a^\infty g(x) \sum_{k=1}^\infty f_k(x)\,dx$$

(A.4.19)

---

[12] More precisely, $\int_I F(x)\,dx \leq \underline{\lim} \int_I F_n(x)\,dx$, in which $\underline{\lim}$ denotes the smallest limit value.

provided that either the integral

$$\int_a^\infty |g(x)| \sum_{k=1}^\infty |f_k(x)|\, dx \qquad or \ the \ series \qquad \sum_{k=1}^\infty \int_a^\infty |g(x)|\ |f_n(x)|\, dx \quad (A.4.20)$$

converges.

This theorem is not applicable in a case such as $\sum (\sin nx)/(xn^p)$ which can be integrated term by term between 1 and ∞ if $p > 1$. Here one might apply the following result:

Theorem A.4.14 *Let $F_k = \int_a^x f_k(\xi)\, d\xi$ and suppose that the series $\sum_k f_k$ is uniformly convergent in any fixed interval (a, b), while the series $\sum_k F_k$ converges uniformly in the infinite interval $a \le x < \infty$. Then both*

$$\sum_k \int_a^\infty f_k(x)\, dx \qquad and \qquad \int_a^\infty \left(\sum_k f_k \right) dx \qquad (A.4.21)$$

converge and are equal.

If in Theorem A.4.13 we relax the hypothesis from convergence everywhere to almost everywhere, we have (see e.g. Riesz & Sz.-Nagy 1955, p. 36):

Theorem A.4.15 *If the series of integrable functions $\sum_k f_k(x)$ is such that*

$$\sum_k \int_a^b |f_k(x)|\, dx \ < \infty \qquad (A.4.22)$$

then it converges almost everywhere to an integrable function and can be integrated term by term.

The result also holds for an infinite interval.

For term-by-term differentiation we have the following result known as the "little Fubini theorem" (see e.g. Kolmogorov & Famin 1957, p. 324; Kharazishvili 2006, p. 112):

Theorem A.4.16 *Let $\{f_n(x)\}$ be a set of monotonic non-decreasing functions on a finite interval $[a, b]$ such that the series $\sum_n f_n$ converges for each point $x \in [a, b]$. Define*

$$f(x) = \sum_n f_n(x). \qquad (A.4.23)$$

Then, for almost all points $x \in [a, b]$, the series can be differentiated term by term.

A.5 Curves

Readers aware of the existence of space-filling curves (e.g. of the Peano or Cantor variety) will appreciate that appeal to the intuitive notion of "curve" may be insufficient for a precise statement of many of the results mentioned in Chapter 17 on analytic functions. For simplicity it is desirable to limit the class of curves to which the theorems refer – luckily with little damage to their practical application.

In view of our specific interests in analytic function theory, here we refer explicitly to curves in the complex plane, although many of the ideas readily generalize to spatial curves in any number of dimensions.

An *arc* or *path* joining a point $z_a = (x_a, y_a)$ to a point $z_b = (x_b, y_b)$ is a line parameterized as

$$z(t) = \phi(t) + i\psi(t), \quad \text{or} \quad (x = \phi(t), y = \psi(t)), \qquad a \le t \le b, \qquad \text{(A.5.1)}$$

with ϕ and ψ continuous functions. If $z(a) = z_a$ and $z(b) = z_b$, the arc is oriented from z_a to z_b.

Definition A.5.1 [Jordan curve] If ϕ and ψ are continuous functions and, for $t_1 \ne t_2$, the points $z(t_1)$ and $z(t_2)$ are distinct, (A.5.1) defines a *simple arc* or *Jordan arc*. If $z(a) = z(b)$ but no other value of t gives the same point, the arc is a *simple closed curve* or *simple Jordan curve*.

One may think of a Jordan curve as a line which can be smoothly put into a one-to-one correspondence with a circle.

The following theorem appears pretty obvious at first sight, but it is actually a deep topological result and surprisingly hard to prove (see e.g. Henrici 1974, p. 235; Burckel 1979, p. 102):

Theorem A.5.1 [Jordan curve theorem] *The complement of any closed Jordan curve $\mathscr{C}$ (i.e., the set of all points of the plane not belonging to $\mathscr{C}$) has exactly two components, with $\mathscr{C}$ as their common boundary. One of these components, called the* interior *of $\mathscr{C}$, is bounded, while the other, the* exterior *of $\mathscr{C}$, is unbounded*

A domain is called *simply connected* if and only if it contains all the points interior to all the simple Jordan curves that it contains.

Jordan arcs and curves are still fairly general concepts, as they do not have to be differentiable at any point or to have a finite length. We can further restrict the concept by introducing the notion of *continuously differentiable* or *smooth* arcs and curves, which are parameterized by functions which possess continuous derivatives ϕ' and ψ' in the ordinary sense at each point $a < t < b$, and in the one-sided sense at $t = a$ and $t = b$. The arc is *regular* if $z'(t) \ne 0$ for any t. A regular differentiable arc has a well-defined tangent at every point oriented in the direction of increasing t. The arc is piecewise regular or differentiable when it can be divided into a finite number of sub-arcs each one of which enjoys the property (see e.g. Conway 1978, p. 62; Henrici 1974, p. 166).

The arc is *rectifiable* if the functions ϕ and ψ have bounded variation (see e.g. Conway 1978, p. 62). A piecewise smooth arc is *rectifiable* and its length ℓ_{ab} is given by

$$\ell_{ab} = \int_a^b |z'(t)| \, dt . \tag{A.5.2}$$

Definition A.5.2 [Winding number] If C is a closed rectifiable curve in the complex plane and ζ a point not belonging to $\mathscr{C}$, the integral

$$W(C, \zeta) = \frac{1}{2\pi i} \oint_C \frac{dz}{z - \zeta} \tag{A.5.3}$$

is an integer called the *winding number* or *index* of C around ζ.

Clearly W equals the "number of turns" that the curve makes around ζ; in particular, $W = 0$ if ζ belongs to the region exterior to C. With this concept we may write a more general statement of Cauchy's integral formula (17.3.21) p. 428 valid for N curves, provided $\sum_k W(C_k, \zeta) = 0$ for any ζ in the domain exterior to all the curves:

$$w(\zeta) \sum_{k=1}^N W(C_k, \zeta) = \frac{1}{2\pi i} \sum_{k=1}^N \oint_{C_k} \frac{w(z)}{z - \zeta} \, dz . \tag{A.5.4}$$

In this form the result is applicable, e.g., to an annulus, in which case there are (at least) two nested curves (se e.g. Conway 1978, p. 85).

A.6 Bounds and limits

Several times in this book the notions of supremum, infimum, maximum etc. were used. We provide here precise definitions for them.

Let S be a set of real numbers, such as the values of a function or the elements of a sequence. If b is larger than any number of S, we say that b is an *upper bound* for the set S.

The *least upper bound* is the *supremum*, denoted by sup S, which may or may not belong to the set in question. When it does, we refer to it as the *maximum*, max S. As an example, if

$$f(x) = \frac{\sin x}{x} \qquad \text{then} \qquad \sup_{-1 \leq x \leq 1} f(x) = 1, \tag{A.6.1}$$

because there is no value of x such that the function equals 1 (as written, the function is undefined at 0). However if we define a new function $\tilde{f}(x) = f(x)$ for $x \neq 0$, $\tilde{f}(0) = 1$, then 1 becomes the maximum of $\tilde{f}(x)$.

Analogous definitions hold for the *greatest lower bound*, or *infimum*, inf S, which is the *minimum*, min S, when it belongs to the set. It is clear that

$$\inf S = -\sup(-S). \tag{A.6.2}$$

If, given an infinite sequence $\{a_n\}$, there is a number a such that the interval $(a - \varepsilon, a + \varepsilon)$ contains an infinite number of elements of the sequence for any $\varepsilon > 0$, the number a is a *limit point* of the sequence. The smallest limit point A is the *lower limit*, or *limes inferior* or *limit inferior* of the sequence:

$$\underline{\lim}_{n\to\infty} u_n = A \qquad \text{or} \qquad \liminf_{n\to\infty} a_n = A. \tag{A.6.3}$$

The greatest limit point B is the *upper limit, limes superior* or *limit superior* of the sequence:

$$\overline{\lim}_{n\to\infty} a_n = B \qquad \text{or} \qquad \limsup_{n\to\infty} a_n = B. \tag{A.6.4}$$

Evidently $A \leq B$, with $A = B$ only if the sequence is convergent. The example $a_n = (-1)^n (n+1)/n$ $(n \geq 1)$, for which $A = -1$, $B = 1$, shows that it is possible for an infinity of elements of the sequence to lie below and above the lower and upper limits, with no element in between. Note also that the infimum of this sequence is $-2 < A$, and the supremum is $3/2 > B$.

If the limit inferior and superior are allowed to be infinite, one may say that they always exist for any sequence of real numbers. When either limit is finite, one can extract from the given sequence a subsequence converging to it. For example, for the sequence $a_n = \sin n$, $\liminf = -1$, $\limsup = +1$ and one can extract subsequences converging to these values.

A characterization is the following (see e.g. Knopp 1966, p. 93): The (finite) number A (or B) is the lower (or upper) limit of the sequence $\{a_n\}$ if and only if, given an arbitrary $\varepsilon > 0$, we have still for an infinite number of n

$$a_n < A + \varepsilon \qquad \text{(respectively } a_n > B - \varepsilon) \tag{A.6.5}$$

while, for at most a finite number of n's,

$$a_n < A - \varepsilon \qquad \text{(respectively } a_n > B + \varepsilon). \tag{A.6.6}$$

In a similar vein one can define the *essential upper bound* of a measurable function $f(x)$ over an interval. A constant F such that the measure of the set where $f(x) > F$ is 0, although the measure of the set where $f(x) > F - \varepsilon$ is positive for any $\varepsilon > 0$, is an an essential upper bound for f over the interval. A similar definition applies to the *essential lower bound*.

References

Ablowitz, M. J. & Fokas, A. S. 1997 *Complex Variables*. Cambridge: Cambridge University Press.

Abramowitz, M. & Stegun, I. A. 1964 *Handbook of Mathematical Functions*. Washington: National Bureau of Standards, reprinted by Dover, New York, 1965. See also Oliver, W. J. *et al.* 2010 *NIST Handbook of Mathematical Functions*. Cambridge: Cambridge University Press.

Adams, R. A. & Fournier, J. J. F. 2003 *Sobolev Spaces*, 2nd edn. New York: Academic Press.

Ahlfors, L. V. 1979 *Complex Analysis*. New York: McGraw-Hill.

Ahlfors, L. V. 1987 *Lectures on Quasiconformal Mappings*. Monterey CA: Wadsworth & Brooks.

Akhiezer, N. I. & Glazman, I. M. 1961 *Theory of Linear Operators in Hilbert Space*. New York: Ungar, reprinted by Dover, New York, 1993.

Apostol, T. M. 1974 *Mathematical Analysis*, 2nd edn. Reading: Addison-Wesley.

Arendt, W., Batty, C. J. K., Hieber, M. & Neubrander, F. 2001 *Vector-Valued Laplace Transforms and Cauchy Problems*. Basel: Birkhäuser.

Barber, J. R. 1992 *Elasticity*. Dordrecht: Kluwer.

Bary, N. K. 1964 *A Treatise on Trigonometric Series*. New York: Pergamon Press.

Batchelor, G. K. 1999 *An Introduction to Fluid Mechanics*, 2nd edn. Cambridge: Cambridge University Press.

Bender, C. M. & Orszag, S. A. 2005 *Advanced Mathematical Methods for Scientists and Engineers*. New York: Springer.

Beltrami, E. J. & Wohlers, M. R. 1966 *Distributions and the Boundary Values of Analytic Functions*. New York: Academic Press.

Berezanski, Y. M., Sheftel, Z. G. & Us, G. F. 1996 *Functional Analysis*. Basel: Birkhäuser.

Bieberbach, L. 1953 *Conformal Mapping*. New York: Chelsea Publishing Co., reprinted by AMS Chelsea Publishing, New York, 2000.

Bleistein, N. & Handelsman, R. A. 1975 *Asymptotic Expansion of Integrals*. Orlando FL: Holt, Reinhart & Winston, reprinted by Dover, New York, 1986.

Boas, R. P. 1987 *Invitation to Complex Analysis*. New York: Random House.

Boccara, N. 1990 *Functional Analysis*. Boston: Academic Press.

Bracewell, R. N. 1986 *The Fourier Transform and Its Applications*. New York: McGraw-Hill.

Bremermann, H. 1965 *Distributions, Complex Variables and Fourier Transforms*. Reading: Addison-Wesley.

Bromwich, T. J. l'A. 1926 *Introduction to the Theory of Infinite Series*, 2nd edn. London: Macmillan & Co., reprinted by Chelsea Publishing Co., New York, 1991.

de Bruijn, N. G. 1958 *Asymptotic Methods in Analysis*. Amsterdam: North-Holland, reprinted by Dover, New York, 1981.

Burckel, R. B. 1979 *An Introduction to Classical Complex Analysis*. New York: Academic Press.

Canuto, C., Hussaini, M. Y., Quarteroni, A. & Zang, T. A. 1988 *Spectral Methods in Fluid Dynamics*. Berlin: Springer.

Carmichael, R. D. & Mitrović, D. 1989 *Distributions and Analytic Functions*. Harlow: Longman.

Chaikin, P. M. & Lubensky, T. C. 1995 *Principles of Condensed Matter Physics*. Cambridge: Cambridge University Press.

Champeney, D. C. 1987 *A Handbook of Fourier Theorems*. Cambridge: Cambridge University Press.

Chandrasekhar, S. 1961 *Hydrodynamic and Hydromagnetic Stability*. Oxford: Clarendon Press, reprinted by Dover, New York, 1981.

Ciarlet, P. G. 1989 *Introduction to Numerical Linear Algebra and Optimization*. Cambridge: Cambridge University Press.

Coddington, E. A. & Levinson, N. 1955 *Theory of Ordinary Differential Equations*. New York: McGraw-Hill.

Constanda, C. 2000 *Direct and Indirect Boundary Integral Equation Methods*. Boca Raton: Chapman & Hall/CRC.

Conway, J. B. 1978 *Functions of One Complex Variable*, 2nd edn. New York: Springer.

Conway, J. B. 1985 *A Course in Functional Analysis*. New York: Springer.

Copson, E. T. 1955 *An Introduction to the Theory of Functions of a Complex Variable*. Oxford: Clarendon Press.

Courant, R. & Hilbert, D. 1953 *Methods of Mathematical Physics*. New York: Interscience, reprinted by Wiley, 1991.

Cullen, C. G. 1972 *Matrices and Linear Transformations*, 2nd edn. Reading: Addison-Wesley.

Dassios, G. & Kleinman, R. E. 1989 On Kelvin inversion and low-frequency scattering. *SIAM Review* **31**, 565–585.

Davis, H. F. 1963 *Fourier Series and Orthogonal Functions*. Boston: Allyn & Bacon, reprinted by Dover, New York, 1989.

Davies, B. 2002 *Integral Transforms and their Applications*, 3rd edn. New York: Springer.

Davies, E. B. 1995 *Spectral Theory and Differential Operators*. Cambridge: Cambridge University Press.

Debnath, L. 1995 *Integral Transforms and Their Applications*. Boca Raton: CRC Press.

Debnath, L. & P., Mikusiński. 1990 *Introduction to Hilbert Spaces with Applications*. Boston: Academic Press.

Doetsch, G. 1943 *Theorie und Anwendung der Laplace-Transformation*. New York: Dover.

Doetsch, G. 1950 *Handbuch der Laplace-Transformation*. Basel: Birkhäuser.

Doetsch, G. 1974 *Introduction to the Theory and Applications of the Laplace Transformation*. New York: Springer.

Doob, J. L. 1994 *Measure Theory*. New York: Springer.

Driscoll, T. A. & Trefethen, L. N. 2002 *Schwarz-Christoffel Mapping*. Cambridge: Cambridge University Press.

Dunford, N. & Schwartz, J. T. 1963 *Linear Operators*. New York: Wiley-Interscience.

Edmunds, D. E. & Evans, W. D. 1987 *Spectral Theory and Differential Operators*. Oxford: Clarendon Press.

Erdélyi, A. Magnus, W. & Oberhettinger, F. 1953 *Higher Transcendental Functions*. New York: McGraw-Hill.

Erdélyi, A. 1956 *Asymptotic Expansions*. New York: Dover.

Erdélyi, A., Magnus, W., Oberhettinger, F. & Tricomi, F. G. 1954 *Tables of Integral Transforms*. New York: McGraw-Hill.

Farkas, H. M. & Kra, I. 1980 *Riemann Surfaces*, 2nd edn. Berlin: Springer.

Fedoriouk, M. 1987 *Méthodes Asymptotiques pour les Équations Différenielles Ordinaires Linéaires*. Moscow: Mir.

Fornberg, B. 1996 *A Practical Guide to Pseudospectral Methods*. Cambridge: Cambridge University Press.

Forsyth, A. R. 1956 *A Treatise on Differential Equations*, 6th edn. London: McMillan & Co., reprinted by Dover, New York, 1996.

Friedlander, G. & Joshi, M. 1998 *Introduction to the Theory of Distributions*, 2nd edn. Cambridge: Cambridge University Press.

Friedman, B. 1956 *Principles and Techniques of Applied Mathematics*. New York: Wiley.

Friedman, A. 1963 *Generalized Functions and Partial Differential Equations*. Englewood Cliffs NJ: Prentice-Hall.

Gantmacher, F. R. 1959 *Matrix Theory*. New York: Chelsea Publishing Co.

Garabedian, P. R. 1964 *Partial Differential Equations*. New York: Wiley, reprinted by AMS Chelsea Publishing, New York, 1998.

Gelfand, I. M. & Shilov, G. E. 1964 *Generalized Functions*, vol. 1. New York: Academic Press.

Godunov, S. K. 1998 *Modern Aspects of Linear Algebra*. Providence: American Mathematical Society.

Golub, G. H. & van Loan, C. F. 1989 *Matrix Computations*, 2nd edn. Baltimore: Johns Hopkins University Press.

Goldstein, H. 1980 *Classical Mechanics*, 2nd edn. Reading: Addison-Wesley.

González, M. O. 1991 *Complex Analysis – Selected Topics*. New York: Dekker.

Gradshteyn, I. S., Ryzhik, I. M., Jeffrey, A. & Zwillinger, D., eds. 2007 *Table of Integrals, Series, and Products*, 7th edn. Orlando: Academic Press.

Griffel, D. H. 2002 *Applied Functional Analysis*. Mineola: Dover.

Gurtin, M. E. 1984 The linear theory of elasticity. In *Mechanics of Solids* (ed. C. Truesdell), vol. 2, pp. 1–295. Berlin: Springer.

Haaser, N. B. & Sullivan, J. A. 1971 *Real Analysis*. New York: Van Nostrand/Reinhold, revised edition reprinted by Dover, New York, 1991.

Hadamard, J. 1923 *Lectures on Cauchy's Problem in Linear Partial Differential Equations*. New Haven: Yale University Press, reprinted by Dover, New York, 1952.

Hajmirzaahmad, M. & Krall, A. M. 1992. Singular second-order operators: the maximal and minimal operators, and selfadjoint operators in between. *SIAM Rev.* **34**, 614–634.

Hajnal, A. & Hamburger, P. 1999 *Set Theory*. Cambridge: Cambridge University Press.

Halmos, P. R. 1958 *Finite-Dimensional Vector Spaces*. Princeton: van Nostrand, reprinted by Springer, 1993.

Halmos, P. R. 1960 *Naive Set Theory*. Princeton: Van Nostrand.

Halmos, P. R. 1974 *Measure Theory*. New York: Van Nostrand.

Hardy, G. H. 1956 *Divergent Series*. Oxford: Clarendon Press.

Hardy, G. H. & Rogosinski, W. W. 1956 *Fourier Series*. Cambridge: Cambridge University Press.

Helmberg, G. 1969 *Introduction to Spectral Theory in Hilbert Space*. Amsterdam: North-Holland.

Henrici, P. 1974 *Applied and Computational Complex Analysis*, vol. 1. New York: Wiley, reprinted, 1993.

Henrici, P. 1986 *Applied and Computational Complex Analysis*, vol. 3. New York: Wiley.

Hetnarski, R. B. & Ignaczak, J. 2004 *Mathematical Theory of Elasticity*. New York: Taylor & Francis.

Hille, E. 1959 *Analytic Function Theory*. New York: Ginn & Co., reprinted by Chelsea Publishing Co., New York, 1982.

Hille, E. & Phillips, R. S. 1965 *Functional Analysis and Semi-Groups*, revised edition edn. Providence: American Mathematical Society

Hinch, E. J. 1991 *Perturbation Methods*. Cambridge: Cambridge University Press.

Hobson, E. W. 1931 *The Theory of Spherical and Ellipsoidal Harmonics*. Cambridge: Cambridge University Press, reprinted by Chelsea Publishing Co., New York, 1965.

Hochstadt, H. 1971 *The Functions of Mathematical Physics*. New York: Wiley-Interscience, reprinted by Dover, New York, 1987.

Holmes, M. H. 1995 *Introduction to Perturbation Methods*. New York: Springer.

Hörmander, L. 1983 *The Analysis of Linear Partial Differential Operators*. Berlin: Springer.

Horn, R. A. & Johnson, C. R. 1994 *Topics in Matrix Analysis*. Cambridge: Cambridge University Press.

Hoskins, R. F. 1999 *Delta Functions*. Chichester: Horwood Publishing.

Hrbacek, K. & Jech, T. 1999 *Introduction to Set Theory*, 3rd edn. New York: Dekker.

Ince, E. L. 1926 *Ordinary Differential Equations*. London: Longmans, Green and Co., reprinted by Dover, New York, 1956.

Isaacson, E. & Keller, H. B. 1966 *Analysis of Numerical Methods*. New York: Wiley, reprinted by Dover, New York, 1994.

Ivanov, V. I. & Trubetskov, M. K. 1995 *Handbook of Conformal Mapping with Computer Aided Visualization*. Boca Raton: CRC Press.

Jackson, J. D. 1998 *Classical Electrodynamics*, 3rd edn. New York: Wiley.

Jeffreys, H. 1962 *Asymptotic Approximations*. Oxford: Clarendon Press.

Jeffreys, H. & Jeffreys, B. 1999 *Methods of Mathematical Physics*, 3rd edn. Cambridge: Cambridge University Press.

Jerry, A. J. 1998 *The Gibbs Phenomenon in Fourier Analysis, Splines and Wavelet Approxmations*. Dordrecht: Kluwer.

John, F. 1955 *Plane Waves and Spherical Means Applied to Partial Differential Equations*. New York: Interscience.

Jones, D. S. 1966 *Generalised Functions*. New York: McGraw-Hill.

Jost, J. 2006 *Compact Riemann Surfaces*. Berlin: Springer.

Kanwal, R. P. 2004 *Generalized Functions*, 3rd edn. Boston: Birkhäuser.

Kellogg, O. D. 1967 *Foundations of Potential Theory*. Berlin: Springer.

Kemmer, N. 1977 *Vector Analysis: A Physicist's Guide to the Mathematics of Fields in Three Dimensions*. Cambridge: Cambridge University Press.

Kharazishvili. A. B. 2006 *Strange Functions in Real Analysis*. Boca Raton: Chapman & Hall.

Knopp, K. 1966 *Theory and Applications of Infinite Series*. New York: Hafner, reprinted by Dover, New York 1990.

Kober, H. 1957 *Dictionary of Conformal Representations*. New York: Dover.

Kolmogorov, A. N. & Fomin, S. V. 1957 *Elements of the Theory of Functions and Functional Analysis*. Rochester NY: Graylock Press, reprinted by Dover, 1999.

Korevaar, J. 2004 *Tauberian Theory: A Century of Development*. Berlin: Springer.

Körner, T. W. 1989 *Fourier Analysis*. Cambridge: Cambridge University Press.

Krantz, S. G. 1999 *Handbook of Complex Variables*. Basel: Birkhäuser.

Kress, R. 1989 *Linear Integral Equations*. Berlin: Springer.

Kubrusly, C. A. 2001 *Elements of Operator Theory*. Boston: Birkhäuser.

Kythe, P. K. 1998 *Computational Conformal Mapping*. Boston: Birkhäuser.

Ladyzhenskaya, O. A. 1985 *The Boundary Value Problems of Mathematical Physics*. New York: Springer.

Lamb, H. 1932 *Hydrodynamics*, 6th edn. Cambridge: Cambridge University Press, reprinted 1993.

Lancaster, P. & Tismenetsky, M. 1985 *The Theory of Matrices*, 2nd edn. Academic Press.

Lanczos, C. 1970 *The Variational Principles of Mechanics*. Toronto: University of Toronto Press, reprinted by Dover, New York, 1986.

Landau, L. D. & Lifshitz, E. M. 1981 *Quantum Mechanics: Non-Relativistic Theory*, 3rd edn. New York: Pergamon, reprinted by Butterworth-Heineman, Oxford, 2003.

Landau, L. D. & Lifshitz, E. M. 1987a *Fluid Mechanics*, 2nd edn. New York: Pergamon, reprinted by Butterworth-Heineman, Oxford, 2003.

Landau, L. D. & Lifshitz, E. M. 1987b *Theory of Elasticty*, 3rd edn. Oxford: Pergamon.

Lang, S. 1983 *Real Analysis*, 2nd edn. Reading: Addison-Wesley.

Lang, S. 1985 *Complex Analysis*, 2nd edn. New York: Springer.

Lebedev, N. N. 1965 *Special Functions and Their Applications*. Englewood Cliffs NJ: Prentice-Hall, reprinted by Dover, New York, 1972.

Levitan, B. M. & Sargsjan, I. S. 1975 *Introduction to Spectral Theory: Selfadjoint Ordinary Differential Operators*. Providence: American Mathematical Society.

Levitan, B. M. & Sargsjan, I. S. 1991 *Sturm-Liouville and Dirac Operators*. Dordrecht: Kluwer.

Lieb, E. H. & Loss, M. 1997 *Analysis*. Providence: American Mathematical Society.

Lighthill, M. J. 1958 *Fourier Analysis and Generalised Functions*. Cambridge: Cambridge University Press.

Lighthill, M. J. 1978 *Waves in Fluids*. Cambridge: Cambridge University Press.

Liseikin, V. D. 1999 *Grid Generation Methods*. New York: Springer.

Liseikin, V. D. 2007 *A Computational Differential Geometry Approach to Grid Generation*. Berlin: Springer.

Liusternik, L. & Sobolev, V. 1961 *Elements of Functional Analysis*. New York: Ungar.

Locker, J. 2000 *Spectral Theory of Non-Self-Adjoint Two-Point Differential Operators*. Providence: American Mathematical Society.

Love, A. E. H. 1927 *A Treatise on the Mathematical Theory of Elasticity*, 4th edn. Cambridge: Cambridge University Press, reprinted by Dover, New York, 1944.

MacCluer, C. R. 2004 *Boundary Value Problems and Fourier Expansions*. New York: Dover.

Markusevich, A. I. 1965 *Theory of Functions*, 2nd edn. New York: Chelsea Publishing Co.

Mathews, J. & Walker, R. L. 1971 *Mathematical Methods of Physics*, 2nd edn. Meno Park: Benjamin.

Medvedev, F. A. 1991 *Scenes from the History of Real Functions*. Basel: Birkhäuser.

Mitrinović, D. S. & Kečkić, J. D. 1984 *The Cauchy Method of Residues. Theory and Applications*. Dordrecht: Reidel.

Mlak, W. 1991 *Hilbert Spaces and Operator Theory*. Dordrecht: Kluwer.

Moler, C. & van Loan, C. F. 1978 19 dubious ways to compute exponential of a matrix. *SIAM Rev.* **20**, 801–836.

Moon, P. & Spencer, D. E. 1961 *Field Theory Handbook*. Berlin: Springer.

Morse, P. M. & Feshbach, H. 1953 *Methods of Theoretical Physics*. New York: McGraw-Hill.

Murdock, J. A. 1991 *Perturbations*. New York: Wiley.

Naimark, M. A. 1967 *Linear Differential Operators*. New York: Ungar.

Narasimhan, R. & Nievergelt, Y. 2001 *Complex Analysis in One Variable*, 2nd edn. Boston: Birkhäuser.

Natanson, I. P. 1961 *Theory of Functions of a Real Variable*. New York: Ungar.

Naylor, A. W. & Sell, G. R. 2000 *Linear Operator Theory in Engineering and Science*, 2nd edn. New York: Springer.

Nussenzweig, H. M. 1972 *Causality and Dispersion Relations*. New York: Academic Press.

Ortega, J. M. 1987 *Matrix Theory, A Second Course*. New York: Plenum.

Paliouras, J. D. & Meadows, D. S. 1990 *Complex Variables for Scientists and Engineers*, 2nd edn. New York: Macmillan.

Pandey, J. N. 1996 *The Hilbert Transform of Schwartz Distributions and Applications*. New York: Wiley.

Panofsky, W. K. H. & Phillips, M. 1956 *Classical Electricity and Magnetism*. Reading: Addison Wesley.

Pathak, R. S. 1997 *Integral Transforms of Generalized Functions and Their Applications*. Amsterdam: Gordon & Breach.

Pease, M. C. 1965 *Methods of Matrix Algebra*. New York: Academic Press.

Pécseli, H. L. 2000 *Fluctuations in Physical Systems*. Cambridge: Cambridge University Press.

Piziak, R. & Odell, P. L. 2007 *Matrix Theory: From Generalized Inverses to Jordan Form.* Boca Raton: Chapman & Hall.

Pomp, A. 1998 *The Boundary-Domain Integral Method for Elliptic Systems.* New York: Springer.

Pozrikidis, C. 1992 *Boundary Integral and Singularity Methods for Linearized Viscous Flow.* Cambridge: Cambridge University Press.

Rao, M. M. 2004 *Measure Theory and Integration,* 2nd edn. New York: Dekker.

Renardy, M. & Rogers, R. C. 1993 *An Introduction to Partial Differential Equations.* New York: Springer.

Richards, I. & Youn, H. K. 1990 *Theory of Distributions: A Non-Technical Introduction.* Cambridge: Cambridge University Press.

Richtmyer, R. D. 1978 *Principles of Advanced Mathematical Physics.* Berlin: Springer.

Riesz, F. & Sz.-Nagy, B. 1955 *Functional Analysis.* New York: Ungar, reprinted by Dover, New York, 1990.

Robinson, J. C. 2001 *Infinite-Dimensional Dynamical Systems.* Cambridge: Cambridge University Press.

Roman, P. 1975 *Some Modern Mathematics for Physicists and Other Outsiders.* New York: Pergamon Press.

Roos, B. W. 1969 *Analytic Functions and Distributions in Physics and Engineering.* New York: Wiley.

Rudin, W. 1987 *Real and Complex Analysis.* New York: McGraw-Hill.

Saichev, A. I. & Woyczyński, W. A. 1997 *Distributions in the Physical and Engineering Sciences.* Boston: Birkhäuser.

Sakurai, J. J. 1994 *Modern Quantum Mechanics,* revised edn. Reading: Addison-Wesley.

Samko, S. G., Kilbas, A. A. & Marichev, O. I. 1993 *Fractional Integrals and Derivatives.* Amsterdam: Gordon & Breach.

Sansone, G. 1959 *Orthogonal Functions.* New York: Interscience, reprinted by Dover, New York, 1991.

Schinzinger, R. & Laura, P. A. A. 1991 *Conformal Mapping: Methods and Applications.* Amsterdam: Elsevier, reprinted by Dover, New York.

Schwartz, L. 1966a *Mathematics for the Physical Sciences.* Reading: Addison-Wesley.

Schwartz, L. 1966b *Théorie des Distributions.* Paris: Hermann.

Schwartz, L. 1970 *Analyse.* Paris: Hermann.

Smirnov, V. I. 1964 *A Course of Higher Mathematics.* Oxford: Pergamon Press.

Smith, M. G. 1966 *Laplace Transform Theory.* London: van Nostrand.

Sneddon, I. N. 1951 *Fourier Transforms.* New York: McGraw-Hill, reprinted by Dover, New York, 1995.

Sneddon, I. N. 1955 Functional analysis. In *Encyclopedia of Physics* (ed. S. Flügge), vol. 2, pp. 198–348. Berlin: Springer.

Sneddon, I. N. 1966 *Mixed Boundary Value Problems in Potential Theory.* Amsterdam: North-Holland.

Sommerfeld, A. 1950 *Partial Differential Equations in Physics.* New York: Academic Press.

Srednicki, M. 2007 *Quantum Field Theory.* Cambridge: Cambridge University Press.

Stakgold, I. 1967 *Boundary Value Problems of Mathematical Physics*. New York: Macmillan.

Stakgold, I. 1997 *Green's Functions and Boundary Value Problems*, 2nd edn. New York: Wiley-Interscience.

Strichartz, R. S. 1994 *A Guide to Distribution Theory and Fourier Transforms*. Boca Raton: CRC Press.

Stroock, D. W. 1999 *A Concise Introduction to the Theory of Integration*, 3rd edn. Boston: Birkhäuser.

Sveshnikov, A. G. & Tikhonov, A. N. 1978 *The Theory Functions of a Complex Variable*. Moscow: Mir Publishers.

Taylor, A. E. & Lay, D. C. 1980 *Introduction to Functional Analysis*. New York: Wiley.

Titchmarsh, A. C. 1948 *Introduction to the Theory of Fourier Integrals*, 2nd edn. Oxford: Clarendon Press.

Titchmarsh, E. C. 1962 *Eigenfunction Expansions*. Oxford: Clarendon Press.

Tranter, C. J. 1966 *Integral Tranforms in Mathematical Physics*, 3rd edn. London: Methuen & Co.

Vallée Poussin, Ch.-J. de La 1949 *Cours d'Analyse Infinitésimale*, 8th edn. Louvain/Paris: Librairie Universitaire/Gauthier-Villars.

van der Pol, B. & Bremmer, H. 1959 *Operational Calculus Based on the Two-Sided Laplace Integral*. Cambridge: Cambridge University Press.

Vladimirov, V. S. 2002 *Methods of the Theory of Generalized Functions*. London: Taylor & Francis.

Watson, G. N. 1944 *A Treatise on the Theory of Bessel Functions*, 2nd edn. Cambridge: Cambridge University Press.

Weinberg, S. 1995 *The Quantum Theory of Fields*. Cambridge: Cambridge University Press.

Weyl, H. 1964 *The Concept of a Riemann Surface*. Reading: Addison-Wesley.

Whitham, G. B. 1974 *Linear and Nonlinear Waves*. New York: Wiley, reprinted, 1999.

Whittaker, E. T. & Watson, G. N. 1927 *Modern Analysis*, 4th edn. Cambridge: Cambridge University Press.

Widder, D. V. 1946 *The Laplace Transform*. Princeton: Princeton University Press.

Widder, D. V. 1961 *Advanced Calculus*, 2nd edn. Englewood Cliffs NJ: Prentice-Hall, reprinted by Dover, New York, 1989.

Widder, D. V. 1971 *An Introduction to Transform Theory*. New York: Academic Press.

Wigner, E. P. 1960 The unreasonable effectiveness of mathematics in the natural sciences. *Comm. Pure Appl. Math.* **13**, 1–14.

Wilkinson, J. H. 1988 *The Algebraic Eigenvealue Problem*, 2nd edn. Oxford: Clarendon Press.

Wouk, A. 1979 *A Course of Applied Functional Analysis*. New York: Wiley.

Zauderer, E. 1989 *Partial Differential Equations of Applied Mathematics*. New York: Wiley.

Zemanian, A. H. *Distribution Theory and Transform Analysis*. New York: McGraw-Hill.

Zwillinger, D. 2002 *CRC Standard Mathematical Tables and Formulae*, 31st edn. Boca Raton: CRC Press.

Index

Bold entries refer to tables

Printed in the United States
by Baker & Taylor Publisher Services